Principles and Practice of Automatic Process Control

Third Edition

Carlos A. Smith, Ph.D., P.E.

Department of Chemical Engineering
University of South Florida

Armando B. Corripio, Ph.D., P.E.

Gordon A. and Mary Cain Department of Chemical Engineering
Louisiana State University

WILEY

John Wiley & Sons, Inc.

ISBN-13 978-0-471-66141-2
ISBN-10 0-471-66141-4

Printed in Asia

10 9 8 7 6 5 4 3 2 1

This work is dedicated with all our love to The Lord our God; all His daily blessings made this work possible once more.

The Smiths:
To the "new generation": Sophia Cristina and Steven Christopher Livingston, Carlos Alejandro Smith, and all others that will hopefully come.

To the "old generation": Tim and Cristina Livingston, and Carlos and Jennifer Smith.

To the "very old generation": Rene & Terina Smith and family who have always been there for us.

To the "dearest generation": Mimi, with my eternal love and thank you for being the best thing that ever happened to me.

To the Corripios' new generation: Nicholas, Robert, Garrett, David, and Roman.

To the memory of our mentor Charles E. Jones, Jr.

Finally, to our dearest homeland, Cuba

Preface

This third edition presents some changes from the second edition, including additions, reorganization, and deletions. The objective of the book, however, remains unchanged, to present the fundamental principles of control theory, and the practice of automatic process control.

WITH THIS TEXT, STUDENTS WILL:

- Develop dynamic mathematical process models that will help in the analysis, design, and operation of control systems.
- Understand how the basic components of control systems work.
- Design and tune feedback controllers
- Apply a variety of techniques that enhance feedback control, including cascade control, ratio control, override control, and selective control, feedforward control, multivariable control, and loop interaction.
- Master the fundamentals of dynamic simulation of process control systems using MATLAB™ and Simulink™.

NEW TO THIS EDITION

- An extended discussion on the development of dynamic balances (Chapter 3). We show "How to" develop dynamic models for physical systems by relating the steady-state balances that students are familiar with to dynamic balances. This discussion helps demystify differential equations and build students' confidence.
- Safety and product quality are emphasized in the examples and problems, especially for Chapters 9, 10, 11 and Appendix E Operating Case Studies.
- Computer problems and modeling examples are integrated throughout the text. Problems with simulation are marked with a computer icon for easy identification.
- A chapter on dynamic simulation (Chapter 13). The principles of simulation of processes and control and instrumentation components are presented in this chapter. Most problems at the end of the chapter provide alternative simulation solutions to the problems marked with icons throughout the book
- Realistic tuning exercises (Appendix D), with Control Tuning Labs available on the book website.
- A discussion and guidelines for plantwide control, with two new case studies (Chapter 12 and Appendix B).
- Two new design case studies (Appendix B).
- Operating case studies (Appendix E).
- More extensive discussion of distributive control systems (DCSs) (Chapter 10).

- The use of the Instrumentation, Systems and Automation Society (ISA) symbols for conceptual diagrams, which eliminate the need to differentiate between pneumatic, electronic, or computer implementation of the various control schemes.

- The material in the second edition related to computer control (Chapters 14 and 15) including the z-transforms, stability of sample-data control systems, and discrete controllers, has been deleted from this edition because they were considered outside the scope of an undergraduate control course.

We feel that these changes are more in focus to what is the practice and implementation of the overwhelming majority of industrial control strategies.

CONTROL TUNING LABS

Appendix D describes several processes (Control Tuning Labs, available on the web at www.wiley.com/college/smith) to practice the tuning of feedback controllers, cascade controllers, feedforward controllers, and the pairing and tuning of controllers in a multivariable control system. These Control Tuning Labs have been used by the authors for many years in their university courses, and in training of practicing engineers.

DYNAMIC SIMULATION

Dynamic simulation is a convenient method for analyzing the performance of processes control systems. Problems throughout the text are marked with a computer icon, indicating that they may be solved through simulation.

Chapter 13 presents the basics for learning how to develop dynamic simulations, based on the dynamic modeling technique presented in Chapters 3 and 4, and the instrument models in Chapter 5.

Two methods are presented:

- Simulation of linear systems through transfer function representation, and
- Simulation of control systems from basic mechanistic models.

The examples presented use the program MATLAB™ with Simulink™ because it is one of the most powerful block-oriented simulation programs that is commonly available to students and engineers. All simulations in the text, as well as simulations using VisSim™, are available from the book website at www.wiley.com/college/smith.

TOPICAL ORGANIZATION

Chapter 1 presents the need for process control, the definitions of terms used in this topic, and the concepts of feedback and feedforward control with their advantages and disadvantages.

Chapter 2 presents the mathematical tools used in the analysis and design of process control systems, Laplace transforms and linearization. The emphasis is on the determination of the quantitative characteristics of the process response—settling time, frequency of oscillation, and damping ratio—rather than the exact determination of the analytical response. This approach allows the students to analyze the response of a dynamic system without having to carry out the time-consuming evaluation of the coefficients in the partial fraction expansion. The chapter also presents the responses of first-, second-, and higher-order systems.

Chapters 3 and 4 discuss the general area of process dynamics. The practice of process control requires, first of all, a complete understanding of the dynamic and steady state behavior of the process; *this is the first objective of these chapters*. Development of dynamic mathematical process models—*modeling*—is an important tool for the analysis, design, and operation of control systems; *modeling is the second objective of Chapters 3 & 4*. A section has been added to Chapter 3 to show students that the development of dynamic, or unsteady-state, models is based on the steady-state balances that they have learned to develop since early in their education. The development of these models requires the use of all aspects of process engineering.

Chapter 5 provides a brief presentation of sensors, transmitters, and control valves, and a detailed presentation of feedback controllers. Chapter 5 should be studied together with Appendix C where practical operating principles of some common sensors, transmitters, and control valves are presented.

The design and tuning of feedback controllers are the subjects of Chapters 6 and 7. Chapter 6 presents the analysis of the stability of feedback control loops. We stress the direct substitution method for determining both the ultimate gain and ultimate period of the loop. In keeping with the spirit of Chapter 2, the examples and problems deal with the determination of the characteristics of the response of the closed loop, not with the exact analytical response of the loop. Computer simulation is used in some of the examples and problems.

Chapter 8 gives a brief introduction to the root locus technique. A detailed presentation on frequency response and its use to analyze and study control systems stability is also presented in the chapter.

Chapters 9 through 12 present in detail a number of control techniques that supplement and enhance feedback control. Chapter 9 presents a thorough treatment of cascade control, including stability considerations, tuning of the master controller, and considerations for its successful implementation. This technique is commonly used as part of other techniques.

Chapter 10 presents ratio control, override control, and selective control. These techniques are presented using several proven industrial examples. The chapter clearly shows that override and selective control are usually used as protective strategies to maintain process variables within limits that must be enforced to *ensure the safety of personnel and equipment, and product quality*. One example of ratio control is also presented to demonstrate safety concerns in combustion process All of the problems given at the end of the chapter are based on our experience as full-time practitioners, or part-time practitioners. Distributed control systems (DCSs) and the different ways to program, or configure, these systems are also presented.

Chapter 11 presents a complete treatment of feedforward control. Discussions of linear and nonlinear feedforward controllers are detailed. Many industrially proven examples are presented.

Multivariable control and loop interaction are the subjects of Chapter 12. The calculation of the relative gain matrix (RGM) and the design of decouplers are kept from the previous edition. Section 12-6 presents the need for *plantwide control* along with several guidelines as an attempt to reduce the task of designing control systems on a plantwide basis; an example is proved to show the application of the guidelines. Several other processes are provided in Appendix B to practice the use of the guidelines in designing control systems for entire processes.

Chapter 13 presents the fundamentals of dynamic simulation of process control systems using MATLAB. Two methods are presented, the simulation of the linear transfer functions of the process and controllers, and the more fundamental simulation of the process mathematical models in the time domain. In this second approach we emphasize the importance of setting the correct initial conditions so that the process response starts from steady state. Although many instructors of process control use the course as a means to train the students in the elegant and interesting algebraic manipulations afforded by Laplace transforms, others prefer to concentrate on the just as interesting aspects of control systems analysis and design while avoiding the algebra. The instructors in the second set may find it convenient to teach Chapter 13 first and then use simulation rather than algebra to study the response of the processes of Chapters 3 and 4, and of the control systems of Chapters 7, 10, 11, and 12. To facilitate this approach, the examples and problems in Chapter 13 deal with the simulation of the processes in examples and problems throughout the text. All of these simulations are available from the website for this text (www.wiley.com/college/smith). The problems with simulation are marked with an icon for easy identification.

As in the first two editions, Appendix A presents some symbols, labels, and other notations commonly used in instrumentation and control diagrams. We have adopted throughout the book the Instrumentation Society of America (ISA) symbols for conceptual diagrams which eliminate the need to differentiate between pneumatic, electronic, or computer implementation of the various control schemes. In keeping with this spirit, we express all instrument signals in percent of range rather than in mA or psig.

Appendix B continues with the topic of *plantwide control* that was discussed in Section 12-6. The appendix presents an example detailing the application of the guidelines presented in Section 12-6, followed by several case studies to practice design control systems for entire processes.

Appendix D, along with the book's website (www.wiley.com/college/smith), presents very practical processes to practice many of the techniques presented in Chapters 7, 9, 11, and 12—examples to tune feedback, cascade, and feedforward controllers and the multivariable techniques. Control Tuning Labs, developed using LavView™, are available on the book website.

Appendix E presents five cases that integrate the material presented in Chapters 5, 6, and 7. The cases are written to resemble a possible industrial situation that a junior engineer may face.

STUDENT RESOURCES

The following resources are available from the book website at www.wiley.com/college/smith. Visit the Student section of the website.

- **Simulations**—all simulations of examples in the text, as well as simulations using VisSim™.

- **Control Tuning Labs**—the authors have developed several processes (described in Appendix D) to practice the tuning of feedback controllers, cascade controllers, feedforward controllers, and the pairing and tuning of controllers in a multivariable control system. These Control Tuning Labs have been used by the authors for many years in their university courses, and in training of practicing engineers. The Control Tuning Labs were developed using LabVIEW™, a trademark of National Instruments™.

If you do not have LabVIEW, there is a free LabVIEW Player™ available on the website, so that students can view the Control Tuning Labs.

INSTRUCTOR RESOURCES

All simulations in the text, examples and problem solutions, as well as simulations using VisSim™ are available from the book website at www.wiley.com/college/smith.

Also, available only to instructors who adopt the text:

- **Solutions Manual**
- **Image Gallery of Text Figures**
- **Text Figures in PowerPoint format**
- **Control Tuning Labs**

These resources are password-protected. Visit the Instructor section of the book website to register for a password to access these materials.

ACKNOWLEDGMENTS

In formulating this edition we have been very fortunate to have received the help and encouragement of several wonderful individuals.

CAS is extremely grateful for the friendship of Dr. Marco E. Sanjuan of the Universidad del Norte in Barranquilla, Columbia. Dr. Sanjuan is the author of Appendix E and coauthor, with the late Dr. Daniel Palomares, of Appendix D. Dr. Sanjuan came to the University of South Florida as a doctoral student; during this time he was a role model for everyone, including his advisor who is now his friend and student.

CAS is also grateful to Renee Dockendorf for her help in developing the new design case studies, and to Dr. Russ Smith, of the Dow Chemical Company, for providing input in different chapters and appendices.

ABC will never forget the encouragement of his students, especially of Olufemi Adebiyi and Craig Plaisance. Craig developed the simple simulation of the proportional-derivative module shown in Section 13-4 while working on a class project. In so doing he taught his teacher.

We are also grateful to the following reviewers, who provided valuable feedback and encouragement as we developed the manuscript for this third edition:

D. John Griffith, Jr., McNeese State University
Ted Huddleston, University of South Alabama
Franklin G. King, North Carolina A&T State University
Michael V. Minnick, West Virginia University Institute of Technology
James M. Munro, South Dakota School of Mines and Technology
Laurent Simon, New Jersey Institute of Technology
Massoud Soroush, Drexel University
Marvin Stone, Oklahoma State University
Don H. Weinkauf, New Mexico Institute of Mining and Technology
Richard L. Zollars, Washington State University

In the preface to the first two editions, we wrote: "To serve as agents in the training and development of young minds is certainly a most rewarding profession." This is still our conviction, and we feel blessed to be able to do so. This aim continues to guide us in this edition.

Carlos A. Smith, Ph.D., P.E.   *Armando B. Corripio, Ph.D., P.E.*
Tampa, Florida   *Baton Rouge, Louisiana*
2005   *2005*

Contents

Chapter 1

Essentials

The purpose of this chapter is to present you with the need for automatic process control and to motivate you, the reader, to study it. Automatic process control is concerned with maintaining process variables, temperatures, pressures, flows, compositions, and the like at some desired operating value. As we shall see, processes are dynamic in nature. Changes are always occurring, and if actions are not taken in response, then the important process variables—those related to safety, product quality, and production rates—will not achieve design conditions.

This chapter also introduces two control systems, takes a look at some of their components, and defines some terms used in the field of process control. Finally, the background needed for the study of process control is discussed.

In writing this book, we have been constantly aware that to be successful, the engineer must be able to apply the principles learned. Consequently, the book covers the principles that underlie the successful practice of automatic process control. The book is full of actual cases drawn from our years of industrial experience as full-time practitioners or part-time consultants. We sincerely hope that you get excited about studying automatic process control. It is a very dynamic, challenging, and rewarding area of process engineering.

1-1 A PROCESS CONTROL SYSTEM

To illustrate process control, let us consider a heat exchanger in which a process stream is heated by condensing steam; the process is sketched in Fig. 1-1.1. The purpose of this unit is to heat the process fluid from some inlet temperature $T_i(t)$ up to a certain desired outlet temperature $T(t)$. The energy gained by the process fluid is provided by the latent heat of condensation of the steam.

In this process there are many variables that can change, causing the outlet temperature to deviate from its desired value. If this happens, some action must be taken to correct the deviation. The objective is to maintain the outlet process temperature at its desired value.

One way to accomplish this objective is by first measuring the temperature $T(t)$, comparing it to its desired value, and on the basis of this comparison, deciding what to do to correct any deviation. The steam valve can be manipulated to correct for the deviation. That is, if the temperature is above its desired value, the steam valve can be throttled back to cut the steam flow (energy) to the heat exchanger. If the temperature is below its desired value, the steam valve could be opened more to increase the steam flow to the exchanger. All of this can be done manually by the operator, and since the procedure is fairly straightforward, it should present no problem. However, there are several problems with this *manual control*. First, the job requires that the operator be frequently looking at the temperature to take corrective action whenever it deviates

Principles and Practice of Automatic Process Control/Third Edition, by C. A. Smith and A. B. Corripio
ISBN 0-471-66141-4   Copyright © 2006 John Wiley & Sons (Asia) Pte. Ltd.

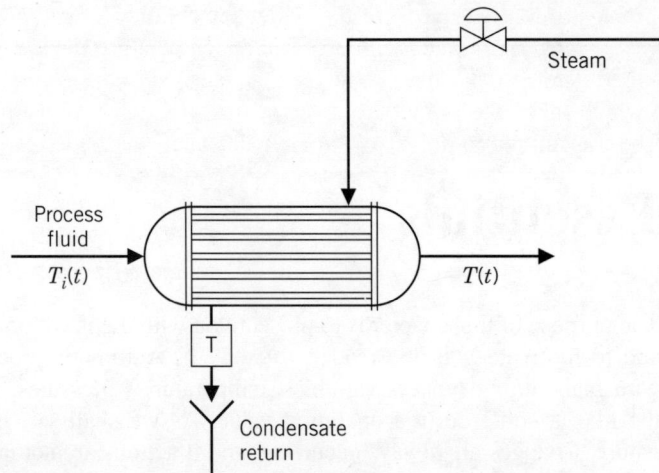

Figure 1-1.1 Heat exchanger.

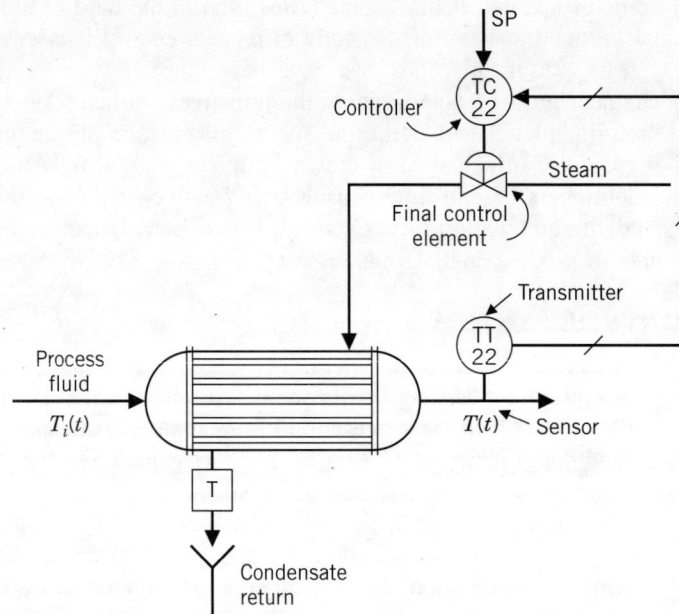

Figure 1-1.2 Heat exchanger control loop.

from its desired value. Second, different operators would make different decisions on how to move the steam valve, resulting in a not consistent operation. Third, since in most process plants there are hundreds of variables that must be maintained at some desired value, this correction procedure would require a large number of operators. Consequently, we would like to accomplish this control automatically. That is, we would like to have systems that control the variables without requiring intervention from the operator. This is what is meant by *automatic process control*.

To achieve automatic process control a *control system* must be designed and implemented. A possible control system and its basic components are shown in Fig. 1-1.2. (Appendix A presents the symbols and identifications for different devices.) The first thing to do is to measure the outlet temperature of the process stream. This is done by a *sensor* (thermocouple, resistance temperature device, filled system thermometer, thermistor, or the like). Usually this sensor is physically connected to a *transmitter*, which takes the output from the sensor and converts it to a signal strong enough to be transmitted to a *controller*. The controller then receives the signal, which is related to the temperature, and compares it with the desired value. Depending on the results of this comparison, the controller decides what to do to maintain the temperature at its desired value. On the basis of this decision, the controller sends a signal to the *final control element*, which in turn manipulates the steam flow. This type of control strategy is known as *feedback control*.

The preceding paragraph presented the three basic components of all control systems. They are

1. *Sensor-Transmitter.* Also often called the primary and secondary elements.
2. *Controller.* The "brain" of the control system.
3. *Final Control Element.* Often a control valve but not always. Other common final control elements are variable-speed pumps, conveyors, electric motors, and electric heaters.

The importance of these components is that they perform the three basic operations that *must* be present in *every* control system. These operations are

1. *Measurement (M).* Measuring the variable to be controlled is usually done by the combination of sensor and transmitter. In some systems the signal from the sensor can be fed directly to the controller and there is no need for the transmitter.
2. *Decision (D).* On the basis of the measurement, the controller decides what to do to maintain the variable at its desired value.
3. *Action (A).* As a result of the controller's decision, the system must then take an action. This is usually accomplished by the final control element.

These three operations, M, D, and A, are always present in every type of control system, and it is *imperative* that they be in a loop. That is, on the basis of the measurement a decision is made, and on the basis of this decision an action is taken. *The action taken must come back and affect the measurement; otherwise, it is a major flaw in the design, and control will not be achieved.* When the action taken does not affect the measurement, an open-loop condition exists and control will not be achieved. The decision-making in some systems is rather simple, whereas in others it is more complex; we will look at many of them in this book.

1-2 IMPORTANT TERMS AND OBJECTIVE OF AUTOMATIC PROCESS CONTROL

At this time it is necessary to define some terms used in the field of automatic process control. The *controlled variable* is the variable that must be maintained, or controlled, at some desired value. In our example of the heat exchanger, the process outlet temperature, $T(t)$, is the controlled variable. Sometimes the term *process variable* is also used to refer to the controlled variable. The *set point* (SP) is the desired value of the controlled variable. Thus, the job of a control system is to maintain the controlled variable at its set point. The *manipulated variable* is the variable used to maintain the controlled variable at its set point. In the example, the steam valve position is the manipulated variable. Finally, any variable that causes the controlled variable to deviate from set point is known as a *disturbance*, or *upset*. In most processes there are a number of different disturbances. In the heat exchanger shown in Fig. 1-1.2, possible disturbances are the inlet process temperature, $T_i(t)$, the process flow, $f(t)$, the energy content of the steam, ambient conditions, process fluid composition, fouling, and so on. It is important to understand that disturbances are always occurring in processes. Steady state is not the rule, and transient conditions are very common. It is because of these disturbances that automatic process control is needed. If there were no disturbances, design operating conditions would prevail and there would be no necessity of continuously "monitoring" the process.

The following additional terms are also important. *Manual control* is the condition in which the controller is disconnected from the process. That is, the controller is not making the decision of how to maintain the controlled variable at set point. It is up to the operator to manipulate the signal to the final control element to maintain the controlled variable at set point. *Closed-loop control* is the condition in which the controller is connected to the process, comparing the set point to the controlled variable, and determining and taking corrective action.

Now that we defined these terms, we can express the objective of an automatic process control meaningfully: *The objective of an automatic process control system is to adjust the manipulated variable to maintain the controlled variable at its set point in spite of disturbances.*

Control is important for many reasons. Those that follow are not the only ones, but we feel they are the most important. They are based on our industrial experience and we would like to pass them on. Control is important to

1. Prevent injury to plant personnel, protect the environment by preventing emissions and minimizing waste, and prevent damage to the process equipment. *SAFETY* must always be in everyone's mind; it is the single most important consideration.

2. Maintain product quality (composition, purity, color, etc.) on a continuous basis and with minimum cost.

3. Maintain plant production rate at minimum cost.

Thus, processes are automated to provide a safe environment and at the same time maintain desired product quality, high plant throughput, and reduced demand on human labor.

1-3 REGULATORY AND SERVO CONTROL

In some processes, the controlled variable deviates from set point because of disturbances. *Regulatory control* refers to systems designed to compensate for these disturbances. In some other instances, the most important disturbance is the set point itself. That is, the set point may be changed as a function of time (typical of this is a batch reactor where the temperature must follow a desired profile), and therefore the controlled variable must follow the set point. *Servo control* refers to control systems designed for this purpose.

Regulatory control is much more common than servo control in the process industries. However, the same basic approach is used in designing both. Thus the principles learned in this book apply to both cases.

1-4   TRANSMISSION SIGNALS, CONTROL SYSTEMS, AND OTHER TERMS

There are three principal types of signals in use in the process industries. The *pneumatic signal*, or air pressure, ranges normally between 3 and 15 psig. The usual representation for pneumatic signals in piping and instrument diagrams (P&IDs) is ——————//————————//——————————//——————. The *electrical signal* ranges normally between 4 and 20 mA. Less often 10 to 50 mA, 1 to 5 V, or 0 to 10 V are used. The usual representation for this signal in P&IDs is a series of dashed lines such as _ __ __ __ __ __ __ _. The third type of signal is the *digital, or discrete, signal* (zeros and ones); a common representation is O————O————O————. In this book we will show signals as ——/————/—— (as shown in Fig. 1-1.2), which is the representation proposed by the Instrumentation Society of America (ISA) when a control concept is shown without concern for specific hardware. The reader is encouraged to read Appendix A where different symbols and labels are presented. Most times we will refer to signals as percent, 0–100 %, as opposed to psig or mA. That is, 0–100 % is equivalent to 3–15 psig or 4–20 mA.

It will help in understanding control systems to realize that signals are used by devices—transmitters, controllers, final control elements, and the like—to communicate. That is, signals are used to *convey information*. The signal from the transmitter to the controller is used by the transmitter to inform the controller of the value of the controlled variable. The signal is not the measurement in engineering units but rather it is a mA, psig, volt, or any other signal that is proportional to the measurement. The relationship to the measurement depends on the calibration of the sensor-transmitter. The controller uses its output signal to tell the final control element what to do: how much to open if it is a valve, how fast to run if it is a variable-speed pump, and so on. Thus, every signal is related to some physical quantity that makes sense from an engineering point of view. The signal from the temperature transmitter in Fig. 1-1.2 is related to the outlet temperature, and the signal from the controller is related to the steam valve position.

It is often necessary to change one type of signal into another. This is done by a *transducer or converter*. For example, there may be a need to change from an electrical signal in mA to a pneumatic signal in psig. This is done by the use of a current (I) to pneumatic (P) transducer (I/P); this is shown graphically in Fig. 1-4.1. The input signal may be 4 to 20 mA and the output 3 to 15 psig. An analog-to-digital converter (A to D) changes from a mA, or volt, signal to a digital signal. There are many other types of transducers: digital-to-analog (D to A), pneumatic-to-current (P/I), voltage-to pneumatic (E/P), pneumatic-to voltage (P/E), and so on.

The term *analog* refers to the controller, or any other instrument, which is either pneumatic or electrical. Most controllers however, are *computer-based*, or *digital*. By computer-based we do not necessarily mean a mainframe computer but rather, anything starting from a microprocessor. In fact, most controllers are microprocessor-based. Chapter 5 presents different types of controllers, and defines some terms related to controllers and control systems.

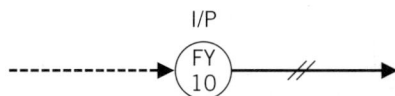

I/P

FY
10

Figure 1-4.1 I/P transducer.

1-5 CONTROL STRATEGIES

1-5.1 Feedback Control

The control scheme shown in Fig. 1-1.2 is referred to as *feedback control* and is also called a *feedback control loop.* One must understand the working principles of feedback control to recognize its advantages and disadvantages; the heat exchanger control loop shown in Fig. 1-1.2 is presented to foster this understanding.

If the inlet process temperature decreases, thus creating a disturbance, its effect must propagate through the heat exchanger before the outlet temperature decreases. Once this temperature changes, the signal from the transmitter to the controller also changes. It is then that the controller becomes aware that a deviation from set point has occurred and that it must compensate for the disturbance by manipulating the steam valve. The controller then signals the valve to increase its opening and thus increase the steam flow. Fig. 1-5.1 shows graphically the effect of the disturbance and the action of the controller.

It is instructive to note that at first the outlet temperature decreases, because of the decrease in inlet temperature, but it then increases even above set point and continues to oscillate around set point until the temperature finally stabilizes. This oscillatory response is typical of feedback control, and shows that it is essentially a trial-and-error operation. That is, when the controller "notices" that the outlet temperature has decreased below the set point, it signals the valve to open, but the opening is more than required. Therefore, the outlet temperature increases above the set point. Noticing this, the controller signals the valve to close again somewhat to bring the temperature back down. This trial and error continues until the temperature reaches and remains at set point.

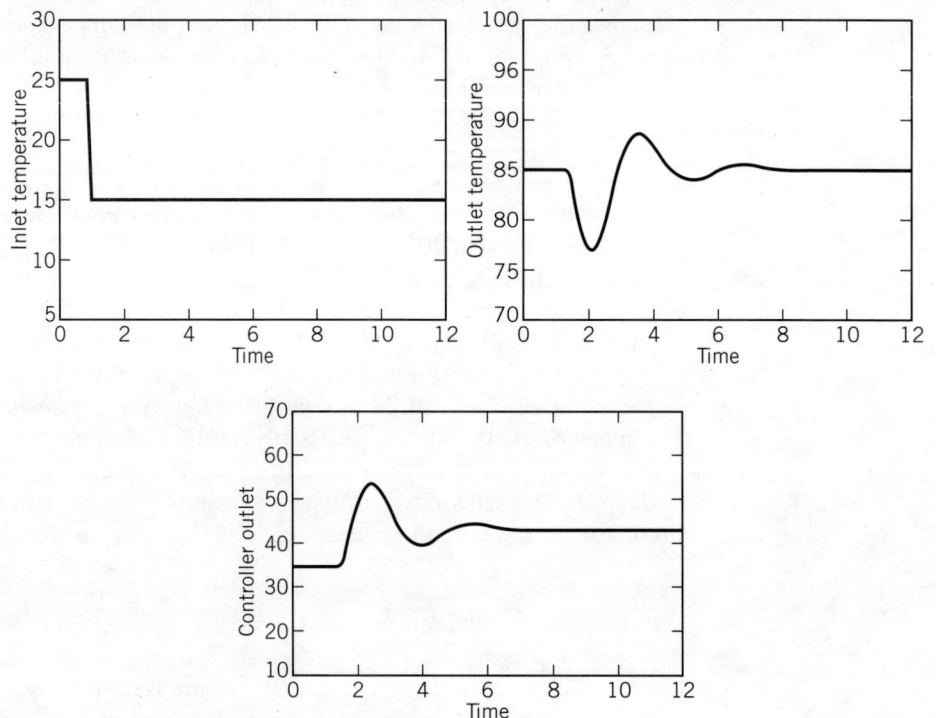

Figure 1-5.1 Response of feedback control.

The *advantage* of feedback control is that it is a very simple technique that compensates for all disturbances. Any disturbance affects the controlled variable, and once this variable deviates from set point, the controller changes its output in such a way as to return the temperature to set point. The feedback control loop does not know, nor does it care, which disturbance enters the process. It tries only to maintain the controlled variable at set point and in so doing compensates for all disturbances. The feedback controller works with minimum knowledge of the process. In fact, the only information it needs is in which direction to move. How much to move is usually adjusted by trial and error. The *disadvantage* of feedback control is that it can compensate for a disturbance only after the controlled variable has deviated from set point. That is, the disturbance must propagate through the entire process before the feedback control scheme can initiate action to compensate for it.

The job of the engineer is to design a control scheme that will maintain the controlled variable at its set point. Once this is done, the engineer must then adjust, or tune, the controller so that it minimizes the trial-and-error operation required to control. Most controllers have up to three terms (also known as parameters) used to tune them. To do a creditable job, the engineer must first know the characteristics of the process to be controlled. Once these characteristics are known, the control system can be designed, and the controller tuned. Process characteristics are explained in Chapters 3 and 4. Chapter 5 presents the meaning of the tuning parameters, and Chapter 7 presents different methods to tune controllers.

1-5.2 Feedforward Control

Feedback control is the most common control strategy in the process industries. Its simplicity accounts for its popularity. In some processes, however, feedback control may not provide the required control performance. For these processes other types of control may have to be designed. Chapters 9, 10, 11, and 12 present additional control strategies that have proved profitable. One such strategy is feedforward control. The objective of feedforward control is to measure the disturbances and compensate for them *before* the controlled variable deviates from set point. If applied correctly, the controlled variable deviation would be minimum.

A concrete example of feedforward control is the heat exchanger shown in Fig. 1-1.2. Suppose that "major" disturbances are the inlet temperature, $T_i(t)$, and the process flow, $f(t)$. To implement feedforward control, these two disturbances must first be measured and then a decision be made about how to manipulate the steam valve to compensate for them. Fig. 1-5.2 shows this control strategy. The feedforward controller makes the decision about how to manipulate the steam valve to maintain the controlled variable at set point, depending on the inlet temperature and process flow.

In Section 1-2 we learned that there are a number of different disturbances. The feedforward control system shown in Fig. 1-5.2 compensates for only two of them. If any of the others enter the process, this strategy will not compensate for it, and the result will be a permanent deviation of the controlled variable from set point. To avoid this deviation, some feedback compensation must be added to feedforward control; this is shown in Fig. 1-5.3. Feedforward control now compensates for the "major" disturbances, while feedback control compensates for all other disturbances. Chapter 11 presents the development of the feedforward controller. Actual industrial cases are used to discuss this important strategy in detail.

It is important to notice that the three basic operations, M, D, A, are still present in this more "advanced" control strategy. Measurement is performed by the sensors and

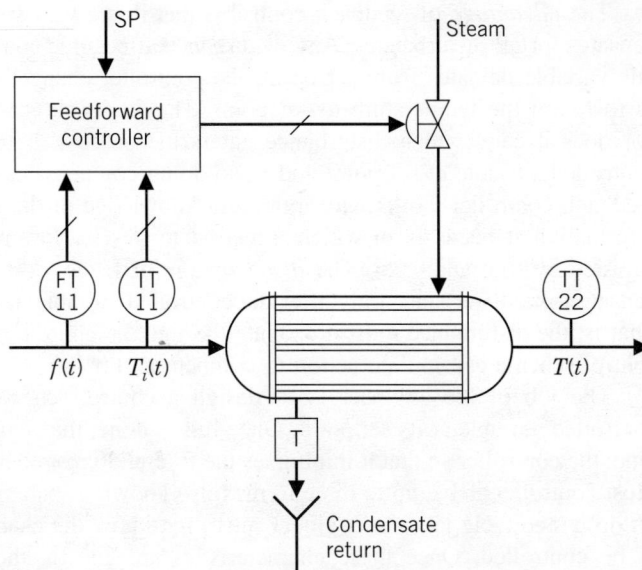

Figure 1-5.2 Heat exchanger feedforward control system.

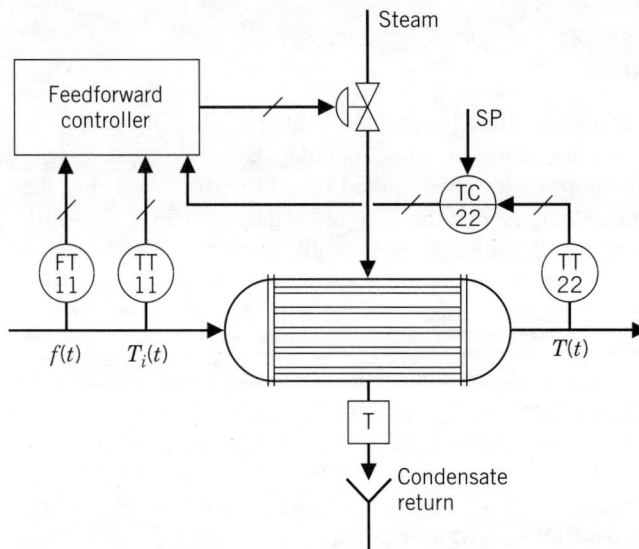

Figure 1-5.3 Heat exchanger feedforward control with feedback compensation.

transmitters. Decision is made by both the feedforward and the feedback controllers. Action is taken by the steam valve.

The advanced control strategies are usually more costly than feedback control in hardware, computing power, and manpower necessary to design, implement, and maintain them. Therefore, the expenses must be justified (safety and/or economics) before they can be implemented. The best procedure is to first design and implement a simple control strategy, keeping in mind that if it does not prove satisfactory, then a more advanced strategy may be justifiable. It is important, however, to recognize that these advanced strategies still require some feedback compensation.

1-6   BACKGROUND NEEDED FOR PROCESS CONTROL

To be successful in the practice of automatic process control, *the engineer must first understand the principles of process engineering*. Therefore, this book assumes that the reader is familiar with the basic principles of thermodynamics, fluid flow, heat transfer, separation processes, reaction processes, and the like.

For the study of process control it is also fundamental to understand how processes behave dynamically. Unfortunately, the overwhelming majority of the Chemical Engineering curriculum up to this course has emphasized the steady-state assumption. Thus, the steady-state behavior of processes is fairly well understood by the students, but the dynamic behavior is not understood, or barely. Although we analyze and design processes assuming steady state, processes hardly run at this condition.

We would like for processes to run at the designed steady state, however, if left to them, processes would not run at the desired steady state. We must force the processes to accomplish this state. Consider again the heat exchanger presented in Section 1-1. This heat exchanger was sized using the steady-state heat transfer equations we learned in the "heat transfer course." If the inlet temperature, the inlet flow, the physical properties of the fluid, the steam conditions, and the cleanliness of the exchanger were to continue constant at the design condition, then the calculated steam flow is exactly what we need to maintain the outlet temperature at the desired value. However, as we discussed in Section 1-2, often the inlet temperature, or the inlet flow, or the physical properties, or the steam conditions, or the ambient conditions, or the fouling may change and, therefore, we must change the steam flow to *force* the process back to the desired steady state (outlet temperature).

How much new steam is needed, and how fast we must change the steam flow to reach the desired temperature depends on the process. Steady-state balances tell us how much steam is needed. However, unsteady-state balances are the ones needed to decide how fast to change the steam flow. Consequently, it is necessary to develop the set of equations that describe the dynamic (time-dependent or transient) behavior of the processes. This is called *modeling*. To do this, the knowledge of the basic principles mentioned in the previous paragraphs, and of mathematics through differential equations is needed. In process control Laplace transforms are used heavily. This greatly simplifies the dynamic analysis of processes and their control systems. Chapter 2 of this book is devoted to the development and usage of the Laplace transforms along with a review of complex number algebra.

Another important "tool" for the study and practice of process control is computer simulation. Many of the equations developed to describe processes are nonlinear in nature and, consequently, the most exact way to solve them is by numerical methods; this means computer solution. The computer solution of process models is called *simulation*. Chapters 3 and 4 present an introduction to the modeling of some processes, and Chapter 13 extends the modeling, and presents simulation.

1-7   SUMMARY

In this chapter we discussed the need for automatic process control. *Industrial processes are not static but rather very dynamic; they are continuously changing because of many types of disturbances.* It is principally because of this dynamic nature that control systems are needed to continuously and automatically watch over the variables that must be controlled.

The working principles of a control system can be summarized with the three letters M, D, and A. M refers to the measurement of process variables. D refers to the decision to be made on the basis of the measurements of the process variables. Finally, A refers to the action to be taken on the basis of that decision.

The basic components of a process control system

were also presented: sensor-transmitter, controller, and final control element. The most common types of signals—pneumatic, electrical, and digital—were introduced along with the purpose of transducers.

Two control strategies were presented: feedback and feedforward control. The advantages and disadvantages of both strategies were briefly discussed. Chapters 6 and 7 present the design and analysis of feedback control loops.

PROBLEMS

1-1. For the following automatic control systems commonly encountered in daily life, identify the devices that perform the measurement (M), decision (D), and action (A) functions, and classify the action function as "On/Off" or "Regulating." Draw also a process and instrumentation diagram using the standard ISA symbols given in Appendix A and determine whether the control is feedback or feedforward.

(a)   House air-conditioning/heating

(b)   Cooking oven

(c)   Toaster

(d)   Automatic sprinkler system for fires

(e)   Automobile cruise speed control

(f)   Refrigerator

1-2. *Instrumentation Diagram: Automatic Shower Temperature Control.* Sketch the process and instrumentation diagram for an automatic control system to control the temperature of the water from a common shower, that is, a system that will automatically do what you do when you adjust the temperature of the water when you take a shower. Use the standard ISA instrumentation symbols given in Appendix A. Identify the measurement (M), decision (D), and action (A) devices of your control system.

Chapter 2

Control Systems Analysis: Mathematical Tools

This chapter presents two mathematical tools that are particularly useful for analyzing process dynamics and designing automatic control systems, Laplace transforms and linearization. Combined, these two techniques allow us to gain insight into the dynamic responses of a wide variety of processes and instruments. In contrast, the technique of computer simulation provides us with a more accurate and detailed analysis of the dynamic behavior of specific systems, but seldom allows us to generalize our findings to other processes.

Laplace transforms are used to convert the differential equations that represent the dynamic behavior of process output variables into algebraic equations. It is then possible to isolate in the resulting algebraic equations what is characteristic of the process, the *transfer function*, from what is characteristic of the input forcing functions. Since the differential equations that represent most processes are nonlinear, linearization is required to approximate nonlinear differential equations with linear ones that can then be treated by the method of Laplace transforms.

The material of this chapter is not just a simple review of Laplace transforms, but a presentation of the tool in the way it is used to analyze process dynamics and design control systems. Also presented are the responses of some common process transfer functions to some common input functions. These responses are related to the parameters of the process transfer functions so that the important characteristics of the responses can be inferred directly from the transfer functions without having to reinvert them each time. We firmly believe that knowledge of Laplace transforms is essential for understanding the fundamentals of process dynamics and control systems design.

2-1 THE LAPLACE TRANSFORM

This section reviews the definition of the Laplace transform and its properties.

2-1.1 Definition of the Laplace Transform

In the analysis of process dynamics, the process variables and control signals are functions of time, t. The Laplace transform of a function of time, $f(t)$, is defined by the following formula:

$$F(s) = \mathscr{L}[f(t)] = \int_0^\infty f(t)e^{-st}\,dt \qquad (2\text{-}1.1)$$

Principles and Practice of Automatic Process Control/Third Edition, by C. A. Smith and A. B. Corripio
ISBN 0-471-66141-4   Copyright © 2006 John Wiley & Sons (Asia) Pte. Ltd.

where

$F(s)$ = the Laplace transform of $f(t)$

s = the Laplace transform variable, time^{-1}

The Laplace transform changes the function of time, $f(t)$, into a function in the Laplace transform variable, $F(s)$. The limits of integration, show that the Laplace transform contains information on the function $f(t)$ for positive time only. This is perfectly acceptable because in process control, as in life, nothing can be done about the past (negative time); control action can only affect the process in the future. The following example uses the definition of the Laplace transform to develop the transforms of a few common forcing functions.

EXAMPLE 2-1.1

The four signals shown in Fig. 2-1.1 are commonly applied as inputs to processes and instruments to study their dynamic responses. We now use the definition of the Laplace transform to derive their transforms.

(A) UNIT STEP FUNCTION

This is a sudden change of unit magnitude as sketched in Fig. 2-1.1a. Its algebraic representation is

$$u(t) = \begin{cases} 0 & t < 0 \\ 1 & t \geq 0 \end{cases}$$

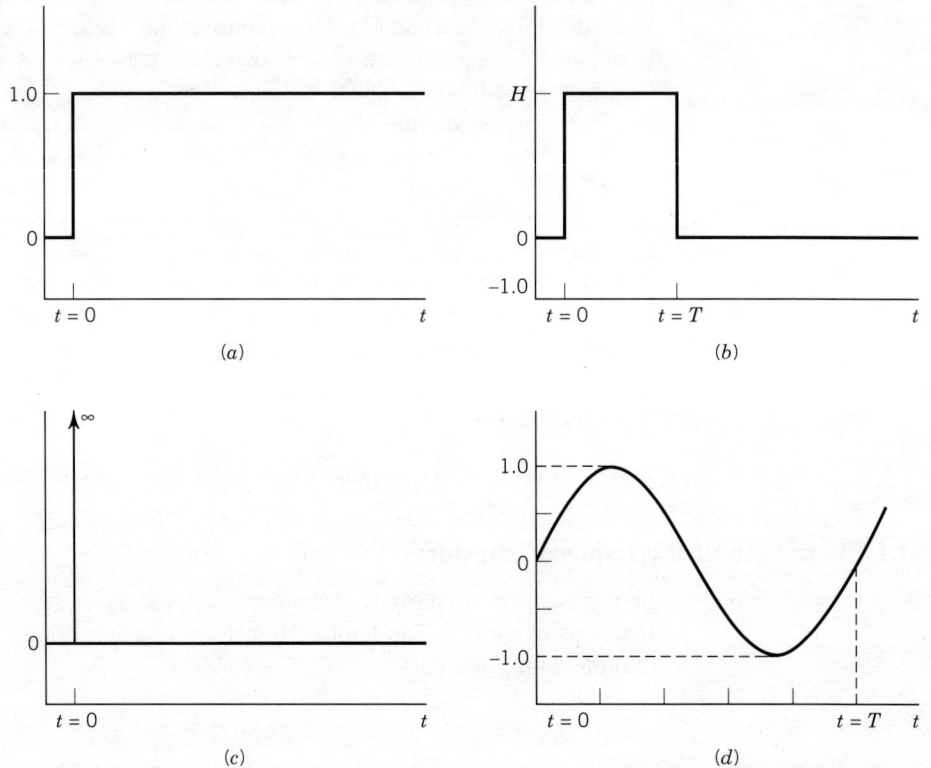

Figure 2-1.1 Common input signals for the study of control system response. (a) Unit step function, $u(t)$. (b) Pulse. (c) Unit impulse function, $\delta(t)$. (d) Sine wave, $\sin \omega t\, (\omega = 2\pi/T)$.

Substituting into Eq. 2-1.1:

$$\mathscr{L}[u(t)] = \int_0^\infty u(t)e^{-st}\,dt = -\frac{1}{s}e^{-st}\Big|_0^\infty = -\frac{1}{s}(0-1) = \frac{1}{s}$$

(B) A PULSE OF MAGNITUDE H AND DURATION T

The pulse sketched in Fig. 2-1.1b is represented by

$$f(t) = \begin{cases} 0 & t < 0, t \geq T \\ H & 0 \leq t \leq T \end{cases}$$

Substituting into Eq. 2-1.1 yields

$$\mathscr{L}[f(t)] = \int_0^\infty f(t)e^{-st}\,dt = \int_0^T He^{-st}\,dt = -\frac{H}{s}e^{-st}\Big|_0^T = -\frac{H}{s}(e^{-sT}-1) = H\frac{1-e^{-sT}}{s}$$

(C) A UNIT IMPULSE FUNCTION

This function, also known as the *Dirac delta function* and represented by $\delta(t)$, is sketched in Fig. 2-1.1c. It is an ideal pulse with zero duration and unit area. All of its area is concentrated at time zero. Since the function is zero at all times except at zero, and the term e^{-st} in Eq. 2-1.1 is equal to unity at $t = 0$, the Laplace transform is

$$\mathscr{L}[\delta(t)] = \int_0^\infty \delta(t)e^{-st}\,dt = 1$$

Notice that the result of the integration, 1, is the area of the impulse. The same result can be obtained by substituting $H = 1/T$ in the result of part (b), so that $HT = 1$, and then taking limits as T goes to zero.

(D) A SINE WAVE OF UNITY AMPLITUDE AND FREQUENCY ω

The sine wave is sketched in Fig. 2-1.1d and is represented in exponential form by

$$\sin \omega t = \frac{e^{i\omega t} - e^{-i\omega t}}{2i}$$

where $i = \sqrt{-1}$ is the unit of imaginary numbers. Substituting into Eq. 2-1.1 yields

$$\mathscr{L}[\sin \omega t] = \int_0^\infty \frac{e^{i\omega t} - e^{-i\omega t}}{2i}e^{-st}\,dt$$

$$= \frac{1}{2i}\int_0^\infty \left[e^{-(s-i\omega)t} - e^{-(s+i\omega)t}\right]dt$$

$$= \frac{1}{2i}\left[-\frac{e^{-(s-i\omega)t}}{s-i\omega} + \frac{e^{-(s+i\omega)t}}{s+i\omega}\right]_0^\infty$$

$$= \frac{1}{2i}\left[-\frac{0-1}{s-i\omega} + \frac{0-1}{s+i\omega}\right]$$

$$= \frac{1}{2i}\frac{2i\omega}{s^2+\omega^2}$$

$$= \frac{\omega}{s^2+\omega^2}$$

The preceding example illustrates some algebraic manipulations required to derive the Laplace transform of various functions using its definition. Table 2-1.1 contains a short list of the Laplace transforms of some common functions.

Table 2-1.1   Laplace transforms of common functions

$f(t)$	$F(s) = \mathscr{L}[f(t)]$
$\delta(t)$	1
$u(t)$	$\dfrac{1}{s}$
t	$\dfrac{1}{s^2}$
t^n	$\dfrac{n!}{s^{n+1}}$
e^{-at}	$\dfrac{1}{s+a}$
te^{-at}	$\dfrac{1}{(s+a)^2}$
$t^n e^{-at}$	$\dfrac{n!}{(s+a)^{n+1}}$
$\sin \omega t$	$\dfrac{\omega}{s^2+\omega^2}$
$\cos \omega t$	$\dfrac{s}{s^2+\omega^2}$
$e^{-at}\sin \omega t$	$\dfrac{\omega}{(s+a)^2+\omega^2}$
$e^{-at}\cos \omega t$	$\dfrac{s+a}{(s+a)^2+\omega^2}$

2-1.2   Properties of the Laplace Transform

This section presents the properties of Laplace transforms in order of their usefulness in analyzing process dynamics and designing control systems. Linearity and the real differentiation and integration theorems are essential for transforming differential equations into algebraic equations. The final value theorem is useful for predicting the final steady-state value of a time function from its Laplace transform, and the real translation theorem is useful for dealing with functions delayed in time. Other properties are useful for deriving the transforms of complex functions from those of simpler functions such as those listed in Table 2-1.1.

Linearity

It is very important to realize that the Laplace transform is a linear operation. This means that, if a is a constant,

$$\mathscr{L}[af(t)] = a\mathscr{L}[f(t)] = aF(s) \tag{2-1.2}$$

The distributive property of addition also follows from the linearity property:

$$\mathscr{L}[af(t) + bg(t)] = a\mathscr{L}[f(t)] + b\mathscr{L}[g(t)] = aF(s) + bG(s) \tag{2-1.3}$$

where a and b are constants. You can easily derive both formulas by application of Eq. 2-1.1, the definition of the Laplace transform.

Real Differentiation Theorem

This theorem, which establishes a relationship between the Laplace transform of a function and that of its derivatives, is most important in transforming differential equations into algebraic equations. It states that

$$\mathscr{L}\left[\frac{df(t)}{dt}\right] = sF(s) - f(0) \tag{2-1.4}$$

Proof. From the definition of the Laplace transform, Eq. 2-1.1,

$$\mathscr{L}\left[\frac{df(t)}{dt}\right] = \int_0^\infty \frac{df(t)}{dt}e^{-st}\,dt$$

Integrate by parts:

$$u(t) = e^{-st} \quad dv = \frac{df(t)}{dt}\,dt$$

$$du = -se^{-st}\,dt \quad v = f(t)$$

$$\mathscr{L}\left[\frac{df(t)}{dt}\right] = [f(t)e^{-st}]_0^\infty - \int_0^\infty f(t)(-se^{-st}\,dt)$$

$$= [0 - f(0)] + s\int_0^\infty f(t)e^{-st}\,dt$$

$$= sF(s) - f(0)$$

The extension to higher derivatives is straightforward.

$$\mathscr{L}\left[\frac{d^2f(t)}{dt^2}\right] = \mathscr{L}\left[\frac{d}{dt}\left(\frac{df(t)}{dt}\right)\right]$$

$$= s\mathscr{L}\left[\frac{df(t)}{dt}\right] - \frac{df}{dt}\bigg|_{t=0}$$

$$= s[sF(s) - f(0)] - \frac{df}{dt}\bigg|_{t=0}$$

$$= s^2F(s) - sf(0) - \frac{df}{dt}\bigg|_{t=0}$$

In general

$$\mathscr{L}\left[\frac{d^nf(t)}{dt^n}\right] = s^nF(s) - s^{n-1}f(0) - \cdots - \frac{d^{n-1}f}{dt^{n-1}}\bigg|_{t=0} \tag{2-1.5}$$

In process control it is normally assumed that the initial conditions are at steady state (time derivatives are zero), and that the variables are deviations from initial conditions (initial value is zero). For this very important case the preceding expression reduces to

$$\mathscr{L}\left[\frac{d^nf(t)}{dt^n}\right] = s^nF(s) \tag{2-1.6}$$

This means that, for the case of zero initial conditions at steady state, the Laplace transform of the derivative of a function is obtained by simply replacing variable s for the "d/dt" operator, and $F(s)$ for $f(t)$.

Real Integration Theorem

This theorem establishes the relationship between the Laplace transform of a function and that of its integral. It states that

$$\mathscr{L}\left[\int_0^t f(t)\,dt\right] = \frac{1}{s}F(s) \qquad (2\text{-}1.7)$$

The proof of this theorem is carried out integrating the definition of the Laplace transform by parts, similar to the previous one, and is left as an exercise. The Laplace transform of the nth integral of a function is the transform of the function divided by s^n.

Real Translation Theorem

This theorem deals with the translation of a function in the time axis, as shown in Fig. 2-1.2. The translated function is the original function delayed in time. As we shall see in Chapter 3, time delays are caused by transportation lag, a phenomenon also known as *dead time*. The theorem states that

$$\mathscr{L}[f(t - t_0)] = e^{-st_0}F(s) \qquad (2\text{-}1.8)$$

Because the Laplace transform does not contain information about the original function for negative time, the delayed function must be zero for all times less than the time delay (see Fig. 2-1.2). This condition is satisfied if the process variables are expressed as deviations from initial steady-state conditions.

Proof. From the definition of the Laplace transform, Eq. 2-1.1,

$$\mathscr{L}[f(t - t_0)] = \int_0^\infty f(t - t_0)e^{-st_0}\,dt$$

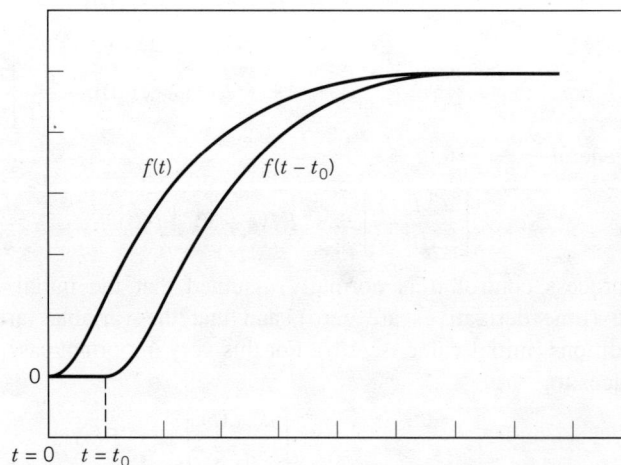

Figure 2-1.2 Function delayed in time is zero for all times less than the time delay t_0.

Let $\tau = t - t_0$ (or $t = t_0 + \tau$) and substitute.

$$\mathscr{L}[f(t - t_0)] = \int_{\tau=-t_0}^{\infty} f(\tau)e^{-s(t_0+\tau)}\, d(t_0 + \tau)$$

$$= \int_{\tau=0}^{\infty} f(\tau)e^{-st_0}e^{-s\tau}\, d\tau$$

$$= e^{-st_0}\int_{0}^{\infty} f(\tau)e^{-s\tau}\, d\tau$$

$$= e^{-st_0}F(s) \quad \text{q.e.d.}$$

Notice that in this proof we made use of the fact that $f(\tau) = 0$ for $\tau < 0$ ($t < t_0$).

Final Value Theorem

This theorem allows us to figure out the final or steady-state value of a function from its transform. It is also useful in checking the validity of derived transforms. If the limit of $f(t)$ as $t \to \infty$ exists, it can be found from its Laplace transform as follows:

$$\lim_{t\to\infty} f(t) = \lim_{s\to 0} sF(s) \tag{2-1.9}$$

The proof of this theorem adds little to its understanding.

The last three properties of the Laplace transform, to be presented next without proof, are not used as often in process dynamic analysis as the ones presented above.

Complex Differentiation Theorem

This theorem is useful for evaluating the transforms of functions involving powers of the independent variable, t. It states that

$$\mathscr{L}[tf(t)] = -\frac{d}{ds}F(s) \tag{2-1.10}$$

Complex Translation Theorem

This theorem is useful for evaluating transforms of functions involving exponential functions of time. It states that

$$\mathscr{L}[e^{at}f(t)] = F(s - a) \tag{2-1.11}$$

Initial Value Theorem

This theorem allows the calculation of the initial value of a function from its transform. It would provide another check of the validity of derived transforms were it not for the fact that, in process dynamic analysis, the initial conditions of the variables are usually zero. The theorem states that

$$\lim_{t\to 0} f(t) = \lim_{s\to\infty} sF(s) \tag{2-1.12}$$

The following examples illustrate the use of the properties of Laplace transforms we have just presented.

EXAMPLE 2-1.2   Derive the Laplace transform of the following differential equation

$$9\frac{d^2y(t)}{dt^2} + 6\frac{dy(t)}{dt} + y(t) = 2x(t)$$

with initial conditions of zero at steady state, that is, $y(0) = 0$, and $dy/dt(0) = 0$.

SOLUTION   By application of the linearity property, Eq. 2-1.3, take the Laplace transform of each term.

$$9\mathscr{L}\left[\frac{d^2y(t)}{dt^2}\right] + 6\mathscr{L}\left[\frac{dy(t)}{dt}\right] + \mathscr{L}[y(t)] = 2\mathscr{L}[x(t)]$$

Then apply the real differentiation theorem, Eq. 2-1.6,

$$9s^2Y(s) + 6sY(s) + Y(s) = 2X(s)$$

Finally, solve for $Y(s)$.

$$Y(s) = \frac{2}{9s^2 + 6s + 1}X(s)$$

The preceding example shows how the Laplace transform converts the original differential equation into an algebraic equation which can then be rearranged to solve for the dependent variable $Y(s)$. Herein lies the great usefulness of the Laplace transform, since algebraic equations are a lot easier to manipulate than differential equations.

EXAMPLE 2-1.3   Obtain the Laplace transform of the following function:

$$c(t) = u(t - 3)[1 - e^{-(t-3)/4}]$$

Note: The term $u(t - 3)$ in this expression shows that the function is zero for $t < 3$. We recall, from Example 2-1.1(a), that $u(t - 3)$ is a change from zero to one at $t = 3$, which means that the expression in brackets is multiplied by zero until $t = 3$, and multiplied by unity after that. Thus, the presence of the unit step function does not alter the rest of the function for $t > 3$.

SOLUTION   Let

$$c(t) = f(t - 3) = u(t - 3)[1 - e^{-(t-3)/4}]$$

then

$$f(t) = u(t)[1 - e^{-t/4}] = u(t) - u(t)e^{-t/4}$$

Apply Eq. 2-1.3, the linearity property, and use entries from Table 2-1.1, with $a = 1/4$.

$$F(s) = \frac{1}{s} - \frac{1}{s + \frac{1}{4}} = \frac{1}{s(4s + 1)}$$

Next, apply the real translation theorem, Eq. 2-1.8,

$$C(s) = \mathscr{L}[f(t - 3)] = e^{-3s}F(s) = \frac{e^{-3s}}{s(4s + 1)}$$

We can check the validity of this answer by using the final value theorem, Eq. 2-1.9.

$$\lim_{t \to \infty} c(t) = \lim_{t \to \infty} u(t - 3)[1 - e^{-(t-3)/4}] = 1$$

$$\lim_{s \to 0} sC(s) = \lim_{s \to 0} s\frac{e^{-3s}}{s(4s + 1)} = 1 \qquad \text{Check!}$$

2-2   SOLUTION OF DIFFERENTIAL EQUATIONS USING THE LAPLACE TRANSFORM

This section presents the use of the Laplace transform to solve the differential equations that represent the dynamics of processes and their control systems. Because our objective is to find out how the output signals respond to input forcing functions, we will always assume that the initial conditions are at steady state (zero time derivatives). We will also define all variables as deviations from their initial values. This forces the initial values of the deviation variables to be also zero.

2-2.1   Laplace Transform Solution Procedure

The procedure for solving a differential equation by Laplace transforms consists of three steps:

1. Transform the differential equation into an algebraic equation in the Laplace transform variable s.

2. Solve for the transform of the output (or dependent) variable.

3. Invert the transform to obtain the response of the output variable with time, t.

Consider the following second-order differential equation:

$$a_2 \frac{d^2 y(t)}{dt^2} + a_1 \frac{dy(t)}{dt} + a_0 y(t) = bx(t) \qquad \textbf{(2-2.1)}$$

The problem of solving this equation can be stated as follows: Given the constant coefficients a_0, a_1, a_2, and b, the initial conditions, $y(0)$ and $dy/dt|_{t=0}$, and the function $x(t)$, find the function $y(t)$ that satisfies the differential equation.

We call the function $x(t)$ the "forcing function" or input variable, and $y(t)$ the "output" or dependent variable. In process control systems, a differential equation like Eq. 2-1.1 usually represents how a particular process or instrument relates its output signal, $y(t)$, to its input signal, $x(t)$. Our approach is that of British inventor James Watt (1736–1819) who considered process variables as signals and processes as signal processors.

The first step is to take the Laplace transform of Eq. 2-2.1. We do this by applying the linearity property of Laplace transforms, Eq. 2-1.3, which allows us to take the Laplace transform of each term separately:

$$a_2 \mathscr{L}\left[\frac{d^2 y(t)}{dt^2} \right] + a_1 \mathscr{L}\left[\frac{dy(t)}{dt} \right] + a_0 \mathscr{L}[y(t)] = b \mathscr{L}[x(t)] \qquad \textbf{(2-2.2)}$$

Assuming for the moment the initial conditions are not zero, the indicated Laplace transforms are obtained using the real differentiation theorem, Eq. 2-1.5.

$$\mathscr{L}\left[\frac{d^2 y(t)}{dt^2} \right] = s^2 Y(s) - sy(0) - \frac{dy}{dt}\bigg|_{t=0}$$

$$\mathscr{L}\left[\frac{dy(t)}{dt} \right] = sY(s) - y(0)$$

Next we substitute these terms into Eq. 2-2.2 and rearrange it to obtain

$$(a_2 s^2 + a_1 s + a_0) Y(s) - (a_2 s + a_1) y(0) - a_2 \frac{dy}{dt}\bigg|_{t=0} = bX(s)$$

The second step is to manipulate this algebraic equation to solve for the transform of the output variable, $Y(s)$.

$$Y(s) = \frac{bX(s) + (a_2 s + a_1)y(0) + a_2 \left.\dfrac{dy}{dt}\right|_{t=0}}{a_2 s^2 + a_1 s + a_0} \tag{2-2.3}$$

This equation shows the effect of the input variable, $X(s)$, and the initial conditions, on the output variable. Since our objective is to study how the output variable responds to the input variable, the presence of the initial conditions complicates our analysis. To avoid this unnecessary complication, we assume that the initial conditions are at steady state, $\left.dy/dt\right|_{t=0} = 0$, and define the output variable as the *deviation* from its initial value, thus forcing $y(0) = 0$. We will show in the next section how this can be done without loss of generality. With zero initial conditions, the equation is reduced to

$$Y(s) = \left[\frac{b}{a_2 s^2 + a_1 s + a_0} \right] X(s) \tag{2-2.4}$$

The form of Eq. 2-2.4 allows us to break the transform of the output variable into the product of two terms, the term in brackets, known as the *transfer function*, and the transform of the input variable, $X(s)$. The transfer function and its parameters characterize the process or device and determine how the output variable responds to the input variable. The concept of transfer function is described in more detail in Chapter 3.

The third and final step is to invert the transform of the output to obtain the time function $y(t)$ which is the response of the output. Inversion is the opposite operation to the taking of the Laplace transform. Before we can invert, we must select a specific input function for $x(t)$. A common function, because of its simplicity, is the unit step function, $u(t)$, which was introduced in Example 2-1.1. From that example, or Table 2-1.1, for $x(t) = u(t)$, $X(s) = 1/s$. Substitute into Eq. 2-2.4 and invert to obtain

$$y(t) = \mathscr{L}^{-1} \left[\frac{b}{a_2 s^2 + a_1 s + a_0} \frac{1}{s} \right] \tag{2-2.5}$$

where the symbol $\mathscr{L}^{-1}$ stands for the inverse Laplace transform. The response to a step input is called the *step response* for short.

The inversion could easily be carried out if we could find the expression within the brackets in Table 2-1.1 or in a more extensive table of Laplace transforms. Obviously, we will not be able to find complex expressions in such a table. The mathematical technique of partial fractions expansion, to be introduced next, is designed to expand the transform of the output into a sum of simpler terms. We can then invert these simpler terms separately by matching entries in Table 2-1.1.

2-2.2   Inversion by Partial Fractions Expansion

British physicist Oliver Heaviside (1850–1925) introduced the mathematical technique of partial fractions expansion as part of his revolutionary "operational calculus." The first step in expanding the transform, Eq. 2-2.5, into a sum of fractions, is to factor its denominator, as follows:

$$(a_2 s^2 + a_1 s + a_0)s = a_2 (s - r_1)(s - r_2)s \tag{2-2.6}$$

where r_1 and r_2 are the roots of the quadratic term, that is, the values of s that satisfy the equation

$$a_2 s^2 + a_1 s + a_0 = 0$$

For a quadratic or second-degree polynomial, the roots can be calculated by the standard quadratic formula:

$$r_{1,2} = \frac{-a_1 \pm \sqrt{a_1^2 - 4a_0 a_2}}{2a_2} \tag{2-2.7}$$

For higher-degree polynomials the reader is referred to any numerical methods text for a root-finding procedure. Most electronic calculators are now able to find the roots of third- and higher-degree polynomials. Computer programs such as Mathcad[1] and MATLAB[2] provide functions for finding the roots of polynomials of any degree.

Once the denominator is factored into first-degree terms, the transform is expanded into partial fractions as follows:

$$Y(s) = \frac{A_1}{s - r_1} + \frac{A_2}{s - r_2} + \frac{A_3}{s} \tag{2-2.8}$$

provided that the roots, r_1, r_2, and $r_3 = 0$, are not equal to each other. For this case of unrepeated roots, the constant coefficients are found by the formula

$$A_k = \lim_{s \to r_k} (s - r_k) Y(s) \tag{2-2.9}$$

We can now carry out the inversion of Eq. 2-2.8 by matching each term to entries in Table 2-1.1; in this case the first two terms match the exponential function with $a = -r_k$, while the third term matches the unit step function. The resulting inverse function is

$$y(t) = A_1 e^{r_1 t} + A_2 e^{r_2 t} + A_3 u(t)$$

Repeated Roots

For the case of *repeated roots*, say $r_1 = r_2$, the expansion is carried out as follows:

$$Y(s) = \frac{A_1}{(s - r_1)^2} + \frac{A_2}{s - r_1} + \frac{A_3}{s} \tag{2-2.10}$$

Coefficient A_3 is calculated as before, but coefficients A_1 and A_2 must be calculated by the following formulas:

$$A_1 = \lim_{s \to r_1} (s - r_1)^2 Y(s)$$

$$A_2 = \lim_{s \to r_1} \frac{1}{1!} \frac{d}{ds} [(s - r_1)^2 Y(s)] \tag{2-2.11}$$

Again, we carry out the inversion of Eq. 2-2.10 by matching terms in Table 2-1.1. The first term matches the sixth term in the table with $a = -r_1$, to give the inverse.

$$y(t) = A_1 t e^{r_1 t} + A_2 e^{r_1 t} + A_3 u(t)$$

---

[1] *Mathcad User's Guide*, by MathSoft, Inc., 201 Broadway, Cambridge, MA, 02139, 2001.

[2] *MATLAB User's Guide*, The MathWorks, Inc., 24 Prime Park Way, Natick, MA, 01760, 2001.

In general, if root r_1 is repeated m times, the expansion is carried out as follows:

$$Y(s) = \frac{A_1}{(s - r_1)^m} + \frac{A_2}{(s - r_1)^{m-1}} + \cdots + \frac{A_m}{s - r_1} + \cdots \qquad (2\text{-}2.12)$$

The coefficients are calculated by

$$A_1 = \lim_{s \to r_1} (s - r_1)^m Y(s)$$

$$A_k = \lim_{s \to r_1} \frac{1}{(k-1)!} \frac{d^{k-1}}{ds^{k-1}} [(s - r_1)^m Y(s)] \qquad (2\text{-}2.13)$$

for $k = 2, \ldots, m$. The inverse function is then:

$$y(t) = \left[\frac{A_1 t^{m-1}}{(m-1)!} + \frac{A_2 t^{m-2}}{(m-2)!} + \cdots + A_m \right] e^{r_1 t} + \cdots \qquad (2\text{-}2.14)$$

The following example is designed to numerically illustrate the partial fractions expansion procedure and the entire inversion process. Three cases are considered: unrepeated real roots, repeated roots, and complex conjugate roots.

EXAMPLE 2-2.1

Given the quadratic differential equation considered in the preceding discussion, Eq. 2-2.1, with zero steady-state initial conditions, we will obtain the unit step response of the output variable $y(t)$ for three different sets of parameters.

(A) UNREPEATED REAL ROOTS

Let $a_2 = 9$, $a_1 = 10$, $a_0 = 1$, and $b = 2$, in Eq. 2-2.1. Then, the unit step response is, from Eq. 2-2.5,

$$y(t) = \mathscr{L}^{-1} \left[\frac{2}{9s^2 + 6s + 1} \frac{1}{s} \right]$$

The roots, from the quadratic equation, are $r_1 = -1/9$, $r_2 = -1$. The denominator is factored as follows:

$$Y(s) = \frac{2}{9 \left(s + \dfrac{1}{9} \right) (s + 1)} \frac{1}{s}$$

$$= \frac{A_1}{s + \dfrac{1}{9}} + \frac{A_2}{s + 1} + \frac{A_3}{s}$$

The coefficients are calculated using Eq. 2-2.9.

$$A_1 = \lim_{s \to -1/9} \frac{2}{9(s + 1)s} = -2.25$$

$$A_2 = \lim_{s \to -1} \frac{2}{9 \left(s + \dfrac{1}{9} \right) s} = 0.25$$

$$A_3 = \lim_{s \to 0} \frac{2}{9 \left(s + \dfrac{1}{9} \right) (s + 1)} = 2$$

Invert by matching entries in Table 2-1.1, to obtain the step response.

$$y(t) = -2.25e^{-t/9} + 0.25e^{-t} + 2u(t)$$

(B) REPEATED ROOTS

Let $a_1 = 6$, and the other parameters be as before. The roots, from the quadratic formula, are $r_1 = r_2 = -1/3$, and the Laplace transform of the output response is

$$Y(s) = \frac{2}{9\left(s + \dfrac{1}{3}\right)^2 s}$$

$$= \frac{A_1}{\left(s + \dfrac{1}{3}\right)^2} + \frac{A_2}{s + \dfrac{1}{3}} + \frac{A_3}{s}$$

The coefficients are, from Eq. 2-2.13,

$$A_1 = \lim_{s \to -1/3} \frac{2}{9s} = -\frac{2}{3}$$

$$A_2 = \lim_{s \to -1/3} \frac{1}{1!}\frac{d}{ds}\left[\frac{2}{9s}\right] = \lim_{s \to -1/3} -\frac{2}{9s^2} = -2$$

and $A_3 = 2$, as before. The step response is then obtained by matching entries in Table 2-1.1.

$$y(t) = \left(-\frac{2}{3}t - 2\right)e^{-t/3} + 2u(t)$$

(C) PAIR OF COMPLEX CONJUGATE ROOTS

Let $a_1 = 3$, and the other parameters be as before. The roots, from the quadratic formula, are $r_{1,2} = -0.167 \pm i0.289$, where $i = \sqrt{-1}$ is the unit of the imaginary numbers. The transform of the output is then

$$Y(s) = \frac{2}{9(s + 0.167 - i0.289)(s + 0.167 + i0.289)s}$$

$$= \frac{A_1}{s + 0.167 - i0.289} + \frac{A_2}{s + 0.167 + i0.289} + \frac{A_3}{s}$$

Once more the coefficients are calculated by Eq. 2-2.9.

$$A_1 = \lim_{s \to -0.167 + i0.289} \frac{2}{9(s + 0.167 + i0.289)s} = -1 + i0.577$$

$$A_2 = \lim_{s \to -0.167 - i0.289} \frac{2}{9(s + 0.167 - i0.289)s} = -1 - i0.577$$

and $A_3 = 2$, as before. The inverse response is again obtained by matching entries in Table 2-1.1. Notice that the fact that the numbers are complex does not affect this part of the procedure.

$$y(t) = (-1 + i0.577)e^{(-0.167 + i0.289)t} + (-1 - i0.577)e^{(-0.167 - i0.289)t} + 2u(t)$$

It is evident from the preceding example that the calculation of the coefficients of the partial fraction expansion can be difficult, especially when the factors of the transform are complex numbers. As we shall see in the next section, the roots of the denominator of the transfer function contain most of the significant information about the response. Consequently, in analyzing the response of process control systems, it is seldom necessary to calculate the coefficients of the partial fraction expansion. This is indeed fortunate.

2-2.3 Handling Time Delays

The technique of partial fractions expansion is restricted for use with Laplace transforms that can be expressed as the ratio of two polynomials. When the response contains time delays, by the real translation theorem, Eq. 2-1.8, an exponential function of s appears in the transform. Since the exponential is a transcendental function, we must appropriately modify the inversion procedure.

If the denominator of the transform contains exponential functions of s, it cannot be factored because the exponential function introduces an infinite number of factors. On the other hand, we can handle exponential terms in the numerator of the transform, as we shall now see.

Consider the case in which there is a single exponential term that can be factored as follows:

$$Y(s) = Y_1(s)e^{-st_0} \tag{2-2.15}$$

The correct procedure is to expand in partial fractions the portion of the transform that does not contain the exponential term.

$$Y_1(s) = \frac{A_1}{s - r_1} + \frac{A_2}{s - r_2} + \cdots + \frac{A_n}{s - r_n} \tag{2-2.16}$$

Then invert this expression.

$$y_1(t) = A_1 e^{r_1 t} + A_2 e^{r_2 t} + \cdots + A_n e^{r_n t} \tag{2-2.17}$$

Now, invert Eq. 2-2.15 making use of the real translation theorem, Eq. 2-1.8.

$$\begin{aligned} y(t) = \mathscr{L}^{-1}\lfloor e^{-st_0} Y_1(s)\rfloor &= y_1(t - t_0) \\ &= A_1 e^{r_1(t-t_0)} + A_2 e^{r_2(t-t_0)} + \cdots + A_n e^{r_n(t-t_0)} \end{aligned} \tag{2-2.18}$$

It is important to realize that the exponential term must be excluded from the partial fractions expansion procedure. Although inclusion of the exponential term in the partial fractions expansion may give the correct result in some special cases, doing so is fundamentally *incorrect*.

Next, let us consider the case of multiple delays. When there is more than one delay term in the numerator of the transform, proper algebraic manipulation will convert the transform into a sum of terms, each having a single exponential function:

$$Y(s) = Y_1(s)e^{-st_{01}} + Y_2(s)e^{-st_{02}} + \cdots \tag{2-2.19}$$

Then expand each of the sub-transforms—$Y_1(s)$, $Y_2(s)$, and so on—in partial fractions and invert them separately, leaving out the exponential terms. Finally, apply Eq. 2-2.18 to each term to produce the result:

$$y(t) = y_1(t - t_{01}) + y_2(t - t_{02}) + \cdots \tag{2-2.20}$$

The following example illustrates this procedure.

EXAMPLE 2-2.2   Given the differential equation

$$\frac{dc(t)}{dt} + 2c(t) = f(t)$$

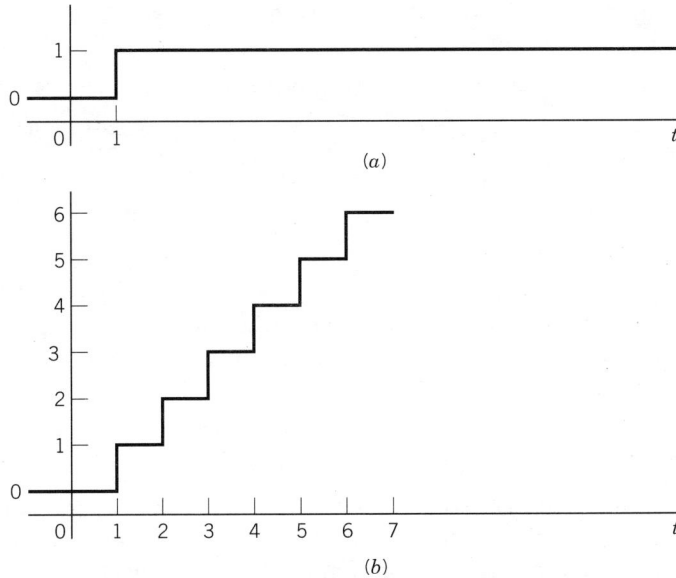

Figure 2-2.1 Input functions for Example 2-2.2. (*a*) Delayed unit step, $u(t-1)$. (*b*) Staircase of unit steps.

with $c(0) = 0$, find the response of the output for

1. A unit step change at $t = 1$: $f(t) = u(t-1)$

2. A staircase function of unit steps at every unit of time

$$f(t) = u(t-1) + u(t-2) + u(t-3) + \cdots$$

The functions are sketched in Fig. 2-2.1.

SOLUTION   **(a)** Transform the differential equation, solve for $C(s)$, and substitute $F(s) = (1/s)e^{-s}$.

$$C(s) = \frac{1}{s+2} F(s) = \frac{1}{s+2} \frac{1}{s} e^{-s}$$

Let $C(s) = C_1(s)e^{-s}$, then invert $C_1(s)$.

$$C_1(s) = \frac{1}{s+2} \frac{1}{s} = \frac{-0.5}{s+2} + \frac{0.5}{s}$$

Invert by matching entries in Table 2-1.1.

$$c_1(t) = -0.5e^{-2t} + 0.5u(t) = 0.5u(t)[1 - e^{-2t}]$$

Apply Eq. 2-2.18.

$$c(t) = \mathscr{L}^{-1}[C_1(s)e^{-s}] = c_1(t-1) = 0.5u(t-1)[1 - e^{-2(t-1)}]$$

Note that the unit step $u(t-1)$ must multiply the exponential term to show that $c(t) = 0$ for $t < 1$.

(b) For the staircase function,

$$C(s) = \frac{1}{s+2} \left(\frac{e^{-s}}{s} + \frac{e^{-2s}}{s} + \frac{e^{-3s}}{s} + \cdots \right)$$

$$= \frac{1}{(s+2)s}(e^{-s} + e^{-2s} + e^{-3s} + \cdots)$$

$$= C_1(s)e^{-s} + C_1(s)e^{-2s} + C_1(s)e^{-3s} + \cdots$$

We note that $C_1(s)$ is the same as for part (a), and therefore $c_1(t)$ is the same. Applying Eq. 2-2.18 to each term results in

$$c(t) = c_1(t-1) + c_1(t-2) + c_1(t-3) + \cdots$$
$$= 0.5u(t-1)[1 - e^{-2(t-1)}] + 0.5u(t-2)[1 - e^{-2(t-2)}]$$
$$+ 0.5u(t-3)[1 - e^{-2(t-3)}] + \cdots$$

The preceding example illustrates how to handle time delays in the input function. The same procedure can be applied when the time delay appears in the transfer function of the system. This situation arises in Chapter 3 in the models of processes with transportation lag.

2-3 CHARACTERIZATION OF PROCESS RESPONSE

In the preceding section we learned that we can express the Laplace of the process output variable as the product of two terms, a transfer function, which is characteristic of the process, and the transform of the input signal. A major objective of this section is to relate the characteristics of the response of the output variable to the parameters of the process transfer function and, in particular, to the roots of the denominator of the transfer function. We will see that most of the important information about the process response can be obtained from these roots, and it is not in general necessary to obtain the exact solution to each problem.

The relevant questions about the output response are the following:

- Is the response stable? That is, will it remain bound when forced by a bound input?
- If stable, what will be its final steady-state value?
- Is the response monotonic or oscillatory?
- If monotonic and stable, how long will it take for the transients to die out?
- If oscillatory, what is the period of oscillation and how long will it take for the oscillations to die out?

We will see that we can obtain the answers to all of these questions from the parameters of the transfer function of the system. But first, let us formally define deviation variables and see how their use, combined with the assumption of steady-state initial conditions, allows us to eliminate the effect of the initial conditions on the response.

2-3.1 Deviation Variables

In Section 2-2 we saw that the response of the output variable is affected not only by the input variables, but also by its initial conditions. As we are interested in studying the response of processes and their control systems to the input variables (disturbances and manipulated variables, defined in Chapter 1), we want to eliminate the effect of the initial conditions on the response. To do this we assume that the initial conditions are at steady state. This makes the initial values of the time derivatives equal to zero, but not the initial value of the output itself. To eliminate the initial value of the output, we replace the output variable with its deviation from the initial value. This gives rise to *deviation variables*, which we defined as

$$Y(t) = y(t) - y(0) \tag{2-3.1}$$

where

$Y(t)$ = deviation variable

$y(t)$ = total value of the variable

In the balance of this book deviation variables will be represented by capital letters and absolute variables by lower-case letters, whenever possible. From the definition of a deviation variable its initial value is always zero: $Y(0) = y(0) - y(0) = 0$.

To illustrate the simplifications that result from the use of deviation variables, consider the nth order linear differential equation:

$$a_n \frac{d^n y(t)}{dt^n} + a_{n-1} \frac{d^{n-1} y(t)}{dt^{n-1}} + \cdots + a_0 y(t) = b_m \frac{d^m x(t)}{dt^m}$$

$$+ b_{m-1} \frac{d^{m-1} x(t)}{dt^{m-1}} + \cdots + b_0 x(t) + c \tag{2-3.2}$$

where $n > m$, $y(t)$ is the output variable, $x(t)$ is the input variable, and c is a constant. At the initial steady state, all the time derivatives are zero, and we can write

$$a_0 y(0) = b_0 x(0) + c \tag{2-3.3}$$

Subtracting Eq. 2-3.3 from Eq. 2-3.2, results in

$$a_n \frac{d^n Y(t)}{dt^n} + a_{n-1} \frac{d^{n-1} Y(t)}{dt^{n-1}} + \cdots + a_0 Y(t)$$

$$= b_m \frac{d^m X(t)}{dt^m} + b_{m-1} \frac{d^{m-1}(t)}{dt^{m-1}} + \cdots + b_0 X(t) \tag{2-3.4}$$

where $Y(t) = y(t) - y(0)$, $X(t) = x(t) - x(0)$, and the deviation variables can be directly substituted for the respective variables in the derivative terms because they only differ by a constant bias:

$$\frac{d^k Y(t)}{dt^k} = \frac{d^k [y(t) - y(0)]}{dt^k} = \frac{d^k y(t)}{dt^k} - \frac{d^k y(0)}{dt^k} = \frac{d^k y(t)}{dt^k}$$

Notice that Eq. 2-3.4 in the deviation variables is essentially the same as Eq. 2-3.2 in the original variables except for constant c that cancels out. This result is general.

2-3.2   Output Response

To show the relationship between the output response and the roots of the denominator of the transfer function, let us Laplace transform the nth-order differential equation in the deviation variables, Eq. 2-3.4, and solve for the transform of the output.

$$Y(s) = \left[\frac{b_m s^m + b_{m-1} s^{m-1} + \cdots + b_0}{a_n s^n + a_{n-1} s^{n-1} + \cdots + a_0} \right] X(s) \tag{2-3.5}$$

where we have made use of the fact that all the initial conditions are zero. The expression in the brackets is the transfer function; its denominator can be factored into n first-degree terms, one for each of its roots.

$$Y(s) = \left[\frac{b_m s^m + b_{m-1} s^{m-1} + \cdots + b_0}{a_n (s - r_1)(s - r_2) \cdots (s - r_n)} \right] X(s) \tag{2-3.6}$$

where $r_1, r_2, \ldots, r_n$, are the roots of the denominator polynomial. Besides the n factors shown in Eq. 2-3.6 there are additional factors introduced by the input variable $X(s)$

which depend on the type of input (step, pulse, ramp, etc.). Next, we expand the transform in partial fractions.

$$Y(s) = \frac{A_1}{s - r_1} + \frac{A_2}{s - r_2} + \cdots + \frac{A_n}{s - r_n} + [\text{terms of } X(s)] \qquad \textbf{(2-3.7)}$$

Finally, we invert the transform by matching entries in Table 2-1.1 to obtain the response as a function of time. If there are no repeated roots, the inverse is

$$Y(t) = A_1 e^{r_1 t} + A_2 e^{r_2 t} + \cdots + A_n e^{r_n t} + (\text{terms of } X) \qquad \textbf{(2-3.8)}$$

The first n terms on the right-hand side come from the transfer function, and the rest of the terms differ depending on the input function $X(t)$.

If any of the roots is repeated p times, its coefficient is replaced by a polynomial in t of degree $p - 1$, as shown in the preceding section. The total number of terms will of course be n, counting the terms in the polynomial of t.

Let us next answer the questions posed at the beginning of this section by analyzing Eq. 2-3.8. We consider first the case in which all the roots are real, and then the possibility of complex conjugate pairs of roots.

All Real Roots

If all the roots are real, the terms of Eq. 2-3.8 are simple exponential functions of time that can only grow with time if the root is positive or decay to zero if the root is negative. Therefore, real roots cannot cause the response to oscillate. Furthermore, if any of the roots are positive, the response will grow exponentially without bounds, so it will be *unstable*. You might ask, what if the coefficient of the term with the positive root is zero? The system is just as unstable; a zero coefficient just means that for a particular input it may not run away, like a pencil standing on its sharpened point, but the slightest deviation from equilibrium will cause it to run away from that position.

So, in answer to our initial questions, if all the roots of the denominator of the transfer function are real,

- The response is monotonic (nonoscillatory).
- It is stable only if all the roots are negative.

Figures 2-3.1a and b are, respectively, examples of stable and unstable monotonic responses.

Regarding the time it takes for the transients to die out, we can see that each exponential term starts at unity ($e^0 = 1$) and, if the root is negative, decays to zero with time. Theoretically an exponential never reaches zero, so we have to define a threshold below which the transient can be considered gone. Let us say we define the threshold for each term of the response as less than 1 % of its initial value. To use a good round number, let $e^{rt} = e^{-5} = 0.0067$, or 0.67 %, which is less than 1 %. Then the time required for the kth exponential term to reach 0.67 % of its initial value is

$$t_k = \frac{-5}{r_k} \qquad \textbf{(2-3.9)}$$

So, the root with the smallest absolute value (least negative) will take the longest to die out. Such a root is called the *dominant root* of the response.

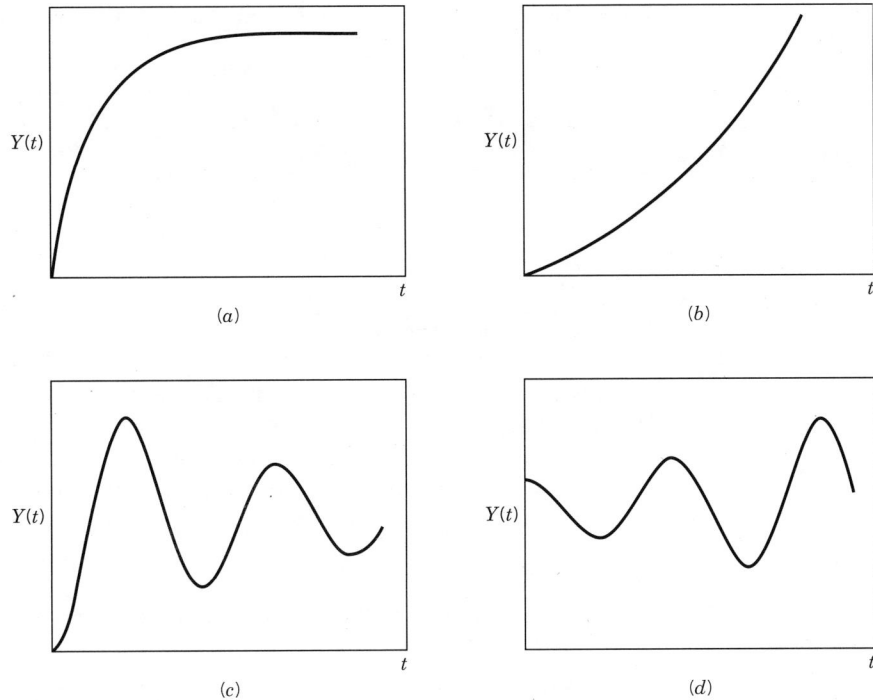

Figure 2-3.1 Examples of responses. (*a*) Stable, negative real root. (*b*) Unstable, positive real root. (*c*) Oscillatory stable, complex roots with negative real part. (*d*) Oscillatory unstable, complex roots with positive real part.

Pair of Complex Conjugate Roots

Complex roots of a real polynomial come in complex conjugate pairs, such as

$$r_1 = \rho + i\omega \quad r_2 = \rho - i\omega$$

where ρ is the real part and ω is the imaginary part. For the case of one such pair of roots, the expanded transform of the output is

$$
\begin{aligned}
Y(s) &= \frac{A_1}{s - \rho - i\omega} + \frac{A_2}{s - \rho + i\omega} + \cdots \\
&= \frac{(A_1 + A_2)(s - \rho)}{(s - \rho)^2 + \omega^2} + \frac{i(A_1 - A_2)\,\omega}{(s - \rho)^2 + \omega^2} + \cdots \\
&= \frac{B(s - \rho)}{(s - \rho)^2 + \omega^2} + \frac{C\omega}{(s - \rho)^2 + \omega^2} + \cdots
\end{aligned}
\qquad \textbf{(2-3.10)}
$$

where

$$B = A_1 + A_2$$
$$C = i(A_1 - A_2)$$

It can be shown that A_1 and A_2 are complex numbers and conjugates of each other. Consequently, B and C are real numbers. We can now invert Eq. 2-3.10 by matching the last two entries in Table 2-1.1, with $a = -\rho$.

$$
\begin{aligned}
Y(t) &= Be^{\rho t} \cos \omega t + Ce^{\rho t} \sin \omega t + \cdots \\
&= e^{\rho t}(B \cos \omega t + C \sin \omega t) + \cdots
\end{aligned}
$$

This equation can be further simplified using the trigonometric identity

$$\sin(\omega t + \theta) = \sin\theta\cos\omega t + \cos\theta\sin\omega t$$

The result is

$$Y(t) = De^{\rho t}\sin(\omega t + \theta) + \cdots \qquad (2\text{-}3.11)$$

where

$D = \sqrt{B^2 + C^2}$, is the initial amplitude

$\theta = \tan^{-1}\dfrac{B}{C}$, is the phase angle, in radians[3]

This result shows that the response is oscillatory, because it contains the sine wave. The amplitude of the sine wave varies with time according to the exponential term $e^{\rho t}$ which is initially unity, but can grow with time if ρ is positive, or decay to zero if negative. So, for the case of one or more pairs of complex conjugate roots, we can further answer the questions at the beginning of this section as follows:

- The response is oscillatory.
- The oscillations grow with time (unstable) if any of the pairs of complex roots has a positive real part.

Figures 2-3.1c and d show, respectively, examples of stable and unstable oscillatory responses.

Equation 2-3.11 shows that the frequency of the sine wave is equal to the imaginary part of the roots, ω, in radians per unit time. The period of the oscillations is the time it takes for a complete cycle, that is, the time it takes for the argument of the sine wave, $\omega t + \theta$, to increase by 2π radians. Thus, the period is

$$T = \frac{2\pi}{\omega} \qquad (2\text{-}3.12)$$

The SI unit for frequency is the hertz (Hz), which is the number of cycles per second, or the reciprocal of the period in seconds. Our formulas, however, require that the frequency be in radians per unit time.

While the period of the oscillations is determined by ω, the imaginary part of the roots, the time it takes for the oscillations to die out is controlled by the real part of the roots, ρ. As with the real roots, the time it takes for the oscillations to decay to less than 1 % of the initial amplitude, specifically $e^{-5} = 0.0067$, or 0.67 %, is

$$t_S = \frac{-5}{\rho} \qquad (2\text{-}3.13)$$

where t_S is approximately the 1 % *settling time*.

Perhaps a better measure of the decay of the oscillations is the *decay ratio*, or the ratio by which the amplitude of the oscillations decays in one period. This number is

$$\text{Decay ratio} = e^{\rho T} = e^{2\pi\rho/\omega} \qquad (2\text{-}3.14)$$

[3] For the formulas derived here, the argument of the trigonometric functions must be in *radians*, which means that your calculator must be in that mode when invoking the trigonometric functions. The trigonometric functions of most computer languages (FORTRAN, Pascal, C, Visual Basic, etc.), and of spreadsheets, work in radians.

Table 2-3.1 Relationship between the Laplace transform $Y(s)$ and its inverse $Y(t)$

Denominator of $Y(s)$	Partial fraction term	Term of $Y(t)$
Unrepeated real root	$\dfrac{A}{s-r}$	Ae^{rt}
Pair of complex conjugate roots where $D = \sqrt{B^2 + C^2}$ $\theta = \tan^{-1}\dfrac{B}{C}$	$\dfrac{Bs + C}{(s-\rho)^2 + \omega^2}$	$De^{\rho t}\sin(\omega t + \theta)$
Real root repeated m times	$\displaystyle\sum_{j=1}^{m}\dfrac{A_j}{(s-r)^j}$	$\displaystyle e^{rt}\sum_{j=1}^{m}\dfrac{A_j t^{j-1}}{(j-1)!}$

Final Steady-State Value

The only question left to be answered is the determination of the final steady-state or equilibrium value of the output after the transients die out. For a final steady state to exist the input variable, $X(t)$, must remain steady for some time. The easiest way to analyze the response to find this final steady-state value is to use the final value theorem of Laplace transforms and assume a step input, $X(t) = \Delta x u(t)$, or $X(s) = \Delta x/s$. Substituting into Eq. 2-3.5 and applying the final value theorem, Eq. 2-1.9, yields

$$\Delta Y = \lim_{s\to 0} s\left[\frac{b_m s^m + b_{m-1}s^{m-1} + \cdots + b_0}{a_n s^n + a_{n-1}s^{n-1} + \cdots + a_0}\right]\frac{\Delta x}{s} = \frac{b_0}{a_0}\Delta x \qquad (2\text{-}3.15)$$

In this section we have derived several formulas for computing important parameters of the output response. Excepting the formula for the final steady-state value, Eq. 2-3.15, all other formulas are based on the roots of the denominator polynomial of the transfer function. None of these parameters depends on the values of the coefficients of the partial fraction expansion.

Table 2-3.1 summarizes the relationships we have established in this section between the output response and the roots of the denominator polynomial in the transfer function. The following example illustrates the application of the ideas discussed in this section.

EXAMPLE 2-3.1

Characterize the responses described by the differential equations given below. Assume the time is measured in minutes and the variables are deviations from initial steady-state conditions.

(a)
$$30\frac{d^3Y(t)}{dt^3} + 43\frac{d^2Y(t)}{dt^2} + 14\frac{dY(t)}{dt} + Y(t) = 2.5X(t)$$

Laplace transform and solve for $Y(s)$.

$$Y(s) = \frac{2.5}{30s^3 + 43s^2 + 14s + 1}X(s)$$

The roots of the denominator are: -0.1, -0.333, and -1.0. Since the roots are all real and negative, the response is monotonic and stable. It is

$$Y(t) = A_1 e^{-0.1t} + A_2 e^{-0.333t} + A_3 e^{-t} + (\text{terms of } X)$$

The time required for each term to decay to 0.67 % of its initial value are, respectively 50 ($=$ $-5/-0.1$), 15, and 5 min, so the first term dominates the response. For a step change in $X(t)$, the final steady-state value is $2.5/1.0 = 2.5$ times the amplitude of the step.

(b)
$$\frac{d^3Y(t)}{dt^3} + 5\frac{d^2Y(t)}{dt^2} + 11\frac{dY(t)}{dt} + 15Y(t) = 12X(t)$$

Laplace transform and solve for $Y(s)$.

$$Y(s) = \frac{12}{s^3 + 5s^2 + 11s + 15}X(s)$$

The roots of the denominator are: $-1 \pm i2$, and -3. Since there is a pair of complex conjugate roots, the response is oscillatory with a frequency of 2 radians/minute and a period of $2\pi/2 = 3.14$ min. The response is stable because the real part of the complex roots is negative and so is the real root. It is

$$Y(t) = De^{-t}\sin(2t + \theta) + A_3e^{-3t} + \text{(terms of } X)$$

The sine wave decays to 0.67 % of its initial amplitude in $-5/-1 = 5$ min, while the term with the real root decays in $-5/-3 = 1.67$ min. Therefore, the sine wave term is the dominant term. The decay ratio of the sine wave is

$$\text{Decay ratio} = e^{2\pi(-1)/2} = e^{-3.14} = 0.043$$

This means that the amplitude is reduced to 4.3 % of its value during one cycle. The final steady-state change in $Y(t)$ is $0.8 (= 12/15)$ times the size of the sustained change in $X(t)$.

EXAMPLE 2-3.2

Response of a Pendulum

SOLUTION

Grandfather and cuckoo clocks use a pendulum for a timing device, that is, a weight suspended by a rod that can oscillate around its equilibrium value that is the vertical position. Determine which parameters of the pendulum (weight, length, shape, etc.) determine its period of oscillation.

A horizontal force balance on the pendulum, neglecting for the moment the resistance of the air and assuming the angle of oscillation is small, results in the following differential equation:

$$M\frac{d^2x(t)}{dt^2} = -Mg\frac{x(t)}{L} + f(t)$$

where $x(t)$ is the horizontal position of the weight in meters (m) from the equilibrium position, M is the mass of the weight in kg, L is the length of the rod in m, $g = 9.8$ m/s^2, is the acceleration of gravity, and $f(t)$ is the force in newtons (N) required to start the pendulum in motion, usually a short pulse or impulse. Assuming the pendulum is originally at equilibrium, $x(0) = 0$, Laplace transform the equation and solve for $X(s)$ to obtain

$$X(s) = \left[\frac{1}{Ms^2 + \dfrac{Mg}{L}}\right]F(s)$$

The roots of the denominator are pure imaginary numbers.

$$r_{1,2} = \pm i\sqrt{\frac{g}{L}}$$

from Eq. 2-3.11, the response is

$$x(t) = D\sin\left(\sqrt{\frac{g}{L}}t + \theta\right) + \text{(terms of } f)$$

That means that the pendulum will oscillate forever with a frequency that is independent of its weight and shape, and a function only of its length and the local acceleration of gravity. The period of oscillation is

$$T = 2\pi \sqrt{\frac{L}{g}}$$

So, if your grandfather clock is gaining time, you must lower the weight along the rod, and if it is losing time, you must raise the weight. For example, a pendulum with a length of 1.0 m will have a period of $2\pi(1/9.8)^{0.5} = 2.0$ s.

The pendulum, of course, does not oscillate forever, because of the resistance of the air. The weight and shape of the pendulum affect the air resistance. Clocks are equipped with a weight or spring mechanism to overcome the resistance of the air. This action could be incorporated into the external force $f(t)$. How could we incorporate the resistance of the air in the equation of motion to show that, left to itself, a pendulum will eventually stop?

Notice that the preceding example was solved without having to evaluate the coefficients of the partial fractions expansion or specify the input function.

2-3.3 Stability

Stability is the ability of the response to remain bound (remain within limits) when subjected to bound inputs. From the discussion in the preceding section, we conclude that the roots of the denominator of the transfer function of a process or device determine the stability of its response to input signals. That discussion can be summarized by the following condition of stability for linear systems: *A system is stable if all the roots of the denominator of its transfer function are either negative real numbers or complex numbers with negative real parts.*

This condition of stability will be discussed further in Chapter 6, where we will see that stability is a very important constraint on the operation and tuning of feedback control loops.

2-4 RESPONSE OF FIRST-ORDER SYSTEMS

As we shall see in Chapter 3, the dynamic response of many processes and control system components can be represented by linear first-order differential equations. We refer to these processes as *first-order systems*. This section presents the response of first-order systems to three different types of input signals, a step function, a ramp, and a sine wave. Our objective is to learn how the parameters of first-order systems affect their response so that later we can infer the important characteristics of the response of a system by simply examining its transfer function. First-order systems are also important because many higher-order systems can be treated as combinations of first-order systems in series and parallel.

Consider the linear first-order differential equation:

$$a_1 \frac{dy(t)}{dt} + a_0 y(t) = bx(t) + c \qquad \textbf{(2-4.1)}$$

where $y(t)$ is the output or dependent variable, $x(t)$ is the input variable, t is time, the independent variable, and the parameters a_1, a_0, b, and c, are constant. We can write the equation at the initial steady state, that is, before any change in input $x(t)$ takes place.

$$a_0 y(0) = bx(0) + c \qquad \textbf{(2-4.2)}$$

Note that this equation establishes a relationship between the initial values of x and y. Subtraction of Eq. 2-4.2 from Eq. 2-4.1 results in

$$a_1 \frac{dY(t)}{dt} + a_0 Y(t) = bX(t) \tag{2-4.3}$$

where

$$Y(t) = y(t) - y(0)$$
$$X(t) = x(t) - x(0)$$

are the deviation variables, and we have made use of the fact that $dy(t)/dt = dY(t)/dt$, because they differ by only the constant bias $y(0)$. Notice that the constant c cancels out.

Equation 2-4.3 is the general linear first-order differential equation in terms of the deviations of the input and output variables from their initial steady-state values. It has three coefficients, a_1, a_0, and b, but, without loss of generality, we can divide the equation by one of the three so that we can characterize the equation by just two parameters. In process control it is customary to divide by the coefficient of the output variable, a_0, provided it is not zero. Such an operation results in the following equation, which we shall call the *standard form* of the linear first-order differential equation

$$\boxed{\tau \frac{dY(t)}{dt} + Y(t) = KX(t)} \tag{2-4.4}$$

where

$\tau = \dfrac{a_1}{a_0}$ is the *time constant*

$K = \dfrac{b}{a_0}$ is the *steady-state gain*

The reason for these names will become apparent as we develop the responses to various types of inputs. Notice that, in order for Eq. 2-4.4 to be dimensionally consistent, τ must have dimension of time, and K must have dimension of Y over dimension of X.

Any linear first-order differential equation can be transformed into the standard form of Eq. 2-4.4, as long as the dependent variable $Y(t)$ appears in the equation. We can then obtain the transfer function of a first-order system by taking the Laplace transform of Eq. 2-4.4. To do this we apply the linearity property, Eq. 2-1.3, and the real differentiation theorem, Eq. 2-1.4, noticing that the initial condition of the deviation variable $Y(t)$ is zero. The result is

$$\tau s Y(s) + Y(s) = KX(s) \tag{2-4.5}$$

Solving for $Y(s)$ yields

$$\boxed{Y(s) = \left[\frac{K}{\tau s + 1} \right] X(s)} \tag{2-4.6}$$

The term in brackets is the transfer function of the first-order system in standard form. Notice that what is characteristic of this form is that the second term in the denominator is unity. When the transfer function is in this form, the numerator term is the gain and the coefficient of s is the time constant.

The root of the denominator of the transfer function is $r = -1/\tau$. From what we learned in the preceding section, we can see that the response of a first-order system is monotonic (one real root), and it is stable if its time constant is positive. Furthermore, the time required for the transients to be reduced to less than 1 % of their initial value, specifically $e^{-5} = 0.0067$, or 0.67 %, is $-5/r = 5\tau$ or five times the time constant. The final steady-state change in the output, obtained by letting $s = 0$ in the transfer function, is K times the sustained change in input, which is precisely why K is the gain; the definition of the gain is the steady-state change in output divided by the sustained change in input.

Having established the general characteristics of the response of first-order systems, we next look at the actual responses to three typical input signals.

2-4.1 Step Response

To obtain the step response of magnitude Δx, we let $X(t) = \Delta x\, u(t)$, where $u(t)$ stands for the unit step function at time zero [see Example 2-1.1(a)]. From Table 2-1.1, the transform of the input is $X(s) = \Delta x/s$. Substitute this into Eq. 2-4.6 and expand in partial fractions to obtain

$$Y(s) = \frac{K}{\tau s + 1}\frac{\Delta x}{s} = \frac{-K\Delta x}{\tau s + 1} + \frac{K\Delta x}{s}$$

Invert by matching entries in Table 2-1.1, with $a = 1/\tau$,

$$\boxed{Y(t) = K\Delta x(1 - e^{-t/\tau})} \tag{2-4.7}$$

This is a very important result. Figure 2-4.1 gives a graph of the response and Table 2-4.1 tabulates the normalized response versus t/τ. Notice that the response starts at the maximum rate of change right after the step is applied and then the rate of change decreases so that the final steady-state value of $K\Delta x$ is approached exponentially. After one time constant the response reaches 63.2 % of its final change, and in five time constants it reaches over 99 % of the change. In other words, the response is essentially complete after five time constants.

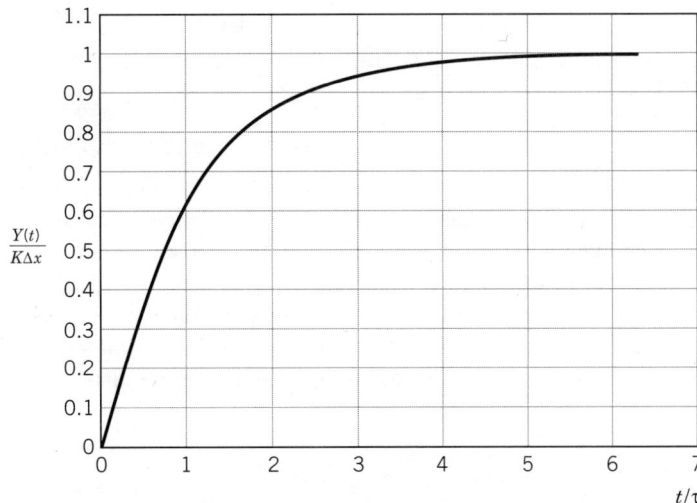

Figure 2-4.1 First-order step response.

Table 2-4.1 First-order step response

$\dfrac{t}{\tau}$	$\dfrac{Y(t)}{K\,\Delta x}$
0	0
1.0	0.632
2.0	0.865
3.0	0.950
4.0	0.982
5.0	0.993
⋮	⋮
∞	1.000

2-4.2 Ramp Response

A ramp is a linear increase in the input with time starting at time zero. The input function is given by $X(t) = rt$, where r is the slope (or rate) of the ramp. From Table 2-1.1, the Laplace transform is $X(s) = r/s^2$. Substitute into Eq. 2-4.6 and expand in partial fractions.

$$Y(s) = \frac{K}{\tau s + 1}\frac{r}{s^2} = \frac{A_1}{s + \dfrac{1}{\tau}} + \frac{A_2}{s^2} + \frac{A_3}{s}$$

Coefficient A_1 is obtained from Eq. 2-2.9, and A_2 and A_3 from Eq. 2-2.13.

$$A_1 = \lim_{s \to -1/\tau}\left(s + \frac{1}{\tau}\right)\frac{Kr}{(\tau s + 1)s^2} = Kr\tau$$

$$A_2 = \lim_{s \to 0} s^2 \frac{Kr}{(\tau s + 1)s^2} = Kr$$

$$A_3 = \lim_{s \to 0}\frac{1}{1!}\frac{d}{ds}\left[\frac{Kr}{\tau s + 1}\right] = -Kr\tau$$

Substitute into the transform and invert by matching entries in Table 2-1.1.

$$\begin{aligned}
Y(t) &= Kr\tau e^{-t/\tau} + (Krt - Kr\tau)u(t) \\
&= Kr\tau e^{-t/\tau} + Kr(t - \tau)u(t)
\end{aligned} \tag{2-4.8}$$

The ramp response, after the exponential term dies out in approximately five time constants, becomes a ramp with slope Kr and delayed by one time constant. To illustrate the way the output is delayed by exactly one time constant relative to the input, Fig. 2-4.2 superimposes plots of $X(t)$ and $Y(t)/K$ versus t; in this manner the two ramps are parallel because they both have slopes (or rates) of r. It is obvious from the plots that the output ramp "lags" the input ramp by one time constant τ. This is why systems represented by a first-order transfer function are also referred to as *first-order lags*.

2-4.3 Sinusoidal Response

To obtain the response of a first-order system to a sine wave, we let the input function be $X(t) = A \sin \omega t$, where A is the amplitude and ω is the frequency in radians/time. From

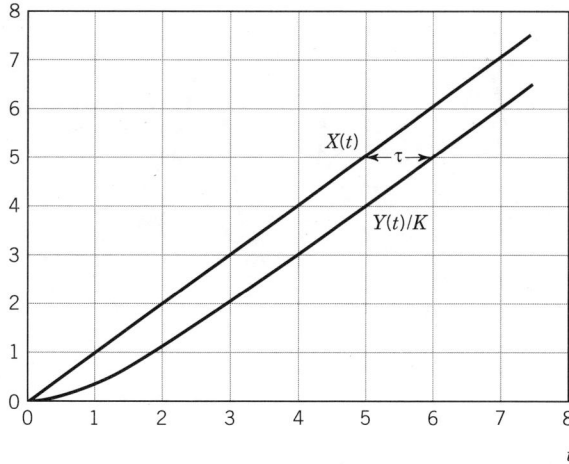

Figure 2-4.2 First-order response to a ramp. The normalized output lags the input by exactly one time constant.

Table 2-1.1, the Laplace transform is $X(s) = A\omega/(s^2 + \omega^2)$. Substitute into Eq. 2-4.6 and expand in partial fractions.

$$Y(s) = \frac{K}{\tau s + 1} \frac{A\omega}{s^2 + \omega^2} = \frac{A_1}{s + \dfrac{1}{\tau}} + \frac{A_2}{s - i\omega} + \frac{A_3}{s + i\omega}$$

where we have made use of $(s^2 + \omega^2) = (s - i\omega)(s + i\omega)$. The coefficients are obtained using Eq. 2-2.9.

$$A_1 = \lim_{s \to -1/\tau} \left(s + \frac{1}{\tau} \right) \frac{KA\omega}{(\tau s + 1)(s^2 + \omega^2)} = \frac{KA\tau\omega}{1 + \tau^2 \omega^2}$$

$$A_2 = \lim_{s \to i\omega} \frac{KA\omega}{(\tau s + 1)(s + i\omega)} = \frac{KA(-\tau\omega - i)}{2(1 + \tau^2 \omega^2)}$$

$$A_3 = \lim_{s \to -i\omega} \frac{KA\omega}{(\tau s + 1)(s - i\omega)} = \frac{KA(-\tau\omega + i)}{2(1 + \tau^2 \omega^2)}$$

Substitute into the transform, invert and, using Eq. 2-3.11 with $\rho - 0$, after some nontrivial manipulations, we obtain

$$Y(t) = \frac{KA\omega\tau}{1 + \tau^2\omega^2} e^{-t/\tau} + \frac{KA}{\sqrt{1 + \tau^2\omega^2}} \sin(\omega t + \theta) \qquad \textbf{(2-4.9)}$$

where $\theta = \tan^{-1}(-\omega\tau)$.

This sinusoidal response of a first-order system is plotted in Fig. 2-4.3. After the exponential term dies out in about five time constants, the response becomes a sine wave of the same frequency ω as the input sine wave. The amplitude of the output sine wave depends on the frequency. At very low frequencies it is just the product of the steady-state gain and the amplitude of the input, but as the frequency of the input sine wave increases, the amplitude of the output sine wave decreases. There is also a phase shift, a lag, θ, which is a function of frequency. This dependence of the response on the frequency of the input sine wave forms the basis for a method for analyzing process dynamics and control systems known as *frequency response*, which is the subject of Chapter 8.

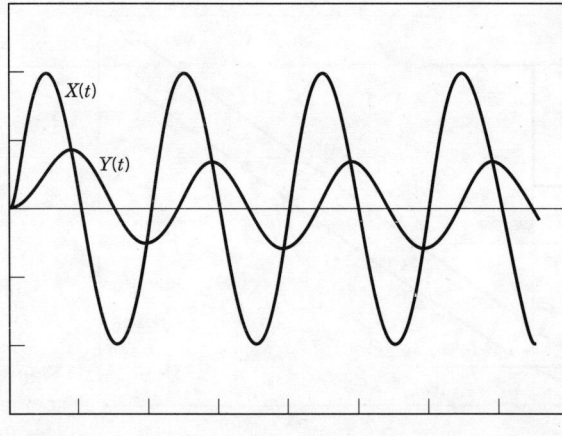

Figure 2-4.3 First-order response to a sine wave. The output sine wave, $Y(t)$, has the same frequency as the input, $X(t)$.

2-4.4 Response with Time Delay

As we shall see in Chapter 3, some process responses exhibit time delays (also known as transportation lag or dead time). By the real translation theorem, the time delay will modify the standard first-order transfer function of Eq. 2-4.6 as follows:

$$Y(s) = \left[\frac{Ke^{-st_0}}{\tau s + 1}\right] X(s) \tag{2-4.10}$$

The term in brackets is an important transfer function used to approximate the response of higher-order processes. We call it a *first-order-plus-dead-time (FOPDT)* transfer function.

The effect of the time delay on the three responses presented in this section is as follows:

Step Response

$$Y(t) = K\,\Delta x u(t - t_0)(1 - e^{(t-t_0)/\tau}) \tag{2-4.11}$$

where the presence of the factor $u(t - t_0)$ shows that the response is zero for $t < t_0$. A plot of this response is shown in Fig. 2-4.4.

Ramp Response

$$Y(t) = u(t - t_0)[Kr\tau e^{-(t-t_0)/\tau} + Kr(t - t_0 - \tau)] \tag{2-4.12}$$

Notice that the effect of the time delay in the long-term response is that the output ramp lags the input ramp by the sum of the time delay and the time constant.

Sinusoidal Response

$$Y(t) = u(t - t_0)\left\{\frac{KA\omega\tau}{1 + \tau^2\omega^2}e^{-(t-t_0)/\tau} + \frac{KA}{\sqrt{1 + \tau^2\omega^2}}\sin[\omega(t - t_0) + \theta]\right\}$$

$$\tag{2-4.13}$$

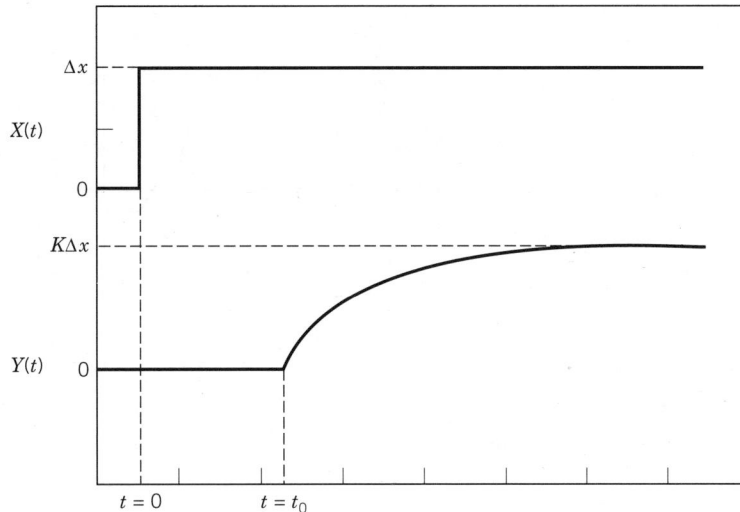

Figure 2-4.4 First-order step response with time delay t_0.

The only effect of the time delay on the long-term response is to increase the phase lag by ωt_0. This increase in phase lag is proportional to the frequency of the input sine wave. The phase lag θ is the same as in Eq. 2-4.9.

2-4.5   Response of a Lead-Lag Unit

A device that is commonly used for dynamic compensation in feedforward controllers, known as a *lead-lag unit*, has the following transfer function:

$$Y(s) = \frac{\tau_{ld}s + 1}{\tau_{lg}s + 1} X(s) \tag{2-4.14}$$

where τ_{ld} is the time constant of the lead term, and τ_{lg} is the time constant of the lag term. Notice that a "lead" is a first-order term in the numerator, while a "lag," as we saw earlier, is a first-order term in the denominator. The step and ramp responses of lead-lag units are helpful in understanding how to tune them, that is, how to adjust the lead and lag time constants to achieve optimum dynamic compensation.

Step Response

The response to a unit step response, $X(s) = 1/s$, is

$$\boxed{Y(t) = u(t) + \left(\frac{\tau_{ld}}{\tau_{lg}} - 1\right) e^{-t/\tau_{lg}}} \tag{2-4.15}$$

Figure 2-4.5 shows a plot of this response for various ratios of τ_{ld}/τ_{lg}. Notice that the initial change in output is controlled by the ratio of the time constants while the time required for the transient to die out is determined by the lag time constant (about five lag time constants). When the lead-to-lag ratio is greater than unity, the response overshoots its final steady state, while if the ratio is less than unity, it undershoots it.

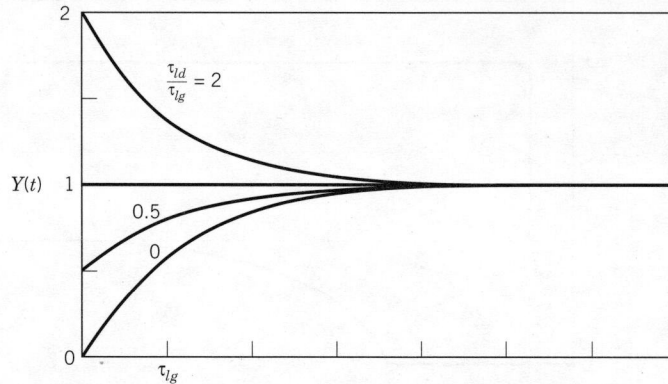

Figure 2-4.5 Lead-lag response to a unit step change.

Ramp Response

The response to a ramp of unity rate, $X(s) = 1/s^2$, is

$$Y(t) = \left(\tau_{ld} - \tau_{lg}\right) e^{-t/\tau_{lg}} + t + \tau_{ld} - \tau_{lg} \qquad \text{(2-4.16)}$$

The ramp response is plotted in Fig. 2-4.6 for two cases, one in which the lead is greater than the lag, and the other in which the lag is greater than the lead, along with the input ramp. Notice that, after the transient term dies out, the response is a ramp that either leads or lags the input ramp by the difference between the lead and the lag, depending on which is longer. It is this response that gives the names "lead" and "lag" to the numerator and denominator terms of the transfer function.

A physical device cannot have more leads than lags, so, in tuning lead-lag units we must keep in mind that, although the lead time constant can be set to zero, the lag time constant cannot be set to zero.

The application of lead-lag units to the dynamic compensation of feedforward controllers is discussed in detail in Section 11-3.

Besides the responses presented in this section, there are other responses of interest, such as responses to impulse functions and pulses. These are proposed as exercises in the problems at the end of this chapter. Another interesting problem proposed as

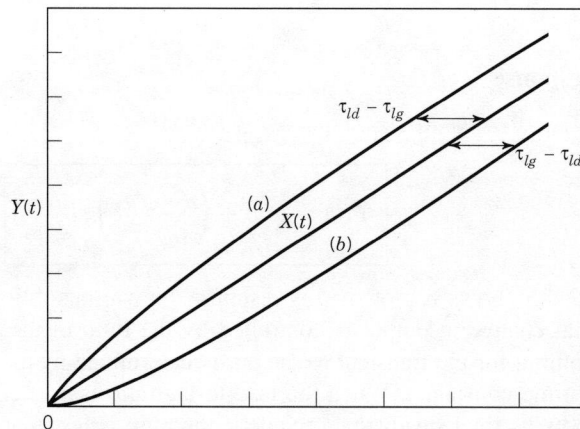

Figure 2-4.6 Lead-lag response to input ramp $X(t)$. (a) Net lead, $\tau_{ld} > \tau_{lg}$. (b) Net lag, $\tau_{lg} > \tau_{ld}$.

an exercise is the response of integrating processes, which are processes that do not contain the term a_0 in Eq. 2-4.3. Section 4-4.1 contains an example of an integrating process.

2-5 RESPONSE OF SECOND-ORDER SYSTEMS

This section presents the response of linear second-order systems to the same three types of input signals for which the response of first-order processes was presented in the preceding section. We will see that the responses are quite different depending on whether the roots of the denominator of the transfer function are real or a complex conjugate pair. When both roots are real the response is said to be *overdamped*, while if the roots are complex the response is said to be *underdamped*. The response of overdamped processes is generalized to systems of order higher than two.

A linear second-order system is one represented by a linear second-order differential equation. A general form of such an equation is

$$a_2 \frac{d^2 y(t)}{dt^2} + a_1 \frac{dy(t)}{dt} + a_0 y(t) = bx(t) + c \qquad \textbf{(2-5.1)}$$

where $y(t)$ is the output variable, $x(t)$ is the input variable, and parameters a_2, a_1, a_0, b, and c, are constant. Assuming the initial conditions are at steady state, the equation at the initial conditions is

$$a_0 y(0) = bx(0) + c \qquad \textbf{(2-5.2)}$$

Subtract Eq. 2-5.2 from Eq. 2-5.1 to obtain

$$a_2 \frac{d^2 Y(t)}{dt^2} + a_1 \frac{dY(t)}{dt} + a_0 Y(t) = bX(t) \qquad \textbf{(2-5.3)}$$

where

$Y(t) = y(t) - y(0)$

$X(t) = x(t) - x(0)$

are the deviation variables. By definition, the initial conditions of the deviation variables are zero. Notice that the constant c cancels out.

The four parameters in Eq. 2-5.3, a_2, a_1, a_0, and b, can be reduced to three by dividing the entire equation by any of them, provided it is not zero. In process control we obtain what we will call the *standard form* of the second-order equation by dividing by coefficient a_0, provided it is not zero. The resulting equation in the standard form is

$$\boxed{\tau^2 \frac{d^2 Y(t)}{dt^2} + 2\zeta\tau \frac{dY(t)}{dt} + Y(t) = KX(t)} \qquad \textbf{(2-5.4)}$$

where

$\tau = \sqrt{\dfrac{a_2}{a_0}}$ is the characteristic time[4]

$\zeta = \dfrac{a_1}{2\tau a_0} = \dfrac{a_1}{2\sqrt{a_0 a_2}}$ is called the *damping ratio*

$K = b/a_0$ is the steady-state gain.

---

[4] In some control textbooks, most notably those written by electrical engineers, the second-order response is characterized by the *natural frequency*, ω_n, defined as the reciprocal of the characteristic time τ.

The reason the parameters τ and ζ are defined as they are will become evident momentarily. In the definition of the characteristic time, we assumed that a_2 and a_0 have the same sign; otherwise the characteristic time would be an imaginary number and lose its usefulness.

Next we take the Laplace transform of Eq. 2-5.4, apply the linearity property and the real differentiation theorem, and solve for the transform of the output to obtain

$$Y(s) = \left[\frac{K}{\tau^2 s^2 + 2\zeta\tau s + 1}\right] X(s) \qquad \textbf{(2-5.5)}$$

where the term in the brackets is the second-order transfer function in standard form. To find the roots of the denominator polynomial we apply the quadratic formula:

$$r_{1,2} = \frac{-\zeta \pm \sqrt{\zeta^2 - 1}}{\tau} \qquad \textbf{(2-5.6)}$$

We see now that the damping ratio determines whether the roots are real or complex. If the absolute value of the damping ratio is one or greater than one, the roots are real, while if the damping ratio is less than unity, the roots are a pair of complex conjugate numbers. When the damping ratio is unity, the two roots are equal to each other and equal to $-1/\tau$. If the damping ratio is zero, the real part of the complex roots is zero, that is, the roots are pure imaginary numbers and equal to $\pm i/\tau$, where i is the square root of -1.

As we learned in Section 2-3, if the roots of the denominator of the transfer function are real numbers, the response is monotonic, while complex roots result in an oscillatory response. Furthermore, for the response to be stable, if the roots are real, they must both be negative, or, if they are complex, the real part must be negative. You are invited to verify that, for the second-order transfer function of Eq. 2-5.5, the condition of stability is satisfied if and only if the damping ratio is positive. We can now see that the term "damping ratio" refers to the damping of oscillations; the behavior of the response is summarized as follows:

For	The response is
$\zeta \geq 1$	overdamped = monotonic and stable
$0 < \zeta < 1$	underdamped = oscillatory and stable
$\zeta = 0$	undamped = sustained oscillations
$-1 < \zeta < 0$	unstable = growing oscillations
$\zeta \leq -1$	runaway = monotonic unstable

The case of $\zeta = 1$ is sometimes called *critically damped*, but this is only the borderline case. Its response is monotonic and stable, just as the overdamped response.

For our purposes, we need only consider the two cases of real and complex roots, which we will call overdamped and underdamped, respectively. The following sections present the specific response equations for step, ramp, and sinusoidal inputs for each of these two cases.

2-5.1   Overdamped Responses

When the damping ratio is greater than unity, the roots given by Eq. 2-5.6 are real numbers. In this case it is better to factor the denominator of the transfer function into two first-order terms containing a time constant each, as follows:

$$\tau^2 s^2 + 2\zeta\tau s + 1 = \tau^2(s - r_1)(s - r_2) = (\tau_{e1}s + 1)(\tau_{e2}s + 1) \quad \textbf{(2-5.7)}$$

where τ_{e1} and τ_{e2} are the *effective time constants*, defined as the negative reciprocals of the roots. For the second-order system, from Eq. 2-5.6, the effective time constants are

$$\tau_{e1} = -\frac{1}{r_1} = \frac{\tau}{\zeta - \sqrt{\zeta^2 - 1}}$$
$$\tau_{e2} = -\frac{1}{r_2} = \frac{\tau}{\zeta + \sqrt{\zeta^2 - 1}} \quad \textbf{(2-5.8)}$$

Next, substitute Eq. 2-5.7 into Eq. 2-5.5 to obtain the transfer function in terms of the effective time constants.

$$Y(s) = \left[\frac{K}{(\tau_{e1}s + 1)(\tau_{e2}s + 1)}\right] X(s) \quad \textbf{(2-5.9)}$$

This is a more convenient transfer function for representing second-order systems when the roots are real numbers. We will use it to develop the various responses.

Step Response

As in the preceding section, we assume the input is a step change of magnitude Δx. Substitute then $X(s) = \Delta x/s$ into Eq. 2-5.9, and expand in partial fractions.

$$Y(s) = \frac{K}{(\tau_{e1}s + 1)(\tau_{e2}s + 1)}\frac{\Delta x}{s} = \frac{A_1}{s + \dfrac{1}{\tau_{e1}}} + \frac{A_2}{s + \dfrac{1}{\tau_{e2}}} + \frac{A_3}{s}$$

Evaluate the coefficients and invert to obtain the output response.

$$Y(t) = K\Delta x\left[u(t) - \frac{\tau_{e1}}{\tau_{e1} - \tau_{e2}}e^{-t/\tau_{e1}} - \frac{\tau_{e2}}{\tau_{e2} - \tau_{e1}}e^{-t/\tau_{e2}}\right] \quad \textbf{(2-5.10)}$$

For the critically damped case, $\zeta = 1$, the two roots are equal to each other, $\tau_{e1} = \tau_{e2} = \tau$, and the response is given by

$$Y(t) = K\Delta x\left[u(t) - \left(\frac{t}{\tau} + 1\right)e^{-t/\tau}\right] \quad \textbf{(2-5.11)}$$

Figure 2-5.1 shows two typical step responses, one overdamped and the other one critically damped. Both of them are monotonic (nonoscillatory). Notice that the initial rate of change of the response is zero, it then increases to a maximum, and finally decreases to exponentially approach its final steady-state change of $K\Delta x$. This differs from the first-order step response of Fig. 2-4.1 in which the maximum rate of change occurred right after the step change was applied (at time zero).

The S-shaped step responses shown in Fig. 2-5.1 are characteristic of many processes.

Ramp Response

We obtain the response to a ramp of rate r, $X(t) = rt$, by substituting its transform, $X(s) = r/s^2$, into Eq. 2-5.9, expanding in partial fractions, and inverting. When the

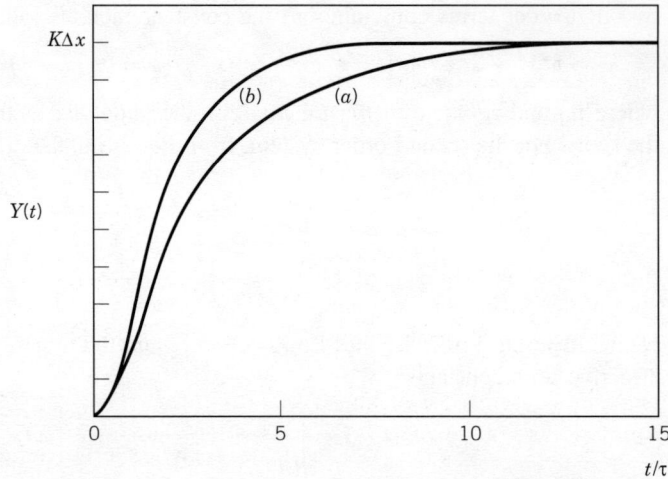

Figure 2-5.1 Second-order step responses. (*a*) Overdamped ($\zeta = 1.5$). (*b*) Critically damped ($\zeta = 1.0$).

two time constants are different, the ramp response is

$$Y(t) = Kr\left[\frac{\tau_{e1}^2}{\tau_{e1} - \tau_{e2}}e^{-t/\tau_{e1}} + \frac{\tau_{e2}^2}{\tau_{e2} - \tau_{e1}}e^{-t/\tau_{e2}} + t - (\tau_{e1} + \tau_{e2})\right] \quad \textbf{(2-5.12)}$$

When the two time constants are equal to τ, the ramp response is

$$Y(t) = Kr[(t + 2\tau)e^{-t/\tau} + t - 2\tau] \quad \textbf{(2-5.13)}$$

The important characteristic that is common to both of these responses is that, after the exponential terms die out, the response becomes a ramp of rate Kr. This output ramp lags the input ramp by the sum of the two time constants. We can extend this result to higher-order systems, if all the roots of the denominator are negative real numbers. The output ramp for an nth-order system lags the input ramp by the sum of all n effective time constants, where the effective time constants are defined as the negative reciprocals of the roots.

Sinusoidal Response

To obtain the response to a sine wave of amplitude A and frequency ω radian/time, $X(t) = A\sin\omega t$, we substitute its transform, $X(s) = A\omega/(s^2 + \omega^2)$, into Eq. 2-5.6, expand in partial fractions, and invert. Chapter 8 presents a formal procedure for carrying out this operation. We present here the resulting response.

$$Y(t) = A_1 e^{-t/\tau_{e1}} + A_2 e^{-t/\tau_{e2}} + \frac{KA}{\sqrt{1 + \tau_{e1}^2\,\omega^2}\sqrt{1 + \tau_{e2}^2\,\omega^2}}\sin(\omega t + \theta) \quad \textbf{(2-5.14)}$$

where

$$\theta = \tan^{-1}(-\omega\tau_{e1}) + \tan^{-1}(-\omega\tau_{e2})$$

Since the exponential terms die out and the sinusoidal term does not, the coefficients of the exponential terms are not important. For the sinusoidal response, the two important characteristics are that the amplitude of the output sine wave decreases as the frequency of the input sine wave increases, and that the phase angle becomes more negative as the frequency increases.

It is even more interesting to note that the effect of the two time lags on the amplitude is multiplicative. This means that the reduction in the output amplitude is the product of the reductions that each lag would cause if it were acting alone. Similarly, the effect of the two lags on the phase angle is additive, the effect is the sum of the effects that each individual lag would cause if it were alone. This result can be extended to a nth-order system, if all the roots of the denominator of the transfer function are negative real numbers. The reduction in the amplitude of the output wave is the product of the reductions that each of the n lags would cause if acting separately. Similarly, the phase angle is the sum of the phase angles that each of the n lags would separately cause. Chapter 8 presents this concept in more detail.

All of the response equations presented in this section apply for both the damping ratio being greater than or equal to unity and less than or equal to minus unity. The difference is that, for positive damping ratios, both effective time constants are positive (both roots are negative) and the response is stable. On the other hand, when the damping ratio is negative, the effective time constants are negative (roots are positive) and the response is monotonically unstable, that is, the output runs away from its initial condition exponentially.

2-5.2 Underdamped Responses

The study of underdamped or oscillatory responses is important because it is the most common response of feedback control systems. Many common devices also exhibit oscillatory behavior, such as pendulums, playground swings, yo-yos, car suspension systems, and doors at department stores.

Second-order systems represented by Eq. 2-5.4 are underdamped when the damping ratio is between -1 and $+1$. We can see from Eq. 2-5.6 that the roots of the denominator of the transfer function form a complex conjugate pair.

$$r_{1,2} = \frac{-\zeta \pm \sqrt{-1(1 - \zeta^2)}}{\tau} = -\frac{\zeta}{\tau} \pm i\frac{\sqrt{1 - \zeta^2}}{\tau} \qquad \textbf{(2-5.15)}$$

From what we learned in Section 2-3, these roots result in a response containing a sine wave with the frequency equal to the imaginary part and a decay rate equal to the real part. If the damping ratio is positive, $0 < \zeta < 1$, the amplitude of the oscillations decays with time and the response is stable, while for a negative damping ratio the amplitude increases with time and the response is unstable; for a damping ratio of zero the oscillations are sustained and the response is said to be *undamped*. Having looked at the generalities of the response, let us next present the specific responses to different input signals.

Step Response

To obtain the step response, let the input signal be a step of magnitude Δx, $X(t) = \Delta x\, u(t)$, and substitute its Laplace transform, $X(s) = \Delta x/s$, into Eq. 2-5.5. After expansion in partial fractions, making use of Eq. 2-3.11, obtain the response

$$Y(t) = K\,\Delta x \left[u(t) - \frac{1}{\sqrt{1-\zeta^2}} e^{-(\zeta/\tau)t} \sin(\psi t + \phi) \right] \qquad \textbf{(2-5.16)}$$

where

$\psi = \dfrac{\sqrt{1-\zeta^2}}{\tau}$ is the frequency in radians/time

$\phi = \tan^{-1} \dfrac{\sqrt{1-\zeta^2}}{\zeta}$ is the phase angle in radians

Figure 2-5.2 presents a plot of this response. Notice that just as for the overdamped responses of Fig. 2-5.1, the maximum rate of change does not occur right after the step change is applied, as in the first-order response. In fact, the initial rate of change is zero. Unlike the overdamped response, the underdamped response oscillates around its final steady state, $K\,\Delta x$, which is also the steady state for the other step responses.

The underdamped step response is so important that it has been characterized by several terms. As for any sine wave, the period of oscillation is the time it takes to complete an entire cycle or 2π radians:

$$T = \frac{2\pi}{\psi} = \frac{2\pi\tau}{\sqrt{1-\zeta^2}} \qquad \textbf{(2-5.17)}$$

As shown in Fig. 2-5.2, the period can be measured in the response by the time between two successive peaks in the same direction. Other terms are defined below.

Although the SI unit for frequency is the hertz (Hz), which is the reciprocal of the period T in seconds, or the number of cycles in one second, the formulas presented here require that the frequencies be in radians per unit time; they also require that the angles be in radians and not in degrees or other units.

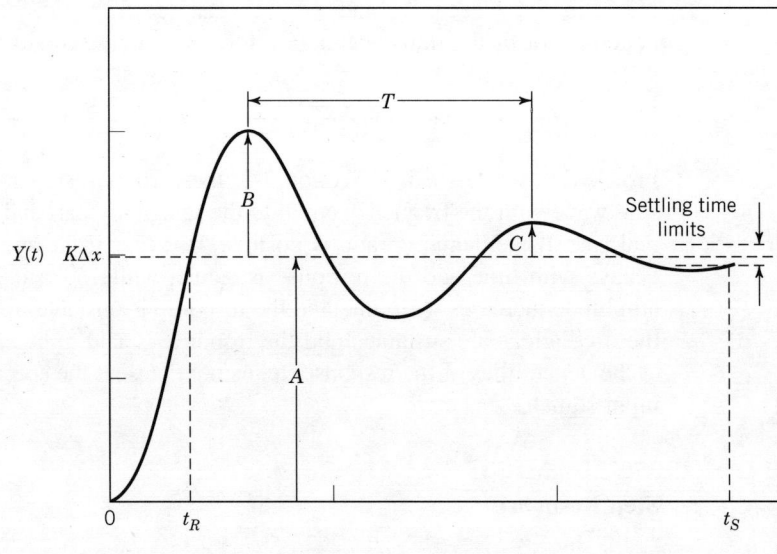

Figure 2-5.2 Second-order underdamped step response ($\zeta = 0.215$).

Decay Ratio   The decay ratio is the ratio by which the amplitude of the sine wave is reduced during one complete cycle. It is defined as the ratio of two successive peaks in the same direction, C/B in Fig. 2-5.2.

$$\text{Decay ratio} = e^{-(\zeta/\tau)T} = e^{-2\pi\zeta/\sqrt{1-\zeta^2}} \qquad \textbf{(2-5.18)}$$

The decay ratio is an important term because it serves as a criterion for establishing satisfactory response of feedback controllers.

Rise Time   This is the time it takes for the response to first reach its final steady-state value, t_R in Fig. 2-5.2. It can be approximated as one-fourth of the period T.

Settling Time   This is the time it takes for the response to come within some prescribed band of the final steady-state value and remain in this band. Typical band limits are $\pm5\%$, $\pm3\%$, or $\pm1\%$ of the total change. The settling time is t_S in Fig. 2-5.2. As discussed in Section 2-3, the real part of the roots of the denominator of the transfer function controls the settling time. For band limits of $\pm1\%$, it is approximately $5\tau/\zeta$.

Overshoot   The overshoot is the fraction (or percent) of the final steady-state change by which the first peak exceeds this change. On the assumption that the first peak occurs approximately half a cycle from the application of the step change, it is

$$\text{Overshoot} = e^{-(\zeta/\tau)(T/2)} = e^{-(\pi\zeta/\sqrt{1-\zeta^2})} \qquad \textbf{(2-5.19)}$$

Figure 2-5.2 shows how to determine the overshoot from a plot of the step response; it is the ratio of B/A, where $A = K\,\Delta x$.

You can see from the number of terms presented so far that the step response of underdamped systems is an important topic. Table 2-5.1 shows the numerical values of some of these terms for several values of the damping ratio. Figure 2-5.3 contains plots of underdamped step responses for several values of the damping ratio.

Ramp Response

To obtain the response of an underdamped second-order system to a ramp of rate r, $X(t) = rt$, substitute its transform, $X(s) = r/s^2$, into Eq. 2-5.5. Then expand in

Table 2-5.1   Second-order underdamped step response

Damping ratio (ζ)	Decay ratio	Overshoot %	$\dfrac{t_R}{\tau}$	$\dfrac{t_S}{\tau}, \pm1\%$
1.0	0	0	—	7.0
0.707	1/500	4.3	2.2	7.1
0.344	1/10	29.3	1.8	14.5
0.215	1/4	50.0	1.6	23.2
0	1/1	100	$\pi/2$	∞

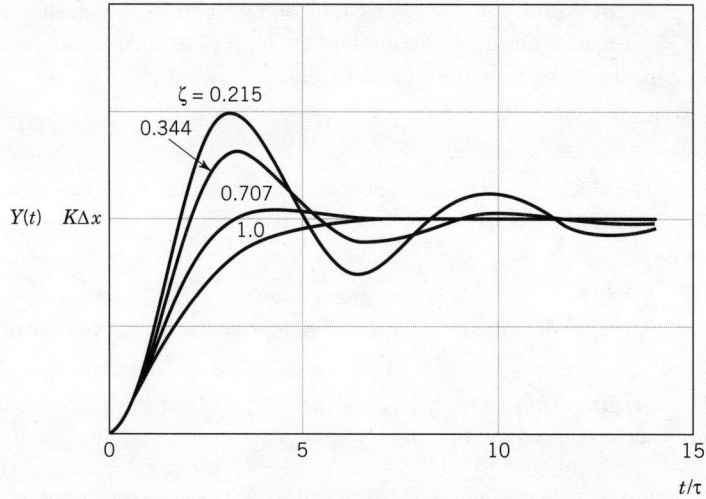

Figure 2-5.3 Effect of damping ratio on the second-order underdamped step response.

partial fractions, evaluate the coefficients, and invert. The result is

$$Y(t) = Kr \left[\frac{\tau}{\sqrt{1 - \zeta^2}} e^{-(\zeta/\tau)t} \sin(\psi t + \phi) + t - 2\zeta\tau \right] \tag{2-5.20}$$

where

$$\phi = \tan^{-1} \left(\frac{2\zeta\sqrt{1 - \zeta^2}}{2\zeta^2 - 1} \right)$$

and ψ is the frequency, which is the same as for the step change. The important characteristic of this response is that, after the sinusoidal term dies out, the output becomes a ramp of rate Kr that lags the input ramp by a time that decreases as the damping ratio decreases. For the undamped response, $\zeta = 0$, the output response is a sustained oscillation around the input ramp.

Sinusoidal Response

To obtain the underdamped response of the second-order system to a sine wave of amplitude A and frequency ω, $A \sin \omega t$, substitute its transform, $X(s) = A\omega/(s^2 + \omega^2)$, into Eq. 2-5.5. Then expand into partial fractions, evaluate the coefficients, and invert. The result is

$$Y(t) = KADe^{-(\zeta/\tau)t} \sin(\psi t + \phi) + \frac{KA}{\sqrt{(1 - \tau^2\omega^2)^2 + (2\zeta\tau\omega)^2}} \sin(\omega t + \theta) \tag{2-5.21}$$

where

$$\theta = \tan^{-1} \left(\frac{2\zeta\tau\omega}{1 - \tau^2 \omega^2} \right)$$

The amplitude D and phase angle ϕ in the first sine term are not important because this is the term that decays with time. After this first term decays, the output response is a sine wave with frequency equal to the frequency of the input signal. The amplitude and the phase angle of the output are functions of the frequency. An interesting effect in the sinusoidal response of underdamped systems is what happens when the input frequency is the system *resonant frequency*, equal to $1/\tau$. According to Eq. 2-5.21, at the resonant frequency the ratio of the amplitude of the output sine wave to that of the input is $K/2\zeta$, that is, inversely proportional to the damping ratio. This phenomenon, known as *resonance*, can result in very high output amplitudes when the damping ratio is small. In the 1940s a bridge at Tacoma Narrows, Washington, collapsed when the wind drove it at its resonant frequency.

2-5.3   Higher-Order Responses

The response of systems represented by differential equations of order higher than two can be thought of as a combination of first-order lags and second-order under-damped responses. When all the roots of the denominator of the transfer function are real, an nth-order system becomes a combination of n first-order lags. We can easily extend the results for the second-order overdamped responses to higher-order overdamped responses. For example, consider the following nth-order overdamped system:

$$Y(s) = \left[\frac{K}{\prod\limits_{k=1}^{n} (\tau_k s + 1)} \right] X(s) \tag{2-5.22}$$

where K is the gain and τ_k are the n effective time constants, or negative reciprocals of the n roots of the denominator polynomial. The response of this system to a step change of magnitude Δx, $X(s) = \Delta x / s$, if all the time constants are different from each other, is given by

$$Y(t) = K \Delta x \left[u(t) - \sum_{k=1}^{n} \frac{\tau_k^{n-1}}{\prod\limits_{\substack{j=1 \\ j \neq k}}^{n} (\tau_k - \tau_j)} e^{-t/\tau_k} \right] \tag{2-5.23}$$

Notice that Eq. 2-5.10 is a special case of this equation for $n = 2$. When all the n time constants are equal to each other, the step response is

$$Y(t) = K \Delta x \left[u(t) - e^{-t/\tau} \sum_{k=1}^{n} \frac{1}{(n-k)!} \left(\frac{t}{\tau} \right)^{n-k} \right] \tag{2-5.24}$$

where τ is the time constant that is repeated n times. Notice that Eq. 2-5.11 is a special case of this equation for $n = 2$.

If the transfer function of the nth-order system contains lead terms,

$$Y(s) = \left[\frac{K \prod\limits_{j=1}^{m} (\tau_{ldj} s + 1)}{\prod\limits_{k=1}^{n} (\tau_{lgk} s + 1)} \right] X(s) \tag{2-5.25}$$

where $n \geq m$, the step response for all the lag time constants being different from each other is

$$Y(t) = K \Delta x \left[u(t) - \sum_{k=1}^{n} \frac{\tau_{lgk}^{n-m-1} \prod_{j=1}^{m} (\tau_{lgk} - \tau_{ldj})}{\prod_{\substack{j=1 \\ j \neq k}}^{n} (\tau_{lgk} - \tau_{lgj})} e^{-t/\tau_{lgk}} \right] \tag{2-5.26}$$

The effect of the lead terms is to speed up the response if the lead time constants are positive, or slow it down if they are negative.

For higher-order underdamped systems, the second-order step response terms defined in this section also apply. However, the formulas presented to calculate the characteristic terms are only valid for estimating the contribution of individual pairs of complex conjugate roots to the overall response. The accuracy of the estimates of the overshoot, rise time, and decay ratio of the total response depend on how dominant is the pair of complex conjugate roots with respect to the other roots. Recall, from Section 2-3, that the dominant roots are those with the least negative real parts, that is, the terms of the response taking the longest to decay.

2-6 LINEARIZATION

A major difficulty in analyzing the dynamic response of many processes is that they are nonlinear, that is, they cannot be represented by linear differential equations. A linear differential equation consists of a sum of terms each of which contains no more than one variable or derivative, which must appear to the first power. In the preceding sections we learned that the method of Laplace transforms allows us to relate the response characteristics of a wide variety of physical systems to the parameters of their transfer functions. Unfortunately, only linear systems can be analyzed by Laplace transforms. There is no comparable technique that can analyze the dynamics of a nonlinear system and generalize the results to represent similar physical systems.

This section presents the technique known as *linearization* to approximate the response of nonlinear systems with linear differential equations that can then be analyzed by Laplace transforms. The linear approximation to the nonlinear equations is valid for a region near some base point around which the linearization is made. To facilitate the manipulation of the linearized equations, we will select the initial steady state as the base point for linearization, and will use *deviation (or perturbation) variables*, as defined in Section 2-3.1.

The following is a list of common nonlinear functions that appear in process dynamic models:

- Enthalpy, H, as a function of temperature, T:

$$H[T(t)] = H_0 + a_1 T(t) + a_2 T^2(t) + a_3 T^3(t) + a_4 T^4(t) \tag{2-6.1}$$

where H_0, a_1, a_2, a_3, and a_4 are constants.

- Antoine equation for the vapor pressure of a pure substance, p^o, as a function of temperature, T:

$$p^o[T(t)] = e^{A - B/[T(t)+C]} \tag{2-6.2}$$

where A, B, and C are constants.

- Equilibrium vapor mole fraction, y, as a function of liquid mole fraction, x:

$$y[x(t)] = \frac{\alpha x(t)}{1 + (\alpha - 1)x(t)} \qquad \text{(2-6.3)}$$

where α is the relative volatility, usually assumed constant.

- Fluid flow, f, as a function of pressure drop, Δp:

$$f[\Delta p(t)] = k\sqrt{\Delta p(t)} \qquad \text{(2-6.4)}$$

where k is a constant conductance coefficient.

- Radiation heat transfer rate, q, as a function of temperature, T:

$$q[T(t)] = \varepsilon \sigma A T^4(t) \qquad \text{(2-6.5)}$$

where $\in, \sigma$, and A are constants.

- Arrhenius equation for the dependence of reaction rate coefficient, k, on temperature, T:

$$k[T(t)] = k_0 e^{-E/RT(t)} \qquad \text{(2-6.6)}$$

where k_0, E, and R are constants.

- Reaction rate, r, as a function of temperature, T, and reactants concentration, $c_A, c_B, \ldots$:

$$r[T(t), c_A(t), c_B(t), \ldots] = k[T(t)]c_A^a(t)c_B^b(t) \cdots \qquad \text{(2-6.7)}$$

where $k[T(t)]$ is given by Eq. 2-6.6, and $a, b, \ldots$ are constant.

All of the nonlinear functions given above except the last one are functions of a single variable. Next, we will introduce the linearization procedure for functions of one variable and then extend it to functions of two or more variables.

2-6.1   Linearization of Functions of One Variable

Any function can be expanded in a Taylor series about a base point, as follows:

$$f[x(t)] = f(\overline{x}) + \left.\frac{df}{dx}\right|_{\overline{x}} [x(t) - \overline{x}] + \frac{1}{2!}\left.\frac{d^2 f}{dx^2}\right|_{\overline{x}} [x(t) - \overline{x}]^2 + \cdots \qquad \text{(2-6.8)}$$

where $\overline{x}$ is the base value of x around which the function is expanded. The linearization of function $f[x(t)]$ consists of approximating it with only the first two terms of the Taylor series expansion.

$$f[x(t)] \approx f(\overline{x}) + \left.\frac{df}{dx}\right|_{\overline{x}} [x(t) - \overline{x}] \qquad \text{(2-6.9)}$$

This is the basic linearization formula. Since $\overline{x}$ is a constant, the right-hand side of the equation is linear in the variable $x(t)$.

Figure 2-6.1 presents a graphical interpretation of the linearization formula, Eq. 2-6.9. The linear approximation is a straight line passing through the point $[\overline{x}, f(\overline{x})]$ with slope $df/dx\big|_{\overline{x}}$. This line is by definition the tangent to $f(x)$ at $\overline{x}$. Notice that the difference between the nonlinear function and its linear approximation is small near the base point $\overline{x}$ and becomes larger the farther $x(t)$ is from $\overline{x}$. The width of the range in which the linear approximation is accurate depends on the function. Some functions are more curved than others and thus have a narrower range over which the linear approximation is accurate.

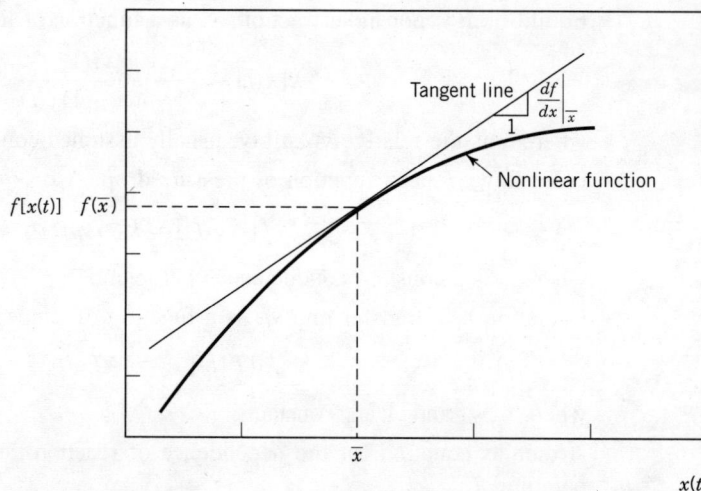

Figure 2-6.1 The linear approximation is the tangent to the nonlinear function at the base point $\overline{x}$.

It is important to realize that what affects the parameters of the transfer function of a linearized system is the slope, $df/dx\big|_{\overline{x}}$, not the value of the function itself, $f(\overline{x})$. This will become obvious when we show how to apply the linearization technique to nonlinear differential equations. The following example illustrates the application of the linearization formula.

EXAMPLE 2-6.1

Linearize the Arrhenius equation, Eq. 2-6.6, for the temperature dependence of chemical reaction rate coefficients. For a reaction with a coefficient $k(\overline{T}) = 100$ s^{-1} and an energy of activation $E = 22{,}000$ kcal/kmole, estimate the error in the slope of the function in the range $\pm10°$C around $\overline{T} = 300°$C (573 K).

SOLUTION

Apply the linearization formula, Eq. 2-6.9, to Eq. 2-6.6.

$$k[T(t)] \approx k(\overline{T}) + \frac{dk}{dT}\bigg|_{\overline{T}}[T(t) - \overline{T}]$$

where

$$\frac{dk}{dT}\bigg|_{\overline{T}} = \frac{d}{dT}\left[k_0 e^{-E/(RT(t))}\right]_{T=\overline{T}} = k_0 e^{-E/(R\overline{T})}\frac{E}{R\overline{T}^2} = k(\overline{T})\frac{E}{R\overline{T}^2}$$

For the numerical values given, with $R = 1.987$ kcal/kmole-K (ideal gas law constant), the base value of the slope is

$$\frac{dk}{dT}\bigg|_{300°C} = (100 \text{ s}^{-1})\frac{(22{,}000 \text{ kcal/kmole})}{(1.987 \text{ kcal/kmole-K})(300 + 273 \text{ K})^2} = 3.37\frac{\text{s}^{-1}}{°\text{C}}$$

and the linear approximation of the function is

$$k[T(t)] \approx 100 \text{ s}^{-1} + \left(3.37\frac{\text{s}^{-1}}{°\text{C}}\right)[T(t) - \overline{T}]$$

In the range 290 to 310°C, the actual function and slope are

$$\text{At } T = 290°\text{C}, \quad k(T) = 70.95 \text{ s}^{-1}, \quad \frac{dk}{dT} = 2.48\frac{\text{s}^{-1}}{°\text{C}}$$

$$\text{At } T = 310°\text{C}, \quad k(T) = 139.3 \text{ s}^{-1}, \quad \frac{dk}{dT} = 4.54\frac{\text{s}^{-1}}{°\text{C}}$$

In comparison, the linear approximation of the function predicts $k(290°C) = 100 + 3.37(290 - 300) = 66.3$ s^{-1} which is -6.6 % in error, and $k(310°C) = 133.7$ s^{-1}, which is -4 % in error. As for the slope, it varies from 2.48 to 4.54 s^{-1}/°C, which is from -26.4 % to $+34.7$ % of the linear approximation, 3.37 s^{-1}/°C.

This example shows that, for the Arrhenius formula, the linear approximation is accurate over a wider range for the function than it is for its slope. Unfortunately, it is the slope that affects the parameters of the transfer function. However, the error of $\pm35\%$ in the parameters is usually satisfactory for many control system calculations.

2-6.2   Linearization of Functions of Two or More Variables

We can use the Taylor series expansion to derive the linearization formula for functions of two or more variables, just as we did to derive Eq. 2-6.9. Here we keep the first partial derivative term for each of the variables. The resulting linear approximation is

$$f[x_1(t), x_2(t), \ldots] \approx f(\overline{x}_1, \overline{x}_2, \ldots) + \frac{\overline{\partial f}}{\partial x_1}[x_1(t) - \overline{x}_1] + \frac{\overline{\partial f}}{\partial x_2}[x_2(t) - \overline{x}_2] + \cdots$$

$$(2\text{-}6.10)$$

where $\dfrac{\overline{\partial f}}{\partial x_k} = \dfrac{\partial f}{\partial x_k}\bigg|_{(\overline{x}_1, \overline{x}_2, \ldots)}$, and $\overline{x}_1, \overline{x}_2, \ldots$, are the base values of each variable. Recall, from calculus, that the partial derivative is the change with respect to one variable keeping all other variables constant. The following examples illustrate the use of Eq. 2-6.10, the linearization formula for functions of more than one variable.

EXAMPLE 2-6.2

A common nonlinear function occurring in component and energy balances is the product of two variables (flow and composition, flow and specific enthalpy, etc.). For this function, the linearization is so simple that sometimes is difficult to grasp. As an example of this simple function, consider the area a of a rectangle as a function of its width, w, and its height, h:

$$a[w(t), h(t)] = w(t)h(t)$$

Linearization, using Eq. 2-6.10, results in

$$a[w(t), h(t)] \approx a(\overline{w}, \overline{h}) + \frac{\partial a}{\partial w}\bigg|_{(\overline{w}, \overline{h})}[w(t) - \overline{w}] + \frac{\partial a}{\partial h}\bigg|_{(\overline{w}, \overline{h})}[h(t) - \overline{h}]$$

$$\approx a(\overline{w}, \overline{h}) + \overline{h}[w(t) - \overline{w}] + \overline{w}[h(t) - \overline{h}]$$

Figure 2-6.2 shows a graphical representation of the area. The figure shows that the error in the approximation is the area of the small rectangle in the upper right-hand corner, $[w(t) - \overline{w}][h(t) - \overline{h}]$. This error is small for small relative increments in the width and height. For example, assume the base values are $\overline{w} = 2$ m, $\overline{h} = 1$ m, and they increment to 2.2 and 1.1 m, respectively. Then the error between the actual area, 2.42 m^2, and the approximation, $2.0 + 1(0.2) + 2(0.1) = 2.40$ m^2, is the area of the rectangle, $(0.2)(0.1) = 0.02$ m^2. Again the accuracy in the function is good (-0.8% error), but the slopes are each off by 10 %.

EXAMPLE 2-6.3

The density of an ideal gas is a function of pressure and temperature:

$$\rho[p(t), T(t)] = \frac{Mp(t)}{RT(t)}$$

where M is the molecular weight, $p(t)$ is the absolute pressure, $T(t)$ is the absolute temperature, and R is the ideal gas law constant. Obtain the linear approximation and evaluate it for air

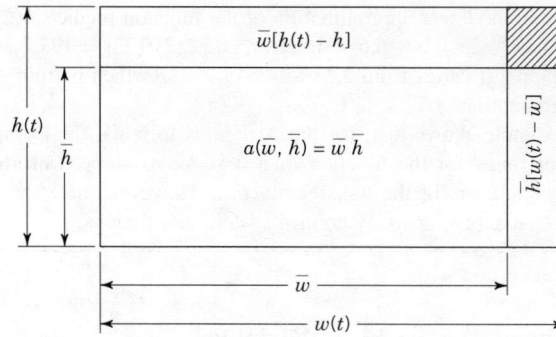

Figure 2-6.2 The cross-hatched area is the error of the linear approximation to the function $a[w(t), h(t)] = w(t)h(t)$.

$(M = 29)$, at 300 K and atmospheric pressure, 101.3 kPa. In these (SI) units the ideal gas law constant is 8.314 kPa-m^3/kmole-K.

SOLUTION

Application of Eq. 2-6.10 results in

$$\rho[p(t), T(t)] \approx \rho(\overline{p}, \overline{T}) + \frac{\partial \rho}{\partial p}\bigg|_{(\overline{p},\overline{T})} [p(t) - \overline{p}] + \frac{\partial \rho}{\partial T}\bigg|_{(\overline{p},\overline{T})} [T(t) - \overline{T}]$$

$$\approx \frac{M\overline{p}}{R\overline{T}} + \frac{\partial}{\partial p}\left[\frac{Mp(t)}{RT(t)}\right]_{(\overline{p},\overline{T})} [p(t) - \overline{p}] + \frac{\partial}{\partial T}\left[\frac{Mp(t)}{RT(t)}\right]_{(\overline{p},\overline{T})} [T(t) - \overline{T}]$$

$$\approx \frac{M\overline{p}}{R\overline{T}} + \frac{M}{R\overline{T}}[p(t) - \overline{p}] - \frac{M\overline{p}}{R\overline{T}^2}[T(t) - \overline{T}]$$

Numerically, the values are

$$\rho[p(t), T(t)] \approx \left(1.178 \frac{\text{kg}}{\text{m}^3}\right) + \left(0.01163 \frac{\text{kg}}{\text{m}^3\text{-kPa}}\right)[p(t) - 101.3 \text{ kPa}]$$

$$- \left(0.00393 \frac{\text{kg}}{\text{m}^3\text{-K}}\right)[T(t) - 300 \text{ K}]$$

where ρ is in kg/m^3, p is in kPa, and T is in K.

One thing to observe in the preceding examples is that in the linear approximations, the coefficient of each variable is constant. So, although we may sometimes show them as functions of the base values of the variables, the equations would not be linear if these base values were not assumed constant. Whenever we can express the coefficients as functions of the base values, we can calculate them at different base values. Let us next apply our newly acquired skill to the linearization of nonlinear differential equations.

2-6.3   Linearization of Differential Equations

The following procedure to linearize nonlinear differential equations assumes that the equations can be expressed as first-order equations. This is not a significant restriction because, as we shall see in Chapters 3 and 4, process dynamic models usually consist of developing a set of first-order differential equations. Only after linearization and Laplace transformation should the first-order equations be combined to form higher-order equations. This is because it is much easier to manipulate the algebraic equations relating the transforms than the original differential equations. Most computer simulation programs also require that the differential equations be first order.

In the following procedure we assume, as we have done so far, that the initial conditions are at steady state. Furthermore, we select the base point for the linearization as the initial steady state, because this greatly simplifies the linearized equations.

Consider the following first-order differential equation with one input:

$$\frac{dy(t)}{dt} = g[x(t), y(t)] + b \qquad (2\text{-}6.11)$$

where $g[x(t), y(t)]$ is a nonlinear function of the input variable, $x(t)$, and the output variable, $y(t)$, and b is a constant. At the initial steady-state conditions, Eq. 2-6.11 can be written as

$$0 = g(\overline{x}, \overline{y}) + b \qquad (2\text{-}6.12)$$

where we have chosen the base point for the linearization to be the initial conditions, $\overline{x} = x(0)$, $\overline{y} = y(0)$. Notice that the time derivative is zero because of the initial steady-state assumption. Subtract Eq. 2-6.12 from Eq. 2-6.11 to obtain

$$\frac{dy(t)}{dt} = g[x(t), y(t)] - g(\overline{x}, \overline{y}) \qquad (2\text{-}6.13)$$

If we now approximate Eq. 2-6.13 using the formula for the linearization of multivariable functions, Eq. 2-6.10, the result is

$$\frac{dy(t)}{dt} \approx \left.\frac{\partial g}{\partial x}\right|_{(\overline{x}, \overline{y})} [x(t) - \overline{x}] + \left.\frac{\partial g}{\partial y}\right|_{(\overline{x}, \overline{y})} [y(t) - \overline{y}] \qquad (2\text{-}6.14)$$

The terms in brackets are the deviation variables that were introduced in Section 2-3.1, because $\overline{x}$ and $\overline{y}$ are the initial conditions. Substitute the deviation variables $X(t) = x(t) - \overline{x}$, $Y(t) = y(t) - \overline{y}$, to obtain

$$\frac{dY(t)}{dt} \approx a_1 X(t) + a_2 Y(t) \qquad (2\text{-}6.15)$$

where $a_1 = \left.\partial g/\partial x\right|_{(\overline{x}, \overline{y})}$ and $a_2 = \left.\partial g/\partial y\right|_{(\overline{x}, \overline{y})}$.

We can generally apply the preceding procedure to any equation for any number of variables. Equation 2-6.15 is the linear approximation of Eq. 2-6.11; compare the two and notice the following:

- The constant b in Eq. 2-6.11 drops out. There should not be any constant terms in the equation relating deviation variables.
- The linearized equation, Eq. 2-6.15, replaces the right-hand side of the original nonlinear equation by a sum of terms each of which consists of a constant times a deviation variable for each variable appearing on the right-hand side of the original differential equation.
- The initial condition of the deviation variable is zero, $Y(0) = y(0) - y(0) = 0$.

We will next illustrate the linearization procedure with an example.

EXAMPLE 2-6.4

The following differential equation results from a reactant mass balance in a stirred tank reactor (the complete model of the reactor is developed in Chapter 4).

$$\frac{dc_A(t)}{dt} = \frac{1}{V} f(t) c_{Ai}(t) - \frac{1}{V} f(t) c_A(t) - k[T(t)] c_A(t)$$

where $k[T(t)]$ is the Arrhenius dependence of the reaction rate with temperature, Eq. 2-6.6. We linearized this function in Example 2-6.1. We assume that V, the reactor volume, is constant. The input variables are $f(t)$, the reactants flow, and $c_{Ai}(t)$, the inlet reactant concentration, while the output variables are $T(t)$, the reactor temperature, and $c_A(t)$, the reactant concentration. Obtain the linear approximation to the equation and the expressions for the time constant and gains of the transfer function of the reactant concentration.

SOLUTION

Each term of the equation is nonlinear: the first two terms each consist of the product of two variables, and the third term is a product of two variables one of which is the nonlinear Arrhenius function. We could linearize each of these terms in turn, substitute the linear approximations into the equation, and then subtract the initial steady-state equation and express in terms of deviation variables. However, we can simplify the algebraic manipulations by the following procedure.

Define the right-hand side of the equation as a nonlinear function of the four variables that appear in it.

$$\frac{dc_A(t)}{dt} = g[f(t), c_{Ai}(t), T(t), c_A(t)]$$

$$= \frac{1}{V} f(t) c_{Ai}(t) - \frac{1}{V} f(t) c_A(t) - k[T(t)] c_A(t)$$

Then, by comparison with Eq. 2-6.11 and its linear approximation, Eq. 2-6.15, we know the linear approximation is of the form

$$\frac{dC_A(t)}{dt} = a_1 F(t) + a_2 C_{Ai}(t) + a_3 \Gamma(t) + a_4 C_A(t)$$

where $C_A(t) = c_A(t) - \overline{c_A}$, $F(t) = f(t) - \overline{f}$, $C_{Ai}(t) = c_{Ai}(t) - \overline{c}_{Ai}$, $\Gamma(t) = T(t) - \overline{T}$ are the deviation variables, and the constants are evaluated by taking the partials of function g.

$$a_1 = \overline{\frac{\partial g}{\partial f}} = \frac{\overline{c}_{Ai} - \overline{c}_A}{V} \qquad\qquad a_2 = \overline{\frac{\partial g}{\partial c_{Ai}}} = \frac{\overline{f}}{V}$$

$$a_3 = \overline{\frac{\partial g}{\partial T}} = -k(\overline{T}) \frac{E}{R\overline{T}^2} \overline{c}_A \qquad a_4 = \overline{\frac{\partial g}{\partial c_A}} = -\frac{\overline{f}}{V} - k(\overline{T})$$

where the line over the partials is shorthand to indicate that they are evaluated at the base point. We invite you to verify the expressions for the constants by taking the derivatives of the nonlinear function g.

The next step is to put the linearized first-order equation in the standard form. To do this we move the term with $C_A(t)$ to the left side of the equal sign and divide by its coefficient, $-a_4$.

$$\tau \frac{dC_A(t)}{dt} + C_A(t) = K_1 F(t) + K_2 C_{Ai}(t) + K_3 \Gamma(t)$$

where the parameters are

$$\tau = -\frac{1}{a_4} = \frac{V}{\overline{f} + Vk(\overline{T})} \qquad\qquad K_1 = -\frac{a_1}{a_4} = \frac{\overline{c}_{Ai} - \overline{c}_A}{\overline{f} + Vk(\overline{T})}$$

$$K_2 = -\frac{a_2}{a_4} = \frac{\overline{f}}{\overline{f} + Vk(\overline{T})} \qquad K_3 = -\frac{a_3}{a_4} = -\frac{Vk(\overline{T})E\overline{c}_A}{R\overline{T}^2[\overline{f} + Vk(\overline{T})]}$$

Finally, Laplace transform and solve for $C_A(s)$:

$$C_A(s) = \frac{K_1}{\tau s + 1} F(s) + \frac{K_2}{\tau s + 1} C_{Ai}(s) + \frac{K_3}{\tau s + 1} \Gamma(s)$$

This model is not complete. It requires another equation for the temperature, $\Gamma(t)$, which is not an independent input. We will present the complete model of the reactor in Chapter 4.

The preceding example shows that the parameters of the transfer function of a linearized equation depend on the values of the variables at the base point. Notice that, as was pointed out earlier, the parameters depend on the partials of the nonlinear function, a_1, a_2, a_3, and a_4, rather than on the value of the function itself.

This section has shown how to linearize nonlinear differential equations so that the powerful technique of Laplace transforms can be applied. Once the transfer function of the linearized equations is developed, the response characteristics can be related to its parameters by the methods discussed in previous sections of this chapter. The important characteristic of nonlinear systems is that their response depends on the operating point. It is convenient to think of the parameters of the linearized system as being valid at the base point rather than in a region of parameter values. In most situations the gains and time constants do not vary enough to significantly affect the performance of control systems, but we must always keep in mind that the parameters do vary and should make allowances for their variation in the design of control systems for nonlinear systems.

2-7 SUMMARY

This chapter presented the techniques of Laplace transforms and linearization in the way they are applied for analyzing the dynamic response of processes and their control systems. The characteristics of the process response to input signals were related to the roots of the denominator of the process transfer function, and the responses and transfer functions of first- and second-order systems were presented. In the chapters that follow, the transfer functions of specific processes will be related to the physical process parameters through the application of the fundamental laws of conservation. Later chapters will use Laplace transforms for designing and analyzing process control systems.

PROBLEMS

2-1. Using the definition of the Laplace transform, derive the transforms $F(s)$ of the following functions.

(a) $f(t) = t$

(b) $f(t) = e^{-at}$ where a is a constant.

(c) $f(t) = \cos \omega t$ where ω is a constant.

(d) $f(t) = e^{-at} \cos \omega t$ where a and ω are constant.

Note: In parts c and d you will need the trigonometric identity

$$\cos x \equiv \frac{e^{ix} + e^{-ix}}{2}$$

Check your answers against the entries in Table 2-1.1.

2-2. Using a table of Laplace transforms and the properties of the transform, find the transforms $F(s)$ of the following functions.

(a) $f(t) = u(t) + 2t + 3t^2$

(b) $f(t) = e^{-2t}[u(t) + 2t + 3t^2]$

(c) $f(t) = u(t) + e^{-2t} - 2e^{-t}$

(d) $f(t) = u(t) - e^{-t} + te^{-t}$

(e) $f(t) = u(t-2)[1 - e^{-2(t-2)} \sin(t-2)]$

2-3. Check the validity of your results to Problem 2-2 by application of the initial and final value theorems. Do these theorems apply in all of the cases?

2-4. In Example 2-1.1(b), the Laplace transform of a pulse was obtained by application of the definition of the transform. Show that the same transform can be obtained by application of the real translation theorem. Notice that the pulse is the difference between two identical step changes of size H with the second one delayed by the duration of the pulse, T.

$$f(t) = Hu(t) - Hu(t - T)$$

2-5. In the statement of the real translation theorem we pointed out that for the theorem to apply, the delayed function has to be zero for all times less than the delay time. Prove this by calculating the Laplace transform of the function

$$f(t) = e^{-(t-t_0)/\tau}$$

where t_0 and τ are constants,

(a) Assuming that it holds for all times greater than zero, that is, that it can be rearranged as

$$f(t) = e^{t_0/\tau} e^{-t/\tau}$$

(b) If it is zero for $t \leq t_0$, that is, that it should be properly written as

$$f(t) = u(t - t_0)e^{-(t-t_0)/\tau}$$

Sketch the graph of the two functions. Are the two answers the same? Which one agrees with the result of the real translation theorem?

2-6. Obtain the solution $Y(t)$, as a deviation from its initial steady-state condition $y(0)$, of the following differential equations. Use the method of Laplace transforms and partial fractions expansion. The forcing function is the unit step function, $x(t) = u(t)$.

(a) $\dfrac{dy(t)}{dt} + 2y(t) = 5x(t) + 3$

(b) $9\dfrac{d^2y(t)}{dt^2} + 18\dfrac{dy(t)}{dt} + 4y(t) = 8x(t) - 4$

(c) $9\dfrac{d^2y(t)}{dt^2} + 9\dfrac{dy(t)}{dt} + 4y(t) = 8x(t) - 4$

(d) $9\dfrac{d^2y(t)}{dt^2} + 12\dfrac{dy(t)}{dt} + 4y(t) = 8x(t) - 4$

(e) $2\dfrac{d^3y(t)}{dt^3} + 7\dfrac{d^2y(t)}{dt^2} + 21\dfrac{dy(t)}{dt} + 9y(t) = 3x(t)$

2-7.   Repeat Problem 2-6(d) using as the forcing function

(a) $x(t) = e^{-t/3}$

(b) $x(t) = u(t - 1)e^{-(t-1)/3}$

2-8.   For the differential equations given in Problem 2-6, determine whether the response is stable or unstable, oscillatory or monotonic. Find also the dominant root, the period of the oscillations and the decay ratio if oscillatory; the time required for the slowest term in the response, or the amplitude of the oscillations, to decay to within less than 1 % (0.67 %) of its initial value; and the final steady-state value of the output. *Note:* It is not necessary to solve Problem 2-6 to answer the questions in this problem.

2-9.   *Second-Order Response: Bird Mobile.* The bird mobile shown in Figure P2-1 has a mass of 50 g and the spring that holds it extends 27 cm when the weight of the bird is applied to it. Neglecting resistance of the air to the motion of the bird, we can derive the following equation by writing a dynamic force balance on the bird:

$$M\dfrac{d^2y(t)}{dt^2} = -Mg - ky(t) + f(t)$$

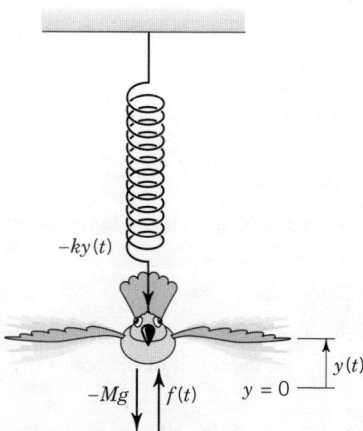

Figure P2-1 Bird mobile for Problem 2-9.

where $y(t)$, is the vertical position of the bird in m, $f(t)$, is the force required to start the bird in motion in N, M, is

the mass of the bird in kg, k, is the spring constant in N/m, and g is the local acceleration of gravity, 9.8 m/s^2. Find the period of oscillation of the bird. Does the solution predict that the bird will oscillate forever? What term must be added to the model equation to more accurately reflect the actual motion of the bird? What is the physical significance of this added term? *Note:* Simulation of this system is the subject of Problem 13-1.

2-10.   For the general first-order differential equation, Eq. 2-4.4, obtain the response to

(a)   An impulse, $X(t) = \delta(t)$.

(b)   The pulse sketched in Fig. 2-1.1b.

Sketch the graph of the response, $Y(t)$, for each case.

2-11.   *Response of an Integrating Process.* The response of the liquid level in a tank is given by the first-order differential equation

$$A\dfrac{dh(t)}{dt} = f(t)$$

where $h(t)$, m, is the level in the tank, $f(t)$, m^3/s, is the flow of liquid into the tank, and A, m^2, is the constant area of the tank. Obtain the transfer function for the tank and the response of the level to a unit step in flow, $F(t) = u(t)$. Sketch the graph of the level response, $H(t)$. Why do you think we call this result the response of an *integrating process*?

2-12.   For the second-order differential equations given in Problem 2-6, find the time constant and damping ratio, and classify them as overdamped or underdamped. For the overdamped equations, figure out the effective time constants, and for the underdamped equations, find the frequency and period of oscillation, the decay ratio, and the percent overshoot, rise time, and settling time on a step input.

2-13.   Evaluate the coefficients of the partial fraction expansion that lead to Eqs. 2-5.10, 2-5.11, 2-5.12, and 2-5.13.

2-14.   Derive Eq. 2-5.23 from Eq. 2-5.22.

2-15.   A common transfer function that models many second-order interacting systems (see Chapter 4), is

$$Y(s) = \dfrac{k_1}{(\tau_1 s + 1)(\tau_2 s + 1) - k_2}X(s)$$

where τ_1, τ_2, k_1, and k_2 are constants. Derive the relationships between the gain, time constant, and damping ratio of the second-order transfer function and the four constants appearing in the transfer function. Assuming that all four constants are positive real numbers,

(a)   Show that the response is overdamped.

(b)   Show that the response is stable if $k_2 < 1$.

(c)   Derive the relationship between the two effective time constants and the four constants of the transfer function.

2-16. The transfer function of a feedback control loop is given by

$$C(s) = \frac{K_c}{(3s+1)(s+1)+K_c} R(s)$$

where K_c is the controller gain. Derive the relationships between the gain, time constant, and damping ratio of the second-order transfer function to the controller gain. Find the ranges of the controller gain for which the response is (i) overdamped, (ii) underdamped, and (iii) undamped. Can the response be unstable for any positive values of the controller gain?

2-17. Linearize the following nonlinear functions and express your results in terms of deviations from the base point.

(a) The equation for enthalpy as a function of temperature, Eq. 2-6.1.

(b) The Antoine equation for the vapor pressure, Eq. 2-6.2.

(c) The equation for vapor mole fraction at equilibrium as a function of liquid mole fraction, Eq. 2-6.3.

(d) The equation for fluid flow as a function of pressure drop, Eq. 2-6.4.

(e) The equation for radiation heat transfer rate as a function of temperature, Eq. 2-6.5.

2-18. As pointed out in the text, the error of the linear approximation usually increases as the variable deviates from its base value. The error in the slope is the one that is important. For the rate of radiation heat transfer as a function of temperature, Eq. 2-6.5, find the range of the temperature for which the slope of the function, $dq/dT|_T$, remains within $\pm 5\%$ of its base value. Calculate also the temperature range for which the linear approximation of the heat transfer rate, q, is within $\pm 5\%$ of its true value. Consider two base values, $\overline{T} = 400$ K and 600 K. Discuss briefly how the range of applicability of the linear approximation, based on its ability to match the slope of the function, varies with the base value of the temperature.

2-19. Solve Problem 2-18 for the formula of the equilibrium vapor mole fraction as a function of liquid mole fraction, Eq. 2-6.3. Calculate the range of values of the liquid mole fraction x, for which the slope, $dy/dx|_{\overline{x}}$, is within $\pm 5\%$ of its base value. Calculate also the range of values of x for which the linear approximation to the equilibrium mole fraction y remains within $\pm 5\%$ of its true value. Consider the following cases.

(a) $\alpha = 1.10$ $\overline{x} = 0.10$ (c) $\alpha = 5.0$ $\overline{x} = 0.10$

(b) $\alpha = 1.10$ $\overline{x} = 0.90$ (d) $\alpha = 5.0$ $\overline{x} = 0.90$

Discuss briefly why you think the range of applicability of the linear approximation, based on its ability to match the slope of the function, varies with both the parameter and the base value for this function.

2-20. The rate of a chemical reaction is given by the following expression.

$$r[c_A(t), c_B(t)] = kc_A^2(t)c_B(t)$$

where $k = 0.5$ m^6/(kmole^2-h), is a constant (isothermal operation). Obtain the linear approximation of this function at $\overline{c}_A = 2$ kmole/m^3, $\overline{c}_B = 1$ kmole/m^3, and find the error in the parameters of the approximation (i.e., the partial derivatives) when each of the concentrations changes, independently, by 1 kmole/m^3. Express the linear approximation in terms of deviation variables.

2-21. Raoult's law gives the vapor mole fraction $y(t)$ at equilibrium as a function of the temperature, $T(t)$, pressure, $p(t)$, and liquid mole fraction, $x(t)$:

$$y[T(t), p(t), x(t)] = \frac{p^o[T(t)]}{p(t)} x(t)$$

where $p^o[T(t)]$ is the vapor pressure of the pure component, given by Antoine's equation, Eq. 2-6.2. Obtain the linear approximation for the vapor mole fraction and express it in terms of deviation variables. Evaluate the parameters of the approximation for benzene at atmospheric pressure (760 mm Hg), 95°C, and a liquid mole fraction of 50 %. The Antoine constants for benzene are $A = 15.9008$, $B = 2788.51$°C, and $C = 220.80$°C, for the vapor pressure in mm Hg.

2-22. Evaluate the parameters of the transform obtained in Example 2-6.4, using the parameters of the Arrhenius formula given in Example 2-6.1 and the following reactor parameters:

$$V = 2.6 \text{ m}^3, \quad \overline{f} = 0.002 \text{ m}^3/\text{s}, \quad \overline{c}_{Ai} = 12 \text{ kmoles/m}^3.$$

Notice that the initial value of the reactant concentration, $\overline{c}_A$, can be found by the condition that the initial condition is at steady state, $\overline{g} = 0$.

2-23. A stray bullet fired by a careless robber punctures the compressed air tank at a gas station. The mass balance of air in the tank is

$$V\frac{d\rho(t)}{dt} = w_i(t) - A_o\sqrt{2\rho(t)[p(t) - p_o]}$$

where

$$\rho(t) = \frac{M}{RT} p(t)$$

$w_i(t)$ kg/s, is the inlet flow from the air compressor, $V = 1.5$ m^3, is the volume of the tank, $A_o = 0.785$ cm^2, is the area of the bullet hole, $M = 29$ kg/kmole, is the molecular weight of air, $R = 8.314$ kPa-m^3/kmole-K, is the ideal gas law constant, and the temperature T is assumed constant at 70°C.

Obtain a linear approximation to the differential equation around the initial pressure of 500 kPa gauge. Obtain also the Laplace transform for the pressure in the tank and evaluate the time constant and gain of the transfer function. *Note*: Simulation of this system is the subject of Problem 13-3.

2-24.   The temperature of a turkey in an oven, assumed uniform throughout the bird, and neglecting the heat absorbed by the cooking reactions, is given by the differential equation

$$Mc_v \frac{dT(t)}{dt} = \sigma \varepsilon A[T_s^4(t) - T^4(t)]$$

where M, is the mass of the turkey in pounds (lb), c_v is the specific heat in Btu/lb-°R, $T(t)$ is the temperature of the turkey in °R, $\sigma = 0.1714 \cdot 10^{-8}$ Btu/h-ft²-°R⁴, is the Stephan–Boltzman constant, ϵ is the emissivity of the skin of the turkey, A is the area of the turkey in square feet (ft²), and $T_s(t)$ is the temperature of the oven in °R.

Obtain a linear approximation to the differential equation. Obtain also the Laplace transform for the temperature of the turkey and write the expressions for the time constant and gain of the transfer function. What is the input variable for this problem? *Note*: Simulation of this system is the subject of Problem 13-4.

2-25.   The temperature of a slab being heated by an electric heater is given by the differential equation

$$C \frac{dT(t)}{dt} = q(t) - \alpha[T^4(t) - T_s^4]$$

where $T(t)$ is the temperature of the slab in °R, assumed uniform, $q(t)$ is the rate of heat input in Btu/h, $C = 180$ Btu/°R, is the heat capacity of the slab, $T_s = 540°$R, is the surrounding temperature (constant), and $\alpha = 5 \cdot 10^{-8}$ Btu/h-°R⁴, is the coefficient of heat radiation. Obtain the linear approximation of the differential equation around the initial steady-state temperature of 700°R. Obtain also the Laplace transform of the temperature of the slab and find the gain and time constant of the transfer function.

Chapter 3

Dynamic Systems: Simple Processes

As briefly presented in Chapter 1, the dynamic response of processes is of prime consideration in the design, analysis, and implementation of process control systems. An interesting and important characteristic of chemical processes is that their dynamics change from one process to another. For instance, the response of temperature is different from the response of level. Further, the response of temperature in a heat exchanger is different from the response of temperature in a furnace. *The principal objective of this chapter is to show how to describe the dynamic response of simple processes using mathematical models, transfer functions, and block diagrams.* Though simple, these processes are taken from actual industrial applications. Chapter 4 presents more complex processes.

3-1 PROCESSES AND IMPORTANCE OF PROCESS CHARACTERISTICS

It is important to start this chapter, and indeed to launch the entire subject of process control, by explaining what a "process" is and describing its characteristics from a process control point of view. To do this, let us consider the heat exchanger of Chapter 1, shown again in Fig. 3-1.1.

The controller's job is to control the process. In the example at hand, the controller is to take action that keeps the outlet temperature, $T(t)$, at a specified value, its set point. The controller receives a signal from the transmitter. It is through the transmitter that the controller "sees" the controlled variable. *Thus, as far as the controller is concerned, the real controlled variable is the transmitter's output.* The relation between the transmitter's output and the physical variable to control, $T(t)$, is given by the transmitter calibration as presented in Chapter 5. The outlet temperature and the transmitter's output should be correctly related. If they are not, for whatever reason, the controller still reacts to the transmitter's output, not to $T(t)$.

In this example the controller is to manipulate the steam valve position to maintain the controlled variable at set point. Note, however, that the way the controller manipulates the valve position is by changing its output signal to the valve. *Thus, as far as the controller is concerned, the real manipulated variable is its own output.* The controller does not manipulate the valve position directly; it only manipulates its output signal.

We can now define the *process* as *anything between the controller's output and the controller's input.* Most often the controller's input is provided by the transmitter

Principles and Practice of Automatic Process Control/Third Edition, by C. A. Smith and A. B. Corripio
ISBN 0-471-66141-4   Copyright © 2006 John Wiley & Sons (Asia) Pte. Ltd.

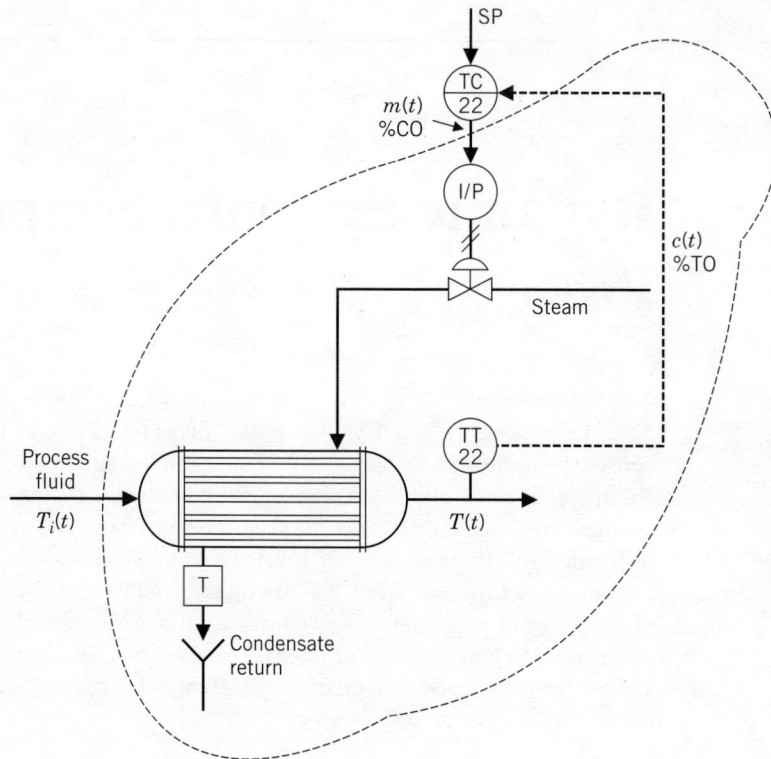

Figure 3-1.1*a* Heat exchanger control system.

output. There are some instances however, in which this may not be the case, such as when a mathematical manipulation, such as filtering, is done on the signal from the transmitter before the controller receives it. Because the transmitter usually provides the input to the controller directly, we can say that *controller's input is equal to the transmitter's output.* Referring to Fig. 3-1.1*a* the process is anything within the area delineated by the dashed curve. The process includes the I/P transducer, the valve, the heat exchanger with associated piping, the sensor, and the transmitter. Note that in the figure we have used the term $c(t)$ for the output signal from the transmitter. This is to remind that this signal is the controlled variable; the unit of $c(t)$ is percent of transmitter's output, or just simply %TO. In the figure we have also used the term $m(t)$ for the output signal from the controller to indicate that this signal is the manipulated variable; the unit of $m(t)$ is percent of controller's output, or just simply %CO. We will use this notation of $c(t)$ and $m(t)$ in this book.

For further understanding of what we have just discussed, consider Fig. 3-1.1*b*. The diagram shows all the parts of the process and how they are related. It shows that the output signal from the controller, $m(t)$, enters the I/P transducer, producing a pneumatic signal. This signal then goes to the valve, producing a steam flow. This flow enters the heat exchanger and, along with other process inputs, produces an output temperature $T(t)$. This temperature is measured by a sensor, and the output signal from the sensor, maybe in millivolts, is received by the transmitter, which produces a signal, $c(t)$, to the controller. Thus, the diagram clearly shows that as far as the controller is concerned, the controlled variable is the transmitter's output and the manipulated variable is the controller's output.

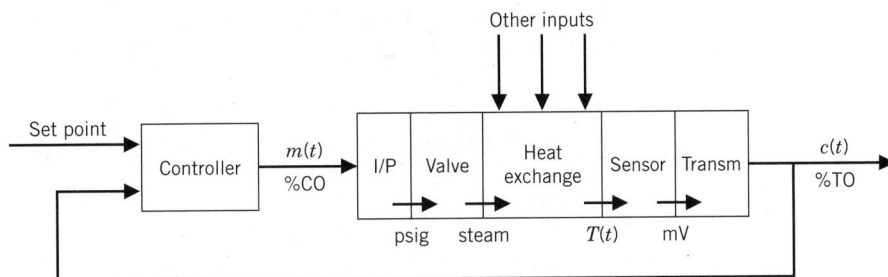

Figure 3-1.1*b* Heat exchanger temperature control.

Why is it necessary to understand the characteristics of the process to be controlled? As we noted in Chapter 1, the control performance provided by the controller depends on the adjustment or specification of different terms in the controller. Setting these terms is referred to as *tuning the controller*. The optimum controller tuning depends on the process to be controlled and on the tuning criterion. Every controller must be tuned specifically for the process it controls. Consequently, to tune a controller, we must first understand the characteristics, or behavior, of the process to be controlled.

Another way to explain the need to understand the characteristics of the process is to realize that in tuning the controller, what we are doing is "adapting" the controller to the process. Thus, it makes sense to first obtain the process characteristics, and then tune the controller, or adapt the controller characteristics, to that of the process. If this is done correctly, the complete closed-loop control system, *process plus controller*, will perform as required.

The present chapter and Chapter 4 discuss processes and their characteristics. Chapter 5 briefly presents some terms related to transmitters and also discusses control valves and controllers and their characteristics. Finally, Chapter 6 puts everything together; it "closes the loop." Chapter 7 shows how to tune the feedback controller once the process characteristics are known. Herein, then, lies the importance of knowing, understanding, and obtaining the process characteristics. *We can tune the controller only after the process steady-state and dynamic characteristics are known.*

3-2   MATHEMATICAL PROCESS MODELING

We stated in the introduction that a principal objective of this chapter is to describe the dynamic response of simple processes using mathematical models, transfer functions, and block diagram. Transfer functions and block diagrams are developed from mathematical models. Mathematical models are sets of equations that describe a process, or system. Actually, engineering students have been writing mathematical models since their physics courses. At that time the models were simple, and most often composed of a single equation. As we progress through engineering, the models become a bit more complex and composed of several equations. The reader may think back to the "material and energy balance" course, and later to the other thermodynamics courses, the transport courses, the unit operations courses, and the reactor design course. In all these courses the models are more complex than those in the physics courses; they are composed of many equations, and often require the use of computers for convenience to obtain a solution. The fact is, engineering students have been developing mathematical models almost from day one!

The mathematical models that we will develop in this, and the next chapters, start from a balance on a conserved quantity: mass or energy. The balance can be written as

$$\begin{array}{c} \text{Rate of mass/energy} \\ \text{into control volume} \end{array} - \begin{array}{c} \text{Rate of mass/energy} \\ \text{out of control volume} \end{array} = \begin{array}{c} \text{Rate of change of} \\ \text{mass/energy accumulated} \\ \text{in control volume} \end{array} \quad \textbf{(3-2.1)}$$

In processes where chemical reactions are not present, moles are also conserved. Thus, in these processes we may substitute the term moles for mass in the balance equation. Section 3-7 discusses processes where chemical reactions are present. Equation 3-2.1 is referred to as an *unsteady-state balance*.

For most of our education, the balances we have written are based on the steady-state assumption, yielding steady-state models. The steady-state assumption states that

$$\begin{array}{c} \text{Rate of change of mass/energy} \\ \text{accumulated in control volume} \end{array} = 0$$

Thus, if the rate of change of mass/energy accumulated in the control volume is zero, the mass/energy in the control volume does not change. There is accumulation of mass/energy, but this quantity is constant! The *steady-state balance* is then expressed by Eq. 3-2.2.

$$\begin{array}{c} \text{Rate of mass/energy} \\ \text{into control volume} \end{array} - \begin{array}{c} \text{Rate of mass/energy out of} \\ \text{control volume} \end{array} = 0 \quad \textbf{(3-2.2)}$$

Although, without a doubt, the reader has seen and applied this expression in previous courses, a brief review here is worthwhile. We take time to do this to show the reader that the models we will be developing are not much different from the ones the reader has already developed in previous courses.

Consider the process shown in Fig. 3-2.1. In this process two gas streams are entering and mixing, and a gas stream leaves. Stream 1 is a mixture of components A and B and Stream 2 is pure A. The contents of the process are assumed to be well mixed. Taking the tank as the control volume, we can write several balances, namely, a total mole balance, a mole balance on component A, and an energy balance.

A steady-state total mole balance gives

$$\dot{n}_1 + \dot{n}_2 - \dot{n}_3 = 0 \quad \textbf{(3-2.3a)}$$

This algebraic equation is typical and we have written it many times in previous courses. Let us understand what this equation is telling us about the physics of the

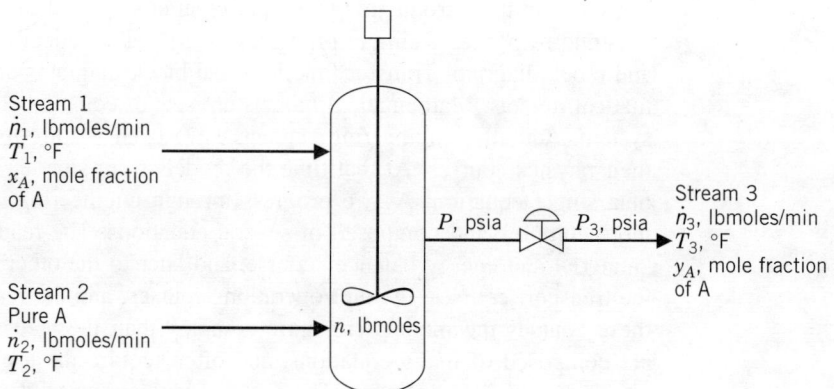

Stream 1
$\dot{n}_1$, lbmoles/min
T_1, °F
x_A, mole fraction
of A

Stream 2
Pure A
$\dot{n}_2$, lbmoles/min
T_2, °F

P, psia P_3, psia

Stream 3
$\dot{n}_3$, lbmoles/min
T_3, °F
y_A, mole fraction
of A

n, lbmoles

Figure 3-2.1 Process.

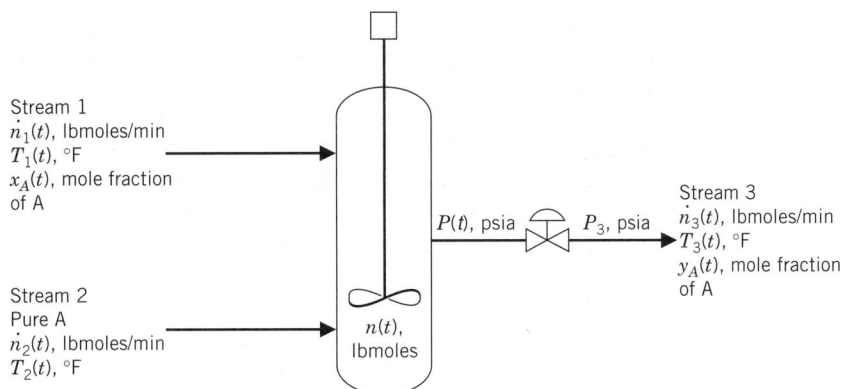

Figure 3-2.2 Process showing variables as a function of time.

process. The equation tells us that the flow rate of stream 3, $\dot{n}_3$, is equal to the sum of the flow rates of streams 2 and 3, $\dot{n}_1 + \dot{n}_2$; obviously we know that under steady state the outlet flow equals the summation of the two inlet flows. It tells us that if either $\dot{n}_1$ or $\dot{n}_2$ change, what is the necessary value of $\dot{n}_3$ to reach the new steady-state condition. What the equation does not tell us is *how long* it takes the process to reach the new condition. We know that if any entering stream changes, the flow of stream 3 may not change instantaneously. Before this flow changes, the pressure in the tank must increase to increase the pressure drop across the valve and thus, increasing the flow; it takes some amount of time for the pressure in the tank to build up. As we discussed in Chapter 1, most of our engineering education has not considered the amount of the time it takes for changes to occur; this is a consequence of the steady-state assumption. In actual practice it takes time for changes to happen. To show that a variable varies with time, is a function of time, we use the notation (t) after the variable; look at Fig. 3-2.2 and Eq. 3-2.3b.

$$\dot{n}_1(t) + \dot{n}_2(t) - \dot{n}_3(t) = 0 \qquad \textbf{(3-2.3b)}$$

Variables that are constant, not function of time, do not have (t) following them.

On the basis of our discussion, Eq. 3-2.3b does not completely describe how changes occur with time; the equation describes the process only at steady state. A more complete description is provided by the application of Eq. 3-2.1, resulting in an unsteady-state balance, or *dynamic model*. Equation 3-2.1 results in

$$\dot{n}_1(t) + \dot{n}_2(t) - \dot{n}_3(t) = \frac{dn(t)}{dt} \qquad \textbf{(3-2.4)}$$

Equation 3-2.4 is a differential equation and its solution describes how the moles accumulated in the tank, $n(t)$, change with time; the term $dn(t)/dt$ is the *rate of change of the total moles accumulated* in the tank. As the total moles of gas increase, the pressure of the gas in the tank, $P(t)$, increases and the outlet flow, $\dot{n}_3(t)$, also increases. Once the outlet flow becomes equal to the sum of the input flows, the derivative term $dn(t)/dt$ becomes zero and a new steady-state condition is reached. Certainly, differential equations look more complex than algebraic equations, but they describe much better the physics of the systems, and often are really not that difficult to solve analytically as shown in Chapter 2, or using a numerical solution as shown in Chapter 13.

There are a couple of things we need to discuss about Eq. 3-2.4. First, note that in writing the unsteady-state mole balance we just added the rate of change of the accumulation term to the familiar steady-state balance; we build in what we know. Second, look at the units of each term. The units of the $\dot{n}(t)$ terms are lbmoles/min, and the same is true for $dn(t)/dt$.

As we previously mentioned, the differential term is the rate of change of the moles accumulated in the tank with respect to time. If $\dot{n}_1(t) + \dot{n}_2(t) - \dot{n}_3(t)$ is not zero, then the moles inside the tank change and $dn(t)/dt$ indicates how fast the moles are changing (positive if the moles are increasing—more moles are entering than leaving—and negative if the moles are decreasing—more moles are leaving than entering). Once the amount of exiting moles becomes equal to the amount entering, the amount of moles inside the tank is constant and $dn(t)/dt = 0$.

A steady-state mole balance on component A gives

$$\dot{n}_1(t)x_A(t) + \dot{n}_2(t) - \dot{n}_3(t)y_A(t) = 0 \tag{3-2.5}$$

Equation 3-2.5 tells us that if any of the inlet operating condition changes, what is the necessary amount of moles of A exiting, given by $\dot{n}_3(t)y_A(t)$, to reach the new steady-state condition. Similarly to the total mole balance, what this equation does not tell us is *how long* it takes the process to reach the new condition. A better description of this process is given by writing an unsteady-state mass balance on component A.

$$\dot{n}_1(t)x_A(t) + \dot{n}_2(t) - \dot{n}_3(t)y_A(t) = \frac{dn_A(t)}{dt} = \frac{d[n(t)y_A(t)]}{dt} \tag{3-2.6}$$

The solution of Eq. 3-2.6 describes how the moles of component A accumulated in the tank, given by $n_A(t) = n(t)y_A(t)$, change with time; the term $dn_A(t)/dt = d[n(t)y_A(t)]/dt$ is the *rate of change of the moles of component A accumulated* in the tank. Please note again that in writing this unsteady-state mole balance on component A we just added the rate of change of the accumulation term to the familiar steady-state balance; once gain we build in what we know.

It is interesting to note that there may be a steady-state operation from a total mole balance point of view, that is, the summation of the total moles entering is equal to the total moles leaving, or $\dot{n}_1(t) + \dot{n}_2(t) - \dot{n}_3(t) = 0$. However, this may not be the case from a mole balance on component A point of view, as when $\dot{n}_1(t)$ and $\dot{n}_2(t)$ remain constant but the mass fraction of A in stream 1, $x_A(t)$, changes.

A steady-state energy balance, assuming negligible heat losses, gives

$$\dot{n}_1(t)\hat{h}_1(t) + \dot{n}_2(t)\hat{h}_2(t) - \dot{n}_3(t)\hat{h}_3(t) = 0 \tag{3-2.7}$$

where $\hat{h}_1(t)$, $\hat{h}_2(t)$, and $\hat{h}_3(t)$ are the molal enthalpies of each stream in Btu/lbmole. Selecting the gas phase at temperature of $0°F$ and 1 atm as the reference state, we can express the enthalpies as a function of temperatures; we assume for the sake of simplicity constant heat capacities. Then,

$$\dot{n}_1(t)\hat{C}_{p1}T_1(t) + \dot{n}_2(t)\hat{C}_{p2}T_2(t) - \dot{n}_3(t)\hat{C}_{p3}T_3(t) = 0 \tag{3-2.8}$$

where $\hat{C}_{p1}$, $\hat{C}_{p2}$, and $\hat{C}_{p3}$ are the molal heat capacities in Btu/lbmole·°F. We can now state a similar argument to the one related to the total balance and to the balance on component A. That is, if any of the inlet operating condition changes, Eq. 3-2.8 tells us the necessary amount of energy exiting, given by $\dot{n}_3(t)\hat{C}_{p3}T_3(t)$, to reach the new steady-state condition. What the equation does not tell us is *how long* it takes the process to reach the new condition. A better description of this process is given by writing an unsteady-state energy balance.

$$\dot{n}_1(t)\hat{h}_1(t) + \dot{n}_2(t)\hat{h}_2(t) - \dot{n}_3(t)\hat{h}_3(t) = \frac{d[n(t)\hat{u}_3(t)]}{dt} \tag{3-2.9}$$

where $u_3(t)$ is the internal energy of the gas in the tank. Equation 3-2.9 can be written in terms of temperature as

$$\dot{n}_1(t)\hat{C}_{p1}T_1(t) + \dot{n}_2(t)\hat{C}_{p2}T_2(t) - \dot{n}_3(t)\hat{C}_{p3}T_3(t) = \frac{d[n(t)\hat{C}_{v3}T_3(t)]}{dt} \qquad \textbf{(3-2.10)}$$

The solution of Eq. 3-2.10 describes how the energy accumulated in the tank, given by $n(t)u_3(t) = n(t)\hat{C}_{v3}T_3(t)$, changes with time; the term $d[n(t)u_3(t)]/dt = d[n(t)\hat{C}_{v3}T_3(t)]/dt$ is the *rate of change of the energy of the gas accumulated* in the tank. Please note once more that in writing the unsteady-state energy balance we just added the rate of change of the accumulation term to the familiar steady-state balance; once gain we build in what we know.

In this section we have explained why steady-state balances, or steady-state models, do not completely describe the behavior of processes. *Unsteady-state balances, or dynamic models*, provide a more complete description; these balances *provide the dynamic and steady-state behavior of processes*. Once the process reaches steady state, the solution of the unsteady-state balance gives exactly the same information as the steady-state balance. Steady-state balances usually result in algebraic equations; unsteady-state balances result in differential equations with time as independent variable. The development of unsteady-state balances just requires the addition of the rate of change of the accumulation term to the usual steady-state balances. This section has shown three common balances; Section 3-7 discusses the balances in processes where reactions occur. The complete mathematical model not only requires the balances but also other equations, as we shall see in the ensuing presentation. In writing the balances, and all other auxiliary equations, we integrate our knowledge of process engineering, such as thermodynamics, heat transfer, fluid flow, mass transfer, and reaction engineering. This makes the modeling of industrial processes most interesting and challenging!

3-3   THERMAL PROCESS EXAMPLE

Consider the tank shown in Fig. 3-3.1. In this process, constant and equal inlet and outlet volumetric flows, liquid densities, and heat capacities are assumed; all of these properties are known. The liquid in the tank is assumed to be well mixed, and the tank is well insulated, that is, there are negligible heat losses to the surroundings. Finally, the energy input by the stirrer is assumed negligible.

We are interested in developing the mathematical model and transfer function that describe how the outlet temperature, $T(t)$, responds to changes in inlet temperature, $T_i(t)$. Taking the contents of the tank as the control volume, an unsteady-state

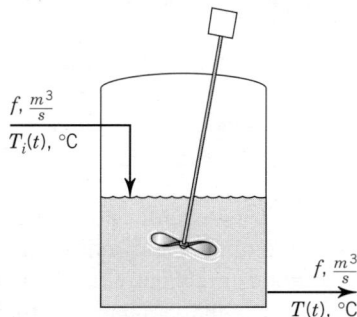

$f, \frac{m^3}{s}$

$T_i(t)$, °C

$f, \frac{m^3}{s}$

$T(t)$, °C   **Figure 3-3.1** Thermal process.

energy balance gives us the desired relation between the inlet and outlet temperatures. Applying Eq. 3-2.1 we get,

$$\text{Rate of energy into} \atop \text{control volume} - \text{Rate of energy out of} \atop \text{control volume} = \text{Rate of change of} \atop \text{energy accumulated in} \atop \text{control volume}$$

or, in terms of an equation,

$$f\rho_i h_i(t) - f\rho h(t) = \frac{d[V\rho u(t)]}{dt}$$

where

$$f = \text{volumetric flow, m}^3\text{/s}$$
$$\rho_i, \rho = \text{inlet and outlet liquid densities, respectively, kg/m}^3$$
$$V = \text{volume of liquid in tank, m}^3$$
$$h_i(t), h(t) = \text{inlet and outlet liquid enthalpies, J/kg}$$
$$u(t) = \text{internal energy of liquid in tank, J/kg}$$

In terms of temperatures, using as reference state for $u(t)$ and $h(t)$ the pure component in the liquid state at $0°\text{F}$ and the pressure of the system, the foregoing equation can be written as

$$f\rho_i C_{p_i} T_i(t) - f\rho C_p T(t) = \frac{d[V\rho C_v T(t)]}{dt} \qquad (3\text{-}3.1)$$

where

$$C_{p_i}, C_p = \text{inlet and outlet liquid heat capacities at constant pressure, respectively,} \atop \text{J/kg-}°\text{C}$$
$$C_v = \text{liquid heat capacity at constant volume, J/kg-}°\text{C}$$
$$T_i(t), T(t) = \text{inlet and outlet temperatures, respectively, }°\text{C}$$

Because the densities and the heat capacities are assumed constant and equal over the operating temperature range, the last equation can be written as

$$f\rho C_p T_i(t) - f\rho C_p T(t) = V\rho C_v \frac{dT(t)}{dt} \qquad (3\text{-}3.2)$$

This equation is a first-order linear ordinary differential equation that provides the relationship between the inlet and outlet temperatures. It is important to note that in this equation there is only one unknown, $T(t)$. The inlet temperature, $T_i(t)$, is an input variable and the one that forces the outlet temperature to change. In this example, we want to study how $T_i(t)$ affects $T(t)$, so it is up to us to decide how this inlet temperature will change. Thus $T_i(t)$ is not considered an unknown. In this chapter and the following one, input variables are not considered unknowns because we have the freedom to change them as we wish.

To show that there is one equation with one unknown, we explicitly write

$$f\rho C_p T_i(t) - f\rho C_p T(t) = V\rho C_v \frac{dT(t)}{dt} \qquad (3\text{-}3.3)$$

$$1 \text{ eq., 1 unk. } [T(t)]$$

Equation 3-3.3 is the mathematical model for this process. The solution of this differential equation yields the response of the outlet temperature as a function of time. As just mentioned, the inlet temperature is the input variable, which is sometimes referred to as the forcing function because it is the variable that forces the outlet temperature to change. The outlet temperature is the output variable, which is sometimes referred to as the responding variable because it is the variable that responds to the forcing function, or input variable.

As stated in the beginning of this example, we are interested in obtaining the transfer function relating $T(t)$ to $T_i(t)$. To do so, we follow a series of steps that yield the desired transfer function; after this example, we will formalize the procedure. We begin by making a variable change that simplifies development of the required transfer function.

Write a steady-state energy balance on the contents of the tank at the initial conditions

$$f\rho C_p \overline{T}_i - f\rho C_p \overline{T} = 0 \tag{3-3.4}$$

where

$\overline{T}, \overline{T}_i$ = initial steady-state values of outlet and inlet temperatures, respectively, $^\circ$C

Subtracting this last equation from Eq. 3-3.3 yields

$$f\rho C_p [T_i(t) - \overline{T}_i] - f\rho C_p [T(t) - \overline{T}] = V\rho C_v \frac{d[T(t) - \overline{T}]}{dt} \tag{3-3.5}$$

Note that the derivative of the temperature is also equal to

$$\frac{d[T(t) - \overline{T}]}{dt} = \frac{dT(t)}{dt} - \frac{d\overline{T}}{dt} = \frac{dT(t)}{dt} - 0$$

which is the result of subtracting the right-hand side of Eqs. 3-3.3 and 3-3.4. This is only a trick that proves helpful in the definition of deviation variables and the development of transfer functions.

As presented in Chapter 2, we now define the following deviation variables

$$\Gamma(t) = T(t) - \overline{T} \tag{3-3.6}$$
$$\Gamma_i(t) = T_i(t) - \overline{T}_i \tag{3-3.7}$$

where

Γ, Γ_i = deviation variables of outlet and inlet temperatures, respectively, $^\circ$C

Substituting Eqs. 3-3.6 and 3-3.7 into 3-3.5 yields

$$f\rho C_p \Gamma_i(t) - f\rho C_p \Gamma(t) = V\rho C_v \frac{d\Gamma(t)}{dt} \tag{3-3.8}$$

Eq. 3-3.8 is the same as Eq. 3-3.3 except that it is written in terms of deviation variables. The solution of this equation yields the deviation variable $\Gamma(t)$ versus time for a certain input $\Gamma_i(t)$. If the actual outlet temperature, $T(t)$, is desired, the steady-state value $\overline{T}$ must be added to $\Gamma(t)$ in accordance with Eq. 3-3.6.

Deviation variables are almost exclusively used throughout control theory. Thus the meaning and importance of deviation variables in the analysis and design of process control systems must be well understood. As explained in Chapter 2, their value indicates the degree of deviation from some initial steady-state value. In practice, this steady-state value may be the desired value of the variable. Another advantage in the use of these variables is that their initial value, assuming we start from the initial steady state, is zero, which simplifies the solution of differential equations such as Eq. 3-3.8 by the Laplace transform.

Equation 3-3.8 can now be rearranged as follows:

$$\frac{V\rho C_v}{f\rho C_p} \frac{d\Gamma(t)}{dt} + \Gamma(t) = \Gamma_i(t)$$

and we let

$$\tau = \frac{V\rho C_v}{f\rho C_p} \tag{3-3.9}$$

so

$$\tau \frac{d\Gamma(t)}{dt} + \Gamma(t) = \Gamma_i(t) \tag{3-3.10}$$

Section 2-4 defined τ as *time constant*, with units of time. From Eq. 3-3.9 we see that for this example,

$$\tau = \frac{[m^3][kg/m^3][J/kg\text{-}°C]}{[m^3/s][kg/m^3][J/kg\text{-}°C]} = \text{seconds}$$

Because Eq. 3-3.10 is a linear differential equation, the use of Laplace transform yields

$$\tau s\Gamma(s) - \tau\Gamma(0) + \Gamma(s) = \Gamma_i(s)$$

But the initial value of the temperature, $T(0)$, is at $\overline{T}$, so $\Gamma(0) = 0$. Performing some simple algebraic manipulations gives

$$\Gamma(s) = \frac{1}{\tau s + 1}\Gamma_i(s) \tag{3-3.11}$$

or

$$\frac{\Gamma(s)}{\Gamma_i(s)} = \frac{1}{\tau s + 1} \tag{3-3.12}$$

Equation 3-3.12 is the desired transfer function. It is a first-order transfer function because it is developed from a first-order differential equation. As we saw in Chapter 2, processes described by this type of transfer function are called *first-order processes, first-order systems,* or *first-order lags.* Equation 2-4.6 presented the general form of this type of transfer function. In the present example the term K is unity.

The term *transfer function* arises from the fact that the solution of the equation translates, or transfers, the input, $\Gamma_i(t)$, to the output, $\Gamma(t)$. Transfer functions are further discussed in Section 3-5.

As a brief review of Chapter 2, let us assume that the inlet temperature to the tank increases by $M°C$. That is, the inlet temperature experiences a step change of M degrees in magnitude. Mathematically, this is written as follows:

$$T_i(t) = \overline{T}_i \qquad\qquad t < 0$$
$$T_i(t) = \overline{T}_i + M \qquad t \geq 0$$

or, in terms of deviation variables,

$$\Gamma_i(t) = Mu(t)$$

where $u(t)$, as shown in Chapter 2, represents a step change of unit magnitude.

Taking the Laplace transform of $\Gamma_i(t)$, we obtain

$$\Gamma_i(s) = \frac{M}{s}$$

Substituting this expression for $\Gamma_i(s)$ into Eq. 3-3.11 results in

$$\Gamma(s) = \frac{M}{s(\tau s + 1)}$$

Using the method of partial fractions presented in Chapter 2 yields

$$\Gamma(s) = \frac{M}{s(\tau s + 1)} = \frac{A}{s} + \frac{B}{\tau s + 1}$$

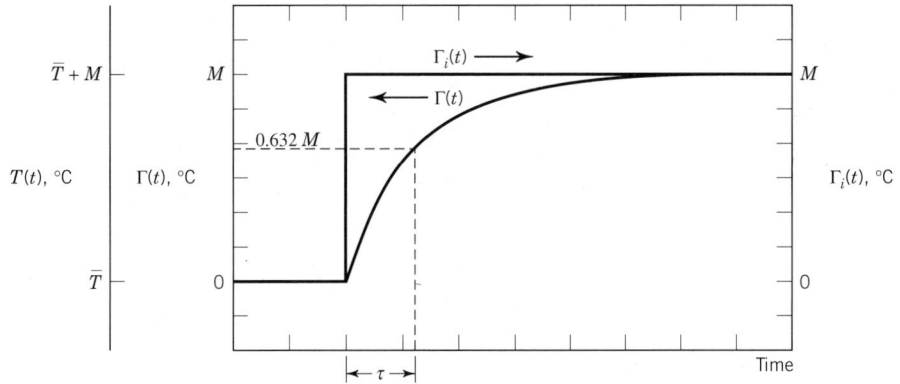

Figure 3-3.2 Response of a first-order process to a step change in input variable.

Obtaining the values of A and B and inverting back to the time domain, we get

$$\Gamma(t) = M(1 - e^{-t/\tau}) \tag{3-3.13}$$

or

$$T(t) = \overline{T} + M(1 - e^{-t/\tau}) \tag{3-3.14}$$

The solutions of Eqs. 3-3.13 and 3-3.14 are shown graphically in Fig. 3-3.2. The steepest slope of the response curve occurs at the beginning of the response; *this is the typical response of first-order systems to a step change in input.*

Section 2-4 presented the significance of the process time constant, τ. However, to review again, let $t = \tau$ in Eq. 3-3.13, which yields

$$\Gamma(\tau) = M(1 - e^{-\tau/\tau}) = M(1 - e^{-1})$$
$$\Gamma(\tau) = 0.632M$$

That is, *for a step change in input variable, the time constant indicates the time it takes the output variable to reach 63.2 % of its total change*; this is shown graphically in Fig. 3-3.2. In five time constants, 5τ, the process reaches 99.3 % of its total change; essentially the response is completed. Therefore, *the time constant is related to the speed of response of the process.* The slower a process responds to an input, the larger the value of τ. The faster the process responds to an input, the smaller the value of τ.

It is important to realize that the time constant is composed of the different physical properties and operating parameters of the process, as shown by Eq. 3-3.9. That is, the time constant depends on the volume of liquid in the tank (V), the heat capacities (C_p and C_v), and the process flow (f). If any of these characteristics changes, the behavior of the process also changes and this change is reflected in the speed of response of the process, or the time constant.

Up to now, we have assumed that the tank is well insulated, that is, there are negligible heat losses to the surroundings. Consequently, there is not a heat loss term in the energy balance. Let us remove this assumption and develop the mathematical model and the transfer functions relating the outlet temperature, $T(t)$, to the inlet temperature, $T_i(t)$, and to the surrounding temperature, $T_s(t)$.

As before, using the same reference state for enthalpies and internal energy, we start with an unsteady-state energy balance:

$$f\rho C_p T_i(t) - q(t) - f\rho C_p T(t) = V\rho C_v \frac{dT(t)}{dt}$$

or

$$f\rho C_p T_i(t) - UA[T(t) - T_s(t)] - f\rho C_p T(t) = V\rho C_v \frac{dT(t)}{dt} \tag{3-3.15}$$

1 eq., 1 unk. [$T(t)$]

where

$q(t)$ = heat transfer rate to the surroundings, J/s
U = overall heat transfer coefficient, J/m^2-K-s
A = heat transfer area, m^2
$T_s(t)$ = temperature of surroundings, °C, an input variable

The overall heat transfer coefficient, U, is a function of several things, one of them being temperature. However, in this particular example we assume it to be constant. Because the mass of liquid in the tank and its density are also assumed to be constant, then the height of liquid is constant, and consequently, the heat transfer area, A, is also constant.

Equation 3-3.15 provides the mathematical model of the process. There is still one equation with one unknown, $T(t)$. The new variable is the surrounding temperature $T_s(t)$, which is another input. As this temperature changes, it affects the heat losses and consequently the process liquid temperature.

To obtain the transfer functions, we start by introducing the deviation variables. This is done by first writing a steady-state energy balance for the process at the initial conditions.

$$f\rho C_p \overline{T}_i - UA[\overline{T} - \overline{T}_s] - f\rho C_p \overline{T} = 0 \tag{3-3.16}$$

Subtracting Eq. 3-3.16 from Eq. 3-3.15 yields

$$f\rho C_p[T_i(t) - \overline{T}_i] - UA[(T(t) - \overline{T}) - (T_s(t) - \overline{T}_s)]$$
$$- f\rho C_p[T(t) - \overline{T}] = V\rho C_v \frac{d[T(t) - \overline{T}]}{dt} \tag{3-3.17}$$

Please note that the trick with the differential term (accumulation) has been done again.

Define a new deviation variable as

$$\Gamma_s(t) = T_s(t) - \overline{T}_s \tag{3-3.18}$$

Substituting Eqs. 3-3.6, 3-3.7, and 3-3.18 into Eq. 3-3.17 yields

$$f\rho C_p \Gamma_i(t) - UA[\Gamma(t) - \Gamma_s(t)] - f\rho C_p \Gamma(t) = V\rho C_v \frac{d\Gamma(t)}{dt} \tag{3-3.19}$$

Equation 3-3.19 is the same as Eq. 3-3.15 except that it is written in terms of deviation variables. This equation is also a first-order linear ordinary differential equation. Equation 3-3.19 can be arranged as follows:

$$\frac{V\rho C_v}{f\rho C_p + UA}\frac{d\Gamma(t)}{dt} + \Gamma(t) = \frac{f\rho C_p}{f\rho C_p + UA}\Gamma_i(t) + \frac{UA}{f\rho C_p + UA}\Gamma_s(t)$$

or

$$\tau\frac{d\Gamma(t)}{dt} + \Gamma(t) = K_1\Gamma_i(t) + K_2\Gamma_s(t) \tag{3-3.20}$$

where

$$\tau = \frac{V\rho C_v}{f\rho C_p + UA}, \text{seconds} \tag{3-3.21}$$

$$K_1 = \frac{f\rho C_p}{f\rho C_p + UA}, \text{dimensionless} \tag{3-3.22}$$

$$K_2 = \frac{UA}{f\rho C_p + UA}, \text{dimensionless} \tag{3-3.23}$$

The right-hand side of Eq. 3-3.20 shows the two input variables, $\Gamma_i(t)$ and $\Gamma_s(t)$, acting on the output variable, $\Gamma(t)$.

Taking the Laplace transform of Eq. 3-3.20 gives

$$\tau s \Gamma(s) - \tau \Gamma(0) + \Gamma(s) = K_1 \Gamma_i(s) + K_2 \Gamma_s(s)$$

But the initial value of the temperature, $T(0)$, is at $\overline{T}$, so $\Gamma(0) = 0$. Rearranging this equation yields

$$\Gamma(s) = \frac{K_1}{\tau s + 1}\Gamma_i(s) + \frac{K_2}{\tau s + 1}\Gamma_s(s) \tag{3-3.24}$$

If the surrounding temperature remains constant, $T_s(t) = \overline{T}_s$ then, $\Gamma_s(t) = 0$, and the transfer function relating the process temperature to the inlet temperature is

$$\frac{\Gamma(s)}{\Gamma_i(s)} = \frac{K_1}{\tau s + 1} \tag{3-3.25}$$

If the inlet liquid temperature remains constant, $T_i(t) = \overline{T}_i$ then $\Gamma_i(t) = 0$, and the transfer function relating the process temperature to the surrounding temperature is

$$\frac{\Gamma(s)}{\Gamma_s(s)} = \frac{K_2}{\tau s + 1} \tag{3-3.26}$$

If both the inlet liquid temperature and the surrounding temperature change, then Eq. 3-3.24 provides the complete relationship, however, it is not a transfer function.

Equations 3-3.25 and 3-3.26 are the typical first-order transfer functions presented in Section 2-4. In this case, however, the *steady-state gains* (sometimes also called *process gains*), K_1 and K_2, are not unity, as was the case in Eq. 3-3.12. To review briefly the significance of the steady-state gains, let us assume that the inlet temperature to the tank increases, in a step fashion, by $M°C$, that is,

$$T_i(t) = \overline{T}_i \qquad t < 0$$
$$T_i(t) = \overline{T}_i + M \qquad t \geq 0$$

or

$$\Gamma_i(t) = Mu(t)$$

and

$$\Gamma_i(s) = \frac{M}{s}$$

The response of the temperature to this forcing function is given by

$$\Gamma(s) = \frac{K_1 M}{s(\tau s + 1)}$$

from which

$$\Gamma(t) = K_1 M(1 - e^{-t/\tau}) \tag{3-3.27}$$

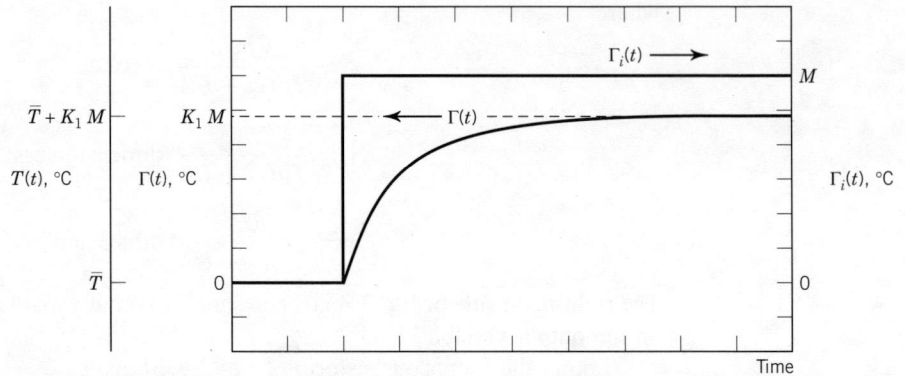

Figure 3-3.3 Response of a first-order process to a step change in input variable.

or

$$T(t) = \overline{T} + K_1 M (1 - e^{-t/\tau}) \qquad (3\text{-}3.28)$$

The output response is shown graphically in Fig. 3-3.3. The total amount of change in $T(t)$ is given by $K_1 M$, the gain times the change in input. Thus *the gain tells us how much the output changes per unit change in input*, or *how much the input affects the output*. That is, the gain defines the *sensitivity* relating the output and input variables! It can also be defined mathematically as follows:

$$K = \frac{\Delta O}{\Delta I} = \frac{\Delta \text{ output variable}}{\Delta \text{ input variable}} \qquad (3\text{-}3.29)$$

The gain is another parameter that describes the characteristics of the process. Consequently, it depends on the physical properties and operating parameters of the process, as shown by Eqs. 3-3.22 and 3-3.23. The gains in this process depend on the flow (f), density (ρ), and heat capacity (C_p) of the process liquid, on the overall heat transfer coefficient (U), and on the heat transfer area (A). If any of these changes, the behavior of the process changes and is reflected in the gain.

There are two gains in this example. The first one, K_1, relates the outlet temperature to the inlet temperature. The other gain, K_2, relates the outlet temperature to the surrounding temperature. The units of the gain term must be the units of the output variable divided by the units of the input variable; this is obvious from Eq. 3-3.29.

Note that the gain clearly indicates the process characteristics. In the first part of this example we assumed the tank to be well insulated, and the gain, given by Eq. 3-3.12, was unity. That is, in steady state all the energy entering with the inlet stream exits with the outlet stream, and the inlet and outlet temperatures are the same. This is not the case when the well-insulated assumption is removed and the tank is permitted to transfer energy with the surroundings. Note that K_1, given by Eq. 3-3.22, is less than unity, indicating that when the inlet temperature increases by M degrees the outlet temperature does not increase by that much. That is, if the energy in the inlet stream increases, the energy in the outlet stream does not increase as much because there is some energy transfer to the surroundings; this of course makes sense. It also makes sense that if $UA \ll f\rho C_p$, the inlet temperature will have a greater effect on the outlet temperature than will the surrounding temperature, that is, $K_2 \ll K_1$.

Equation 3-3.24 shows that there is only one time constant in this process. That is, the time it takes the outlet temperature to reach a certain percentage of its total change due to a change in inlet temperature is equal to the time it takes to reach the same percentage of its total change when the surrounding temperature changes.

It is always important during the analysis of any process to stop at some point to check the development for possible errors; Eq. 3-3.20 provides a convenient point. A quick check can be made by examining the signs of the equation to see whether they make sense in the real world. In Eq. 3-3.20 both gains are positive. The equation indicates that if the inlet temperature increases, the outlet temperature also increases; which makes sense for this process. The equation also shows that if the surrounding temperature increases, the outlet temperature increases. This makes sense because if the surrounding temperature increases, the rate of heat losses from the tank decreases, thereby increasing the temperature of the contents of the tank. Another check consists in examining the units of τ and K. We know what each of them should be, and the defining equations, Eqs. 3-3.21 through 3-3.23 in this example, should confirm these expectations. This quick check builds our confidence and permits us to proceed with the analysis with a renewed hope of success.

Before finishing with this section, let us summarize the procedure we followed to develop the transfer functions.

1. Write the set of unsteady-state equations that describe the process. This is called modeling.
2. Write the steady-state equations at the initial conditions.
3. Subtract the two set of equations, and define deviation variables.
4. Obtain the Laplace transforms of the model in deviation variables.
5. Obtain the transfer functions by solving the Laplace transform explicitly for the transformed output variable(s).

We have followed these five steps in our thermal example. They constitute an organized procedure that yields the transfer functions.

3-4   DEAD TIME

Consider the process shown in Fig. 3-4.1. This is essentially the same process as the one shown in Fig. 3-3.1. In this case, however, we are interested in knowing how $T_1(t)$ responds to changes in inlet and surrounding temperatures.

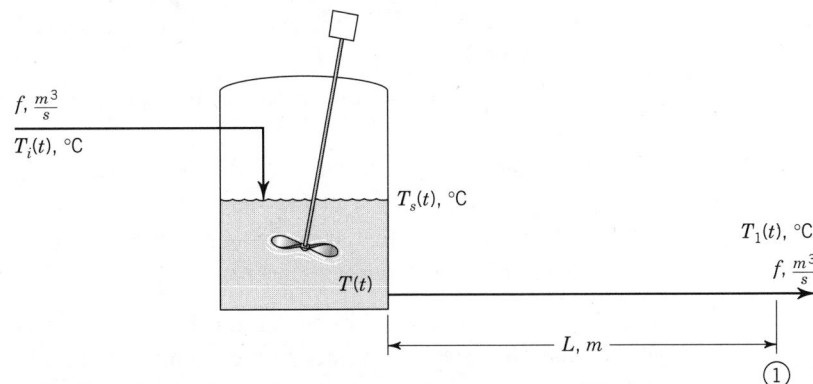

Figure 3-4.1 Thermal process.

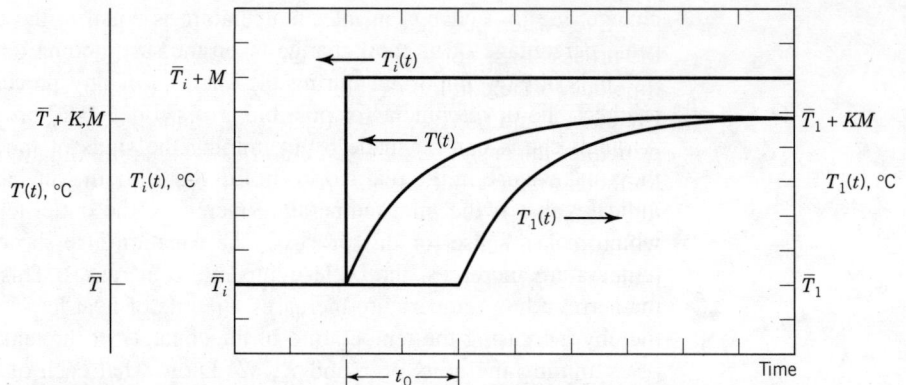

Figure 3-4.2 Response of a thermal process to a step change in inlet temperature.

Let us make the following two assumptions about the exit pipe between the tank and point 1. First, the pipe is well insulated. Second, the flow of liquid through the pipe is ideal plug flow (highly turbulent) with no energy diffusion or dispersion so that there is essentially no backmixing of the liquid in the pipe.

Under these assumptions, the response of $T_1(t)$ to the disturbances $T_i(t)$ and $T_s(t)$ will be the same as $T(t)$ except that it will be delayed by some amount of time. That is, there will be a finite amount of time between the initial response of $T(t)$ and the change of $T_1(t)$; this delay is shown graphically in Fig. 3-4.2. This finite amount of time has developed because of the time it takes the liquid to move from the exit of the tank to point 1 and is called a pure dead time, transportation delay, or time delay. It is represented by t_o and in this case can be easily estimated from

$$t_o = \frac{\text{distance}}{\text{velocity}} = \frac{L}{f/A_p} = \frac{A_p L}{f} \tag{3-4.1}$$

where

f = volumetric flow, m³/s
A_p = cross-sectional area of pipe, m²
L = length of pipe, m

Different physical variables travel at different velocities:

- Electric voltage and current travel at the speed of light: 300,000 km/s, or 984,106 ft/s.
- Pressure and liquid flow travel at the speed of sound in the fluid: 340 m/s or 1100 ft/s.
- Temperature, composition, and other fluid properties travel at the velocity of the fluid: typically, up to 5 m/s (15 ft/s) for liquids, and 60 m/s (200 ft/s) for gases.
- Solid properties travel at the velocity of the solid, for example, coal in a conveyor, cake in a filter bed, and paper in a paper machine.

From this information, we can see that for the distances which are typical of industrial process control systems, pure dead time is only significant for temperature, composition, and other fluid and solid properties which are propagated through space by the moving fluid or solid.

Even when pure dead time (dead time due to transportation) is negligible relative to the process time constant, the response of many processes may appear to exhibit dead time due to the combination of several first-order processes in series, as we shall see in Chapters 4 and 6. This pseudo-dead time cannot be easily evaluated from fundamental principles and must be obtained empirically by approximation of the process response. Methods to carry out such empirical evaluation will be presented in Chapter 7.

Because dead time is an integral part of processes, it must be accounted for in the transfer functions. Equation 2-1.8 indicates that the Laplace transform of a delayed function is equal to the Laplace transform of the nondelayed function times the term $e^{-t_o s}$; the term $e^{-t_o s}$ is the Laplace transform of dead time. Thus, if the transfer functions relating $T_1(t)$ to $T_i(t)$ and $T_s(t)$ are required, using the assumptions stated at the beginning of the section, the transfer functions given by Eqs. 3-3.25 and 3-3.26 are multiplied by $e^{-t_o s}$ or

$$\frac{\Gamma_1(s)}{\Gamma_i(s)} = \frac{K_1 e^{-t_o s}}{\tau s + 1} \tag{3-4.2}$$

and

$$\frac{\Gamma_1(s)}{\Gamma_s(s)} = \frac{K_2 e^{-t_o s}}{\tau s + 1} \tag{3-4.3}$$

At this point, it must be recognized that the dead time is another term that helps define the characteristics of the process. Equation 3-4.1 shows that t_o depends on some physical properties and operating characteristics of the process, similar to K and τ. If any condition of the process changes, this change may be reflected in a change in t_o.

Before concluding this section, we must stress that one of the worst thing that can happen to a feedback control loop is a significant amount of dead time in the loop. The performance of feedback control loops is severely affected by dead time as will be shown in Chapters 6 and 8. Thus, processes and control systems should be designed to keep the dead time to a minimum. Some steps we can take to minimize dead time include putting the measurements as close to the equipment as possible, selecting rapidly responding sensors and final control elements, and using electronic instead of pneumatic instrumentation for processes with short time constants.

3-5 TRANSFER FUNCTIONS AND BLOCK DIAGRAMS

3-5.1 Transfer Functions

Chapter 2 presented the concept of transfer functions. This concept is so fundamental to the study of process dynamics and automatic process control that at this time we briefly consider, once more, some of its important properties and characteristics.

We have already defined a transfer function as the ratio of the Laplace-transformed output variable to the Laplace-transformed input variable. Transfer functions are usually represented by

$$G(s) = \frac{Y(s)}{X(s)} = \frac{K(a_m s^m + a_{m-1} s^{m-1} + \cdots + a_1 s + 1) e^{-t_o s}}{(b_n s^n + b_{n-1} s^{n-1} + \cdots + b_1 s + 1)} \tag{3-5.1}$$

where

$G(s)$ = general representation of a transfer function

$Y(s)$ = Laplace transform of the output variable

$X(s)$ = Laplace transform of the input variable

K, a's, b's = constants

$\quad t_o$ = dead time

Equation 3-5.1 shows the most general and best way to write a transfer function. When written in this way (the coefficient of s^0 is 1 in both the numerator and denominator polynomials) K represents the gain of the system and will have as units the units of $Y(t)$ over the units of $X(t)$. The other constants, a's and b's, will have as units $(time)^i$, where i is the power of the Laplace variable, s, multiplied by the particular constant; this will render a dimensionless term inside the parentheses because the unit of s is 1/time.

Note: In general, the unit of s is the reciprocal of the unit of the independent variable used in the definition of Laplace transform, Eq. 2-1.1. In process dynamics and control the independent variable is time, and so the unit of s is 1/time.

The transfer function completely defines the steady-state and dynamic characteristics, or the total response, of a system described by a linear differential equation. It is characteristic of the system, and its terms determine whether the system is stable or unstable and whether its response to a nonoscillatory input is oscillatory. The system, or process, is said to be stable when its output remains bound (finite) for all times for a bound input. Chapter 2 presented some discussions on stability and how it is related to terms in the transfer function. Chapters 6 and 8 treat in more detail the subject of stability of process systems.

The following are some important properties of transfer functions:

1. In the transfer functions of real physical systems, the highest power of s in the numerator is never higher than that in the denominator. In other words, $n \geq m$.

2. The transfer function relates the transforms of the deviation of the input and output variables from some initial steady state. Otherwise, the nonzero initial conditions would contribute additional terms to the transform of the output variable. As an example, if we had decided to obtain the Laplace transform of Eq. 3-3.3, without subtracting from it the steady-state energy balance to obtain Eq. 3-3.8, the result would have been

$$\frac{V\rho C_v}{f\rho C_p}\frac{dT(t)}{dt} + T(t) = T_i(t)$$

$$\tau\frac{dT(t)}{dt} + T(t) = T_i(t)$$

$$\tau s T(s) - sT(0) + T(s) = T_i(s)$$

because at steady state $T(0) = \overline{T}$,

$$\tau s T(s) - \tau\overline{T} + T(s) = T_i(s)$$

$$\tau s T(s) + T(s) = T_i(s) + \tau\overline{T}$$

$$T(s) = \frac{1}{\tau s + 1}T_i(s) + \frac{\tau\overline{T}}{\tau s + 1}$$

and because of the last term (due to the nonzero initial condition), it is impossible to obtain a transfer function from this equation.

3. For stable systems, the steady-state relationship between the change in output variable and the change in input variable can be obtained by

$$\lim_{s \to 0} G(s)$$

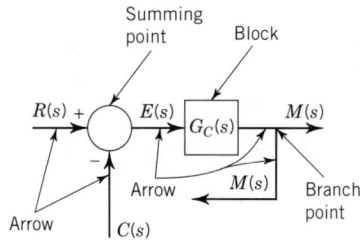

Figure 3-5.1 Elements of a block diagram.

This stems from the final value theorem, presented in Chapter 2

$$\frac{Y(s)}{X(s)} = G(s)$$

$$Y(s) = G(s)X(s)$$

$$\lim_{t \to \infty} Y(t) = \lim_{s \to 0} sY(s)$$

$$\lim_{t \to \infty} Y(t) = \lim_{s \to 0} sG(s)X(s)$$

$$\lim_{t \to \infty} Y(t) = [\lim_{s \to 0} G(s)][\lim_{s \to 0} sX(s)]$$

$$\lim_{t \to \infty} Y(t) = [\lim_{s \to 0} G(s)][\lim_{t \to \infty} X(t)]$$

This means that the change in the output variable after a very long time, if bound, can be obtained by multiplying the transfer function evaluated at $s = 0$ times the final value of the change in input.

3-5.2 Block Diagrams

A very useful tool in process control is the graphical representation of transfer functions by means of block diagrams. This section offers an introduction to block diagrams and block diagram algebra.

All block diagrams are formed by a combination of four basic elements: arrows, summing points, branch points, and blocks; Fig. 3-5.1 shows these elements. The arrows indicate flow of information; they represent process variables or control signals. Each arrowhead indicates the direction of the flow of information. The summing points represent the algebraic summation of the input arrows, $E(s) = R(s) - C(s)$. A branch point is the position on an arrow at which the information branches out and goes concurrently to other summing points or blocks. The blocks represent the mathematical operation, in transfer function form such as $G(s)$, which is performed on the input to produce the output. The arrows and block shown in Fig. 3-5.1 represent the mathematical expression

$$M(s) = G_c(s)E(s) = G_c(s)[R(s) - C(s)]$$

Any block diagram can be handled, or manipulated, algebraically. Table 3-5.1 shows some rules of block diagram algebra. These rules are important any time a complicated block diagram is simplified. Let us look at some examples of block diagram algebra.

EXAMPLE 3-5.1   Draw the block diagram depicting Eqs. 3-3.12 and 3-3.24.
Equation 3-3.12 is shown in Fig. 3-5.2. Equation 3-3.24 may be drawn in two different ways, as shown in Fig. 3-5.3. Often the one with fewer blocks is preferred because it is simpler.

Table 3-5.1 Rules for block diagram algebra

1. $Y(s) = X_1(s) - X_2(s) - X_3(s)$

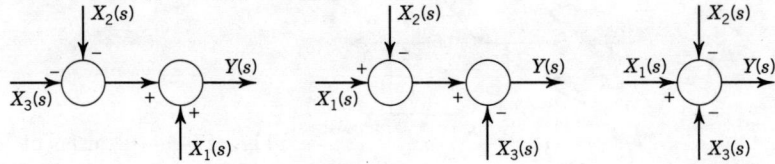

2. Associative and commutative properties:

$$Y(s) = G_1(s)G_2(s)X(s) = G_2(s)G_1(s)X(s)$$

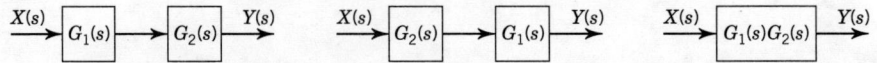

3. Distributive property:

$$Y(s) = G_1(s)[X_1(s) - X_2(s)] = G_1(s)X_1(s) - G_1(s)X_2(s)$$

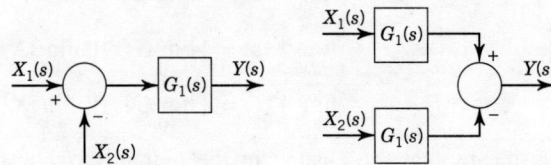

4. Blocks in parallel:

$$Y(s) = [G_1(s) + G_2(s)]X(s) = G_1(s)X(s) + G_2(s)X(s)$$

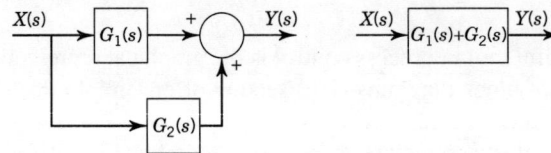

5. Positive feedback loop:

$$Y(s) = G_1(s)[X(s) + G_2(s)Y(s)] = \frac{G_1(s)}{1 - G_1(s)G_2(s)}X(s)$$

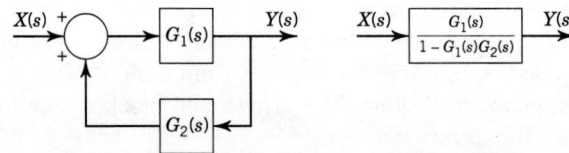

6. Negative feedback loop:

$$Y(s) = G_1(s)[X(s) - G_2(s)Y(s)] = \frac{G_1(s)}{1 + G_1(s)G_2(s)}X(s)$$

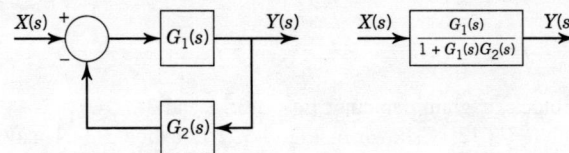

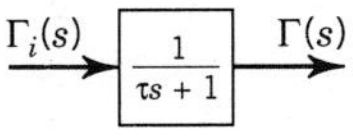

Figure 3-5.2 Block diagram of Eq. 3-3.12.

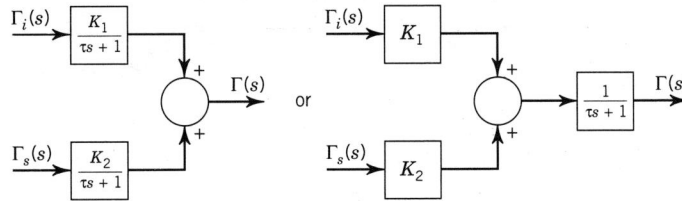

Figure 3-5.3 Block diagram of Eq. 3-3.24.

The block diagrams of Eq. 3-3.24 show graphically that the total response of the system is obtained by algebraically adding the response due to a change in inlet temperature to the response due to a change in surrounding temperature. This algebraic addition of responses due to several inputs to obtain the final response is a property of linear systems and is called the principle of superposition. This principle also serves as the basis for defining linear systems. That is, we say that a system is linear if it obeys the principle of superposition.

EXAMPLE 3-5.2

Determine the transfer functions relating $Y(s)$ to $X_1(s)$ and $X_2(s)$ from the block diagram shown in Fig. 3-5.4a. That is, obtain

$$\frac{Y(s)}{X_1(s)} \quad \text{and} \quad \frac{Y(s)}{X_2(s)}$$

Using rule 4, the block diagram shown in Fig. 3-5.4a can be reduced to that of Fig. 3-5.4b (please note that reduction is used in this context to mean simplification and that it consists of reducing the number of blocks). Using rule 2, Fig. 3-5.4b can be further reduced to that of Fig. 3-5.4c. Then

$$Y(s) = G_3(G_1 - G_2)X_1(s) + (G_4 - 1)X_2(s)$$

from which the two desired transfer functions can be determined

$$\frac{Y(s)}{X_1(s)} = G_3(G_1 - G_2)$$

and

$$\frac{Y(s)}{X_2(s)} = G_4 - 1$$

Example 3-5.2 has shown a procedure to reduce a block diagram to a transfer function. This reduction of block diagrams is necessary in the study of process control, as will be clear in later chapters. In these chapters, numerous examples of block diagrams of feedback, cascade, feedforward, and multivariable control systems are developed. Let us look at the reduction to transfer functions of some of these block diagrams.

EXAMPLE 3-5.3

Figure 3-5.5a shows the block diagram of a typical feedback control system. From this diagram determine

$$\frac{C(s)}{L(s)} \quad \text{and} \quad \frac{C(s)}{C^{\text{set}}(s)}$$

(a)

(b)

(c)

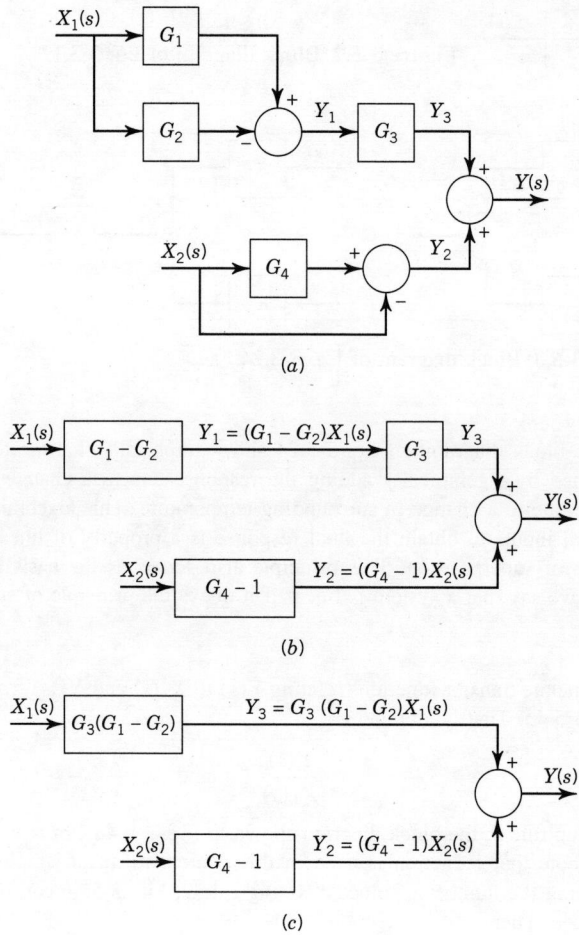

Figure 3-5.4 Block diagram for Example 3-5.2.

Figures 3-5.5b through d show the different reduction steps using the rules of Table 3-5.1. Finally from Fig. 3-5.5d we obtain the transfer functions as

$$\frac{C(s)}{C^{\text{set}}(s)} = \frac{G_1 G_C G_2 G_3 G_4}{1 + G_C G_2 G_3 G_4 G_6} \tag{3-5.2}$$

and

$$\frac{C(s)}{L(s)} = \frac{G_5 G_4}{1 + G_C G_2 G_3 G_4 G_6} \tag{3-5.3}$$

Example 3-5.3 shows how to reduce a simple feedback control system block diagram to transfer functions. These types of block diagrams and transfer functions will become useful in Chapters 6, 7, and 8, when feedback control is discussed.

The transfer functions given by Eqs. 3-5.2 and 3-5.3 are referred to as closed-loop transfer functions. The reason for this term will become evident in Chapter 6. Looking at Eq. 3-5.2 note that the numerator is the multiplication of all of the transfer functions in the forward path between the two variables related by the transfer function, $C(s)$ and $C^{\text{set}}(s)$. The denominator of this equation is one (1) plus the multiplication of all the transfer functions in the control loop shown in Fig. 3-5.5a. Inspection of Eq. 3-5.3

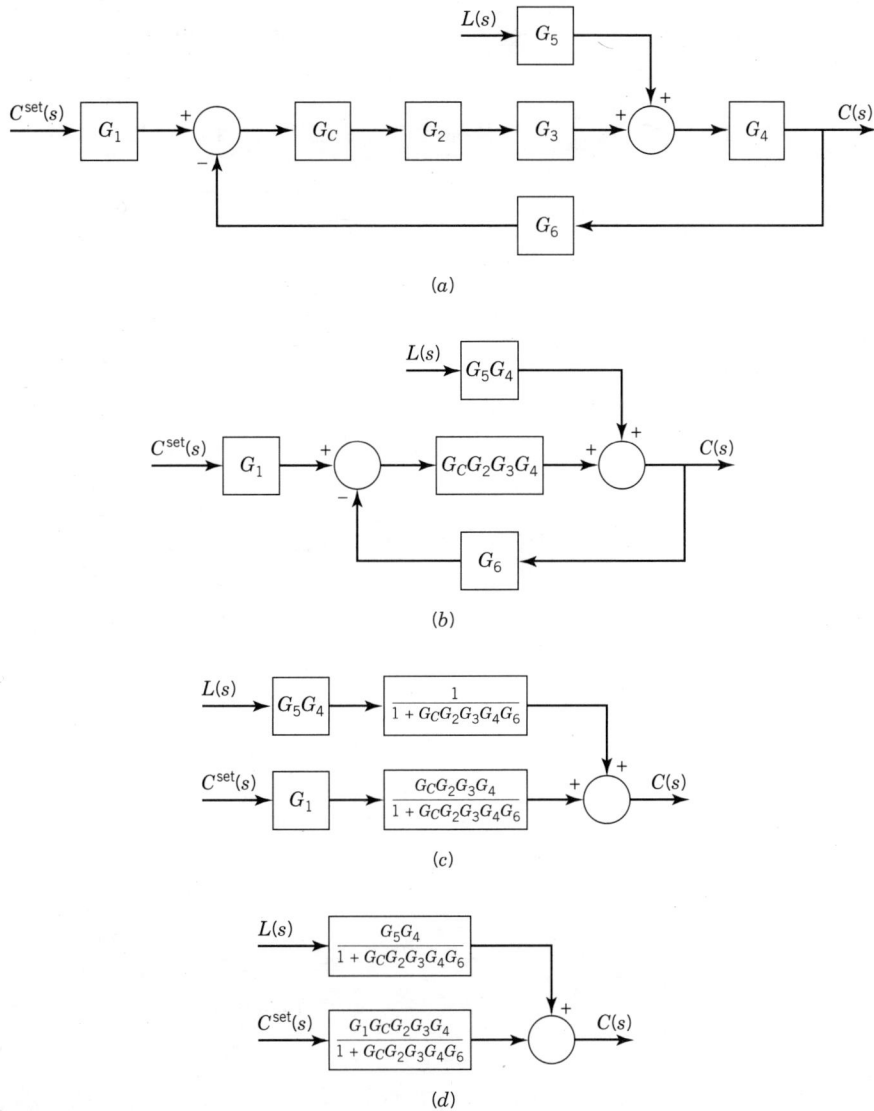

(a)

(b)

(c)

(d)

Figure 3-5.5 Block diagram of a feedback control system.

shows that the numerator is again the multiplication of the transfer functions in the forward path between $L(s)$ and $C(s)$. The denominator is the same as the one of Eq. 3-5.2. If there had been more than one forward path between the input and output, the development would have shown the numerator to be the algebraic summation of the product of the transfer functions in each forward path.

EXAMPLE 3-5.4

Consider another typical block diagram as shown in Fig. 3-5.6a. Chapter 9 shows that this block diagram depicts a cascade control system. For now, simply determine the following transfer functions:

$$\frac{C(s)}{R(s)} \quad \text{and} \quad \frac{C(s)}{L(s)}$$

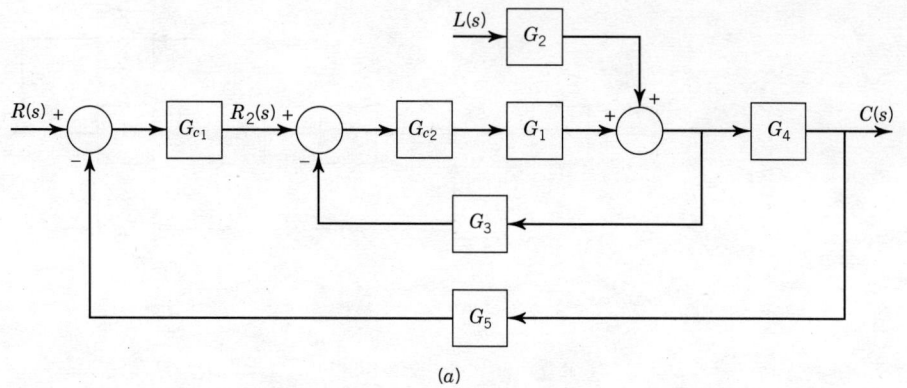

(a)

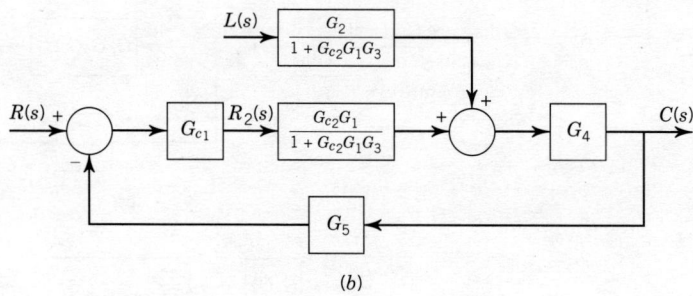

(b)

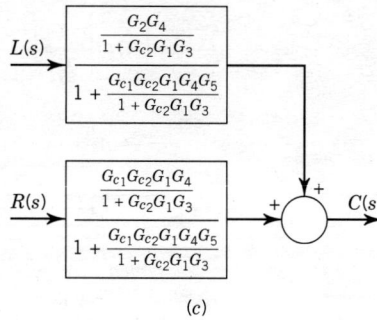

(c)

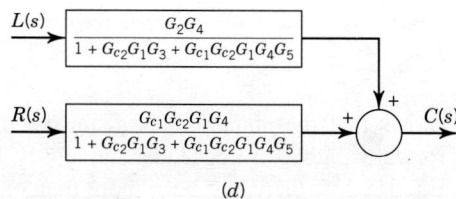

(d)

Figure 3-5.6 Block diagram of a cascade control system.

The block diagram of Fig. 3-5.6*a* can be thought of as being composed of two closed-loop systems, one inside the other (in practice, this is exactly what it is). Figure 3-5.6*b* and *c* show the steps to reduce the block diagram of Fig. 3-5.6*a*; rule 6 is applied twice. From Fig. 3-5.6*d* the following transfer functions are obtained:

$$\frac{C(s)}{R(s)} = \frac{G_{c_1} G_{c_2} G_1 G_4}{1 + G_{c_2} G_1 G_3 + G_{c_1} G_{c_2} G_1 G_4 G_5} \tag{3-5.4}$$

and

$$\frac{C(s)}{L(s)} = \frac{G_2 G_4}{1 + G_{c_2} G_1 G_3 + G_{c_1} G_{c_2} G_1 G_4 G_5} \qquad \text{(3-5.5)}$$

We have learned how to develop several transfer functions, Eqs. 3-5.2, 3-5.3, 3-5.4, and 3-5.5, from block diagrams. We have not intended, however, to give their significance; this will be done in the chapters where control systems are presented.

A useful recommendation is to write next to each arrow the units of the process variable or control signal that the arrow represents. If this is done, then it is fairly simple to recognize the units of the gain of a block, which are the units of the output arrow over the units of the input arrow. This procedure also helps in avoiding the algebraic summation of arrows with different units; it is extensively illustrated in Chapters 6, 9, 10, 11, and 12.

As mentioned at the beginning of this section, block diagrams are a very helpful tool in process control. They show the flow of information in a graphical way, identify the input and output signals (or variables) in a system, and show the occurrence of loops and parallel paths. We will learn and practice more about the logic of drawing block diagrams as we continue in our study of process dynamics and control. Later chapters make use of them to help analyze and design control systems.

3-6   GAS PROCESS EXAMPLE

Consider the gas tank shown in Fig. 3-6.1. A fan blows air into a tank, and from the tank the air flows out through a valve. For purposes of this example, let us suppose that the air flow delivered by the fan is given by

$$f_i(t) = 0.16 m_i(t)$$

where

$f_i(t)$ = gas flow in scf/min, where scf is cubic feet at standard conditions of 60°F and 1 atm.

$m_i(t)$ = signal to fan, %

The flow through the valve is expressed by

$$f_o(t) = 0.00506 m_o(t) \sqrt{p(t)[p(t) - p_1(t)]}$$

where

$f_o(t)$ = gas flow, scf/min

$m_o(t)$ = signal to valve, %

$p(t)$ = pressure in tank, psia

$p_1(t)$ = downstream pressure from valve, psia

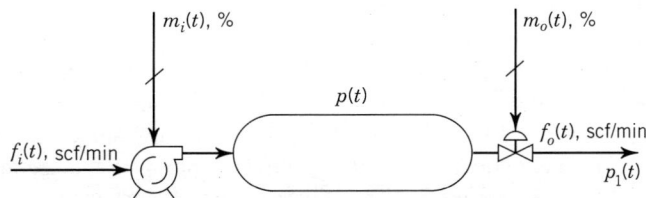

Figure 3-6.1 Gas vessel.

The volume of the tank is 20 ft^3, and it can be assumed that the process occurs isothermally at 60°F. The initial steady-state conditions are

$$\overline{f}_i = \overline{f}_o = 8 \text{ scfm}; \quad \overline{p} = 40 \text{ psia}; \quad \overline{p}_1 = 1 \text{ atm}; \quad \overline{m}_i = \overline{m}_o = 50 \%$$

We want to develop the mathematical model, transfer functions, and block diagram that relate the pressure in the tank to changes in the signal to the fan, $m_i(t)$; in the signal to the valve, $m_o(t)$; and in the downstream pressure, $p_1(t)$.

We must first develop the mathematical model. An unsteady-state mole balance around the control volume, defined as the fan, tank, and outlet valve, provides the starting relation. That is,

$$\begin{array}{ccc} \text{Rate of moles into} & & \text{Rate of moles out of} \\ \text{control volume} & - & \text{control volume} \end{array} = \begin{array}{c} \text{Rate of change of moles} \\ \text{accumulated in control} \\ \text{volume} \end{array}$$

or, in equation form,

$$\overline{\rho} f_i(t) - \overline{\rho} f_o(t) = \frac{dn(t)}{dt} \qquad \textbf{(3-6.1)}$$

1 eq., 3 unk. $[f_i(t), f_o(t), n(t)]$

where

$\overline{\rho}$ = molal density of gas at standard conditions, 0.00263 lbmoles/scf

$n(t)$ = moles of gas in tank, lbmoles

The fan provides another equation:

$$f_i(t) = 0.16 m_i(t) \qquad \textbf{(3-6.2)}$$

2 eq., 3 unk.

Note that because $m_i(t)$ is an input variable, it is up to us to decide how it will change. Thus it is not considered an unknown.

The valve provides still another equation:

$$f_o(t) = 0.00506 m_o(t) \sqrt{p(t)[p(t) - p_1(t)]} \qquad \textbf{(3-6.3)}$$

3 eq., 4 unk. $[p(t)]$

The signal $m_o(t)$ and downstream pressure $p_1(t)$ are also input variables and thus are not considered unknowns.

Because the pressure in the tank is low, the ideal gas equation of state can be used to relate the moles in the tank to the pressure.

$$p(t)V = n(t)RT \qquad \textbf{(3-6.4)}$$

4 eq., 4 unk.

The set of Eqs. 3-6.1 through 3-6.4 constitutes the mathematical model for this process. The solution of this set of equations describes, considering the assumptions taken, how the pressure in the tank (the output) responds to changes in $m_i(t)$, $m_o(t)$, and $p_1(t)$ (the inputs).

So far we have completed the first step of the procedure outlined at the end of Section 3-3. Before proceeding to the second step, we must realize that the expression for $f_o(t)$, Eq. 3-6.3, is a nonlinear equation. The Laplace transformation can be applied only to linear equations. Thus, before continuing to the second step, we must linearize all the nonlinear terms. This linearization is done using Taylor series expansion as presented in Chapter 2.

Because $f_o(t) = f_o[m_o(t), p(t), p_1(t)]$, its linearization is done with respect to $m_o(t)$, $p(t)$, and $p_1(t)$ about their steady-state values $\overline{m}_o$, $\overline{p}$, and $\overline{p}_1$.

$$f_o(t) \approx \overline{f}_o + \left.\frac{\partial f_o(t)}{\partial m_o(t)}\right|_{ss} [m_o(t) - \overline{m}_o] + \left.\frac{\partial f_o(t)}{\partial p(t)}\right|_{ss} [p(t) - \overline{p}]$$
$$+ \left.\frac{\partial f_o(t)}{\partial p_1(t)}\right|_{ss} [p_1(t) - \overline{p}_1]$$

or

$$f_o(t) \approx \overline{f}_o + C_1[m_o(t) - \overline{m}_o] + C_2[p(t) - \overline{p}] + C_3[p_1(t) - \overline{p}_1] \qquad \textbf{(3-6.5)}$$

where

$$C_1 = \left.\frac{\partial f_o(t)}{\partial m_o(t)}\right|_{ss} = 0.00506\sqrt{\overline{p}(\overline{p} - \overline{p}_1)} \qquad \textbf{(3-6.6)}$$

$$C_2 = \left.\frac{\partial f_o(t)}{\partial p(t)}\right|_{ss} = 0.00506\overline{m}_o(1/2)[\overline{p}(\overline{p} - \overline{p}_1)]^{-1/2}(2\overline{p} - \overline{p}_1) \qquad \textbf{(3-6.7)}$$

$$C_3 = \left.\frac{\partial f_o(t)}{\partial p_1(t)}\right|_{ss} = 0.00506\overline{m}_o(1/2)[\overline{p}(\overline{p} - \overline{p}_1)]^{-1/2}(-\overline{p}) \qquad \textbf{(3-6.8)}$$

and

$$\overline{f}_o = f_o[\overline{m}_o, \overline{p}, \overline{p}_1] \qquad \textbf{(3-6.9)}$$

Now there is a set of linear equations, Eqs. 3-6.1, 3-6.2, 3-6.4, and 3-6.5, that describes the process around the linearization values of $\overline{m}_o, \overline{p}$, and $\overline{p}_1$; there are still four equations with four unknowns.

To simplify this set somewhat, solve for $n(t)$ in Eq. 3-6.4 and substitute it in Eq. 3-6.1

$$\overline{\rho}f_1(t) - \overline{\rho}f_o(t) = \frac{V}{RT}\frac{dp(t)}{dt} \qquad \textbf{(3-6.10)}$$

With this simple substitution the set of equation is reduced to three equations, Eqs. 3-6.10, 3-6.2, and 3-6.5, with three unknowns, $f_i(t)$, $f_o(t)$, and $p(t)$.

We can now proceed with the next two steps of the procedure, which call to write the steady-state equations, subtract them from their respective counterparts, and defining the required deviation variables.

First we write a steady-state mole balance around the tank.

$$\overline{\rho}\overline{f}_i - \overline{\rho}\overline{f}_o = 0$$

Subtracting this equation from Eq. 3-6.10 gives

$$\overline{\rho}[f_i(t) - \overline{f}_i] - \overline{\rho}[f_o(t) - \overline{f}_o] = \frac{V}{RT}\frac{d[p(t) - \overline{p}]}{dt} \qquad \textbf{(3-6.11)}$$

Defining the following deviation variables

$$F_i(t) = f_i(t) - \overline{f}_i; \qquad F_o(t) = f_o(t) - \overline{f}_o; \qquad P(t) = p(t) - \overline{p}$$

and substituting these variables into Eq. 3-6.11 yields

$$\overline{\rho}F_i(t) - \overline{\rho}F_o(t) = \frac{V}{RT}\frac{dP(t)}{dt} \qquad \textbf{(3-6.12)}$$

Writing the steady-state equation for the fan and subtracting it from Eq. 3-6.2 gives

$$F_i(t) = 0.16 M_i(t) \tag{3-6.13}$$

where $M_i(t) = m_i(t) - \overline{m}_i$.

From Eq. 3-6.5, after subtraction of the steady-state value of $\overline{f}_o$ from both sides of the equation

$$F_o(t) = C_1 M_o(t) + C_2 P(t) + C_3 P_1(t) \tag{3-6.14}$$

where

$$M_o(t) = m_o(t) - \overline{m}_o$$
$$P_1(t) = p_1(t) - \overline{p}_1$$

Recapping what has been done, there are now three equations, Eqs. 3-6.12, 3-6.13, and 3-6.14 and three unknowns, $F_i(t)$, $F_o(t)$, and $P(t)$. All of these equations and variables are in deviation form.

We now proceed with the last two steps of the procedure. Substituting Eqs. 3-6.13, and 3-6.14 into Eq. 3-6.12, taking the Laplace transform, and rearranging yield

$$P(s) = \frac{K_1}{\tau s + 1} M_i(s) - \frac{K_2}{\tau s + 1} M_o(s) - \frac{K_3}{\tau s + 1} P_1(s) \tag{3-6.15}$$

where

$$K_1 = \frac{0.16}{C_2}, \frac{psi}{\%}; \quad K_2 = \frac{C_1}{C_2}, \frac{psi}{\%}; \quad K_3 = \frac{C_3}{C_2}, \frac{psi}{psi}; \quad \tau = \frac{V}{RT\overline{\rho}C_2}, \min \tag{3-6.16}$$

The desired transfer functions can now be obtained

$$\frac{P(s)}{M_i(s)} = \frac{K_1}{\tau s + 1} \tag{3-6.17}$$

$$\frac{P(s)}{M_o(s)} = \frac{-K_2}{\tau s + 1} \tag{3-6.18}$$

and

$$\frac{P(s)}{P_1(s)} = \frac{-K_3}{\tau s + 1} \tag{3-6.19}$$

Because the steady-state values and other process information are known, all gains and the time constant can be evaluated as

$$K_1 = 0.615 \frac{psi}{\%}; \quad K_2 = 0.619 \frac{psi}{\%}; \quad K_3 = -0.611 \frac{psi}{psi}; \quad \tau = 5.242 \min$$

All of the transfer functions are of first order. Figure 3-6.2 shows a block diagram for this process.

After considering the presentation in Chapter 2 about transfer functions and their response to inputs, and after what has been presented in this chapter, we should have a good feeling for the complete response of any first-order system. We know by analyzing Eq. 3-6.17 that, if the signal to the fan increases by 10 %, the pressure in the tank will ultimately change by $+(10)(K_1)$ psi. We also know that 63.2 % of the change, or $0.632(10)(K_1)$, will occur in one time constant. This response is shown graphically in Fig. 3-6.3. Remember that K_1 is the gain that $M_i(t)$ has on $P(t)$, and that τ gives how fast $P(t)$ responds to a change in $M_i(t)$.

Equation 3-6.18 indicates that if the signal to the valve increases by 5 %, the pressure in the tank will decrease by $(5)K_2$ psi. The negative sign in front of the gain

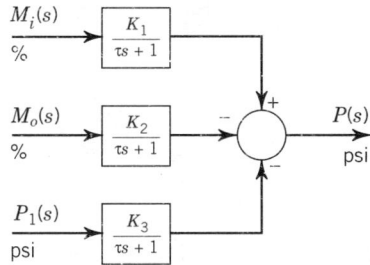

Figure 3-6.2 Block diagrams for gas process.

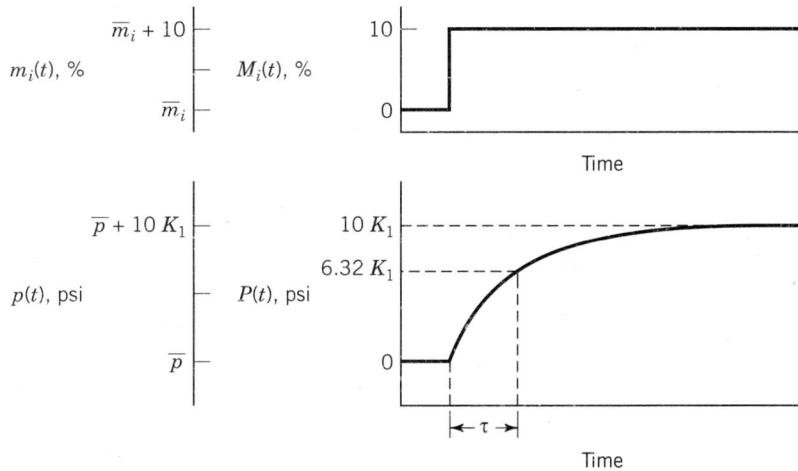

Figure 3-6.3 Response of pressure to signal to fan.

indicates this type of response. Certainly it makes sense that if the signal to the valve increases, opening the valve and thus extracting more gas from the tank, the pressure in the tank should fall.

Equation 3-6.19 indicates that if the downstream pressure from the valve increases by 3 psi, the pressure in the tank will decrease by $(3)K_3$ psi. That is, if $P_1(t)$ changes by $+3$ psi, $P(t)$ will change by $-(3)K_3$. From a physical point of view, however, this does not make any sense. If the downstream pressure increases, then the flow through the valve decreases, increasing the pressure in the tank. So, where is the discrepancy? Reviewing the definition of K_3 we see that it depends on C_3 and, on the basis of Eq. 3-6.8, it is obvious that C_3 is negative. Thus K_3 is negative ($K_3 = -0.611$), and consequently, the pressure in the tank actually increases.

At this point, we should reformulate the procedure for obtaining the transfer functions. This is needed because we now realize that linearization of nonlinear terms is an important step in the procedure.

1. Write the set of unsteady-state equations that describe the process. This is called modeling.

2. Linearize the model if necessary.

3. Write the steady-state equations at the initial conditions.

4. Subtract the two sets of equations, and define the deviation variables.

5. Obtain the Laplace transform of the linear model in deviation variables.

6. Obtain the transfer functions by solving the Laplace transform explicitly for the transformed output variable(s).

3-7   CHEMICAL REACTORS

3-7.1   Introductory Remarks

The example presented in this section involves a chemical reaction. Because the stoichiometry of the reactions are given in moles, the balances done in chemical reactors are usually mole balances, either on a specific component i, or on total moles. The problem, however, is that mole balances cannot be written using the equations presented in the introduction to this chapter. That is, taking the reactor as control volume,

$$\begin{array}{ccc} \text{Rate of component } i & \text{Rate of component } i & \text{Rate of change of} \\ & - & \neq \quad \text{component } i \\ \text{into reactor} & \text{out of reactor} & \text{accumulated in reactor} \end{array}$$

Moles are not necessarily conserved in chemical reactions. Consider, for example, the reaction $2A + B \rightarrow 3S + P$. Under steady-state operation the moles of A exiting the reactor are not the same as entering (reactant A is consumed!). Similarly, 3 moles of reactants are used, whereas 4 moles of products are formed, so the total moles are not conserved, either. Remember, however, the total mass is always conserved.

Therefore, the mole balance equations must account for the production or depletion of moles due to reaction. The unsteady-state component mole balance that accounts for this production or depletion is written as

$$\begin{array}{ccccc} \text{Rate of} & \text{Rate of} & \text{Rate of} & & \text{Rate of} \\ & & \text{production} & & \text{change of} \\ \text{component } i \text{ into} - & \text{component } i \text{ out} + & \text{of} & = & \text{component } i \\ \text{reactor} & \text{of reactor} & \text{component } i & & \text{accumulated} \\ & & & & \text{in reactor} \end{array}$$

The rate of production of component i in the reactor is usually given by

$$\text{Rate of production of component } i = \nu_i r_k V, \text{ moles of component } i/\text{time}$$

where

ν_i = stoichiometric coefficient of component i in the reaction

V = volume of reacting mixture

r_k = rate of reaction of key component in the reaction. This rate (always positive) is usually given as $\dfrac{\text{mole of key component formed or reacted}}{(\text{volume of reacting mixture})(\text{time})}$.

An important term in the above definitions is key component. The key component may be any component—reactant or product—in the reaction. The stoichiometric coefficient, ν_i, of the chosen component is made equal to 1. A positive ν_i indicates production of component i, a negative ν_i indicates depletion of component i. Thus the rate of change of any component i is expressed as the multiplication of the rate of reaction of the key component, the volume of the reacting mixture, and the number of moles of component i changing per mole of key component reacting.

To demonstrate further the application of this component mole balance consider the reaction previously given. Assume that for this reaction the rate is experimentally determined to be $r_B = kc_A(t)c_B(t)$, moles of B/(volume)(time), where B is the key component. Therefore, $v_A = -2$, $v_B = -1$, $v_P = 1$, $v_S = 3$.

An unsteady-state mole balance on component A is written as

$$\begin{matrix} \text{Rate of moles of} \\ \text{A into reactor} \end{matrix} - \begin{matrix} \text{Rate of moles of} \\ \text{A out of reactor} \end{matrix} + (-2)r_B V = \frac{dn_A(t)}{dt}$$

where $n_A(t)$ is the moles of A accumulated in the reactor.

An unsteady-state mole balance on component S is written as

$$\begin{matrix} \text{Rate of moles of} \\ \text{S into reactor} \end{matrix} - \begin{matrix} \text{Rate of moles of} \\ \text{S out of reactor} \end{matrix} + 3r_B V = \frac{dn_S(t)}{dt}$$

where $n_S(t)$ is the moles of S accumulated in the reactor.

The unsteady-state total mole balance is written as

$$\begin{matrix} \text{Rate of total} \\ \text{moles into reactor} \end{matrix} - \begin{matrix} \text{Rate of total} \\ \text{moles out of} \\ \text{reactor} \end{matrix} + \begin{matrix} \text{Rate of} \\ \text{production of} \\ \text{total moles} \end{matrix} = \begin{matrix} \text{Rate of change} \\ \text{of total moles} \\ \text{accumulated in} \\ \text{reactor} \end{matrix}$$

and

$$\text{Rate of production of total moles} = v_T r_k V, \text{moles/time}$$

where

$v_T = \sum_i v_i$. For the particular reaction at hand,

$v_T = v_S + v_P - v_A - v_B = 3 + 1 - 2 - 1 = 1$

Similarly, the energy balance must also account for the energy given off or taken in by the reaction. The energy balance equation is usually written as

$$\begin{matrix} \text{Rate of energy} \\ \text{into reactor} \end{matrix} - \begin{matrix} \text{Rate of energy} \\ \text{out of reactor} \end{matrix} - \begin{matrix} \text{Rate of energy} \\ \text{associated} \\ \text{with reaction} \end{matrix} = \begin{matrix} \text{Rate of change of} \\ \text{energy accumulated} \\ \text{in reactor} \end{matrix}$$

A usual reference state for the enthalpies and internal energy is the pure components in the phase (liquid, gas, or solid) in which the reaction takes place, temperature of $25°C$, and the pressure of the system. Using this reference state, we can write

$$\text{Rate of energy associated with reaction} = V r_k \Delta H_r, \text{energy/time}$$

where ΔH_r is the enthalpy of reaction evaluated at the temperature of the reaction in energy/mole of key component. Note that by convention, ΔH_r for exothermic reactions is negative, and for endothermic reactions is positive.

3-7.2   Chemical Reactor Example

Consider the chemical reactor system shown in Fig. 3-7.1. The reactor is a vessel where the "well-known" reaction A $\rightarrow$ B occurs. Let us assume that the reaction occurs at constant volume and temperature. In addition, let us assume constant physical properties and that the reactor is well mixed. The rate of reaction is given by the expression

$$r_A(t) = kc_A^2(t)$$

where

$r_A(t)$ = rate of reaction of component A, kmoles of A/m³-s

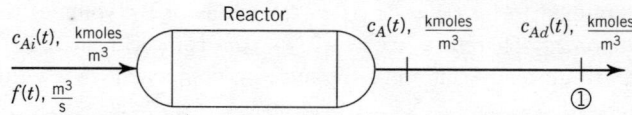

Figure 3-7.1 Isothermal well-mixed chemical reactor.

k = constant of reaction, $\text{m}^3/\text{kmoles-s}$

$c_A(t)$ = concentration of component A in reactor, kmoles of A/m^3

The objective is to develop the mathematical model, transfer functions, and draw the block diagram, relating $c_A(t)$ and $c_{Ad}(t)$ to the inputs $f(t)$ and $c_{Ai}(t)$.

Our procedure calls for first developing the mathematical model. Remember, in our way of doing things, those input variables, $f(t)$ and $c_{Ai}(t)$ in this case, are not considered unknowns. The control volume is the reactor, specifically the contents of the reactor. For this process, an unsteady-state mole balance on component A, of the type presented at the beginning of this section, provides the first equation:

| Rate of moles of component A into control volume | − | Rate of moles of component A out of control volume | + | Rate of change of component A in control volume | = | Rate of change of moles of component A accumulated in control volume |

or, in equation form,

$$f(t)c_{Ai}(t) - f(t)c_A(t) + (-1)Vr_A(t) = V\frac{dc_A(t)}{dt} \qquad (3\text{-}7.1)$$

1 eq., 2 unk. $[c_A(t), r_A(t)]$

The rate of reaction expression provides another equation

$$r_A(t) = kc_A^2(t) \qquad (3\text{-}7.2)$$

2 eq., 2 unk.

Equations 3-7.1 and 3-7.2 constitute the mathematical model for this process. Writing this model constitutes the first step in our procedure. The second step calls for linearizing the nonlinear terms in the model.

Linearizing the first two terms of Eq. 3-7.1 and Eq. 3-7.2 around the initial steady-state values of $\bar{f}, \bar{c}_{Ai}$, and $\bar{c}_A$ yields

$$f(t)c_{Ai}(t) \approx \overline{f}\,\overline{c}_{Ai} + \overline{c}_{Ai}(f(t) - \overline{f}) + \overline{f}(c_{Ai}(t) - \overline{c}_{Ai}) \qquad (3\text{-}7.3)$$

$$f(t)c_A(t) \approx \overline{f}\,\overline{c}_A + \overline{c}_A(f(t) - \overline{f}) + \overline{f}(c_A(t) - \overline{c}_A) \qquad (3\text{-}7.4)$$

$$r_A(t) \approx \overline{r}_A + 2k\overline{c}_A(c_A(t) - \overline{c}_A) \qquad (3\text{-}7.5)$$

Substituting Eqs. 3-7.3, 3-7.4, and 3-7.5 into Eq. 3-7.1 yields

$$\overline{f}\,\overline{c}_{Ai} + \overline{c}_{Ai}(f(t) - \overline{f}) + \overline{f}(c_{Ai}(t) - \overline{c}_{Ai}) - \overline{f}\,\overline{c}_A - \overline{c}_A(f(t) - \overline{f})$$
$$- \overline{f}(c_A(t) - \overline{c}_A) - V\overline{r}_A - 2Vk\overline{c}_A(c_A(t) - \overline{c}_A) = V\frac{dc_A(t)}{dt} \qquad (3\text{-}7.6)$$

Equation 3-7.6 is the equation that describes the process around the linearization values. We can now proceed to obtain the transfer functions. Writing a mole balance at the initial steady state and subtracting it from Eq. 3-7.6 yields

$$\overline{c}_{Ai}F(t) + \overline{f}C_{Ai}(t) - \overline{c}_A F(t) - \overline{f}C_A(t) - 2k\overline{c}_A V C_A(t) = V\frac{dC_A(t)}{dt} \qquad (3\text{-}7.7)$$

where

$$F(t) = f(t) - \overline{f}; \quad C_{Ai}(t) = c_{Ai}(t) - \overline{c}_{Ai}; \quad C_A(t) = c_A(t) - \overline{c}_A$$

From Eq. 3-7.7,

$$C_A(s) = \frac{K_1}{\tau s + 1} F(s) + \frac{K_2}{\tau s + 1} C_{Ai}(s) \tag{3-7.8}$$

where

$$K_1 = \frac{\overline{c}_{Ai} - \overline{c}_A}{\overline{f} + 2k\overline{c}_A V}, \text{ kmoles-s/(m}^3)^2$$

$$K_2 = \frac{\overline{f}}{\overline{f} + 2k\overline{c}_A V}, \text{ dimensionless}$$

$$\tau = \frac{V}{\overline{f} + 2k\overline{c}_A V}, \text{ seconds}$$

From Eq. 3-7.8 the desired transfer functions can be obtained. They are

$$\frac{C_A(s)}{F(s)} = \frac{K_1}{\tau s + 1} \tag{3-7.9}$$

and

$$\frac{C_A(s)}{C_{Ai}(s)} = \frac{K_2}{\tau s + 1} \tag{3-7.10}$$

To obtain the relationships for $c_{Ad}(t)$, assuming ideal plug flow and no reaction occurring in the outlet pipe, we can state

$$c_{Ad}(t) = c_A(t - t_o)$$

or in terms of deviation variables

$$C_{Ad}(t) = C_A(t - t_o) \tag{3-7.11}$$

and

$$t_o = \frac{LA_p}{\overline{f}} \tag{3-7.12}$$

where

t_o = dead time between the reactor outlet and point (1), seconds
L = distance between the reactor outlet and point (1), m
A_p = cross-sectional area of pipe, m^2

The Laplace transform of Eq. 3-7.11 gives

$$C_{Ad}(s) = e^{-t_o s} C_A(s) \tag{3-7.13}$$

Thus, from Eqs. 3-7.9, 3-7.10, and 3-7.13 the final desired transfer functions are

$$\frac{C_{Ad}(s)}{F(s)} = \frac{K_1 e^{-t_o s}}{\tau s + 1} \tag{3-7.14}$$

and

$$\frac{C_{Ad}(s)}{C_{Ai}(s)} = \frac{K_2 e^{-t_o s}}{\tau s + 1} \tag{3-7.15}$$

Figure 3-7.2 shows two different ways to draw the block diagram for this reactor.

(a)

(b)

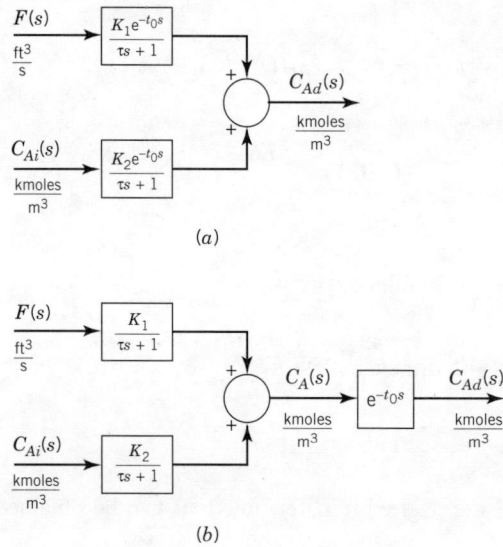

Figure 3-7.2 Block diagram for well-mixed isothermal chemical reactor.

3-8   EFFECTS OF PROCESS NONLINEARITIES

A most important characteristic of processes is their linear or nonlinear behavior. To understand what these terms mean and appreciate their significance, consider the thermal process presented in Section 3-3. In this particular process, because the flow f is considered constant, the gains, K_1 and K_2, are constants over the complete operating range. That is, their numerical value, given by Eqs. 3-3.22 and 3-3.23, do not ever change, no matter what the process operating condition. The value of the time constant τ, Eq. 3-3.21, is also constant for this system. The fact that the parameters that describe the characteristics of this process are constants means that the behavior of the process is also constant. That is, the process will behave in the same manner, with the same sensitivity and speed of response, at any operating condition. Processes that exhibit this characteristic are called linear processes.

In Section 3-1 we noted that the controller must be tuned, or adapted, to the process to obtain adequate control performance. Because the behavior of a linear process is the same over the complete operating range, if the controller is optimally tuned at one operating condition, it is also optimum at any other operating condition. This is certainly an ideal operation and the one we could hope for.

However, consider now the gas process presented in Section 3-6. In this process the gains K_1, K_2, and K_3 as given by Eq. 3-6.16 depend on C_1, C_2, and C_3, and the numerical value of these terms depend on the values of $\bar{p}$, $\bar{p}_1$, and $\bar{m}_o$ around which the linearization of the nonlinear function $f_o(t)$ was done. Therefore, the numerical values of K_1, K_2, and K_3 also depend on where the linearization was performed. The numerical value of the time constant, as also given by Eq. 3-6.16, also depends on C_2. This means that the values of the terms that describe the process characteristics, and thus the process behavior itself, depend on the operating condition. The process behavior changes as the operating conditions change! Processes that exhibit these characteristics are called *nonlinear processes*. Nonlinearity is a characteristic of most chemical processes.

To demonstrate graphically the effect of the process nonlinearities, two different cases are shown. In the first case, the pressure in the tank was allowed to vary between 25 psia and 70 psia, while keeping the process flow and other process conditions at their

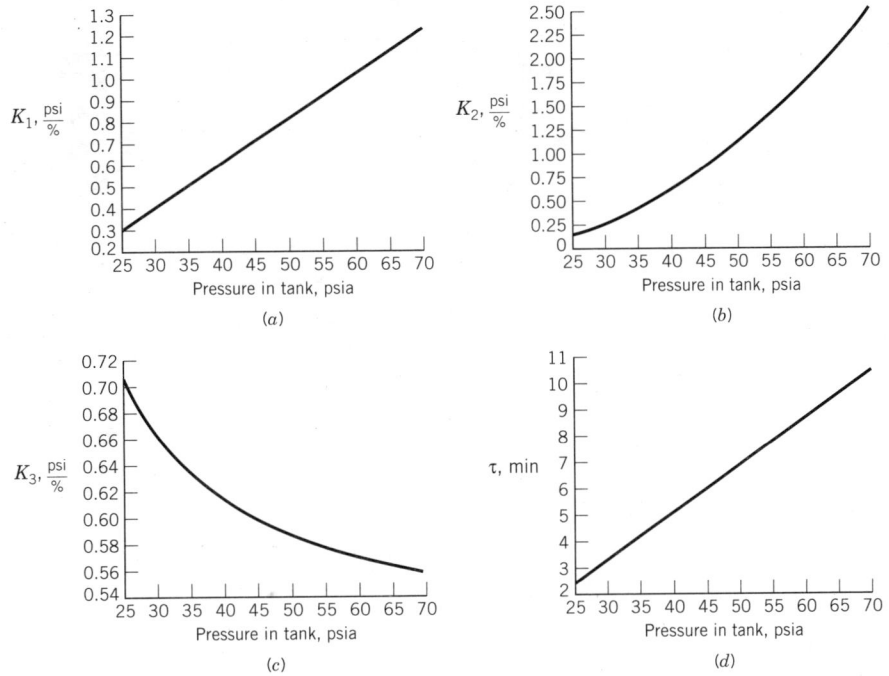

Figure 3-8.1 Gains and time constant as a function of pressure in tank.

steady state. The signal to the outlet valve was allowed to vary to keep the process flow constant. This is the case when it is desired to run the process at different pressures even though the process flow has not changed. Figure 3-8.1 shows how K_1, K_2, K_3, and τ vary as the pressure in the tank varies. Figure 3-8.1a shows that K_1 varies by a factor of 4, and Fig. 3-8.1b shows that K_2 varies by a factor of 10. Similarly, Fig. 3-8.1d shows that τ varies by a factor of 5. K_3 is not affected as much as shown in Fig. 3-8.1c.

Another interesting case occurs when the signal to the fan varies, thus varying the process flow through the tank, while keeping the pressure constant. This could be the case when a pressure control system is having to react to upsets, the upset being the signal to the fan in this case. The signal to the outlet valve was allowed to change to match the outlet gas flow to the inlet flow provided by the fan, and to keep the pressure in the tank constant. Figure 3-8.2 shows how K_1, K_2, K_3, and τ vary as the signal to the fan varies. All of these figures show the nonlinear characteristics of this simple process.

The nonlinear behavior of processes is very detrimental to their control. As the process behavior changes with operating conditions, the controller should be re-tuned, or re-adapted, to maintain optimum control performance. Often, the best we can do is to tune the controller so that its performance is best at the design operating point and acceptable over the expected range of operating conditions; tuning methods are presented in Chapter 6. Techniques have been developed to permit the controller to re-tune itself, automatically, as the process characteristics change. These techniques are referred to as self-tuning, or adaptive tuning and the reader is referred to the previous edition of this book for a brief presentation (Smith and Corripio, 1997).

Although it is not previously presented, the dead time also depends on the operating conditions. Equation 3-4.1 shows that if the process flow varies, the dead time will also vary. Thus, all the terms that describe the process behavior are functions of the operating conditions.

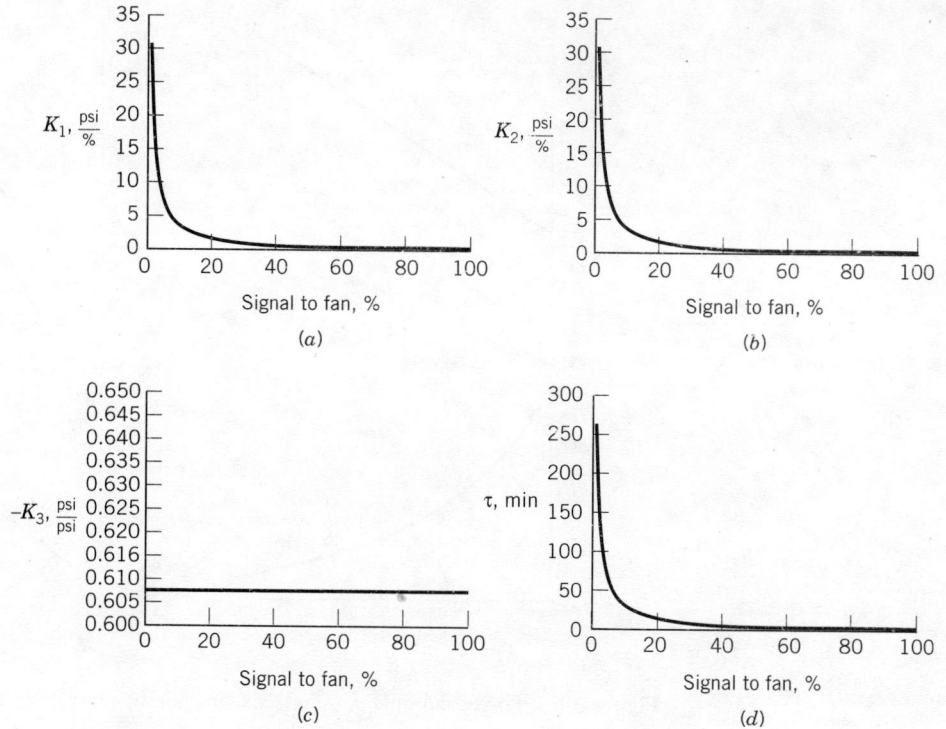

Figure 3-8.2 Gains and time constant as a function of signal to fan.

Process nonlinearities are certainly not a desirable characteristic, but they are unfortunately a realistic and very common one.

3-9   ADDITIONAL COMMENTS

It is now important to analyze what we have done from a more general point of view. If we look at the form of the transfer functions that have been developed in the different examples, Eqs. 3-3.12, 3-3.25, 3-3.26, 3-6.17, 3-6.18, 3-6.19, 3-7.9, and 3-7.10, they are all of the form

$$\frac{Y(s)}{X(s)} = \frac{K}{\tau s + 1} \tag{3-9.1}$$

where

$Y(s)$ = Laplace transform of the output variable

$X(s)$ = Laplace transform of the input variable

In Section 2-4, Eq. 3-9.1 was defined as the standard form of the transfer function for a first-order system. The distinguishing characteristic of this form is that the second term of the denominator is unity. This is the form of all first-order systems regardless whether they are thermal, fluid, reacting, mechanical, or electrical systems. This is important because it says that the behavior of any system, no matter what type, described by Eq. 3-9.1 is the same; they all respond the same way to forcing functions. The meaning of gain, K, and time constant, τ, is the same for all of them.

Sometimes dead time is present, and in this case the transfer function becomes

$$\frac{Y(s)}{X(s)} = \frac{Ke^{-t_o s}}{\tau s + 1} \tag{3-9.2}$$

Equation 3-9.2 is more general than Eq. 3-9.1.

One of the most important terms in the study of automatic control is the time constant, τ. We have developed several expressions for τ, Eqs. 3-3.9, 3-3.21, and 3-6.16. These equations are all analogous, that is, they are all of the following form

$$\tau = \frac{\text{capacitance}}{\text{conductance}} \tag{3-9.3}$$

The capacitance is a measure of the ability of the process to accumulate the quantity conserved (mass or energy). The conductance is a measure of the process ability to regulate itself.

For example, for the thermal system of Section 3-3, the time constant, Eq. 3-3.9, is

$$\tau = \frac{V\rho C_v}{f\rho C_p} = \frac{\text{capacitance}}{\text{conductance}}, \frac{\frac{\text{Joules}}{°C}}{\frac{\text{Joules/sec}}{°C}}$$

We can also write this expression for τ, assuming that $C_v = C_p$, which is a good assumption for liquids, as

$$\tau = \frac{V}{f}$$

This expression clearly shows the accumulation, V, and flow, f, terms. Table 3-9.1 presents the analogy for the processes shown in this chapter. A process that has not been presented here, but is given as an exercise in Problem 3-1, is that of mixing, or blending. This process is similar to a reacting process in which no reaction occurs. This is expressed by assigning the reaction constant, k, a value of zero in the time constant in Section 3-7. In this case

$$\tau = \frac{V}{\overline{f} + 2k\overline{C}_A V} = \frac{V}{\overline{f}}, \frac{m^3}{m^3/s}$$

Another comment we wish to make concerns the method used to obtain the desired transfer functions and block diagrams. As you have undoubtedly noticed, the procedure

Table 3-9.1 Time constant analogy for different processes

Process	Variable	Time constant	Capacitance	Conductance
Thermal	Temperature	$\frac{V\rho C_v}{f\rho C_p}$ (Eq. 3-2.9)	$V\rho C_v, \frac{J}{°C}$	$f\rho C_p, \frac{J/s}{°C}$
Thermal	Temperature	$\frac{V\rho C_v}{f\rho C_p + UA}$ (Eq. 3-2.21)	$V\rho C_v, \frac{J}{°C}$	$f\rho C_p + UA, \frac{J/s}{°C}$
Gas	Gas pressure, flow	$\frac{V}{RT\overline{\rho}C_2}$ (Eq. 3-5.16)	$\frac{V}{RT}, \frac{\text{Ibmole}}{\text{psia}}$	$\overline{\rho}C_2, \frac{\text{Ibmole}}{\text{min-psia}}$
Reacting	Concentration	$\frac{V}{\overline{f} + 2k\overline{c}_A V}$ (Eq. 3-6.16)	V, m^3	$\overline{f} + 2k\overline{c}_A V, m^3/s$

first of all requires a good knowledge of process engineering. The steps followed to obtain the transfer functions were outlined in Section 3-6.

You must have also noted that most of the time, the equations that describe the process are nonlinear. We have linearized them to be able to obtain the desired transfer functions. These transfer functions describe the process in a region close to the linearization values. Outside this region the linearization will "break down" and give erroneous results. The size of the region where the transfer functions are valid depends on the degree of nonlinearity of the process. For a very nonlinear process, the valid region will be very close to the linearization values. The region "opens" as the degree of nonlinearity of the process lessens. The only way to obtain an accurate solution from the set of equations, the *mathematical model*, over the complete operating range is by numerical methods, or computer solution. However, this technique does not allow a general analysis of the process dynamics. Chapter 13 presents in more detail process simulation.

Finally, a comment about the response of first-order systems to different types of forcing functions; the responses to step function, ramp function, and sinusoidal functions are presented in Chapter 2. The response to a step function is particularly important in process control studies and thus it has also been shown in this chapter. It is clear from Fig. 3-3.2 that the steepest slope occurs at the beginning of the response. This characteristic is typical of first-order systems.

All the responses shown in this chapter reach a new operating value. That is, the responses to a bounded input are also bounded; the system "regulates" itself to a new value. The majority of processes are of this type and are sometimes referred to as self-regulating processes. There are some processes, however, that do not regulate themselves to a new value before they reach an extreme operating condition. These processes are referred to as nonself-regulating processes and examples of them are given in Chapter 4.

3-10   SUMMARY

This chapter began by explaining that from a controls point of view, a "process" is everything except the controller. That is, the process consists of the sensor, transmitter, process unit, valve, and transducer, if present. We noted that as far as the controller is concerned, its controlled variable is the signal it receives from the transmitter. Its manipulated variable is its own output signal, that is, the signal the controller sends out to the final control element. A discussion of why it is necessary to study the process characteristic was included.

The chapter then continued with an introduction to dynamic process modeling, and particularly why steady-state balances do not result in useful models for process control. It was shown that the unsteady-state balances, also called dynamic balances, are an extension of the steady-state balances when the rate of change of the accumulation term is added. Several simple processes were used to illustrate the development of mathematical models, transfer functions, and block diagrams. All of the processes studied in the chapter are described by first-order ordinary differential equations. The starting point is usually a balance equation. In order to develop the set of equations, we must end up with the same number of independent equations as unknowns. This is why

we have stressed the unknowns next to each equation. This should help us keep track of the equations needed to describe the process and develop the model.

Several other concepts were reviewed and further explained in this chapter. Transfer functions were defined as the ratio of the Laplace-transformed output variable to the Laplace-transformed input variable. The meaning of transfer functions was explained: they fully describe the steady-state and dynamic behavior of the system. Transfer functions indicate how much and how fast processes change. The variables used in the transfer functions are in deviation form.

The transfer functions developed in this chapter are of the general form

$$\frac{Y(s)}{X(s)} = \frac{Ke^{-t_o s}}{\tau s + 1} \qquad \textbf{(3-10.1)}$$

These transfer functions are called first-order-plus-dead-time (FOPDT) transfer functions, or first-order lags. This transfer function contains three parameters: process gain, K, process time constant, τ, and process dead time, t_o. Understanding these parameters is fundamental to the study of process control. The process gain, K, specifies the amount of change of the output variable per unit change in the input variable; it is defined mathematically as follows:

$$K = \frac{\Delta Y}{\Delta X} = \frac{\Delta \text{ output variable}}{\Delta \text{ input variable}} \qquad \textbf{(3-10.2)}$$

The process time constant, τ, is related to the speed of response of the process once the process starts to respond to an input. The time constant was shown to be, in Chapter 2 and in Eq. 3-3.13, the time required by the output variable to reach 63.2 % of the total change when the input variable changes in a step fashion. The slower a process is to respond to an input, the larger the value of τ. The process dead time, t_o, is the time interval between the change in input variable and the time the output variable starts to respond. Therefore, the process time constant and process dead time are the terms

that describe the dynamics of the system; the process gain describes the steady-state characteristics of the system.

We must remember that these three parameters, K, τ, and t_o, are functions of the physical parameters of the process. It was shown that for a linear system, these parameters are constant over the complete operating range of the process. For a nonlinear system, the parameters were shown to be functions of the operating conditions and, consequently, not constant over the operating range. This nonlinear characteristic of processes is a most important consideration for their control. This chapter showed how to obtain the aforementioned parameters starting from balance equations. Chapter 7 shows how to evaluate them from process data.

The next chapter shows the development of transfer functions and block diagrams for more complex processes.

REFERENCES

1. SMITH, C. A., and A. B. CORRIPIO, *Principles and Practice of Automatic Process Control*, 2nd ed., New York: John Wiley & Sons, 1997.

PROBLEMS

3-1. Consider the mixing process shown in Fig. P3-1. You may assume that the density of the input and output streams are very similar and that the flow rates f_1 and f_2 are constant. It is desired to understand how each inlet concentration affects the outlet concentration. Develop the mathematical model, determine the transfer functions, and draw the block diagram for this mixing process. Show the units of all gains and time constants.

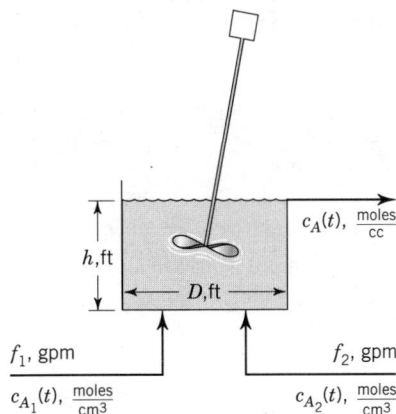

Figure P3-1 Sketch for Problem 3-1.

Note: The simulation of this process is the subject of Problems 13-6 and 13-7.

3-2. Consider the isothermal reactor shown in Fig. P3-2. The rate of reaction is given by

$$r_A(t) = kc_A(t), \text{ moles of A}/(ft^3 - min)$$

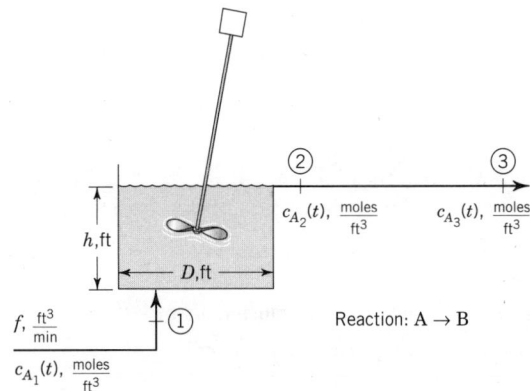

Figure P3-2 Sketch for Problem 3-2.

where k is constant. You may assume that the density and all other physical properties of products and reactants are similar. You may also assume that the flow regime between points 2 and 3 is very turbulent (plug flow), minimizing backmixing. Develop the mathematical model, and obtain the transfer functions relating

(a) The concentration of A at point 2 to the concentration of A at point 1.

(b) The concentration of A at point 3 to the concentration of A at point 2.

(c) The concentration of A at point 3 to the concentration of A at point 1.

Note: The simulation of this process is the subject of Problem 13-8.

3-3. A storage tank has a diameter of 20 ft and a height of 10 ft. The output volumetric flow from this tank is given by

$$f_{out}(t) = 2h(t)$$

where $h(t)$ is the height of liquid in the tank. At a particular time the tank is at steady state with an input flow of 10 ft^3/min.

(a) What is the steady-state liquid height in the tank?

(b) If the input flow is ramped up at the rate of 0.1 ft^3/min, how many minutes will it take for the tank to overflow?

(Copyright 1992 by the American Institute of Chemical Engineers; reproduced by permission of Center for Chemical Process Safety of AIChE.)

3-4. Consider the temperature sensor sketched in Fig. P3-3. The bulb and its surrounding thermowell are at a uniform temperature, $T_b(t)$, and the surroundings are at a uniform temperature, $T_S(t)$. The exchange of heat between the surroundings and the bulb is given by

$$q(t) = hA[T_S(t) - T_b(t)]$$

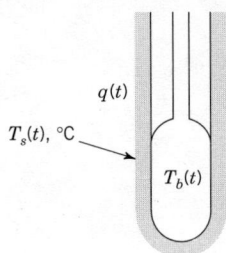

Figure P3-3 Sketch for Problem 3-4.

where

$q(t)$ = heat transfer rate, J/s

$\quad h$ = film coefficient of heat transfer, J/s-m^2-$^\circ$C

$\quad A$ = contact area between the bulb and its surroundings, m^2

Let M, kg, be the mass of the bulb and thermowell, and C_v, J/kg-$^\circ$C, its heat capacity. Obtain the transfer function that describes the response of the temperature of the bulb when the surrounding temperature changes. List all assumptions and draw the block diagram for the bulb. Express the time constant and the gain in terms of the bulb parameters. *Note*: the transfer function derived here generally represents the dynamic response of most temperature sensors, regardless of their type.

3-5. Hot water at a rate of 2 liters/min (constant) and temperature $T_h(t)$ is mixed with cold water at a constant rate of 3 liters per minute and a constant temperature of 20°C. Both streams flow into a bathtub, but because of carelessness, the water is overflowing and keeping the bathtub full of water. The volume of the bathtub is 100 liters. Assuming the water in the bathtub is perfectly mixed, derive the differential equation relating the temperature in the bathtub, $T(t)$, to the temperature of the hot water, $T_h(t)$. Obtain the transfer function $\Gamma(s)/\Gamma_h(t)$, and calculate its gain and time constant.

3-6. Process wastewater (density = 1000 kg/m^3) flows at 500,000 kg/h into a holding pond with a volume of 5000 m^3 and then flows from the pond to a river. Initially, the pond is at steady state with a negligible concentration of pollutants [$x(0) = 0$]. Because of a malfunction in the wastewater treating process, the concentration of pollutants in the inlet stream suddenly increases to 500 mass ppm (kg pollutant per million kg of water) and stays constant at that value (step change).

(a) Assuming a perfectly mixed pond, obtain the transfer function of the pollutant concentration in the outlet stream to the concentration of the inlet stream, and determine for how long can the process malfunction go undetected before the outlet concentration of pollutants exceeds the regulated maximum value of 350 ppm.

(b) Repeat part (a) assuming that the water flows in plug flow (without mixing) through the pond. Notice that this means the pond behaves as a pipe and the response of the concentration is a pure transportation lag.

(c) In both parts (a) and (b), it is assumed that the entire volume of the pond is active. How would your answers be affected if portions of the pond were stagnant and not affected by the flow of water in and out?

3-7. In Dr. Corripio's home, the hot water line between the water heater and his shower is 1/2-in. copper tubing (cross-sectional area = 0.00101 ft^2) and about 30 ft long. On a cold Baton Rouge morning, Dr. Corripio turned the hot water valve on the shower fully opened and got a flow of 2 gallons per minute. How long did he have to wait for the hot water to reach the shower (and probably burn him)? Write the transfer function $\Gamma_s(s)/\Gamma_h(s)$ for the hot water line, where $\Gamma_s(t)$ is the deviation temperature at the shower, and $\Gamma_h(t)$ is the deviation temperature in the hot water heater, when the hot water valve is opened. Draw the block diagram for the hot water line. What is the transfer function when the hot valve is closed? Could you predict this from your previous answer?

3-8. Brine from a pond is pumped at 100 ft^3/min to a process through a line that has two different diameters, before and after the pump. The inside diameters and lengths of the pipes are as follows:

	Before the pump	After the pump
Inside diameter, in.	6.00	5.25
Length, ft	1000	2000

You may assume that the brine does not mix in the pipe. When the concentration in the pond changes, how long does it take for the concentration of the stream entering the process to change? Write the transfer function for the concentration out of the pipe to the concentration in the pond.

3-9. It is desired to model the response of the temperature, $T(t)$, °C, in a fish tank to changes in the heat input from the electric heater, $q(t)$, W; ambient temperature, $T_s(t)$, °C; and ambient partial pressure of water in the air, $p_s(t)$, Pa, under the following assumptions:

(a) The water in the tank is perfectly mixed.

(b) Transfer of heat and mass to the surroundings is only from the free surface of the water (transfer of heat through the glass sides is negligible).

(c) The overall heat transfer coefficient to the surroundings, U, W/m² − °C, and overall mass transfer coefficient of water vapor, K_y, kg/s − m² − Pa, are constant.

(d) The physical properties of water, specific heat, C_p, J/kg − °C, and latent heat, λ, J/kg, are constant.

(e) The rate of vaporization of water from the tank is proportional to the difference in partial pressures

$$w(t) = K_y A[p^o(T) - p_s(t)], \text{ kg/s}$$

where $p^o(T)$, Pa is the vapor pressure of water and is given by Antoine's equation.

A, m², is the area of the free surface of the water.

(f) The rate of vaporization is so small that the total mass of water in the tank, M, kg, may be assumed constant.

Obtain the transfer functions that represent the response of the tank temperature when the heat input from the electric heater, the surrounding temperature, and the surrounding water partial pressure change. Draw the block diagram for this system.

3-10. Water is poured at a rate $f_i(t)$, cm³/s, into a cup measuring 6.5 cm in diameter and 10 cm high. The cup has a circular hole in the bottom measuring 0.2 cm in diameter. The velocity of the water through the hole is given, from Bernoulli's equation, by

$$v(t) = \sqrt{2gh(t)}$$

where g is the local acceleration of gravity, 980 cm/s², and $h(t)$, cm, is the level of the water in the cup. Obtain the transfer function between the level of the water in the cup, $H(s)$, and the inlet flow $F_i(s)$, when the cup is half full of water ($h = 5$ cm).

3-11. Consider the flash drum shown in Fig. P3-4. $z(t)$, $x(t)$, and $y(t)$ are the mole fractions of the most volatile component in the feed, liquid, and vapor streams, respectively. The total mass of liquid and vapor accumulated in the drum, the temperature, and the pressure can all be assumed constant. If equilibrium between the vapor and liquid phases

leaving the drum is assumed, the following relationship between $y(t)$ and $x(t)$ can be established

$$y(t) = \frac{\alpha x(t)}{1 + (\alpha - 1)x(t)}$$

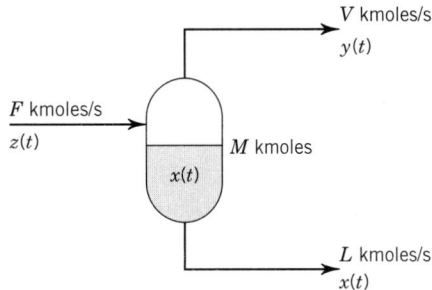

Figure P3-4 Sketch for Problem 3-11.

The steady state, and other process information is

$$M = 500 \text{ kmoles}; \quad F = 10 \text{ kmoles/s};$$
$$L = 5 \text{ kmoles/s}; \quad \alpha = 2.5; \quad x(0) = 0.4$$

Obtain the transfer function that relate the outlet liquid composition, $x(t)$, to the feed composition, $z(t)$. Determine also the numerical value of all the terms in the transfer function. *Note:* The simulation of this process is the subject of Problem 13-9.

3-12. Figure P3-5 shows a tray of a distillation column. The flow from the tray is given by the Francis weir formula (adapted from Perry's Chemical Engineering Handbook, 6th edition, 1984, p. 5–19).

$$f_o(t) = 0.415wh^{1.5}(t)\sqrt{2g}$$

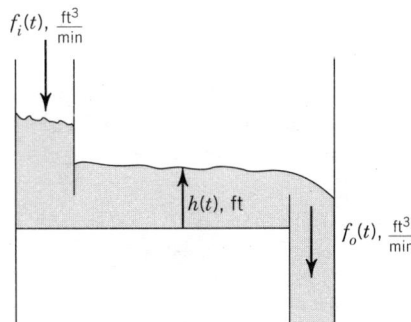

Figure P3-5 Sketch for Problem 3-12.

where

$h(t) = $ the liquid level on the tray above the top of the weir, ft

$w = $ width of the weir over which the liquid overflows, ft

$g = $ local acceleration of gravity (32.2 ft/s²).

The steady-state inlet flow and process parameters are as follows: tray cross-sectional area $= 11.2$ ft^2; $w = 3.0$ ft; $f_i(0) = 30$ ft^3/min.

Obtain the transfer functions that relate the height of the water above the weir and the flow from the tray to the inlet flow to the tray. State all assumptions, and calculate the numerical values of the tray time constant and gain. Also draw the complete block diagram relating the variables. *Note*: The simulation of this process is the subject of Problem 13-10.

3-13. Consider an adiabatic, exothermic, perfectly mixed (what else?) chemical reactor where the reaction $A + B \rightarrow C$ (what else?) takes place. Let

ρ = density of reactants and product (constant), kmole/m^3

f = flow of inlet and outlet streams (constant), m^3/s

$T_i(t)$ = inlet temperature, K

$T(t)$ = temperature in reactor, K

ΔH_r = heat of reaction (constant and negative), J/kmole

c_p, c_v = heat capacities (constants), J/kmole-K

V = volume of liquid in tank (constant), m^3

The kinetics for the reaction is expressed by the following zeroth-order expression

$$r_A = -k_o e^{-E/RT(t)}$$

where

k_o = frequency factor, kmole/m^3-s

E = activation energy, J/kmole

R = ideal gas constant, J/kmole-K

Determine the transfer function $\Gamma(s)/\Gamma_i(s)$ for the reactor. Express the time constant and gain in terms of the physical parameters. Under what conditions can the time constant be negative? What would be the consequences of a negative time constant?

3-14. Consider the process shown in Fig. P3-6. The tank is spherical with a radius of 4 ft. The nominal mass flow into and out of the tank is 30,000 lbm/hr, the density of the liquid is 70 lbm/ft^3, and the steady-state level is 5 ft. The volume of a sphere is given by $4\pi r^3/3$. The relation between volume and height is given by

$$V(t) = V_T \left[\frac{h^2(t)[3r - h(t)]}{4r^3} \right]$$

and the flows through the valves by

$$\dot{m}(t) = 500 C_v vp(t) \sqrt{G_f \Delta P(t)}$$

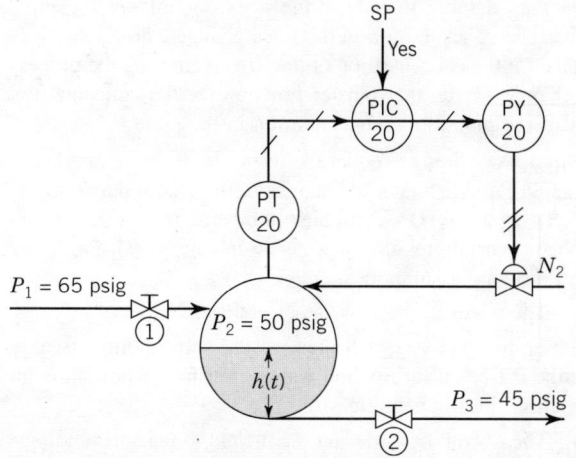

Figure P3-6 Sketch for Problem 3-14.

where

r = radius of sphere, ft

$V(t)$ = volume of liquid in tank, ft^3

V_T = total volume of tank, ft^3

$h(t)$ = height of liquid in tank, ft

$\dot{m}(t)$ = mass flow rate, lbm/hr

C_v = valve coefficient, gpm/(psi$^{1/2}$)

     $C_{v1} = 20.2$ gpm/(psi$^{1/2}$)   and

     $C_{v2} = 28.0$ gpm/(psi$^{1/2}$)

$\Delta P(t)$ = pressure drop across valve, psi

G_f = specific gravity of fluid

$vp(t)$ = valve position, a fraction of valve opening

The pressure above the liquid level is maintained constant at a value of 50 psig. Obtain the transfer functions that relate the level of liquid in the tank to changes in valve positions of valves 1 and 2. Also, plot the gains and time constants versus different operating levels while keeping the valve positions constant.

3-15. Consider the heating tank shown in Fig. P3-7. A process fluid is being heated in the tank by an electrical heater. The rate of heat transfer, $q(t)$, to the process fluid is related to the signal, $m(t)$, by

$$q(t) = am(t)$$

You may assume that the heating tank is well insulated, that the fluid is well mixed in the tank, and the heat capacity and density of the fluid are constant. Develop the mathematical model that describes how the inlet temperature, $T_i(t)$; the process flow, $f(t)$; and the signal, $m(t)$, affect the outlet temperature $T(t)$. Then determine the transfer functions, and draw the block diagram for this process.

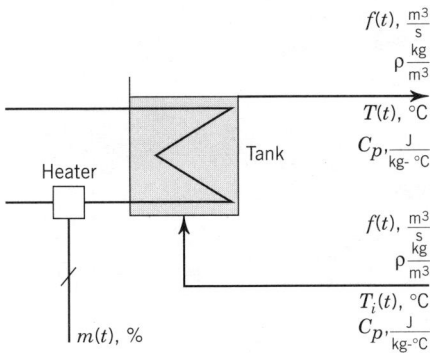

Figure P3-7 Sketch for Problem 3-15.

3-16. Consider the mixing process shown in Fig. P3-8. The purpose of this process is to blend a stream, weak in component A, with another stream, pure A. The density of stream 1, ρ_1, can be considered constant because the amount of A in this stream is small. The density of the outlet stream is, of course, a function of the concentration and is given by

$$\rho_3(t) = a_3 + b_3 c_{A3}(t)$$

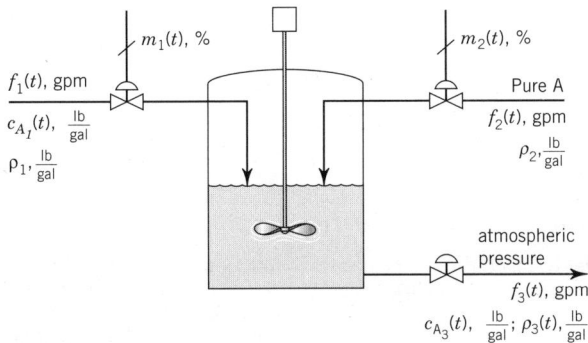

Figure P3-8 Sketch for Problem 3-16.

The flow through valve 1 is given by

$$f_1(t) = C_{v1} vp_1(t)\sqrt{\frac{\Delta P_1}{G_1}}$$

The flow through valve 2 is given by

$$f_2(t) = C_{v2} vp_2(t)\sqrt{\frac{\Delta P_2}{G_2}}$$

Finally, the flow through valve 3 is given by

$$f_3(t) = C_{v3}\sqrt{\frac{\Delta P_3(t)}{G_3(t)}}$$

The relationship between the valve position and the signal is given by

$$vp_1(t) = a_1 + b_1(m_1(t) - d_1)$$

and

$$vp_2(t) = a_2 + b_2(m_2(t) - d_2)$$

where

$a_1, b_1, d_1, a_2, b_2, d_2, a_3, b_3 =$ known constants

$C_{v1}, C_{v2}, C_{v3} =$ valve coefficients of valve 1, 2, and 3, respectively, $m^3/(s\text{-}psi^{1/2})$

$vp_1(t), vp_2(t) =$ valve position of valves 1 and 2, respectively, a dimensionless fraction

$\Delta P_1, \Delta P_2 =$ pressure drop across valves 1 and 2, respectively (constants), psi

$\Delta P_3(t) =$ pressure drop across valve 3, psi

$G_1, G_2 =$ specific gravity of streams 1 and 2, respectively (constants), dimensionless

$G_3(t) =$ specific gravity of stream 3, dimensionless

Develop the mathematical model that describes how the forcing functions $m_1(t)$, $m_2(t)$, and $c_{Ai}(t)$ affect $h(t)$, and $c_{A3}(t)$; determine the transfer functions; and draw the block diagram. Be sure to show the units of all the gains, and time constants.

3-17. Consider the tank shown in Fig. P3-9. A 10 % (± 0.2 %) by weight NaOH solution is being used for a caustic washing process. In order to smooth variations in flow rate and concentration, a 8000-gal tank is being used as surge tank.

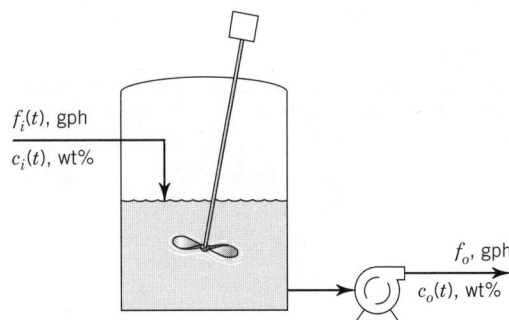

Figure P3-9 Sketch for Problem 3-17.

The steady-state conditions are as follows:

$$\overline{V} = 4000 \text{ gal}; \qquad \overline{f}_i = \overline{f}_o = 2500 \text{ gph};$$
$$\overline{c}_i = \overline{c}_o = 10 \text{ wt\%}$$

The tank contents are well mixed, and the density of all streams is 8.8 lbm/gal.

(a)   An alarm will sound when the outlet concentration drops to 9.8 wt% (or rises to 10.2 wt%). Assume that the flows are constant.

(i)   Obtain the transfer function relating the outlet concentration to the inlet concentration. Obtain the numerical values of all gains and time constants.

(ii)   Because of an upset, the inlet concentration, $c_i(t)$, drops to 8 % NaOH instantaneously. Determine how long it will take before the alarm sounds.

(b)   Consider now that the inlet flow, $f_i(t)$, can vary whereas the outlet flow is maintained constant at 2500 gph. Therefore, the volume in the tank can also vary.

(iii)   Develop the differential equation that relates the volume in the tank to the flows in and out.

(iv)   Develop the differential equation that relates the mass of NaOH in the tank to the inlet flow and inlet concentration.

(v)   Obtain the transfer function relating the volume in the tank to the inlet flow.

(vi)   Obtain the transfer function relating the outlet concentration to the inlet flow and the inlet concentration. Obtain the numerical values of all gains and time constants.

(vii)   Suppose now that the inlet flow to the tank drops to 1000 gph. Determine how long it takes to empty the tank.

3-18.   The blending tank shown in Fig. P3-10 is perfectly mixed. The input variables are the solute concentrations and flows of the inlet streams, $c_1(t), c_2(t)$ [kg/m^3], $f_1(t)$, and $f_2(t)$ [m^3/min]. The volume of liquid in the tank, V [m^3], can be assumed constant, and variation of stream densities with composition may be neglected.

(a)   Obtain the transfer functions for the outlet composition $C(s)$, kg/m^3, and outlet flow $F(s)$, m^3/min, to the four input variables, and write the expressions for the time constant and gains of the blender in terms of the parameters of the system.

(b)   Draw the block diagram for the blender, showing all transfer functions.

(c)   Calculate the numerical values of the time constants and gains for a blender that is initially mixing a stream containing 80 kg/m^3 of solute with a second stream containing 30 kg/m^3

of the solute to produce 4.0 m^3/min of a solution containing 50 kg/m^3 of the solute. The volume of the blender is 40 m^3.

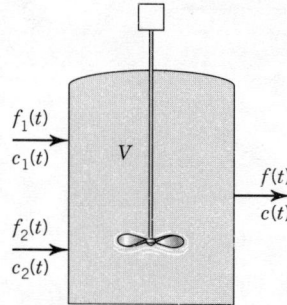

Figure P3-10 Sketch for Problem 3-18.

Note: The simulation of this process is the subject of Problem 13-11.

3-19.   Draw the block diagram representing the following transfer functions. In each case, do not do any algebraic manipulations to simplify the transfer functions, but use the rules of block diagram algebra to simplify the diagram if possible.

(a) $Y(s) = \dfrac{K_1}{\tau_1 s + 1} X(s) + \dfrac{K_2}{\tau_2 s + 1} X(s)$

(b) $Y(s) = \dfrac{1}{\tau s + 1}[K_1 F_1(s) - K_2 F_2(s)]$

(c) $Y_1(s) = G_1(s)X(s) + G_3(s)Y_2(s)$
$Y_2(s) = G_2(s)Y_1(s)$

3-20.   Determine the transfer function $C(s)/R(s)$ for the system shown in Fig. P3-11.

3-21.   Determine the transfer function $C(s)/R(s)$ for the system shown in Fig. P3-12.

3-22.   Determine the transfer function $C(s)/L(s)$ for the system shown in Fig. P3-13.

3-23.   Obtain the response of a process described by a first-order-plus-dead-time transfer function to the forcing function shown in Fig. P3-14.

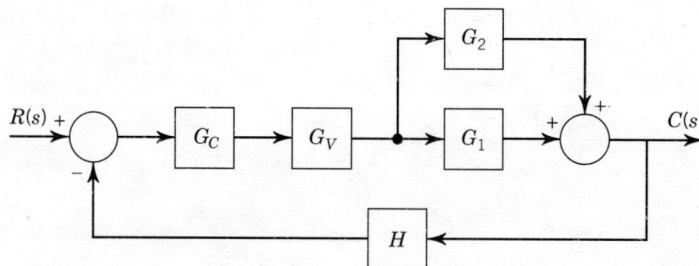

Figure P3-11 Sketch for Problem 3-20.

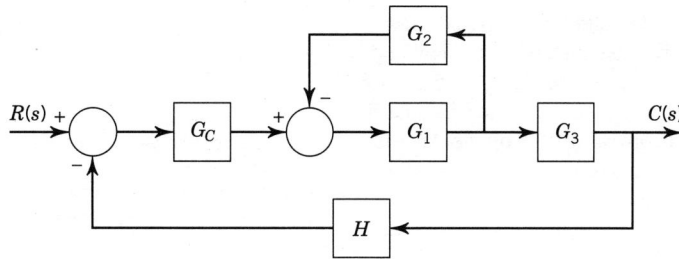

Figure P3-12 Sketch for Problem 3-21.

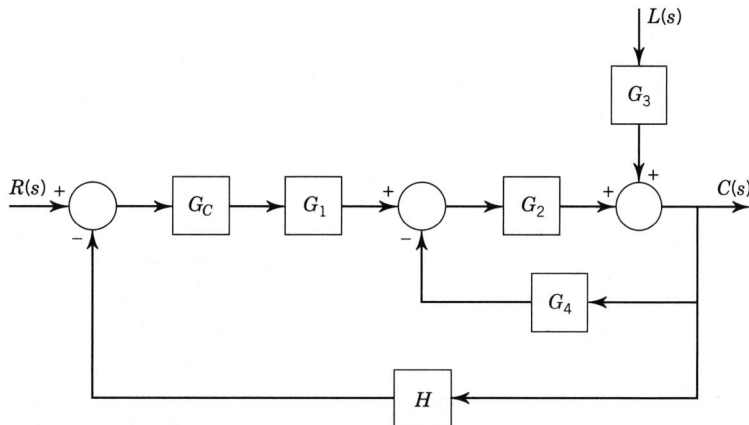

Figure P3-13 Sketch for Problem 3-22.

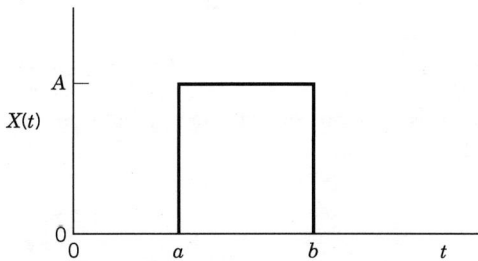

Figure P3-14 Sketch for Problem 3-23.

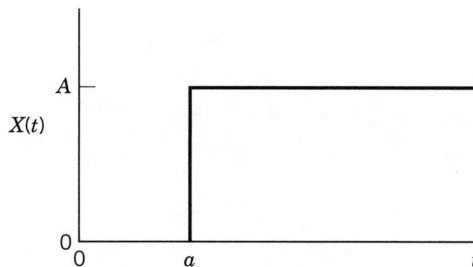

Figure P3-15 Sketch for Problem 3-24.

3-24. Assume that the following equation describes a certain process

$$\frac{Y(s)}{X(s)} = \frac{3e^{-0.5s}}{5s + 0.2}$$

(a) Obtain the steady-state gain, time constant, and dead time of this process.

(b) The initial condition of the variable y is $y(0) = 2$. For a forcing function as shown in Fig. P3-15, what is the final value of $y(t)$?

3-25. Obtain the response of a process described by a first-order transfer function to an impulse forcing function.

3-26. A gas detector is used to determine the concentration of a flammable gas in a gas stream. Normally the gas concentration is 1 % by volume, well below the alarm limit of 4 % and the lower flammability limit of 5 %. If the gas concentration is above the lower flammability limit it is flammable. A particular gas detector demonstrates first-order behavior with a time constant of 5 s. At a particular time, the gas stream is flowing at 1 m^3/s through a duct with a cross-sectional area of 1 m^2. If the gas concentration suddenly increases from 1 to 7 % by volume, how many cubic meters of flammable gas pass the sensor before the alarm is sounded? Is it possible for a plug of flammable gas to pass the detector without the alarm ever being sounded? (Copyright 1992 by the American Institute of Chemical Engineers; reproduced by permission of Center for Chemical Process Safety of AIChE.)

3-27. Consider the chemical reactor shown in Fig. P3-16. In this reactor, an endothermic reaction of the type $A + 2B \rightarrow C$ takes place. The rate of appearance of A, is given by

$$r_A(t) = k_o e^{-E/RT(t)} c_A(t) c_B(t)$$

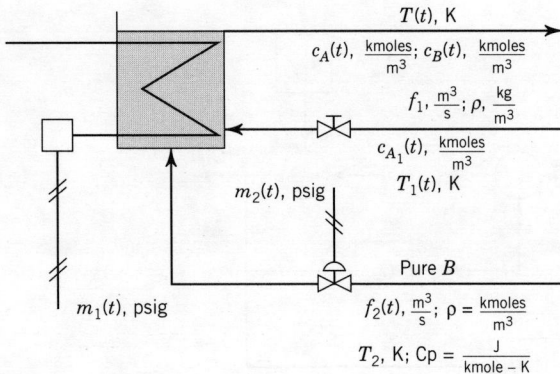

Figure P3-16 Sketch for Problem 3-27.

where

$r_A(t)$ = rate of disappearance of A, kmole of A/(m^3-s)

k_o = frequency factor (constant) m^3/(kmole-s)

E = energy of activation (constant) cal/gmole

R = gas law constant, 1.987 cal/(gmole-K)

$T(t)$ = temperature in reactor, K

$c_A(t)$ = concentration of A in reactor, kmole/m^3

$c_B(t)$ = concentration of B in reactor, kmole/m^3

ΔH_r = heat of reaction, J/kmole

The heat input to the reactor is related to the signal to the heater by the following expression

$$q(t) = r m_1(t)$$

where

$q(t)$ = heat input to reactor, J/s

r = constant

The flow of pure B through the valve is given by

$$f_2(t) = C_{v2} v p_2(t) \sqrt{\frac{\Delta P_2}{G_2}}$$

where

C_{v2} = valve coefficient (constant) m^3/(s-psi$^{1/2}$)

ΔP_2 = pressure drop across valve (constant) psi

G_2 = specific gravity of B (constant) dimensionless

$vp_2(t)$ = valve position, a fraction

You may assume that the reactor is well insulated, and that the physical properties of the reactants and products are similar. The flow rate f_1 can be assumed to be constant. The valve position $vp_2(t)$ is linearly related to the signal $m_2(t)$. Develop the mathematical model that describes the interactions between the input variables $m_1(t), m_2(t), c_{Ai}(t)$, and the outlet temperature $T(t)$; determine the transfer functions; and draw the block diagram. Show the units of all gains and time constants.

Chapter 4

Dynamic Systems: Higher-Order Processes

The previous chapter investigated the steady-state and dynamic response of simple processes which were all described by first-order ordinary differential equations. The objective of this chapter is to investigate the steady-state and dynamic characteristics of processes described by higher-order ordinary differential equations. Thus the processes shown in this chapter are more complex; however, they are also more representative of those found in industry.

It is important to remember why we are going through this modeling and analysis procedure. Do not get lost in the mathematics; that is not the reason. Before a control system is designed and implemented, it is imperative to understand the characteristics and behavior of processes. Instilling this understanding is the objective of both chapters. Mathematical methods enable us to quantify the process characteristics.

4-1 NONINTERACTING SYSTEMS

Higher-order processes and systems are classified as either noninteracting or interacting. This section presents two examples of noninteracting systems, and Section 4-2 presents three examples of interacting systems. These terms are explained in the respective sections.

4-1.1 Noninteracting Level Process

Consider the set of tanks shown in Fig. 4-1.1. In this process all the tanks are open to the atmosphere, and the liquid temperature is constant. The openings of the valves remain constant and the flow of liquid through the valves is given by

$$f(t) = C_v \sqrt{\frac{\Delta P(t)}{G_f}}$$

where

$f(t)$ = flow through valve, m^3/s

C_v = valve coefficient, m^3/s-$kPa^{1/2}$

$\Delta P(t)$ = pressure drop across valve, kPa

G_f = specific gravity of liquid, dimensionless

Because the tanks are open to the atmosphere and the valves discharge to atmospheric pressure, the pressure drop across each valve is given by

$$\Delta P(t) = P_u(t) - P_d = P_a + \rho g h(t) - P_a = \rho g h(t)$$

Principles and Practice of Automatic Process Control/Third Edition, by C. A. Smith and A. B. Corripio
ISBN 0-471-66141-4   Copyright © 2006 John Wiley & Sons (Asia) Pte. Ltd.

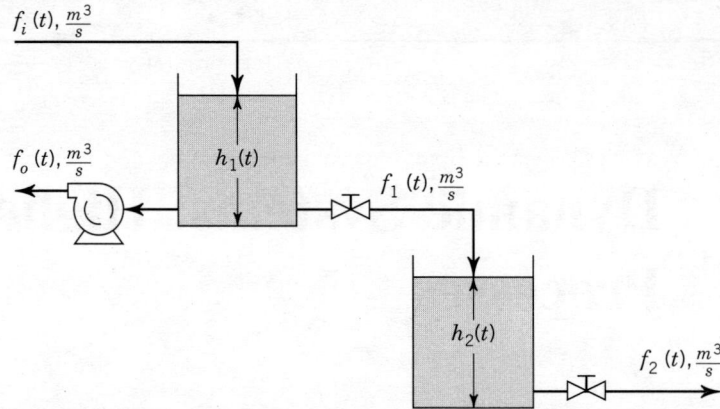

Figure 4-1.1 Tanks in series—noninteracting system.

where

$P_u(t)$ = upstream pressure from valve, kPa

P_d = downstream pressure from valve, kPa

P_a = atmospheric pressure, kPa

ρ = density of liquid, kg/m^3

g = acceleration due to gravity, 9.8 m/s^2

$h(t)$ = liquid level in tank, m

Thus, the valve equation for this process becomes

$$f(t) = C_v \sqrt{\frac{\Delta P(t)}{G_f}} = C_v \sqrt{\frac{\rho g h(t)}{G_f}} = C_v' \sqrt{h(t)}$$

where $C_v' = C_v \sqrt{\dfrac{\rho g}{G_f}}$

It is desired to know how the level in the second tank, $h_2(t)$, is affected by the inlet flow into the first tank, $f_i(t)$, and by the pump flow, $f_o(t)$. The objective is to develop the mathematical model, determine the transfer functions relating $h_2(t)$ to $f_i(t)$ and $f_o(t)$, and draw the block diagram.

Writing an unsteady-state mass balance around the first tank (control volume) gives

Rate of mass into tank − Rate of mass out of tank = Rate of change of mass accumulated in tank

or, in equation form

$$\rho f_i(t) - \rho f_1(t) - \rho f_o(t) = \frac{dm_1(t)}{dt}$$

where $m_1(t)$ = mass of liquid accumulated in the first tank, kg. This mass is given by

$$m_1(t) = \rho A_1 h_1(t)$$

where

A_1 = cross-sectional area of first tank, uniform throughout, m^2

$h_1(t)$ = liquid level in first tank, m

Then substituting the expression for $m_1(t)$ into the mass balance yields

$$\rho f_i(t) - \rho f_1(t) - \rho f_o(t) = \rho A_1 \frac{dh_1(t)}{dt} \qquad (4\text{-}1.1)$$

1 eq., 2 unk. $[f_1(t), h_1(t)]$

As was done in the previous chapter, we do not consider the input variables, $f_i(t)$ and $f_o(t)$, unknowns; it is up to us to specify how they will change.

The valve expression provides another equation

$$f_1(t) = C'_{v_1} \sqrt{h_1(t)} \qquad (4\text{-}1.2)$$

2 eq., 2 unk.

Equations 4-1.1 and 4-1.2 describe the first tank. We now proceed to the second tank. An unsteady-state mass balance around the second tank gives

$$\rho f_1(t) - \rho f_2(t) = \rho A_2 \frac{dh_2(t)}{dt} \qquad (4\text{-}1.3)$$

3 eq., 4 unk. $[f_2(t), h_2(t)]$

Again, the valve expression provides another equation

$$f_2(t) = C'_{v_2} \sqrt{h_2(t)} \qquad (4\text{-}1.4)$$

4 eq., 4 unk.

The set of Eqs. 4-1.1 through 4-1.4 describes the process; this set is the mathematical model of the process.

We now proceed to obtain the transfer functions. Because Eqs. 4-1.2 and 4-1.4 are nonlinear, they must first be linearized. This yields

$$f_1(t) \approx \overline{f}_1 + C_1[h_1(t) - \overline{h}_1] \qquad (4\text{-}1.5)$$

and

$$f_2(t) \approx \overline{f}_2 + C_2[h_2(t) - \overline{h}_2] \qquad (4\text{-}1.6)$$

where

$$C_1 = \left.\frac{\partial f_1(t)}{\partial h_1(t)}\right|_{SS} = \frac{1}{2} C'_{v_1} (\overline{h}_1)^{-1/2}, \frac{\mathrm{m}^3/\mathrm{s}}{\mathrm{m}}, \text{ and}$$

$$C_2 = \left.\frac{\partial f_2(t)}{\partial h_2(t)}\right|_{SS} = \frac{1}{2} C'_{v_2} (\overline{h}_2)^{-1/2}, \frac{\mathrm{m}^3/\mathrm{s}}{\mathrm{m}}$$

Equations 4-1.1, 4-1.3, 4-1.5, and 4-1.6 provide a set of linear equations that describes the process around the linearization values $\overline{h}_1$ and $\overline{h}_2$. Substituting Eq. 4-1.5 into Eq. 4-1.1, substituting Eq. 4-1.6 into Eq. 4-1.3, writing the steady-state mass balances, defining the deviation variables, and rearranging yield

$$\tau_1 \frac{dH_1(t)}{dt} + H_1(t) = K_1 F_i(t) - K_1 F_o(t) \qquad (4\text{-}1.7)$$

and

$$\tau_2 \frac{dH_2(t)}{dt} + H_2(t) = K_2 H_1(t) \qquad (4\text{-}1.8)$$

where

$$H_1(t) = h_1(t) - \overline{h}_1; \; H_2(t) = h_2(t) - \overline{h}_2; \; F_i(t) = f_i(t) - \overline{f}_i;$$
$$F_o(t) = f_o(t) - \overline{f}_0; \; F_1(t) = f_1(t) - \overline{f}_1$$

and

$$\tau_1 = \frac{A_1}{C_1}, \text{ seconds}; \ \tau_2 = \frac{A_2}{C_2}, \text{ seconds}; \ K_1 = \frac{1}{C_1}, \frac{m}{m^3/s}; \ K_2 = \frac{C_1}{C_2}, \text{ dimensionless}$$

Equation 4-1.7 relates the level in the first tank to the inlet and pump flows. Equation 4-1.8 relates the level in the second tank to the level in the first tank.

Taking the Laplace transform of Eqs. 4-1.7 and 4-1.8 and rearranging, we get

$$H_1(s) = \frac{K_1}{\tau_1 s + 1} F_i(s) - \frac{K_1}{\tau_1 s + 1} F_o(s) \tag{4-1.9}$$

$$H_2(s) = \frac{K_2}{\tau_2 s + 1} H_1(s) \tag{4-1.10}$$

To determine the desired transfer functions, we substitute Eq. 4-1.9 into Eq. 4-1.10 which yields

$$H_2(s) = \frac{K_1 K_2}{(\tau_1 s + 1)(\tau_2 s + 1)}[F_i(s) - F_o(s)] \tag{4-1.11}$$

from which the individual desired transfer functions can be obtained

$$\frac{H_2(s)}{F_i(s)} = \frac{K_1 K_2}{(\tau_1 s + 1)(\tau_2 s + 1)} \tag{4-1.12}$$

and

$$\frac{H_2(s)}{F_o(s)} = \frac{-K_1 K_2}{(\tau_1 s + 1)(\tau_2 s + 1)} \tag{4-1.13}$$

When the denominator of these two transfer functions is expanded into a polynomial form, the power on the s operator is two. Thus these transfer functions are called *second-order transfer functions* or *second-order lags*. Their development shows that they are "formed" by two first-order transfer functions, or differential equations, in series.

The block diagram for this system can be represented in different forms as shown in Fig. 4-1.2. The block diagram of Fig. 4-1.2a is developed by "chaining" Eqs. 4-1.9 and 4-1.10. The diagram shows that the inlet and pump flows initially affect the level in the first tank. A change in this level then affects the level in the second tank.

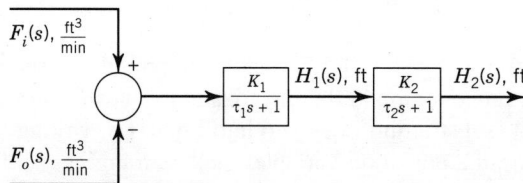

(a)

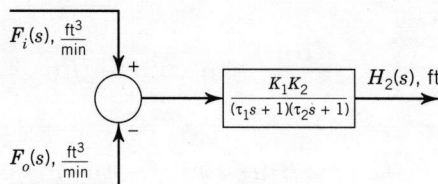

(b)

Figure 4-1.2 Block diagram for two noninteracting tanks.

Figure 4-1.2*b* shows a more compact diagram. Even though the block diagram of Fig. 4-1.2*a* provides a better description of the physics involved (how things really happen), both diagrams are used without any preference.

Now let us extend the process shown in Fig. 4-1.1 by one more tank, as shown in Fig 4-1.3. For this new process, the objective is to develop the mathematical model, determine the transfer functions relating the level in the third tank to the inlet flow and to the pump flow, and draw the block diagram.

Because the first two tanks have already been modeled, Eqs. 4-1.1 through 4-1.4, the third tank is now modeled. Writing an unsteady-state mass balance around the third tank results in

$$\rho f_2(t) - \rho f_3(t) = \rho A_3 \frac{dh_3(t)}{dt} \qquad \textbf{(4-1.14)}$$

$$5 \text{ eq., } 6 \text{ unk. } [f_3(t), h_3(t)]$$

The valve expression provides the next required equation:

$$f_3(t) = C'_{v_3}\sqrt{h_3(t)} \qquad \textbf{(4-1.15)}$$

$$6 \text{ eq., } 6 \text{ unk.}$$

The new process, Fig. 4-1.3, is now modeled by Eqs. 4-1.1 through 4-1.4, 4-1.14, and 4-1.15.

Proceeding as before, we get from Eq. 4-1.14 and the linearized form of Eq. 4-1.15 the equation

$$\tau_3 \frac{dH_3(t)}{dt} + H_3(t) = K_3 H_2(t) \qquad \textbf{(4-1.16)}$$

where

$$H_3(t) = h_3(t) - \overline{h}_3; \quad C_3 = \left.\frac{\partial f_3(t)}{\partial h_3(t)}\right|_{SS} = \frac{1}{2}C'_{v_3}(\overline{h}_3)^{-1/2}, \frac{m^3/s}{m}$$

$$\tau_3 = \frac{A_3}{C_3}, \text{ seconds}; \quad K_3 = \frac{C_2}{C_3}, \text{ dimensionless}$$

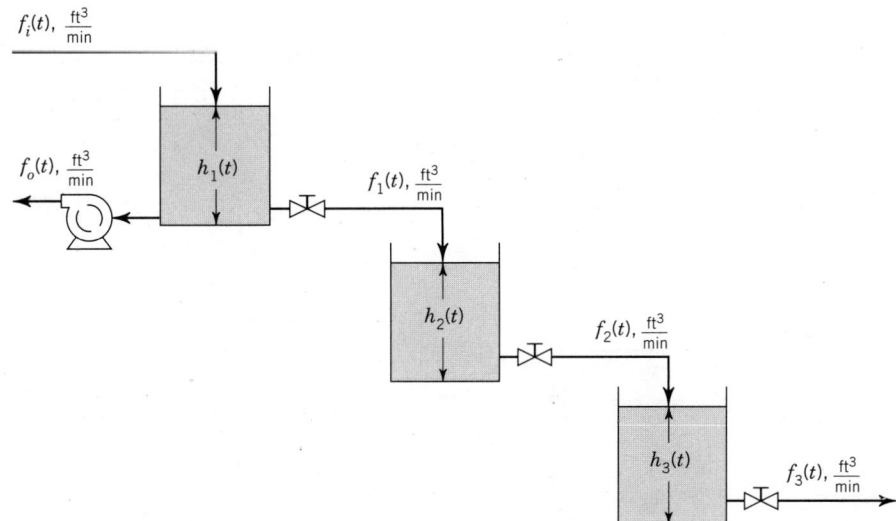

Figure 4-1.3 Tanks in series—noninteracting system.

Taking the Laplace transform of Eq. 4-1.16 and rearranging, we obtain

$$H_3(s) = \frac{K_3}{\tau_3 s + 1} H_2(s) \tag{4-1.17}$$

Finally, substituting Eq. 4-1.11 into the above equation gives

$$H_2(s) = \frac{K_1 K_2 K_3}{(\tau_1 s + 1)(\tau_2 s + 1)(\tau_3 s + 1)}[F_i(s) - F_o(s)] \tag{4-1.18}$$

from which the following transfer functions are determined:

$$\frac{H_3(s)}{F_i(s)} = \frac{K_1 K_2 K_3}{(\tau_1 s + 1)(\tau_2 s + 1)(\tau_3 s + 1)} \tag{4-1.19}$$

and

$$\frac{H_2(s)}{F_o(s)} = \frac{-K_1 K_2 K_3}{(\tau_1 s + 1)(\tau_2 s + 1)(\tau_3 s + 1)} \tag{4-1.20}$$

When the denominator of these two transfer functions is expanded into a polynomial form, the power on the s operator is three. Thus they are referred to as *third-order transfer functions* or *third order-lags*. Figure 4-1.4 shows a block diagram for this process.

The processes shown on Figs. 4-1.1 and 4-1.3 are referred to as *noninteracting systems* because there is no full interaction between the variables. That is, the level in the first tank affects the level in the second tank, but the level in the second tank does not in turn affect the level in the first tank. The level in the second tank does not "feed back" to affect the level in the first tank. The cause-and-effect relationship is a one-way path. The same is true for the levels in the second and third tanks.

It is important to remember what we said about transfer functions in Chapters 2 and 3. Transfer functions completely describe the characteristics of linear processes and those around the linearization values for nonlinear processes. Equation 4-1.20, for example, shows that if the pump flow increases by 10 m³/s, then the level in the third tank will change by $-10K_1 K_2 K_3$ m; that is, it will decrease by $10K_1 K_2 K_3$ m. The dynamics of the change will depend on τ_1, τ_2, and τ_3. These dynamics are discussed in detail in Section 4-3. Transfer functions quantify the process characteristics, or behavior.

Note that the transfer functions presented in this section were obtained by multiplying first-order transfer functions in series. For example,

$$\frac{H_2(s)}{F_i(s)} = \frac{H_1(s)}{F_i(s)} \cdot \frac{H_2(s)}{H_1(s)}$$

and

$$\frac{H_3(s)}{F_o(s)} = \frac{H_1(s)}{F_o(s)} \cdot \frac{H_2(s)}{H_1(s)} \cdot \frac{H_3(s)}{H_2(s)}$$

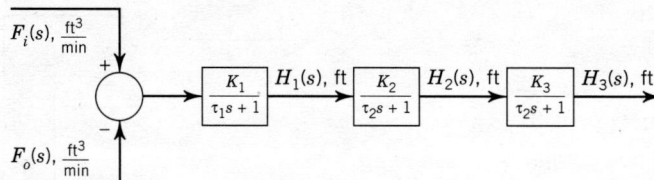

Figure 4-1.4 Block diagram for three noninteracting tanks.

In general this is the case for noninteracting systems only. It can be generalized by writing

$$G(s) = \prod_{i=1}^{n} G_i(s) \qquad \textbf{(4-1.21)}$$

where

n = number of noninteracting systems in series

$G(s)$ = transfer function relating the output from the last system, the nth system, to the input to the first system

$G_i(s)$ = individual transfer function of each system

Remember, Eq. 4-1.21 is true only for noninteracting systems.

4-1.2   Thermal Tanks in Series

Consider the set of tanks shown in Fig. 4-1.5. The first tank provides some mixing and residence time to stream A. Tank 2 provides mixing of streams A and B. Let us assume that the volumetric flows of each stream, f_A and f_B, are constant; that the density and heat capacity, ρ and C_p, of each stream are equal to each other and constants; that since the fluids are liquid $C_p = C_v$; that the two tanks are next to each other; and finally, that the heat losses to the surroundings and the paddle work are negligible. The volume of tanks 1 and 2 are V_1 and V_2, respectively.

It is desired to know how the outlet liquid temperature from the second tank, $T_4(t)$, is affected by the inlet temperature of stream A, $T_1(t)$, and by that of stream B, $T_3(t)$. For this process let us develop the mathematical model, determine the transfer functions that relate $T_4(t)$ to $T_1(t)$ and $T_3(t)$, and draw the block diagram.

The flows and densities are constants, so the mass accumulated in each tank is also constant. Therefore, a total mass balance around both tanks indicates that the total flow out of the second tank is equal to the sum of the individual inlet streams, or $f_A + f_B$.

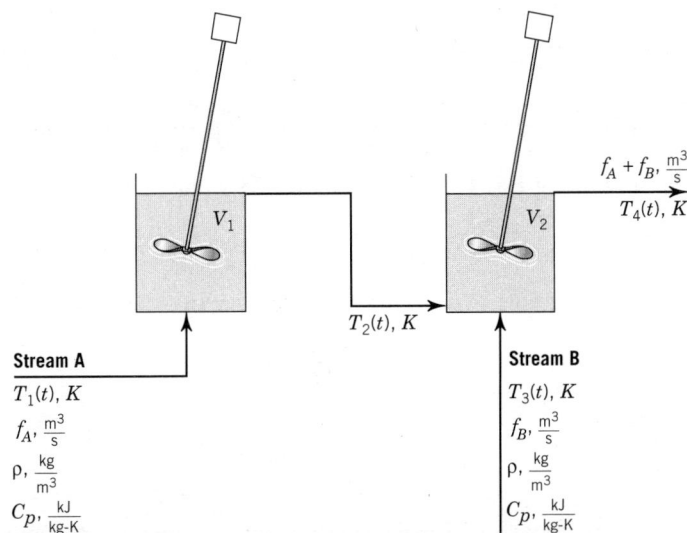

Figure 4-1.5 Thermal tanks in series—noninteracting system.

We start by writing an unsteady-state energy balance on the contents of the first tank:

$$f_A \rho h_1(t) - f_A \rho h_2(t) = V_1 \rho \frac{du_2(t)}{dt}$$

where

$h(t) = $ specific enthalpy, kJ/kg

$u(t) = $ specific internal energy, kJ/kg

Or, in terms of temperature, using as reference state for $h(t)$ and $u(t)$ the components in the liquid phase at 0 K,

$$f_A \rho C_p T_1(t) - f_A \rho C_p T_2(t) = V_1 \rho C_p \frac{dT_2(t)}{dt} \tag{4-1.22}$$

$$\text{1 eq., 1 unk. } [T_2(t)]$$

Another unsteady-state energy balance on the contents of the second tank yields

$$f_A \rho C_p T_2(t) + f_B \rho C_p T_3(t) - (f_A + f_B) \rho C_p T_4(t) = V_2 \rho C_p \frac{dT_4(t)}{dt} \cdot \tag{4-1.23}$$

$$\text{2 eq., 2 unk. } [T_4(t)]$$

Equations 4-1.22 and 4-1.23 are the mathematical model that relate the output variable, $T_4(t)$, to the inputs of interest, $T_1(t)$ and $T_3(t)$.

To develop the transfer functions and block diagrams, we first realize that this model is a set of linear equations and that, accordingly, there is no need for linearization. Thus we proceed by writing the steady-state energy balances, defining deviation variables, taking Laplace transforms, and rearranging to yield from Eq. 4-1.22

$$\Gamma_2(s) = \frac{1}{\tau_1 s + 1} \Gamma_1(s) \tag{4-1.24}$$

and from Eq. 4-1.23

$$\Gamma_4(s) = \frac{K_1}{\tau_2 s + 1} \Gamma_2(s) + \frac{K_2}{\tau_2 s + 1} \Gamma_3(s) \tag{4-1.25}$$

where

$$K_1 = \frac{f_A}{f_A + f_B}, \text{ dimensionless} \qquad K_2 = \frac{f_B}{f_A + f_B}, \text{ dimensionless}$$

$$\tau_1 = \frac{V_1}{f_A}, \text{ seconds} \qquad\qquad \tau_2 = \frac{V_2}{f_A + f_B}, \text{ seconds}$$

Substituting Eq. 4-1.24 into Eq. 4-1.25 yields

$$\Gamma_4(s) = \frac{K_1}{(\tau_1 s + 1)(\tau_2 s + 1)} \Gamma_1(s) + \frac{K_2}{\tau_2 s + 1} \Gamma_3(s) \tag{4-1.26}$$

from which the two required transfer functions are obtained:

$$\frac{\Gamma_4(s)}{\Gamma_1(s)} = \frac{K_1}{(\tau_1 s + 1)(\tau_2 s + 1)} \tag{4-1.27}$$

and

$$\frac{\Gamma_4(s)}{\Gamma_3(s)} = \frac{K_2}{\tau_2 s + 1} \tag{4-1.28}$$

Equation 4-1.27 is the transfer function relating the outlet temperature to the inlet temperature of stream A; it is a second-order transfer function. Equation 4-1.28 is the transfer function relating the outlet temperature to the inlet temperature of stream B; it is a first-order transfer function. The block diagram is shown in Fig. 4-1.6.

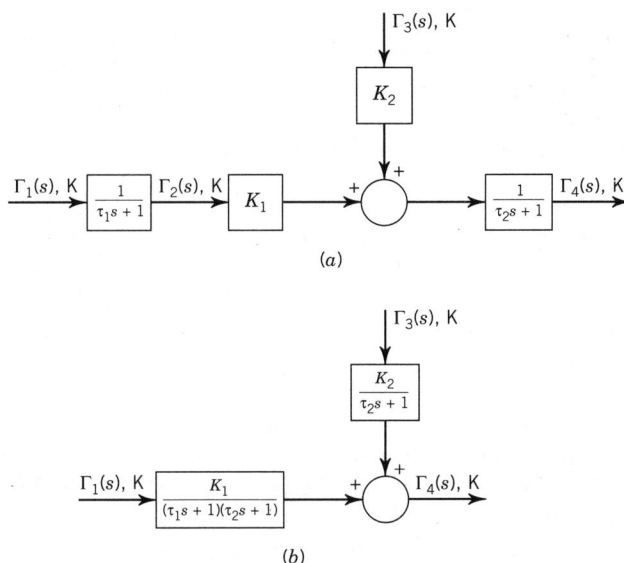

(a)

(b)

Figure 4-1.6 Block diagrams for thermal tanks in series.

On the basis of what we learned in Chapters 2 and 3 we know that the dynamic response of $\Gamma_4(t)$ to changes in $\Gamma_1(t)$, expressed by Eq. 4-1.27, is different to the response of $\Gamma_4(t)$ to changes in $\Gamma_3(t)$, expressed by Eq. 4-1.28. The reader must also try to understand this difference by "looking at" the physical system. If $\Gamma_1(t)$ changes it must affect $\Gamma_2(t)$ before $\Gamma_4(t)$ starts to feel the effect; this is shown in Fig. 4-1.6a. However, if $\Gamma_3(t)$ changes, it will start affecting $\Gamma_4(t)$ right away. $\Gamma_4(t)$ will respond more slowly to changes in $\Gamma_1(t)$ than to changes in $\Gamma_3(t)$. How much slower the response is given by the time constant τ_1, which, as shown by its definition, is related to the tank volume and the flow.

We have simplified this process by the assumptions taken. However, you may want to think how the development—and indeed the final form—of the transfer functions, would be affected by removing some of the assumptions. For example, what if a long pipe exists between the tanks? And what if we allow variations in the flows of streams A and B?

4-2   INTERACTING SYSTEMS

Interacting systems are more frequently encountered in industry than noninteracting systems; this section presents three examples. The differences in the dynamic response between the noninteracting and interacting systems are also presented.

4-2.1   Interacting Level Process

Let us rearrange the tanks of Fig. 4-1.1 to result in the new process shown in Fig. 4-2.1. In this case the pressure drop, $\Delta P(t)$, across the valve between the two tanks is given by

$$\Delta P(t) = P_u(t) - P_d(t) = [P_a + \rho g h_1(t)] - [P_a + \rho g h_2(t)] = \rho g[h_1(t) - h_2(t)]$$

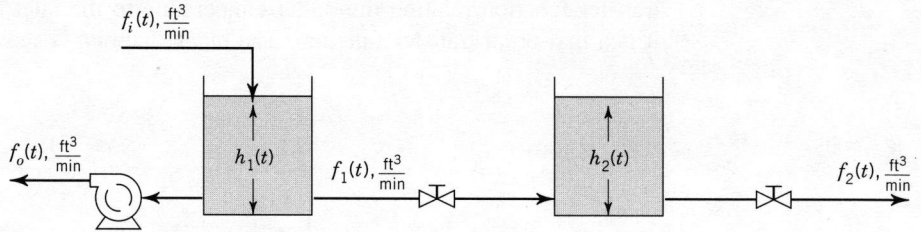

Figure 4-2.1 Tanks in series—interacting system.

Substituting this pressure drop expression into the valve equation results in

$$f_1(t) = C_{v_1}\sqrt{\frac{\Delta P(t)}{G_f}} = C_{v_1}\sqrt{\frac{\rho g[h_1(t) - h_2(t)]}{G_f}}$$

$$f_1(t) = C'_{v_1}\sqrt{h_1(t) - h_2(t)}$$

This new process is referred to as an interacting system. The valve equation shows that the flow between the two tanks depends on the levels in *both* tanks, each affecting the other. That is, the level in the first tank affects the level in the second tank and, at the same time, the level in the second tank affects that in the first tank. Each element of the process affects each other. The cause-and-effect relation is a two-way path.

We are still interested in determining how the level in the second tank is affected by the flow into the first tank and by the pump flow. Let us develop the mathematical model, determine the transfer functions, and draw the block diagram for this new process.

We start by writing an unsteady-state mass balance around the first tank, this is given by Eq. 4-1.1.

$$\rho f_i(t) - \rho f_1(t) - \rho f_o(t) = \rho A_1 \frac{dh_1(t)}{dt} \tag{4-1.1}$$

1 eq., 2 unk. $[f_1(t), h_1(t)]$

The valve equation provides the next equation

$$f_1(t) = C'_{v_1}\sqrt{h_1(t) - h_2(t)} \tag{4-2.1}$$

2 eq., 3 unk. $[h_2(t)]$

Another independent equation is still needed. An unsteady-state mass balance around the second tank, Eq. 4-1.3, provides the needed equation:

$$\rho f_1(t) - \rho f_2(t) = \rho A_2 \frac{dh_2(t)}{dt} \tag{4-1.3}$$

3 eq., 4 unk. $[f_2(t)]$

The expression for the flow $f_2(t)$ is given by Eq. 4-1.4.

$$f_2(t) = C'_{v_2}\sqrt{h_2(t)} \tag{4-1.4}$$

4 eq., 4 unk.

Equations 4-1.1, 4-2.1, 4-1.3, and 4-1.4 constitute the mathematical model.

We continue with the usual procedure to obtain the transfer functions and block diagram. Because Eq. 4-2.1 is nonlinear it is linearized as

$$f_1(t) \approx \overline{f}_1 + C_4[h_1(t) - \overline{h}_1] - C_4[h_2(t) - \overline{h}_2] \tag{4-2.2}$$

where

$$C_4 = \left.\frac{\partial f_1(t)}{\partial h_1(t)}\right|_{SS} = -\left.\frac{\partial f_1(t)}{\partial h_2(t)}\right|_{SS} = \frac{1}{2}C'_{v_1}(\overline{h}_1 - \overline{h}_2)^{-1/2}, \frac{m^3/s}{m}$$

Equation 4-1.4 is linearized as given by Eq. 4-1.6.

$$f_2(t) \approx \overline{f}_2 + C_2[h_2(t) - \overline{h}_2] \tag{4-1.6}$$

Equations 4-1.1, 4-1.3, 4-2.2, and 4-1.6 provide the set of linear equations that describe the process around the linearization values $\overline{h}_1$ and $\overline{h}_2$.

Substituting Eq. 4-2.2 into Eq. 4-1.1, writing the steady-state mass balance around the first tank, defining deviation variables, taking Laplace transforms, and rearranging yield

$$H_1(s) = \frac{K_4}{\tau_4 s + 1}[F_i(s) - F_o(s)] + \frac{1}{\tau_4 s + 1}H_2(s) \tag{4-2.3}$$

where

$$K_4 = \frac{1}{C_4}, \quad \frac{m}{m^3/s}; \quad \tau_4 = \frac{A_1}{C_4}, \text{ seconds}$$

Following the same procedure for the second tank gives

$$H_2(s) = \frac{K_5}{\tau_5 s + 1}H_1(s) \tag{4-2.4}$$

where

$$K_5 = \frac{C_4}{C_4 + C_2}, \text{ dimensionless} \quad \tau_5 = \frac{A_2}{C_4 + C_2}, \text{ seconds}$$

Finally, substituting Eq. 4-2.3 into Eq. 4-2.4

$$H_2(s) = \frac{K_4 K_5}{(\tau_4 s + 1)(\tau_5 s + 1)}[F_i(s) - F_o(s)] + \frac{K_5}{(\tau_4 s + 1)(\tau_5 s + 1)}H_2(s)$$

and rearranging

$$H_2(s) = \frac{\dfrac{K_4 K_5}{1 - K_5}}{\left(\dfrac{\tau_4 \tau_5}{1 - K_5}\right)s^2 + \left(\dfrac{\tau_4 + \tau_5}{1 - K_5}\right)s + 1}[F_i(s) - F_o(s)] \tag{4-2.5}$$

from which the desired transfer functions are obtained:

$$\frac{H_2(s)}{F_i(s)} = \frac{\dfrac{K_4 K_5}{1 - K_5}}{\left(\dfrac{\tau_4 \tau_5}{1 - K_5}\right)s^2 + \left(\dfrac{\tau_4 + \tau_5}{1 - K_5}\right)s + 1} \tag{4-2.6}$$

and

$$\frac{H_2(s)}{F_o(s)} = \frac{-\dfrac{K_4 K_5}{1 - K_5}}{\left(\dfrac{\tau_4 \tau_5}{1 - K_5}\right)s^2 + \left(\dfrac{\tau_4 + \tau_5}{1 - K_5}\right)s + 1} \tag{4-2.7}$$

These transfer functions are of second order. Block diagrams depicting this interacting process are shown in Figure 4-2.2. Figure 4-2.2a develops directly from Eq. 4-2.5. Figure 4-2.2b develops by "chaining" Eqs. 4-2.3 and 4-2.4. Note also that Fig. 4-2.2a

(a)

(b)

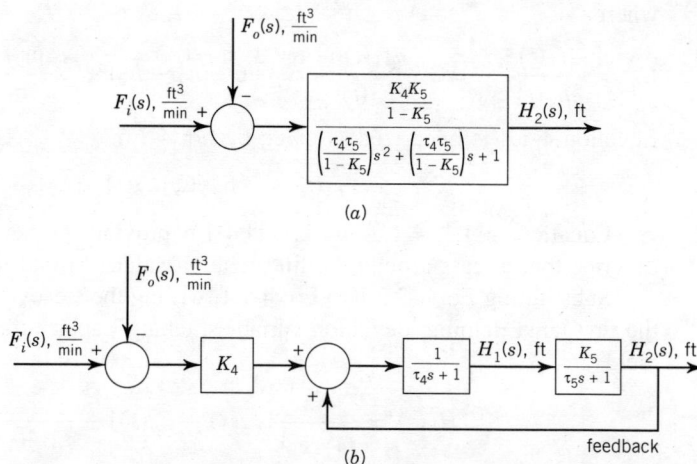

Figure 4-2.2 Block diagrams for an interacting two-tank system.

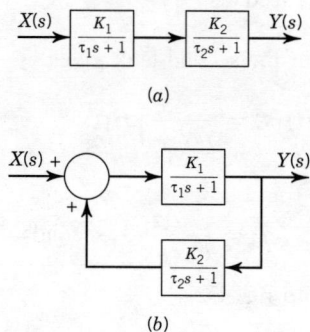

(a)

(b)

Figure 4-2.3 (a) Block diagram of a noninteracting system. (b) Block diagram of an interacting system.

can be obtained directly from Fig. 4-2.2b using the "positive feedback" rule of block diagrams presented in Chapter 3. The interacting nature of this process is clearly shown in Fig. 4-2.2b. The figure shows that $H_1(s)$ is the input to obtain $H_2(s)$ but also $H_2(s)$ is another input to obtain $H_1(s)$, as indicated by the "feedback path." Oftentimes we referred to this type of system as "interacting lags."

At this time, there are several things we can learn by comparing the transfer functions of the interacting and noninteracting systems. Consider Fig. 4-2.3 where a block diagram of a noninteracting system and one of an interacting system are shown. For the noninteracting system the transfer function is

$$\frac{Y(s)}{X(s)} = \frac{K_1 K_2}{(\tau_1 s + 1)(\tau_2 s + 1)} \tag{4-2.8}$$

As presented in Section 2-5, the "effective" time constants are the negative of the reciprocal of the roots of the denominator of the transfer function. For the foregoing transfer function, the effective time constants are equal to the individual τ values; that is, $\tau_{1_{eff}} = \tau_1$ and $\tau_{2_{eff}} = \tau_2$.

For the interacting system the transfer function is

$$\frac{Y(s)}{X(s)} = \frac{\dfrac{K_1}{(\tau_1 s + 1)}}{1 - \dfrac{K_1 K_2}{(\tau_1 s + 1)(\tau_2 s + 1)}} \tag{4-2.9}$$

or

$$\frac{Y(s)}{X(s)} = \frac{K_1(\tau_2 s + 1)}{(\tau_1 s + 1)(\tau_2 s + 1) - K_1 K_2} = \frac{K_1(\tau_2 s + 1)}{\tau_1 \tau_2 s^2 + (\tau_1 + \tau_2)s + (1 - K_1 K_2)} \quad \textbf{(4-2.10)}$$

The roots of the denominator are

$$\text{Roots} = \frac{-(\tau_1 + \tau_2) \pm \sqrt{(\tau_1 + \tau_2)^2 - 4\tau_1 \tau_2 (1 - K_1 K_2)}}{2\tau_1 \tau_2} \quad \textbf{(4-2.11)}$$

or, making use of the assumption $\tau_1 = \tau_2 = \tau$,

$$\text{Roots} = \frac{-\left(1 + \sqrt{K_1 K_2}\right)}{\tau}, \quad \frac{-\left(1 - \sqrt{K_1 K_2}\right)}{\tau}$$

from which the "effective" time constants for the interacting system can be obtained as

$$\tau_{1_{eff}} = \frac{\tau}{1 + \sqrt{K_1 K_2}} \text{ and } \tau_{2_{eff}} = \frac{\tau}{1 - \sqrt{K_1 K_2}}$$

The ratio of these two terms is

$$\frac{\tau_{2_{eff}}}{\tau_{1_{eff}}} = \frac{1 + \sqrt{K_1 K_2}}{1 - \sqrt{K_1 K_2}}$$

which is a number greater than one even though $\tau_1 = \tau_2$! This result clearly shows that the larger τ the interacting system "experiences," $\tau_{2_{eff}}$, is larger than any individual τ.

The following observations, conclusions, and comments are related to the above analysis and to the general subject of higher-order systems.

1. Most times the "effective" time constants are real, yielding a nonoscillatory response to step change in input. The roots are real if, in Eq. 4-2.11,

$$(\tau_1 + \tau_2)^2 - 4\tau_1 \tau_2 (1 - K_1 K_2) > 0$$

or

$$\tau_1^2 + 2\tau_1 \tau_2 + \tau_2^2 - 4\tau_1 \tau_2 + 4\tau_1 \tau_2 K_1 K_2 > 0$$

or

$$\tau_1^2 - 2\tau_1 \tau_2 + \tau_2^2 + 4\tau_1 \tau_2 K_1 K_2 > 0$$

or

$$(\tau_1 - \tau_2)^2 + 4\tau_1 \tau_2 K_1 K_2 > 0$$

and this is true if $\tau_1 \tau_2 K_1 K_2 > 0$. Because for most cases $\tau_1 > 0$, $\tau_2 > 0$, and $K_1 K_2 > 0$ then the roots are real.

The exception to the above statement is the exothermic continuous stirred tank reactor where sometimes one of the τ values is negative. Refer to Section 4-2.3 where a reactor is presented and a τ is negative. Section 4-4.2 also presents another reactor and shows the oscillatory response.

Shinskey (1988) points out that, for interacting systems, the higher the interaction, the more different the two effective time constants are, and therefore the more controllable the process.

2. In Chapter 2 and in the present chapter, we have defined and used several times the term "effective" time constant. Let us discuss this term a bit further.

Chapters 2 and 3 showed that when the input to a first-order system changes in a step fashion, the time constant (τ) is the time required for the system to reach 63.2 % of its total change. This definition applies to first-order systems. In higher-order systems there is no one time constant as defined above. That is, we cannot say that any one of the τ values in a higher-order system represents the time to reach 63.2 % of the total change. However, the τ values in the transfer functions of these systems are still an indication of the dynamics of the system. The slower the system, the larger the τ values, and the faster the system, the smaller they are. This is the reason why we use the term *effective time constant* instead of just *time constant*; we still use the same representation. Often, in everyday conversation, we drop the word *effective* and use only time constant. What is important to remember is that τ is a parameter of the system related to its dynamics; that for first-order systems it has a definite definition, and that for higher-order systems it is only an indication.

4-2.2   Thermal Tanks with Recycle

Consider the process shown in Fig. 4-2.4. This process is essentially the same one described in Section 4-1.2 except that a recycle stream to the first tank has been added. Let us suppose that this recycle stream is a constant 20 % of the total flow out from the process. In addition, let us accept the same assumptions as in Section 4-1.2.

It is required to know how the outlet temperature from the second tank, $T_4(t)$, responds to changes in the inlet temperatures of streams A and B. Develop the mathematical model, determine the transfer functions that relate $T_4(t)$ to $T_1(t)$ and $T_3(t)$, and draw the block diagram for this process.

As in Section 4-1.2, we start by writing an unsteady-state energy balance on the contents of the first tank.

$$f_A \rho C_p T_1(t) + 0.2(f_A + f_B)\rho C_p T_4(t) - [f_A + 0.2(f_A + f_B)]\rho C_p T_2(t)$$

$$= V_1 \rho C_p \frac{dT_2(t)}{dt} \tag{4-2.12}$$

$$\text{1 eq., 2 unk. } [T_2(t), T_4(t)]$$

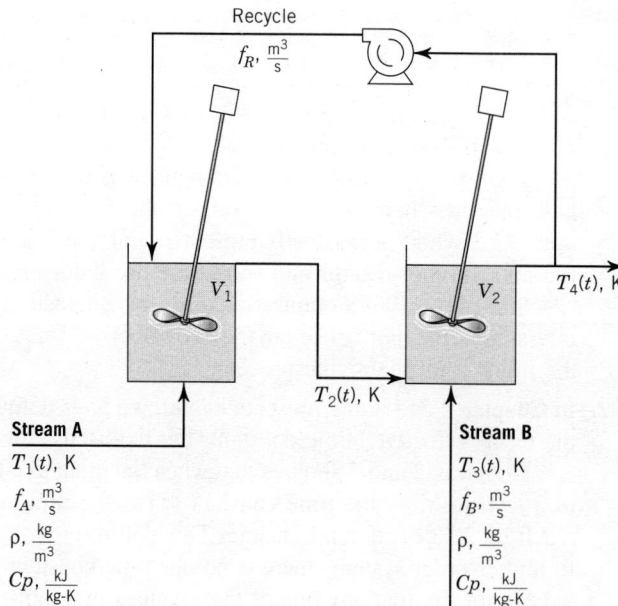

Figure 4-2.4 Thermal tanks with recycle.

Then we write an unsteady-state energy balance on the contents of the second tank.

$$[f_A + 0.2(f_A + f_B)]\rho C_p T_2(t) + f_B \rho C_p T_B(t) - 1.2(f_A + f_B)\rho C_p T_4(t) \quad \text{(4-2.13)}$$

$$= V_2 \rho C_p \frac{dT_4(t)}{dt} \qquad \qquad \text{2 eq., 2 unk.}$$

The mathematical model for this process is given by Equations 4-2.12 and 4-2.13.

To obtain the required transfer functions and block diagram we proceed in the usual way and obtain from Eq. 4-2.12

$$\Gamma_2(s) = \frac{K_1}{\tau_1 s + 1}\Gamma_1(s) + \frac{K_2}{\tau_2 s + 1}\Gamma_4(s) \qquad \text{(4-2.14)}$$

and from Eq. 4-2.13

$$\Gamma_4(s) = \frac{K_3}{\tau_2 s + 1}\Gamma_2(s) + \frac{K_4}{\tau_2 s + 1}\Gamma_3(s) \qquad \text{(4-2.15)}$$

where

$$K_1 = \frac{f_A}{f_A + 0.2(f_A + f_B)}, \text{ dimensionless;}$$

$$K_2 = \frac{0.2(f_A + f_B)}{f_A + 0.2(f_A + f_B)}, \text{ dimensionless}$$

$$K_3 = \frac{f_A + 0.2(f_A + f_B)}{1.2(f_A + f_B)}, \text{ dimensionless;} \quad K_4 = \frac{f_B}{1.2(f_A + f_B)}, \text{ dimensionless}$$

$$\tau_1 = \frac{V_1}{f_A + 0.2(f_A + f_B)}, \text{ seconds;} \quad \tau_2 = \frac{V_2}{1.2(f_A + f_B)}, \text{ seconds}$$

Substituting Eq. 4-2.14 into Eq. 4-2.15 and solving for $T_4(t)$ give

$$\Gamma_4(s) = \frac{K_3 K_1}{(\tau_1 s + 1)(\tau_2 s + 1) - K_2 K_3}\Gamma_1(s) + \frac{K_4(\tau_1 s + 1)}{(\tau_1 s + 1)(\tau_2 s + 1) - K_2 K_3}\Gamma_3(s)$$
$$\text{(4-2.16)}$$

from which the two required transfer functions can be obtained:

$$\frac{\Gamma_4(s)}{\Gamma_1(s)} = \frac{K_3 K_1}{(\tau_1 s + 1)(\tau_2 s + 1) - K_2 K_3} \qquad \text{(4-2.17)}$$

and

$$\frac{\Gamma_4(s)}{\Gamma_3(s)} = \frac{K_4(\tau_1 s + 1)}{(\tau_1 s + 1)(\tau_2 s + 1) - K_2 K_3} \qquad \text{(4-2.18)}$$

Figure 4-2.5 shows two different ways to draw the block diagram. Figure 4-2.5a is developed by chaining Eqs. 4-2.14 and 4-2.15. Figure 4-2.5b is the graphical representation of Eq. 4-2.16. The feedback path in Fig. 4-2.5a shows graphically the interactive nature of this process.

The transfer functions given by Eqs. 4-2.17 and 4-2.18 are of second order, as expressed by the denominator terms. Even though both denominators are the same, the dynamic response of $\Gamma_4(t)$ to a change in $\Gamma_1(t)$ is different from the response to a change in $\Gamma_3(t)$. The fact that the term $(\tau_1 s + 1)$ appears in the numerator of Eq. 4-2.18, and not in that of Eq. 4-2.17, shows this difference. The presence of this term, as will be shown in Section 4-4.1, results in a faster dynamic response. Thus, Eqs. 4-2.17 and 4-2.18 tell us that $\Gamma_4(t)$ responds faster to a change in $\Gamma_3(t)$ than to a change in $\Gamma_1(t)$. From a physical point of view this makes sense. Looking at Fig. 4-2.4 we notice that a change in $\Gamma_1(t)$ affects first the temperature in the first tank, $\Gamma_2(t)$,

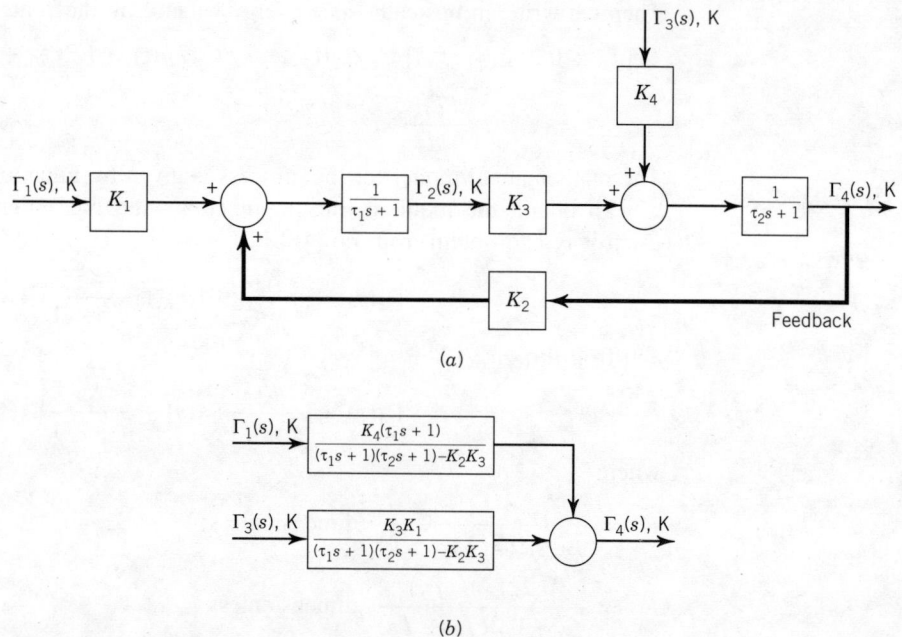

Figure 4-2.5 Block diagrams for thermal tanks with recycle.

and then affects the temperature in the second tank, $\Gamma_4(t)$. A change in $\Gamma_3(t)$, however, affects $\Gamma_4(t)$ directly.

The previous paragraph brings up a most important consideration. We must never, during any mathematical analysis, forget the physics of the process. If our analysis does not describe what happens in reality, then it is no good. Mathematics is a "tool" to describe nature.

4-2.3   Nonisothermal Chemical Reaction

Consider the reactor shown in Fig. 4-2.6. The reactor is a stirred tank where the exothermic reaction A $\rightarrow$ B occurs. To remove the heat of reaction the reactor is surrounded by a jacket through which a cooling liquid flows. Let us assume that the heat losses to the surroundings are negligible, and that the thermodynamic properties, densities, and heat capacities of the reactants and products are both equal and constant. The heat of reaction is constant and is given by ΔH_r in Btu/lbmole of A reacted. Let us also assume that the level of liquid in the reactor tank is constant; that is, the rate of mass into the tank is equal to the rate of mass out of the tank. Finally, the rate of reaction is given by

$$r_A(t) = k_o e^{-E/RT(t)} c_A^2(t), \qquad \frac{\text{lbmoles of A produced}}{\text{ft}^3 - \text{min}}$$

where the frequency factor, k_o, and energy of activation, E, are constants. Table 4-2.1 gives the steady-state values of the variables and other process specifications.

It is desired to find out how the outlet concentration of A, $c_A(t)$, and the outlet temperature, $T(t)$, respond to changes in the inlet concentration of A, $c_{A_i}(t)$; the inlet temperature of the reactant, $T_i(t)$; the inlet temperature of the cooling liquid, $T_{c_i}(t)$; and the flows $f(t)$ and $f_c(t)$. The objective, therefore, is to develop the mathematical

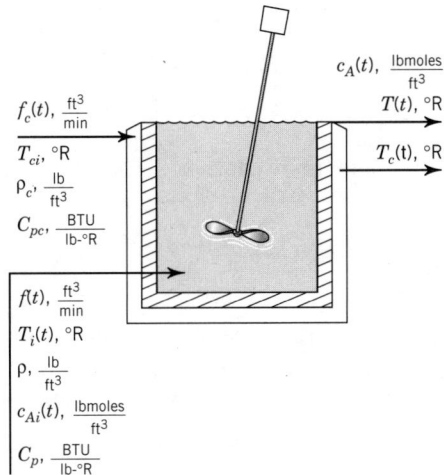

$f_c(t)$, $\frac{ft^3}{min}$
T_{ci}, °R
ρ_c, $\frac{lb}{ft^3}$
C_{pc}, $\frac{BTU}{lb\text{-}°R}$

$c_A(t)$, $\frac{lbmoles}{ft^3}$
$T(t)$, °R
$T_c(t)$, °R

$f(t)$, $\frac{ft^3}{min}$
$T_i(t)$, °R
ρ, $\frac{lb}{ft^3}$
$c_{Ai}(t)$, $\frac{lbmoles}{ft^3}$
C_p, $\frac{BTU}{lb\text{-}°R}$

Figure 4-2.6 Nonisothermal chemical reactor.

Table 4-2.1   Process information and steady-state values

Process information

$V = 13.26$ ft^3 $k_o = 8.33 \times 10^8$ ft^3/(lbmole-min)
$E = 27{,}820$ Btu/lbmole $R = 1.987$ Btu/(lbmole-°R)
$\rho = 55$ lbm/ft^3 $C_p = 0.88$ Btu/(lbm-°F)
$\Delta H_r = -12{,}000$ Btu/lbmole $U = 75$ Btu/(h-ft^2-°F)
$A = 36$ ft^2; $\rho_c = 62.4$ lbm/ft^3 $C_{pc} = 1.0$ Btu/(lbm-°F) $V_c = 1.56$ ft^3

Steady-state values

$c_{Ai}(t) = 0.5975$ lbmole/ft^3 $T_i(t) = 635$°R
$T_c = 602.7$°R $f = 1.3364$ ft^3/min
$c_A(t) = 0.2068$ lbmole/ft^3 $T(t) = 678.9$°R
$T_{ci}(t) = 540$°R $f_c(t) = 0.8771$ ft^3/min

model, determine the transfer functions relating $c_A(t)$ and $T(t)$ to $c_{A_i}(t)$, $T_i(t)$, $T_{c_i}(t)$, $f(t)$, and $f_c(t)$, and draw the block diagram for this process.

Before we accomplish the objectives, it might be wise to discuss why the interest in learning how the outlet temperature responds to the different inputs. This temperature is most often economically unimportant; however, it is related to safety, production rate, yield, and other operational objectives. Because temperature is easy to measure, it is usually controlled as a way to control the reactor performance.

Returning to our objectives, we start by considering the reactor as the control volume and writing an unsteady-state mole balance on component A as discussed in Section 3-7.1

| Rate of component A into reactor | − | Rate of component A out of reactor | + | Rate of production of component A | = | Rate of change of component A accumulated in reactor |

or

$$f(t)c_{A_i}(t) - f(t)c_A(t) - Vr_A(t) = V\frac{dc_A(t)}{dt} \qquad \textbf{(4-2.19)}$$

1 eq., 2 unk. $[r_A(t), c_A(t)]$

where V = volume of liquid in reactor, ft^3. The rate expression provides another equation

$$r_A(t) = k_o e^{-E/RT(t)} c_A^2(t) \qquad \textbf{(4-2.20)}$$

2 eq., 3 unk. $[T(t)]$

We still need another equation, specifically, an equation to obtain temperature. Usually, an energy balance provides this necessary equation. Thus, writing an unsteady-state energy balance on the contents of the reactor, as also presented in Section 3-7.1, gives

$$\begin{array}{ccccc} \text{Rate of energy} & - & \text{Rate of energy} & - & \text{Rate of energy} \\ \text{into reactor} & & \text{out of reactor} & & \text{associated} \\ & & & & \text{with reaction} \end{array} = \begin{array}{c} \text{Rate of change of} \\ \text{energy accumulated} \\ \text{in reactor} \end{array}$$

$$f(t)\rho C_p T_i(t) - UA[T(t) - T_c(t)] - f(t)\rho C_p T(t) - Vr_A(\Delta H_r)$$
$$= V\rho C_v \frac{dT(t)}{dt} \qquad \textbf{(4-2.21)}$$

3 eq., 4 unk. $[T_c(t)]$

where

$\quad U$ = overall heat transfer coefficient, assumed constant, Btu/ft^2-$°$R-min

$\quad A$ = heat transfer area, ft^2

ΔH_r = heat of reaction, Btu/lbmole of A reacted

$\quad C_v$ = heat capacity at constant volume, Btu/lb-$°$R

Writing an unsteady-state energy balance in the contents of the cooling jacket—a new control volume—provides another equation

$$f_c(t)\rho_c C_{p_c} T_{c_i}(t) + UA[T(t) - T_c(t)] - f_c(t)\rho_c C_{p_c} T_c(t)$$
$$= V_c \rho_c C_{v_c} \frac{dT_c(t)}{dt} \qquad \textbf{(4-2.22)}$$

4 eq., 4 unk.

where

V_c = volume of cooling jacket, m^3

C_{v_c} = heat capacity at constant volume of cooling liquid, assumed constant, BTU/lb-$°$R

Equations 4-2.19 through 4-2.22 constitute the model of the process.

To obtain the transfer functions and the block diagram, realize that this set of equations is nonlinear, so the nonlinear terms must first be linearized. Doing so, and defining the following deviation variables

$$C_{A_i}(t) = c_{A_i}(t) - \overline{c}_{A_i}; \; C_A(t) = c_A(t) - \overline{c}_A; \; \Gamma(t) = T(t) - \overline{T}; \; F(t) = f(t) - \overline{f}$$
$$\Gamma_i(t) = T_i(t) - \overline{T}_i; \; F_c(t) = f_c(t) - \overline{f}_c; \; \Gamma_c(t) = T_c(t) - \overline{T}_c; \; \Gamma_{c_i}(t) = T_{c_i}(t) - \overline{T}_{c_i}$$

we get, from Eq. 4-2.19,

$$C_A(s) = \frac{K_1}{\tau_1 s + 1} C_{A_i}(s) + \frac{K_2}{\tau_1 s + 1} F(s) - \frac{K_3}{\tau_3 s + 1} \Gamma(s) \qquad \textbf{(4-2.23)}$$

where

$$\tau_1 = \frac{V}{\overline{f} + 2Vk_o e^{-E/R\overline{T}}\overline{c}_A} = 2.07 \, \text{min}; \quad K_1 = \frac{\overline{f}}{\overline{f} + 2Vk_o e^{-E/R\overline{T}}\overline{c}_A} = 0.209$$

$$K_2 = \frac{\overline{c}_{A_i} - \overline{c}_A}{\overline{f} + 2Vk_o e^{-E/R\overline{T}}\overline{c}_A} = 0.0612 \frac{\text{lbmole/ft}^3}{\text{ft}^3/\text{min}}; \quad K_3 = \frac{V\dfrac{E}{R\overline{T}^2}\overline{r}_A}{\overline{f} + 2Vk_o e^{-E/R\overline{T}}\overline{c}_A}$$

$$= 0.00248 \frac{\text{lbmole/ft}^3}{^\circ\text{R}}$$

From Eq. 4-2.21,

$$\Gamma(s) = \frac{K_4}{\tau_2 s + 1} F(s) + \frac{K_5}{\tau_2 s + 1} \Gamma_i(s) - \frac{K_6}{\tau_2 s + 1} C_A(s) + \frac{K_7}{\tau_2 s + 1} \Gamma_c(s) \qquad \textbf{(4-2.24)}$$

where

$$\tau_2 = \frac{V\rho C_v}{V(\Delta H_r)\overline{r}_A \dfrac{E}{R\overline{T}^2} + UA + \overline{f}\rho C_p} = -7.96 \, \text{min}$$

$$K_4 = \frac{\rho C_p(\overline{T}_i - \overline{T})}{V(\Delta H_r)\overline{r}_A \dfrac{E}{R\overline{T}^2} + UA + \overline{f}\rho C_p} = 26.35 \frac{^\circ\text{R}}{\text{ft}^3/\text{min}}$$

$$K_5 = \frac{\overline{f}\rho C_p}{V(\Delta H_r)\overline{r}_A \dfrac{E}{R\overline{T}^2} + UA + \overline{f}\rho C_p} = -0.802$$

$$K_6 = \frac{2V(\Delta H_r)k_o e^{-E/R\overline{T}}\overline{c}_A}{V(\Delta H_r)\overline{r}_A \dfrac{E}{R\overline{T}^2} + UA + \overline{f}\rho C_p} = 751.48 \frac{^\circ\text{R}}{\text{lbmoles/ft}^3}$$

$$K_7 = \frac{UA}{V(\Delta H_r)\overline{r}_A \dfrac{E}{R\overline{T}^2} + UA + \overline{f}\rho C_p} = -0.558$$

Finally, Eq. 4-2.22 yields

$$\Gamma_c(s) = \frac{K_8}{\tau_3 s + 1} F_c(s) + \frac{K_9}{\tau_3 s + 1} \Gamma_{c_i}(s) + \frac{K_{10}}{\tau_3 s + 1} \Gamma(s) \qquad \textbf{(4-2.25)}$$

where

$$\tau_3 = \frac{V_c \rho_c C_{p_c}}{UA + \overline{f}_c \rho_c C_{p_c}} = 0.976 \, \text{min} \quad K_8 = \frac{\rho_c C_{p_c}(\overline{T}_{c_i} - \overline{T}_c)}{UA + \overline{f}_c \rho_c C_{p_c}} = -39.23 \frac{^\circ\text{R}}{\text{ft}^3/\text{min}}$$

$$K_9 = \frac{\overline{f}_c \rho_c C_{p_c}}{UA + \overline{f}_c \rho_c C_{p_c}} = 0.5488 \quad K_{10} = \frac{UA}{UA + \overline{f}_c \rho_c C_{p_c}} = 0.4512$$

Substituting Eq. 4-2.25 into Eq. 4-2.24 gives

$$\Gamma(s) = \frac{(\tau_3 s + 1)}{(\tau_2 s + 1)(\tau_3 s + 1) - K_7 K_{10}}[K_4 F(s) + K_5 \Gamma_i(s) - K_6 C_A(s)]$$

$$+ \frac{K_7}{(\tau_2 s + 1)(\tau_3 s + 1) - K_7 K_{10}}[K_8 F_c(s) + K_9 \Gamma_{c_i}(s)] \qquad \textbf{(4-2.26)}$$

Substituting Eq. 4-2.26 into Eq. 4-2.23 yields

$$C_A(s) = \frac{K_1[(\tau_2 s + 1)(\tau_3 s + 1) - K_7 K_{10}]}{(\tau_1 s + 1)(\tau_2 s + 1)(\tau_3 s + 1) - K_7 K_{10}(\tau_1 s + 1) - K_3 K_6(\tau_3 s + 1)} C_{A_i}(s)$$

$$+ \frac{K_2[(\tau_2 s + 1)(\tau_3 s + 1) - K_7 K_{10}] - K_3 K_4(\tau_3 s + 1)}{(\tau_1 s + 1)(\tau_2 s + 1)(\tau_3 s + 1) - K_7 K_{10}(\tau_1 s + 1) - K_3 K_6(\tau_3 s + 1)} F(s)$$

$$- \frac{K_3 K_5(\tau_3 s + 1)}{(\tau_1 s + 1)(\tau_2 s + 1)(\tau_3 s + 1) - K_7 K_{10}(\tau_1 s + 1) - K_3 K_6(\tau_3 s + 1)} \Gamma_i(s)$$

$$- \frac{K_3 K_7}{(\tau_1 s + 1)(\tau_2 s + 1)(\tau_3 s + 1) - K_7 K_{10}(\tau_1 s + 1) - K_3 K_6(\tau_3 s + 1)}$$

$$\times [K_8 F_c(s) + K_9 \Gamma_{c_i}(s)] \qquad \textbf{(4-2.27)}$$

and from Eq. 4-2.27, the following required transfer functions can be obtained.

$$\frac{C_A(s)}{C_{A_i}(s)} = \frac{K_1[(\tau_2 s + 1)(\tau_3 s + 1) - K_7 K_{10}]}{(\tau_1 s + 1)(\tau_2 s + 1)(\tau_3 s + 1) - K_7 K_{10}(\tau_1 s + 1) - K_3 K_6(\tau_3 s + 1)}$$

$$\textbf{(4-2.28)}$$

or

$$\frac{C_A(s)}{C_{A_i}(s)} = \frac{0.427(0.95s + 1)(1 - 6.54s)}{26.27 s^3 + 36.31 s^2 + 10.14 s + 1} \qquad \textbf{(4-2.29)}$$

$$\frac{C_A(s)}{F(s)} = \frac{K_2[(\tau_2 s + 1)(\tau_3 s + 1) - K_7 K_{10}] - K_3 K_4(\tau_3 s + 1)}{(\tau_1 s + 1)(\tau_2 s + 1)(\tau_3 s + 1) - K_7 K_{10}(\tau_1 s + 1) - K_3 K_6(\tau_3 s + 1)}$$

$$\textbf{(4-2.30)}$$

or

$$\frac{C_A(s)}{F(s)} = \frac{0.0182(0.95s + 1)(1 - 44.75s)}{26.27 s^3 + 36.31 s^2 + 10.14 s + 1} \qquad \textbf{(4-2.31)}$$

$$\frac{C_A(s)}{\Gamma_i(s)} = \frac{-K_3 K_5(\tau_3 s + 1)}{(\tau_1 s + 1)(\tau_2 s + 1)(\tau_3 s + 1) - K_7 K_{10}(\tau_1 s + 1) - K_3 K_6(\tau_3 s + 1)}$$

$$\textbf{(4-2.32)}$$

or

$$\frac{C_A(s)}{\Gamma_i(s)} = \frac{-0.0032(0.976s + 1)}{26.27 s^3 + 36.31 s^2 + 10.14 s + 1} \qquad \textbf{(4-2.33)}$$

$$\frac{C_A(s)}{F_c(s)} = \frac{-K_3 K_7 K_8}{(\tau_1 s + 1)(\tau_2 s + 1)(\tau_3 s + 1) - K_7 K_{10}(\tau_1 s + 1) - K_3 K_6(\tau_3 s + 1)}$$

$$\textbf{(4-2.34)}$$

or

$$\frac{C_A(s)}{F_c(s)} = \frac{0.0887}{26.27s^3 + 36.31s^2 + 10.14s + 1} \tag{4-2.35}$$

$$\frac{C_A(s)}{\Gamma_{c_i}(s)} = \frac{-K_3 K_7 K_9}{(\tau_1 s + 1)(\tau_2 s + 1)(\tau_3 s + 1) - K_7 K_{10}(\tau_1 s + 1) - K_3 K_6(\tau_3 s + 1)} \tag{4-2.36}$$

or

$$\frac{C_A(s)}{\Gamma_{c_i}(s)} = \frac{-0.00124}{26.27s^3 + 36.31s^2 + 10.14s + 1} \tag{4-2.37}$$

From Eqs. 4-2.23, 4-2.24, and 4-2.25, we also obtain

$$\frac{\Gamma(s)}{\Gamma_i(s)} = \frac{1.31(2.07s + 1)(0.976s + 1)}{26.27s^3 + 36.31s^2 + 10.14s + 1} \tag{4-2.38}$$

$$\frac{\Gamma(s)}{F(s)} = \frac{-31.79(0.976s + 1)(1 - 2.77s)}{26.27s^3 + 36.31s^2 + 10.14s + 1} \tag{4-2.39}$$

$$\frac{\Gamma(s)}{C_{A_i}(s)} = \frac{256(0.976s + 1)}{26.27s^3 + 36.31s^2 + 10.14s + 1} \tag{4-2.40}$$

$$\frac{\Gamma(s)}{F_c(s)} = \frac{-35.77(2.07s + 1)}{26.27s^3 + 36.31s^2 + 10.14s + 1} \tag{4-2.41}$$

and

$$\frac{\Gamma(s)}{\Gamma_{c_i}(s)} = \frac{0.5(2.07s + 1)}{26.27s^3 + 36.31s^2 + 10.14s + 1} \tag{4-2.42}$$

All of the transfer functions developed are of third order. However, the dynamic behavior of the responding variables varies significantly depending on the forcing function. The differences are due to the terms in the numerator. Equations 4-2.35 and 4-2.37 show that the dynamic behavior of $C_A(t)$ to changes in $F_c(t)$ and $\Gamma_{c_i}(t)$ are the same but different from the behavior due to changes in $C_{A_i}(t)$, $F(t)$, or $\Gamma_i(t)$. Furthermore, Eqs. 4-2.29, 4-2.31, and 4-2.33 also indicate different dynamic behavior. Similarly, Eqs. 4-2.41 and 4-2.42 indicate the same dynamic behavior of $\Gamma(t)$ to changes in $F_c(t)$ and $\Gamma_{c_i}(t)$. Note that the dynamic response of $C_A(t)$ to a change in $\Gamma_i(t)$ is the same as the dynamic response of $\Gamma(t)$ to a change in $C_{A_i}(t)$, as indicated by Eqs. 4-2.33 and 4-2.40. Section 4-3 explains in detail the effect of the term $(\tau s + 1)$ in the numerator of the transfer function, and Section 4-4.2 explains the significance of the similar, but distinctly different, term $(\tau s - 1)$.

Figures 4-2.7a and 4-2.7b show different ways to draw the block diagram for this reactor. Although Fig. 4-2.7b seems to be a bit less complex, Fig. 4-2.7a clearly shows the feedback paths indicating the interactions.

In Chapter 3 the nonlinear characteristics of processes were presented and discussed. Chemical reactors are nonlinear in their behavior, so it is appropriate to use this reactor to demonstrate once more the nonlinear characteristics of processes. Figure 4-2.8 shows how four of the terms that describe the process vary as the concentration in the reactor, $c_A(t)$, is operated at different conditions. To obtain these different conditions, the coolant flow, $f_c(t)$, was varied, which also resulted in a variation of the temperatures, $T(t)$ and $T_c(t)$, in the reactor. Figure 4-2.8a shows how the gain in Eq. 4-2.29 varies. This gain is calculated from Eq. 4-2.28 as $(K_1 - K_1 K_7 K_{10})/$

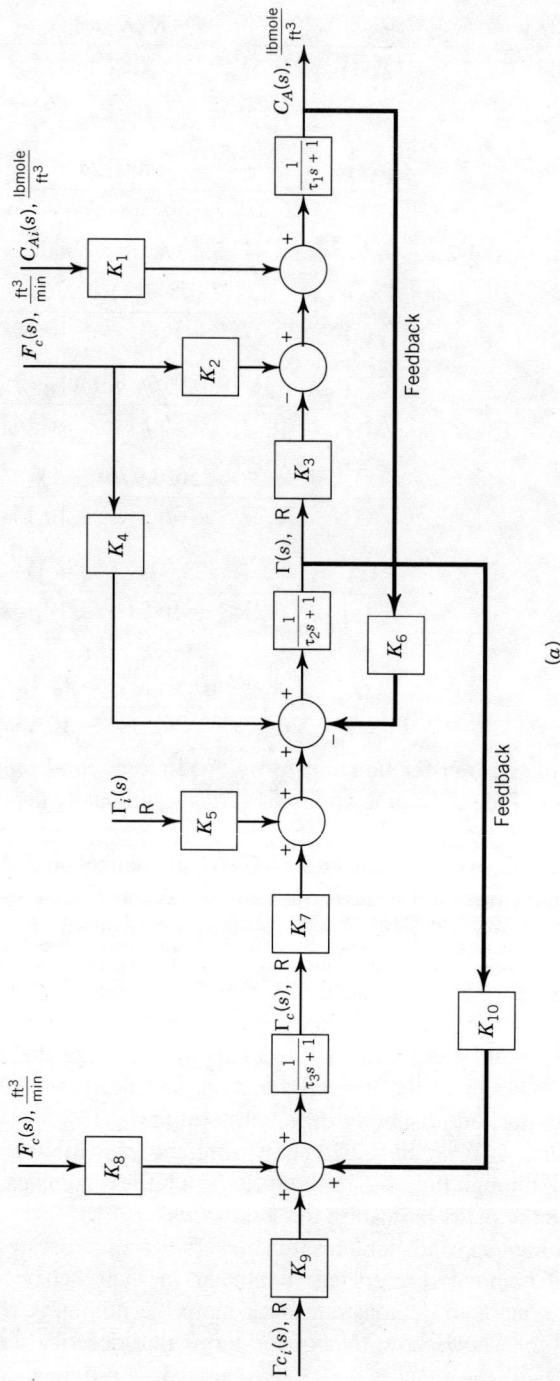

Figure 4-2.7a Block diagram for nonisothermal chemical reactor.

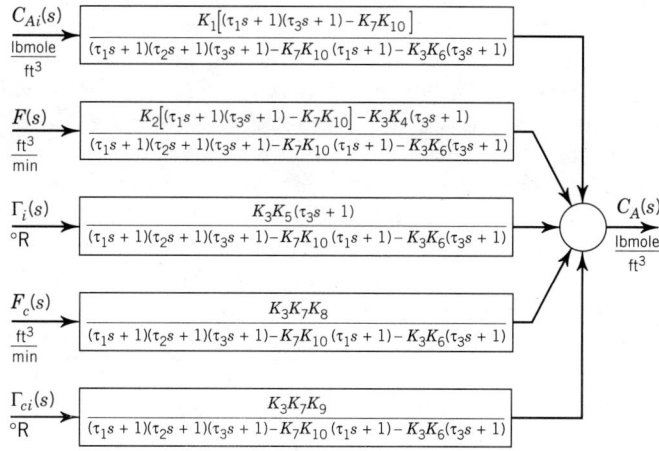

(b)

Figure 4-2.7b Block diagram for nonisothermal chemical reactor.

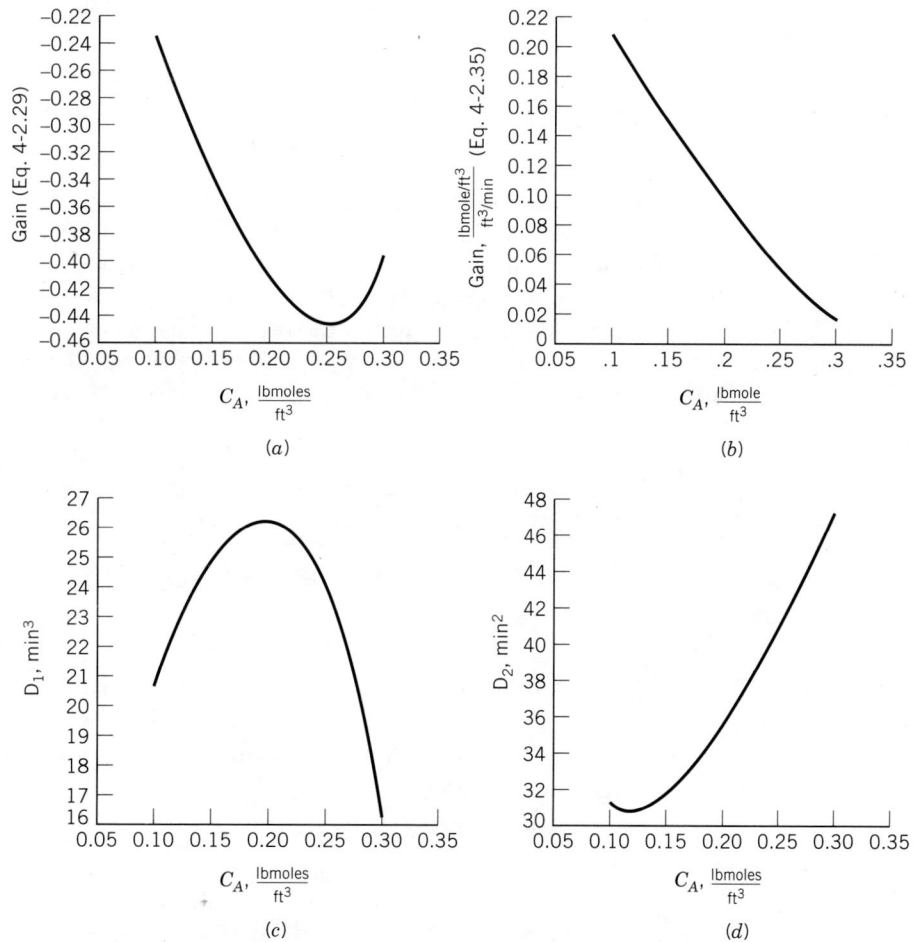

Figure 4-2.8 Variations of process parameters with operating conditions.

$(1 - K_7 K_{10} - K_3 K_6)$. At the original steady-state condition the value of this gain is -0.427. *[Note: Eq. 4-2.29 shows a value of $+0.427$ for the constant where the gain usually goes. However, remember that the gain of a transfer function is calculated by $\lim_{s \to 0} G(s)$. When this limit is applied to Eq. 4-2.29 the gain results in -0.427. The term $(6.54\,s - 1)$, which is explained in Section 4-4.2, changes the sign.]* The figure shows that sometimes the variation is as much as a factor of 2. The figure also shows that after some minimum the gain starts to increase again. Figure 4-2.8*b* shows the gain of Eq. 4-2.35, which is calculated from Eq. 4-2.34 as $(-K_3 K_7 K_8)/(1 - K_7 K_{10} - K_3 K_6)$. The figure shows a change of greater than a factor of 10. Figure 4-2.8*c* shows how the first term of the denominator, referred to as D_1, of all the transfer functions developed varies. This term is calculated as $(\tau_1 \tau_2 \tau_3)/(1 - K_7 K_{10} - K_3 K_6)$, and has a value of 26.27 min^3 at the original steady state. Figure 4-2.8*d* shows the second term of the denominator, referred to as D_2, of the transfer functions. This term is calculated as $(\tau_1 \tau_2 + \tau_1 \tau_3 + \tau_2 \tau_3)/(1 - K_7 K_{10} - K_3 K_6)$, and has a value of 36.31 min^2 at the original steady state. As we discussed in Chapter 3, the variations in process behavior, the nonlinearities, have a significant effect on the control of the process.

In this reactor example, the heat transfer rate expression $UA[T(t) - T_c(t)]$ has been used. This expression says that once the cooling water temperature, $T_c(t)$, changes, the contents of the reactor immediately feel a change in heat transfer. Thus the dynamics of the wall have been neglected. In reality, however, these dynamics may be significant. When the cooling water temperature changes, then the heat transfer to the wall changes. As the wall temperature changes, the heat transfer from the wall to the reactants then changes. Thus it is only after the wall temperature feels the change that the heat transfer to the reacting mass starts to change. Therefore, the wall represents another capacitance in the system, the magnitude of which depends on thickness, density, heat capacity, and other physical properties of the material of construction of the wall.

Taking the wall into consideration gives a better understanding of the capacitance. We will assume that the wall is at a uniform temperature, $T_m(t)$, because the heat transfer resistance of the wall is small compared to the resistances of the films on each side. Sometimes, one of the two resistances is much larger than the other. In this case, the capacitance of the wall can be lumped with the capacitance of the side of the smaller resistance, and they are assumed to be at the same temperature.

When we consider the reactor wall, the unsteady-state mole balance on component A and the rate of reaction remain the same thus providing two equations, Eqs. 4-2.19 and 4-2.20, with three unknowns, $r_A(t)$, $c_A(t)$, and $T(t)$. The unsteady-state energy balance on the contents of the reactor is changed to

$$f(t)\rho C_p T_i(t) - h_i A_i[T(t) - T_m(t)] - f(t)\rho C_p T(t) - V r_A(t)(\Delta H_r) = V\rho C_p \frac{dT(t)}{dt}$$

(4-2.43)

3 eq., 4 unk. $[T_m(t)]$

where

h_i = inside film heat transfer coefficient, assumed constant, Btu/ft^2-min-°R

A_i = inside heat transfer area, ft^2

$T_m(t)$ = temperature of metal wall,°R

Proceeding with an unsteady-state energy balance on the wall, we can write

$$h_i A_i[T(t) - T_m(t)] - h_o A_o[T_m(t) - T_c(t)] = V_m \rho_m C_{v_m} \frac{dT_m(t)}{dt}$$ **(4-2.44)**

4 eq., 5 unk. $[T_c(t)]$

where

h_o = outside film heat transfer coefficient, assumed constant, Btu/ft^2- min -$^\circ$R

A_o = outside heat transfer area, ft^2

V_m = volume of the metal wall, ft^3

ρ_m = density of the metal wall, lb/ft^3

C_{v_m} = capacity at constant volume of the metal wall, Btu/lb-$^\circ$R

Finally, an unsteady-state energy balance on the cooling water gives the other required equation:

$$f_c(t)\rho_c C_{p_c} T_{c_i}(t) + h_o A_o[T_m(t) - T_c(t)] - f_c(t)\rho_c C_{p_c} T_c(t)$$
$$= V_c \rho_c C_{v_c} \frac{dT_c(t)}{dt} \qquad \textbf{(4-2.45)}$$

5 eq., 5 unk.

Five equations are now required to describe the reactor. Equation 4-2.44 is the new equation describing the dynamics of the wall.

From Eqs. 4-2.19 and 4-2.20, Eq. 4-2.23 is obtained as previously shown. We write this equation again for convenience.

$$C_A(s) = \frac{K_1}{\tau_1 s + 1} C_{A_i}(s) + \frac{K_2}{\tau_1 s + 1} F(s) - \frac{K_3}{\tau_3 s + 1} \Gamma(s) \qquad \textbf{(4-2.23)}$$

From Eq. 4-2.43 and using the procedure previously learned, we obtain

$$\Gamma(s) = \frac{K_{11}}{\tau_4 s + 1} F(s) + \frac{K_{12}}{\tau_4 s + 1} \Gamma_i(s) - \frac{K_{13}}{\tau_4 s + 1} C_A(s) + \frac{K_{14}}{\tau_4 s + 1} \Gamma_m(s) \qquad \textbf{(4-2.46)}$$

where

$$\tau_4 = \frac{V\rho C_v}{V(\Delta H_r)\bar{r}_A \dfrac{E}{R\bar{T}^2} + h_i A_i + \bar{f}\rho C_p}, \text{ minutes}$$

$$K_{11} = \frac{\rho C_p(\bar{T}_i - \bar{T})}{V(\Delta H_r)\bar{r}_A \dfrac{E}{R\bar{T}^2} + h_i A_i + \bar{f}\rho C_p}, \quad \frac{^\circ R}{\text{ft}^3/\min}$$

$$K_{12} = \frac{\bar{f}\rho C_p}{V(\Delta H_r)\bar{r}_A \dfrac{E}{R\bar{T}^2} + h_i A_i + \bar{f}\rho C_p}, \text{ dimensionless}$$

$$K_{13} = \frac{V(\Delta H_r)k_o e^{-E/R\bar{T}}}{V(\Delta H_r)\bar{r}_A \dfrac{E}{R\bar{T}^2} + h_i A_i + \bar{f}\rho C_p}, \quad \frac{^\circ R}{\text{lbmole/ft}^3}$$

$$K_{14} = \frac{h_i A_i}{V(\Delta H_r)\bar{r}_A \dfrac{E}{R\bar{T}^2} + h_i A_i + \bar{f}\rho C_p}, \text{ dimensionless}$$

From Eq. 4-2.44 and the usual procedure, we obtain

$$\Gamma_m(s) = \frac{K_{15}}{\tau_5 s + 1} \Gamma(s) + \frac{K_{16}}{\tau_5 s + 1} \Gamma_c(s) \qquad \textbf{(4-2.47)}$$

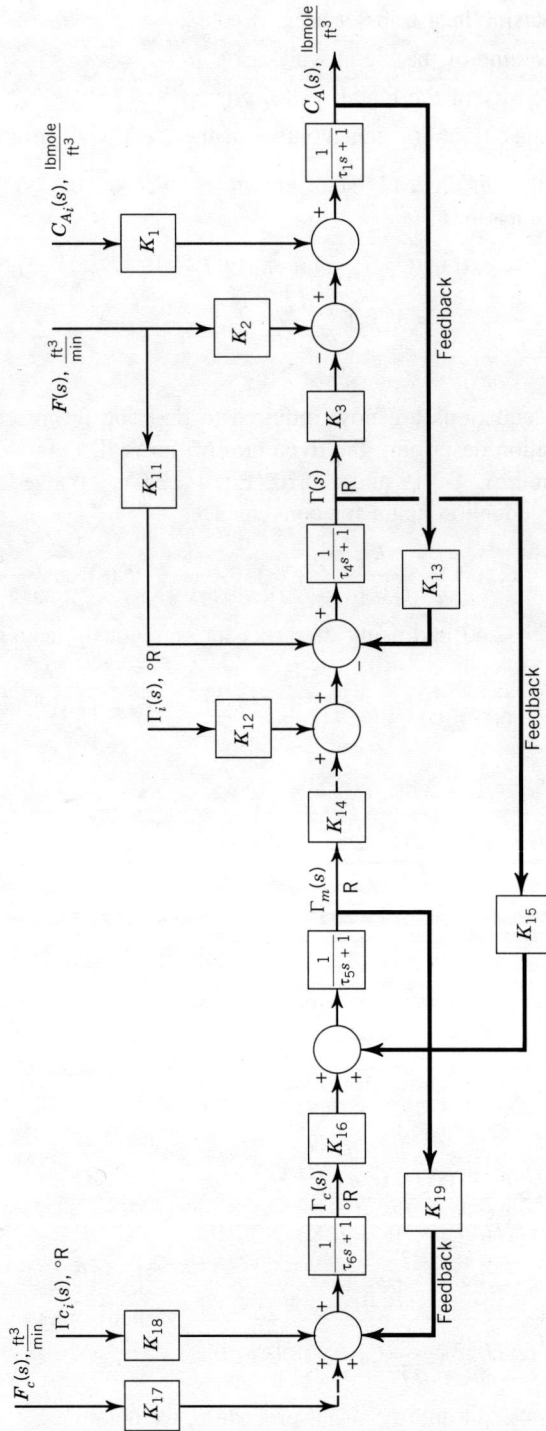

Figure 4-2.9 Block diagram for nonisothermal chemical reactor—wall considered.

where

$$\tau_5 = \frac{V_m \rho_m C_{v_m}}{h_i A_i + h_o A_o}, \quad \text{minutes} \quad K_{15} = \frac{h_i A_i}{h_i A_i + h_o A_o}, \quad \text{dimensionless}$$

$$K_{16} = \frac{h_o A_o}{h_i A_i + h_o A_o}, \quad \text{dimensionless}$$

Finally, from Eq. 4-2.45

$$\Gamma_c(s) = \frac{K_{17}}{\tau_6 s + 1} F_c(s) + \frac{K_{18}}{\tau_6 s + 1} \Gamma_{c_i}(s) + \frac{K_{19}}{\tau_6 s + 1} \Gamma_m(s) \qquad \textbf{(4-2.48)}$$

where

$$\tau_6 = \frac{V_c \rho_c C_{v_c}}{h_o A_o + \overline{f}_c \rho_c C_{p_c}}, \text{minutes} \quad K_{17} = \frac{\rho_c C_{p_c}(\overline{T}_{c_i} - \overline{T}_c)}{h_o A_o + \overline{f}_c \rho_c C_{p_c}}, \frac{°\text{R}}{\text{ft}^3/\text{min}}$$

$$K_{18} = \frac{\overline{f}_c \rho_c C_{p_c}}{h_o A_o + \overline{f}_c \rho_c C_{p_c}}, \text{dimensionless} \quad K_{19} = \frac{h_o A_o}{h_o A_o + \overline{f}_c \rho_c C_{p_c}}, \text{dimensionless}$$

With Eqs. 4-2.23, 4-2.46, 4-2.47, and 4-2.48, the block diagram for this process can be developed. This block diagram, shown in Fig. 4-2.9, shows that now there are three feedback paths indicating the interactive nature of the process.

Finally, as the reader has undoubtedly noticed, the development of any desired transfer function for this system is more complex (even though only algebraic manipulation is required) than for the previous case. As any good textbook would say, the development of these transfer functions from the foregoing equations is "left to the reader as an exercise."

4-3   RESPONSE OF HIGHER-ORDER SYSTEMS

Several types of higher-order transfer functions were developed in the previous sections. Two of the most common ones are

$$G(s) = \frac{Y(s)}{X(s)} = \prod_{i=1}^{n} G_i(s) = \frac{K}{\displaystyle\prod_{i=1}^{n}(\tau_i s + 1)} \qquad \textbf{(4-3.1)}$$

and

$$G(s) = \frac{Y(s)}{X(s)} = \frac{K \displaystyle\prod_{j=1}^{m}(\tau_{ld_j} s + 1)}{\displaystyle\prod_{i=1}^{n}(\tau_{lg_i} s + 1)}; \quad n > m \qquad \textbf{(4-3.2)}$$

A third type of transfer function developed, the one with the term $(\tau s - 1)$ in the numerator, is discussed in Section 4-4.3.

Chapter 2 presented the response of higher-order systems. This section presents a brief review of the response of Eqs. 4-3.1 and 4-3.2 to a step change in input. We believe this brief presentation makes it easy to understand the difference in dynamic response between the different systems studied in this and the previous chapter.

Consider the transfer function given by Eq. 4-3.1 with real and distinct roots. In the time domain, the response to a step change of unit magnitude is given by Eq. 4-3.3.

$$Y(t) = K \left[1 - \sum_{i=1}^{n} \frac{\tau_i^{n-1} e^{-t/\tau_i}}{\prod\limits_{\substack{j=1 \\ j \neq i}}^{n} (\tau_i - \tau_j)} \right] \tag{4-3.3}$$

The general method for solving transfer functions with other types of roots is presented in Chapter 2.

Figure 4-3.1 shows the response of systems of Eq. 4-3.1 with $n = 2$ through $n = 6$ to a unit step change in forcing function, $X(s) = 1/s$, where all the time constants are equal to 1 minute (Eq. 4-3.3 does not apply in this case because the roots are not distinct). From the figure, it is clear that as the order of the system increases, the initial response of the system is slower and slower. That is, there is an "apparent" dead time that also seems to increase. This is important in the study of automatic process control because most industrial processes are composed of first-order systems in series. Analyzing Fig. 4-3.1 in more detail, we may realize that the response of systems which are third-order and higher-order systems looks similar to the response of a second-order overdamped system with some amount of dead time. Because of this similarity, the response of these systems can be approximated by that of a second-order-plus-dead-time (SOPDT). Mathematically this is shown as follows

$$\frac{Y(s)}{X(s)} = \frac{K}{\prod\limits_{i=1}^{n} (\tau_i s + 1)} \approx \frac{K e^{-t_o s}}{(\tau_a s + 1)(\tau_b s + 1)} \tag{4-3.4}$$

for $n > 2$.

The response of processes described by Eq. 4-3.2, with real and distinct roots, to a step change of unit magnitude in forcing function is given by Eq. 4-3.5.

$$Y(t) = K \left[1 - \sum_{i=1}^{n} \frac{\prod\limits_{j=1}^{m} (\tau_{lg_i} - \tau_{ld_j}) \tau_{lg_i}^{n-m-1}}{\prod\limits_{\substack{j=1 \\ j \neq i}}^{n} (\tau_{lg_i} - \tau_{ld_j})} e^{-\frac{t}{\tau_{lg_i}}} \right] \tag{4-3.5}$$

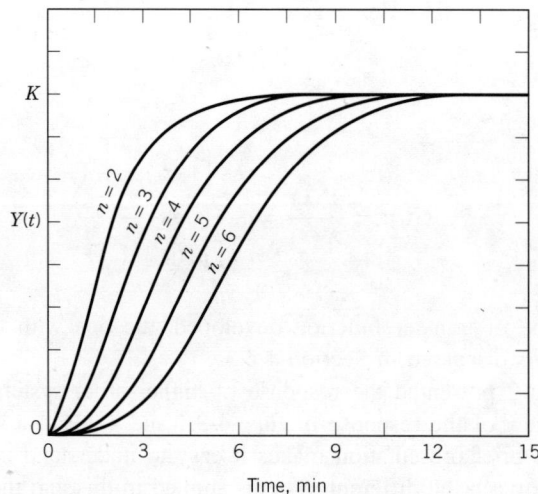

Figure 4-3.1 Response of overdamped higher-order systems to a unit step change in input.

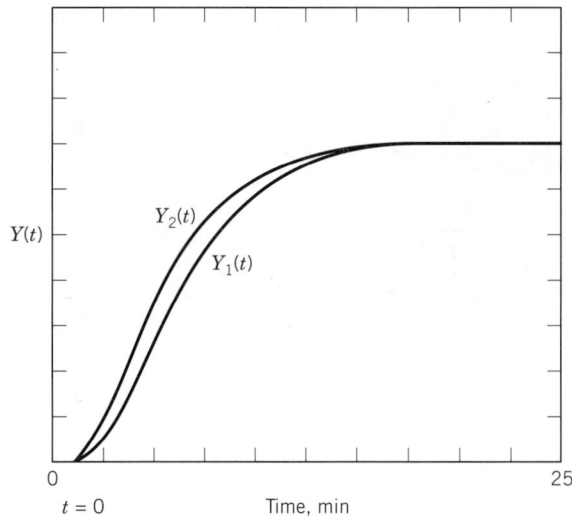

Figure 4-3.2 Comparison of responses of Eqs. 4-3.6 and 4-3.7.

To obtain a better understanding of the term $(\tau_{ld}s + 1)$, let us compare the response of the following two processes to a step change of unit magnitude in $X(s)$.

$$\frac{Y_1(s)}{X(s)} = \frac{1}{(s + 1)(2s + 1)(3s + 1)} \tag{4-3.6}$$

$$\frac{Y_2(s)}{X(s)} = \frac{(0.5s + 1)}{(s + 1)(2s + 1)(3s + 1)} \tag{4-3.7}$$

Figure 4-3.2 shows the two responses. The effect of the $(\tau_{ld}s + 1)$ term is to "speed up" the response of the process. This is opposite to the effect of $1/(\tau_{lg}s + 1)$. In Chapter 3, the term $1/(\tau_{lg}s + 1)$ was referred to as a *first-order lag*. Consequently, we refer to the term $(\tau_{ld}s + 1)$ as a *first-order lead*. This is why the notation τ_{lg}, indicating a "lag" time constant, and τ_{ld}, indicating a "lead" time constant, is used. Note that when τ_{ld} becomes equal to τ_{lg}, the transfer function becomes of one order less. Chapter 2 also presented the concepts of lead and lag using the results of a response to a ramp function.

A common characteristic of all the responses presented so far is that they all reach a new steady state, or operating condition. Processes that show this characteristic—that is, those processes that after a step change in input reach a new steady state—are sometimes classified as *self-regulating processes*; most processes are of this type. Section 4-4 presents two examples of *non-self-regulating processes*.

4-4   OTHER TYPES OF PROCESS RESPONSES

This section presents some systems that cannot be classified as any of the types presented so far. The first two systems presented are sometimes classified under the general heading of non-self-regulating; this section explains the reason for this term. The third system presented, though of the self-regulating type, has a different response from the systems presented in this and the previous chapter.

4-4.1   Integrating Processes: Level Process

Consider the process tank shown in Fig. 4-4.1. An input stream enters the tank freely, whereas the output stream depends on the speed of the pump. The pump speed is

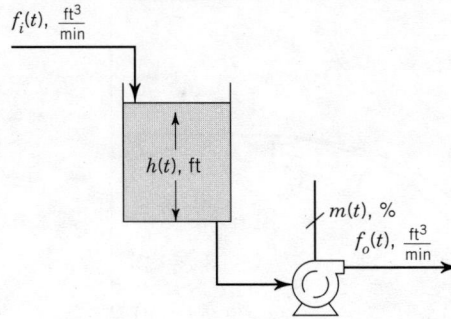

Figure 4-4.1 Process tank with pump manipulating outlet flow.

regulated by the signal $m(t)$, %. The relation between the output flow and the signal is given by

$$\tau_p \frac{df_o(t)}{dt} + f_o(t) = K_p m(t)$$

That is, the pump flow does not respond instantaneously to a change in signal but rather as a first-order response with time constant τ_p (min) and gain $K_p (\text{ft}^3/\min/\%)$. It is important to realize that the pump flow does not depend on the level in the tank but only on the input signal $m(t)$.

Develop the mathematical model, obtain the transfer functions, and draw the block diagrams, that relate the tank level, $h(t)$, to the input flow, $f_i(t)$, and the input signal, $m(t)$.

An unsteady-state mass balance around the tank provides the first equation needed:

$$\rho f_i(t) - \rho f_o(t) = \rho A \frac{dh(t)}{dt} \tag{4-4.1}$$

1 eq., 2 unk. $[f_o(t), h(t)]$

where

ρ = liquid density, assumed constant, lbm/ft^3
A = cross-sectional area of tank, ft^2

The pump provides the second equation.

$$\tau_p \frac{df_o(t)}{dt} + f_o(t) = K_p m(t) \tag{4-4.2}$$

2 eq., 2 unk.

Only two equations are required to model this simple process.
Following the usual procedure, we obtain from Eq. 4-4.1

$$H(s) = \frac{1}{As}[F_i(s) - F_o(s)] \tag{4-4.3}$$

where the deviation variables are

$$H(t) = h(t) - \overline{h}; \quad F_i(t) = f_i(t) - \overline{f}_i; \quad F_o(t) = f_o(t) - \overline{f}_o$$

From Eq. 4-4.2 we obtain

$$F_o(s) = \frac{K_p}{\tau_p s + 1} M(s) \tag{4-4.4}$$

where the new deviation variable is $M(t) = m(t) - \overline{m}$

Substituting Eq. 4-4.4 into Eq. 4-4.3 yields

$$H(s) = \frac{1}{As} F_i(s) - \frac{K_p}{As(\tau_p s + 1)} M(s) \qquad \textbf{(4-4.5)}$$

from which we can write the following transfer functions:

$$\frac{H(s)}{F_i(s)} = \frac{1}{As} \qquad \textbf{(4-4.6)}$$

and

$$\frac{H(s)}{M(s)} = \frac{-K_p}{As(\tau_p s + 1)} \qquad \textbf{(4-4.7)}$$

These two transfer functions are different from the ones developed so far in this and the previous chapter. The single s term in the denominator indicates the "integrating" nature of the process. Let us develop the response of the system to a change of $-B\%$ in the signal $m(t)$. That is,

$$M(t) = -Bu(t)$$
$$M(s) = -\frac{B}{s}$$

and using the techniques learned in Chapter 2, we find that

$$H(s) = \frac{K_p B}{As^2(\tau_p s + 1)}$$

and inverting this equation back to the time domain yields

$$H(t) = \frac{K_p B}{A} \left(t - \tau_p + \tau_p e^{-t/\tau_p} \right) \qquad \textbf{(4-4.8)}$$

This equation shows that, as time increases, the exponential term decays to zero but the first term continues to increase; this results in a ramp-type level response. In theory the level should continue to increase, "integrating," without bounds. Realistically, the level will stop increasing when it overflows, an extreme operating condition. If the signal had increased, increasing the pump speed, the analysis would have shown the same type of response but in the opposite direction. That is, the level would decrease theoretically without bounds. Realistically, the level would stop decreasing when it reaches a very low level, or when the pump starts to cavitate. In practice, however, tanks are usually instrumented with high/low level alarms and switches designed to avoid these extreme operating conditions. These safety controls are required in any well-designed process. Figure 4-4.2 shows the response of the system, and Fig. 4-4.3 shows the block diagram.

The integrating nature of this system develops because the outlet flow, $f_o(t)$, is not a function of the level in the tank but is only a function of the signal to the pump as expressed by Eq. 4-4.2. That is, there is no "process feedback" to provide regulation. Very often, control valves are used to manipulate the outlet stream. Figure 4-4.4 shows two possible arrangements. Figure 4-4.4a is essentially the same as Fig. 4-4.1; that is, the valve's upstream pressure is provided by the pump and therefore is independent of the level in the tank. In Fig. 4-4.4b the upstream pressure is dependent on the level and therefore the flow is also dependent on the level. This dependency provides the "process feedback" necessary for self-regulation. In this case, the transfer functions would have been

$$\frac{H(s)}{F_i(s)} = \frac{K_1}{\tau s + 1} \qquad \textbf{(4-4.9)}$$

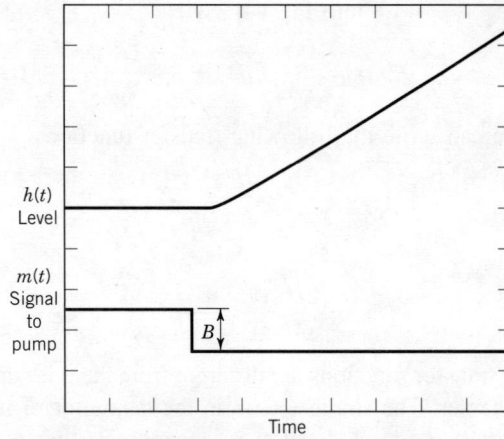

Figure 4-4.2 Response of tank level to a change in signal to pump.

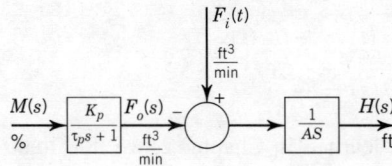

Figure 4-4.3 Block diagram for process tank.

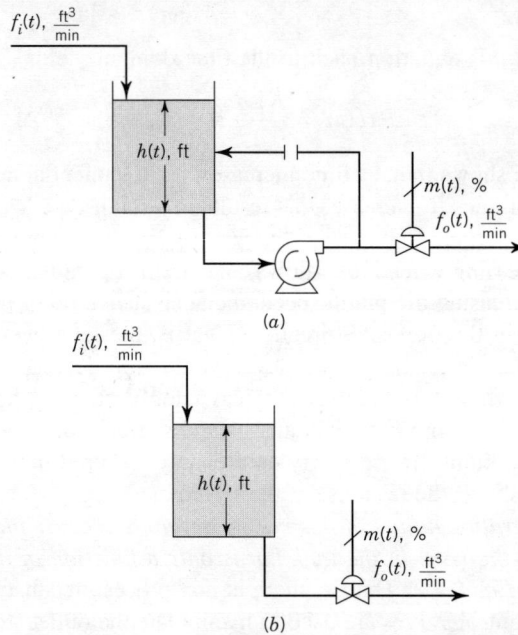

(a)

(b)

Figure 4-4.4 Process tank with control valve manipulating outlet flow.

and

$$\frac{H(s)}{M(s)} = \frac{-K_2}{(\tau s + 1)(\tau_v s + 1)} \qquad (4\text{-}4.10)$$

where

τ = time constant of the tank

τ_v = time constant of the valve

Sometimes, however, the level does not strongly affect the outlet flow, which results in a similar type of response as the one shown in Fig. 4-4.2. To explore this point further, assume that the signal to the valve changes by −B %, closing the valve some amount. In this case, the level rises, and in so doing it increases the outlet flow due to the liquid head. This process continues until the outlet flow balances the inlet flow, reaching a new steady state. But suppose that the necessary increase in level to reach steady state puts it above the maximum tank height, thus overflowing the tank. In this case, even though the process is trying to balance itself, the operation still results in a response similar to integrating.

Any system described by a transfer function containing an isolated s term in the denominator is referred to as an "integrating" system. The response of these systems to a step change in input is such that, in theory, they will not reach a new steady-state value, or operating condition. That is, they do not regulate themselves to a new steady-state condition and thus are sometimes classified as *non-self-regulating systems*. In practice, as shown in this section, they reach an "extreme" steady-state condition. The most common example of an integrating system is a level process. Level control is discussed in more detail in Chapter 7.

4-4.2 Open-Loop Unstable Process: Chemical Reactor

Consider a chemical reactor, shown in Fig. 4-4.5, where the exothermic reaction A → B takes place. To remove the heat of reaction, a jacket surrounds the reactor where a cooling liquid is maintained at 100°F as a result of a high recirculation rate.

It is desired to develop the set of equations that describe this process and to write the transfer functions relating the outlet reactor temperature and concentration to the inlet temperature and inlet concentration. Assume that the reactor contents are well mixed, that the reactor is well insulated, and that the heat capacities and densities of the reactant and product are equal to each other. Table 4-4.1 presents all the necessary process information and steady-state values.

We start by writing an unsteady-state mole balance on the reactant component A:

$$f c_{A_i}(t) - f c_A(t) - V r_A(t) = V \frac{d c_A(t)}{dt}$$ **(4-4.11)**

$$1 \text{ eq., 2 unk. } [r_A(t), c_A(t)]$$

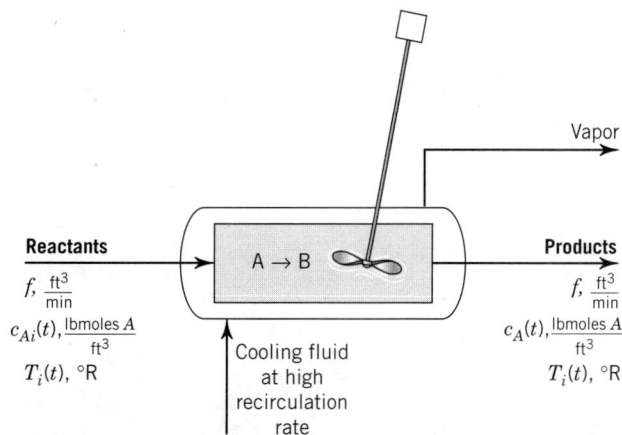

Figure 4-4.5 Chemical reactor.

Table 4-4.1 Process information and steady-state values

Process information

$V = 13.26$ ft^3; $A = 36$ ft^2
$E = 27{,}820$ Btu/lbmole; $R = 1.987$ Btu/lbmole-°R
$\rho = 55$ lbm/ft^3; $C_p = 0.88$ Btu/lbm-°F
$\Delta H_r = -12{,}020$ Btu/lbmole; $U = 75$ Btu/(hr-ft^2-°F)
$k_o = 1.73515 \times 10^{13}$ min^{-1}

Steady-state values

$c_{Ai}(t) = 0.8983$ lbmole/ft^3; $T_i(t) = 578$°R
$T_c = 560.0$°R; $f = 1.3364$ ft^3/min
$c_A(t) = 0.08023$ lbmole/ft^3; $T(t) = 690.0$°R

---

where

$r_A(t)$ = rate of reaction, lbmoles of A reacted/ft^3-min
$\quad V$ = volume of reactor, ft^3

The rate of reaction gives

$$r_A(t) = k_o \, e^{-E/RT(t)} c_A(t) \qquad \textbf{(4-4.12)}$$

2 eq., 3 unk. [$T(t)$]

An energy balance on the contents of the reactor provides another equation:

$$f\rho C_p T_i(t) - UA[T(t) - T_c] - f\rho C_p T(t) - V r_A(t)(\Delta H_r) = V \rho C_v \frac{dT(t)}{dt} \qquad \textbf{(4-4.13)}$$

3 eq., 3 unk.

where

T_c = temperature of boiling liquid in cooling jacket, °R
ΔH_r = heat of reaction, Btu/lbmole of A reacted, assumed constant

Following the usual procedure, we obtain

$$\frac{C_A(s)}{C_{A_i}(s)} = \frac{\dfrac{K_1}{1 - K_2 K_4}(\tau_2 s + 1)}{\dfrac{\tau_1 \tau_2}{1 - K_2 K_4} s^2 + \dfrac{\tau_1 + \tau_2}{1 - K_2 K_4} s + 1} \qquad \textbf{(4-4.14)}$$

$$\frac{C_A(s)}{\Gamma_i(s)} = \frac{\dfrac{-K_2 K_3}{1 - K_2 K_4}}{\dfrac{\tau_1 \tau_2}{1 - K_2 K_4} s^2 + \dfrac{\tau_1 + \tau_2}{1 - K_2 K_4} s + 1} \qquad \textbf{(4-4.15)}$$

$$\frac{\Gamma(s)}{C_{A_i}(s)} = \frac{\dfrac{-K_1 K_4}{1 - K_2 K_4}}{\dfrac{\tau_1 \tau_2}{1 - K_2 K_4} s^2 + \dfrac{\tau_1 + \tau_2}{1 - K_2 K_4} s + 1} \qquad \textbf{(4-4.16)}$$

and

$$\frac{\Gamma(s)}{\Gamma_i(s)} = \frac{\dfrac{K_3}{1 - K_2 K_4}(\tau_1 s + 1)}{\dfrac{\tau_1 \tau_2}{1 - K_2 K_4}s^2 + \dfrac{\tau_1 + \tau_2}{1 - K_2 K_4}s + 1} \qquad \textbf{(4-4.17)}$$

where

$$K_1 = \frac{f}{f + V k_o e^{-E/R\overline{T}}}, \text{ dimensionless; } K_2 = \frac{V k_o E \overline{c}_A e^{-E/R\overline{T}}}{R\overline{T}^2(f + V k_o e^{-E/R\overline{T}})}, \frac{\text{lbmoles A/ft}^3}{°\text{R}}$$

$$K_3 = \frac{f \rho C_p}{\dfrac{V k_o E \overline{c}_A \Delta H_r e^{-E/R\overline{T}}}{R\overline{T}^2} + f \rho C_p + UA}, \text{ dimensionless}$$

$$K_4 = \frac{V k_o \Delta H_r e^{-E/R\overline{T}}}{\dfrac{V k_o E \overline{c}_A \Delta H_r e^{-E/R\overline{T}}}{R\overline{T}^2} + f \rho C_p + UA}, \frac{°\text{R}}{\text{lbmoles A/ft}^3}$$

$$\tau_1 = \frac{V}{f + V k_o e^{-E/R\overline{T}}}, \text{ min; } \tau_2 = \frac{V \rho C_v}{\dfrac{V k_o E \overline{c}_A \Delta H_r e^{-E/R\overline{T}}}{R\overline{T}^2} + f \rho C_p + UA}, \text{ min}$$

and the deviation variables are

$$C_A(t) = c_A(t) - \overline{c}_A; \quad C_{A_i}(t) = c_{A_i}(t) - \overline{c}_{A_i};$$
$$\Gamma(t) = T(t) - \overline{T}; \quad \Gamma_i(t) = T_i(t) - \overline{T}_i$$

As discussed in Chapter 2, for a system to be stable, all the roots of the denominator of the transfer function must have negative real parts. Thus for the present chemical reactor, the roots are given by

$$\text{Roots} = \frac{-(\tau_1 + \tau_2) \pm [(\tau_1 + \tau_2)^2 - 4\tau_1 \tau_2(1 - K_2 K_4)]^{1/2}}{2\tau_1 \tau_2} \qquad \textbf{(4-4.18)}$$

As we have learned in this and the previous chapter, for nonlinear systems the numerical values of the process parameters, τ_1, τ_2, K_2, and K_4, vary as the operating conditions, $\overline{c}_A$ and $\overline{T}$, vary. Thus the location of the roots and the stability itself also varies. Table 4-4.2 shows the roots as the operating conditions change. To generate this information, the energy removed from the reactor by the cooling flow was varied by adjusting T_c to obtained the desired operating conditions of $\overline{c}_A$ and $\overline{T}$.

Let us analyze further the response of this chemical reactor. Figure 4-4.6a shows the temperature and concentration responses to a change in $-5°$R in inlet temperature; these responses are oscillatory around a temperature of $684°$R and a concentration of 0.102 lbmole/ft^3. Table 4-4.2 shows that at the above temperature and concentration the roots are at $-0.0819 \pm 0.2341i$, indicating stable (negative real parts) and oscillatory (imaginary parts) responses. Figure 4-4.6b shows the responses to a change of $-10°$F in inlet temperature. In this case the temperature starts to decrease, and the concentration to increase, with apparently no bounds. Table 4-4.2 shows that at a temperature around $668°$R the roots have positive real parts indicating an unstable behavior, or what is commonly referred to as an *open-loop unstable response*. In theory, this decrease in temperature, and increase in concentration, should continue. However,

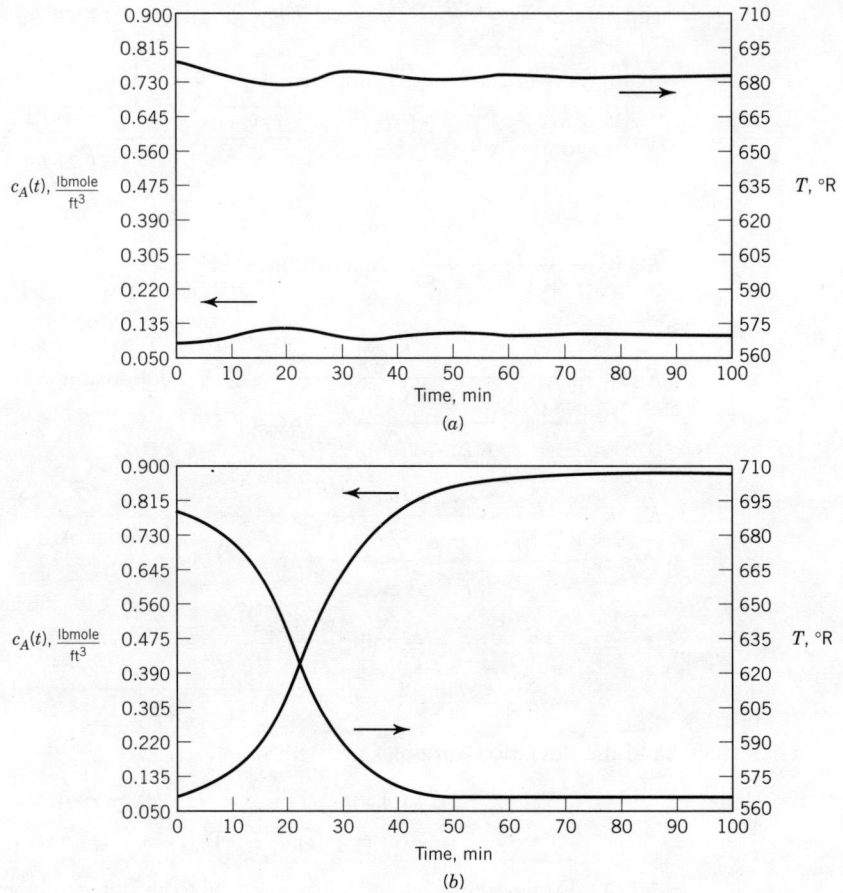

Figure 4-4.6 (*a*) Response of temperature and concentration to a change of $-5°R$ in inlet temperature. (*b*) Response of temperature and concentration to a change of $-10°R$ in inlet temperature.

Table 4-4.2   Roots versus operating conditions

$\bar{c}_A$, lbmoles/ft^3	$\bar{T}$,$°$R	Roots		$\bar{c}_A$, lbmoles/ft^3	$\bar{T}$,$°$R	Roots	
0.8905	560.08	$-0.1018,$	-0.1577	0.4159	644.00	$-0.0805,$	$+0.3032$
0.8868	566.00	$-0.1026,$	-0.1515	0.3499	650.00	$-0.0749,$	$+0.3275$
0.8815	572.00	$-0.1040,$	-0.1425	0.2526	660.00	$-0.0564,$	$+0.3063$
0.8738	578.00	$-0.1080,$	-0.1279	0.1894	668.00	$-0.1249,$	$+0.1987$
0.8671	582.00	-0.1135	$\pm0.0096i$	0.1629	672.00	-0.0647	$\pm0.0669i$
0.8588	586.00	-0.1079	$\pm0.0165i$	0.1020	684.00	-0.0819	$\pm0.2341i$
0.8485	590.00	-0.1012	$\pm0.0203i$	0.0802	690.00	-0.1980	$\pm0.2502i$
0.8287	596.00	-0.0886	$\pm0.0211i$	0.0630	696.00	-0.3506	$\pm0.1767i$
0.7270	614.00	$-0.0882,$	$+0.0316$	0.0536	700.00	$-0.6491,$	-0.3037
0.6764	620.00	$-0.0884,$	$+0.0871$	0.0331	712.00	$-1.7969,$	-0.2095
0.6381	624.00	$-0.0881,$	$+0.1265$	0.0241	720.00	$-2.8438,$	-0.1941
0.5530	632.00	$-0.0864,$	$+0.2061$	0.0176	728.00	$-4.2452,$	-0.1859

Fig. 4-4.6*b* shows that eventually, the temperature stabilizes at 566°R, and the concentration at 0.8393 lbmole/ft^3. Table 4-4.2 shows that at these conditions, the roots regain their negative real components. At 566°R the reaction is "quenched" indicating no conversion. At the final temperature of 566°R the transfer functions are written as

$$\frac{\Gamma(s)}{\Gamma_i(s)} = \frac{K_3(9.79s + 1)}{(9.75s + 1)(6.6s + 1)} \text{ and } \frac{C_A(s)}{\Gamma_i(s)} = \frac{-K_2K_3}{(9.75s + 1)(6.6s + 1)}$$

However, at $T = 620$°R, which is an unstable operating condition, the transfer functions are written as

$$\frac{\Gamma(s)}{\Gamma_i(s)} = \frac{K_3(7.47s + 1)}{(11.3s + 1)(1 - 11.47s)} \text{ and } \frac{C_A(s)}{\Gamma_i(s)} = \frac{-K_2K_3}{(11.3s + 1)(1 - 11.47s)}$$

Recall that the τ values are the negative reciprocal of the roots.

To explore this behavior further, consider the following transfer function

$$\frac{Y(s)}{X(s)} = \frac{1}{1 - \tau s}$$

The response to a step change in input, $X(s) = 1/s$, is given in the time domain by

$$Y(t) = 1 - e^{t/\tau}$$

The positive exponential term indicates an unbounded response.

Textbooks in reactor design (Levenspiel, 1972; Fogler, 1992) commonly discuss the important concept of stability from a steady-state point of view; Fig. 4-4.7*a* shows a typical graph presented. The figure, generated using the steady-state values of Table 4-4.1, shows a plot of the heats generated and removed versus the temperature in the reactor. There are three points (A$_1$, B$_1$, and C$_1$) where both heats are equal (balanced). Only two of these points, A$_1$ and C$_1$, represent stable operating conditions; point B$_1$ is an unstable condition. At any point to the left of B$_1$ the heat generated is less than the heat removed, so the temperature in the reactor will decrease until both heats are equal, which occurs at point A$_1$. At any point to the right of B$_1$ the heat generated is greater than the heat removed, so the temperature will increase until both heats are equal, which occurs at point C$_1$. Around point B$_1$ the temperature moves quickly away from that operating condition. Temperatures 573 and 690°R are the stable conditions. Figure 4-4.7*b* shows the same graph except that now the inlet temperature has been changed by -5 to 573°R. The figure shows that the new stable conditions are at 569 and 684°R. Points A$_2$ and C$_2$ are the stable conditions, whereas point B$_2$ represents the unstable condition; the temperature moves quickly away from point B$_2$. Figure 4-4.7*c* shows the graph when the inlet temperature is changed by -10 to 568°R. In this case there are only two conditions where the heat generated and removed are equal. Point A$_3$ represents the only stable condition; at point B$_3$ the curves just touch each other, which indicate an unstable condition. The three figures indicate that for this reactor the stable operating conditions "move away" from a temperature range from 615 to 670°R; stable conditions occur below 615°R or above 670°R. This same indication of stability can be obtained observing the roots given in Table 4-4.2; at any temperature between 615 and 670°R there is always one positive root, indicating unstable behavior. The roots also indicate the oscillatory nature of the behavior; this characteristic is not indicated in the steady-state analysis.

In the example at hand, the unstable behavior occurred when the temperature decreased. In this case, the extreme operating condition reached was that of "quenching" the reaction. In some other reactions, however, the unstable behavior may occur when the temperature increases. That is, in these reactions if the positive real

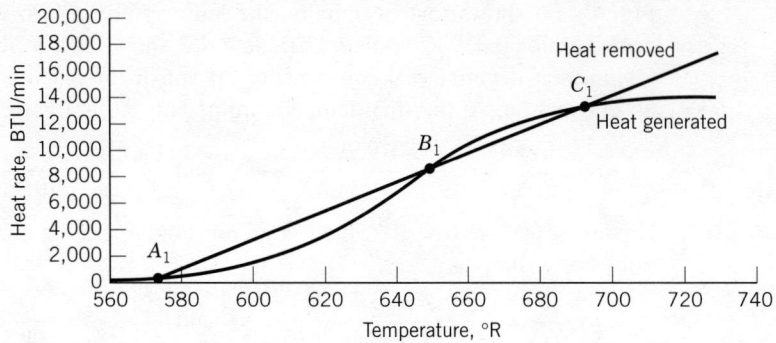

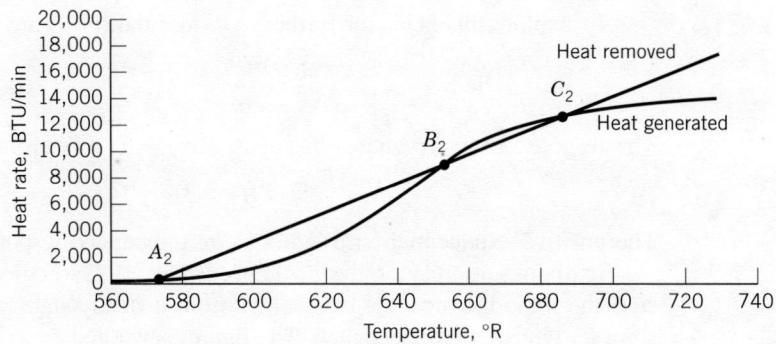

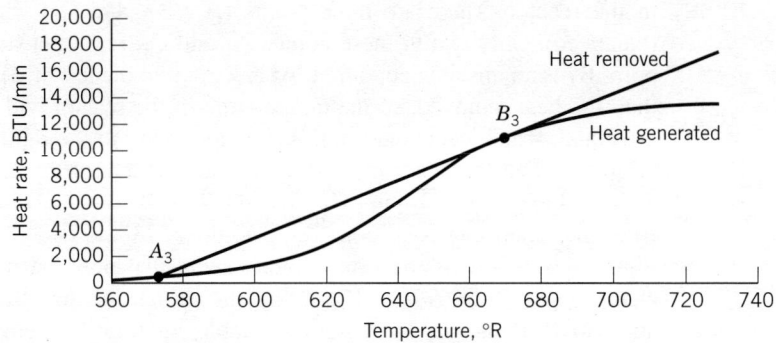

Figure 4-4.7 Steady-state analysis of heat generated and heat removed in chemical reactor. (*a*) Inlet temperature 578°R. (*b*) Inlet temperature 573°R. (*c*) Inlet temperature 568°R.

part(s) persist(s), then the temperature in the reactor will continue to increase, theoretically without bounds, and the concentration to decrease. Very often, engineers refer to this rapid increase in temperature as a runaway reaction. Several things may happen if this occurs. If the temperature increases beyond the maximum safe temperature limit for which the reactor was designed, an explosion, a meltdown, or the like may occur. *To prevent this unsafe operation, safety overrides must be triggered soon enough and must have the capacity to stop the process.* If the temperature does not reach the maximum safe limit, and it is left alone, it may actually reach a new steady state. This occurs because as the reactants are depleted, the heat generated reaches a limit; it does not continue to increase. At this moment, the heat removal may become equal to the heat generated by the reaction. If so, a new steady state is obtained; this condition is essentially points C_1 and C_2 in Figs. 4-4.7a and b. A type of reactor where this depletion does not easily occur is a nuclear reactor. The nuclear rods—the fuel—do not deplete easily, and thus, there are plenty of reactants available.

Systems described by transfer functions with a $(\tau s - 1)$ or $(1 - \tau s)$ term in the denominator are referred to as *open-loop unstable*. They are also sometimes classified as *non-self-regulating* systems, because as long as the roots with positive real parts persist, these systems will not reach a steady-state condition. The most common example of open-loop unstable behavior is an exothermic reaction.

The design of chemical reactors where exothermic reactions occur is most important, and it affects their controllability. These reactors must have enough cooling capacity to avoid "runaway reactions," and the material of construction must be able to sustain high temperatures for safe operation. However, other nonsafe operations may develop sometimes even before a high temperature is reached. Suppose, for example, that beyond a certain temperature, a new reaction starts to produce a toxic chemical compound.

4-4.3   Inverse Response Processes: Chemical Reactor

In Section 4-2.3 we considered a nonisothermal chemical reactor. Several transfer functions, Eqs. 4-2.28 through 4-2.42, were developed. Chapter 2 presented the response of transfer functions, similar to Eqs. 4-2.35, 4-2.37, 4-2.38, and 4-2.40 through 4-2.42, to a step change in forcing function. This section presents and analyzes the response of Eqs. 4-2.29, 4-2.31, and 4-2.39 to the same type of forcing function. That is, we wish to look at the response of systems described by transfer functions with the term $(\tau s - 1)$ in the numerator.

Figure 4-4.8 shows the response of the temperature, Eq. 4-2.39, and of the concentration, Eq. 4-2.31, to a step change in process flow. It is interesting to note that both initial responses are in the opposite directions to the final ones. That is, the concentration initially starts to increase and then decreases. The temperature first tends to

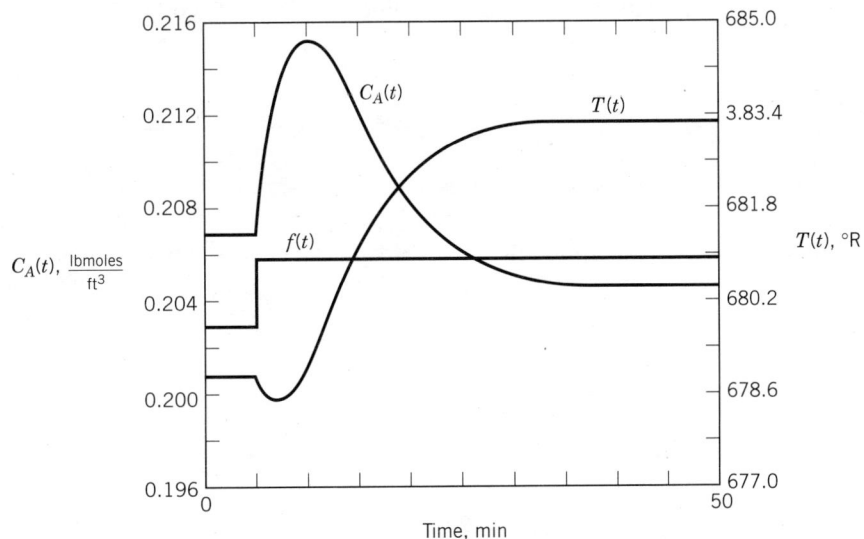

Figure 4-4.8 Inverse response of concentration and temperature to a change in process flow of 0.15 ft^3/min.

decrease and then increases. This type of response is called *inverse response*; certainly there is an explanation for it. Realizing that the temperature of the inlet stream is colder than the contents of the reactor, we should not be surprised that when the inlet flow increases, its initial effect is to reduce the temperature in the reactor. Similarly, because the concentration of the inlet stream is greater than the one in the reactor, the initial effect of an increase in inlet flow is to increase the concentration in the reactor. The effect of a lower temperature in the reactor is to reduce the rate of reaction, whereas the effect of a higher concentration is to increase it. Thus the lower temperature and the higher concentration represent two opposing effects. The final response is the net result of these opposing effects. As the figure shows, the temperature eventually increases and the concentration decreases, indicating that the rate of reaction increases until a new steady state is reached.

Mathematically, the inverse response behavior is represented by a positive root in the numerator of the transfer function. As will be presented in Chapter 8, roots of the numerator of transfer functions are called *zeros*. Equation 4-2.39 has a zero at $+0.361$, due to $(1 - 2.77s)$, and Eq. 4-2.31 has a zero at $+0.0223$, due to $(1 - 44.75s)$. Similarly, by inspection of Eq. 4-2.29 we realize that the outlet concentration, $c_A(t)$, will exhibit an inverse response when the inlet concentration, $c_{A_i}(t)$, changes.

As we have said, the inverse response can be thought of as the net result of two opposing effects. This phenomenon can be expressed mathematically as two parallel first-order systems with gains of opposite signs; this is shown in Fig. 4-4.9. From the figure the following transfer function can be obtained:

$$\frac{\Gamma(s)}{F(s)} = \frac{K_2}{\tau_2 s + 1} - \frac{K_1}{\tau_1 s + 1}$$

$$\frac{\Gamma(s)}{F(s)} = \frac{K_2 \tau_1 s + K_2 - K_1 \tau_2 s - K_1}{(\tau_2 s + 1)(\tau_1 s + 1)}$$

$$\frac{\Gamma(s)}{F(s)} = \frac{(K_2 \tau_1 - K_1 \tau_2)s + (K_2 - K_1)}{(\tau_2 s + 1)(\tau_1 s + 1)}$$

This equation provides an inverse response when process 1 reacts faster than process 2, that is, $\tau_1 < \tau_2$. In addition, the gain of process 2 must be larger than that of process 1, that is, $|K_2| > |K_1|$. Under these conditions, the numerator of the transfer function will have its root at

$$s = -\frac{K_2 - K_1}{K_2 \tau_1 - K_1 \tau_2}$$

Not all chemical reactors exhibit the inverse response behavior. Other common processes that exhibit this type of response are fluidized coal gasifiers, where increased combustion air flow first expands the bed and then consumes the material at a faster rate; distillation columns (Buckley *et al.*, 1975); and the water level in a boiler drum.

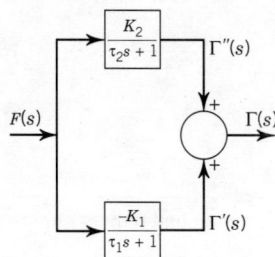

Figure 4-4.9 Explanation of inverse response.

The control of processes exhibiting inverse response presents a challenge to the control engineer (Iinoya, 1962). Chapter 11 presents the control of water level in a boiler drum and it shows the extra control sometimes used to "counteract" the response and provide the control performance required.

4-5   SUMMARY

This chapter has investigated the steady-state and dynamic characteristics of processes described by higher-order differential equations. It presented the development of the mathematical models, transfer functions, and block diagrams of these processes. We found that most processes are described by higher-order transfer functions. One of the most important and significant facts presented, as far as process control is concerned, is that as the order of the system increases, the apparent dead time also increases. This fact was clearly shown in Section 4-3, Fig. 4-3.1. This is one of the two most common reasons for the occurrence of dead time in processes; the other being due to transportation delays, as shown in Chapter 3. Figure 4-3.1 also shows an important difference between the response of first-order and higher-order systems. For first-order systems, the steepest slope on the response curve, to a step change in input, is the initial one. For higher-order systems this is not the case; the steepest slope occurs later on the response curve.

The chapter also introduced and explained the significance of noninteracting and interacting systems.

Another important concept presented was the meaning of the different terms in the transfer functions. The term $(\tau s + 1)$ in the denominator of a transfer function increases the order of the system and slows the response, as shown in Fig. 4-3.1. In this case we refer to $(\tau s + 1)$ as a first-order lag, or just simply as a lag. However, when $(\tau s + 1)$ is present in the numerator of a transfer function, it indicates a faster response, as shown in Fig. 4-3.2. In this case we refer to $(\tau s + 1)$ as a first-order lead, or just simply as a lead. The presence of an s term in the denominator of a transfer function indicates an integrating system, as shown in Fig. 4-4.1. When the term $(\tau s - 1)$ is present in the denominator of a transfer function it indicates an open-loop unstable system, as shown in Section 4-4.2. When the term $(\tau s - 1)$ is present in the numerator of a transfer function it indicates an inverse response behavior, as shown in Section 4-4.3.

4-6   OVERVIEW OF CHAPTERS 3 AND 4

Chapters 3 and 4 complete our presentation of the types, behavior, and characteristics of processes. All of the characteristic terms, gains, time constants, and dead times were obtained starting from first principles, usually mass and energy balances. Sometimes, however, it is difficult to obtain them as we have done in these two chapters. This is mainly due to the complexity of the processes, or to the lack of knowledge or understanding of some physical or chemical properties. In these cases we must use empirical means to obtain these terms. Some of these methods are presented in Chapter 7.

As indicated in Chapter 3, we must first understand the processes before control systems can be designed. Thus, now that we have completed our study of processes we are ready to control them. Chapter 5 discusses some aspects of the sensor and transmitter combination, control valves, and different types of feedback controllers. Finally Chapter 6, and subsequent chapters, put everything together. Prepare yourself for fun and challenging chapters. The subject presented in these chapters show you how to design control systems that will ensure that your processes be safe to operate, and at the same time produce the desired quality product at the design rate.

REFERENCES

1. SHINSKEY, F. G., *Process Control Systems*, 3rd ed., New York: McGraw-Hill, 1988, p. 47.

2. LEVENSPIEL, O., *Chemical Reaction Engineering*, New York: John Wiley & Sons, 1972, Chapter 8.

3. FOGLER, H. S., *Elements of Chemical Reaction Engineering*, Englewoods Cliff, New Jersey: Prentice-Hall, 1992, Section 8-7.

4. BUCKLEY, P. S., R. K. COX, and D. L. ROLLINS, "Inverse Response in a Distillation Column." *Chemical Engineering Progress*, June 1975.

5. IINOYA, K., and R. J. ALTPETER, "Inverse Response in Process Control." *Industrial and Engineering Chemistry*, July 1962.

PROBLEMS

4-1. Consider the process shown in Fig. P4-1. The mass flow rate of liquid through the tanks is constant at 250 lbm/min. The liquid inlet temperature to the first tank is 75°F; the density of the liquid may be assumed constant at 50 lbm/ft^3; the heat capacity may also be assumed constant at 1.3 Btu/lbm-°F; and the volume of each tank is 10 ft^3. You may neglect heat losses to the surroundings.

It is desired to know how the inlet temperature $T_i(t)$ and the heat transfer $q(t)$ affect the outlet temperature $T_3(t)$. For this process develop the mathematical model, determine the transfer functions relating $T_3(t)$ to $T_i(t)$ and $q(t)$, and draw the block diagram. Give the numerical values and units of each parameter in all transfer functions.

4-2. Consider the process described in Section 3-6. In that problem the relation expressing the flow provided by the fan and the signal to the fan is algebraic. This means that the fan does not have any dynamics, that is, the fan is instantaneous. In reality this is not the case. Let us assume that the fan has some dynamics such that the flow responds to a change in

signal as a first-order response with a time constant of 10 s. Obtain the same information required in Section 3-5.

4-3. Several streams are mixed in the process shown in Fig. P4-2. Streams 5, 2, and 7 are solutions of water and component A; stream 1 is pure water. The steady-state value for each stream is given in Table P4-1. Determine the following transfer functions, with the numerical values for every term

$$\frac{X_6(s)}{X_5(s)}, \quad \frac{X_6(s)}{X_2(s)} \quad \text{and} \quad \frac{X_6(s)}{F_1(s)}$$

4-4. The following irreversible elementary reaction takes place in the reactor shown in Fig. P4-3

$$A + B \rightarrow \text{Product}$$

The rate of consumption of reactant A is given by

$$r_A(t) = -kc_A(t)c_B(t)$$

where $r_A(t)$ is the reaction rate, lbmoles/gal-min.

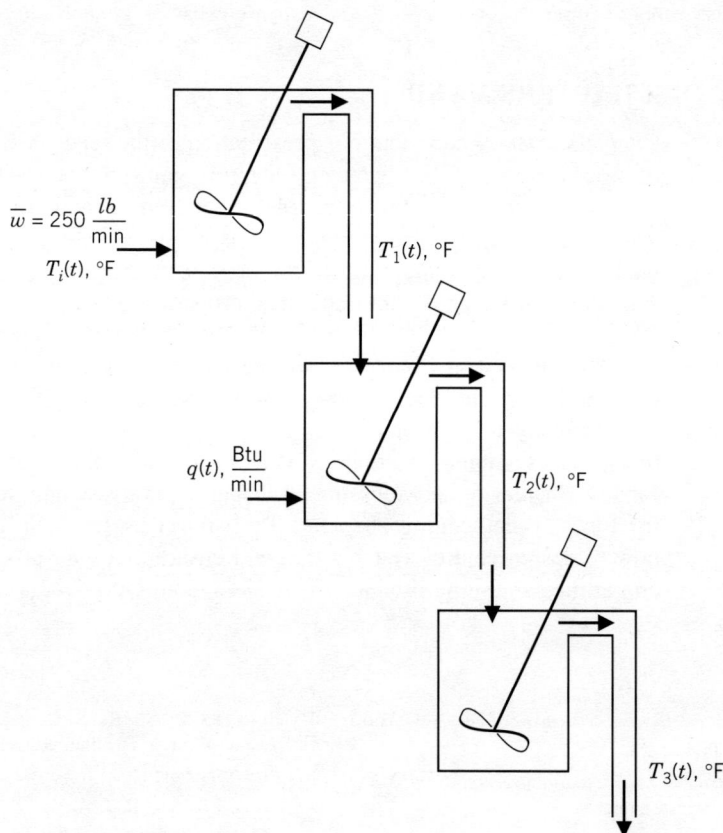

Figure P4-1 Tanks for Problem 4-1.

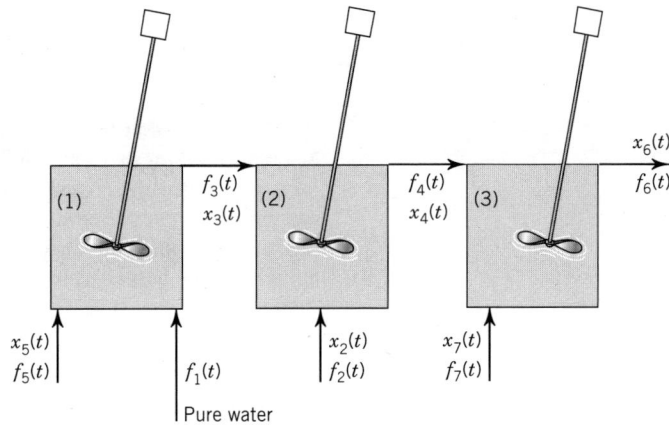

Figure P4-2 Mixing tanks for Problem 4-3.

Table P4-1 Process Information and Steady-State Values for Problem 4-3

Information

Tank Volumes: $V_1 = V_2 = V_3 = 7000$ gal
The density of all streams can be considered similar and constant.

Steady-state values

Stream	Flow, gpm	Mass fraction
1	1900	0.000
2	1000	0.990
3	2400	0.167
4	3400	0.409
5	500	0.800
6	3900	0.472
7	500	0.900

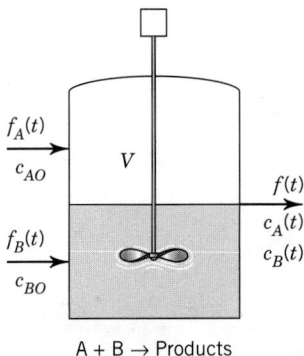

A + B → Products

Figure P4-3 Reactor for Problem 4-4.

The reactor's contents are perfectly mixed and the temperature, volume, and density of the reacting mixture may be assumed constant.

Assuming that the inlet reactant concentrations, c_{A_o} and c_{B_o} [kmoles/m^3], are constant. Obtain the transfer functions for the outlet concentrations, $C_A(s)$ and $C_B(s)$, to changes in the inlet flows, $f_A(t)$ and $f_B(t)$ [m^3/min], and draw the block diagram for the reactor showing all transfer functions. Obtain also the overall transfer function for $C_A(s)$ in terms of the input variables only [not including $C_B(s)$, but taking into consideration the interaction between the two concentrations]. Factor the denominator of the transfer function and obtain the effective time constants (negative reciprocals of the roots), and the steady-state gain.

4-5. Consider the tank, shown in Fig. P-4.4, where a fluid is mixed with saturated steam at 1 atm. The steam condenses in the liquid, and the tank is full all the time. The steady-state values and some process information are as follows:

$$\bar{f}_1 = 25 \text{ gpm}; \bar{T}_1 = 60°\text{F};$$

$$\bar{m}_2 = 3.09 \frac{\text{lbm}}{\text{min}}; \bar{T}_3 = 80°\text{F}; \bar{m} = 50\%$$

$$\rho_1 = \rho_3 = 7\frac{\text{lbm}}{\text{gal}}; C_{p_1} = C_{p_3} = 0.8\frac{\text{Btu}}{\text{lbm-}°\text{F}};$$

Volume of tank = 5 gal

The flow through the valve is given by

$$w_2(t) = 1.954vp(t)\sqrt{\Delta P}$$

The pressure drop across the valve is a constant 10 psi. The valve position, $vp(t)$, is linearly related to the signal $m(t)$. As the signal goes between 0 and 100 %, the valve position goes between 0 and 1. The valve's dynamics can be described with a first-order time constant of 4 s. Develop the mathematical model, and obtain the transfer functions relating the temperature

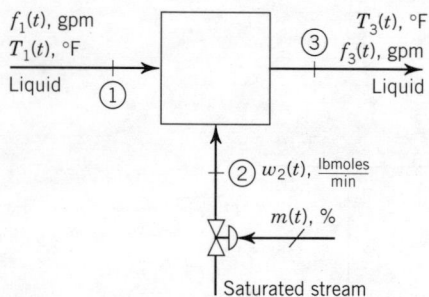

Figure P4-4 Mixing tank for Problem 4-5.

$T_3(t)$ to $f_1(t)$, $T_1(t)$, and $m(t)$. Be sure to provide the numerical values, and units of all gains, and time constants.

4-6. Figure P4-5 shows a tank used for continuous extraction of a solute from a liquid solution to a solvent. One way to model the extractor is, as shown in the sketch to the right, by assuming two perfectly mixed phases, the extract and the raffinate, separated by an interface across which the solute diffuses at a rate given by

$$n(t) = K_a V[c_1(t) - c_1^*(t)]$$

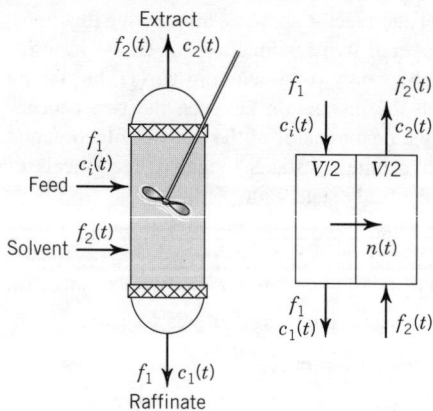

Figure P4-5 Extraction unit for Problem 4-6.

where $n(t)$, kmoles/s, is the rate of solute mass transfer across the interface; K_a, s^{-1}, is the coefficient of mass transfer; V, m^3, is the contact volume; $c_1(t)$, kmoles/m^3, is the solute concentration in the raffinate phase; and $c_1^*(t)$, kmoles/m^3, is the raffinate concentration that would be in equilibrium with the extract phase. The equilibrium relationship can be expressed as a straight line

$$c_1^*(t) = mc_2(t)$$

where m, is the slope of the equilibrium line, and $c_2(t)$ is the concentration of the solute in the extract phase. For simplicity, you may assume that the volume of each phase is one-half the total contact volume, and that the feed flow, f_1,

is constant. The two input variables are the feed concentration, $c_1(t)$, and the flow of pure solvent, $f_2(t)$. You may also assume that the variation of the densities of the streams with concentration can be neglected.

Derive the transfer functions of the extractor, draw the block diagram for the extractor, and obtain the overall transfer functions for the composition of each phase in terms of the input variables. Factor the denominator of the overall transfer functions for the extractor and express the roots in terms of the process parameters. Can the response of the concentrations be oscillatory? Can it be unstable? Justify your answers by analyzing the expressions for the roots.

4-7. A jacketed stirred tank is used to cool a process stream by flowing cooling water through the jacket as shown in Fig. P4-6. The process input variables to be considered are the flow of cooling water, $f_c(t)$, m^3/min, and the inlet temperature of the process stream, $T_i(t)$, °C. The process output variables of interest are the outlet temperatures of the process and water, $T(t)$ and $T_c(t)$, °C, respectively.

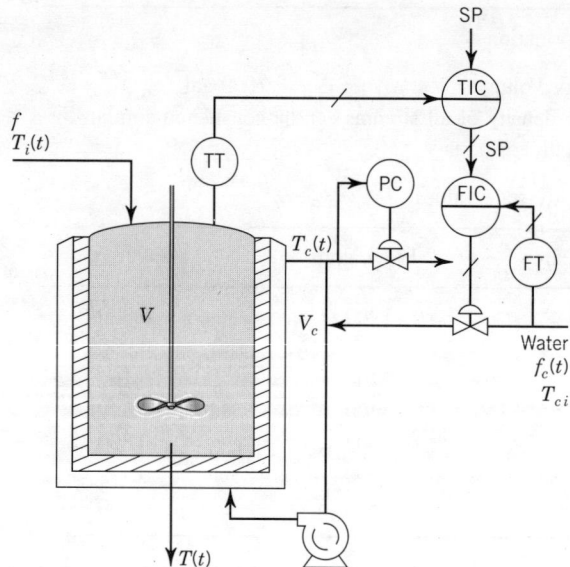

Figure P4-6 Jacketed stirred tank for Problem 4-7.

(a) List the necessary assumptions and derive, from basic principles, the following differential equations that represent the dynamic response of the process

$$\frac{dT(t)}{dt} = \frac{f}{V}[T_i(t) - T(t)] - \frac{UA}{V\rho C_v}[T(t) - T_c(t)]$$

$$\frac{dT_c(t)}{dt} = \frac{f_c(t)}{V_c}[T_{c_i} - T_c(t)] + \frac{UA}{V_c\rho_c C_{v_c}}[T(t) - T_c(t)]$$

where U, J/min-m^2-°C, is the overall heat transfer coefficient and A, m^2, is the area of heat transfer to the jacket.

(b) Laplace-transform the equations (after linearizing them) and derive the transfer functions of the process. Draw the block diagram and obtain the overall transfer function for the temperature of the fluid leaving the tank. Factor the denominator of the overall transfer function and determine the expressions for the roots as functions of the process parameters. Can the response of the concentrations be oscillatory? Can it be unstable? Justify your answers by analyzing the expressions for the roots.

4-8. One way to model imperfect mixing in a stirred tank is to divide the tank into two or more perfectly mixed sections with recirculation between them. Assume we divide the blending tank of Problem 3-18 into two perfectly mixed volumes V_1 and V_2 (so that $V_1 + V_2 = V$), as shown in Fig. P4-7, where $c_i(t)$ and $c(t)$ are the concentrations of the solute in each of the two sections. For simplicity, assume that the inlet flows and the recirculation flow f_R, volumes, and density are constant. Show that the transfer function of the outlet concentration to either inlet concentration (use stream 1 as an example), is given by

$$\frac{C(s)}{C_1(s)} = \frac{K_1}{(\tau_1 s + 1)(\tau_2 s + 1) - K_R}$$

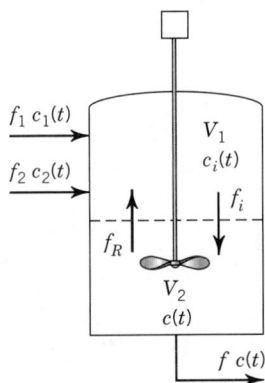

Figure P4-7 Mixing tank for Problem 4-8.

where

$$K_1 = \frac{f_1}{f + f_R}, \ \tau_1 = \frac{V_1}{f + f_R}, \ \tau_2 = \frac{V_2}{f + f_R}, \text{ and}$$
$$K_R = \frac{f_R}{f + f_R}$$

Calculate the parameters of the transfer function using the numbers given in Problem 3-18 assuming that the two volumes are equal to each other and that the recirculation flow is (a) zero, (b) f, and (c) 5f. For each of these cases calculate the gain and the effective time constants of the transfer function (the negative reciprocals of the roots of the denominator). To what values do the effective time constants go when the recirculation flow becomes very large? How does this result compare to the result of Problem 3-18.

4-9. Consider the two stirred tank reactors in series with recycle shown in Fig. P4-8. You may assume the following:

- Each reactor is perfectly mixed and the temperature is constant.
- The reactor volumes, V_1 and V_2, are constant, and so is the density of the reacting mixture.
- The flow into the first reactor, f_o, and the recycle flow, f_R, are constant.
- The chemical reaction is elementary first order, so that its rate of reaction is given by the expression:

$$r_A = kc_A(t), \text{ lbmole/ft}^3\text{- min}$$

where

$c_A(t)$ = concentration of reactant A, lbmole/ft^3.

k = constant reaction rate coefficient, min^{-1}.

- The reactors are initially at steady state with an inlet concentration c_{A_o}.
- The transportation lag between the reactors and in the recycle line is negligible.

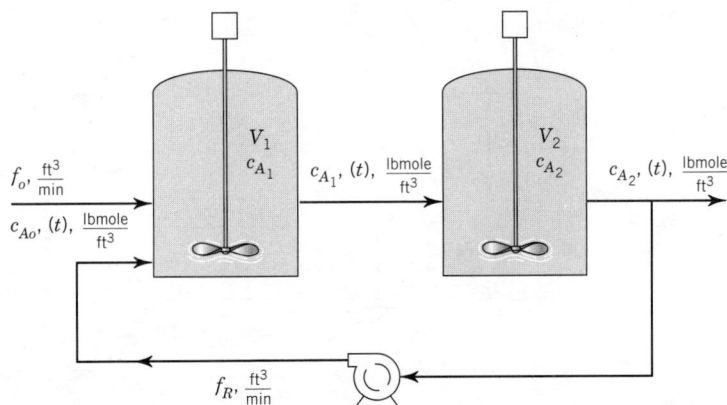

Figure P4-8 Reactors in series for Problem 4-9.

(a) Determine the process transfer functions.

(b) Draw the block diagram for the two reactors.

(c) Use block diagram algebra to determine the transfer function $\dfrac{C_{A_2}(s)}{C_{A_o}(s)}$ for the two reactors.

(d) Determine the gain and the effective time constants of this transfer function, in terms of the parameters of the system: V_1, V_2, f_o, f_R, and k.

(e) Answer the following questions:

(i) Can the system be unstable (negative effective time constants)?

(ii) Can the response of the composition be underdamped (complex conjugate effective time constants)?

(iii) What do the effective time constants become as the recycle flow, f_R, becomes much larger than the inlet flow f_o?

4-10. Consider the two gas tanks shown in Fig. P4-9. The gas may be assumed to be isothermal and behave as an ideal gas, so that the density in each tank is proportional to the pressure in that tank by the following formula:

$$\rho(t) = \frac{Mp(t)}{RT}$$

where

$\rho(t)$ = gas density, lb/ft^3

M = gas molecular weight, lb/lbmole

R = ideal gas constant, 10.73 ft^3-psia/lbmole-°R

T = gas temperature, °R

$p(t)$ = pressure in the tank, psia

The gas completely fills the volumes of the tanks, which are constant.

(a) *Critical (choked) flow through the valves.*

If the flow (in lbm/min) through the valves is assumed to be critical or "choked," it is proportional to the upstream pressure for each valve

$$w_1(t) = k_{v_1} p_1(t) \quad w_2(t) = k_{v_2} p_2(t)$$

where k_{v_1} and k_{v_2} are constant valve coefficients (lb/min)/psia, which depend on the valve capacity, gas specific gravity, temperature, and valve design. Choked flow is discussed in detail in Appendix C.

Obtain the transfer functions relating the pressure in each tank to the inlet flow to the first tank. Draw a block diagram showing the pressures. If there were n identical tanks in series, all having the same volumes, and all valves having the same valve coefficients, what would be the transfer function for the pressure in the last tank to the flow into the first tank, $P_n(s)/W_i(s)$?

(b) *Subcritical flows through the valves.*

If the flows through the valves are subcritical, they are given by

$$w_1(t) = k_{v_1}\sqrt{p_1(t)[p_1(t) - p_2(t)]}$$
$$w_2(t) = k_{v_2}\sqrt{p_2(t)[p_2(t) - p_3]}$$

where the valve coefficients k_{v_1} and k_{v_2} are not numerically the same as for critical flow, and the discharge pressure p_3 may be assumed constant.

Obtain the transfer functions relating the pressure in each tank to the inlet flow to the first tank. Draw the block diagram for the tanks showing the transfer function of each block. Write the overall transfer function, $P_2(s)/W_i(s)$, and

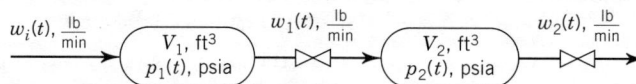

Figure P4-9 Gas tanks for Problem 4-10.

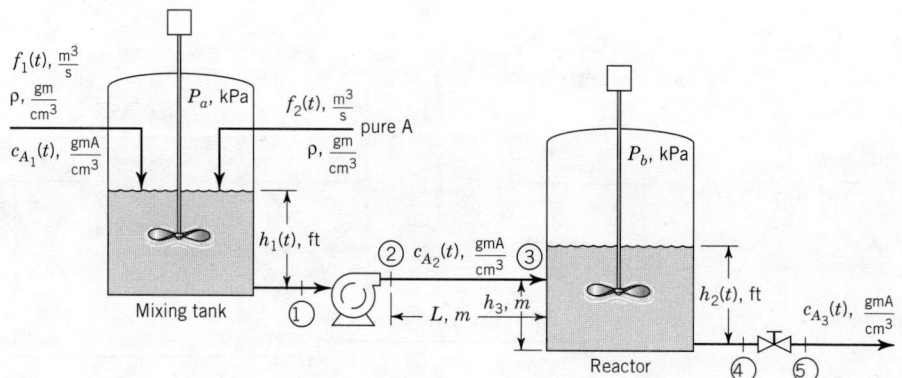

Figure P4-10 Process for Problem 4-11.

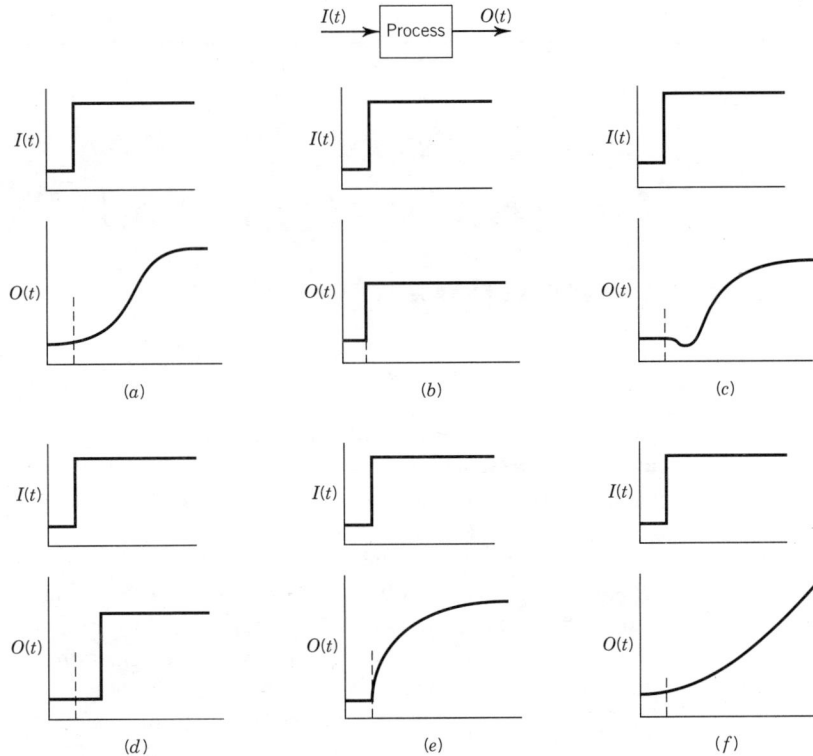

Figure P4-11 Process responses for Problem 4-12.

the formulas for the effective time constants and for the steady-state gains of the transfer functions in terms of the process parameters. *Note*: the effective time constants are defined as the negative reciprocal roots of the denominator of the overall transfer function.

4-11. Consider the process shown in Fig. P4-10. The following is known about the process:

- The density of all streams is approximately equal.
- The flow through the constant speed pump is given by

$$f(t) = A[1 + B\{p_1(t) - p_2(t)\}^2], \, \text{m}^3/\text{s}$$

where A and B are constants.

- The pipe between points 2 and 3 is rather long with a length of L, m. The flow through this pipe is highly turbulent (plug flow). The diameter of the pipe is D, m. The pressure drop between points 2 and 3 is constant, ΔP, kPa.
- We may assume that the energy effects associated with the reaction (A $\rightarrow$ B) are negligible and, consequently, the

reaction occurs at a constant temperature. The rate of reaction is given by

$$r_A(t) = kc_A(t), \, \text{kg/m}^3\text{-s}$$

- The flow through the outlet valve is given

$$f(t) = C_v\sqrt{h_2(t)}$$

Develop the mathematical model, and obtain the block diagram that shows the effect of the forcing functions $f_2(t)$, and $c_{A_1}(t)$ on the responding variables $h_1(t)$, $h_2(t)$, and $c_{A_3}(t)$.

4-12. Figure P4-11 shows the response of different processes to a step change in input. Give an indication of the form(s) of possible transfer function(s) for each process.

4-13. Consider the chemical reactor presented in Section 4-4.2. The section shows, in Fig. 4-4.6, the response of the temperature in the reactor to changes (decreases) in inlet temperature. What would be the response—stable or not, oscillatory or not—to increases in inlet temperature?

Chapter 5

Control Systems and Their Basic Components

Chapter 1 presented the three basic components of control systems: sensors-transmitters, controllers, and final control elements. In that chapter, we learned that these components perform the three basic operations of every control system: measurement (M), decision (D), and action (A).

The present chapter takes a brief look at the sensor-transmitter combination, the M component, followed by a more detailed study of control valves, the A component, and feedback controllers, the D component. Appendix C presents numerous diagrams, schematics, and other figures to aid in the presentation of different types of sensors, transmitters, and control valves. Thus Appendix C complements this chapter, and the reader is encouraged to read it together with this chapter.

5-1 SENSORS AND TRANSMITTERS

The sensor produces a phenomenon—mechanical, electrical, or the like—related to the process variable it measures. The transmitter in turn converts this phenomenon into a signal that can be transmitted. Thus, the purpose of the sensor-transmitter combination is to generate a signal, the transmitter output, which is related to the process variable. Ideally this relationship should be linear; that is, the transmitter output signal should be proportional to the process variable. Often this is the case, as, for example, with pressure, level, and some temperature transmitters, such as resistance temperature devices (RTDs). In other situations the transmitter output is a known nonlinear function of the process variable, as, for example, thermocouples and orifice flowmeters.

There are three important terms related to the sensor-transmitter combination. The *range* of the instrument is given by the low and high values of the process variable that is measured. Consider a pressure sensor-transmitter that has been calibrated to measure a process pressure between the values of 20 psig and 50 psig. We say that the range of this sensor-transmitter combination is 20 to 50 psig. The *span* of the instrument is the difference between the high and low values of the range. For the pressure transmitter we have described, the span is 30 psi. The low value of the range is often referred to as the *zero* of the instrument. This value does not have to be zero in order to be called the zero of the instrument. For our example, the zero of the instrument is 20 psig.

Appendix C presents some of the most common industrial sensors: pressure, flow, temperature, and level. That appendix also briefly discusses the working principles of an electrical, and of a pneumatic transmitter.

Principles and Practice of Automatic Process Control/Third Edition, by C. A. Smith and A. B. Corripio
ISBN 0-471-66141-4   Copyright © 2006 John Wiley & Sons (Asia) Pte. Ltd.

$$\frac{PV(s)}{\text{Process variable}} \longrightarrow \boxed{H(s)} \xrightarrow{\begin{array}{c} C(s) \\ \text{Transmitter output} \end{array}}$$

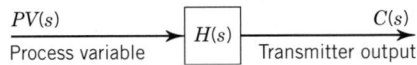

Figure 5-1.1 Block diagram of a sensor-transmitter combination.

The transfer function of the sensor-transmitter combination relates its output signal to its input, which is the process variable; this is shown in Fig. 5-1.1. The simplest form of the transfer function is a first-order lag

$$H(s) = \frac{C(s)}{PV(s)} = \frac{K_T}{\tau_T s + 1} \qquad (5\text{-}1.1)$$

where

K_T = transmitter gain

τ_T = transmitter time constant

When the relationship between the transmitter output, $C(s)$ in %TO, and the process variable, PV, is linear, the transmitter gain is simple to obtain once the span is known. Consider an electronic pressure transmitter with a range of 0 to 200 psig. A diagram showing the output versus the process variable is shown in Fig. 5-1.2. From the definition of gain in Chapter 3, the transmitter gain can be obtained by considering the entire change in output over the entire change in input, which is the span of the transmitter,

$$K_T = \frac{(20 - 4)\ \text{mA}}{(200 - 0)\ \text{psig}} = \frac{16\ \text{mA}}{200\ \text{psig}} = 0.08\ \text{mA/psig}$$

or, in percent transmitter output (%TO),

$$K_T = \frac{(100 - 0)\ \%\text{TO}}{(200 - 0)\ \text{psig}} = 0.5\ \%\text{TO/psig}$$

Thus, the gain of a sensor-transmitter is the ratio of the span of the output signal to the span of the measured variable.

The preceding example assumed that the gain of the sensor-transmitter is constant over the complete operating range. For most sensor-transmitters this is the case, but there are some instances, such as a differential pressure sensor used to measure flow, when this is not so. A differential pressure sensor measures the differential pressure, h, across an orifice. Ideally, this differential pressure is proportional to the square of the volumetric flow rate, f. That is,

$$f^2 \propto h$$

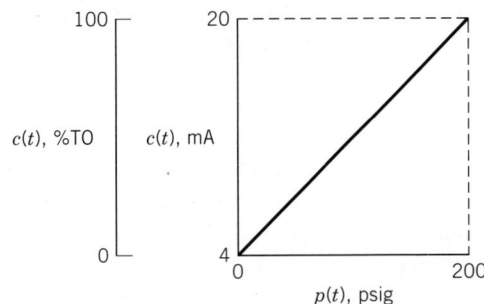

Figure 5-1.2 Linear electronic pressure transmitter.

The equation that describes the output signal, in %, from a differential pressure transmitter when used to measure volumetric flow with a range of $0 - f_{max}$ gpm is

$$c = \frac{100}{(f_{max})^2} f^2$$

where

c = output signal, %TO

f = volumetric flow

From this equation, the local gain of the transmitter is obtained as follows:

$$K_T = \frac{dc}{df} = \frac{2(100)}{(f_{max})^2} f$$

If the transmitter were linear, its gain would be

$$K_T' = \frac{100}{f_{max}}$$

The expression for K_T shows that the gain is not constant but is rather a function of flow. The greater the flow, the greater the gain. Specifically,

$$\text{at} \left(\frac{f}{f_{max}} \right) \quad 0 \quad 0.1 \quad 0.5 \quad 0.75 \quad 1.0$$

$$\left(\frac{K_T}{K_T'} \right) \quad 0 \quad 0.2 \quad 1.0 \quad 1.50 \quad 2.0$$

So the actual gain varies from zero to twice what the gain would be if the transmitter were linear. This fact results in nonlinearity in flow control systems. Most manufacturers offer differential pressure transmitters with built-in square root extractors yielding a linear transmitter. Also, most modern distributed control systems offer automatic square root extraction of signals. This makes the combination of sensor-transmitter-square root linear with a gain of $100/f_{max}$. Chapter 10 discusses in more detail the use of square root extractors.

The dynamic parameters are generally obtained empirically using methods similar to the ones shown in Chapters 6 and 9, or are provided by some manufacturers. For an example in which the time constant can be estimated from basic principles, see Problem 3-4. Some analyzer sensor-transmitters, such as chromatographs, present a dead time due to their analysis time and sampling operation.

5-2 CONTROL VALVES

Control valves are the most common final control elements. They perform the action (A) function of the control system by adjusting the flows that affect the controlled variables. This section presents the most important aspects of control valves: the selection of their action and fail position, their capacity and sizing, their flow characteristics, their gain, and their transfer function. Appendix C presents different types of valves and their accessories. The reader is strongly encouraged to read Appendix C along with this section.

A control valve acts as a variable restriction in a process pipe. By changing its opening it changes the resistance to flow and, thus, the flow itself. Throttling flows is what control valves are all about. The controller output signal positions the valve, determining the valve position that in turn determines the degree of restriction to flow. Therefore, the controller output signal is the input to the valve, and the flow is the output of the valve.

5-2.1   The Control Valve Actuator

Figure 5-2.1a shows the instrumentation schematic of a control valve. Even with electronic instrumentation, an air pressure actuator is the most common means of adjusting the position of control valves because of its high reliability and low maintenance requirements. When the signal from the controller is a 4- to 20-mA signal, a current-to-pressure transducer, labeled I/P in Fig. 5-2.1a, is required to convert the current to an air pressure. The transducer, however, does not change the signal and it can be omitted in a conceptual diagram, Fig. 5-2.1b. In this diagram, the controller signal $m(t)$ is in percent controller output (%CO), as opposed to mA or psig.

The control valve actuator consists of a diaphragm and a spring, with the diaphragm attached to the stem, which positions the flow restriction in the valve body. Figure 5-2.1 shows these parts of the valve and Appendix C presents pictures of several types of control valves with additional details of their parts.

The actuator, as shown in Fig. 5-2.1b, converts the controller output, $m(t)$, into the valve position, $vp(t)$. The valve position is usually expressed as a fraction that varies between zero and unity. When the valve position is zero, the valve is closed and the flow is zero. At the other extreme, the valve position is unity, the valve is fully opened, and the flow is maximum. For a full-range valve actuator, a 1 % change in controller output results in a 0.01 change in fraction valve position. Most control valves use a full-range actuator.

Control Valve Action

The first question the engineer must answer when specifying a control valve is: what do I want the valve to do when the energy supply fails? This question concerns the "fail position" or "action" of the valve. The main consideration in answering this question is, or should be, safety. When the safest position of the valve is the closed position, the engineer must specify a "fail-closed" (FC) valve. Such a valve requires energy to open and is also called "air-to-open" (AO). The other possibility is a "fail-open" (FO) valve. Fail-open valves require energy to close and are called "air-to-close" (AC) valves.

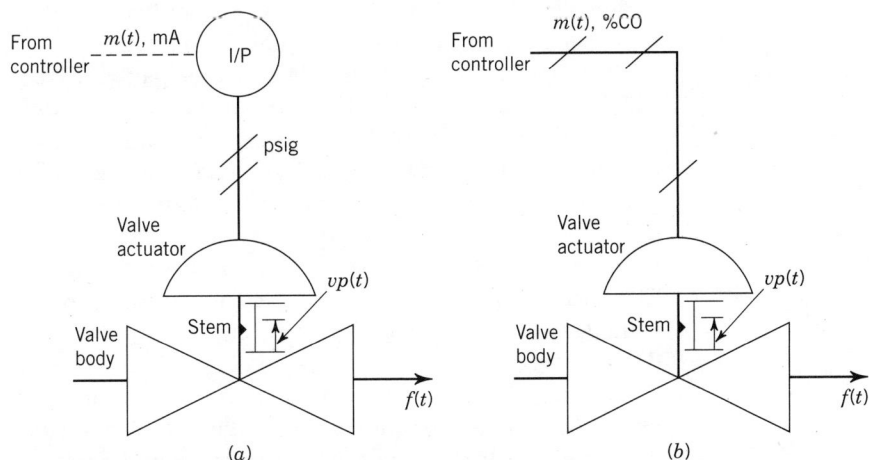

Figure 5-2.1 Instrumentation schematic for a control valve. (a) Detailed. (b) Conceptual simplification.

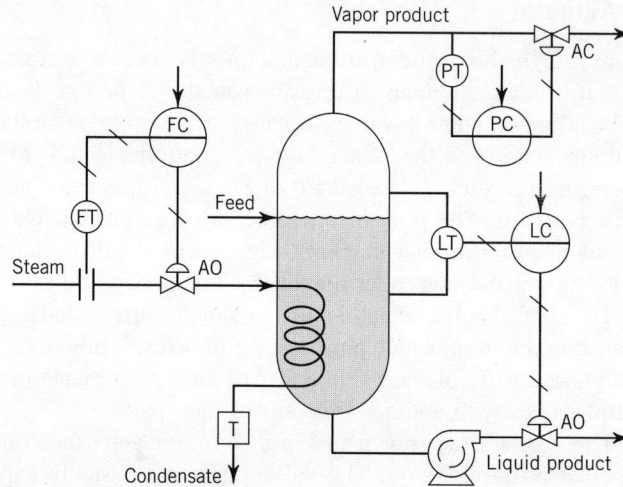

Figure 5-2.2 Fail positions of control valves on a flash drum.

To illustrate the selection of the action of control valves, let us consider the flash drum sketched in Fig. 5-2.2. Steam condenses in a coil to partially vaporize the liquid feed and separate its components into the vapor and liquid products. There are three valves in this example: one on the steam line to the coil, and one on each of the liquid and vapor products. The valve on the liquid product controls the level in the tank, and the one on the vapor product controls the pressure in the tank. The question is: what do we want each of these valves to do if the electrical or air supply were to fail? As previously explained, each valve must move to its safest position when either the electric power or the air pressure fail. The safest position for the steam valve is closed because this prevents a high steam flow that could vaporize all of the liquid and overheat the coil. Therefore, we select a fail-closed (FC) or air-to-open (AO) valve for the steam valve. For the liquid product valve, a fail-closed or air-to-open valve would keep the liquid stored in the tank. This action gives the operator time to shut down the feed to the tank and correct the cause of the failure. It is seldom safe for the liquid product to flow uncontrolled to the process downstream. Finally, a fail-open (FO) or air-to-close (AC) valve on the vapor product line would allow the vapor to flow out of the tank and prevent the tank from pressurizing.

It is important to note that in this example we have considered only the safety conditions around the flash drum. This may not necessarily be the safest operation of the process. The safety of the entire process requires that we also consider the effect of each flow on the downstream and upstream equipment. For example, when the vapor product valve fails opened, an unsafe condition may result in the process that receives the vapor. If this is so, the valve must fail closed. The engineer must then provide a separate pressure relief valve to route the vapors to an appropriate disposal system. The selection of the fail position of control valves is part of the procedure known as Hazard Analysis (HazAn). Teams of engineers perform such a procedure at process design time.

It is important to realize that safety is the only consideration in selecting the action of the control valve. As we shall see in the next section, the action of the control valve directly affects the action of the feedback controller.

The action of the valve determines the sign of the gain of the valve. An air-to-open valve has a positive gain, and an air-to-close valve has a negative gain. This is easy to see from the following formulas relating the valve position to the controller output.

$$\text{Air-to-open}: \quad \overline{vp} = \frac{\overline{m}}{100}$$

$$\text{Air-to-close}: \quad \overline{vp} = 1 - \frac{\overline{m}}{100}$$

These formulas relate the steady-state values of the variables. They do not consider the dynamic response of the actuator.

5-2.2   Control Valve Capacity and Sizing

The purpose of the control valve is to regulate the manipulated flow in the control system. To regulate flow, the flow capacity of the control valve varies from zero when the valve is closed to a maximum when the valve is fully opened, that is, when the fraction valve position is one. This subsection looks at the formulas provided by valve manufacturers to estimate the flow capacity of control valves and to size a valve for a given service. We will use the formulas provided by valve manufacturers to estimate the flow capacity of control valves and to size a valve for a given service. The next subsection presents the dependence of valve capacity on valve position.

Following a convention adopted by all control valve manufacturers, the flow capacity of a control valve is determined by its capacity factor or flow coefficient, C_v, introduced in 1944 by Masoneilan International, Inc. (Reference 2). By definition, the C_v coefficient is "the flow in U.S. gallons per minute (gpm) of water that flows through a valve at a pressure drop of 1 psi across the valve." For example, a valve with a C_v coefficient of 25 can deliver 25 gpm of water when it has a 1 psi pressure drop. Valve catalogs list the C_v coefficient of valves by type and size. Figures C-10.1a through c contain samples of valve catalog entries.

Liquid Service

A control valve is simply an orifice with a variable area of flow. The C_v coefficient and the basic principles regulating flow through an orifice provide the following formula for the liquid flow through the valve

$$f = C_v \sqrt{\frac{\Delta p_v}{G_f}} \tag{5-2.1}$$

where

f = liquid flow, U.S. gpm

Δp_v = pressure drop across the valve, psi

G_f = specific gravity of liquid at flowing conditions

Simple conversion of units in Eq. 5-2.1 gives the mass flow through the valve in lb/h:

$$w = \left(f \frac{\text{gal}}{\text{min}} \right) \left(60 \frac{\text{min}}{\text{h}} \right) \left(8.33 G_f \frac{\text{lb}}{\text{gal}} \right) = 500 C_v \sqrt{G_f \Delta p_v} \tag{5-2.2}$$

where w is the mass flow in lb/h, and 8.33 lb/gal is the density of water.

There are several other considerations, such as corrections for very viscous fluids, flashing, and cavitation, in determining the flow through control valves for liquid service. These considerations are discussed in Appendix C.

Compressible Flow

Different manufacturers have developed different formulas to model the flow of compressible fluids—gases, vapors, and steam—through their control valves. Although we present the compressible flow formulas proposed by Masoneilan (Reference 2) there are several manufacturers that market good quality valves, including Fisher Controls, Crane Company, DeZurik, Foxboro, Honeywell, and others. Masoneilan is chosen because their equations and methods are typical of the industry.

Although the equations for compressible flow look quite different from the equation for liquids, it is important to realize that they derive from the equation for liquids. They simply contain the units conversion factors and the corrections for temperature and pressure, which affect the density of a gas. It is important to realize that the C_v coefficient of a valve is the same whether the valve is used for liquid or gas service.

Masoneilan (Reference 2) proposes the following equation for gas or vapor flow in cubic feet per hour, at the standard conditions of one atm and 60°F,

$$f_s = 836 C_v C_f \frac{p_1}{\sqrt{GT}} (y - 0.148 y^3) \qquad (5\text{-}2.3)$$

where

f_s = gas flow in scfh (scfh = ft^3/h at standard conditions of 14.7 psia and 60°F)

G = gas specific gravity with respect to air, calculated by dividing the molecular weight of the gas by 29, the average molecular weight of air

T = temperature at the valve inlet, °R(= °F + 460)

C_f = critical flow factor. The numerical value for this factor ranges between 0.6 and 0.95. Figure C-10.4 shows this factor for different valve types.

p_1 = pressure at valve inlet, psia

At the standard conditions of 1 atm and 60°F, the density of a gas or vapor is given by its average molecular weight in lb divided by 380 ft^3. So, the mass rate through the valve is

$$w = \left(f_s \frac{\text{scf}}{\text{h}} \right) \left(M_w \frac{\text{lb}}{\text{lbmole}} \right) \left(\frac{1 \text{ lbmole}}{380 \text{ scf}} \right) \qquad (5\text{-}2.4)$$

where w is the gas flow in lb/h and M_w lb/lbmole is the average molecular weight of the gas.

The term y expresses the compressibility effects on the flow, and is defined by

$$y = \frac{1.63}{C_f} \sqrt{\frac{\Delta p_v}{p_1}} \qquad (5\text{-}2.5)$$

where

$\Delta p_v = p_1 - p_2$, pressure drop across the valve, psi

p_2 = pressure at valve exit, psia

and y has a maximum value of 1.5. At low ratios of the pressure drop to the inlet pressure, the gas flow is approximately incompressible and proportional to the square root of the pressure drop across the valve. The formulas match this fact because, at low values of y, the function $y - 0.148 y^3$ becomes very close to y. As the ratio of the pressure drop to the inlet pressure increases, the flow through the valve becomes choked because the velocity of the gas approaches the velocity of sound, which is the maximum it can reach. Under this condition, known as critical flow, the flow becomes independent of the exit pressure and of the pressure drop across the valve. The formulas also match this fact because, as y approaches its maximum value of 1.5, the function $y - 0.148y^3$ approaches 1.0. When this happens, the flow becomes proportional to the upstream pressure, p_1.

The critical flow factor C_f is an empirical factor that accounts for the pressure profile in the valve when the flow becomes critical. Notice that it cancels out at low ratios of the pressure drop to the inlet pressure when the term $0.148y^3$ becomes negligible. As shown in Fig. C-10.4, the C_f factor depends on the type of valve and even on the direction of flow. This is because the flow patterns in the valve affect the pressure profile and consequently the density of the gas.

EXAMPLE 5-2.1

Liquid and Gas Flow through a Control Valve

From Fig. C-10-1a, a 3-in. Masoneilan valve with full trim has a capacity factor of 110 gpm/(psi)$^{1/2}$ when fully opened. The pressure drop across the valve is 10 psi.

(a) Calculate the flow of a liquid solution with density 0.8 g/cm^3 (the density of water is 1 g/cm^3),

(b) Calculate the flow of a gas with average molecular weight of 35 when the valve inlet conditions are 100 psig and 100°F.

SOLUTION

(a) For the liquid solution, using Eq. 5-2.1,

$$f = 110\sqrt{\frac{10}{0.8}} = 389 \text{ gpm}$$

or, in mass units,

$$w = (500)(110)\sqrt{(0.8)(10)} = 155{,}600 \text{ lb/h}$$

(b) For the gas, with $M_w = 35$, $G = 35/29 = 1.207$, inlet pressure $p_1 = 100 + 14.7 = 114.7$ psia, $T = 100 + 460 = 560°$R, assuming $C_f = 0.9$, and using Eqs. 5-2.5 and 5-2.3,

$$y = \frac{1.63}{0.9}\sqrt{\frac{10}{114.7}} = 0.535$$

$$f_s = 836(110)(0.9)\frac{114.7}{\sqrt{(1.207)(560)}} = [0.535 - 0.148(0.535)^3]$$

$$= 187{,}000 \text{ scfh}$$

In mass rate units, using Eq. 5-2.4,

$$w = \left(187{,}000 \frac{\text{scf}}{\text{h}}\right)\left(35 \frac{\text{lb}}{\text{lbmole}}\right)\left(\frac{1 \text{ lbmole}}{380 \text{ scf}}\right) = 17{,}200 \frac{\text{lb}}{\text{hr}}$$

Sizing of Control Valves

Part of the job of a control engineer is to size control valves for a given service. The formulas presented thus far in this section, although useful for estimating the flow through a control valve, were developed for sizing control valves. To size a control valve for liquid service we must know the flow through the valve, the pressure drop across the valve, and the specific gravity of the liquid. For compressible flow, the additional data required are the inlet pressure and temperature, and the average molecular weight of the fluid. With this information, the engineer should use the appropriate formula provided by the specific valve manufacturer to calculate the C_v coefficient. The formulas will be very similar to those presented here, Eqs. 5-2.1 through 5-2.5. Once the C_v coefficient is known, the engineer selects, from the manufacturer's catalog, a valve that is large enough for the service. Generally, the calculated C_v falls between two different valve sizes, in which case the larger of the two should be selected. Tables provided by valve manufacturers are very similar to those presented in Fig. C-10.1 (Appendix C).

When sizing the valve for a brand new service, the flow is obtained from the process steady-state design conditions. This is the flow through the valve at the nominal production rate of the process. We will call this flow the nominal flow through the valve, and denote it $\overline{f}$. The pressure drop across the valve at nominal flow is the one to use in sizing the valve. For example, the formula to size a valve for liquid service is, from Eq. 5-2.1,

$$\overline{C_v} = \overline{f}\sqrt{\frac{G_f}{\overline{\Delta p_v}}}$$

where $\overline{\Delta p_v}$ is the pressure drop across the valve in (psi) when the flow is the nominal flow, $\overline{f}$, in gpm. Obviously, the valve coefficient must be greater than the one calculated from the above equation. This is because, if the valve is to regulate the flow, it must be able to increase the flow beyond the nominal flow. We call the ratio of the valve coefficient when the valve is fully opened to the valve coefficient at nominal flow, $C_{v,\max}/\overline{C_v}$, the overcapacity factor of the valve. Typical overcapacity factors are 1.5, for 50 % overcapacity, and 2.0, for 100 % overcapacity.

Sometimes the control engineer must also choose the pressure drop across the valve at nominal flow, a decision often made in cooperation with the process engineer. The pressure drop across the valve represents an energy loss to the process and should be kept as low as possible, but seldom less than 5 psi. Higher pressure drops are required when the pressure drop in the line and equipment in series with the valve is high, as we shall see in the subsection on valve characteristics.

EXAMPLE 5-2.2

Sizing a Steam Control Valve

A control valve is to regulate the flow of steam into a distillation column reboiler with a design heat transfer rate of 15 million Btu/h. The supply steam is saturated at 20 psig. Size the control valve for a pressure drop of 5 psi and 100 % overcapacity.

SOLUTION

From the steam tables we find that the steam saturation temperature is 259°F and its latent heat of condensation is 930 Btu/lb. This means that the nominal flow of steam is $15,000,000/930 = 16,100$ lb/h. The valve inlet pressure is $20 + 14.7 = 34.7$ psia, the molecular weight of steam is M_w is 18 lb/lbmole, and its specific gravity is $G = 18/29 = 0.621$. Assuming a Masoneilan valve with $C_f = 0.8$, from Eqs. 5-2.5, 5-2.4, and 5-2.3,

$$y = \frac{1.63}{0.8}\sqrt{\frac{5}{34.7}} = 0.773$$

$$y - 0.148y^3 = 0.705$$

$$\overline{f}_s = \left(16,100\ \frac{\text{lb}}{\text{h}}\right)\left(\frac{1\ \text{lbmole}}{18\ \text{lb}}\right)\left(\frac{380\ \text{scf}}{\text{lbmole}}\right) = 340,000\ \text{scfh}$$

$$\overline{C}_v = \frac{(340,000)\sqrt{(0.621)(259+460)}}{(836)(0.8)(34.7)(0.705)} = 440\ \frac{\text{gpm}}{\sqrt{\text{psi}}}$$

For 100 % overcapacity, the valve coefficient when fully opened is

$$C_{v,\max} = 2.0\overline{C}_v = 880\ \frac{\text{gpm}}{\sqrt{\text{psi}}}$$

From Fig. C-10.1a, a 10-in. Masoneilan valve, with a coefficient of 1000, is the smallest valve with enough capacity for this service.

EXAMPLE 5-2.3

Sizing a Valve for Liquid Service

Figure 5-2.3 shows a process for transferring an oil from a storage tank to a separation tower. The tank is at atmospheric pressure and the tower works at 25.9 in. Hg absolute (12.7 psia). Nominal oil flow is 700 gpm, its specific gravity is 0.94, and its vapor pressure at the flowing temperature of 90°F is 13.85 psia. The pipe is 8-in. schedule 40 commercial steel pipe, and the efficiency of the pump is 75 %. Size a valve to control the flow of oil. From fluid flow correlations, the frictional pressure drop in the line is found to be 6 psi.

SOLUTION

Before we can size this valve we must decide where to place it in the line and the pressure drop across the valve at nominal flow. The placement of the valve is important here because there is a possibility that the liquid will flash as its pressure drops through the valve. This would require a larger valve, since the density of the flashing mixture of liquid and vapor will be much less than that of the liquid. Notice that if we place the valve at the entrance to the tower, the liquid will flash because the exit pressure, 12.7 psia, is less than the vapor pressure at the flowing temperature, 13.85 psia. A better location for the valve is at the discharge of the pump, where the exit pressure is higher due to the hydrostatic pressure of the 60 ft of elevation plus most of the 6 psi of friction drop. The hydrostatic pressure is $(62.3 \text{ lb/ft}^3)(0.94)(60 \text{ ft})/(144 \text{ in.}^2/\text{ft}^2) = 24.4$ psi. This means that the pressure at the valve exit will be at least 37.1 psia $(24.4 + 12.7)$, well above the vapor pressure of the oil. There will be no flashing through the valve. The valve should never be placed at the suction of the pump, because there the pressure is lower and flashing would cause cavitation of the pump.

For the pressure drop across the valve we will use 5 psi, or about the same as the friction drop in the line. To get an idea of the cost of this pressure drop, for an electricity cost of \$0.03/kW-h, and 8200 h/year of operation of the pump, the annual cost due to the 5 psi drop across the valve is

$$\left(\frac{700 \text{ gal}}{\text{min}}\right)\left(\frac{1 \text{ ft}^3}{7.48 \text{ gal}}\right)\left(\frac{(5)(144) \text{ lbf}}{\text{ft}^2}\right)\left(\frac{1}{0.75}\right)\left(\frac{1 \text{ kW-min}}{44,250 \text{ ft-lbf}}\right)$$

$$\times \left(\frac{8200 \text{ h}}{\text{yr}}\right)\left(\frac{\$0.03}{\text{kW-h}}\right) = \$500/\text{yr}$$

where 0.75 is the efficiency of the pump. This cost may appear insignificant until one considers that a typical process may require several hundred control valves.

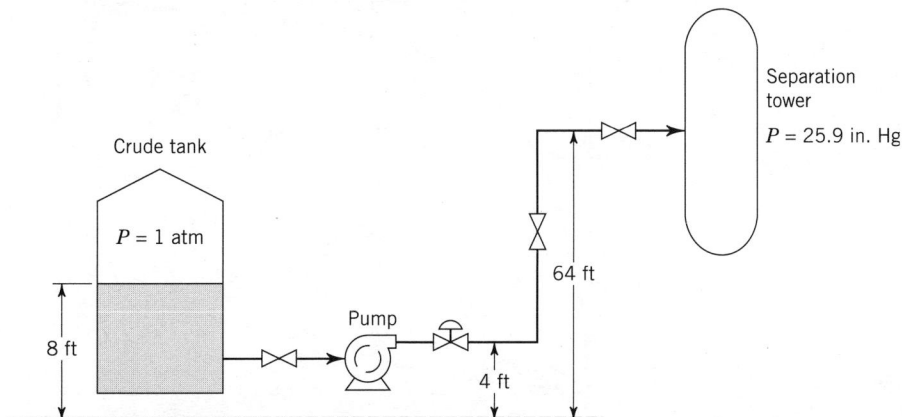

Figure 5-2.3 Process schematic for Example 5-2.3.

The maximum valve coefficient (fully opened) for 100 % overcapacity is

$$C_{v,\max} = 2(700)\sqrt{\frac{0.94}{5}} = 607\ \frac{\text{gpm}}{\sqrt{\text{psi}}}$$

This requires an 8-in. Masoneilan valve (Fig. C-10.1a), which has a C_v of 640. As a comparison, a pressure drop across the valve of 2 psi requires a C_v of 960, corresponding to an 10-in. valve. The annual cost due to a pressure drop of 2 psi is $200/year. A valve pressure drop of 10 psi requires a C_v of 429, corresponding to an 8-in. valve, and represents an annual cost of $1000/year.

5-2.3   Control Valve Characteristics

The C_v coefficient of a control valve depends on the valve position. It varies from zero when the valve is closed, $vp = 0$, to a maximum value, $C_{v,\max}$, when the valve is fully opened, that is, when the fraction valve position is unity. It is this variation in the C_v that allows the valve to continuously regulate the flow. The particular function relating the C_v coefficient to the valve position is known as the inherent valve characteristics. Valve manufacturers can shape the valve characteristics by arranging the way the area of the valve orifice varies with valve position.

Figure 5-2.4 shows three common valve characteristics, the quick-opening, linear, and equal percentage characteristics. As is evident from its shape, the quick-opening characteristic is not suitable for regulating flow, because most of the variation in the valve coefficient takes place in the lower third of the valve travel. Very little variation in coefficient takes place for most of the valve travel. Quick-opening valves are appropriate for relief valves and for on–off control systems. Relief valves must allow a large flow as quickly as possible to prevent overpressuring of process vessels and other equipment. On–off control systems work by providing either full flow or no flow. They do not regulate the flow between the two extremes.

The two characteristics normally used to regulate flow are the linear and equal percentage characteristics. The function for linear characteristics is

$$C_v(vp) = C_{v,\max}vp \tag{5-2.6}$$

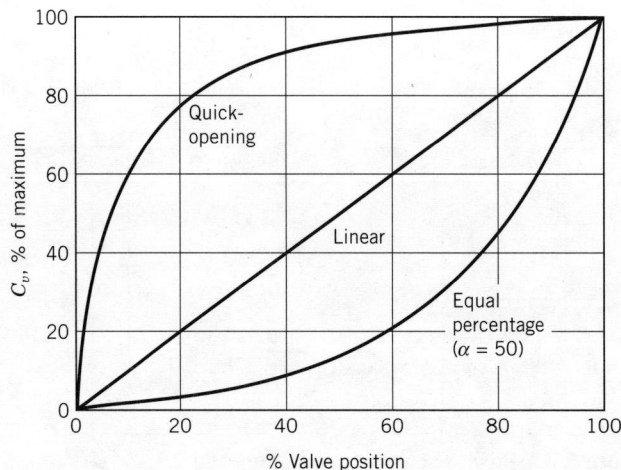

Figure 5-2.4 Inherent valve characteristic curves.

and that for equal percentage characteristics is

$$C_v(vp) = C_{v,\max}\alpha^{vp-1} \tag{5-2.7}$$

where α is the rangeability parameter, having a value of 50, or 100, with 50 being the most common. The actual equal percentage characteristic does not fit Eq. 5-2.7 all the way down to the closed position because the exponential function cannot predict zero flow at zero valve position. In fact, it predicts a coefficient of $C_{v,\max}$ at $vp = 0$. Because of this, the actual characteristic curve deviates from the exponential function in the lower 5 % of the travel.

The linear characteristic produces a coefficient proportional to the valve position. At 50 % valve position the coefficient is 50 % of its maximum coefficient.

The exponential function has the property that equal increments in valve position result in equal relative or percentage increments in the valve coefficient, therefore the name. That is, when increasing the valve position by 1 % in going from 20 to 21 % valve position, the flow increases by 1 % of its value at the 20 % position. If the valve position increases by 1 % in going from 60 to 61 % position, the flow increases by 1 % of its value at the 60 % position.

The reader may ask, what makes an exponential function useful for regulating flow? To achieve uniform control performance, the control loop should have a constant gain. A linear valve characteristic may appear to be the only one that provides a constant gain. However, as seen in Chapters 3 and 4, most processes are nonlinear in nature, and many exhibit a decrease in gain with increasing load. For such processes, the equal percentage characteristic, having a gain that increases as the valve opens (see Fig. 5-2.4), compensates for the decreasing process gain. As far as the controller is concerned, it is the product of the gains of the valve, the process, and the sensor/transmitter, that must remain constant.

The selection of the correct valve characteristics for a process requires a detailed analysis of the characteristics or "personality" of the process. However, several rules of thumb, based on previous experience, help us in making the decision. Briefly, we can say that valves with the linear flow characteristic are used when the process is linear and the pressure drop across the valve does not vary with flow. Equal percentage valves are probably the most common ones. They are generally used when the pressure drop across the valve varies with flow, and with processes in which the gain decreases when the flow through the valve increases.

Valve Rangeability

Closely associated with the valve characteristics is the valve rangeability or turn down ratio. The valve rangeability is the ratio of the maximum controllable flow to the minimum controllable flow. It is therefore a measure of the width of operating flows the valve can control. Because the flow must be under control, these flows cannot be determined when the valve is against one of its travel limits. A common way to define the maximum and minimum flows is at the 95 and 5 % valve positions, that is

$$\text{Rangeability} = \frac{\text{Flow at 95 \% valve position}}{\text{Flow at 5 \% valve position}} \tag{5-2.8}$$

Another definition uses the 90 and 10 % valve positions.

If the pressure drop across the valve is independent of flow, the flow through the valve is proportional to its C_v coefficient. Then we can calculate the valve rangeability from its inherent characteristics. From Eq. 5-2.6, the linear characteristic produces a rangeability of $0.95/0.05 = 19$, while, from Eq. 5-2.7, the equal percentage characteristic has an inherent rangeability of $\alpha^{-0.05}/\alpha^{-0.95} = \alpha^{0.90}$, which is 33.8 for $\alpha = 50$ and 63.1 for $\alpha = 100$. From Fig. 5-2.4 we can see that the rangeability of a quick-opening valve is about 3. This is one reason quick-opening valves are not suitable for regulating flow.

Installed Valve Characteristics

When the pressure drop in the line and equipment in series with a valve is significant compared with the pressure drop across the valve, the valve pressure drop varies with the flow through the valve. This pressure drop variation causes the variation of the flow with valve position to be different from the variation of the C_v coefficient. In other words, the installed flow characteristics of the valve are different from the inherent C_v characteristics. To develop a model for the installed flow characteristics, consider the piping system shown in Fig. 5-2.5. Although in this system the valve is in series with a heat exchanger, any flow resistance in series with the valve will cause the phenomenon we are about to describe and model.

There are two basic assumptions to our model: (1) the pressure drop in the line and equipment in series with the valve, Δp_L, varies with the square of the flow. (2) there is a total pressure drop, Δp_0, that is independent of flow. This total pressure drop provides the total pressure differential available across the valve plus the line and equipment. The first of these assumptions is approximately valid when the flow is turbulent, which is the most common flow regime in industrial equipment. We can always find the total pressure drop, Δp_0, by finding the pressure drop across the valve when it is closed, because then the flow, and consequently the frictional pressure drop in the line and equipment, are zero.

Let

$$\Delta p_L = k_L G_f f^2 \tag{5-2.9}$$

where

Δp_L = frictional pressure drop across the line, fittings, equipment, etc. in series with the control valve, psi

f = flow through the valve and line, gpm

k_L = constant friction coefficient for the line, fittings, equipment, etc., psi/(gpm)^2

G_f = specific gravity of the liquid (water = 1)

The pressure drop across the valve is obtained from Eq. 5-2.1.

$$\Delta p_v = G_f \frac{f^2}{C_v^2} \tag{5-2.10}$$

The total pressure drop is the sum of the two.

$$\Delta p_0 = \Delta p_v + \Delta p_L = \left(\frac{1}{C_v^2} + k_L \right) G_f f^2 \tag{5-2.11}$$

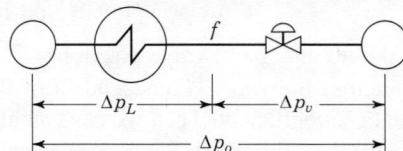

Figure 5-2.5 Valve in series with a heat exchanger. The pressure drop across the valve varies with the flow.

Solving for the flow yields

$$f = \frac{C_v}{\sqrt{1 + k_L C_v^2}} \sqrt{\frac{\Delta p_o}{G_f}} \qquad (5\text{-}2.12)$$

This formula constitutes the model of the installed characteristics for any valve in liquid service. Notice that if the line pressure drop is negligible, $k_L = 0$, $\Delta p_0 = \Delta p_v$, and Eq. 5-2.12 becomes the same as Eq. 5-2.1. In this case the installed characteristics are the same as the inherent characteristics because the pressure drop across the valve is constant. The friction coefficient is calculated from the line pressure drop at nominal flow. From Eq. 5-2.10,

$$k_L = \frac{\overline{\Delta p_L}}{G_f \overline{f}^2} \qquad (5\text{-}2.13)$$

To obtain the installed characteristics in fraction of maximum flow, we first obtain the maximum flow through the valve by substituting the maximum C_v in Eq. 5-2.12:

$$f_{\max} = \frac{C_{v,\max}}{\sqrt{1 + k_L C_{v,\max}^2}} \sqrt{\frac{\Delta p_o}{G_f}} \qquad (5\text{-}2.14)$$

Then, divide Eq. 5-2.12 by Eq. 5-2.14 to obtain

$$\frac{f}{f_{\max}} = \frac{C_v}{C_{v,\max}} \sqrt{\frac{1 + k_L C_{v,\max}^2}{1 + k_L C_v^2}} \qquad (5\text{-}2.15)$$

Notice that the maximum flow through the valve, $f_{\max}$, is independent of the valve characteristics while the normalized installed characteristics, Eq. 5-2.15, are independent of the total pressure drop, Δp_0. In fact, for a valve with a given capacity, the normalized flow characteristics, and consequently its rangeability, depend only on the friction coefficient of the line, k_L, and the inherent characteristics of the valve.

This model—Eqs. 5-2.12, 5-2.14, and 5-2.15—applies only to liquid flow through the valve without flashing. We could develop a similar model for gas flow through the valve. However, such a model must differentiate between the pressure drop in the line upstream of the valve from the pressure drop downstream of the valve. It must also consider whether the flow through the valve is critical or subcritical. Therefore, such a model could not be represented by simple formulas, but by a computer program or workbook. The use of the formulas for installed characteristics in liquid flow is displayed in the following example.

EXAMPLE 5-2.4

Installed Flow Characteristics of a Liquid Valve

SOLUTION

For the valve of Example 5-2.3, find the maximum flow through the valve, the installed flow characteristics, and the rangeability of the valve. Assume both linear and equal percentage characteristics with a rangeability parameter of 50. Analyze the effect of varying the pressure drop across the valve at nominal flow.

Although the pressure rise through the pump of Fig. 5-2.3 is also variable, we will assume that it is constant for simplicity. Alternatively, we could handle the variation of the pressure rise by adding the difference between the pressure rise at zero flow and the pressure rise at nominal flow to the 6-psi pressure drop in the line at nominal flow. In Example 5-2.3 we figured out that for a 5-psi drop across the valve, specific gravity of 0.94, and nominal flow of 700 gpm, the required valve coefficient for 100 % overcapacity is 607 gpm/(psi)$^{1/2}$. However, the smallest valve with this capacity is an 8-in. valve with $C_{v,\max} = 640$ gpm/(psi)$^{1/2}$; we will use this value. The line friction coefficient is

$$k_L = \frac{6 \text{ psi}}{(0.94)(700 \text{ gpm})^2} = 13.0 \times 10^{-6} \frac{\text{psi}}{(\text{gpm})^2}$$

and the total (constant) flow-dependent pressure drop is

$$\Delta p_0 = \Delta p_v + \Delta p_L = 5 + 6 = 11 \text{ psi}$$

The maximum flow is, from Eq. 5-2.14,

$$f_{\max} = \frac{640}{\sqrt{1 + (13 \times 10^{-6})(640)^2}} \sqrt{\frac{11}{0.94}} = 870 \text{ gpm}$$

Had we used the calculated $C_{v,\max}$ of 607, we would have gotten a maximum flow of 862 gpm. Either way, the maximum flow is much less than twice the nominal flow, 1400 gpm, although the valve was sized for 100 % overcapacity. This is because the line resistance limits the flow as the valve opens. It is not possible to select a valve big enough to deliver twice the nominal flow, because even if the valve is removed and the entire 11 psi were across the line, the flow would be $700(11/6)^{1/2} = 947$ gpm.

To obtain the valve rangeability, calculate the flow at 95 % valve position and at 5 % valve position using Eq. 5-2.12. For linear characteristics, at $vp = 0.05$, $C_v = C_{v,\max}vp = (640)(0.05) = 32$, and, from Eq. 5-2.12,

$$f_{0.05} = \frac{32}{\sqrt{1 + (13 \times 10^{-6})(32)^2}} \sqrt{\frac{11}{0.94}} = 109 \text{ gpm}$$

Similarly, at $vp = 0.95$, $C_v = 608$, and $f_{0.95} = 862$ gpm, The rangeability is then $(862)/(109) = 7.9$, which is much less than the nominal rangeability of 19. For the equal percentage characteristics with rangeability parameter $\Delta = 50$, the flows are

$$\text{At } vp = 0.05, \quad C_v = C_{v,\max}\alpha^{vp-1} = (640)(50)^{0.05-1} = 15.6 \quad f = 53.2 \text{ gpm}$$

Similarly, at $vp = 0.95$, $C_v = 526$, $f = 839$ gpm, and the installed rangeability is $839/53.2 = 15.8$, also much lower than the inherent rangeability of 33.8, but about twice the rangeability of the linear valve for the same service.

Table 5-2.1 summarizes the results for pressure drops across the valve of 2 psi, 5 psi, and 10 psi. In each case it is assumed that the line pressure drop does not change, but that the total

Table 5-2.1 Results for Example 5-2.4

	Valve pressure drop, psi		
	2	5	10
Total pressure drop, psi	8	11	16
Calculated $C_{v,\max}$	960	607	429
Required valve size[a]	10-in.	8-in.	8-in.
Actual $C_{v,\max}$[a]	1000	640	640
Maximum flow, gpm	779	870	1049
Linear rangeability	5.4	7.9	7.9
Equal % rangeability	10.8	15.8	15.8

[a]From Fig. C-10.1 Appendix C.

available pressure drop is the sum of the valve and line pressure drops at design flow. This is realistic, because, at process design time, the valve pressure drop is decided on and then the pump is sized to provide the necessary total pressure drop. The table shows that the maximum flow increases with pressure drop even when the valve size decreases. Notice also that the rangeability of the valve does not change when the valve size and the line pressure drop remain the same, even when the total available pressure drop increases.

Figure 5-2.6 shows plots of the normalized installed characteristics corresponding to the three pressure drops. These characteristics were computed using Eq. 5-2.15. For comparison, Fig. 5-2.6 also shows the inherent characteristics of each valve; these are the flow characteristics when there is no pressure drop in the line (constant pressure drop across the valve). The characteristics for valve pressure drops of 5 and 10 psi overlap each other because the valve

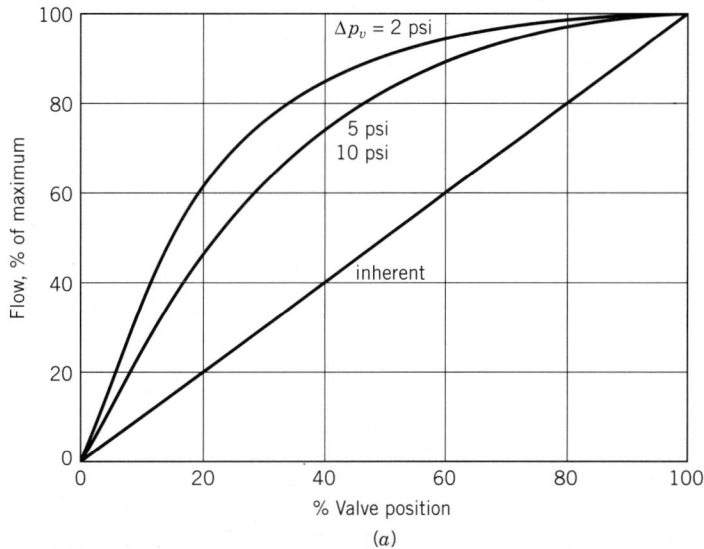

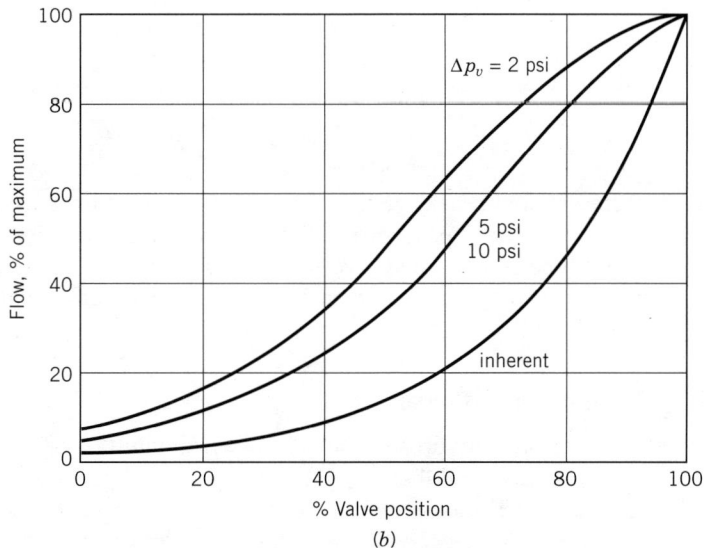

Figure 5-2.6 Installed flow characteristics of the control valve of Example 5-2.4. (*a*) Linear inherent characteristics. (*b*) Equal percentage characteristics with $\alpha = 50$.

size does not change. Notice how the installed characteristics for the linear valve turn into quick-opening characteristics, more so for the larger valve. By contrast, the installed characteristics for the equal percentage valve remain more linear, although they too flatten out at the high flows.

5-2.4   Control Valve Gain and Transfer Function

The gain of the valve, as that of any other device, is the steady-state change in output divided by the change in input. The valve schematic diagram of Fig. 5-2.1 shows that the output of the valve is the flow and its input is the controller output signal in percent controller output (%CO). The gain of the valve is therefore defined by

$$K_v = \frac{df}{dm} \frac{\text{gpm}}{\%\text{CO}} \tag{5-2.16}$$

The valve gain can also be defined in other units such as (lb/h)/(%CO), and scfh/(%CO). Using the chain rule of differentiation, we can show the valve gain as the product of three terms relating the dependence of the valve position on the controller output, the dependence of the C_v on the valve position, and the dependence of the flow on the C_v

$$K_v = \left(\frac{dvp}{dm}\right)\left(\frac{dC_v}{dvp}\right)\left(\frac{df}{dC_v}\right) \tag{5-2.17}$$

The dependence of the valve position is simply the conversion of percent controller output to fraction valve position, but the sign depends on whether the valve fails closed or opened

$$\frac{dvp}{dm} = \pm\frac{1}{100} \frac{\text{frn}.vp}{\%\text{CO}} \tag{5-2.18}$$

where the plus sign is used if the valve fails closed (air-to-open), and the minus sign if the valve fails opened (air-to-close).

The dependence of the C_v on the valve position depends on the valve characteristics. From Eqs. 5-2.6 and 5-2.7, for linear characteristics,

$$\frac{\overline{dC_v}}{dvp} = C_{v,\text{max}} \tag{5-2.19}$$

and for equal percentage,

$$\frac{\overline{dC_v}}{dvp} = (\ln\alpha)C_{v,\text{max}}\alpha^{\overline{vp}-1} = (\ln\alpha)\overline{C_v} \tag{5-2.20}$$

where the nonlinear exponential function has been linearized.

Finally, the dependence of flow on the C_v is a function of the installed characteristics of the control valve. We will consider first the simpler case of constant pressure drop across the valve and then the more complex case, which considers the pressure drop in the line in series with the valve.

Constant Valve Pressure Drop

When the pressure drop in the line in series with the valve is negligible, the inlet and outlet pressures and thus the valve pressure drop, remain constant. For liquid service, from Eq. 5-2.1, the dependence of flow on the C_v coefficient is

$$\frac{df}{dC_v} = \sqrt{\frac{\Delta p_v}{G_f}} \tag{5-2.21}$$

The gain of a valve with linear characteristics is now obtained by substituting Eqs. 5-2.18, 5-2.19, and 5-2.21 into Eq. 5-2.17.

$$K_v = \pm \frac{1}{100} C_{v,\max} \sqrt{\frac{\Delta p_v}{G_f}} = \frac{f_{\max}}{100} \frac{\text{gpm}}{\%\text{CO}} \qquad (5\text{-}2.22)$$

where $f_{\max}$ is the flow through the valve when it is fully opened. Notice that the gain of the linear valve is constant when the pressure drop across the valve is constant. It can be similarly shown, and left as an exercise, that the gain for either liquid or gas flow in mass units is

$$K_v = \pm \frac{w_{\max}}{100} \frac{\text{lb/h}}{\%\text{CO}} \qquad (5\text{-}2.23)$$

when the valve inlet and outlet pressures do not vary with flow.

The gain of a valve with equal percentage characteristics is similarly obtained by substituting Eqs. 5-2.18, 5-2.20, and 5-2.21 into Eq. 5-2.17.

$$K_v = \pm \frac{1}{100} (\ln \alpha) \overline{C_v} \sqrt{\frac{\Delta p_v}{G_f}} = \pm \frac{\ln \alpha}{100} \overline{f} \frac{\text{gpm}}{\%\text{CO}} \qquad (5\text{-}2.24)$$

This formula shows that the gain of an equal percentage valve is proportional to the flow when the pressure drop across the valve is constant. The gain for either liquid or gas flow in mass units is

$$K_v = \pm \frac{\ln \alpha}{100} \overline{w} \frac{\text{lb/h}}{\%\text{CO}} \qquad (5\text{-}2.25)$$

when the valve inlet and outlet pressures do not vary with flow.

Variable Pressure Drop across the Valve

To obtain the dependence of the flow on the C_v coefficient for liquid flow when the pressure drop across the valve is variable, differentiate Eq. 5-2.12 with respect to the C_v using the rules of differential calculus to obtain

$$\frac{\overline{df}}{dC_v} = \frac{\sqrt{1 + k_L \overline{C_v}^2} - \overline{C_v}(1 + k_L \overline{C_v}^2)^{-0.5} k_L \overline{C_v}}{1 + k_L \overline{C_v}^2} \sqrt{\frac{\Delta p_0}{G_f}}$$

$$= \left(1 + k_L \overline{C_v}^2 \right)^{-3/2} \sqrt{\frac{\Delta p_0}{G_f}} \qquad (5\text{-}2.26)$$

For a valve with linear characteristics, the expression in Eq. 5-2.26 is multiplied by the constant $\pm C_{v,\max}/100$ to obtain the valve gain. It is easy to see that the gain of the linear valve decreases as the valve opens, because of the increase in $\overline{C_v}$. For the equal percentage valve, the gain is obtained by substituting Eqs. 5-2.18, 5-2.20, and 5-2.26 into Eq. 5-2.17.

$$K_v = \pm \frac{\ln \alpha}{100} \frac{\overline{C_v}}{\left(1 + k_L \overline{C_v}^2 \right)^{3/2}} \sqrt{\frac{\Delta p_0}{G_f}}$$

$$= \pm \frac{\ln \alpha}{100} \frac{\overline{f}}{1 + k_L \overline{C_v}^2} \frac{\text{gpm}}{\%\text{CO}} \qquad (5\text{-}2.27)$$

where we have substituted Eq. 5-2.12. Notice that this gain is less variable with valve opening, because the flow term in the numerator tends to cancel some of the effect of the C_v term in the denominator, at least until the valve is near fully opened. This near linearity of the installed characteristics of the equal percentage valve can also be observed in the plots of Fig. 5-2.6b.

EXAMPLE 5-2.5

Gain of Steam Valve with Constant Pressure Drop

SOLUTION

Find the gain and the valve position at design conditions for the steam valve of Example 5-2.2. Assume that the 10-in. valve with $C_{v,max} = 1000$ is selected and that the pressures around the valve are independent of flow. Consider both a valve with linear characteristics and an equal percentage valve with rangeability parameter of 50. For the latter, find the gain at the nominal flow of 16,100 lb/h.

Being a steam valve, we will assume that it fails closed to prevent overheating the reboiler. Then, as the controller signal opens the valve, the valve gain is positive.

For the linear valve, the valve position at design flow is found from Eq. 5-2.6.

$$\overline{vp} = \frac{\overline{C_v}}{C_{v,max}} = \frac{440}{1000} = 0.440$$

The gain is obtained from Eq. 5-2.23.

$$K_v = +\frac{w_{max}}{100} = \frac{1}{100}(16,100)\left(\frac{1000}{440}\right) = 366\frac{lb/h}{\%CO}$$

where, since the pressures are constant, we have used the ratio of the C_v values to estimate the maximum flow.

For the equal percentage valve with $\alpha = 50$, the valve position is calculated using Eq. 5-2.7.

$$\alpha^{\overline{vp}-1} = \frac{\overline{C_v}}{C_{v,max}} = \frac{440}{1000} = 0.440$$

$$\overline{vp} = \frac{\ln(0.440)}{\ln(50)} + 1 = 0.79$$

The gain is obtained using Eq. 5-2.25.

$$K_v = +\frac{\ln\alpha}{100}\overline{w} = \frac{\ln(50)}{100}16,100 = 630\frac{lb/h}{\%CO}$$

As expected, the gain of the equal percentage valve at design flow is greater than the constant gain of the linear valve.

EXAMPLE 5-2.6

Valve Gain with Variable Pressure Drop

SOLUTION

Calculate the gain of the valve in Example 5-2.4 at nominal flow. Consider both a linear valve and an equal percentage valve with rangeability parameter $\alpha = 50$.

From Examples 5-2.3 and 5-2.4, we know that $\Delta p_0 = 11$ psi, $\overline{f} = 700$ gpm, $k_L = 13.0 \times 10^{-6}$ psi/(gpm)^2, $\overline{C_v} = 303$, and $C_{v,max} = 640$ gpm/(psi)$^{1/2}$. Since the valve feeds a distillation column, let us assume that it fails closed, so its gain is positive because the controller signal opens it. The gain of the linear valve with variable pressure drop is obtained by substituting Eq. 5-2.18, 5-2.19, and 5-2.26 into Eq. 5-2.17.

$$K_v = +\frac{1}{100}\frac{C_{v,max}}{\left(1 + k_L\overline{C_v}^2\right)^{3/2}}\sqrt{\frac{\Delta p_0}{G_f}}$$

$$= \frac{1}{100}\frac{640}{\left(1 + 13.0 \times 10^{-6}(303)^2\right)^{3/2}}\sqrt{\frac{11}{0.94}} = 6.7\frac{gpm}{\%CO}$$

This gain is less than half the gain of 15 gpm/%CO the valve would have if the pressure drop of 5 psi across the valve remained constant.

The gain of the equal percentage valve is obtained from Eq. 5-2.27.

$$K_v = +\frac{\ln(50)}{100}\frac{700}{1 + 13.0 \times 10^{-6}(303)^2} = 12.5\ \frac{\text{gpm}}{\text{\%CO}}$$

This gain is about half the gain of 27 gpm/%CO the valve would have if the pressure drop across the valve were independent of flow.

Valve Transfer Function

Figure 5-2.7 shows the block diagram for a control valve. It is usually sufficient to model the valve as a first-order lag, resulting in the following transfer function

$$G_v(s) = \frac{K_v}{\tau_v s + 1} \tag{5-2.28}$$

where

K_v = valve gain, gpm/%CO or (lb/h)/%CO or scfh/%CO

τ_v = time constant of valve actuator, minutes

The actuator time constant is usually of the order of a few seconds and can be neglected when the process time constants are of the order of minutes.

The block diagram of Fig. 5-2.7 assumes that the pressure drop across the valve is either constant or a function of flow only. When the pressure drop across the valve is a function of other process variables, as in the control of level or gas pressure, the block diagram must include the effect of these variables on the flow through the valve. Chapters 3 and 4 show examples of block diagrams in which level and pressure variables affect the flow through the valve.

5-2.5 Control Valve Summary

This section has presented some important considerations in the modeling and sizing of control valves. Although there are other considerations that must be taken when specifying a control valve, the formulas presented here allow the modeling of control valves for the purposes of designing and analyzing the complete control system. The reader who wants more details on the complete specification of control valves is directed to read the many fine references given at the end of this chapter.

5-3 FEEDBACK CONTROLLERS

This section presents the most important types of industrial controllers. Specifically, we will consider the different types of algorithms used in analog controllers and the most common ones used in Distributed Control Systems (DCSs) and in "stand-alone controllers," which are also sometimes referred to as "single-loop controllers," or just simply "loop-controllers." As presented in Chapter 1, the DCSs and the "stand-alone" controllers are computer-based, so they do not process the signals on a continuous basis but rather in a discrete fashion. However, the sampling time for these systems is

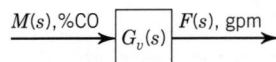

$M(s)$,%CO $\longrightarrow$ $\boxed{G_v(s)}$ $\xrightarrow{F(s),\ \text{gpm}}$

Figure 5-2.7 Block diagram of a control valve.

rather fast, usually ranging from 10 times a second to about once per second. Thus, for all practical purposes these controllers appear to be continuous.

Briefly, the controller is the "brain" of the control loop. As we noted in Chapter 1, the controller performs the decision (D) operation in the control system. To do this, the controller:

1. Compares the process signal it receives, the controlled variable, with the set point.

2. Sends an appropriate signal to the control valve, or any other final control element, in order to maintain the controlled variable at its set point.

The controllers have a series of buttons/windows that permits adjusting the set point, reading the value of the controlled variable, transferring between the automatic and manual modes, reading the output signal from the controller, and adjusting the output signal when in the manual mode.

The auto/manual mode determines the operation of the controller. When in the auto (automatic) mode, the controller decides on the appropriate signal and outputs it to the final control element to maintain the controlled variable at the set point. In the manual mode, the controller stops deciding and allows the operating personnel to change the output manually. In this mode, the controller just provides a convenient (and expensive) way to adjust the final control element. In auto, information from the manual adjustment is ignored, or disabled; only the set point influences the output. In manual, on the other hand, the set point has no effect on the controller output; only the manual output influences the output. When a controller is in manual, there is not much need for it. *Only when the controller is in automatic are the benefits of automatic process control obtained.*

5-3.1   Actions of Controllers

The selection of the controller action is critical. *If the action is not correctly selected, the controller will not control.* Let us see how to select the action and what it means.

Consider the heat exchanger control loop shown in Fig. 5-3.1; the process is at steady state, and the set point is constant. Assume that the signal from the temperature transmitter increases, indicating that the outlet temperature has increased above set point. To return this temperature to set point, the controller must close the steam valve by some amount. Since the valve is fail-closed (FC), the controller must reduce its output signal to the valve (see the arrows in the figure). When an increase in signal to the controller requires a decrease in controller output, the controller must be set to *reverse action*. Often the term *increase/decrease* is also used.

Alternatively, consider the level control loop shown in Fig. 5-3.2; the process is at steady state, and the set point is constant. Assume that the signal from the level transmitter increases, indicating that the level has increased above the set point. To return this level to set point, the controller must open the valve by some amount. Since the valve is fail-closed (FC), the controller must increase its output signal to the valve (see the arrows in the figure). When an increase in signal to the controller requires a increase in controller output, the controller must be set to *direct action*. Often the term *increase/increase* is also used.

In summary, to determine the action of a controller, the engineer must know:

1. The process requirements for control.

2. The fail-safe action of the control valve or other final control element.

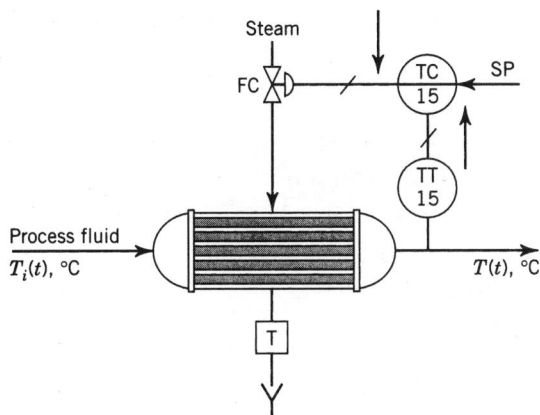

Figure 5-3.1 Heat exchanger control loop.

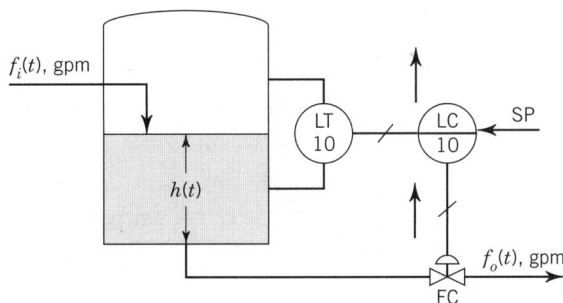

Figure 5-3.2 Liquid level control loop.

Both things must be taken into consideration. What should be the action of the level controller if a fail-open (FO) valve is used? And what should it be if the level is controlled with the inlet flow instead of the outlet flow? In the first case the control valve action changes, whereas in the second case the process requirement for control changes.

The controller action is usually set by a switch, by a configuration bit, or by answering a question while programming/configuring the controller.

5-3.2   Types of Feedback Controllers

The way feedback controllers make a decision is by solving an equation based on the difference between the controlled variable and the set point. In this section, we examine the most common types of controllers by looking at the equations that describe their operation.

As we saw in Chapter 1, the signals entering and exiting the controllers are either electrical or pneumatic. Even in computer systems the signals entering from the field are electrical before they are converted, by an analog-to-digital (A/D) converter, to digital values. Likewise, the signal the computer system sends back to the field is an electrical signal. To help simplify the presentation that follows, we will use all signals in percent. That is, we will speak of 0–100 % rather than 4–20 mA or 3–15 psig, or any other type of signal.

As we have said, feedback controllers decide what to do to maintain the controlled variable at set point by solving an equation based on the difference between the set point and the controlled variable. This difference, or error, is computed as

$$e(t) = r(t) - c(t) \tag{5-3.1}$$

where

$c(t)$ = controlled variable. Most often, the controlled variable is given by the transmitter output (TO) and consequently it has unit of %TO.

$r(t)$ = set point [$r(t)$ is used because originally the term *reference value* was used]. This is the desired value of the controlled variable and thus it has unit of %TO.

$e(t)$ = error, %TO.

The error could have also being computed as $e(t) = c(t) - r(t)$. However, Eq. 5-3.1 will be the convention used in this book.

Equation 5-3.1 is written in deviation variable form as

$$E(t) = R(t) - C(t) \tag{5-3.2}$$

where

$E(t)$ = error in deviation form. Assuming that the error at the initial steady state is zero, which is the convention used in this book, $E(t) = e(t) - 0$.

$R(t)$ = set point in deviation variable form. It is defined as $R(t) = r(t) - \bar{r}$ where $\bar{r}$ is the initial steady-state value of the set point.

$C(t)$ = controlled variable in deviation form. It is defined as $C(t) = c(t) - \bar{c}$ where $\bar{c}$ is the initial steady-state value of the controlled variable.

Taking the Laplace transform of Eq. 5-3.2 yields

$$E(s) = R(s) - C(s) \tag{5-3.3}$$

The conventional block diagram representation for the controller is shown in Fig. 5-3.3. $M(s)$ is the Laplace variable used to denote the controller output and therefore, it has units of percent controller output (%CO). $G_c(s)$ is the transfer function that describes how the controller acts on an error. The following paragraphs present the different controllers along with their transfer function.

Proportional Controller (P)

The proportional controller is the simplest type of controller we will discuss. The equation that describes its operation is

$$m(t) = \bar{m} + K_c e(t) \tag{5-3.4}$$

where

$m(t)$ = controller output, %CO. The term $m(t)$ is used to stress that as far as the controller is concerned, this output is the manipulated variable.

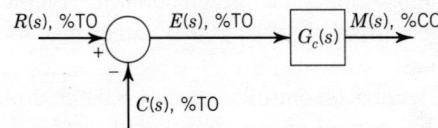

Figure 5-3.3 Block diagram representation of controller.

$$K_c = controller\ gain,\ \frac{\%CO}{\%TO}$$

$$\overline{m} = bias,\ \%CO.$$

$\overline{m}$ is the output from the controller when the error is zero; its value is a constant, and it is set equal to the output when the controller is switched to manual. It is very often initially set at mid-scale, 50 %CO.

Note that since the controlled variable is the signal from the transmitter with unit of %TO, the set point must also have unit of %TO. As the set point is entered in engineering units of the process variable, it is converted by the control system (controller) into %TO. This conversion is done using the transmitter range.

Equation 5-3.4 shows that the output of the controller is proportional to the error between the set point and the controlled variable. The proportionality is given by the controller gain, K_c. As a result of our definition of error, when K_c is positive, an increase in the controlled variable, $c(t)$, results in a decrease in controller output, $m(t)$. Thus, a positive K_c results in a reverse-acting controller. To obtain a direct-acting controller we must either use a negative K_c, or reverse the definition of the error, that is, $e(t) = c(t) - r(t)$. In this text we will use the definition of the error as in Eq. 5-3.1 and use a negative K_c when a direct-acting controller is required. Most industrial feedback controllers, however, do not allow negative gains; in such cases the error computation is reversed. The change in error calculation is done internally by the controller. The user does not have to do anything but select the correct action. Note that whatever definition is used, the effect of the set point on the output is opposite to the effect of the controlled variable.

The controller gain determines how much the output from the controller changes for a given change in error; this is illustrated graphically in Fig. 5-3.4. The figure shows that the larger the K_c value, the more the controller output changes for a given error. Thus K_c establishes the sensitivity of the controller to an error, that is, how much the controller output changes per unit change in error.

Proportional controllers have the advantage of only one tuning parameter, K_c. However, they suffer a major disadvantage: the controlled variable operates with an *offset*. Offset can be described as a *steady-state deviation of the controlled variable from set point*, or simply as a *steady-state error*. To examine the meaning of offset, consider the liquid level control loop shown in Fig. 5-3.2. The design operating conditions are $\overline{f}_i = \overline{f}_o = 150$ gpm, and $\overline{h} = 6$ ft. Let us also assume that in order for the outlet valve to deliver 150 gpm, the signal to it must be 50 %CO. If the inlet flow, $f_i(t)$, increases, then the response of the system with proportional controller looks like Fig. 5-3.5. The controller returns the controlled variable to a steady value, but not to the required set point. The difference between the set point and the new steady state is the offset. The proportional controller is not "intelligent enough" to drive the controlled variable back to set point. The new steady-state value satisfies the controller.

Figure 5-3.5 shows three response curves corresponding to three different values of K_c. This figure shows that the larger the value of K_c the smaller the offset. Why not, then, set a maximum gain to eliminate the offset? Figure 5-3.5 also shows that although the larger K_c reduces the offset, the more oscillatory the process becomes. For most processes there is a maximum value of K_c beyond which the process goes unstable. Thus there is a limit to the value at which we can set K_c while at the same time maintaining stability. Consequently, the offset cannot be completely eliminated. The calculation of this maximum value of the controller gain, referred to as the ultimate gain, K_{cu}, is presented in Chapters 6, 7, and 8.

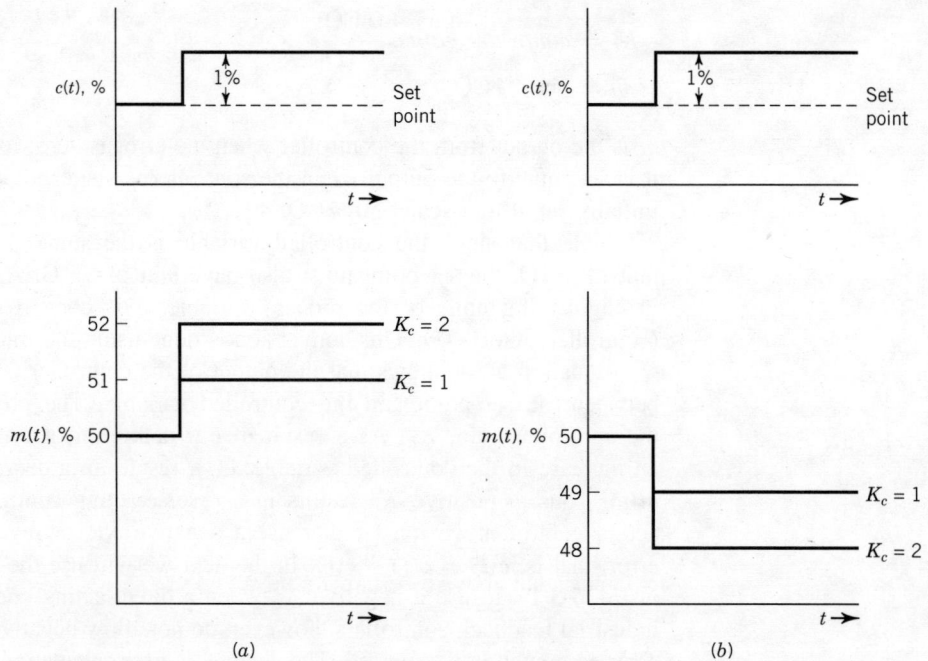

Figure 5-3.4 Effect of controller gain on output of controller. (*a*) Direct-acting controller, (*b*) Reverse-acting controller.

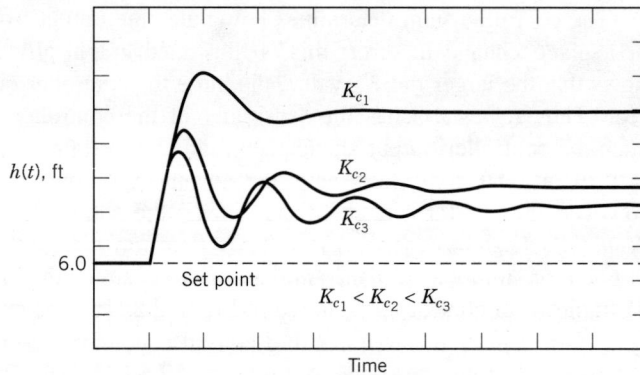

Figure 5-3.5 Response of liquid level process.

Let us now look at a simple explanation why offset exists; a more rigorous proof is given in Chapter 6. Consider the liquid level control system shown in Fig. 5-3.2 with the same operating conditions previously given, that is, $\overline{f}_i = \overline{f}_o = 150$ gpm, and $\overline{h} = 6$ ft. Recall that the proportional controller, direct acting $(-K_c)$, solves the following equation

$$m(t) = 50 \% + (-K_c)e(t) \tag{5-3.5}$$

Assume now that the inlet flow increases up to 170 gpm. When this happens, the liquid level increases and the controller increases its output to open the valve and bring the level back down. In order to reach a steady operation, the outlet flow must now be 170 gpm. To deliver this new flow, the outlet valve must be open more than before, when it needed to deliver 150 gpm. This is a fail-closed valve, so let us assume that the new required signal to the valve to deliver 170 gpm is 60 %. That is, the output from the controller must be 60 %. Looking back at Eq. 5-3.5, we note that the only way for the controller output to be 60 % is for the second term of the right-hand side to have a value of $+10$ % and for this to be so, the error term cannot be zero at steady state. *This required steady-state error is the offset!* Note that a negative error means that the controlled variable is greater than the set point. The actual level in feet can be calculated from the calibration of the level transmitter.

Two points need to be stressed in this example. First, the magnitude of the offset depends on the value of the controller gain. Because the total term must have a value of $+10$ %CO,

K_c	$e(\infty)$, Offset, % TO
1	10
2	5.0
4	2.5

As previously mentioned, the larger the gain, the smaller the offset. The reader must remember that above a certain K_c most processes go unstable. The controller equation does not show this, however; it will be discussed in Chapter 6.

Second, it seems that all a proportional controller is doing is reaching a steady-state operating condition. Once a steady state is reached, the controller is satisfied. The amount of deviation from set point, or offset, depends on the controller gain.

Many controller manufacturers do not use the term K_c for the tuning parameter; they use the term *proportional band*, PB. The relationship between gain and proportional band is given by

$$PB = \frac{100}{K_c} \tag{5-3.6}$$

In these cases, the equation that describes the proportional controller is written as

$$m(t) = \overline{m} + \frac{100}{PB} e(t) \tag{5-3.7}$$

PB is usually referred to as percent proportional band.

Equation 5-3.6 presents a most important fact. A large controller gain is the same as a low, or narrow, proportional band, and a low controller gain is the same as a large, or wide, proportional band. An increase in PB is similar to a decrease in K_c, resulting in a controller less sensitive to an error. A decrease in PB is similar to an increase in K_c, resulting in a more sensitive controller. K_c and PB are reciprocals, so care must be taken when tuning the controller.

Let us offer another definition of proportional band. *The proportional band is the error (expressed in percentage of the range of the controlled variable) required to move the output of the controller from its lowest to its highest value.* Consider the heat exchanger control loop shown in Fig. 5-3.1. The temperature transmitter has a range from 100 to 300°C, and the set point of the controller is at 200°C. Figure 5-3.6 gives a graphical explanation of this definition of PB. The figure shows that a 100 % PB means that as the controlled variable varies by 100 % of its range, the controller output varies by 100 % of its range. A 50 % PB means that as the controlled variable varies by 50 % of its range, the controller output varies by 100 % of its range. Also note that a proportional only controller with a 200 % PB will not move its output the entire range. A 200 % PB means a very small controller gain or controller sensitivity to errors.

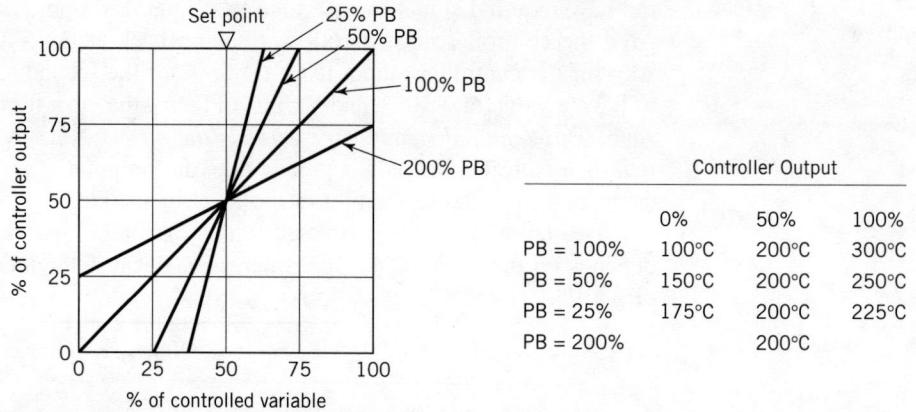

Figure 5-3.6 Definition of proportional band.

To obtain the transfer function for the proportional controller, we can write Eq. 5-3.1 as

$$m(t) - \overline{m} = K_c e(t)$$

or in deviation variable form,

$$M(t) = K_c E(t)$$

where $M(t) = m(t) - \overline{m}$ and $E(t)$ is as previously defined. Taking the Laplace transform yields the following transfer function

$$G_c(s) = \frac{M(s)}{E(s)} = K_c \qquad (5\text{-}3.8)$$

Equation 5-3.8 is the transfer function of a proportional controller and the one to use in Fig. 5-3.3 when this controller is used.

To briefly summarize, proportional controllers are the simplest controllers and offer the advantage of only one tuning parameter, K_c or PB. The disadvantage of these controllers is the operation with an offset in the controlled variable. In some processes, such as the level in a surge tank, the cruise control in a car, or a thermostat in a house, this may not be of any major consequence. In cases in which the process can be controlled within a band from set point, proportional controllers are sufficient. However, when the process variable must be controlled at the set point, not near it, proportional controllers do not provide the required control.

Proportional-Integral Controller (PI)

Most processes cannot be controlled with an offset; that is, they must be controlled at the set point. In these instances, an extra amount of intelligence must be added to the proportional controller to remove the offset. This new intelligence, or new mode of control, is the integral, or reset, action; consequently, the controller becomes a proportional-integral (PI) controller. The describing equation is

$$m(t) = \overline{m} + K_c e(t) + \frac{K_c}{\tau_I} \int e(t)\, dt \qquad (5\text{-}3.9)$$

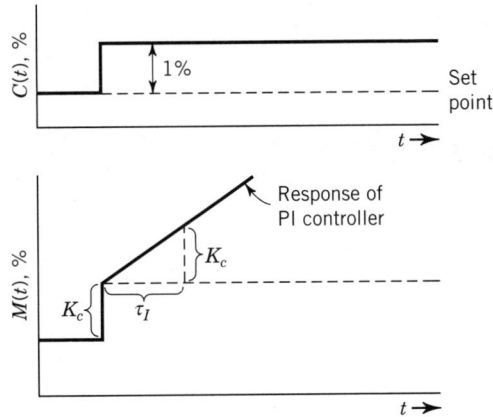

Figure 5-3.7 Response of PI controller (direct response) to a step change in error.

where $\tau_I = $ *integral (or reset) time*. Most often, the time unit used is minutes; less often seconds are used. The unit used depends on the manufacturer. Therefore, the PI controller has two parameters, K_c and τ_I, both of which must be adjusted (tuned) to obtain satisfactory control.

To understand the physical significance of the reset time, τ_I, consider the hypothetical example shown in Fig. 5-3.7. At some time, $t = 0$, a constant error of 1 % in magnitude is introduced in the controller. At this moment the PI controller solves the following equation

$$m(t) = 50\ \% + K_c(1) + \frac{K_c}{\tau_I} \int_0^t (1)\, dt$$

or

$$m(t) = 50\ \% + K_c + \frac{K_c}{\tau_I}\, t$$

When the error is introduced at $t = 0$, the controller output changes immediately by an amount equal to K_c; this is the response due to the proportional mode. As time increases, the output also increases in a ramp fashion as expressed by the equation and shown in the figure. Note that when $t = \tau_I$ the controller's output becomes

$$m(t) = 50\ \% + K_c + K_c$$

Thus, in an amount of time equal to τ_I, the integral mode repeats the action taken by the proportional mode. The smaller the value of τ_I, the faster the controller integrates. Realize that the smaller the value of τ_I, the larger the term in front of the integral, K_c/τ_I, and consequently, the more weight is given to the integral term.

To understand why the PI controller removes the offset, consider the level control system previously used to explain the offset required by a P controller. Figure 5-3.8 shows the response of the level under P and PI controllers to a change in inlet flow from 150 to 170 gpm. The response with a P controller shows the offset, whereas the response with a PI controller shows that the level returns to set point, with no offset. Under PI control, as long as the error is present, the controller keeps changing its output (integrating the error). Once the error disappears (goes to zero), the controller does not change its output anymore (it integrates a function with a value of zero). As shown in the figure, at time t_f the error disappears. The signal to the valve must still be 60 %, requiring the valve to deliver 170 gpm. Let us look at the PI equation at the

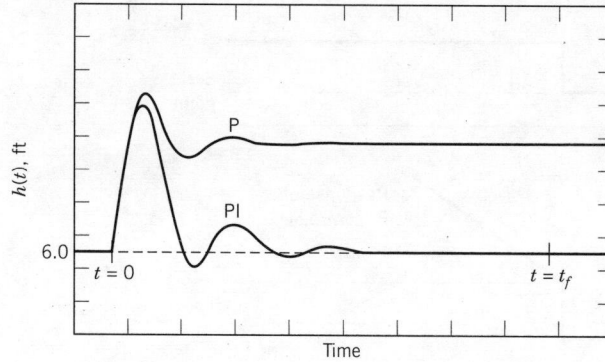

Figure 5-3.8 Response of liquid level process under P and PI controllers.

moment the steady state is reached.

$$m(t) = 50\,\% + K_c(0) + \frac{K_c}{\tau_I} \int (0)\, dt$$

or

$$m(t) = 50\,\% + 0\,\% + 10\,\% = 60\,\%$$

The equation shows that even with a "zero" error, the integral term is not "zero" but rather 10 %, which provides the required output of 60 %. The fact that the error is zero does not mean that the value of the integral term is zero. It means that the integral term remains constant at the last value! Integration means area under the curve, and even though the level is the same at $t = 0$ and at $t = t_f$, the value of the integral is different (a different area under the curve) at these two times. The value of the integral term times K_c/τ_I is equal to 10 %. Once the level returns to set point, the error disappears and the integral term remains constant. *Integration is the mode that removes the offset!*

This has been a brief explanation of why reset action removes the offset; Chapter 6 provides a more rigorous proof.

Some manufacturers do not use the term reset time, τ_I, for their tuning parameter. They use the reciprocal of reset time, which we shall referred to it as *reset rate*, τ_I^R; that is,

$$\tau_I^R = \frac{1}{\tau_I} \tag{5-3.10}$$

The unit of τ_I^R is therefore 1/time, or simply (time)$^{-1}$. Note that when τ_I is used and faster integration is desired, a smaller value must be used in the controller. However, when τ_I^R is used, a larger value must be used. Therefore, before tuning the reset term, the user must know whether the controller uses reset time (time) or reset rate (time^{-1}). τ_I and τ_I^R are reciprocals, so their effects are opposite.

As we learned in the previous section, two terms are used for the proportional mode (K_c and PB), and we have just learned that there are also two terms for the integral mode (τ_I and τ_I^R). This can be confusing, so it is important to keep the differences in mind when tuning a controller. Equations 5-3.9, 5-3.11, 5-3.12, and 5-3.13 show four possible combinations (Eq. 5-3.24 in Section 5-3.3 presents still another combination) of tuning parameters; we refer to Eq. 5-3.9 as the *classical controller*.

$$m(t) = \overline{m} + \frac{100}{PB} e(t) + \frac{100}{PB \cdot \tau_I} \int e(t)\, dt \tag{5-3.11}$$

$$m(t) = \overline{m} + \frac{100}{\text{PB}}e(t) + \frac{100\,\tau_I^R}{\text{PB}} \int e(t)\,dt \qquad (5\text{-}3.12)$$

$$m(t) = \overline{m} + K_c\, e(t) + K_c\tau_I^R \int e(t)\,dt \qquad (5\text{-}3.13)$$

Using the same procedure we followed for the proportional controller, we obtain the following transfer function for the PI controller from Eq. 5-3.9.

$$G_c(s) = \frac{M(s)}{E(s)} = K_c\left(1 + \frac{1}{\tau_I s}\right) = K_c\left(\frac{\tau_I s + 1}{\tau_I s}\right) \qquad (5\text{-}3.14)$$

To summarize, proportional-integral controllers have two tuning parameters: the gain or proportional band, and the reset time or reset rate. *The advantage is that the integration removes the offset.* Close to 85 % of all controllers in use are of this type.

Proportional-Integral-Derivative Controller (PID)

Sometimes another mode of control is added to the PI controller. This new mode of control is the derivative action, which is also called the rate action, or preact. Its purpose is to anticipate where the process is heading by looking at the time rate of change of the error, its derivative. The describing equation is

$$m(t) = \overline{m} + K_c e(t) + \frac{K_c}{\tau_I} \int e(t)\,dt + K_c\tau_D\frac{de(t)}{dt} \qquad (5\text{-}3.15)$$

where $\tau_D = $ *derivative (or rate) time.* Most often the time unit used is minutes, however, some manufacturers use seconds.

The PID controller has three terms, K_c or PB, τ_I or τ_I^R, and τ_D, that must be adjusted (tuned) to obtain satisfactory control. The derivative action gives the controller the capability to anticipate where the process is heading—that is, to "look ahead"—by calculating the derivative of the error. The amount of "anticipation" is decided by the value of the tuning parameter, τ_D.

Let us consider the heat exchanger shown in Fig. 5-3.1 and use it to clarify what is meant by anticipation. Assume that the inlet process temperature decreases by some amount and the outlet temperature starts to decrease correspondingly, as shown in Fig. 5-3.9. At time t_a, the amount of the error is positive and small. Consequently, the amount of control correction provided by the proportional and integral modes is small. However, the derivative of this error, the slope of the error curve, is large and positive, making the control correction provided by the derivative mode large. By looking at the derivative of the error, the controller knows that the controlled variable is heading away from set point rather fast and it uses this fact to help in controlling. At time t_b, the error is still positive and is larger than before. The amount of control correction provided by the proportional and integral modes is also larger than before and still adding to the output of the controller to open the steam valve further. However, the derivative of the error at this time is negative, signifying that the error is decreasing; the controlled variable has started to come back to set point. Using this fact, the derivative mode starts to subtract from the other two modes, because it recognizes that the error is decreasing. This algorithm results in reduced overshoot and oscillations around set point.

PID controllers are recommended to be used in slow processes (processes with long time constants), such as temperature loops, which are usually free of noise. Fast processes (processes with short time constants) are easily susceptible to process noise.

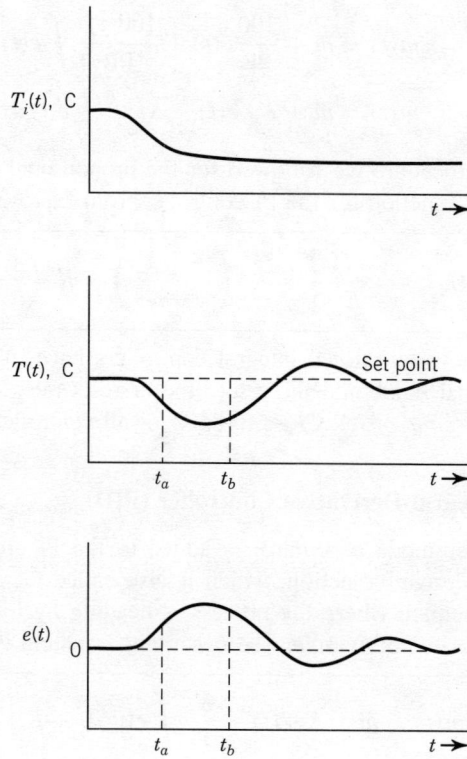

Figure 5-3.9 Heat exchanger control.

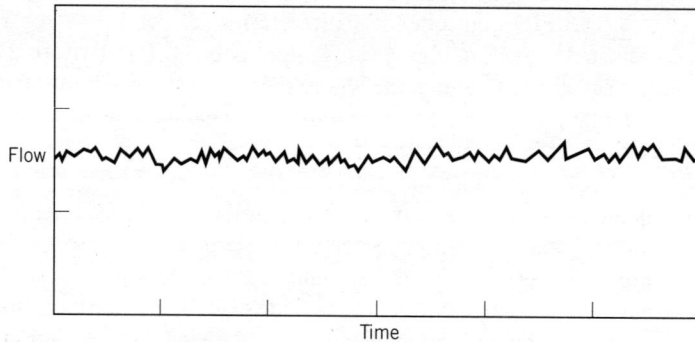

Figure 5-3.10 Recording of flow.

Typical of these fast processes are flow loops and liquid pressure loops. Consider the recording of a flow shown in Fig. 5-3.10. The application of the derivative mode will only result in the amplification of the noise, because the derivative of the fast changing noise is a large value. Processes with long time constants are usually damped and, consequently, are less susceptible to noise.

The transfer function of a PID controller is obtained using the same procedure followed for P and PI controllers

$$G_c(s) = \frac{M(s)}{E(s)} = K_c \left(1 + \frac{1}{\tau_I s} + \tau_D s \right) \tag{5-3.16}$$

Actually, when the PID controller is implemented with Eq. 5-3.16 it does not work very well. To improve the performance of the derivative mode the algorithm is slightly changed to

$$G_c(s) = \frac{M(s)}{E(s)} = K_c \left[1 + \frac{1}{\tau_I s} + \frac{\tau_D s}{\alpha \tau_D s + 1} \right] \qquad \textbf{(5-3.17)}$$

The equation shows that the derivative portion is multiplied by the term $1/(\alpha \tau_D s + 1)$. This term, which can be recognized as the transfer function of a first-order system with gain of unity and a time constant equal to $\alpha \tau_D$, is referred to as a filter. The filter does not usually affect the performance of the controller because its time constant, $\alpha \tau_D$, is small. Typical values of α range between 0.05 and 0.2, depending on the manufacturer.

Equation 5-3.17 can be algebraically rearranged into

$$G_c(s) = \frac{M(s)}{E(s)} = K_c \left[\frac{(\alpha + 1)\tau_D s + 1}{\alpha \tau_D s + 1} + \frac{1}{\tau_I s} \right] \qquad \textbf{(5-3.18)}$$

The term $[(\alpha + 1)\tau_D s + 1]/(\alpha \tau_D s + 1)$ is a lead-lag unit which was introduced in Chapter 2 and is further discussed in Chapter 11. This transfer function shows the PID controller as a lead-lag unit in parallel with integration.

In analog controllers and many computer-based controllers the describing transfer function for the PID controllers used is

$$G_c(s) = \frac{M(s)}{E(s)} = K_c' \left(1 + \frac{1}{\tau_I' s} \right) \left(\frac{\tau_D' s + 1}{\alpha \tau_D' s + 1} \right) \qquad \textbf{(5-3.19)}$$

Figure 5-3.11 shows the block diagram of Eq. 5-3.19. The diagram shows that this PID controller can be considered as a lead-lag unit in series with a PI controller, sometimes referred to as "rate-before-reset."

In Eq. 5-3.19 the prime notation has been used to indicate that the tuning parameters are not the same as those in Eq. 5-3.16 or 5-3.17. Using algebraic manipulations with Eq. 5-3.16 and 5-3.19 the following relations can be obtained

$$K_c' = K_c \left(0.5 + \sqrt{0.25 - \frac{\tau_D}{\tau_I}} \right)$$

$$\tau_I' = \tau_I \left(0.5 + \sqrt{0.25 - \frac{\tau_D}{\tau_I}} \right) \qquad \textbf{(5-3.20)}$$

$$\tau_D' = \frac{\tau_D}{0.5 + \sqrt{0.25 - \frac{\tau_D}{\tau_I}}}$$

Chapter 6 shows how to obtain the tuning parameters K_c, τ_I, τ_D, K_c', τ_I', and τ_D'.

The controller described by Eq. 5-3.16 is sometimes referred to as an "ideal" PID, whereas the controller described by Eq. 5-3.19 is referred to as an actual PID.

To summarize, PID controllers have three tuning parameters: the gain or proportional band, the reset time or reset rate, and the rate time. PID controllers are

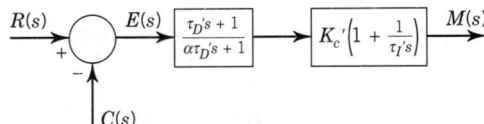

Figure 5-3.11 Block diagram of PID controller—Eq. 5-3.19.

recommended for processes that are free of noise. The advantage of the derivative mode is that it provides anticipation.

Proportional-Derivative Controller (PD)

This controller is used in processes where a proportional controller can be used, where steady-state offset is acceptable, but where some amount of anticipation is desired and no noise is present. The describing equation is

$$m(t) = \overline{m} + K_c e(t) + K_c \tau_D \frac{de(t)}{dt} \qquad (5\text{-}3.21)$$

and the "ideal" transfer function is

$$G_c(s) = \frac{M(s)}{E(s)} = K_c(1 + \tau_D s) \qquad (5\text{-}3.22)$$

whereas the "actual," or implemented, transfer function is

$$G_c(s) = \frac{M(s)}{E(s)} = K_c \left[\frac{(\alpha + 1)\tau_D s + 1}{\alpha \tau_D s + 1} \right] \qquad (5\text{-}3.23)$$

5-3.3   Modifications to the PID Controller and Additional Comments

Section 5-3.2 pointed out the differences in tuning parameters, K_c versus PB, and τ_I versus τ_I^R. It is unfortunate that there is no one single set, but it is a fact, and the engineer must be aware of the differences. There is yet one more set of parameters used by some manufacturers, shown as

$$m(t) = \overline{m} + K_c e(t) + K_I \int e(t)\,dt + K_D \frac{de(t)}{dt} \qquad (5\text{-}3.24)$$

The three tuning parameters are in this case K_c, K_I, and K_D.

There are other common modifications found in some controllers. Figure 5-3.12a shows the usual way to introduce a set point change. When this takes place, a step change in error is also introduced, as shown in Fig. 5-3.12b. Because the derivative calculation is based on the error, this calculation results in a drastic change in controller output (see Fig. 5-3.12c). Such a change is unnecessary and often detrimental to the process operation. The most common way to avoid this problem is to use the negative of the derivative of the controlled variable, $-dc(t)/dt$, instead of the derivative of the error. That is,

$$m(t) = \overline{m} + K_c e(t) + \frac{K_c}{\tau_I} \int e(t)\,dt - K_c \tau_D \frac{dc(t)}{dt} \qquad (5\text{-}3.25)$$

The response of both derivatives is the same when the set point is constant

$$\frac{de(t)}{dt} = \frac{d[r(t) - c(t)]}{dt} = \frac{dr(t)}{dt} - \frac{dc(t)}{dt}$$

Under constant set point the first derivative term on the right hand is zero and thus,

$$\frac{de(t)}{dt} = -\frac{dc(t)}{dt}$$

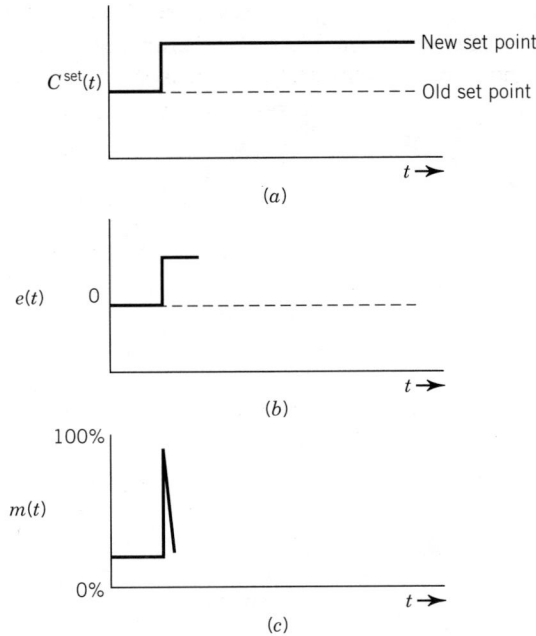

$C^{set}(t)$ —— New set point
- - - - - - - - Old set point
(a)

$e(t)$ 0
(b)

100%
$m(t)$
0%
(c)

Figure 5-3.12 Effect of set point changes in PID controllers.

At the moment the set point change is introduced, the "new" derivative does not produce the drastic response. Shortly after, the responses of the two derivatives become the same again. The Laplace transform of Eq. 5-3.25 is given by Eq. 5-3.26.

$$M(s) = K_c \left(\left(1 + \frac{1}{\tau_I s} \right) E(s) - \tau_D C(s) \right) \qquad (5\text{-}3.26)$$

or

$$M(s) = K_c \left(\left(1 + \frac{1}{\tau_I s} \right) E(s) - \frac{\tau_D s}{\alpha \tau_D s + 1} C(s) \right) \qquad (5\text{-}3.27)$$

or

$$M(s) = K_c' \left(1 + \frac{1}{\tau_I' s} \right) \left[R(s) - \frac{\tau_D' s + 1}{\alpha \tau_D' s + 1} C(s) \right] \qquad (5\text{-}3.28)$$

Figure 5-3.13 shows the block diagrams of Eqs. 5-3.27 and 5-3.28. This modification is commonly referred to as derivative-on-process variable.

The algorithm given in Eq. 5-3.25 drastically reduces the undesirable effect of set point changes on the response of the algorithm. However, the proportional term,

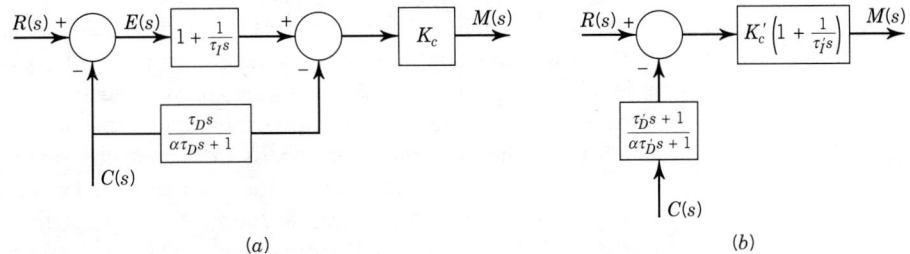

Figure 5-3.13 Block diagrams of derivative-on-process variable PID controller. (a) Eq. 5-3.27, (b) Eq. 5-3.28.

$K_c e(t) = K_c[r(t) - c(t)]$, still provides a sudden response when the set point is changed. This sudden change in response due to the proportional mode is referred to as *proportional kick*. Under some circumstances, such as for large values of K_c, this response may also be detrimental to the process operation. The following modification is sometimes proposed

$$m(t) = \overline{m} - K_c c(t) + \frac{K_c}{\tau_I} \int e(t)\,dt - K_c \tau_D \frac{dc(t)}{dt} \qquad (5\text{-}3.29)$$

The algorithms shown in Eqs. 5-3.15, 5-3.25, and 5-3.29 are different in their response to set point changes; however, their responses are the same for disturbances.

Another modification to the basic PID algorithm is the one in which the control calculation is based on the square of the error, or

$$m(t) = \overline{m} + K_c \left[|e(t)| e(t) + \frac{1}{\tau_I} \int e(t)\,dt + \tau_D \frac{de(t)}{dt} \right] \qquad (5\text{-}3.30)$$

The basic idea is that when the error is small, not much corrective action is needed. When $e(t)$ is small, $|e(t)|\, e(t)$ is smaller and not much action is obtained. However, when $e(t)$ is large, significant corrective action is needed to return to set point; in this case $|e(t)|\, e(t)$ is larger and provides the required action. Error-squared controllers are usually difficult to tune. They have shown some advantages in controlling integrating processes such as level loops.

The PID-gap controller, also referred as the dead-band controller, is not really a modification to the basic PID algorithm. In this controller, as long as the controlled variable is within some prescribed gap, or band, from set point (say ± 1 %, ± 3 %, or the like), no action is taken. The rationale is that these small deviations are due only to noise and are not really process deviations, so there is no need to take corrective action. Outside the prescribed band, the controller works as usual.

Let us now look at another option in controllers. Consider the heat exchanger of Fig. 5-3.1. The temperature controller is in automatic and controlling at set point, say, at $140°$C. That is, both the set point and controlled variable are at $140°$C. Now for some reason, the operator or engineer sets the controller in manual and increases the controller output. This action opens the valve, permitting more steam into the heat exchanger. As a result, the temperature increases to a new value, say, $150°$C. If the controller is then transferred to automatic, it will see an error, because the set point is still at $140°$C and the controlled variable is now at $150°$C. The controller, of course, suddenly closes the valve to correct for the deviation. This sudden change in signal to the valve represents a "bump" to the process and in some cases may be detrimental to the operation. If a *bumpless transfer* is desired when transferring from manual to automatic, the error must be zero; that is, the set point and controlled variable must be equal. The error can be made zero by either manually reducing the controller output to bring the temperature back to $140°$C or by increasing the set point to $150°$C to match the temperature. Once either action is taken, a bumpless transfer results when the controller is transferred to automatic. Computer-based controllers offer a standard option called *tracking*, or specifically in this case *process variable tracking (PV-tracking)*, which allows a bumpless transfer automatically. If this option is selected, whenever the controller is in manual the set point is forced to be equal to the controlled variable; that is, the set point tracks the controlled variable. This action results in a zero error while the controller is in manual and, therefore, at the moment of its transfer back to automatic. Once the controller is in automatic, the set point remains at the new value, $150°$C in the given example, not at the original value, $140°$C. Note that PV-tracking is an option and does not have to be selected when programming/configuring the controller. The tracking options are quite useful in control strategies for safety and improved performance.

5-3.4   Reset Windup and Its Prevention

The problem of reset windup is an important and realistic one in process control. It may occur whenever a controller contains the integral mode of control. Let us use the heat exchanger control loop shown in Fig. 5-3.1 to explain this problem.

Suppose that the process inlet temperature drops by an unusually large amount. This disturbance will reduce the outlet temperature. The controller (PI or PID) will in turn ask the steam valve to open. Because this is a fail-closed valve, the signal from the controller will increase until, because of the reset action, the outlet temperature equals the desired set point. But suppose that in restoring the controlled variable to set point, the controller integrates up to 100 % because the drop in inlet temperature is too large. At this point the steam valve is wide open, so the control loop cannot do any more. Essentially, the process is out of control. This is demonstrated graphically in Fig. 5-3.14, which shows that when the valve is fully open, the outlet temperature is not at set point. Because there is still an error, the controller will try to correct for it by further increasing (integrating the error) its output, even though the valve will not open more after 100 %. The output of the controller can in fact integrate

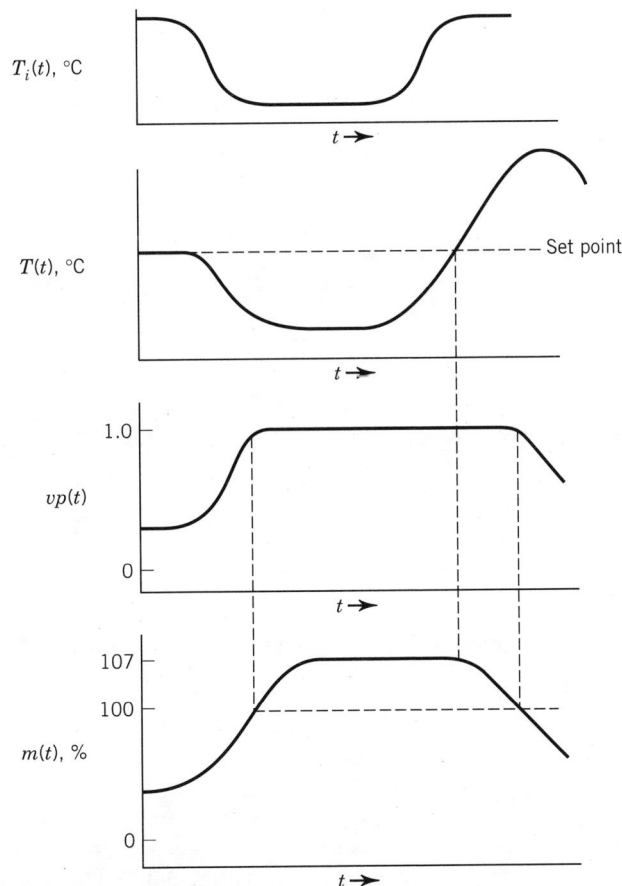

Figure 5-3.14 Heat exchanger control—reset windup.

above 100 %. Some controllers can integrate between -15 and 115 %, others between -7 and 107 %, and still others between -5 and 105 %. Analog controllers can also integrate outside their limits of 3–15 psig, or 4–20 mA. Let us suppose the controller being used can integrate up to 107 %. At that point, the controller cannot increase its output anymore; its output has saturated. This state is also shown in Fig. 5-3.14. This saturation is due to the reset action of the controller and is referred to as *reset windup*.

Suppose now that the inlet temperature goes back up; the outlet process temperature will in turn start to increase, as also shown in Fig. 5-3.14. This figure shows that the outlet temperature reaches and passes the set point and the valve remains wide open when, in fact, it should be closing. The valve is not closing because the controller must integrate from 107 % down to 100 % before it starts to close. By the time this happens, the outlet temperature has overshot the set point by a significant amount.

As we have said, this problem of reset windup may occur whenever integration is present in the controller. It can be avoided if the controller is set in manual as soon as its output reaches 100 % (or 0 %); this action will stop the integration. The controller can be set back to automatic when the temperature starts to decrease (or increase) again. The disadvantage of this operation is that it requires the operator's attention. Note that the prevention of reset windup requires stopping the integration, not limiting the controller output, when the controller reaches the 0 % or 100 % limit. Figure 5-3.15 shows a limiter on the output of the controller that does not prevent windup. Although the output does not go beyond the limits, the controller may still be internally wound up, because it is the integral mode that winds up.

There is an ingenious way to limit the integration when the controller output reaches its limits. Consider the PI controller transfer function

$$M(s) = K_c \left[1 + \frac{1}{\tau_I s} \right] E(s)$$

or

$$M(s) = K_c E(s) + M_I(s) \tag{5-3.31}$$

where

$$M_I(s) = \frac{K_c}{\tau_I s} E(s)$$

or

$$\tau_I s M_I(s) = K_c E(s) \tag{5-3.32}$$

From Eq. 5-3.31 we get

$$K_c E(s) = M(s) - M_I(s) \tag{5-3.33}$$

Equating Eqs. 5-3.32 and 5-3.33 and rearranging yield

$$M_I(s) = \frac{1}{\tau_I s + 1} M(s) \tag{5-3.34}$$

The implementation of Eqs. 5-3.31 and 5-3.34 is shown in Fig. 5-3.16. When the limiter is placed as shown in the figure, $M_I(s)$ will be automatically limited. $M_I(s)$ is

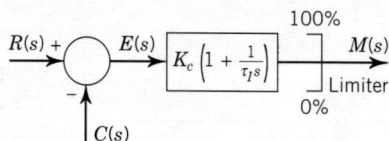

Figure 5-3.15 Limiting controller output.

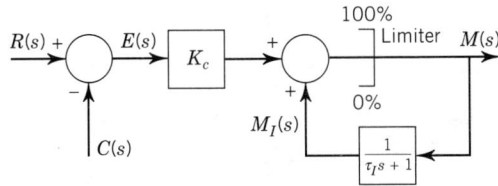

Figure 5-3.16 Reset-feedback implementation block diagram.

always lagging $M(s)$ with a gain of 1 and an adjustable parameter τ_I, so it can never get outside the range within which $M(s)$ is limited. In other words, if $M(s)$ reaches one of its limits $M_I(s)$ will approach that limit, say, 100 %. Then, at the moment the error turns negative the controller output becomes

$$m(t) = 100 + K_c e(t) < 100 \% \qquad as \ e(t) < 0$$

That is, the controller output will come off the limit, closing the valve, the instant the controlled variable crosses the set point!

Note that at steady state, the error is zero

$$M(s) = M_I(s) = M_I(s) + K_c E(s)$$

and for this to be true, $E(s) = 0$. Thus there is no offset. This way to implement this reset windup protection is commonly referred to as *reset feedback* (RFB).

Reset windup protection is an option that must be bought in analog controllers. It is a standard feature in any computer-based controller.

Reset windup occurs any time a controller is not in charge, such as when a manual bypass valve is open or when there is insufficient manipulated variable power. It also typically occurs in batch processes, in cascade control, and when more than one controller, as in override control schemes, drives a final control element. Cascade control is presented in Chapter 9, and override control is presented in Chapter 10.

5-3.5 Feedback Controller Summary

This section has presented the subject of process controllers. The purpose of the controllers is to adjust the manipulated variable to maintain the controlled variable at set point. We considered the significance of, and how to choose, the action of the controller (reverse or direct). The different types of controllers were presented, including the significance of the tuning parameters gain (K_c) or proportional band (PB), reset time (τ_I) or reset rate (τ_I^R), and rate time (τ_D). Finally, the subject of reset windup was presented and its significance discussed.

We have not discussed yet the important subject of obtaining the optimum setting of the tuning parameters. "Tuning the controller" is presented in Chapter 7.

5-4 SUMMARY

In this chapter, we looked at some of the hardware necessary to build a control system. The chapter began with a brief look at some terms related to sensors and transmitters and a discussion of the parameters that describe these devices. It continued with some important considerations about control valves, such as fail action, sizing, and characteristics. The reader is referred to Appendix C for more information on sensors, transmitters, and valves.

A discussion of feedback process controllers followed. The four most common types of controllers were presented, along with some modifications. The physical significance of their parameters was explained. The tuning of these parameters is presented in Chapter 7.

We are now ready to apply what we have learned in the first five chapters of this book to design process control systems.

REFERENCES

1. *Control Valve Handbook*, Marshalltown, Iowa: Fisher Controls Co.

2. *Masoneilan Handbook for Control Valve Sizing*, Norwood, MA: Masoneilan International, Inc.

3. *Fisher Catalog 10*, Marshalltown, Iowa: Fisher Controls Co.

PROBLEMS

5-1. Calculate the gain, in percent transmitter output (%TO) per variable unit (specify units), write the transfer function, and draw the block diagram, for each of the following cases:

(a) A temperature transmitter with a range of 100 to 150°C, and a time constant of 1.2 min.

(b) A temperature transmitter with a range of 100 to 350°F and a time constant of 0.5 min.

(c) A pressure transmitter with a range of 0 to 50 psig and a time constant of 0.05 min.

(d) A level transmitter with a range of 0 to 8 ft and a negligible time constant.

(e) A flow transmitter, consisting of a differential pressure transmitter measuring the pressure drop across an orifice sized for a maximum flow of 750 gpm, when the flow is 500 gpm. The time constant is negligible.

5-2. Liquid levels in storage tanks are frequently determined by measuring the pressure at the bottom of the tank. In one such tank the material stored in the tank was changed and an overflow resulted. Why? (Copyright 1992 by the American Institute of Chemical Engineers; reproduced by permission of Center for Chemical Process Safety of AIChE.)

5-3. An operator was told to control the temperature of a reactor at 60°C. The operator set the set point of the temperature controller at 60. The scale actually indicated 0–100 % of a temperature range of 0–200°C. This caused a runaway reaction which overpressurized the vessel. Liquid was discharged and injured the operator. What was the set point temperature the operator actually set? (Copyright 1992 by the American Institute of Chemical Engineers; reproduced by permission of Center for Chemical Process Safety of AIChE.)

5-4. Specify the proper fail-safe action for the valves in the following services. Specify either fail-open or fail-close.

(a) A flammable solvent is heated by steam in a heat exchanger. The valve manipulates the flow of steam to the exchanger.

(b) A valve manipulates the flow rate of reactant to a reactor vessel. The reaction is exothermic.

(c) A valve manipulates the flow rate of reactant to a reactor vessel. The reaction is endothermic.

(d) A valve manipulates the flow of natural gas (combustible) to a furnace. Another valve manipulates the flow of combustion air to the same furnace.

(Copyright 1992 by the American Institute of Chemical Engineers; reproduced by permission of Center for Chemical Process Safety of AIChE.)

5-5. Size a control valve to regulate the flow of 50 psig saturated steam to a heater. The nominal flow is 1200 lb/h and the outlet pressure is 5 psig.

(a) Obtain the C_v coefficient for 50 % overcapacity (assume $C_f = 0.8$),

(b) Obtain the valve gain in (lb/h)/%CO (assume the valve is linear with constant pressure drop).

5-6. The nominal liquid flow through a control valve is 52,500 lb/h, and the required maximum flow is 160,000 lb/h. Operating conditions call for an inlet pressure of 229 psia and an outlet pressure of 129 psia. At the flowing temperature of 104°F the liquid has a vapor pressure of 124 psia, a specific gravity of 0.92, and a viscosity of 0.2 cp. The critical pressure of the liquid is 969 psia. (See Appendix C for sizing formulas for flashing liquids.) Obtain the C_v coefficient for the valve.

5-7. A control valve is to regulate the flow of a gas with a molecular weight of 44. Process design conditions call for a nominal flow of 45,000 scfh, inlet pressure and temperature of 110 psig and 100°F, respectively, and an outlet pressure of 11 psig.

(a) Obtain the C_v coefficient for 100 % overcapacity (assume $C_f = 0.85$),

(b) Obtain the valve gain in scfh/%CO (assume the valve is linear with constant pressure drop).

5-8. You are asked to design a control valve to regulate the flow of benzene in the line shown in Fig. P5-1. The process design calls for a nominal flow of 100,000 kg/h and a temperature of 155°C. At the design flow, the frictional pressure drop in the line between points 1 and 2 is 100 kPa. The density of benzene at the flowing temperature is 730 kg/m³. Assume that the pressures shown in the diagram do not change with flow.

(a) Recommend a proper location for the control valve.

(b) Size the valve for 100 % overcapacity.

5-9. In the line sketched in Fig. P5-2, ethylbenzene flows at 950 gpm (nominal) and 445°F (density = 42.0 lb/ft³). The frictional pressure drop between points 1 and 2 is 12.4 psi.

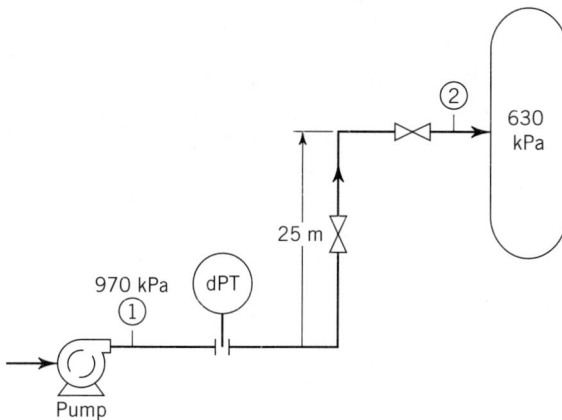

Figure P5-1 Benzene process for Problems 5-8 and 5-12.

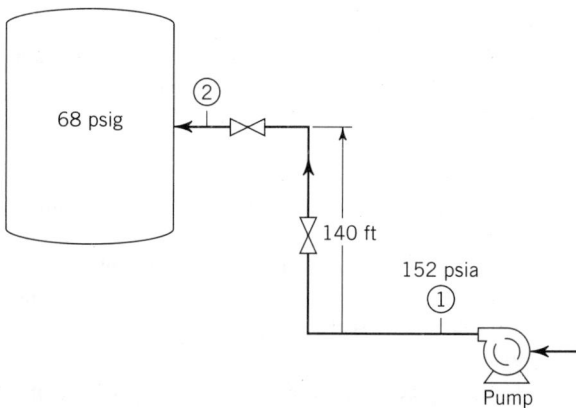

Figure P5-2 Ethylbenzene process for Problems 5-9 and 5-13.

(a)   Recommend a proper location for the control valve.

(b)   Size the valve for 100 % overcapacity.

5-10.   The nominal flow of a liquid through a control valve is 480 gpm. At this flow the frictional pressure drop in the line is 15 psi. The total pressure drop available across the valve and line is 20 psi, independent of flow, and the specific gravity of the liquid is 0.85.

(a)   Size the valve for 100 % overcapacity.

(b)   Find the flow through the valve when it is fully opened. (*Hint*: It is not 960 gpm.)

(c)   Calculate the gain through the valve at design flow assuming it has linear inherent characteristics.

(d)   Obtain the rangeability of the valve.

State your assumptions in solving this problem.

5-11.   Repeat Problem 5-10 if the total available pressure drop is increased to 35 psi to have more pressure drop across the valve. Estimate also the incremental annual cost of running the pump to provide the additional 15 psi of pressure

drop. Use the economic parameters of Example 5-2.3, and a pump efficiency of 70 %.

5-12.   The valve of Problem 5-8 has inherent equal percentage characteristics with rangeability parameter of 50.

(a)   Find the flow through the valve when it is fully opened. (*Hint*: It is not 200,000 kg/h.)

(b)   Obtain the rangeability of the control valve.

(c)   Estimate the gain of the valve at the design flow, in (kg/h)/%CO.

(d)   Plot the normalized installed characteristics.

State your assumptions in solving this problem.

5-13.   The valve of Problem 5-9 has linear inherent characteristics.

(a)   Obtain the flow through the valve when it is fully opened. (*Hint*: It is not 1900 gpm.)

(b)   Calculate the rangeability of the control valve.

(c)   Find the gain of the valve at the design flow, in gpm/%CO.

(d)   Plot the normalized installed characteristics.

State your assumptions in solving this problem.

5-14.   Derive Eqs. 5-2.23 and 5-2.25 for a gas, if inlet and outlet pressures are constant with flow. Would the equation also apply if the mass flow is replaced with the flow in scfh?

5-15.   *Design of Gas Flow Control Loop.* A flow control loop, consisting of an orifice in series with the control valve, a differential pressure transmitter, and a controller, is to be designed for a nominal process flow of 180,000 scfh of air. Valve inlet conditions are 100 psig and 60°F, and the outlet pressure is 80 psig. The valve has linear characteristics and a square root extractor is built into the transmitter so that its output signal is linear with flow. The valve time constant is 0.06 min, and the transmitter time constant is negligible. A proportional-integral (PI) controller controls the flow.

(a)   Obtain the valve capacity factor, C_v, and the gain of the valve. Size it for 100 % overcapacity, and assume $C_f = 0.9$ (Masoneilan).

(b)   Calculate the gain of the transmitter if it is calibrated for a range of 0 to 250,000 scfh.

(c)   Draw the instrumentation diagram and the block diagram of the flow control loop showing the specific transfer functions of the controller, the control valve, and the flow transmitter.

5-16.   Consider the pressure control system shown in Fig. P5-3. The pressure transmitter, PT25, has a range of 0 to 100 psig. The controller, PC25, is a proportional only controller, its bias value is set at mid-scale, and its set point is 10 psig. Obtain the correct action of the controller and the proportional band required so that when the pressure in the tank is 30 psig, the valve will be wide open.

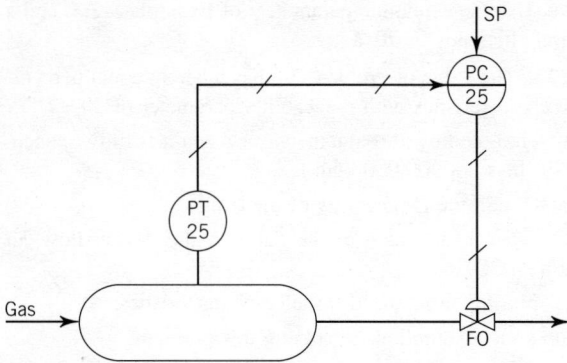

Figure P5-3 Pressure control system for Problem 5-16.

5-17. Let us change the pressure control system of Problem 5-16. The new control scheme is shown in Fig. P5-4. This control scheme is called cascade control; its benefits and principles are explained in Chapter 9. In this scheme, the pressure controller sets the set point of the flow controller. The pressure transmitter has a range of 0 to 100 psig and the flow transmitter range is 0 to 3000 scfh. Both controllers are proportional only. The normal flow rate through the valve is 1000 scfh, and to give this flow the valve must be 33 % opened. The control valve has linear characteristics.

(a) Obtain the action of both controllers.

(b) Choose the bias values ($\overline{m}$) for both controllers so that no offset occurs in either controller.

(c) Obtain the proportional band setting of the pressure controller so that when the tank pressure reaches 40 psig the set point to the flow controller is 1700 scfh. The set point of the pressure controller is 10 psig.

(d) Obtain the action of both controllers if the valve were to be fail-closed (air-to-open).

5-18. Consider the level loop shown in Fig. 5-3.2. The steady-state operating conditions are $\overline{f}_i = \overline{f}_o = 150$ gpm

and $\overline{h} = 6$ ft. For this steady-state the FC valve requires a 50 % signal. The level transmitter has a range of 0 to 20 ft. A proportional only controller, with $K_c = 1$, is used in this process. Calculate the offset if the inlet flow increases to 170 gpm and the valve requires a signal of 57 % to deliver this flow. Report the offset in % of scale and in feet.

5-19. A controller receives a signal from a temperature transmitter with a range of 100 to 150°C. Assume the controller is proportional-integral (PI) with a gain of 3 %CO/%TO and an integral (or reset) time of 5 min.

(a) Write the transfer function of the controller relating the output $M(s)$ to the error signal $E(s)$; assume both signals are in percent of range. Show the numerical values of the controller parameters.

(b) Calculate the gain of the transmitter and write its transfer function assuming that it can be represented by a first-order lag with a time constant of 0.1 min.

(c) Draw a block diagram of the transmitter and controller showing all transfer functions. The input signals to the diagram are the process temperature $T(s)$ and its set point $T^{set}(s)$, both in °C.

(d) Assume a sustained step change in set point of 1°C is applied to the controller, and, because of a loss of the signal to the control valve, the process temperature remains constant and equal to the original set point. Calculate the sustained error in %TO, and the controller output in %CO at the following times: right after the change in set point, 5 min later, and 10 min later. Sketch a plot of the error and the controller output versus time.

5-20. Consider the concentration control loop for the two stirred reactors shown in Fig. P5-5. The rate of consumption of reactant A in each reactor is given by the following formula:

$$r_A(t) = kc_A(t)$$

where $r_A(t)$ is the reaction rate in lbmole/gal-min, and $c_A(t)$ is the concentration of reactant A in the reactor, in

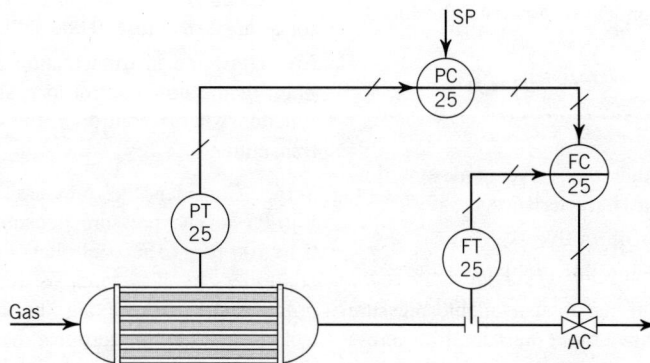

Figure P5-4 Pressure control system for Problem 5-17.

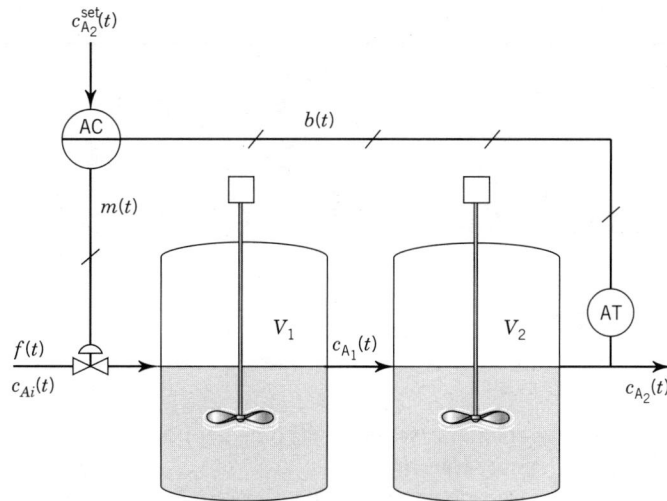

Figure P5-5 Stirred reactors in series for Problem 5-20.

lbmole/gal. Assume the reactor volumes, V_1 and V_2, gal, the rate coefficients, k_1 and k_2, min^{-1}, and the density of the fluid ρ, lb/gal, are constant.

(a)   Obtain the transfer functions for the concentration from the reactors, $C_{A1}(s)$ and $C_{A2}(s)$, to the input variables, the flow $F(s)$, and inlet concentration, $C_{Ai}(s)$. Draw the block diagram for the reactors.

(b)   Size the control valve for 100 % overcapacity and a nominal flow of 100 gpm. The pressure drop across the valve is constant at 9 psi, and the specific gravity of the reactant stream is 0.81. Assume the time constant of the valve actuator is negligible, and the valve is air-to-open with linear characteristics. Calculate the gain of the valve and draw the block diagram for the valve.

(c)   The concentration transmitter has a calibrated range of 0 to 0.2 lbmole/gal and a time constant of 0.5 min. Calculate the transmitter gain and draw the block diagram for the transmitter.

(d)   Draw a block diagram of the concentration control loop showing all transfer functions. Use a proportional-integral-derivative (PID) controller. Should the controller action be direct or reverse?

5-21.   Show how the relationships given in Eq. 5-3.20 were obtained from Eqs. 5-3.16 and 5-3.19.

Analysis of Feedback Control Loops

In previous chapters we have become familiar with the dynamic characteristics of processes, sensor-transmitters, control valves, and controllers. We have also learned how to write linearized transfer functions for each of these components and to recognize the parameters that are significant to the design of automatic control systems, namely, the steady-state gain, the time constants, and the dead time (transportation lag or time delay). In this chapter we will see how these concepts are put together to design single-loop feedback control systems. We will first analyze a simple feedback control loop and learn how to draw a block diagram for it and determine its characteristic equation. Then we will examine the significance of the characteristic equation in terms of how it can be used to determine the stability of the loop.

The methods that we will study in this chapter are most applicable to the design of feedback control loops for industrial processes. Two other design techniques, root-locus and frequency response analysis, which have been traditionally applied to inherently linear systems, will be presented in Chapter 8.

6-1 THE FEEDBACK CONTROL LOOP

The concept of feedback control, although more than 2000 years old, did not find practical application in industry until James Watt applied it to the control of the speed of his steam engine about 200 years ago. Since then, the number of industrial applications have proliferated to the point that today almost all automatic control systems include feedback control. None of the advanced control techniques that have been developed in the last 70 years to enhance the performance of feedback control loops have been able to replace it. We will study these advanced techniques in later chapters.

To review the concept of feedback control, let us again look at the heat exchanger example of Chapter 1. Figure 6-1.1 presents a sketch of the exchanger. Our objective is to maintain the outlet temperature of the process fluid, $T_o(t)$, at its desired value or set point, $T_o^{\text{set}}(t)$, in the presence of variations of the process fluid flow, $W(t)$, and inlet temperature, $T_i(t)$. We select the steam flow $W_s(t)$ as the variable that can be adjusted to control the outlet temperature; the amount of energy supplied to the process fluid is proportional to the steam flow.

Feedback control works as follows: a sensor-transmitter (TT42) measures the outlet temperature or *controlled variable*, $T_o(t)$, generates a signal $C(t)$ proportional to it, and sends it to the controller (TC42) where it is compared to the set point, $T_o^{\text{set}}(t)$. The controller then calculates an output signal or *manipulated variable*, $M(t)$, on the basis

Principles and Practice of Automatic Process Control/Third Edition, by C. A. Smith and A. B. Corripio
ISBN 0-471-66141-4   Copyright © 2006 John Wiley & Sons (Asia) Pte. Ltd.

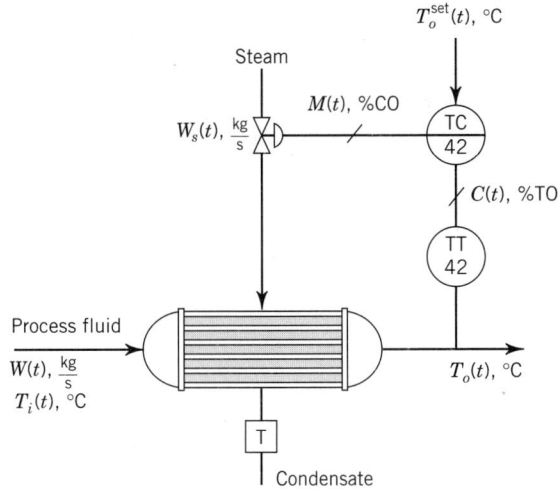

Figure 6-1.1 Feedback control loop for temperature control of a heat exchanger.

of the *error* or difference between the measurement and the set point. This controller output signal is sent to the actuator of the steam control valve. The valve actuator positions the valve in proportion to the controller output signal. Finally, the steam flow, a function of the valve position, determines the energy rate to the exchanger and therefore the controlled outlet temperature.

The term *feedback* derives from the fact that the controlled variable is measured and this measurement is "fed back" to reposition the steam valve. This causes the signal variations to move around the loop as follows:

> *Variations in outlet temperature are sensed by the sensor-transmitter and sent to the controller causing the controller output signal to vary. This in turn causes the control valve position and consequently the steam flow to vary. The variations in steam flow cause the outlet temperature to vary, thus completing the loop.*

This loop structure is what makes feedback control simultaneously simple and effective. When properly tuned, the feedback controller can maintain the controlled variable at or near the set point in the presence of any disturbance (e.g., process flow and inlet temperature), without knowledge of what the disturbance is or of its magnitude.

As we saw in Section 5-3, the most important requirement of the controller is the direction of its action (or simply *action*), direct or reverse. In the case of the temperature controller the correct action is reverse, because an increase in temperature requires a decrease in the controller output signal to close the valve and reduce the steam flow. This assumes that the control valve is air-to-open, so that the steam flow will be cut off in case of loss of electric power or instrument air pressure (fail-closed).

The performance of the control loop can best be analyzed by drawing the block diagram for the entire loop. To do this, we first draw the block for each component and then connect the output signal from each block to the next block. Let us start with the heat exchanger. In Chapters 3 and 4 we learned that the linear approximation to the response of the output of any process can be represented by the sum of a series of blocks, one for each input variable. As Fig. 6-1.2 shows, the block diagram for the heat exchanger consists of three blocks, one for each of its three inputs: the process flow, $W(s)$, inlet temperature, $T_i(s)$, and steam flow, $W_s(s)$. The corresponding transfer functions are $G_w(s)$, $G_T(s)$, and $G_s(s)$.

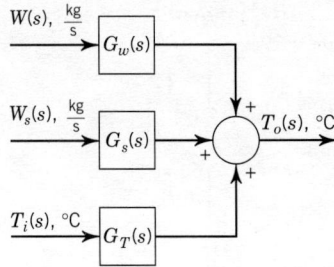

Figure 6-1.2 Block diagram of the heat exchanger of Fig. 6-1.1.

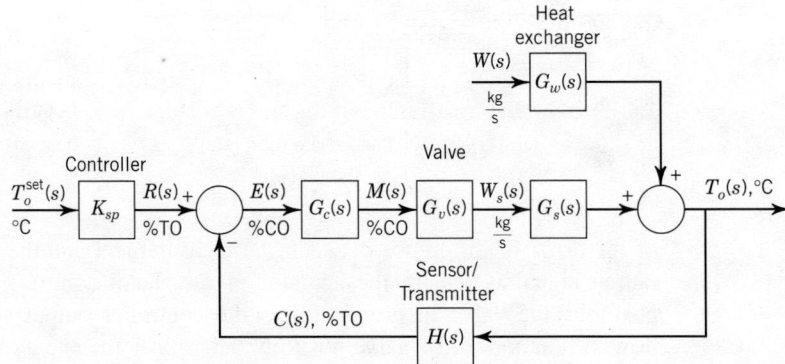

Figure 6-1.3 Block diagram of heat exchanger temperature control loop.

Figure 6-1.3 shows the complete block diagram for the feedback control loop. To simplify the discussion that follows, we have purposely omitted the inlet temperature, $T_i(s)$, as an input signal. This effectively assumes that the inlet temperature is constant and selects the process flow, $W(s)$, as representative of either disturbance. The symbols in Fig. 6-1.3 are as follows:

$E(s)$ = the error, % transmitter output (%TO)

$G_c(s)$ = the controller transfer function, %CO/%TO

$G_v(s)$ = the control valve transfer function, (kg/s)/%CO

$H(s)$ = the sensor-transmitter transfer function, %TO/°C

K_{sp} = the scale factor for the temperature set point, %TO/°C

It is important at this point to notice the correspondence between the blocks (or groups of blocks) in the block diagram, Fig. 6-1.3, and the components of the control loop, Fig. 6-1.1. Matching the symbols used to identify the various signals facilitates this comparison. It is also important to recall from Chapter 3 that the blocks on the diagram represent linear relationships between the input and output signals, and that the signals are deviations from initial steady-state values and are not absolute variable values.

The term K_{sp} is a scale factor that converts the set point, usually calibrated in the same units as the controlled variable, to the same basis as the transmitter signal, that is °C to %TO. It can be shown that for the measurement and the set point to be on the same scale, K_{sp} must be numerically equal to the transmitter gain.

The sign convention used in the block diagram of Fig. 6-1.3 agrees with the convention used in Section 5-3 for the calculation of the error (set point − measurement).

This convention will be used throughout this book. Notice that this makes the sign around the loop negative if the gains of all the blocks and summers in the loop are positive, as they are in this case. *A negative feedback gain is a requirement for stability.* Following this convention, a reverse-acting controller must have a positive gain, while a direct-acting controller must have a negative gain, as you can verify by analyzing the controller section of the block diagram. The convention is not selected this way to confuse you, but to graphically emphasize the negative feedback gain on the block diagram (otherwise the minus sign would be hidden in the controller gain).

6-1.1 Closed-Loop Transfer Function

We can see by the inspection of the closed-loop block diagram of Fig. 6-1.3 that the loop has one output signal, the controlled variable $T_o(s)$, and two input signals, the set point $T_o^{\text{set}}(s)$, and the disturbance, $W(s)$. Since the steam flow is connected to the outlet temperature through the control loop, we might expect that the "closed-loop response" of the system to the various inputs would be different from the response when the loop is "open." Most control loops can be opened by flipping a switch on the controller from the "automatic" to the "manual" position (see Section 5-3). When the controller is in the manual position, its output does not respond to the error signal and it is thus independent of the set point and measurement signals. On the other hand, when in "automatic" the controller output varies when the measurement signal varies.

We can determine the closed-loop transfer function of the loop output with regard to any of its inputs by applying the rules of block diagram algebra learned in Chapter 3 to the diagram of the loop. To review, suppose we want to derive the response of the outlet temperature $T_o(s)$ to the process flow $W(s)$. We first write the equations for each block in the diagram, as follows:

$$E(s) = K_{sp}T_o^{\text{set}}(s) - C(s) \tag{6-1.1}$$

$$M(s) = G_c(s)E(s) \tag{6-1.2}$$

$$W_s(s) = G_v(s)M(s) \tag{6-1.3}$$

$$T_o(s) = G_s(s)W_s(s) + G_w(s)W(s) \tag{6-1.4}$$

$$C(s) = H(s)T_o(s) \tag{6-1.5}$$

Next we assume that the set point does not vary; that is, its deviation variable is zero,

$$T_o^{\text{set}}(s) = 0$$

and eliminate all the intermediate variables by combining Eqs. 6-1.1 through 6-1.5. The result is

$$T_o(s) = G_s(s)G_v(s)G_c(s)[-H(s)T_o(s)] + G_w(s)W(s) \tag{6-1.6}$$

Solving for $T_o(s)$, and dividing by $W(s)$, we get

$$\boxed{\frac{T_o(s)}{W(s)} = \frac{G_w(s)}{1 + H(s)G_s(s)G_v(s)G_c(s)}} \tag{6-1.7}$$

This is the closed-loop transfer function between the process flow and the outlet temperature. Similarly, we let $W(s) = 0$ and combine Eqs. 6-1.1 through 6-1.5, to obtain the closed-loop transfer function between the set point and the outlet temperature.

$$\boxed{\frac{T_o(s)}{T_o^{\text{set}}(s)} = \frac{K_{sp}G_s(s)G_v(s)G_c(s)}{1 + H(s)G_s(s)G_v(s)G_c(s)}} \tag{6-1.8}$$

As we saw in Chapter 3, the denominator is the same for both inputs while the numerator is different for each input. We recall further that the denominator is one plus the product of the transfer functions of all the blocks that are in the loop itself, while the numerator of each transfer function is the product of the blocks that are in the direct path between the specific input and the output of the loop. These results apply to any block diagram that contains a single loop.

It is enlightening to check the units of the product of the blocks in the loop, as follows:

$$H(s) \cdot G_s(s) \cdot G_v(s) \cdot G_c(s) = \left(\frac{\%TO}{°C} \right) \left(\frac{°C}{kg/s} \right) \left(\frac{kg/s}{\%CO} \right) \left(\frac{\%CO}{\%TO} \right) = \text{dimensionless}$$

This shows that the product of the transfer functions of the blocks in the loop is dimensionless, as it should be. We can also verify that the units of the numerator of each of the closed-loop transfer functions are the units of the output variable divided by the units of the corresponding input variable.

Simplified Block Diagram

It is convenient to simplify the block diagram of Fig. 6-1.3 by combining blocks. Applying the rules of block diagram algebra from Chapter 3, we obtain the diagram of Fig. 6-1.4. The transfer functions of the simplified diagram are

$$G_1(s) = G_v(s)G_s(s)H(s) \qquad \qquad \text{(6-1.9)}$$
$$G_2(s) = G_w(s)H(s) \qquad \qquad \text{(6-1.10)}$$

In the simplified diagram the loop signals are in percent of range and the feedback gain is unity, which is the reason the loop in the diagram is sometimes called a *unity feedback loop*. The closed-loop transfer function of the output signal, which is now the transmitter output, is

$$C(s) = \frac{G_c(s)}{1 + G_c(s)G_1(s)} R(s) + \frac{G_2(s)}{1 + G_c(s)G_1(s)} W(s) \qquad \text{(6-1.11)}$$

where $R(s)$ is the reference signal (set point) in %TO. Except for the name of the flow disturbance, the block diagram of Fig. 6-1.4 can represent any feedback control loop.

The following example demonstrates how to develop the closed-loop transfer function from the principles we learned in Chapters 3, 4, and 5.

EXAMPLE 6-1.1

Temperature Control of a Continuous Stirred Tank Heater

The stirred tank sketched in Fig. 6-1.5 is used to heat a process stream so that its premixed components achieve a uniform composition. Temperature control is important because a high temperature tends to decompose the product while a low temperature results in incomplete mixing. The tank is heated by steam condensing inside a coil. A proportional-integral-derivative

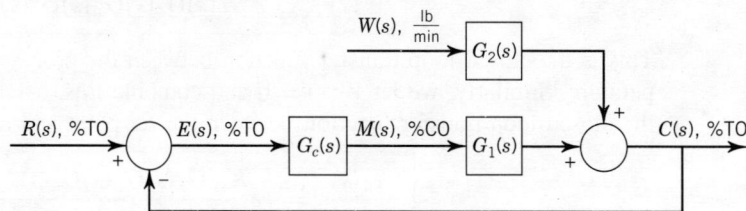

Figure 6-1.4 Simplified block diagram of a feedback control loop.

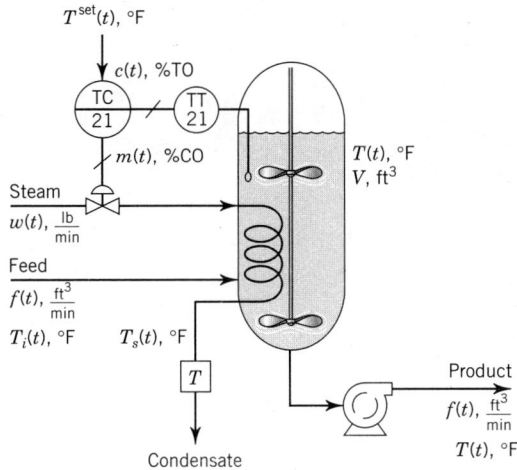

Figure 6-1.5 Temperature control of the stirred tank heater of Example 6-1.1.

(PID) controller is used to control the temperature in the tank by manipulating the steam valve position. Derive the complete block diagram and the closed-loop transfer function from the following design data.

Process. The feed has a density ρ of 68.0 lb/ft^3, a heat capacity c_p of 0.80 Btu/lb-$^\circ$F. The volume V of liquid in the reactor is maintained constant at 120 ft^3. The coil consists of 205 ft of 4-in. schedule 40 steel pipe, weighing 10.8 lb/ft with a heat capacity of 0.12 Btu/lb-$^\circ$F and an outside diameter of 4.500 in. The overall heat transfer coefficient U, based on the outside area of the coil, has been estimated as 2.1 Btu/min-ft^2-$^\circ$F. The steam available is saturated at a pressure of 30 psia; it can be assumed that its latent heat of condensation λ is constant at 966 Btu/lb. It can also be assumed that the inlet temperature, T_i, is constant.

Design conditions. The feed flow f at design conditions is 15 ft^3/min and its temperature T_i is 100°F. The contents of the tank must be maintained at a temperature T of 150°F. Possible disturbances are changes in feed rate and temperature.

Temperature sensor and transmitter. The temperature sensor has a calibrated range of 100 to 200°F and a time constant τ_T of 0.75 min.

Control valve. The control valve is to be designed for 100 % overcapacity, and pressure drop variations can be neglected. The valve is an equal percentage valve with a rangeability parameter α of 50. The actuator has a time constant τ_v of 0.20 min.

SOLUTION

Our approach will be to derive the equations that describe the dynamic behavior of the tank, the control valve, the sensor-transmitter, and the controller. Then we will Laplace-transform them to obtain the block diagram of the loop.

Process. An energy balance on the liquid in the tank, assuming negligible heat losses, perfect mixing, and constant volume and physical properties, results in the equation

$$V\rho c_v \frac{dT(t)}{dt} = f(t)\rho c_p T_i(t) + UA[T_s(t) - T(t)] - f(t)\rho c_p T(t)$$

1 eq., 2 unk. (T, T_s)

where

A = the heat transfer area, ft^2

$T_s(s)$ = the condensing steam temperature, $^\circ$F

and the other symbols have been defined in the statement of the problem. For the liquid contents of the tank, the c_v in the accumulation term is essentially equal to c_p.

An energy balance on the coil, assuming that the coil metal is at the same temperature as the condensing steam, results in

$$C_M \frac{dT_s(t)}{dt} = w(t)\lambda - UA[T_s(t) - T(t)]$$

2 eq., 3 unk. (w)

where

$w(t)$ = the steam rate, lb/min

C_M = heat capacitance of the coil metal, Btu/°F

Since the steam rate is the output of the control valve and an input to the process, our process model is complete.

Linearization and Laplace transformation. By the methods learned in Section 2-6, we obtain the linearized tank model equations in terms of deviation variables.

$$V \rho c_p \frac{d\Gamma(t)}{dt} = \rho c_p (T_i - \overline{T}) F(t) + UA\Gamma_s(t) - (UA + \overline{f}\rho c_p)\Gamma(t)$$

$$C_M \frac{d\Gamma_s(t)}{dt} = \lambda W(t) - UA\Gamma_s(t) + UA\Gamma(t)$$

where $\Gamma(t)$, $F(t)$, and $W(t)$ are the deviation variables.

Taking the Laplace transform of these equations and rearranging, as learned in Chapters 2, 3, and 4, we get

$$\Gamma(s) = \frac{K_F}{\tau s + 1} F(s) + \frac{K_s}{\tau s + 1}\Gamma_s(s)$$

$$\Gamma_s(s) = \frac{1}{\tau_c s + 1}\Gamma(s) + \frac{K_w}{\tau_c s + 1} W(s)$$

where

$$\tau = \frac{V\rho c_p}{UA + \overline{f}\rho c_p} \qquad \tau_c = \frac{C_M}{UA}$$

$$K_F = \frac{\rho c_p (T_i - \overline{T})}{UA + \overline{f}\rho c_p} \qquad K_w = \frac{\lambda}{UA}$$

$$K_s = \frac{UA}{UA + \overline{f}\rho c_p}$$

Control valve. The transfer function for an equal percentage valve with constant pressure drop is, from Section 5-2,

$$G_v(s) = \frac{W(s)}{M(s)} = \frac{K_v}{\tau_v s + 1}$$

3 eq., 4 unk. (M)

where $M(s)$ is the controller output signal in percent controller output (%CO), and the valve gain is, from Section 5-2,

$$K_v = \frac{\overline{w}(\ln \alpha)}{100}$$

Sensor-transmitter (TT21). The sensor-transmitter can be represented by a first-order lag:

$$H(s) = \frac{C(s)}{\Gamma(s)} = \frac{K_T}{\tau_T s + 1}$$

4 eq., 5 unk. (C)

where $C(s)$ is the LaPlace transform of the transmitter output signal, %TO, and the transmitter gain is, from Section 5-1,

$$K_T = \frac{100 - 0}{200 - 100} = 1.0 \frac{\%\text{TO}}{°\text{F}}$$

The transfer function of the PID controller is, from Section 5-3,

$$G_c(s) = \frac{M(s)}{R(s) - C(s)} = K_c \left(1 + \frac{1}{\tau_I s} + \tau_D s\right)$$

5 eq., 5 unk. Solved!

where K_c is the controller gain, τ_I is the integral time, and τ_D is the derivative time. This completes the derivation of the equations for the temperature control loop.

Block diagram of the loop. Figure 6-1.6 shows the complete block diagram for the loop. All of the transfer functions in the diagram have been derived above. Using the rules for block diagram manipulation learned in Chapter 3, we obtain the simpler diagram of Fig. 6-1.7. The transfer functions in the simpler diagram are

$$G_F(s) = \frac{K_F(\tau_c s + 1)}{(\tau s + 1)(\tau_c s + 1) - K_s}$$

$$G_s(s) = \frac{K_w K_s}{(\tau s + 1)(\tau_c s + 1) - K_s}$$

The closed-loop transfer function of the outlet process temperature is

$$\Gamma(s) = \frac{K_{sp} G_c(s) G_v(s) G_s(s)}{1 + H(s) G_c(s) G_v(s) G_s(s)} \Gamma^{\text{set}}(s) + \frac{G_F(s)}{1 + H(s) G_c(s) G_v(s) G_s(s)} F(s)$$

Initial steady-state and parameter values. Table 6-1.1 gives the numerical values of the parameters in the transfer functions, calculated from the data given in the problem statement. The base values for the linearization are the design conditions, assumed to be the initial conditions and at

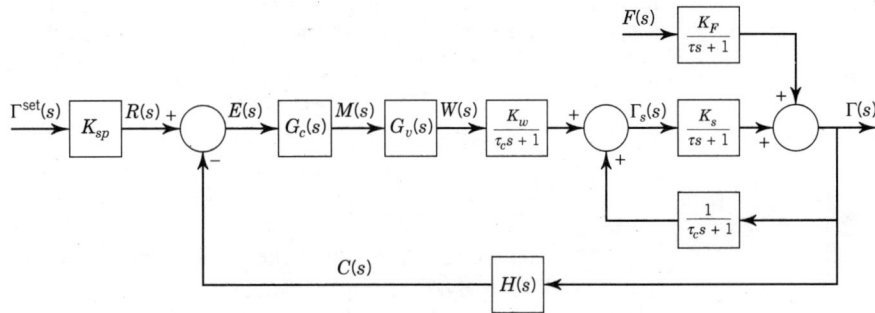

Figure 6-1.6 Block diagram of temperature control loop of stirred tank heater.

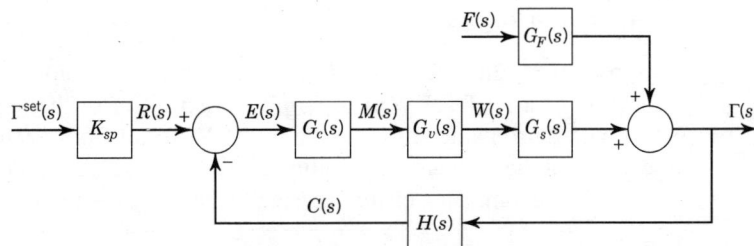

Figure 6-1.7 Simplified block diagram of temperature control loop.

Table 6-1.1   Parameters for Example 6-1.1

$A = 241.5 \text{ ft}^2$	$\tau = 4.93 \text{ min}$
$C_M = 265.7 \text{ Btu/}°\text{F}$	$\tau_c = 0.524 \text{ min}$
$K_F = 2.06°\text{F/(ft}^3\text{/min)}$	$K_w = 1.905°\text{F/(lb/min)}$
$K_s = 0.383°\text{F/}°\text{F}$	$K_{sp} = K_T = 1.0 \text{ %TO/}°\text{F}$
$K_v = 1.652 \text{ (lb/min)/%CO}$	$\tau_v = 0.20 \text{ min}$
$\tau_T = 0.75 \text{ min}$	

steady state. From the model equations for the tank and the coil, we compute the initial steam temperature and steam flow at steady state.

$$\overline{f}\rho c_p T_i + UA(\overline{T}_s - \overline{T}) - \overline{f}\rho c_p \overline{T} = 0$$

$$\overline{w}\lambda - UA(\overline{T}_s - \overline{T}) = 0$$

$$\overline{T}_s = \overline{T} + \frac{\overline{f}\rho c_p(\overline{T} - T_i)}{UA} = 150 + \frac{(15)(68)(0.8)(150 - 100)}{(2.1)(241.5)} = 230°\text{F}$$

$$\overline{w} = \frac{UA(\overline{T}_s - \overline{T})}{\lambda} = \frac{(2.1)(241.5)(230 - 150)}{966} = 42.2 \frac{\text{lb}}{\text{min}}$$

We can see, from Fig. 6-1.6, that the coil and the tank form a set of two interacting lags. This means that we must calculate effective time constants from the parameters in Table 6-1.1. They are 8.34 and 0.502 min, resulting in the following transfer functions:

$$G_1(s) = G_v(s)G_s(s)H(s) = \frac{1.652}{0.2s + 1} \frac{1.183}{(8.34s + 1)(0.502s + 1)} \frac{1.0}{0.75s + 1} \frac{\text{%TO}}{\text{%CO}}$$

$$G_2(s) = G_F(s)H(s) = \frac{-3.34(0.524s + 1)}{(8.34s + 1)(0.502s + 1)} \frac{1.0}{0.75s + 1} \frac{\text{%TO}}{\text{ft}^3/\text{min}}$$

where the gains are $K_w K_s/(1 - K_s) = (1.905)(0.383)/(1 - 0.383) = 1.183°\text{F/(lb/min)}$, and $K_F/(1 - K_s) = (-2.06)/(1 - 0.383) = -3.34°\text{F/(ft}^3\text{/min)}$. The closed-loop transform of the temperature transmitter output is then

$$C(s) = \frac{G_c(s)G_1(s)}{1 + G_c(s)G_1(s)} R(s) + \frac{G_2(s)}{1 + G_c(s)G_1(s)} F(s)$$

These transfer functions match the unity feedback loop of Fig. 6-1.4.

This example illustrates how the basic principles of process engineering can be put to work in analyzing simple feedback control loops. From the closed-loop transfer functions we can calculate the response of the closed loop to various input forcing functions for different values of the controller tuning parameters, K_c, τ_I, and τ_D.

6-1.2   Characteristic Equation of the Loop

As we saw in the preceding discussion, the denominator of the closed-loop transfer function of a feedback control loop is independent of the location of the input to the loop and thus characteristic of the loop. We recall from Section 2-3 that the unforced response of the loop and its stability depend on the roots of the equation that is obtained when the denominator of the transfer function of the loop is set equal to zero.

$$\boxed{1 + H(s)G_s(s)G_v(s)G_c(s) = 0} \tag{6-1.12}$$

This is the *characteristic equation* of the loop. Notice that the controller transfer function is very much a part of the characteristic equation of the loop. *This is why the response of the loop can be shaped by tuning the controller.* The other elements that form part of the characteristic equation are the sensor-transmitter, the control valve, and that part of the process that affects the response of the controlled variable to the manipulated variable, $G_s(s)$. On the other hand, the process transfer function related to the disturbance, $G_w(s)$, is not part of the characteristic equation.

To show that the characteristic equation determines the unforced response of the loop, let us derive the response of the closed loop to a change in process flow by inverting the Laplace transform of the output signal, as we learned to do in Section 2-2. Assume that the characteristic equation can be expressed as an nth-degree polynomial in the Laplace transform variable s

$$1 + H(s)G_s(s)G_v(s)G_c(s) = a_n s^n + a_{n-1}s^{n-1} + \cdots + a_0 = 0 \qquad \textbf{(6-1.13)}$$

where $a_n, a_{n-1}, \ldots, a_0$ are the polynomial coefficients. With an appropriate computer program, we can find the n roots of this polynomial, and factor it as follows:

$$1 + H(s)G_s(s)G_v(s)G_c(s) = a_n(s - r_1)(s - r_2) \cdots (s - r_n) \qquad \textbf{(6-1.14)}$$

where $r_1, r_2 \ldots, r_n$ are the roots of the characteristic equation. These roots can be real numbers or pairs of complex conjugate numbers, and some of them may be repeated, as we saw in Section 2-2.

From Eq. 6-1.7 we obtain

$$T_o(s) = \frac{G_w(s)}{1 + H(s)G_s(s)G_v(s)G_c(s)} W(s) \qquad \textbf{(6-1.15)}$$

Next let us substitute Eq. 6-1.14 for the denominator and assume that other terms will appear because of the input forcing function, $W(s)$.

$$T_o(s) = \frac{\text{numerator terms}}{a_n(s - r_1)(s - r_2) \cdots (s - r_n)(\text{input terms})} \qquad \textbf{(6-1.16)}$$

We then expand this expression into partial fractions

$$T_o(s) = \frac{b_1}{s - r_1} + \frac{b_2}{s - r_2} + \cdots + \frac{b_n}{s - r_n} + (\text{input terms}) \qquad \textbf{(6-1.17)}$$

where $b_1, b_2, \ldots, b_n$ are the constant coefficients that are determined by the method of partial fractions expansion (see Section 2-2.2). Inverting this expression with the help of a Laplace transform table (e.g., Table 2-1.1), we obtain

$$T_o(t) = b_1 e^{r_1 t} + b_2 e^{r_2 t} + \cdots + b_n e^{r_n t} + (\text{input terms})$$
$$\text{UNFORCED RESPONSE} + \text{FORCED RESPONSE} \qquad \textbf{(6-1.18)}$$

We have thus shown that each of the terms of the unforced response contains a root of the characteristic equation. We recall that the coefficients $b_1, b_2, \ldots, b_n$ depend on the actual input forcing function and so does the exact response of the loop. However, the speed with which the unforced response terms die out ($r_i < 0$), diverge ($r_i > 0$), or oscillate (r_i complex) is determined entirely by the roots of the characteristic equation. We will use this concept in the next section to determine the stability of the loop. The following example illustrates the effect of a pure proportional and a pure integral controller on the closed-loop response of a first-order process. We will see that the pure proportional controller will speed up the first-order response and result in an *offset* or steady-state error, as discussed in Section 5-3. On the other hand, the integral controller produces a second-order response that, as the controller gain increases, changes from overdamped to underdamped. As discussed in Section 2-5.2, the underdamped response is oscillatory.

EXAMPLE 6-1.2

Control of a First-Order Process

In the simplified block diagram of Fig 6-1.4, the process can be represented by a first-order lag.

$$G_1(s) = \frac{K}{\tau s + 1}$$

Determine the closed-loop transfer function and the response to a unit step change in set point for

(a) a proportional controller, $G_c(s) = K_c$

(b) a pure integral controller, $G_c(s) = K_I/s$, where K_I is the controller integral gain in min^{-1}.

SOLUTION

By block diagram algebra we obtain the closed-loop transfer function

$$\frac{C(s)}{R(s)} = \frac{G_1(s)G_c(s)}{1 + G_1(s)G_c(s)}$$

(a) We then substitute the process and controller transfer functions and simplify

$$\frac{C(s)}{R(s)} = \frac{KK_c}{\tau s + 1 + KK_c} = \frac{\dfrac{KK_c}{1 + KK_c}}{\dfrac{\tau}{1 + KK_c}s + 1} = \frac{K'}{\tau' s + 1}$$

We can easily see that the closed-loop response is first-order with steady-state gain of $K' = KK_c/(1 + KK_c)$, and time constant of $\tau' = \tau/(1 + KK_c)$. Notice that the closed-loop gain is always less than unity and that the closed-loop time constant is always less than the open-loop time constant τ. In other words, the closed-loop responds faster than the open-loop system but does not quite match the set point at steady state, that is, there will be *offset*.

Figure 6-1.8 shows the closed-loop unit step responses for several positive values of the loop gain, KK_c. These responses are typical first order (see Section 2-4). The response approaches the set point as the loop gain increases, as stated in Section 5-3.

What would be the response if the loop gain, KK_c, were negative? You can easily verify that for loop gains between 0 and −1, the response is stable, but the offset would be greater than if no control action were taken at all ($K_c = 0$). You can also verify that for loop gains less than −1 the response is unstable. In contrast, positive loop gains result in a stable response with decreasing *offset*.

(b) Substitute the integral controller transfer function into the closed-loop transfer to obtain

$$\frac{C(s)}{R(s)} = \frac{\dfrac{KK_I}{s}}{\tau s + 1 + \dfrac{KK_I}{s}} = \frac{KK_I}{\tau s^2 + s + KK_I}$$

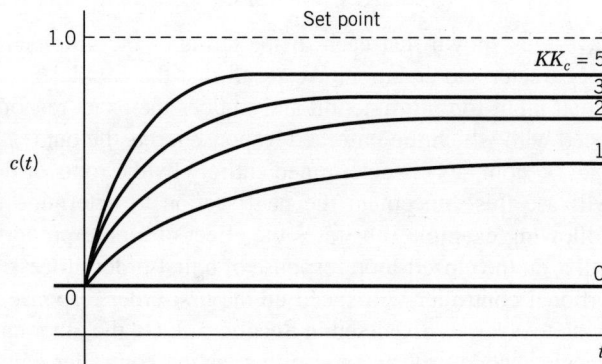

Figure 6-1.8 Unit step response to set point for closed loop of first-order system with proportional controller (Example 6-1.2).

By the extension of the final value theorem to transfer functions (see Section 3-5.1), we substitute $s = 0$ to obtain the steady-state gain.

$$\lim_{s \to 0} \frac{C(s)}{R(s)} = \frac{K K_I}{K K_I} = 1$$

This means that, for the integral controller, the controlled variable will always match the set point at steady state; that is, there will not be an offset.

The characteristic equation of the loop is obtained by setting the denominator of the closed-loop transfer function equal to zero. The roots of this quadratic equation are

$$r_{1,2} = \frac{-1 \pm \sqrt{1 - 4 K K_I \tau}}{2\tau}$$

These roots are real for $0 \le K K_I \tau \le \frac{1}{4}$ and complex conjugates $K K_I \tau > \frac{1}{4}$. As we saw in Section 2-5, when the roots are real the response is overdamped, and when the roots are complex conjugates, the response is underdamped (oscillatory). This means that, for the loop considered here, the response becomes oscillatory when the loop gain increases. This property is common to most feedback loops.

Figure 6-1.9 shows the closed-loop unit step responses for several positive values of the loop gain. By comparing the characteristic equation of the closed loop with that for the standard underdamped second-order system (Eq 2-5.4), we can calculate the damping ratio and the frequency of oscillation as a function of the loop parameters. They are

$$\zeta = \frac{1}{\sqrt{4 K K_I \tau}} \qquad \omega = \frac{\sqrt{4 K K_I \tau - 1}}{2\tau}$$

Table 6-1.2 gives the values of the damping ratio and the frequency of oscillation for several positive values of the loop gain which correspond to the values for the responses of Fig. 6-1.9. This demonstrates how the adjustable controller gain shapes the response of the closed loop.

It can be readily shown that, for this loop, any negative value of the loop gain $K K_I$ results in two real roots, one of which is positive. This means that the response will exponentially run away with time. On the other hand, for positive values of the loop gain, the roots are either negative real numbers or complex numbers with negative real roots. This, coupled with the unity gain, means that the response always converges to the set point when the loop gain is positive.

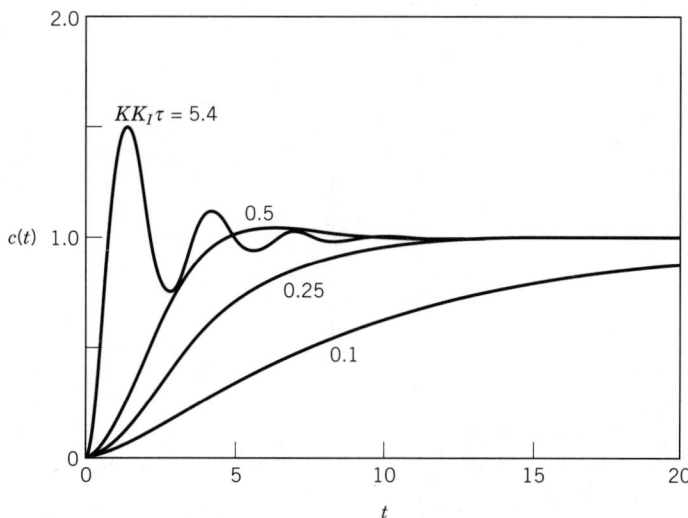

Figure 6-1.9 Unit step response to set point for closed loop of first-order system with integral controller (Example 6-1.2).

Table 6-1.2 Underdamped response for
Example 6-1.2

$KK_I\tau$	ζ	$\omega\tau$	Comments
0.25	1.0	0	Critically damped
0.50	0.707	0.5	5 % overshoot
5.40	0.215	2.3	Quarter decay ratio

The preceding example illustrates the point discussed in Chapter 4 regarding the fact that although most processes are inherently overdamped, their response can be underdamped when forming part of a closed feedback control loop.

EXAMPLE 6-1.3

Flow Control Loop

As we shall see in Chapters 9, 10 and 11, flow control loops are commonly used as the innermost loop in cascade, ratio, and feedforward control systems. Develop the closed-loop transfer function for a flow control loop with a proportional-integral (PI) controller.

SOLUTION

Figure 6-1.10 shows a schematic diagram of a flow control loop and its corresponding block diagram. To concentrate on the response of the flow $F(s)$ to its set point, $F^{set}(s)$, we will assume constant pressure drop across the control valve. However, one of the purposes of the flow controller is to compensate for changes in the pressure drop across the valve (disturbance). Notice that the flow control loop does not have a process! This is because the controlled variable, the flow, is the output of the control valve.

As discussed in Section 5-2, the control valve can be represented by a first-order lag.

$$G_v(s) = \frac{F(s)}{M(s)} = \frac{K_v}{\tau_v s + 1}\frac{\text{gpm}}{\%\text{CO}}$$

Flow transmitters are usually fast and can thus be represented by just a gain. Assuming a linear transmitter, the gain is, from Section 5-1,

$$H(s) = K_T = \frac{100}{f_{\max}}\frac{\%\text{TO}}{\text{gpm}}$$

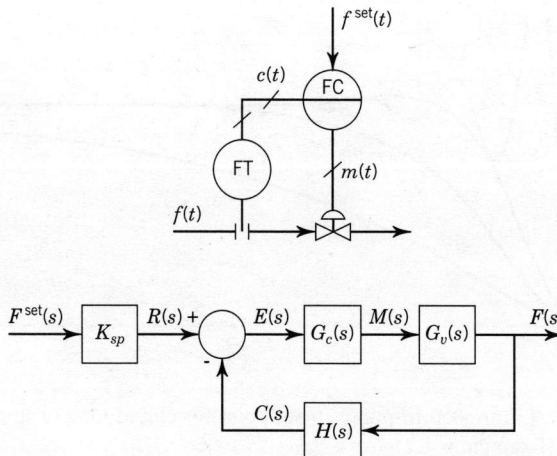

Figure 6-1.10 Schematic and block diagram of a flow control loop.

We apply block diagram algebra (Chapter 3) to the diagram of Fig. 6-1.10 to obtain the transfer function of the closed loop.

$$\frac{F(s)}{F^{set}(s)} = \frac{K_{sp}G_v(s)G_c(s)}{1 + K_T G_v(s)G_c(s)}$$

where $K_{sp} = K_T$. From Section 5-3, the transfer function of the PI controller is

$$G_c(s) = K_c\left(1 + \frac{1}{\tau_I s}\right) = \frac{K_c(\tau_I s + 1)}{\tau_I s}\frac{\%CO}{\%TO}$$

Substitute into the closed-loop transfer function and simplify to obtain

$$\frac{F(s)}{F^{set}(s)} = \frac{K_T K_v K_c(\tau_I s + 1)}{\tau_I s(\tau_v s + 1) + K_T K_v K_c(\tau_I s + 1)}$$

The response is second order. It can be underdamped (oscillatory) or overdamped, depending on the controller parameters. A fast first-order response can be obtained by setting the integral time equal to the valve time constant, $\tau_I = \tau_v$.

$$\frac{F(s)}{F^{set}(s)} = \frac{K_T K_v K_c}{\tau_v s + K_T K_v K_c} = \frac{1}{\tau_{FC}s + 1}$$

where the closed-loop time constant is $\tau_{FC} = \tau_v/K_T K_v K_c$. Notice that the closed-loop response is faster (shorter time constant) as the controller gain increases, and that the steady-state gain is unity, that is, there is no offset.

When the flow control loop is part of a cascade control system, its set point is sometimes in percent of range instead of in engineering units (gpm). In such cases, the input to the loop is $R(s)$ instead of $F^{set}(s)$ (see Fig. 6-1.10). The transfer function is then

$$\frac{F(s)}{R(s)} = \frac{K_v K_c}{\tau_v s + K_T K_v K_c} = \frac{1/K_T}{\tau_{FC}s + 1}\frac{gpm}{\%TO}$$

The gain of the flow control loop is then $1/K_T$.

Notice that this is very similar to the gain of a linear valve with constant pressure drop, except that the maximum flow here is the upper limit of the flow transmitter range (see Section 5-2).

The formulas derived in this example apply to liquid, gas, and steam valves, with the units appropriately adjusted (e.g., gpm, scfm, lb/h). They are independent of the flow characteristics of the valve and of whether the pressure drop is constant or variable.

6-1.3   Steady-State Closed-Loop Response

We have seen in the preceding examples that the final or steady-state value is an important aspect of the closed-loop response. This is because, in industrial process control practice, the presence of steady-state error or *offset* is usually unacceptable. We shall learn in this section how to calculate the offset when it is present. To do this we return to the exchanger of Fig. 6-1.1 and the corresponding block diagram of Fig. 6-1.3. As we learned earlier, this is a linearized representation of the heat exchanger. Our approach is to obtain the steady-state closed-loop relationships between the output variable and each of the inputs to the loop by application of the final value theorem to the closed-loop transfer function. From Eq. 6-1.7, the closed-loop transfer function between the outlet temperature and the process fluid flow is

$$\frac{T_o(s)}{W(s)} = \frac{G_w(s)}{1 + H(s)G_s(s)G_v(s)G_c(s)} \tag{6-1.7}$$

We recall that this expression assumes that the deviation variables for the inlet temperature T_i and the set point T_o^{set} are zero when these other inputs remain constant. We also recall, from Section 3-5, that the steady-state relationship between the output and the input to a transfer function is obtained by setting $s = 0$ in the transfer function. This follows from the final value theorem of Laplace transforms. Applying this method to Eq. 6-1.7, we obtain

$$\frac{\Delta T_o}{\Delta W} = \frac{G_w(0)}{1 + H(0)G_s(0)G_v(0)G_c(0)} \qquad (6\text{-}1.19)$$

where

ΔT_o = the steady-state change in outlet temperature, °C

ΔW = the steady-state change in process fluid flow, kg/s

If we assume, as is usually the case, that the process is stable,

$G_w(0) = K_w$, the process open-loop gain to a change in process fluid flow, °C/(kg/s)

$G_s(0) = K_s$, the process open-loop gain to a change in steam flow, °C/(kg/s)

Similarly, for the valve and the sensor-transmitter,

$G_v(0) = K_v$, the valve gain, (kg/s)/%CO

$H(0) = K_T$, the sensor-transmitter gain, %TO/%CO

Finally, if the controller does not have integral mode,

$G_c(0) = K_c$, the proportional gain, %CO/%TO

Substituting these terms into Eq. 6-1.19 yields

$$\frac{\Delta T_o}{\Delta W} = \frac{K_w}{1 + KK_c} \frac{°C}{kg/s} \qquad (6\text{-}1.20)$$

where $K = K_T K_s K_v$ is the combined gain of the elements of the loop other than the controller, in %TO/%CO. Since the change in set point is zero, the steady-state error or offset is $e = \Delta T_o^{\text{set}} - \Delta T_o = -\Delta T_o$ °C, and, combining this relationship with Eq. 6-1.20, gives

$$\frac{e}{\Delta W} = \frac{-K_w}{1 + KK_c} \frac{°C}{kg/s} \qquad (6\text{-}1.21)$$

Note that the offset decreases as the controller gain, K_c, is increased.

Following an identical procedure for Eq. 6-1.8, we obtain the steady-state relationship to a change in set point at constant process fluid flow.

$$\frac{\Delta T_o}{\Delta T_o^{\text{set}}} = \frac{K_s K_v K_c K_{sp}}{1 + K_T K_s K_v K_c} = \frac{KK_c}{1 + KK_c} \frac{°C}{°C} \qquad (6\text{-}1.22)$$

where ΔT_o^{set} is the steady-state change in set point, °C, and we have used $K_{sp} = K_T$. The offset in this case is $e = \Delta T_o^{\text{set}} - \Delta T_o$ °C. Combining this relationship with Eq. 6-1.22 gives

$$\frac{e}{\Delta T_o^{\text{set}}} = 1 - \frac{KK_c}{1 + KK_c} = \frac{1 + KK_c - KK_c}{1 + KK_c} = \frac{1}{1 + KK_c} \frac{°C}{°C} \qquad (6\text{-}1.23)$$

Again the offset is smaller the higher the controller gain.

Effect of Integral Mode

For a proportional-integral (PI) controller

$$G_c(0) = \lim_{s \to 0} K_c \left(1 + \frac{1}{\tau_I s} \right) = \infty$$

in which case, by substitution into Eq. 6-1.19 or 6-1.23, in place of K_c, we can see that the offset is zero. The same is true for a PID controller and for a pure-integral controller.

EXAMPLE 6-1.4

For the heat exchanger of Fig. 6-1.1, calculate the linearized ratios for the steady-state error in outlet temperature to

(a) A change in process flow.

(b) A change in set point.

The operating conditions and instrument specifications are

Process fluid flow	$w = 12$ kg/s
Inlet temperature	$T_i = 50°C$
Set point	$T_o^{set} = 90°C$
Heat capacity of fluid	$c_p = 3.75$ kJ/kg°C
Latent heat of steam	$\lambda = 2250$ kJ/kg
Capacity of steam valve	$w_{s,max} = 1.6$ kg/s
Transmitter range	50–150°C

SOLUTION

If we assume that heat losses are negligible, we can write the following steady-state energy balance:

$$\overline{w}c_p \left(\overline{T}_o - T_i \right) = \overline{w}_s \lambda$$

and, solving for $\overline{w}_s$, the steam flow required to maintain T_o at 90°C is

$$\overline{w}_s = \frac{\overline{w}c_p(\overline{T}_o - T_i)}{\lambda} = \frac{(12)(3.75)(90 - 50)}{2250} = 0.80 \text{ kg/s}$$

The next step is to calculate the steady-state open-loop gains of each of the elements in the loop.

Exchanger. From the steady-state energy balance, solving for T_o yields

$$\overline{T}_o = T_i + \frac{\overline{w}_s \lambda}{\overline{w}c_p}$$

By linearization, as learned in Section 2-6, we obtain

$$K_w = \frac{\overline{\partial T_o}}{\partial w} = -\frac{\overline{w}_s \lambda}{\overline{w}^2 c_p} = -\frac{(0.8)(2250)}{(12)^2(3.75)} = -3.33 \frac{°C}{\text{kg/s}}$$

$$K_s = \frac{\overline{\partial T_o}}{\partial w_s} = \frac{\lambda}{\overline{w}c_p} = \frac{(2250)}{(12)(3.75)} = 50 \frac{°C}{\text{kg/s}}$$

Control valve. Assuming a linear valve with constant pressure drop, the gain of the valve is, from Section 5-2,

$$K_v = \frac{w_{s,max}}{100} = \frac{1.6}{100} = 0.016 \frac{\text{kg/s}}{\%CO}$$

Sensor-transmitter. From Section 5-1 the gain of the transmitter is

$$K_T = \frac{100 - 0}{150 - 50} = 1.0 \frac{\%TO}{°C}$$

Then $K = K_T K_s K_v = (1.0)(50)(0.016) = 0.80$ %TO/%CO.

Table 6-1.3 Offset for heat exchanger control loop

K_c, $\dfrac{\%\text{CO}}{\%\text{TO}}$	$\dfrac{e}{\Delta w}$, $\dfrac{°\text{C}}{\text{kg/s}}$	$\dfrac{e}{\Delta T_o^{\text{set}}}$, $\dfrac{°\text{C}}{°\text{C}}$
0	3.33	1.00
0.5	2.38	0.714
1.0	1.85	0.556
5.0	0.67	0.200
10.0	0.37	0.111
20.0	0.20	0.059
100.0	0.04	0.012

(a) Substitute into Eq. 6-1.21 to get

$$\frac{e}{\Delta w} = \frac{-K_w}{1 + K K_c} = \frac{-3.33}{1 + 0.80 K_c}\ \frac{°\text{C}}{\text{k/s}}$$

(b) Substitute into Eq. 6-1.23 to get

$$\frac{e}{\Delta T_o^{\text{set}}} = \frac{1}{1 + K K_c} = \frac{1}{1 + 0.80 K_c}\ \frac{°\text{C}}{°\text{C}}$$

The results for different values of K_c are given in Table 6-1.3. We see that the offset in outlet temperature approaches zero as the gain is increased. These results illustrate the point made in Section 5-3 regarding the fact that the offset decreases when the gain of the proportional controller is increased. As pointed out there, the gain of the controller is limited by the stability of the loop. We shall see this in the next section.

6-2 STABILITY OF THE CONTROL LOOP

As defined in Section 2-3.3, a system is stable if its output remains bound for a bound input. Most industrial processes are open-loop stable; that is, they are stable when not a part of a feedback control loop. This is equivalent to saying that most processes are self-regulating; that is, the output will move from one steady state to another when driven by changes in its input signals. As we learned in Chapter 4, a typical example of an open-loop unstable process is an exothermic stirred tank reactor.

Even for open-loop stable processes, stability becomes a consideration when the process becomes a part of a feedback control loop. This is because the signal variations may reinforce each other as they travel around the loop causing the output—and all the other signals in the loop—to become unbound. As noted in Chapter 1, the behavior of a feedback control loop is essentially oscillatory—"trial and error." Under some circumstances the oscillations may increase in magnitude, resulting in an unstable process. The easiest illustration of an unstable feedback loop is the controller whose direction of action is the opposite of what it should be. For example, in the heat exchanger sketched in the preceding section, if the controller output were to increase with increasing temperature (direct-acting controller), the loop would be unstable because the opening of the steam valve would cause a further increase in temperature. What is needed in this case is a reverse-acting controller that decreases its output when the temperature increases so as to close the steam valve and bring the temperature back down. However, even for a controller with the proper action, the system may become unstable because of the lags in the loop. This usually happens as the loop gain is increased. The controller gain at which the loop reaches the threshold of instability is therefore of utmost importance in the design of a feedback control loop. This maximum gain is known as the *ultimate gain*.

In this section we will determine a criterion for the stability of dynamic systems and study the direct substitution method for calculating the ultimate gain. Then we will study the effect of various loop parameters on its stability.

6-2.1   Criterion of Stability

We have seen earlier that the response of a control loop to a given input can be represented (Eq. 6-1.18) by

$$C(t) = b_1 e^{r_1 t} + b_2 e^{r_2 t} + \cdots + b_n e^{r_n t} + \text{(input terms)} \qquad \textbf{(6-2.1)}$$

where $C(t)$ is the controlled variable and $r_1, r_2, \ldots, r_n$ are the roots of the characteristic equation of the loop.

Assuming that the input terms remain bound as time increases, the stability of the loop requires that the unforced response terms also remain bound as time increases. This depends only on the roots of the characteristic equation and can be expressed as follows:

For real roots: If $r < 0$ then $e^{rt} \to 0$ as $t \to \infty$

For complex roots: $r = \sigma + i\omega$ $e^{rt} = e^{\sigma t} \sin(\omega t + \theta)$

If $\sigma < 0$ then $e^{\sigma t} \sin(\omega t + \theta) \to 0$ as $t \to \infty$

In other words, the real part of the complex roots and the real roots must be negative in order for the corresponding terms in the response to decay to zero. This result is not affected by repeated roots, as this only introduces a polynomial of time into the solution which cannot overcome the effect of the decaying exponential term (see Chapter 2). Notice that, if any root of the characteristic equation is a positive real number, or a complex number with a positive real part, that term on the response (Eq. 6-2.1) will be unbound and the entire response will be unbound even though all the other terms may decay to zero. This brings us to the following statement of the criterion for the stability of a control loop:

> *For a feedback control loop to be stable, all of the roots of its characteristic equation must be either negative real numbers or complex numbers with negative real parts.*

If we now define the complex s plane as a two-dimensional graph with the horizontal axis for the real parts of the roots and the vertical axis for the imaginary parts, we can make the following graphical statement of the criterion of stability (see Fig. 6-2.1):

> *For a feedback control loop to be stable, all of the roots of its characteristic equation must fall on the left-hand half of the s plane, also known as the "left-hand plane."*

We must point out that both of these statements of the stability criterion in the Laplace domain apply in general to any physical system, not just feedback control loops. In each case we obtain the characteristic equation by setting the denominator of the transfer function of the system equal to zero.

Having articulated the criterion of stability, let us turn our attention to the determination of the stability of a control loop.

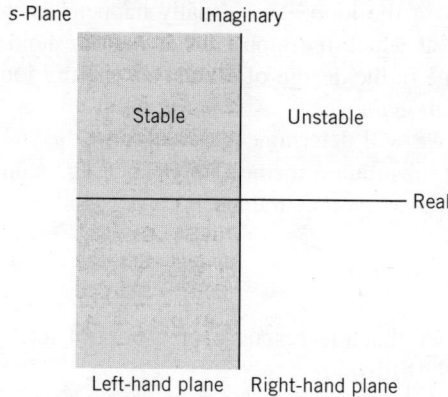

Figure 6-2.1 The s plane, showing the regions of stability and instability for the location of the roots of the characteristic equation.

6-2.2 Direct Substitution Method

Direct substitution is a convenient method for determining the range of controller parameters for which the closed-loop response is stable. The method is based on the fact that the roots of the characteristic equation vary continuously with the loop parameters. Consequently, at the point of instability, at least one and usually two of the roots must lie on the imaginary axis of the complex plane as they cross from the left half-plane to the right. This means that the roots are pure imaginary numbers—zero real parts—at the verge of instability. At this point the loop is said to be marginally stable, and the corresponding term on the loop output is, in the Laplace domain,

$$C(s) = \frac{b_1 s + b_2}{s^2 + \omega_u^2} + \text{(other terms)} \tag{6-2.2}$$

or, inverting, this term, from Table 2-1.1, is a sine wave in the time domain:

$$C(t) = b_1' \sin(\omega_u t + \theta) \tag{6-2.3}$$

where ω_u is the frequency of sine wave, θ is its phase angle, and b_1' is its amplitude (constant). This means that, at the point of marginal stability, the characteristic equation must have a pair of pure imaginary roots at $r_{1,2} = \pm i\omega_u$. The frequency ω_u with which the loop oscillates is the *ultimate frequency*. The controller gain at which this point of marginal instability is reached is called the *ultimate gain*. At a gain just below the ultimate, the loop oscillates with decaying amplitude, while at a gain just above the ultimate gain, the amplitude of the oscillations increases with time. At the point of marginal stability the amplitude of the oscillation remains constant with time. Figure 6-2.2 shows these responses along with the graphical representation of the *ultimate period*, T_u. This is the period of the oscillations at the ultimate gain, and is related to the ultimate frequency, ω_u, in rad/s, by

$$T_u = \frac{2\pi}{\omega_u} \tag{6-2.4}$$

The method of direct substitution consists of substituting $s = i\omega_u$ in the characteristic equation. This results in a complex equation that can be converted into two simultaneous equations.

$$\text{Real part} = 0$$
$$\text{Imaginary part} = 0$$

(a)

$\leftarrow T_u \rightarrow$

(b)

(c)

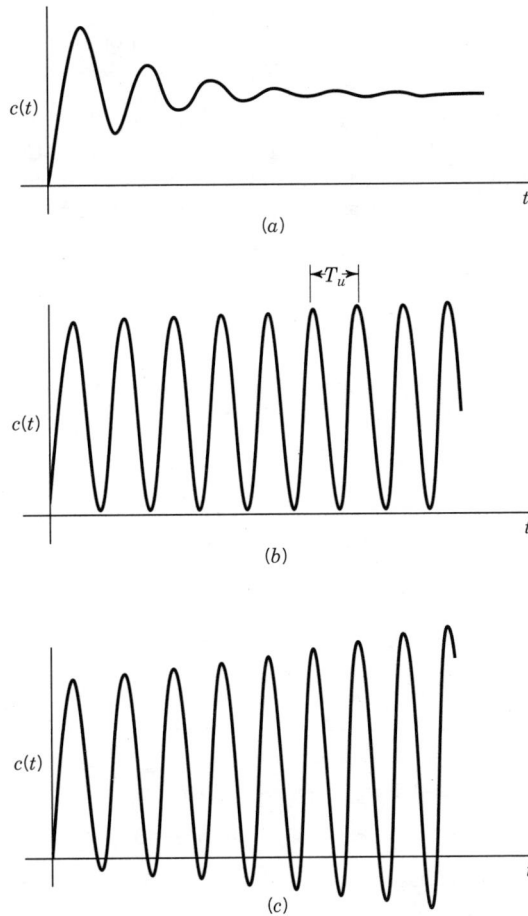

Figure 6-2.2 Response of closed loop with the controller gain less than (a), equal to (b), and greater than (c) the ultimate gain.

From these we can solve for two unknowns: one is the ultimate frequency ω_u, the other is any of the parameters of the loop, usually the controller gain at the point of marginal instability or ultimate gain. Generally, the closed-loop response is unstable when the controller gain is greater than the ultimate gain. The following example shows how the direct substitution method is used to compute the ultimate gain and period of the heat exchanger control loop.

EXAMPLE 6-2.1

Ultimate Gain and Period of Temperature Controller by Direct Substitution

Let us assume that the transfer functions for the various elements of the temperature control loop of Fig. 6-1.3 are as follows:

Exchanger. The exchanger response to the steam flow has a gain of $50°C/(kg/s)$ and a time constant of $30s$.

$$G_s(s) = \frac{50}{30s + 1} \frac{°C}{kg/s}$$

Sensor-transmitter. The sensor-transmitter has a calibrated range of 50 to $150°C$ and a time constant of $10s$.

$$H(s) = \frac{1.0}{10s + 1} \frac{\%TO}{°C}$$

Control valve. The control valve has a maximum capacity of 1.6 kg/s of steam, linear characteristics, a constant pressure drop, and a time constant of $3s$.

$$G_v(s) = \frac{0.016}{3s + 1} \frac{\text{kg/s}}{\%\text{CO}}$$

Controller. The controller is proportional only. $G_c(s) = K_c$ %CO/%TO.

The problem is then to determine the ultimate controller gain, that is, the value of K_c at which the loop becomes marginally stable, and the ultimate period.

SOLUTION

The characteristic equation is given by Eq. 6-1.12:

$$1 + H(s)G_s(s)G_v(s)G_c(s) = 1 + \frac{1.0}{10s + 1} \cdot \frac{50}{30s + 1} \cdot \frac{0.016}{3s + 1} \cdot K_c = 0$$

where we have substituted the transfer function for each element of the loop. We must now rearrange this equation into polynomial form.

$$(10s + 1)(30s + 1)(3s + 1) + 0.80K_c = 0$$
$$900s^3 + 420s^2 + 43s + 1 + 0.80K_c = 0$$

Next we substitute $s = i\omega_u$ at $K_c = K_{cu}$.

$$900i^3\omega_u^3 + 420i^2\omega_u^2 + 43i\omega_u + 1 + 0.80K_{cu} = 0$$

Then we substitute $i^2 = -1$ and separate the real and imaginary parts.

$$(-420\omega_u^2 + 1 + 0.80K_{cu}) + i(-900\omega_u^3 + 43\omega_u) = 0 + i0$$

From this complex equation we obtain the following two equations, as both the real and imaginary parts must be zero.

$$-420\omega_u^2 + 1 + 0.80K_{cu} = 0$$
$$-900\omega_u^3 + 43\omega_u = 0$$

The solution of this set has the following possibilities:

For $\omega_u = 0$	$K_{cu} = -1.25$ %CO/%TO
For $\omega_u = 0.2186$ rad/s	$K_{cu} = 23.8$ %CO/%TO

The first solution corresponds to the monotonic instability caused by having the wrong action on the controller. In this case the system does not oscillate but moves monotonically in one direction or the other. The crossing of the imaginary axis occurs at the origin ($s = 0$). This solution is irrelevant.

The ultimate gain for the second solution is the one that is relevant. At this gain the loop response oscillates with a frequency of 0.2186 rad/s (0.0348 hertz) or a period of

$$T_u = \frac{2\pi}{\omega_u} = \frac{2\pi}{0.2186} = 28.7 \text{ s}$$

We saw in the preceding section that the offset or steady-state error inherent in proportional controllers can be reduced by increasing the controller gain. We see here that stability imposes a limit on how high that gain can be. It is of interest to study how the other parameters of the loop affect the ultimate gain and period.

EXAMPLE 6-2.2

Ultimate Gain and Period of Continuous Stirred Tank Heater

Obtain the ultimate gain and period for the temperature control loop of the continuous stirred tank heater of Example 6-1.1, assuming a proportional controller.

SOLUTION   From the transfer function for the closed loop, developed in Example 6-1.1, the characteristic equation of the loop is

$$1 + G_c(s)G_1(s) = 0$$

with $G_c(s) = K_c$ %CO/%TO (proportional controller), and

$$G_1(s) = \frac{1.954}{(0.2s + 1)(8.34s + 1)(0.502s + 1)(0.75s + 1)}\frac{\%\text{TO}}{\%\text{CO}}$$

where $1.954 = 1.652 \cdot 1.183 \cdot 1.0$. Substitute $G_1(s)$ and $G_c(s)$ into the characteristic equation and clear fractions to get

$$(0.2s + 1)(8.34s + 1)(0.502s + 1)(0.75s + 1) + 1.954K_c = 0$$

or, in polynomial form:

$$0.628s^4 + 5.303s^3 + 12.73s^2 + 9.790s + 1 + 1.954K_c = 0$$

Substitute $s = i\omega_u$ at $K_c = K_{cu}$, and group the real and imaginary parts.

$$(0.628\,\omega_u^4 - 12.73\,\omega_u^2 + 1 + 1.954\,K_{cu}) + i(-5.303\,\omega_u^3 + 9.790\,\omega_u) = 0 + i0$$

Solve for ω_u by setting the imaginary part to zero and for K_{cu} by setting the real part to zero.

$$\omega_u = \sqrt{\frac{9.790}{5.303}} = 1.359\,\frac{\text{rad}}{\text{s}}$$

$$K_{cu} = \frac{-0.628(1.359)^4 + 12.73(1.359)^2 - 1}{1.954} = 10.4\,\frac{\%\text{CO}}{\%\text{TO}}$$

The ultimate periods is $T_u = 2\pi/1.359 = 4.6$ min.

6-2.3   Effect of Loop Parameters on the Ultimate Gain and Period

Let us assume that the calibrated range of the temperature sensor-transmitter in Example 6-2.1 is reduced to 75 to 125°C. The new transmitter gain is

$$K_T = \frac{100 - 0}{125 - 75} = 2.0\frac{\%\text{TO}}{°\text{C}}$$

The characteristic equation of the loop becomes

$$900s^3 + 420s^2 + 43s + 1 + 1.60K_c = 0$$

and the ultimate gain and period are

$$K_{cu} = 11.9\frac{\%\text{CO}}{\%\text{TO}} \qquad T_u = 28.7\text{ s}$$

This is exactly half the ultimate gain for the base case, showing that the ultimate loop gain remains the same. The loop gain is defined as the product of the gains of all of the blocks in the loop.

$$K_L = K_v K_s K_T K_c \qquad\qquad \textbf{(6-2.5)}$$

where K_L is the (dimensionless) loop gain.

Similarly, if we were to double the capacity of the control valve, and thus its gain, the ultimate controller gain would be reduced to half its value for the base case.

Next let us assume that a faster sensor-transmitter with a time constant of 5 s is installed in this service replacing the 10-s instrument. The new transfer function is

$$H(s) = \frac{1.0}{5s + 1}\frac{\%\text{TO}}{°\text{C}}$$

The characteristic equation is then

$$1 + \frac{1.0}{5s+1} \cdot \frac{50}{30s+1} \cdot \frac{0.016}{3s+1} \cdot K_c = 0$$
$$450s^3 + 255s^2 + 38s + 1 + 0.80K_c = 0$$

and the ultimate gain, frequency, and period are

$$\omega_u = 0.2906 \ \frac{\text{rad}}{\text{s}}, \qquad T_u = 21.6 \ \text{s}, \qquad K_{cu} = 25.7 \ \frac{\%\text{CO}}{\%\text{TO}}$$

The reduction of the time constant of the sensor has resulted in a slight increase in the ultimate gain and a decrease in the period of oscillation of the loop. This is because we have reduced the measurement lag on the control loop. A similar result would be obtained if the time constant of the control valve were to be reduced. However, the increase in the ultimate gain would be even less because the valve is not as slow as the sensor-transmitter. You are invited to verify this.

Finally let us consider a case in which a change in exchanger design results in a shorter time constant for the process, namely from 30 to 20 s. The new transfer function is

$$G_s(s) = \frac{50}{20s+1} \ \frac{°\text{C}}{\text{kg/s}}$$

The characteristic equation is then

$$1 + \frac{1.0}{10s+1} \cdot \frac{50}{20s+1} \cdot \frac{0.016}{3s+1} \cdot K_c = 0$$
$$600s^3 + 290s^2 + 33s + 1 + 0.80K_c = 0$$

and the ultimate frequency, gain, and period are

$$\omega_u = 0.2345 \ \frac{\text{rad}}{\text{s}}, \qquad T_u = 26.8 \ \text{s}, \qquad K_{cu} = 18.7 \ \frac{\%\text{CO}}{\%\text{TO}}$$

Surprisingly, the ultimate gain is reduced by a reduction in the process time constant. This is opposite to the effect of reducing the time constant of the sensor-transmitter. The reason is that when the longest or dominant time constant is reduced, the relative effect of the other lags in the loop becomes more pronounced. In other words, in terms of the ultimate gain, reducing the longest time constant is equivalent to proportionately increasing the other time constants in the loop. However, the loop with the shorter process time constant responds faster than the original one, as shown by the shorter ultimate period. The results of the direct substitution method for the four cases considered here are summarized in Table 6-2.1. We note that the loop can oscillate significantly faster when the time constant of the sensor-transmitter is reduced from 10 to 5 s. Also, the loop oscillates slightly faster when the exchanger time constant is reduced from 30 to 20 s, in spite of the significant reduction in ultimate gain. Changing the gains of the blocks on the loop has no effect on the frequency of oscillation, or on the ultimate loop gain.

6-2.4 Effect of Dead Time

We have seen how the direct substitution method allows us to study the effect of various loop parameters on the stability of the feedback control loop. Unfortunately, the method fails when any of the blocks on the loop contains a dead-time (transportation lag or time delay) term. This is because the dead time introduces an exponential function

Table 6-2.1 Direct substitution results for heat exchanger control loop

	K_{cu}	ω_u, rad/s	T_u, s
1. Base case	23.8	0.2186	28.7
2. $H(s) = \dfrac{2.0}{10s+1}$	11.9	0.2186	28.7
3. $H(s) = \dfrac{1.0}{5s+1}$	25.7	0.2906	21.6
4. $G_s(s) = \dfrac{50}{20s+1}$	18.7	0.2345	26.8

of the Laplace transform variable in the characteristic equation. This means that this equation is no longer a polynomial and the methods we have learned in this section no longer apply. An increase in dead time tends to reduce the ultimate loop gain very rapidly. This effect is similar to the effect of increasing the nondominant time constants of the loop in that it is relative to the magnitude of the dominant time constant. We will be able to study the exact effect of dead time on loop stability when we study the method of frequency response in Chapter 8.

We must point out that the exchanger we have used in this chapter is a distributed-parameter system, that is, the temperature of the process fluid is distributed throughout the exchanger. The transfer functions for such systems usually contain a least one dead-time term, which, for simplicity, we have ignored.

An estimate of the ultimate gain and frequency of a loop with dead time may sometimes be obtained by using an approximation to the dead-time transfer function. A popular approximation is the first-order Padé approximation, given by

$$e^{-t_0 s} \approx \frac{1 - \dfrac{t_0}{2}s}{1 + \dfrac{t_0}{2}s} \tag{6-2.6}$$

where t_0 is the dead time. More accurate higher-order approximations are also available, but they are too complex to be practical. The following example illustrates the use of the Padé approximation with the direct substitution method.

EXAMPLE 6-2.3

Ultimate Gain and Frequency of First-Order-Plus-Dead-Time Process

Let the process transfer function of the loop of Fig. 6-1.4 be

$$G_1(s) = \frac{Ke^{-t_0 s}}{\tau s + 1}$$

where K is the gain, t_0 is the dead time, and τ is the time constant. Determine the ultimate gain and frequency of the loop as a function of the process parameters if the controller is a proportional controller, $G_c(s) = K_c$ %CO/%TO.

SOLUTION

From Example 6-1.2, the characteristic equation of the loop is

$$1 + G_1(s)G_c(s) = 1 + \frac{KK_c e^{-t_0 s}}{\tau s + 1} = 0$$

Substitute the first-order Padé approximation, Eq. 6-2.6, to get

$$1 + \frac{KK_c \left(1 - \frac{t_0}{2}s\right)}{(\tau s + 1)\left(1 + \frac{t_0}{2}s\right)} = 0$$

Clear the fraction.

$$\frac{t_0}{2}\tau s^2 + \left(\tau + \frac{t_0}{2} - KK_c\frac{t_0}{2}\right)s + 1 + KK_c = 0$$

By the direct substitution method, $s = i\omega_u$ at $K_c = K_{cu}$, after setting the real and imaginary parts equal to zero and solving the two equations simultaneously, yields

$$\omega_u = \frac{2}{t_0}\sqrt{\frac{t_0}{\tau} + 1}$$

$$KK_{cu} = 1 + 2\left(\frac{\tau}{t_0}\right)$$

These formulas show that the ultimate loop gain goes to infinity—with no stability limit—as the dead time approaches zero, which agrees with the results of Example 6-1.2. However, any finite amount of dead time imposes a stability limit on the loop gain. The ultimate frequency increases with decreasing dead time and becomes very small as the dead time increases. This means that dead time slows down the response of the loop.

6-2.5 Summary

From the results of direct-substitution analysis in the preceding examples we can summarize the following general effects of various loop parameters.

- Stability imposes a limit on the overall loop gain, so that an increase in the gain of the control valve, the transmitter, or the process, results in a decrease in the ultimate controller gain.

- An increase in dead time or in any of the nondominant (smaller) time constants of the loop results in a reduction of the ultimate gain.

- A decrease in the dominant (longest) time constant of the loop results in a decrease in the ultimate loop gain and an increase in the ultimate frequency of the loop.

6-3 SUMMARY

This chapter presented the analysis of feedback control loops. We have learned how to develop the closed-loop transfer function and the characteristic equation of the loop, and how to estimate the closed-loop steady-state gain, the ultimate gain, and the ultimate period of the loop. We have also studied how the various loop parameters affect the ultimate gain and period. The next chapter looks at various important methods for tuning feedback controllers.

PROBLEMS

6-1. A feedback control loop is represented by the block diagram of Fig. 6-1.4. The process can be represented by two lags in series.

$$G_1(s) = \frac{K}{(\tau_1 s + 1)(\tau_2 s + 1)}$$

where the process gain is $K = 0.10$ %TO/%CO and the time constants are

$$\tau_1 = 1 \text{ min}$$
$$\tau_2 = 0.8 \text{ min}$$

The controller is a proportional controller: $G_c(s) = K_c$.

(a) Write the closed-loop transfer function and the characteristic equation of the loop.

(b) For what values of the controller gain is the loop response to a step change in set point overdamped, critically damped, and underdamped? Can the loop be unstable?

(c) Find the effective time constants, or the second-order time constant and the damping ratio, of the closed loop, for $K_c = 0.1$, 0.125, and 0.20 %CO/%TO.

(d) Determine the steady-state offset for each of the gains in part (c) and a unit step change in set point.

6-2. Do Problem 6-1 for a process transfer function of

$$G(s) = \frac{6(1-s)}{(s+1)(0.5s+1)} \frac{\%}{\%}$$

Transfer functions such as this are typical of processes consisting of two lags in parallel with opposite action (see Section 4-4.3). The controller is a proportional controller as in Problem 6-1.

6-3. A feedback control loop is represented by the block diagram of Fig. 6-1.4. The process can be represented by a first-order lag and the controller is proportional-integral (PI):

$$G_1(s) = \frac{K}{\tau s + 1}$$

$$G_c(s) = K_c \left(1 + \frac{1}{\tau_I s} \right)$$

Without loss of generality you can set the process time constant τ equal to 1, and the process gain K equal to 1.

(a) Write the closed-loop transfer function and the characteristic equation of the loop. Is there offset?

(b) Is there an ultimate gain for this loop?

(c) Determine the response of the closed loop to a step change in set point for $\tau_I = \tau$ as the controller gain varies from zero to infinity.

6-4. Given the feedback control loop of Problem 6-1 and a pure integral controller

$$G_c(s) = \frac{K_I}{s}$$

(a) Determine the ultimate controller gain and the ultimate period.

(b) Recalculate the ultimate controller gain for $\tau_2 = 0.10$ and for $\tau_2 = 2$. Are your results what you expected?

6-5. For the feedback control loop of Problem 6-1 and a proportional-integral controller

$$G_c(s) = K_c \left(1 + \frac{1}{\tau_I s} \right)$$

(a) Determine the ultimate loop gain $K K_{cu}$ and the ultimate period of oscillation as functions of the integral τ_I.

(b) Determine the damping ratio and the decay ratio with the controller gain set equal to one-half the ultimate gain, and the integral time set equal to 1.

6-6. *Design of Gas Flow Control Loop.* A flow control loop, consisting of an orifice in series with the control valve, a differential pressure transmitter, and a controller, is to be designed for a nominal process flow of 150 kscf/h (kscf = 1000 cubic feet of gas at standard conditions of 60°F and 1 atm). The upstream conditions are constant at 100 psig and 60°F, the downstream pressure is constant at 80 psig, and the fluid is air (mol. wt. = 29). The valve has equal percentage characteristics with $\alpha = 50$, and a square root extractor is built into the transmitter so that its output signal is linear with flow. The valve time constant is 0.06 min, and the transmitter time constant is negligible. A proportional-integral (PI) controller controls the flow. Draw the block diagram of the flow control loop showing the specific transfer functions of the controller, the control valve, and the flow transmitter. Write the closed-loop transfer function for the loop, and find the time constant of the loop for $K_c = 0.9$ %CO/%TO, and $\tau_I = \tau_v$.

6-7. *Steam Flow Control Loop.* A process heater requires 3500 lb/h of steam to heat a process fluid. A control valve and linear flow transmitter are installed to control the flow of the steam. The conditions are as follows: control valve upstream conditions are 45 psig, superheated 50°F; downstream pressure is 20 psig; critical flow factor is 0.8. A linear valve sized for 100 % overcapacity is proposed. The flow transmitter is sized to measure a maximum flow of 5000 lb/h and its output is linear with flow; that is, it has a built-in square root extractor. Draw the block diagram for the flow control loop showing all transfer functions and write the closed-loop transfer function. Use a proportional-integral (PI) feedback controller with the integral time set equal to the time constant of the control valve. Find the time constant of the loop for $K_c = 0.5$ %CO/%TO.

6-8. For a feedback control loop represented by the block diagram of Fig. 6-1.4, determine the ultimate gain and period for a proportional controller and each of the following process transfer functions:

(a) $G_1(s) = \dfrac{1}{(s+1)^4}$

(b) $G_1(s) = \dfrac{1}{(s+1)^2}$

(c) $G_1(s) = \dfrac{1}{(4s+1)(2s+1)(s+1)}$

(d) $G_1(s) = \dfrac{(0.5s+1)}{(4s+1)(2s+)(s+1)}$

(e) $G_1(s) = \dfrac{1}{(4s+1)(0.2s+1)(0.1s+1)}$

(f) $G_1(s) = \dfrac{e^{-0.6s}}{6s+1}$

6-9. Work Problem 6-8 using a pure integral controller, $G_c(s) = K_I/s$.

6-10. An open-loop unstable process can be represented by the block diagram of Fig. 6-1.4 and the following transfer function:

$$G_1(s) = \frac{K}{(5s-1)(\tau_v s + 1)(\tau_T s + 1)}$$

where τ_v and τ_T are, respectively, the time constants of the control valve and the transmitter. Assuming a proportional controller, find the range of the loop gain KK_c for which the loop is stable if

(a)   the valve and transmitter time constants are negligible.

(b)   the valve time constant is negligible and $\tau_T = 1.0$ min.

(c) $\tau_v = 0.1$ min and $\tau_T = 1.0$ min.

Hint: Parts (a) and (b) can be solved by writing the roots of the characteristic equation as functions of the loop gain. Part (c) requires the direct substitution method. Notice that there is a lower stability limit on the loop gain for each case.

6-11. Calculate the ultimate gain and period of oscillation for a proportional analyzer controller installed on the blending tank of Problem 3-18. The control valve is to be installed on the dilute stream and sized for 100 % overcapacity. The valve has linear characteristics, a constant pressure drop of 5 psi, and can be represented by a first-order lag with a time constant of 0.1 min. The analyzer transmitter has a range of 20 to 70 kg/m³, and can be represented by a first-order lag with a time constant of 3 min. Draw also the block diagram of the loop showing all transfer functions, and calculate the offset caused by a change of 0.1 m³/min in the flow of the concentrated solution when the controller gain is one-half the ultimate gain.

Note: The simulation of this process is the subject of Problem 13-11.

6-12. In Section 4-2.3, a nonisothermal chemical reactor is modeled in detail. Calculate the ultimate gain and period of a proportional temperature controller for the reactor assuming that the control valve is installed on the cooling water line to the jacket to manipulate the cooling water flow $f_c(t)$. The valve is equal percentage with constant pressure drop and $\alpha = 50$, and the temperature transmitter has a range of 640 to 700°R. The time constants of the valve and the transmitter can be neglected. Draw also the block diagram of the temperature control loop showing all transfer functions. *Hint:* For simplicity you may consider the cooling water flow as the only input variable, that is, assume all other input variables are constant.

Note: The simulation of this loop is the subject of Problem 13-21.

6-13. Solve Problem 6-12 assuming the control valve is installed on the reactants line to manipulate the flow of reactants $f(t)$. Assume all other input variables, including the coolant flow, are constant. Use a linear control valve with constant pressure drop and sized for 100 % overcapacity.

6-14. In Problem 4-4 you are asked to model three mixing tanks in series. Find the ultimate gain and period of a proportional controller that is to control the outlet composition from the third tank, $x_6(t)$, by manipulating the flow of water into the first tank, $f_1(t)$. Assume an equal percentage control valve with constant pressure drop and $\alpha = 50$. The analyzer transmitter has a range of 0.30 to 0.70 mass fraction units, and the time constants of the valve and transmitter can be neglected. Draw also the block diagram of the loop and calculate the offset caused by a change of 10 gpm in flow f_2 when the controller gain is set equal to one-half the ultimate gain.

6-15. *Feedback Control of Reactors in Series.* Consider the control of the concentration out of the second of the two reactors in series of Problem 4-9 by manipulating the reactants flow. The reactors have each a volume of 125 ft³, the inlet concentration is initially $c_{Ai}(0) = 7.0$ lbmole/ft³, and the rate coefficient in each reactor is $k = 0.2$ min⁻¹. The initial inlet flow is $f(0) = 10$ ft³/min, and the recycle flow is zero. The control valve is linear, sized for 100 % overcapacity, and has a negligible time lag. The analyzer transmitter has a range of 0 to 5 lbmole/ft³ and can be represented by a first-order lag with a time constant of 0.5 min. The set point of the analyzer controller is initially equal to the initial steady-state value of the concentration from the second reactor. Draw the block diagram of the control loop showing all transfer functions and calculate the ultimate gain and period of the loop for a proportional controller. Calculate also the offset of the controller for a change in inlet concentration from 7.0 to 8.0 lbmole/ft³, assuming the controller gain is set equal to one-half the ultimate gain. What would be the offset if the loop is opened (controller on "manual" or $K_c = 0$)? What would it be if the controller were a proportional-integral (PI) controller?

Note: The simulation of this process is the subject of Problem 13-22.

6-16. In Problem 4-5 you were asked to model a tank in which steam is mixed with a liquid stream. Find the ultimate gain and period for a proportional controller which is to control the temperature of the stream leaving the tank by manipulating the steam valve position. Draw also the block diagram for the temperature control loop showing all transfer functions. Calculate the offset caused by a change 2 gpm in the inlet liquid flow when the controller gain is set equal to one-half the ultimate gain.

6-17. *Composition Control of Three Isothermal Reactors in Series.* Consider the concentration control loop for the three stirred reactors shown in Fig. P6-1. Each reactor has a volume of 1000 gal, and the design flow of reactants is 100 gpm. The initial inlet concentration of reactant A into the first reactor is 4 lb/gal, and the reaction rate is proportional to the concentration of A in each reactor with a constant coefficient of 0.1 min⁻¹. The outlet concentration transmitter has a range of 0 to 1.0 lb/gal, and a negligible time constant. The reactant control valve is linear with constant pressure drop of

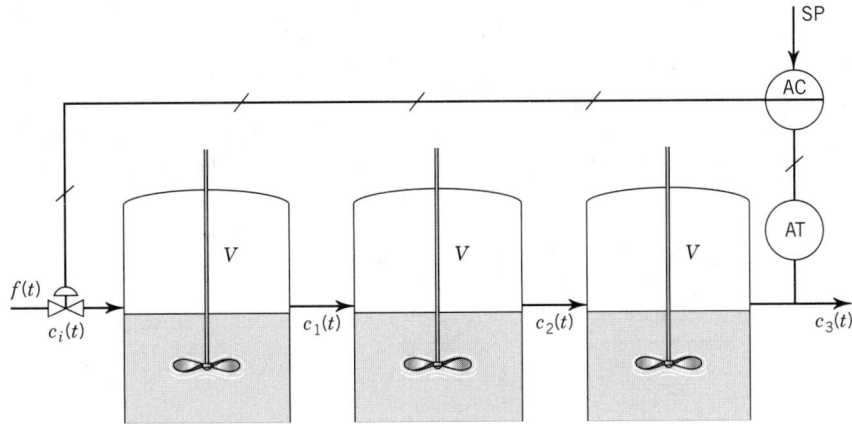

Figure P6-1 Three stirred reactors in series for Problem 6-17.

5 psi and is sized for 100 % overcapacity. The time constant of the valve is 0.1 min.

(a) Draw a block diagram of the loop showing all transfer functions.

(b) If a proportional controller with a gain of 1.0 %CO/%TO is installed on this system, what would be the offset caused by a 1.0 lb/gal change in inlet reactant concentration? What would be the offset if the controller gain were zero? What if the controller were proportional-integral (PI)?

(c) Calculate the ultimate gain and period for the loop assuming a proportional controller.

Note: The simulation of this process is the subject of Problem 13-23.

6-18. *Compressor Suction Pressure Control.* Figure P6-2 shows the schematic of a compressor suction pressure control loop. A mass balance on the suction volume results in the following approximate linear model for the suction pressure:

$$P_s(s) = \frac{0.5}{7.5s + 1}[F_i(s) - F_c(s)] \text{ psi}$$

where $F_i(s)$ and $F_c(s)$ are, respectively, the inlet and compressor flows, kscf/min (1 kscf = 1000 ft^3 at standard conditions of 1 atm and 60°F), and the time constant is in seconds. The response of the compressor flow to the controller output signal, $M(s)$, in %CO, is

$$F_c(s) = \frac{0.36}{2.5s + 1}M(s)$$

The pressure transmitter has a range of 0 to 20 psig and can be represented by a first-order lag with a time constant of 1.2 s.

(a) Draw the block diagram for the loop and write the closed-loop transfer function and the characteristic equation. Must the controller be direct-acting or reverse-acting?

(b) Calculate the ultimate gain and period assuming a proportional controller.

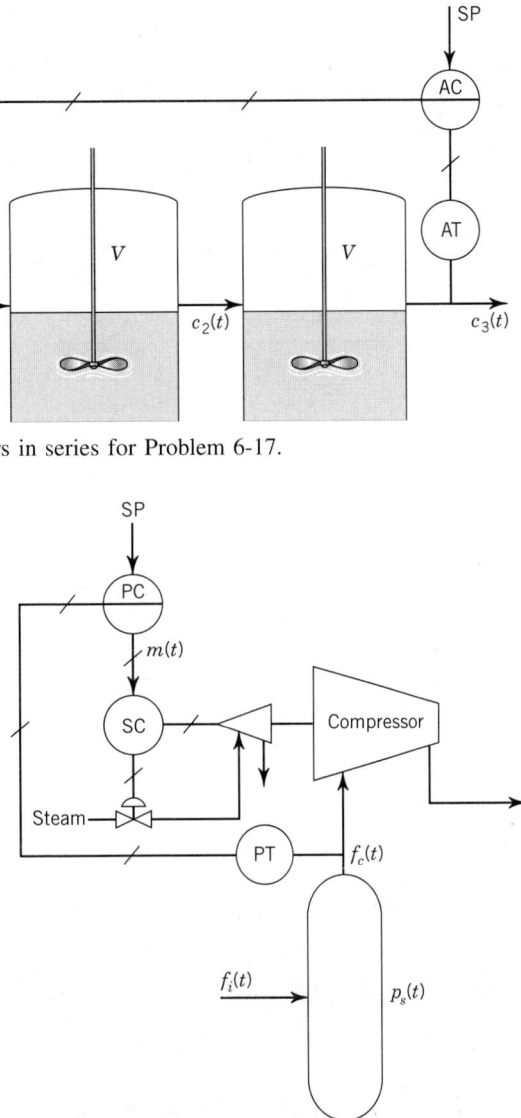

SC = Turbine speed controller

Figure P6-2 Compressor pressure control for Problem 6-18.

(c) Calculate the offset caused by a change of 1.0 kscf/min in the inlet flow when the controller gain is one-half the ultimate gain.

6-19. The parameters for the stirred tank cooler of Problem 4-7 are: $V = 5.0$ m^3, $U = 200$ kJ/min-m^2-°C, $A = 4.0$ m^2, $V_c = 1.1$ m^3, $\rho = 800$ kg/m^3, $c_p = 3.8$ kJ/kg-°C, $\rho_c = 1000$ kg/m^3, $c_{pc} = 4.2$ kJ/kg-°C. The design conditions are as follows: process flow = 0.10 m^3/min, inlet process temperature = 70°C, outlet process temperature = 45°C, and the coolant inlet temperature is 25°C. The temperature transmitter (TT) has a range of 20 to 70°C, and can be represented by a first-order lag with a time constant of 0.6 min. The coolant flow transmitter (FT) has a range of 0 to 0.8 m³/min, and the time constant of the flow control loop (FIC) is 0.1 min.

(a) Draw the complete block diagram for the temperature control loop showing all transfer functions. Should the coolant valve fail open or closed? Must the controller be direct-acting or reverse-acting?

(b) Calculate the ultimate gain and period for the temperature control loop assuming a proportional controller (TIC).

(c) Calculate the offset caused by a 5°C increase in inlet process temperature when the temperature controller gain is one-half the ultimate gain. What is the offset when the controller gain is zero (manual state)? What is the offset if the controller is proportional-integral (PI)?

Note: The simulation of this process is the subject of Problem 13-18.

6-20. A gas storage tank, shown in Fig. P6-3, supplies a gas with a molecular weight of 50 to two processes. The first process receives a normal flow of 500 scf/min (scf = ft³ at 1 atm and 60°F) and operates at a pressure of 30 psig, while the second process operates at a pressure of 15 psig. A process operating at 90 psig supplies gas to the storage tank at a rate of 1500 scf/min. The tank has a capacity of 550,000 ft³ and operates at 45 psig and 350°F. You may assume that

the pressure transmitter responds instantaneously with a calibrated range of 0 to 100 psig.

(a) Size all three valves for 100 % overcapacity. For all valves you can use the factor $C_f = 0.9$.

(b) Draw the complete block diagram for the system. You can consider as disturbances $P_1(t)$, $P_3(t)$, $P_4(t)$, $vp_3(t)$, $vp_4(t)$ and the set point to the controller.

(c) Can the feedback loop go unstable? If yes, what is its ultimate gain?

(d) Using a proportional only controller with a gain of 50 %CO/%TO, what is the offset observed for a set point change of +5 psi?

6-21. Consider the electric heater shown in Fig. P6-4. Two liquid streams with variable mass rates $w_A(t)$ and $w_B(t)$ come together in a tee and pass through the heater where they are thoroughly mixed and heated to temperature $T(t)$. The outlet temperature is controlled by manipulating the current through an electric coil. The outlet mass fraction of component B is also controlled by manipulating the inlet flow of stream B. The following information is known:

• The pressure drop across the valves can be assumed constant so, the flow through the valves is given by

$$w_A(t) = K_{v1}vp_1(t) \qquad w_B(t) = K_{v2}vp_2(t)$$

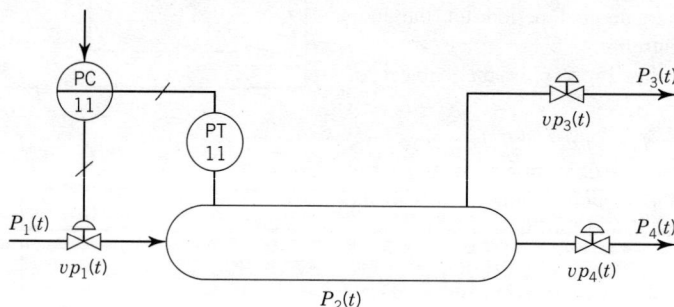

Figure P6-3 Gas storage tank for Problem 6-20.

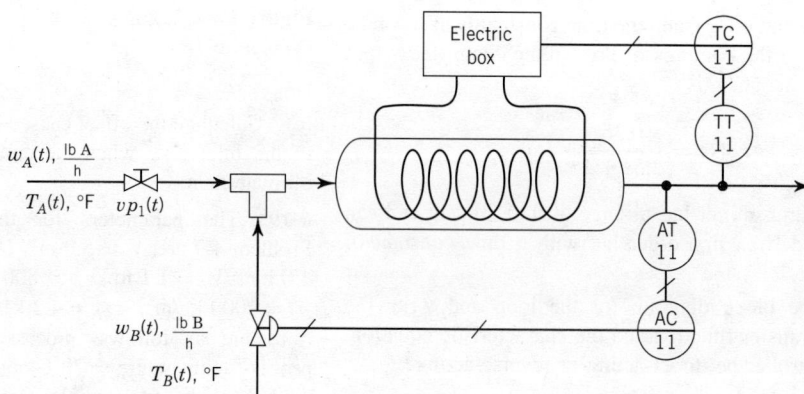

Figure P6-4 Electric heater for Problem 6-21.

These two streams are each pure in components A and B, respectively.

- The mass fraction of B in the outlet stream is related to the electrical conductivity of the stream. The conductivity of this stream is inversely proportional to the mass fraction, x_B, that is,

$$\text{Conductivity} = \frac{\beta}{x_B}$$

where β is a constant, mho-mass fraction/m. The conductivity transmitter has a range of C_L to C_H mho/m.

- You may assume that the heat transfer rate, q, is linear with the output of the controller in the range 0 to q_{max}.

The disturbances to this system are $vp_1(t)$, $T_A(t)$, and $T_B(t)$.

(a) Derive, from basic principles, the set of equations that describe the composition (conductivity) control loop. State all assumptions.

(b) Linearize the equations from part (a) and draw the complete block diagram for the conductivity loop. Show the transfer function of each block. Specify the required action of the composition controller assuming that the control valve is air-to-open.

(c) Derive, from basic principles, the set of equations that describe the temperature control loop. State all assumptions.

(d) Linearize the equations from part (c) and draw the complete block diagram for the temperature loop. Show the transfer function of each block. Specify also the action of the temperature controller.

(e) Write the characteristic equation for each of the control loops. Can either loop be made unstable by increasing the controller gain? Discuss briefly.

6-22. Consider the system shown in Fig. P6-5. In each of the two tanks the reaction A → E takes place. The rate of

reaction is given by

$$r(t) = kc_A(t), \frac{\text{lbmoles}}{\text{gal-min}}$$

where k is the reaction rate coefficient, min^{-1}, and $c_A(t)$ is the concentration, lbmole/gal. The disturbances to this process are $f_i(t)$ and $c_{Ai}(t)$. The concentration out of the second reactor is controlled by manipulating a stream of pure A to the first reactor. The density of this stream is ρ_A in lbmole/gal. The temperature in each reactor can be assumed constant. The following design data are known:

Reactor volumes:	$V_1 = 500$ gal,	$V_2 = 500$ gal
Reaction rate coefficients:	$k_1 = 0.25$ min^{-1},	$k_2 = 0.50$ min^{-1}
Properties of stream A:	$\rho_A = 2.0$ lbmole/gal,	$MW_A = 25$
Design conditions:	$\bar{c}_{Ai} = 0.8$ lbmole/gal,	$\bar{f}_i = 50$ gal/min

$f_A = 50$ gal/min

Control valve: $\Delta p_v = 10$ psi, linear characteristics.

Concentration transmitter range: 0.05 to 0.5 lbmole/gal. The dynamics of this transmitter can be represented by a first-order lag with a time constant of 0.5 min.

(a) Size the control valve for 100 % overcapacity. Report the C_v and the gain of the valve.

(b) Derive, from basic principles, the set of equations that describe the composition control loop. State all assumptions.

(c) Linearize the equations from part (b) and draw the complete block diagram of the composition control loop. Show all transfer functions with the numerical values and units of all gains and time constants, except for the controller.

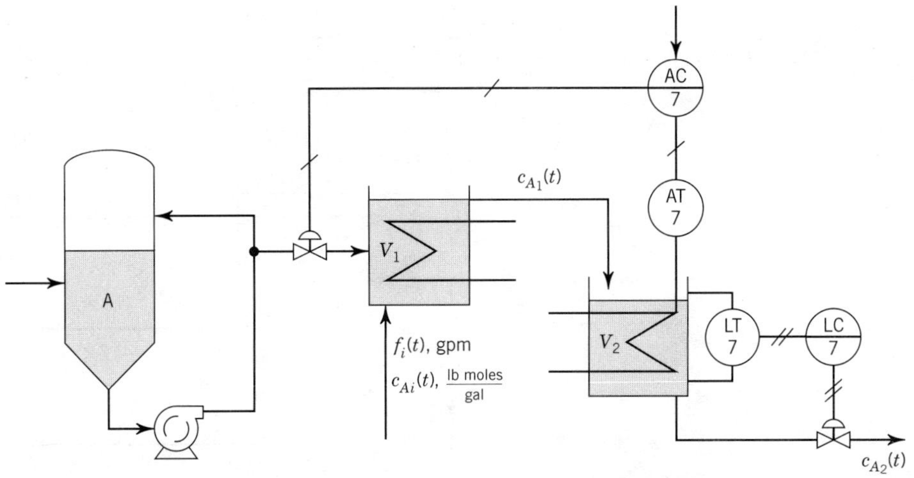

Figure P6-5 Reactors in series for Problem 6-22.

(d)   Obtain the closed-loop transfer functions

$$\frac{C_{A_2}(s)}{C_{A_2}^{set}(s)}; \qquad \frac{C_{A_2}(s)}{F_i(s)}; \qquad \frac{C_{A_2}(s)}{C_{A_i}(s)}$$

(e)   Calculate the ultimate gain and period of the loop.

6-23.   Consider the process shown in Fig. P6-6. In the first tank, two streams of rates $f_1(t)$ and $f_2(t)$ are being mixed and heated. The heating medium flows at such a high rate that its temperature change from inlet to exit is not significant. Thus, the heat transfer rate can be described by $UA[T_{c1}(t) - T_3(t)]$. It can also be assumed that the densities and heat capacities of all streams are not strong functions of temperature or composition. The outlet flow from the first tank flows into the second tank where it is again heated, this time by condensing steam. The rate of heat transfer can be described by $w_s(t)\lambda$, where $w_s(t)$ is the mass flow rate of steam and λ is the latent heat of vaporization of the steam. Assume that the pressure drop across the steam valve is constant and its time constant is τ_v. The temperature transmitter has a range T_L to T_H, and a time constant τ_T. Assuming that the heat losses from both tanks are negligible and that the important disturbances are $T_1(t)$, $T_2(t)$, and $T_{c1}(t)$, obtain the complete block diagram of the temperature control loop and its characteristic equation. Derive the transfer function of each block.

6-24.   Consider the process shown in Fig. P6-7. The process fluid entering the tank is oil with a density of 53 lb/ft^3, a heat capacity of 0.45 Btu/lb-°F, and an inlet temperature of 70°F. This oil is to be heated up to 200°F by saturated steam at 115 psig. The pressure in the tank, above the oil level, is maintained at 40 psia by a blanket of inert gas, N$_2$. Assume that the tank is well insulated, that the physical properties of the oil are not strong functions of temperature, that the liquid is well mixed, and that the level covers the heating coil. The following data are also known:

$\overline{p}_1 = 45$ psig; $\overline{p}_3 = 15$ psig

Heat transfer coefficient = 136 Btu/hr-ft^2-°F

Heating surface area = 127.5 ft^2

Heating coil: $\frac{1}{2}$-in. O.D., 20 BWG tubes, 974 linear ft., mass of tube metal = 0.178 lb/ft, c_p of tube metal = 0.12 Btu/lb-°F.

Tank diameter: 3 ft

Level transmitter: 7 to 10 ft range, 0.01 min time constant.

Temperature transmitter: 100 to 300°F range, 0.5 min time constant.

(a)   Size the control valves for 50 % overcapacity, the nominal oil flow rate is 100 gpm. The pressure drop across the steam valve can be assumed to be constant.

(b)   Obtain the complete block diagram for the level control loop. Use a proportional only controller.

(c)   Obtain the complete block diagram and the characteristic equation of the temperature control loop. Use a PID controller. Show the numerical values of all the gains and time constants in the transfer functions.

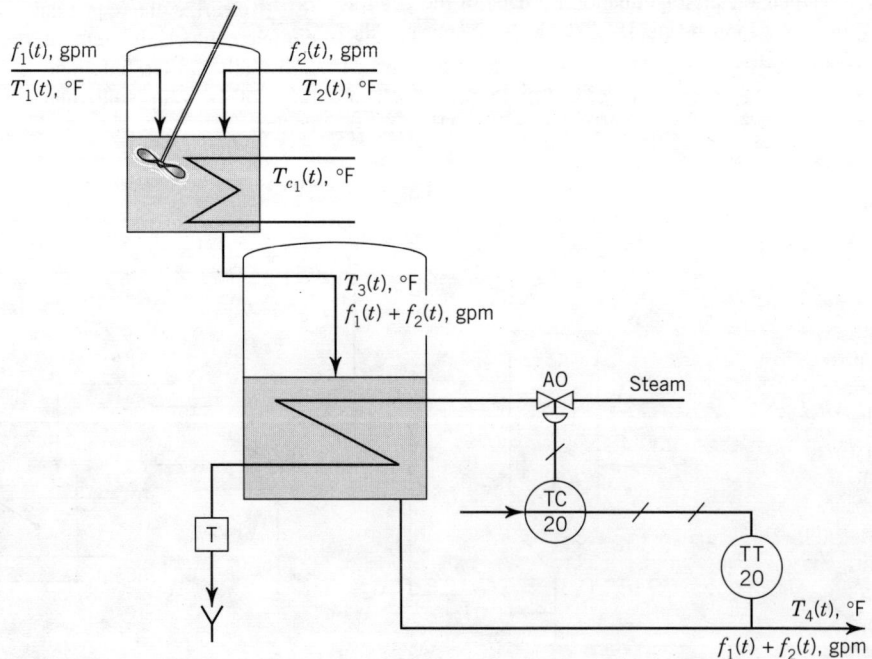

Figure P6-6 Heaters for Problem 6-23.

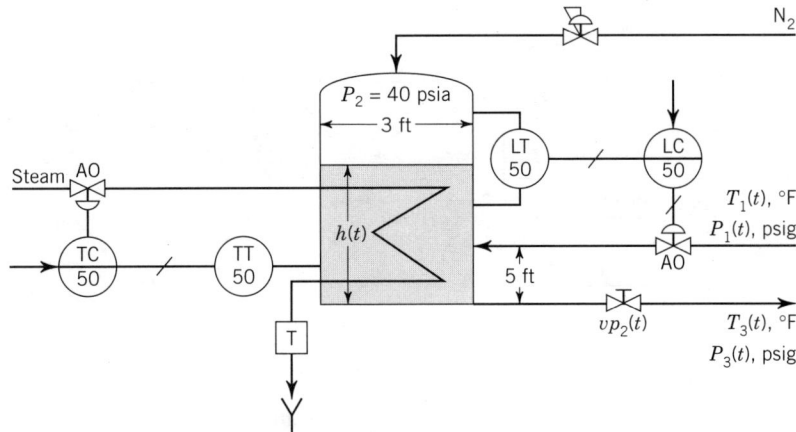

Figure P6-7 Oil heater for Problem 6-24.

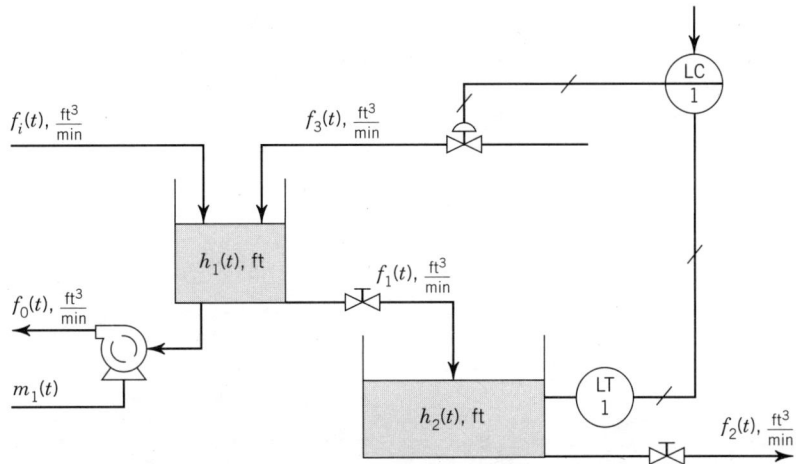

Figure P6-8 Level controller for problem 6-25.

(d) Calculate the ultimate gain and period of oscillation of the temperature control loop.

Note: The simulation of this process is the subject of Problem 13-24.

6-25. Consider the process shown in Fig. P6-8. Assume the two outlet valves remain at constant opening and their downstream pressure is also constant; the pump flow is linear with the controller output in the range 0 to f_{max}; the variable-speed pump has a time constant, relating the flow to the input signal, $m_1(t)$, of τ_p s; the control valve is linear and has a time constant, relating the flow to the pneumatic signal, of τ_v s; the pressure drop across the control valve is constant; the level transmitter has a range 0 to h_{max} and a negligible time constant. The diameters of the tanks are D_1 and D_2. The valve coefficients are C_{v1}, C_{v2}, and C_{v3}.

(a) Draw the block diagram and derive the transfer functions for this control system. The disturbances are $f_i(t)$ and $m_1(t)$.

(b) Write the characteristic equation of the level control loop and determine its ultimate gain and period as functions of the system parameters.

6-26. Consider the process shown in Fig. P6-9. In this process a waste gas is enriched with natural gas to be used as fuel in a small furnace. The enriched waste gas must have a certain heating value to be used as fuel. The control strategy calls for measuring the heating value of the gas leaving the process and manipulating the natural gas flow (using a variable-speed fan) to maintain the heating value set point. The waste gas is composed of methane (CH_4) and some low heating value combustibles. The natural gas composition can be considered constant and is composed mainly of methane and some small amount of other hydrocarbons. The heating value of the enriched waste gas is related to the mole fraction of methane by the following relation:

$$hv(t) = c + gx_3(t)$$

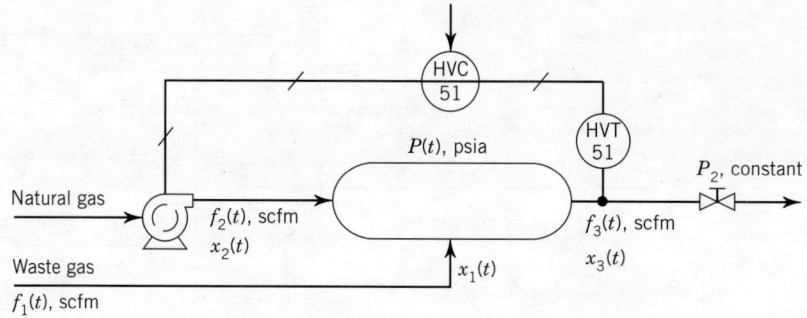

Figure P6-9 Gas enriching tank for Problem 6-26.

where $hv(t)$ is the heating value, $x_3(t)$ is the mole fraction of methane, and c and g are constants. The variable-speed fan is such that at full speed its flow is $f_{2\,max}$. It can be assumed that the relationship between the flow and the input signal to the fan driver is linear. This driver has a time constant τ_F. The outlet valve has a constant opening. A proportional-integral controller is used to control the heating value. The sensor-transmitter has a time constant of τ_T min. The specific gravity of the enriched gas is related to the mole fraction of methane by

$$G(t) = a + bx_3(t)$$

where a and b are constants.

(a) Draw the block diagram for this control system and derive all transfer functions. The possible disturbances are $f_1(t)$, $x_1(t)$, and $x_2(t)$.

(b) Write the characteristic equation of the feedback control loop.

Chapter 7

Adjusting Controller Parameters

In this chapter we will study the tuning of feedback controllers, that is, the adjustment of the controller parameters to match the characteristics (or personality) of the rest of the components of the loop. We will look at two methods for characterizing the process dynamic characteristics: the on-line or closed-loop tuning method, and the step-testing or open-loop method. We will also look at three different specifications of control loop performance: quarter decay ratio response, minimum error integral, and controller synthesis. The latter method, in addition to providing us some simple controller tuning relationships, will give us some insight into the selection of the proportional, integral, and derivative modes for various process transfer functions.

Tuning is the procedure of adjusting the feedback controller parameters to obtain a specified closed-loop response. The tuning of a feedback control loop is analogous to the tuning of an automobile engine, a television set, or a stereo system. In each of these cases the difficulty of the problem increases with the number of parameters that must be adjusted. For example, tuning a simple proportional-only or integral-only controller is similar to adjusting the volume of a stereo sound system. As only one parameter or "knob" needs to be adjusted, the procedure consists of moving it in one direction or the other until the desired response (or volume) is obtained. The next degree of difficulty is the tuning of a two-mode or proportional-integral (PI) controller, which is similar to adjusting the bass and treble on a stereo system. Since two parameters, the gain and the reset time, must be adjusted, the tuning procedure is significantly more complicated than when only one parameter needs to be adjusted. Finally, the tuning of three-mode or proportional-integral-derivative (PID) controllers represents the next higher degree of difficulty since three parameters—the gain, the reset time, and the derivative time—must be adjusted.

Although we have drawn an analogy between the tuning of a stereo system and that of a feedback control loop, we do not want to give the impression that the two tasks have the same degree of difficulty. The main difference lies in the speed of response of the stereo system versus that of a process loop. On the stereo system we get almost immediate feedback on the effect of our tuning adjustments. On the other hand, although some process loops do have relatively fast responses, for many process loops we may have to wait several minutes and maybe even hours to observe the response that results from our tuning adjustments. This makes the tuning of feedback controllers by trial and error a tedious and time-consuming task. Yet this happens to be the most common method used by control and instrument engineers in industry. A number of tuning procedures and formulas have been introduced to aid and give insight into the tuning procedure. We will study some of these procedures in this

Principles and Practice of Automatic Process Control/Third Edition, by C. A. Smith and A. B. Corripio
ISBN 0-471-66141-4   Copyright © 2006 John Wiley & Sons (Asia) Pte. Ltd.

chapter. However, we must keep in mind that no one procedure will give better results than any other for all process control situations.

The values of the tuning parameters depend on the desired closed-loop response and on the dynamic characteristics or personality of the other elements of the control loop, particularly the process. We saw in Chapters 3 and 4 that if the process is non-linear, as is usually the case, its characteristics will change from one operating point to the next. This means that a particular set of tuning parameters can produce the desired response at only one operating point, given that standard feedback controllers are basically linear devices. For operation in a range of operating conditions, a compromise must be reached in arriving at an acceptable set of tuning parameters, as the response will be sluggish at one end of the range and oscillatory at the other.

One characteristic of feedback control that greatly simplifies the tuning procedure is that the performance of the loop is not a strong function of the tuning parameters. In other words, the performance does not vary much with the tuning parameters. Changes of less than 50 % on the values of the tuning parameters seldom have significant effects on the response of the loop. We will emphasize this by never showing the values of the tuning parameters with more than two significant digits. With this in mind, let us look at some of the procedures that have been proposed to tune industrial controllers.

7-1   QUARTER DECAY RATIO RESPONSE BY ULTIMATE GAIN

Ziegler and Nichols proposed this pioneer method, also known as the closed-loop or on-line tuning method, in 1942. It consists of two steps:

Step 1. The determination of the dynamic characteristics or personality of the control loop.

Step 2. The estimation of the controller tuning parameters that produce a desired response for the dynamic characteristics determined in the first step—in other words, the matching of the personality of the controller to that of the other elements in the loop.

In this method the dynamic characteristics of the process are represented by the *ultimate gain* of a proportional controller, and the *ultimate period* of oscillation of the loop. These parameters, introduced in Section 6-2, can be determined by the direct substitution method if the transfer functions of all of the components of the loop are known quantitatively. As this is not usually the case, we must often experimentally determine the ultimate gain and period from the actual process by the following procedure:

1. Switch off the integral and derivative actions of the feedback controller so as to have a proportional controller. In some controllers the integral action cannot be switched off, but can be detuned by setting the integral time to its maximum value or, equivalently, the integral rate to its minimum value.

2. With the controller in *automatic* (i.e., the loop closed), increase the proportional gain (or reduce the proportional band) until the loop oscillates with constant amplitude. Record the value of the gain that produces sustained oscillations as K_{cu}, the ultimate gain. This step is carried out in discrete gain increments, bumping the system by applying a small set point change at each gain setting. To prevent the loop from going unstable, smaller increments in gain are made as the ultimate gain is approached.

3. From a time recording of the controlled variable such as Fig. 7-1.1, the period of oscillation is measured and recorded as T_u, the ultimate period.

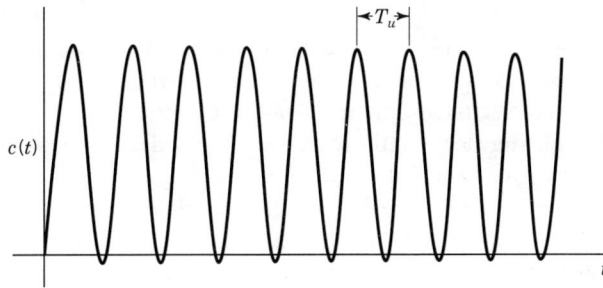

Figure 7-1.1 Response of the loop with the controller gain set equal to the ultimate gain K_{cu}. T_u is the ultimate period.

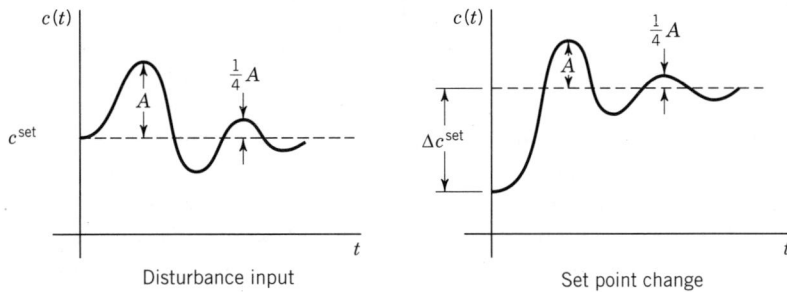

Disturbance input

Set point change

Figure 7-1.2 Quarter decay ratio response to disturbance input and to change in set point.

For the desired response of the closed loop, Ziegler and Nichols (1942) specified a decay ratio of one-fourth. The decay ratio is the ratio of the amplitudes of two successive oscillations. It should be independent of the input to the system and depend only on the roots of the characteristic equation for the loop. Typical quarter decay ratio response for a disturbance input and a set point change are shown in Fig. 7-1.2.

Once the ultimate gain and period are determined, they are used in the formulas of Table 7-1.1 for calculating the controller tuning parameters that produce quarter decay ratio responses.

Table 7-1.1 shows that the introduction of integral action forces a reduction of 10 % in the gain of the PI controller, as compared to the proportional controller gain. Derivative action, on the other hand, allows an increase in both the proportional gain

Table 7-1.1   Tuning formulas for quarter decay ratio response

Controller type	Proportional gain	Integral time	Derivative time
Proportional-only, P	$K_c = \dfrac{K_{cu}}{2}$	—	—
Proportional-integral, PI	$K_c = \dfrac{K_{cu}}{2.2}$	$\tau_I = \dfrac{T_u}{1.2}$	—
Proportional-integral-derivative, PID[a]	$K_c' = \dfrac{K_{cu}}{1.7}$	$\tau_I' = \dfrac{T_u}{2}$	$\tau_D' = \dfrac{T_u}{8}$

[a]The PID formulas are for the actual PID controller, Eq. 5-3.19. To convert to the ideal PID controller, Eq. 5-3.17: $K_c = K_c'(1 + \tau_D'/\tau_I')$; $\tau_I = \tau_I' + \tau_D'$; $\tau_D = \tau_D'\tau_I'/(\tau_I' + \tau_D')$.

and the integral rate (a decrease in integral time) of the PID controller as compared to the PI controller. This is because the integral action introduces a lag in the operation of the feedback controller, while the derivative action introduces an advance or lead. This will be discussed in more detail in Chapter 8.

The quarter decay ratio response is very desirable for disturbance inputs because it prevents a large initial deviation from the set point without being too oscillatory. However, it is not as desirable for step changes in set point, because it causes a 50 % overshoot. This is because the maximum deviation from the new set point in each direction is one-half the preceding maximum deviation in the opposite direction (see Fig. 7-1.2). This difficulty can easily be corrected by reducing the proportional gain from the value predicted by the formulas of Table 7-1.1. In fact, the decay ratio is a direct function of the controller gain, and can be adjusted at any time by simply changing the gain. In other words, if for a given process the quarter decay ratio response is too oscillatory, a reduction of the gain will smooth out the response.

The discussion in the preceding paragraph brings out the main advantage of the closed-loop tuning formulas: they reduce the tuning procedure to the adjustment of a single parameter, the controller gain. On the assumption that a good estimate of the ultimate period can be obtained by observing the closed-loop response, the reset and derivative times can be set based on this value. The response can then be molded by adjusting the proportional gain. Because of the insensitivity of the response to the precise values of the tuning parameters, it is not absolutely necessary to make the closed-loop response oscillate with sustained oscillations to estimate a good value of the ultimate period. Any oscillation caused by the proportional controller can be used to obtain an approximate value of the ultimate period that is usually good enough for tuning.

It has been said that a difficulty of the quarter decay ratio response is that the set of tuning parameters necessary to obtain it is not unique, except for the case of the proportional controller. In the case of PI controllers, we can easily verify that for each value of the integral time, we could find a value of the gain that produces a quarter decay ratio response and vice versa. The same is true for the PID controller. The simple tuning formulas proposed by Ziegler and Nichols give ballpark figures that produce fast response for most industrial loops.

The PID tuning formulas of Table 7-1.1 are for the "actual" (or *series*) PID controller transfer function given by Eq. 5-3.19. We know this because these were the only PID controllers available when Ziegler and Nichols developed their tuning formulas. Today, many computer control packages use the "ideal" (or *parallel*) PID transfer function of Eq. 5-3.18. Table 7-1.1 includes formulas to calculate the parameters of the ideal PID controller.

EXAMPLE 7-1.1

Given the characteristic equation of the continuous stirred tank heater derived in Example 6-1.1, determine the quarter decay ratio tuning parameters for the PID controller by the ultimate gain method. Also calculate the roots of the characteristic equation for the controller tuned with these parameters, and the actual decay ratio.

SOLUTION

In Example 6-2.2 the ultimate gain and period of the loop for a proportional controller were obtained by direct substitution:

$$K_{cu} = 10.4 \frac{\%CO}{\%TO}, \qquad T_u = 4.6 \text{ min}$$

According to Table 7-1.1, the tuning parameters for quarter decay ratio response of a PID controller are

$$K_c' = \frac{K_{cu}}{1.7} = 6.1 \; \frac{\%CO}{\%TO}, \qquad \tau_I' = \frac{4.6}{2} = 2.3 \; \text{min}, \qquad \tau_D' = \frac{4.6}{8} = 0.58 \; \text{min}$$

The transfer function of the PID controller is

$$G_c(s) = 6.1 \left(1 + \frac{1}{2.3s} \right) (1 + 0.58s)$$

Substitute into the characteristic equation from Example 6-2.2, and clear fractions.

$$s(8.34s + 1)(0.502s + 1)(0.2s + 1)(0.75s + 1)$$
$$+ (1.954)(6.1)(s + 0.435)(1 + 0.58s) = 0$$

where $0.435 = 1/2.3$. In polynomial form,

$$0.628s^5 + 5.303s^4 + 12.73s^3 + 16.74s^2 + 12.98s + 5.181 = 0$$

Using a computer program, the roots of this characteristic equation are

$$-0.327 \pm i1.232, \qquad -0.519, \qquad -1.784, \qquad -5.49$$

The response of the closed loop has the following form:

$$T(t) = b_1 e^{-0.327t} \sin(1.232t + \theta_1) + b_2 e^{-0.519t}$$
$$+ b_3 e^{-1.784t} + b_4 e^{-5.49t} + \text{(input terms)}$$

where the parameters b_1, b_2, b_3, b_4, and θ_1, could be evaluated by the technique of partial fractions expansion (see Section 2-2.2) and depend on the particular input applied to the system.

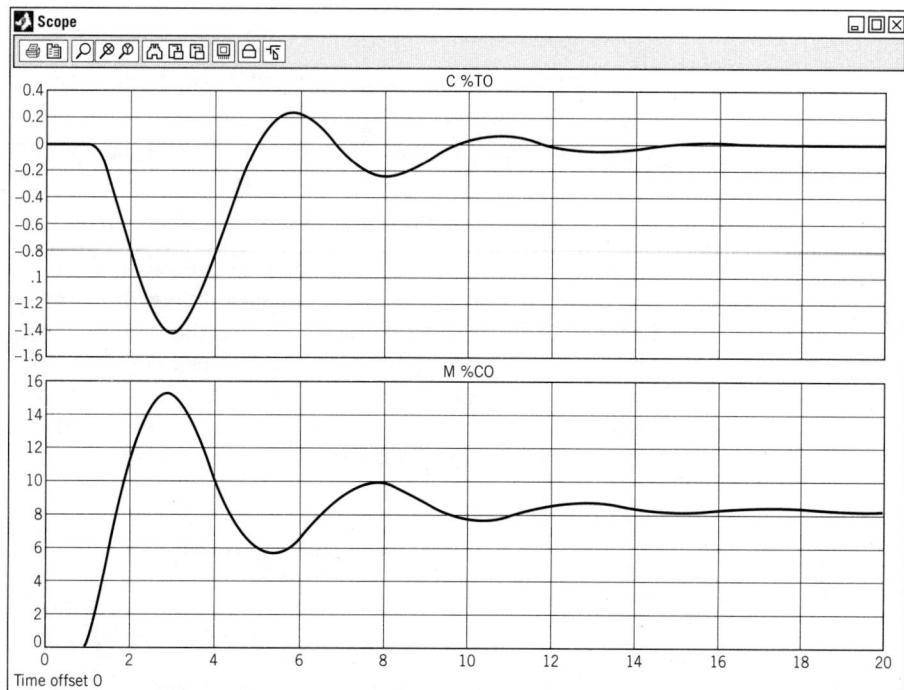

Figure 7-1.3 Responses of the controlled and manipulated variables to a step change of 5 ft^3/min in process flow for the stirred tank heater with a series PID controller.

The pair of complex conjugate roots dominates the response because the corresponding term in the response decays the slowest. It has a period of oscillation of $T = 2\pi/1.232 = 5.1$ min, a 1 % settling time of $-5/-0.327 = 15.3$ min, and a decay ratio of

$$e^{-0.327(5.1)} = 0.19$$

which is close to the theoretical decay ratio of 0.25. This shows that the tuning formulas are not exact in terms of the response specification.

The actual response of the outlet tank temperature to a change in process flow is shown in Fig. 7-1.3. This response was obtained by simulation of the loop (see Chapter 13). Notice that the period of oscillation, settling time, and decay ratio agree with the values calculated.

7-2 OPEN-LOOP PROCESS CHARACTERIZATION

The Ziegler–Nichols on-line tuning method we have just introduced is the only one that characterizes the process by the ultimate gain and the ultimate period. Most of the other controller tuning methods characterize the process by a simple first- or second-order model with dead time. In order to better understand the assumptions involved in such characterization, let us consider the block diagram of a feedback control loop given in Fig. 7-2.1. The symbols shown in the block diagram are

$R(s)$ = the Laplace transform of the set point

$M(s)$ = the Laplace transform of the controller output

$C(s)$ = the Laplace transform of the transmitter output

$E(s)$ = the Laplace transform of the error signal

$U(s)$ = the Laplace transform of the disturbance

$G_c(s)$ = the controller transfer function

$G_v(s)$ = the transfer function of the control valve (or final element)

$G_m(s)$ = the process transfer function between the controlled and manipulated variables

$G_u(s)$ = the process transfer function between the controlled variable and the disturbance

$H(s)$ = the transfer function of the sensor-transmitter

Using the simple block diagram algebra manipulations learned in Chapter 3, we can draw the equivalent block diagram of Fig. 7-2.2. In this diagram there are only two blocks in the control loop, one for the controller and the other for the rest of the components of the loop, plus one block for each disturbance. The advantage of this simplified representation is that it highlights the two signals in the loop that can be usually observed and recorded: the controller output $M(s)$ and the transmitter signal $C(s)$. For simple loops no signal or variable can be observed except these two.

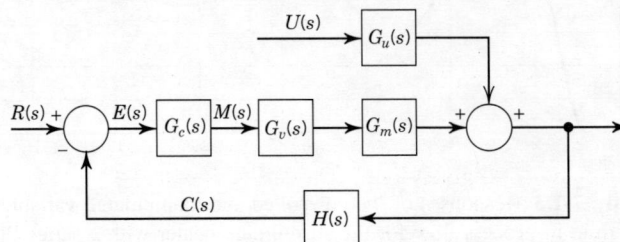

Figure 7-2.1 Block diagram of typical feedback control loop.

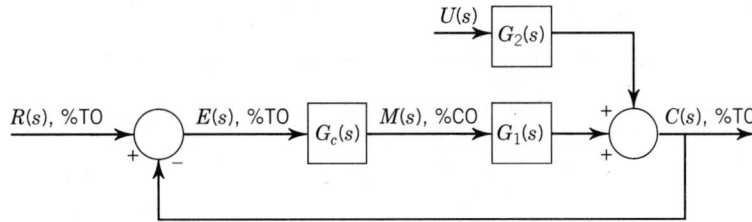

Figure 7-2.2 Equivalent simplified block diagram in which all the field instruments and the process have been lumped into single blocks.

Therefore, the lumping of the transfer functions of the control valve, the process, and the sensor-transmitter into a single block is not just a convenience, but a practical necessity. Let us call these combinations of transfer functions $G_1(s)$ and $G_2(s)$:

$$G_1(s) = G_v(s)G_m(s)H(s)$$
$$G_2(s) = G_u(s)H(s) \tag{7-2.1}$$

The combined transfer function that is in the loop, $G_1(s)$, is precisely what is approximated by low-order models for the purpose of characterizing the dynamic response of the process. The point is that the characterized "process" includes the dynamic behavior of the control valve and of the sensor-transmitter. The two most common models used to characterize the process are

First-order-plus dead-time (FOPDT) model

$$G_1(s) = \frac{K\ e^{-t_0 s}}{\tau s + 1} \tag{7-2.2}$$

Second-order-plus dead-time (SOPDT) model

$$G_1(s) = \frac{K\ e^{-t_0 s}}{(\tau_1 s + 1)(\tau_2 s + 1)} \tag{7-2.3}$$

$$G_1(s) = \frac{K\ e^{-t_0 s}}{\tau^2 s^2 + 2\zeta\tau s + 1} \tag{7-2.4}$$

where

$$K = \text{the process steady-state gain}$$
$$t_0 = \text{the effective process dead time}$$
$$\tau, \tau_1, \tau_2 = \text{the effective process time constants}$$
$$\zeta = \text{the effective process damping ratio}$$

For underdamped processes, $\zeta < 1$.

Of these, the FOPDT model is the one on which most controller tuning formulas are based. This model characterizes the process by three parameters: the gain K, the dead time t_0, and time constant τ. The question then, is how can these parameters be determined for a given loop? The answer is that some dynamic test must be performed on the actual system or on a computer simulation of the process. The simplest test that can be performed is a step test.

7-2.1 Process Step Testing

The step test procedure is carried out as follows:

1. With the controller on "manual" (that is, the loop opened) a step change in the controller output signal $m(t)$ is applied to the process. The magnitude of the change should be large enough for the consequent change in the transmitter signal to be measurable, but not so large that the response will be distorted by the process nonlinearities.

2. The response of the transmitter output signal $c(t)$ is recorded on a strip chart recorder or equivalent device, making sure that the resolution is adequate in both the amplitude and the time scale. The resulting plot of $c(t)$ versus time must cover the entire test period from the introduction of the step test until the system reaches a new steady state. Typically, a step test lasts between a few minutes and several hours, depending on the speed of response of the process.

It is of course imperative that no disturbances enter the system while the step test is performed. A typical test plot, also known as a *process reaction curve*, is sketched in Fig. 7-2.3. As we saw in Section 2-5, the S-shaped response is characteristic of second- and higher-order processes with or without dead time. The next step is to match the process reaction curve to a simple process model in order to determine the model parameters. Let us do this for the first-order-plus dead-time (FOPDT) model.

In the absence of disturbances, and for the conditions of the test, the block diagram of Fig. 7-2.2 can be redrawn as in Fig. 7-2.4. The response of the transmitter output signal is given by

$$C(s) = G_1(s)M(s) \qquad (7\text{-}2.5)$$

For a step change in controller output of magnitude Δm and a FOPDT model, Eq. 7-2.2, we have

$$C(s) = \frac{K\,e^{-t_0 s}}{\tau s + 1} \cdot \frac{\Delta m}{s} \qquad (7\text{-}2.6)$$

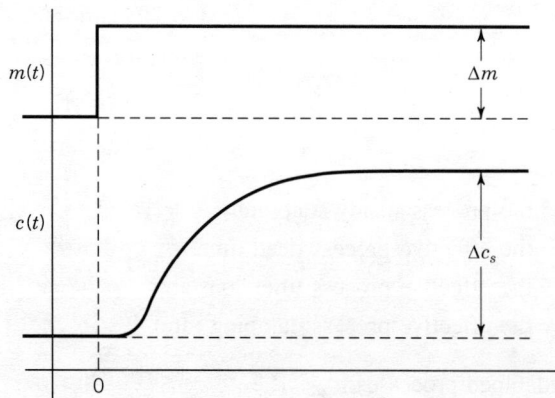

Figure 7-2.3 Process reaction curve or open-loop step response.

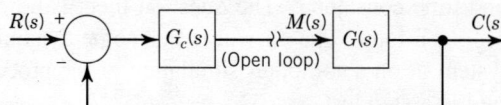

Figure 7-2.4 Block diagram for open-loop step test.

Expanding this expression by partial fractions (see Section 2-2.2), we obtain

$$C(s) = K \ \Delta m \ e^{-t_0 s} \left[\frac{1}{s} - \frac{\tau}{\tau s + 1} \right] \tag{7-2.7}$$

Inverting with the help of a Laplace transform table (Table 2-1.1) and applying the real translation theorem of Laplace transforms (see Section 2-1.2), we get

$$C(t) = K \ \Delta m \ u(t - t_0) \left(1 - e^{-(t-t_0)/\tau} \right) \tag{7-2.8}$$

where the unit step function $u(t - t_0)$ is included to explicitly indicate that

$$C(t) = 0 \qquad \text{for } t < t_0 \tag{7-2.9}$$

Variable C is the perturbation or change of the transmitter output from its initial value, $C(t) = c(t) - c(0)$.

Estimation of the Process Gain

A graph of Eq. 7-2.8 is shown in Fig. 7-2.5. In this figure the term Δc_s is the steady-state change in $c(t)$. From Eq. 7-2.8 we find

$$\Delta c_s = \lim_{t \to \infty} C(t) = K \ \Delta m \tag{7-2.10}$$

From this equation, and realizing that the model response must match the process reaction curve at steady state, we can calculate the steady-state gain of the process, which is one of the model parameters.

$$K = \frac{\Delta c_s}{\Delta m} \tag{7-2.11}$$

This formula, as shown in Chapters 2 and 3, is the definition of the gain.

Estimation of Time Constant and Dead Time

Three methods have been proposed to estimate the dead time t_0 and time constant τ, each of which results in different values.

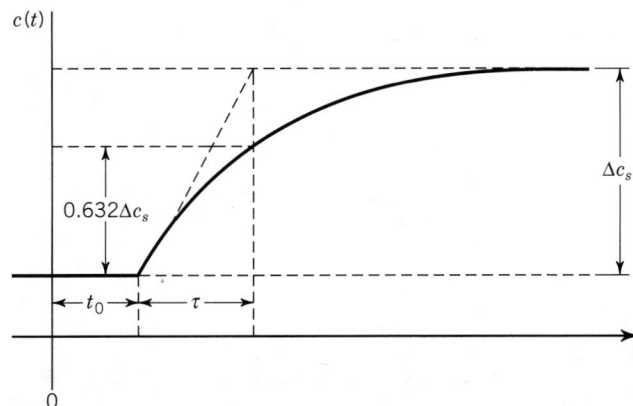

Figure 7-2.5 Step response of first-order-plus-dead-time process, showing the graphical definition of the dead time t_0 and time constant τ.

Fit 1 This method uses the line that is tangent to the process reaction curve at the point of maximum rate of change. For the FOPDT model this happens at $t = t_0$, as is evident from inspecting the model response of Fig. 7-2.5. From Eq. 7-2.8, this initial (maximum) rate of change is

$$\left.\frac{dC}{dt}\right|_{t=t_0} = K \ \Delta m \left(\frac{1}{\tau}\right) = \frac{\Delta c_s}{\tau} \qquad (7\text{-}2.12)$$

From Fig. 7-2.5, we see that this result tells us that the line of maximum rate of change crosses the initial value line at $t = t_0$ and the final value line at $t = t_0 + \tau$. This finding suggests the construction for determining t_0 and τ shown in Fig. 7-2.6a. The line is drawn tangent to the actual process reaction curve at the point of maximum rate of change. The model response using these values of t_0 and τ is shown by the dashed line in the figure. Evidently, the model response obtained with this fit does not match the actual response very well.

Fit 2 In this fit, t_0 is determined in the same manner as in fit 1, but the value of τ is the one that forces the model response to coincide with the actual response at $t = t_0 + \tau$. According to Eq. 7-2.8, this point is

$$C(t_0 + \tau) = K \ \Delta m (1 - e^{-1}) = 0.632 \ \Delta c_s \qquad (7\text{-}2.13)$$

Figure 7-2.6b shows that the model response for fit 2 is much closer to the actual response than for fit 1. The value of the time constant obtained by fit 2 is usually less than that obtained by fit 1, but the dead time is exactly the same.

Fit 3 The least precise step in the determination of t_0 and τ by the previous two methods is the drawing of the line tangent to the process reaction curve at the point of maximum rate of change. Even for fit 2, for which the value of $(t_0 + \tau)$ is independent of the tangent line, the estimated values of both t_0 and τ depend on the line. To

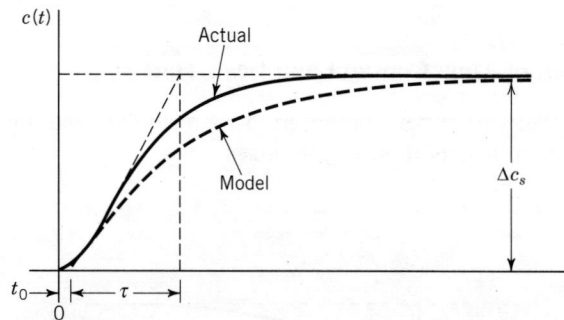

Figure 7-2.6a FOPDT model parameters by fit 1.

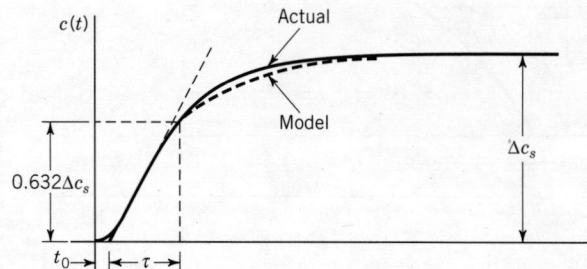

Figure 7-2.6b FOPDT model parameters by fit 2.

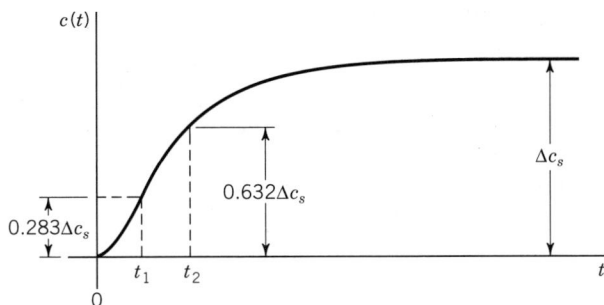

Figure 7-2.6c FOPDT model parameters by fit 3.

eliminate this dependence on the tangent line, Dr. Cecil L. Smith (1972) proposes that the values of t_0 and τ be selected such that the model and actual responses coincide at two points in the region of high rate of change. The two points recommended are $(t_0 + \tau/3)$ and $(t_0 + \tau)$. The second of these points is located as in fit 2, while the first point is located, from Eq. 7-2.8, at

$$C(t) = K \ \Delta m \left(1 - e^{-1/3}\right) = 0.283 \ \Delta c_s \qquad \textbf{(7-2.14)}$$

These two points are labeled t_2 and t_1, respectively, in Fig. 7-2.6c. The value of t_0 and τ can then be easily obtained by the simple solution of the following set of equations:

$$t_0 + \frac{\tau}{3} = t_1$$
$$t_0 + \tau = t_2 \qquad \textbf{(7-2.15)}$$

which reduces to

$$\tau = \frac{3}{2}(t_2 - t_1)$$
$$t_0 = t_2 - \tau \qquad \textbf{(7-2.16)}$$

where

t_1 = time at which $C = 0.283 \ \Delta c_s$

t_2 = time at which $C = 0.632 \ \Delta c_s$

Our past experience shows that results obtained by this method are more reproducible than those obtained by the other two. We therefore recommend this method for the estimation of t_0 and τ from the process reaction curve.

Various models have been proposed in the literature to estimate the parameters of a second-order-plus-dead-time (SOPDT) model to the process reaction curve. Our experience is that the precision of these methods is very low. The reason is that the step test does not provide enough information to extract the additional parameters—second time constant or damping ratio—that is required by the SOPDT. In other words, the increased complexity of the model demands a more sophisticated dynamic test.

Since most of the controller tuning formulas that we are about to introduce are based on FOPDT model parameters, we may find ourselves in some situation in which we have the parameters of a high-order model and need to estimate the equivalent first-order model parameters. Although there is no general procedure to do this, the following rule of thumb might provide a rough estimate for a first approximation:

If one of the time constants of the high-order model is much longer than the others, the effective time constant of the first-order model can be estimated to be equal to the longest time constant. The effective dead time of the first-order model can then be approximated by the sum of all of the smaller time constants plus the dead time of the high-order model.

EXAMPLE 7-2.1

Estimate the FOPDT parameters for the temperature control loop of the exchanger of Example 6-2.1. The combined transfer function for the control valve, exchanger, and sensor-transmitter for that example is

$$G(s) = \frac{0.80}{(10s + 1)(30s + 1)(3s + 1)}$$

SOLUTION

Assuming that the $30s$ time constant is much longer than the other two, we can roughly approximate

$$\tau = 30 \text{ s}$$
$$t_0 = 10 + 3 = 13 \text{ s}$$

and the gain is of course the same, that is, $K = 0.80$. The resulting FOPDT model transfer function is then

$$G(s) = \frac{0.80 e^{-13s}}{30s + 1} \quad \textbf{(Model A)}$$

We will next compare this rough approximation with the experimentally determined FOPDT parameters from the process reaction curve. Figure 7-2.7 shows the process reaction curve for the three first-order lags in series that we have assumed represent the heat exchanger, control valve, and sensor-transmitter. The response of Fig. 7-2.7 was obtained by simulating the three first-order lags on a computer, applying a 5 % step change on the controller output signal, and recording the output of the sensor-transmitter versus time. From this result we can calculate the FOPDT parameters using the three fits presented above.

Process gain: $\quad K = \dfrac{\Delta c_s}{\Delta m} = \dfrac{4°\text{F}}{5 \text{ \%CO}} \cdot \dfrac{100 \text{ \%TO}}{(150 - 50)°\text{F}} = 0.80 \dfrac{\text{\%TO}}{\text{\%CO}}$

Fit 1 From Fig. 7-2.7, $t_0 = 7.2$ s, $\tau = 61.5 - 7.2 = 54.3$ s. Then,

$$G(s) = \frac{0.80 e^{-7.2s}}{54.3s + 1} \quad \textbf{(Model B)}$$

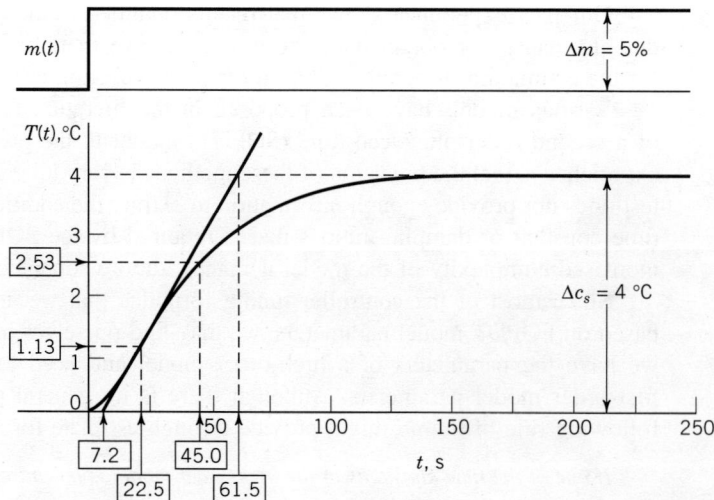

Figure 7-2.7 Step response for the heat exchanger temperature (Example 7-2.1).

Fit 2   From Fig. 7-2.7, $t_0 = 7.2$ s; at $C(t_2) = 0.632\,(4°F) = 2.533°F$, $t_2 = 45$ s. Then, $\tau = 45 - 7.2 = 37.8$ s, and

$$G(s) = \frac{0.80\,e^{-7.2s}}{37.8s + 1} \qquad \textbf{(Model C)}$$

Fit 3   From Fig. 7-2.7, at $C(t_1) = 0.283\,(4°F) = 1.13°F$, $t_1 = 22.5$ s. Then $\tau = (3/2)$ $(45.0 - 22.5) = 33.8$ s, and $t_0 = 45.0 - 33.8 = 11.2$ s. Then

$$G(s) = \frac{0.80\,e^{-11.2s}}{33.8s + 1} \qquad \textbf{(Model D)}$$

As we shall see in the following sections, an important parameter in terms of tuning is the ratio of the dead time to the time constant. The values for the four FOPDT model approximations are as follows:

Model	A (rough)	B (fit 1)	C (fit 2)	D (fit 3)
t_0, s	13.0	7.2	7.2	11.2
τ, s	30.0	54.3	37.8	33.8
t_0/τ	0.43	0.13	0.19	0.33

We see that the ratio t_0/τ is the most sensitive parameter, varying by a factor of slightly over 3:1. Recall that fits 2 and 3 provide the closest approximations to the actual step response.

EXAMPLE 7-2.2

Given a second-order process

$$G(s) = \frac{C(s)}{M(s)} = \frac{K}{(\tau_1 s + 1)(\tau_2 s + 1)} \quad \tau_1 \geq \tau_2$$

determine the parameters of a first-order-plus-dead-time (FOPDT) model

$$G'(s) = \frac{K\,e^{-t_0' s}}{\tau' s + 1}$$

using fit 3, as a function of the ratio of τ_2/τ_1.

SOLUTION

To obtain the unit step response of the actual process, we solve for $C(s)$ and substitute $M(s) = 1/s$ into the transfer function.

$$C(s) = \frac{K}{(\tau_1 s + 1)(\tau_2 s + 1)} \frac{1}{s}$$

Expanding in partial fractions, and inverting with the help of a Laplace transform table (Table 2-1.1), we obtain

$$C(t) = K\left[1 - \frac{\tau_1}{\tau_1 - \tau_2}e^{-t/\tau_1} + \frac{\tau_2}{\tau_1 - \tau_2}e^{-t/\tau_2}\right]$$

As $t \to \infty$, $C(t) \to K$. For fit 3, at $t_1 = t_0' + \tau'/3$,

$$C(t_1) = (1 - e^{-1/3})K = K\left[1 - \frac{\tau_1}{\tau_1 - \tau_2}e^{-t_1/\tau_1} + \frac{\tau_2}{\tau_1 - \tau_2}e^{-t_1/\tau_2}\right]$$

and at $t_2 = t_0' + \tau'$,

$$C(t_2) = (1 - e^{-1})K = K\left[1 - \frac{\tau_1}{\tau_1 - \tau_2}e^{-t_2/\tau_1} + \frac{\tau_2}{\tau_1 - \tau_2}e^{-t_2/\tau_2}\right]$$

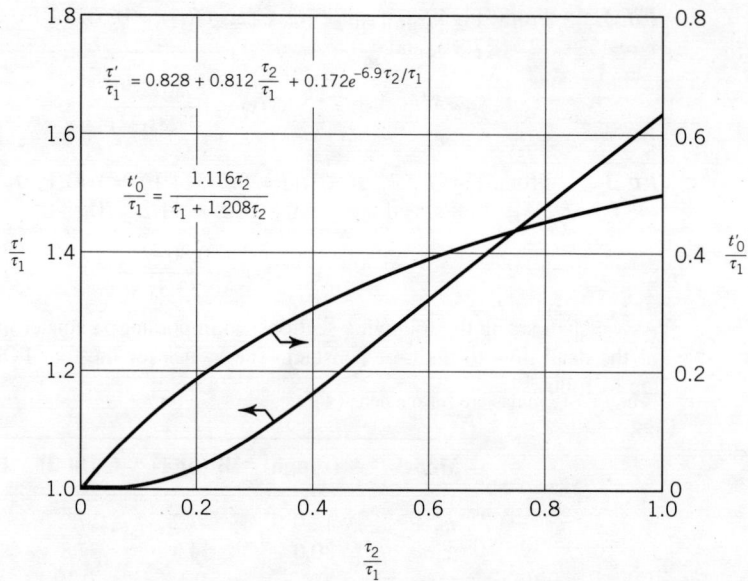

Figure 7-2.8 FPODT model dead time and time constant for fit 3 approximation to overdamped second-order system. (Reproduced by permission of Reference 3.)

or

$$e^{-1/3} = \frac{\tau_1}{\tau_1 - \tau_2}\left(e^{-t_1/\tau_1} - \frac{\tau_2}{\tau_1}e^{-t_1/\tau_2}\right) \qquad \textbf{(A)}$$

$$e^{-1} = \frac{\tau_1}{\tau_1 - \tau_2}\left(e^{-t_2/\tau_1} - \frac{\tau_2}{\tau_1}e^{-t_2/\tau_2}\right) \qquad \textbf{(B)}$$

Similar formulas can be obtained for the special case $\tau_1 = \tau_2$. The values of t_1 and t_2 are obtained by trial and error from these equations. Then, Eq. 7-2.16 gives

$$\tau' = \frac{3}{2}(t_2 - t_1) \qquad t_0' = t_2 - \tau'$$

Martin (1975) used a computer program to solve this problem, and the results are plotted in Fig. 7-2.8. As the figure shows, the maximum effective dead time takes place when the two time constants are equal:

$$\text{For } \tau_1 = \tau_2 \qquad t_0' = 0.505\tau_1 \qquad \tau' = 1.64\tau_1$$
$$\text{For } \tau_2 \ll \tau_1 \qquad t_0' \to \tau_2 \qquad \tau' \to \tau_1$$

This is the basis for the rule of thumb presented earlier. We can use Fig. 7-2.8 to refine this rule of thumb for systems represented by three or more first-order lags in series. For instance, for the heat exchanger example, we can refine the rough approximation model as follows:

Assume $\tau_1 = 30$ s, $\tau_2 = 10 + 3 = 13$ s, so that $\tau_2/\tau_1 = 13/30 = 0.433$. Then, from Fig. 7-2.8, $t_0' = 0.33\tau_1 = 9.9$ s, $\tau' = 1.2\tau_1 = 36$ s. These values are close to the values obtained by fit 3, Model C, in Example 7-2.1. In this case τ_2 is not that much smaller than τ_1.

7-2.2   Tuning for Quarter Decay Ratio Response

In addition to their on-line tuning formulas, Ziegler and Nichols (1942) proposed a set of formulas based on the parameters of a first-order model fit to the process reaction curve. These formulas are given in Table 7-2.1. Although the parameters they

Table 7-2.1   Tuning formulas for quarter decay ratio response

Controller type	Proportional gain	Integral time	Derivative time
Proportional-only, P	$K_c = \dfrac{1}{K}\left(\dfrac{t_0}{\tau}\right)^{-1}$	—	—
Proportional-integral, PI	$K_c = \dfrac{0.9}{K}\left(\dfrac{t_0}{\tau}\right)^{-1}$	$\tau_I = 3.33t_0$	—
Proportional-integral-derivative, PID[a]	$K_c' = \dfrac{1.2}{K}\left(\dfrac{t_0}{\tau}\right)^{-1}$	$\tau_I' = 2.0t_0$	$\tau_D' = \dfrac{t_0}{2}$

[a]The PID formulas are for the actual PID controller, Eq. 5-3.19. To convert to the ideal PID controller, Eq. 5-3.17: $K_c = K_c'(1 + \tau_D'/\tau_I');\ \tau_I = \tau_I' + \tau_D';\ \tau_D = \tau_D'\tau_I'/(\tau_I' + \tau_D')$.

used were not precisely the gain, time constant, and dead time, their formulas can be modified and expressed in terms of these parameters. They used fit 1 to determine the model parameters.

As we can see in Table 7-2.1, the relative magnitudes of the gain, integral time, and derivative time between the P, PI, and PID controllers are the same as for the on-line tuning formulas that are based on the ultimate gain and period (Table 7-1.1). The formulas for the gain show that the loop gain, KK_c, is inversely proportional to the ratio of the effective dead time to the effective time constant.

In using the formulas of Table 7-2.1, we must keep in mind that they are empirical and apply only to a limited range of dead time to time constant ratios. In our experience, they are most applicable for a range of t_0/τ of around 0.10 to 0.5.

As was pointed out on the discussion of on-line tuning, the quarter decay ratio formulas can be adjusted to less oscillatory responses by simply reducing the proportional gain from the value given by the tuning formula. Unfortunately, the formulas in Table 7-2.1 give the reset and derivative times as functions of the process dead time which cannot be estimated as readily as the ultimate period.

EXAMPLE 7-2.3   Compare the values of the tuning parameters for the temperature control of the exchanger of Figure 6-1.1 using the quarter decay ratio on-line tuning and the FOPDT parameters estimated in Example 7-2.1. In earlier examples we determined the following results for the exchanger temperature control loop:

By direct substitution method, from Example 6-2.1:

$$K_{cu} = 23.8\ \%CO/\%TO \qquad T_u = 28.7\ \text{s}$$

By fit 1 approximation, from Example 7-2.1:

$$K = 0.80\ \%CO/\%TO \qquad t_0 = 7.2\ \text{s} \qquad \tau = 54.3\ \text{s}$$

SOLUTION   For a proportional only (P) controller,

On-line tuning (Table 7-1.1)   **Open-loop tuning (Table 7-2.1)**

$$K_c = \frac{1}{2}(23.8) = 12\ \frac{\%CO}{\%TO} \qquad K_c = \frac{1}{0.80}\left(\frac{7.2}{54.3}\right)^{-1} = 9.4\ \frac{\%CO}{\%TO}$$

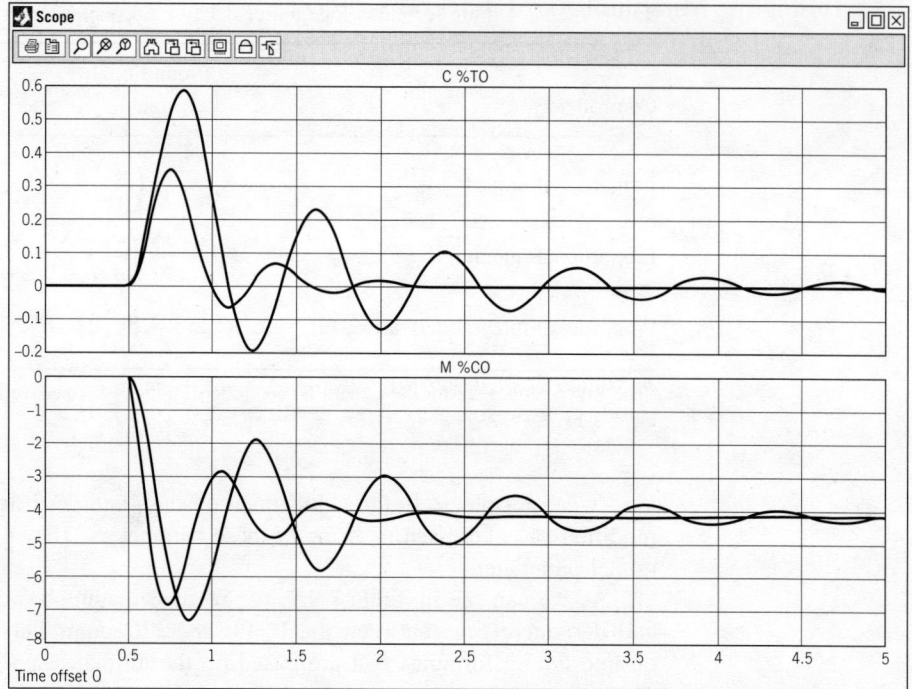

Figure 7-2.9 Comparison of the PI versus PID responses for the heat exchanger of Example 7-2.3. Input is a 2 kg/s step increase in process flow.

For a proportional-integral (PI) controller,

$$K_c = \frac{23.8}{2.2} = 11\,\frac{\%CO}{\%TO} \qquad\qquad K_c = \frac{0.9}{0.80}\left(\frac{7.2}{54.3}\right)^{-1} = 8.5\,\frac{\%CO}{\%TO}$$

$$\tau_I = \frac{28.7}{1.2} = 24 \text{ s } (0.40 \text{ min}) \qquad\qquad \tau_I = 3.33(7.2) = 24 \text{ s } (0.40 \text{ min})$$

For a proportional-integral-derivative (PID) controller,

$$K_c' = \frac{23.8}{1.7} = 14\,\frac{\%CO}{\%TO} \qquad\qquad K_c' = \frac{1.2}{0.80}\left(\frac{7.2}{54.3}\right)^{-1} = 11\,\frac{\%CO}{\%TO}$$

$$\tau_I' = \frac{28.7}{2.0} = 14 \text{ s } (0.24 \text{ min}) \qquad\qquad \tau_I' = 2.0(7.2) = 14 \text{ s } (0.24 \text{ min})$$

$$\tau_D' = \frac{28.7}{8} = 3.6 \text{ s } (0.06 \text{ min}) \qquad\qquad \tau_D' = 0.5(7.2) = 3.6 \text{ s } (0.06 \text{ min})$$

The agreement is evident. Notice, however, that this agreement depends on using the fit 1 model parameters, which happens to be what Ziegler and Nichols used.

Figure 7-2.9 compares the responses of the transmitter and controller outputs using the PI and PID controller parameters given in the right-hand column above. To generate these plots the exchanger was simulated using the transfer functions of Example 6-2.1 and the controller was simulated as described in Section 13-4. The figure shows that on the responses to a change in process flow, the PID controller results in a smaller initial deviation from set point, fewer oscillations, and faster settling time than the PI controller. We conclude that PID controllers should be used anytime that the process model exhibits a high dead time to time constant ratio, say greater than 0.05.

7-2.3   Tuning for Minimum Error Integral Criteria

Because of the nonuniqueness of the quarter decay ratio tuning parameters, a substantial research project was conducted at Louisiana State University under Professors Paul W. Murrill and Cecil L. Smith to develop tuning relationships that were unique. They used the first-order-plus-dead-time (FOPDT) model parameters to characterize the process. Their specification of the closed-loop response is basically a minimum error or deviation of the controlled variable from its set point. Since the error is a function of time for the duration of the response, the sum of the error at each instant of time must be minimized. This is by definition the integral of the error with time, or the shaded area in the responses illustrated in Fig. 7-2.10. Since the tuning relationships are intended to minimize the integral of the error, they are referred to as *minimum error integral tuning*. However, the integral of the error cannot be minimized directly, because a very large negative error would be the minimum. One way we prevent negative values of the performance function is to use the integral of the absolute value of the error (IAE).

$$\text{IAE} = \int_0^\infty |e(t)|\, dt \qquad (7\text{-}2.17)$$

Other formulations include the square of the error (ISE) and time weighted integrals, ITAE and ITSE, but we will use only the IAE here. The reader should consult the original publications by López *et al.* (1967) for the tuning formulas for other error integrals.

The integral extends from the occurrence of the disturbance or change in set point ($t = 0$) to a very long time thereafter ($t = \infty$). This is because the ending of the responses cannot be fixed beforehand. The only problem with this definition of the integral is that it becomes undetermined when the error is not forced to zero. This happens only when the controller does not have integral action, because of offset or steady-state error. In this case, the error in the definition is replaced with the difference between the controlled variable and its final steady-state value.

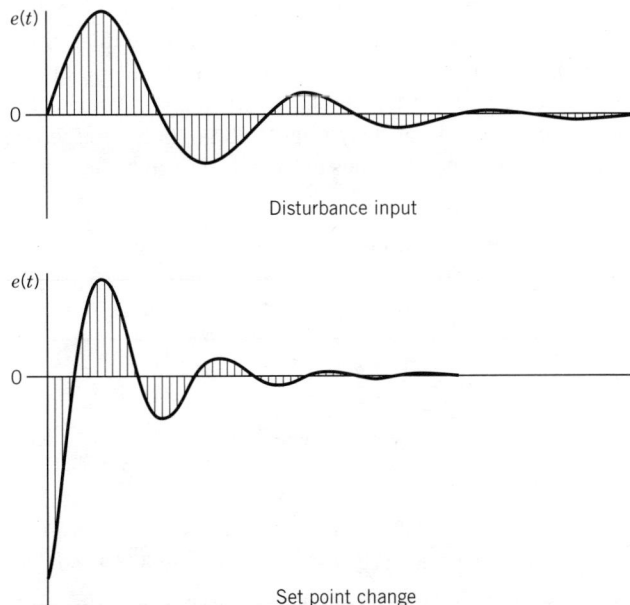

Figure 7-2.10 Definition of error integrals for disturbance and for set point changes.

Equation 7-2.17 constitutes the error integral that can be minimized for a given loop by adjusting the controller parameters. Unfortunately, the optimum set of parameter values is a function of the type of input, that is, disturbance or set point, and of its shape, for example, step change, ramp, and so on. In terms of the shape of the input, the step change is usually selected because it is the most disruptive that can occur in practice, while in terms of the input type, we must select either set point or disturbance input for tuning, according to which one is expected to affect the loop more often. When set point inputs are more important, the purpose of the controller is to have the controlled variable track the set point, and the controller is referred to as a *servo regulator*. When the purpose of the controller is to maintain the controlled variable at a constant set point in the presence of disturbances, the controller is called a *regulator*. The optimum tuning parameters in terms of minimum error integral are different for each case. Most process controllers are considered to be regulators, except for the slave controllers in cascade control schemes, which are servo regulators. We will study cascade control in Chapter 9.

When the controller is tuned for optimum response to a disturbance input, an additional decision must be made regarding the process transfer function to the particular disturbance. This is complicated by the fact that the controller response cannot be optimum for each disturbance if there are more than one major disturbance signals entering the loop. Since the process transfer function is different for each disturbance and for the controller output signal, the optimum tuning parameters are functions of the relative speed of response of the controlled variable to the disturbance input. The slower the response to the disturbance input, the tighter the controller can be tuned, that is, the higher the controller gain can be. At the other extreme, if the controlled variable were to respond instantaneously to the disturbance, the controller tuning would be the least tight it can be, which would be exactly equivalent to the tuning for set point changes.

López *et al.* (1967) developed tuning formulas for minimum error integral criteria based on the assumption that the process transfer function to disturbance inputs is identical to the transfer functions to the controller output signal. In other words, with reference to Fig. 7-2.2, they assumed that $G_2(s) = G_1(s)$. The tuning formulas for minimum IAE are listed in Table 7-2.2.

These formulas indicate the same trend as the quarter decay ratio formulas, except that the integral time depends to some extent on the effective process time constant and less on the process dead time. We must again keep in mind that these formulas are empirical and should not be extrapolated beyond a range of (t_0/τ) of between 0.1 and 1.0. (This is the range of values used by López in his correlations.) As is the case for the

Table 7-2.2 Minimum IAE formulas for disturbance inputs

Controller type	Proportional gain	Integral time	Derivative time
Proportional-only, P	$K_c = \dfrac{0.902}{K}\left(\dfrac{t_0}{\tau}\right)^{-0.985}$	—	—
Proportional-integral, PI	$K_c = \dfrac{0.984}{K}\left(\dfrac{t_0}{\tau}\right)^{-0.986}$	$\tau_I = \dfrac{\tau}{0.608}\left(\dfrac{t_0}{\tau}\right)^{0.707}$	—
Proportional-integral-derivative, PID	$K_c = \dfrac{1.435}{K}\left(\dfrac{t_0}{\tau}\right)^{-0.921}$	$\tau_I = \dfrac{\tau}{0.878}\left(\dfrac{t_0}{\tau}\right)^{0.749}$	$\tau_D = 0.482\tau\left(\dfrac{t_0}{\tau}\right)^{1.137}$

Table 7-2.3 Minimum IAE formulas for set point changes

Controller type	Proportional gain	Integral time	Derivative time
Proportional-integral, PI	$K_c = \dfrac{0.758}{K}\left(\dfrac{t_0}{\tau}\right)^{-0.861}$	$\tau_I = \dfrac{\tau}{1.02 - 0.323\left(\dfrac{t_0}{\tau}\right)}$	—
Proportional-integral-derivative, PID	$K_c = \dfrac{1.086}{K}\left(\dfrac{t_0}{\tau}\right)^{-0.869}$	$\tau_I = \dfrac{\tau}{0.740 - 0.130\left(\dfrac{t_0}{\tau}\right)}$	$\tau_D = 0.348\tau\left(\dfrac{t_0}{\tau}\right)^{0.914}$

quarter decay ratio tuning formulas, these formulas predict that both the proportional and integral actions go to infinity as the process approaches a first-order process without dead time. This behavior is typical of tuning formulas for disturbance inputs.

The set point tuning formulas given in Table 7-2.3 were developed by Rovira (1981) who omitted relationships for pure proportional controllers on the assumption that the minimum error integral criteria is not appropriate for those applications for which a pure proportional controller is indicated, for example, flow averaging by proportional level control (see Section 7-3). These formulas are also empirical and should not be extrapolated beyond the range of (t_0/τ) between 0.1 and 1.0. They predict that for a single-capacitance (first-order) process without dead time, the integral time approaches the time constant of the process while the proportional gain goes to infinity and the derivative time to zero. These results are typical of set point tuning formulas.

EXAMPLE 7-2.4

Minimum IAE Tuning for Heat Exchanger

Determine the tuning parameters that result from the minimum integral of the absolute error for disturbance inputs for the heat exchanger temperature controller using the FOPDT model transfer function of Example 7-2.1. Consider **(a)** a P controller, **(b)** a PI controller, and **(c)** a PID controller.

SOLUTION

The FOPDT model parameters from Example 7-2.1 are, for fit 3,

$$K = 0.80 \ \%/\%; \qquad \tau = 33.8 \ s; \qquad t_0 = 11.2 \ s$$

The minimum error integral tuning parameters for disturbance inputs can be calculated using the formulas from Table 7-2.2.

(a) P CONTROLLER

$$K_c = \frac{0.902}{K}\left(\frac{t_0}{\tau}\right)^{-0.985} = 3.3 \ \frac{\%CO}{\%TO}$$

Compare with the gain for quarter decay ratio, from Example 7-2.3, of 9.4 %CO/%TO.

(b) PI CONTROLLER

$$K_c = \frac{0.984}{0.80}\left(\frac{11.2}{33.8}\right)^{-0.986} = 3.7 \ \frac{\%CO}{\%TO} \qquad \tau_I = \frac{33.8}{0.608}\left(\frac{11.2}{33.8}\right)^{0.707} = 25.5 \ s \ (0.42 \ min)$$

Compare with the quarter decay values, from Example 7-2.3, of $K_c = 8.5$ %CO/%TO, and $\tau_I = 0.40$ min.

(c) *PID CONTROLLER*

$$K_c = \frac{1.435}{0.80}\left(\frac{11.2}{33.8}\right)^{-0.921} = 5.0 \,\frac{\%CO}{\%TO}$$

$$\tau_I = \frac{33.8}{0.878}\left(\frac{11.2}{33.8}\right)^{0.749} = 16.8 \text{ s } (0.28\,\text{min})$$

$$\tau_D = 0.482 \cdot 33.8 \cdot \left(\frac{11.2}{33.8}\right)^{1.137} = 4.64 \text{ s } (0.077\,\text{min})$$

These parameters are for the ideal or parallel PID controller. To compare them to the quarter decay ratio parameters of Example 7-2.3, which are for the actual or series PID controller, we must convert them to the parameters of the ideal PID controller (see Table 7-2.1).

$$K_c = K_c'\left(1 + \frac{\tau_D'}{\tau_I'}\right) = 11\left(1 + \frac{0.06}{0.24}\right) = 14 \,\frac{\%CO}{\%TO}$$

$$\tau_I = \tau_I' + \tau_D' = 0.24 + 0.06 = 0.30\,\text{min}$$

$$\tau_D = \frac{\tau_I'\tau_D'}{\tau_I' + \tau_D'} = \frac{0.24 \cdot 0.06}{0.24 + 0.06} = 0.048\,\text{min}$$

The conclusion we can draw from comparing these tuning parameters with the parameters for quarter decay ratio is that all of these formulas result in values of the same order of magnitude or ballpark. Notice that the quarter decay ratio gains are higher than those for minimum IAE, meaning that they result in more aggressive control.

Figure 7-2.11 shows the PID controller responses of the transmitter and controller outputs using the minimum IAE and quarter decay ratio tuning. The quarter decay ratio responses show a smaller initial deviation from set point for the change in process flow and faster approach to the new set point on the set point change. This better performance is at the expense of a slightly higher oscillatory behavior and higher variation in the controller output.

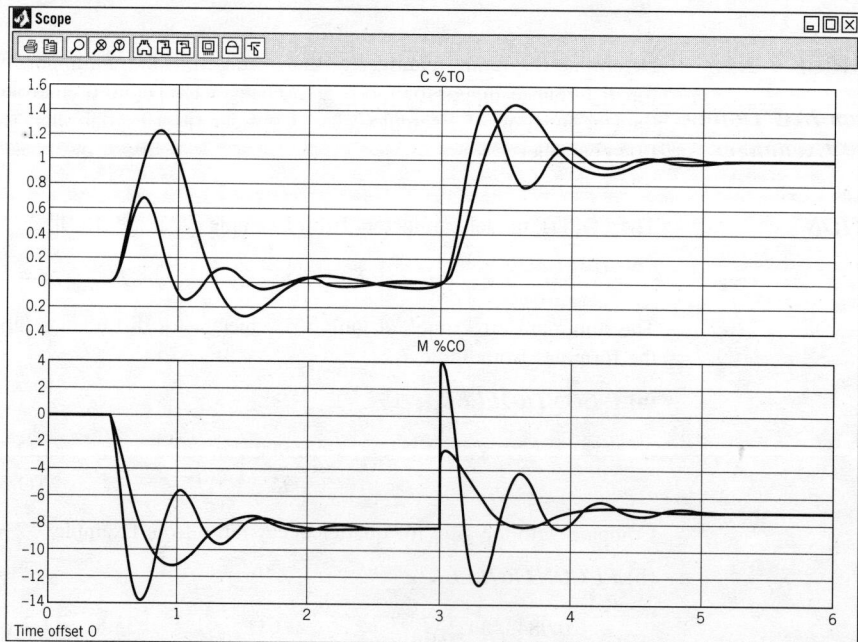

Figure 7-2.11 Comparison of PID controller responses for minimum IAE versus quarter decay ratio tuning for the heat exchanger of Example 7-2.4. Input is a 2 kg/s step increase in process flow followed by a 1°C change in set point.

EXAMPLE 7-2.5

Disturbance versus Set Point Tuning Formulas

Compare the responses of the heat exchanger to unit step changes in disturbance and in set point obtained when a PID controller is tuned for minimum IAE for disturbance inputs with those obtained for the same criteria for set point changes.

SOLUTION

The FOPDT parameters are

$$K = 0.8 \ \%TO/\%CO \qquad \tau = 33.8 \ s \qquad t_0 = 11.2 \ s$$

The minimum IAE parameters for disturbance inputs were determined in Example 7-2.4 as $K_c = 5.0 \ \%CO/\%TO$, $\tau_I = 0.28$ min, $\tau_D = 0.077$ min. For set point changes the parameters of the PID controller are determined from the formulas of Table 7-2.3.

$$K_c = \frac{1.086}{0.80} \left(\frac{11.2}{33.8}\right)^{-0.869} = 3.5 \ \frac{\%CO}{\%TO}$$

$$\tau_I = \frac{33.8}{0.740 - 0.130\left(\frac{11.2}{33.8}\right)} = 48 \ s \ (0.80 \min)$$

$$\tau_D = 0.348 \cdot 33.8 \cdot \left(\frac{11.2}{33.8}\right)^{0.914} = 4.3 \ s \ (0.071 \min)$$

We see that the set point tuning formulas result in a smaller gain and a longer integral time than the disturbance formulas. Figure 7-2.12 compares the responses of the transmitter and controller outputs for the two sets of tuning formulas. The figure shows that for the disturbance in process flow, the disturbance tuning parameters result in a slightly smaller initial deviation

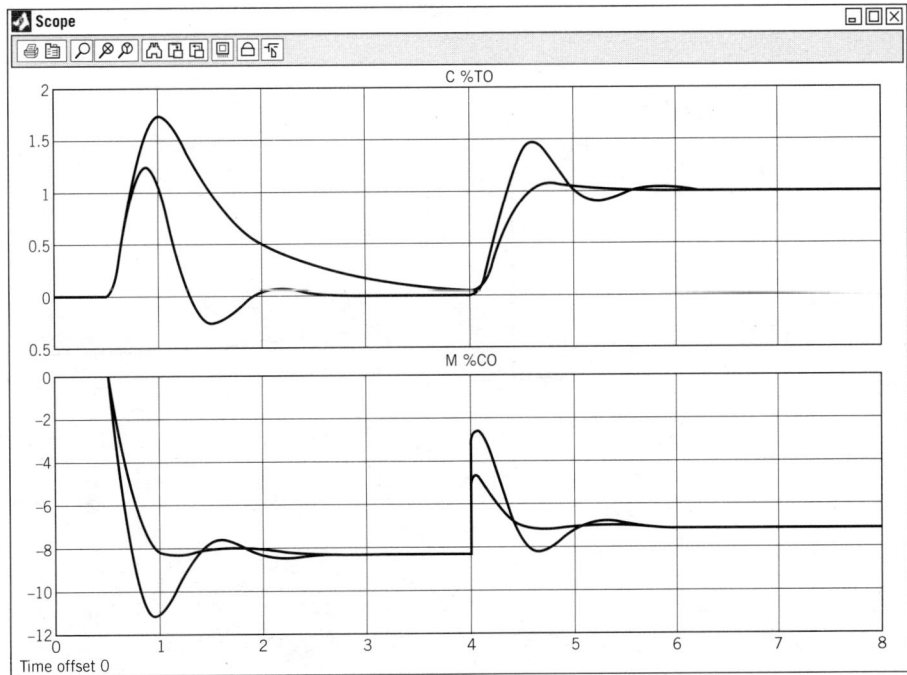

Figure 7-2.12 Comparison of PID controller responses tuned for minimum IAE for disturbance inputs versus minimum IAE for set point changes for the heat exchanger of Example 7-2.5. Input is a 2 kg/s increase in process flow followed by a 1°C change in set point.

from and a faster return to the set point than the set point tuning parameters, while for the set point change the set point tuning parameters result in a significantly smaller overshoot, less oscillatory behavior, and a shorter settling time than the disturbance tuning parameters. Notice that the disturbance formulas result in more variation in the controller output. These variations may be undesirable because they cause disturbances to other control loops.

As would be expected, each set of tuning parameters performs best for the input for which it is designed. The responses obtained are a direct result of the higher gain and shorter reset time obtained with disturbance tuning.

7-2.4 Tuning Sampled-Data Controllers

Today in industry most control functions are implemented using microprocessors (distributed controllers), minicomputers, and digital computers. A common characteristic of these installations is that the control calculations are performed at regular intervals of time T, the *sample time*. This is in contrast to analog (electronic and pneumatic) instruments that perform their functions continuously with time. Sampling is also characteristic of some analyzers, such as on-line gas chromatographs.

The discrete mode of operation characteristic of computers requires that at each sampling instant, the transmitter signals be sampled, the value of the manipulated variable be calculated, and the controller output signal be updated. The output signals are then held constant for a full sample time until the next update. This is illustrated in Fig. 7-2.13. As might be expected, this sampling and holding operation has an effect on the performance of the controller and thus on its tuning parameters.

The sampling time of computer controllers varies from about one-third of a second to several minutes, depending on the application. *A good rule of thumb is that the sample*

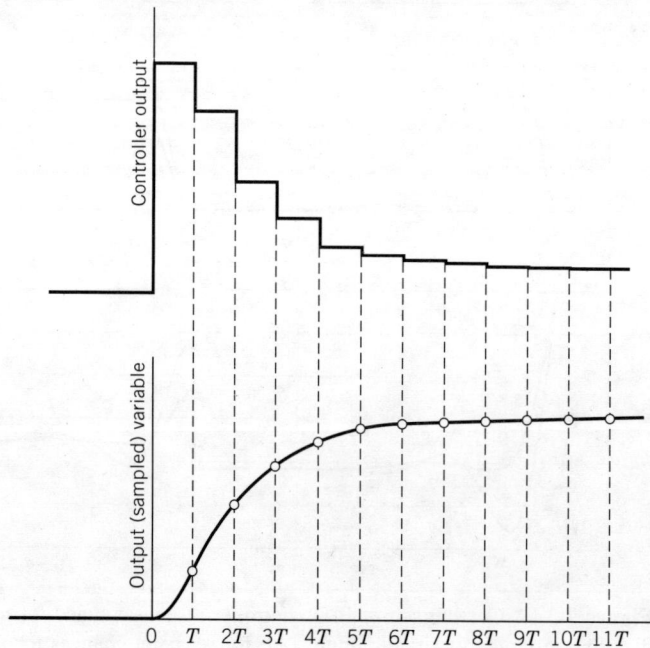

Figure 7-2.13 The output of a sampled-data controller is updated at uniform intervals of time T and held constant between updates.

time should be about one-tenth of the effective process time constant. When the sample time is of this order of magnitude, its effect can be taken into consideration in the tuning formulas by adding one-half the sample time to the process dead time and then using this corrected dead time in the tuning formulas for continuous controllers (Tables 7-2.1, 7-2.2, and 7-2.3). This method, proposed by Moore *et al.* (1969), says that the dead time to use in the tuning formulas is

$$t_{0c} = t_0 + \frac{T}{2} \tag{7-2.18}$$

where

t_{0c} = the corrected dead time

t_0 = the dead time of the process

T = the time interval between samples (sample time)

Notice that the on-line tuning method inherently incorporates the effect of sampling when the ultimate gain and period are determined for the loop with the sampled-data controller in automatic.

Tuning formulas that are specific for sampled-data controllers have been developed by Chiu *et al.* (1973) and reproduced by Corripio (2001).

7-2.5   Summary of Controller Tuning

We have thus far presented two methods for measuring the dynamic characteristics of the process in a feedback control loop: the ultimate gain method, and the step test or process reaction curve. We have also presented one set of tuning formulas for the ultimate gain method and three sets of formulas for the first-order-plus-dead-time model parameters. For a given process, all four sets of tuning formulas result in controller parameters that are in the same ballpark. These tuning parameters are just starting values that must be adjusted in the field so that the true "personality" of the specific process can be matched by the controller. We must reiterate a point made at the beginning of this chapter. As discussed in Chapters 3 and 4, most processes are nonlinear and their dynamic characteristics (e.g., ultimate gain and frequency, FOPDT model parameters) vary from one operating point to another. It follows that the controller parameters arrived at by the tuning procedure are at best a compromise between slow behavior at one end of the operating range and oscillatory behavior at the other. In short, tuning is not an exact science. However, we must also keep in mind that the tuning formulas offer us insight into how the various controller parameters depend on such process parameters as the gain, the time constant, and the dead time.

7-3   TUNING CONTROLLERS FOR INTEGRATING PROCESSES

Integrating processes represent a special tuning case because the standard first- or second-order models presented in the preceding section cannot represent the process. This is because the process is not self-regulating, that is, it will not reach a steady state if driven by a sustained disturbance with the loop opened. As a further consequence, feedback control is absolutely required to operate an integrating process, and open-loop step tests could only be carried out for very brief periods of time. By far the most common integrating process is the control of liquid level.

A special feature of liquid level control is that there are two opposite specifications of control loop performance: *tight level control* and *averaging level control.* Tight level control requires that the level be kept at or very near its set point, as in natural circulation evaporators and reboilers, because of the large sensitivity of the heat transfer rate on the level. Averaging level control is specified for surge tanks and accumulators, where the objective is to average out or attenuate variations in inlet flow so that the outlet flow does not vary suddenly. Reactors and similar equipment would require an intermediate specification where the objective of controlling the level is to keep the volume of liquid in the tank approximately constant. In such applications it would be acceptable to allow the level to vary about ±5 % around its set point, which is looser than for an evaporator or reboiler.

In this section we will use level control as an example of an integrating process. We will derive a model of a simple tank level control and show that, in many cases, it is possible to compute the tuning parameters from the process design parameters.

7-3.1   Model of Liquid Level Control System

Liquid level control is one of the few continuous process that can be treated as an integrating process. Figure 7-3.1 shows a schematic of the level control loop with the control valve on the discharge of the pump that draws the liquid from the tank. In Section 4-1.1 similar level processes were modeled as self-regulating because of the effect of the level on the pressure drop across the valve and, through it, on the outlet flow. However, this self-regulating effect is usually negligible, especially when a pump, as in this case, provides the pressure drop across the valve. A numerical demonstration of this fact is provided as a problem at the end of this chapter.

To model the response of the level in the tank, we write a mass balance around the tank

$$A \frac{dh(t)}{dt} = f_i(t) - f_o(t) \tag{7-3.1}$$

where A is the cross-sectional area of the tank, $h(t)$ is the level in the tank, $f_i(t)$ is the total flow into the tank, $f_o(t)$ is the flow out of the tank, and we are assuming consistent units. Subtracting the initial steady state, and Laplace-transforming the resulting equation in terms of the deviation variables, we obtain

$$H(s) = \frac{1}{As}[F_i(s) - F_o(s)] \tag{7-3.2}$$

1 eq., 2 unk. (H, F_o)

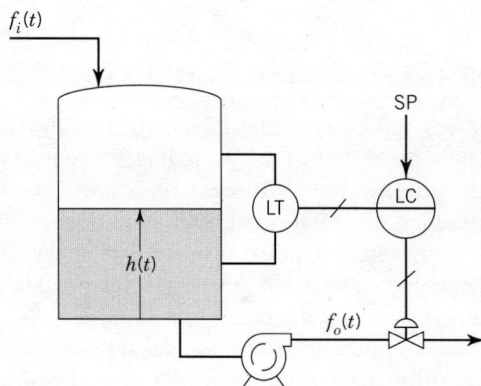

Figure 7-3.1 Schematic of a level control loop with manipulation of the outlet stream.

Assuming the outlet flow is only a function of the valve position, and modeling the valve as a first-order lag (see Section 5-2), we get

$$F_o(s) = \frac{K_v}{\tau_v s + 1} M(s)$$

(7-3.3)

2 eq., 3 unk. (M)

where K_v is the valve gain, τ_v is the valve time constant, and $M(s)$ is the controller output signal, %CO. Level transmitters are usually very fast and can be modeled as simple gains:

$$C(s) = K_T H(s)$$

(7-3.4)

3 eq., 4 unk. (C)

where $C(s)$ is the transmitter output signal, %TO, and K_T is the transmitter gain, from Section 5-1,

$$K_T = \frac{100}{h_{max} - h_{min}}$$

(7-3.5)

where h_{max} and h_{min} are the limits of the transmitter range.

A level controller is usually calibrated in percent of transmitter output (%TO) because this value tells how full the tank is without having to know the actual limits on the level transmitter. Because of this, the set point and the level are displayed in %TO. In terms of these variables, the controller output is

$$M(s) = G_c(s)\lfloor C^{set}(s) - C(s)\rfloor$$

(7-3.6)

4 eq., 4 unk.

where $C^{set}(s)$ is the set point in %TO, $G_c(s)$ is the controller transfer function, and we have followed our standard sign convention in calculating the error.

This completes the model of the level control loop. Figure 7-3.2a shows the detailed block diagram of the loop, while Fig. 7-3.2b shows the simplified diagram. The two process gains shown in the latter diagram are

$$K = \frac{K_v K_T}{A} \frac{\%TO}{\%CO\text{-min}}$$
$$K_u = \frac{K_T}{A} \frac{\%TO}{ft^3}$$

(7-3.7)

Notice that these formulas assume consistent units. For example, for K_v in gpm/%CO, and K_T in %TO/ft, A must be in gal/ft, and the inlet flow must also be in gpm so K_u can be in %TO/gal.

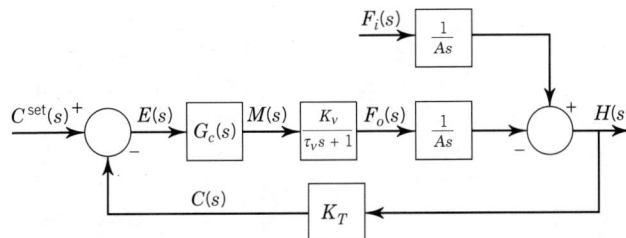

Figure 7-3.2a Block diagram of the level control loop of Fig. 7-3.1.

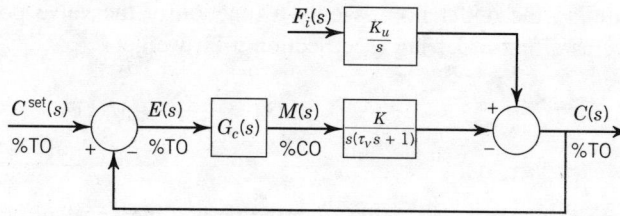

Figure 7-3.2b Simplified block diagram of the level control loop of Fig. 7-3.1.

From the block diagram of Fig. 7-3.2*b*, after some algebraic manipulation, we can derive the following closed-loop transfer function for the level

$$C(s) = \frac{-KG_c(s)}{s(\tau_v s + 1) - KG_c(s)} C^{\text{set}}(s) + \frac{K_u(\tau_v s + 1)}{s(\tau_v s + 1) - KG_c(s)} F_i(s) \qquad \textbf{(7-3.8)}$$

where the minus signs indicate that we have a positive feedback loop, which means that the controller must have a negative gain (direct acting). We can arrive at exactly the same conclusion by analyzing Fig. 7-3.1. An increase in level must cause the controller output to increase to open the outlet valve and increase the outlet flow. This demands a direct acting controller.

7-3.2   Proportional Level Controller

Many level controllers, whether tuned for tight or averaging level control, are proportional controllers. As a matter of fact, we will see in the following section that controller synthesis for an integrating process results in a proportional controller. Let us then look at a proportional controller for the level control loop.

To look at a direct-acting proportional controller, we substitute its transfer function, $G_c(s) = -K_c$, into Eq. 7-3.8 to obtain

$$C(s) = \frac{KK_c}{s(\tau_v s + 1) + KK_c} C^{\text{set}}(s) + \frac{K_u(\tau_v s + 1)}{s(\tau_v s + 1) + KK_c} F_i(s) \qquad \textbf{(7-3.9)}$$

As a proportional controller usually results in offset or steady-state error, let us obtain an expression for the offset before we move into controller tuning. To obtain the offset caused by a sustained change in set point, Δc^{set}, and a sustained change in inlet flow, Δf_i, we let $s = 0$ in Eq. 7-3.9 to obtain

$$\Delta c = \Delta c^{\text{set}} + \frac{K_u}{KK_c} \Delta f_i$$

where Δc is the resulting steady-state change in level in %TO. The offset is then the steady-state error.

$$e = \Delta c^{\text{set}} - \Delta c = -\frac{K_u}{KK_c} \Delta f_i = -\frac{1}{K_v K_c} \Delta f_i \qquad \textbf{(7-3.10)}$$

where we have substituted Eq. 7-3.7. This result shows that for an integrating process there is no offset for changes in set point, only for changes in flow. The offset is inversely proportional to the controller gain K_c.

The characteristic equation of the loop is obtained by setting the denominator of Eq. 7-3.9 equal to zero.

$$\tau_v s^2 + s + KK_c = 0$$

The roots of this equation can be obtained directly by the quadratic formula and rearranged into the following form:

$$r_{1,2} = \frac{-1 \pm \sqrt{1 - 4\tau_v K K_c}}{2\tau_v} \qquad \text{(7-3.11)}$$

This expression tells us that the roots of the characteristic equation are real and negative as long as the gain is limited to

$$0 < K_c < \frac{A}{4\tau_v K_v K_T} \qquad \text{(7-3.12)}$$

where we have substituted the value of K from Eq. 7-3.7. The expression given in Eq. 7-3.12 represents the maximum proportional controller gain that will result in a nonoscillatory response. We notice, from Eq. 7-3.11, that if the controller gain is increased beyond this maximum value, the response becomes oscillatory, but it cannot be unstable, no matter how high the controller gain, because the real part of the complex roots is always negative (see Section 2-3.3).

If the controller gain is set equal to the expression of Eq. 7-3.12, the closed-loop transfer function will have two equal roots equal to $-1/2\tau_v$. The effective time constants of the closed loop will then be equal to each other and to twice the valve time constant. Since most valves are very fast, this means that a proportional level controller can be tuned to give a very fast nonoscillatory response.

EXAMPLE 7-3.1

Calculate the maximum gain of a proportional level controller that results in a nonoscillatory response. The level transmitter has a range of 2 to 10 ft above the bottom of a distillation column 8.0 ft in diameter. The design flow of the bottoms product is 500 gpm and the control valve on that line is linear and sized for 100 % overcapacity. The time constant of the valve is 3 s (0.05 min). Calculate also the effective time constant of the closed loop and the offset caused by a sustained change of 100 gpm in flow.

SOLUTION

Assuming constant pressure drop across the valve, the valve gain, from Section 5-2.4, is

$$K_v = \frac{f_{o,\max}}{100} = \frac{2(500)}{100} = 10.0 \frac{\text{gpm}}{\text{\%CO}}$$

The transmitter gain, from Eq. 7-3.5 is

$$K_T = \frac{100}{10 - 2} = 12.5 \frac{\text{\%TO}}{\text{ft}}$$

The area of the tower is

$$A = \frac{\pi (8.0)^2}{4} \left(7.48 \frac{\text{gal}}{\text{ft}^3} \right) = 376 \frac{\text{gal}}{\text{ft}}$$

The maximum controller gain that will produce nonoscillatory response is, from Eq. 7-3.12

$$K_c = \frac{A}{4\tau_v K_v K_T} = \frac{(376 \text{ gal/ft})}{4(0.05 \text{ min})(10.0 \text{ gpm/\%CO})(12.5 \text{ \%TO/ft})} = 15.0 \frac{\text{\%CO}}{\text{\%TO}}$$

This represents a proportional band of less than 7 %. The effective time constants of the closed loop, when the gain is set equal to this value, are $2(0.05) = 0.10$ min, or 6 s each, which is very fast, considering the size of the column.

Finally, the offset caused by a 100-gpm change in flow, from Eq. 7-3.10,

$$e = -\frac{1}{K_v K_c} \Delta f_i = -\frac{(100 \text{ gpm})}{(10.0 \text{ gpm/\%CO})(15.0 \text{ \%CO/\%TO})} = -0.66 \text{ \%TO}$$

This is an imperceptible change, less than 1 % of the transmitter range.

The preceding example shows that a proportional level controller can be tuned for nonoscillatory response from the design parameters of the process. Obviously this tuning is for tight level control, that is, for reducing variations in level. Let us next look at averaging level control.

7-3.3   Averaging Level Control

The purpose of an average level controller is to average out sudden variations in the disturbance flows so as to produce a smooth varying manipulated flow. For example, if the tank of Fig. 7-3.1 were a surge tank on the feed to a continuous distillation column, it would be very desirable that the column not be subjected to sudden variations in flow, because this could cause flooding and upset the product compositions.

A proportional controller is ideal for averaging level control, but obviously we would like its gain to be as low as possible so that it lets the level in the tank vary and absorb the variations in disturbance flows. How low can the gain be? The minimum controller gain is the one that prevents the level from exceeding the range of the level transmitter at any time. To derive this minimum controller gain, let us recall the formula for a proportional controller, Eq. 5-3.4:

$$m(t) = \overline{m} + K_c \, e(t) \tag{5-3.4}$$

If we set $\overline{m} = 50$ %CO, the control valve will be exactly half opened when the level is at the set point. If we further set the set point to 50 %TO, then the maximum value of the error is ± 50 %TO, because the transmitter can only read from 0 to 100 %TO. From the above equation we see that the minimum gain that prevents the level from exceeding the limits of the transmitter is 1.0 %CO/%TO (100 %PB). If the gain were lower than unity, the level would have to exceed the transmitter range for the valve to either open fully ($m = 100$ %) or close fully ($m = 0$ %). For gains greater than unity the valve can reach one of its limits before the level reaches the corresponding limit on the transmitter range, but then the manipulated flow will vary more than necessary. In summary,

The ideal averaging level controller is a proportional controller with the set point at 50 %TO, the output bias at 50 %CO, and the gain set at 1.0 %CO/%TO.

EXAMPLE 7-3.2

Comparison of Tight and Averaging Level Control

Figure 7-3.3 shows the schematic of a process in which each of three batch reactors dumps its contents into a tank at a rate of 70 ft^3/min for a period of 4 min, starting at different times. The tank, with an area of 20 ft^2, is intended to smooth out the sudden variations in inlet flow, so that the flow out of the tank is maintained relatively constant. This is because sudden variations in the outlet flow would upset the continuous process downstream of the tank.

The level in the tank is measured with sensor-transmitter LT and controlled by proportional controller LC which adjusts the position of the control valve on the outlet stream from the tank. The control valve fails closed and the level controller is direct acting.

Compare, by simulation, the response of the level and the outlet flow using tight level control, $K_c = 100$ %CO/%TO, and averaging level control, $K_c = 1.0$ %CO/%TO.

SOLUTION

The tank is simulated by numerical integration of Eq. 7-3.1 with three inlet flows, one outlet flow, and an area $A = 20$ ft^2. Each inlet flow is simulated with a positive step input of magnitude 70 ft^3/min, followed by a negative step input of the same magnitude 4 min later. Thus each pair of step inputs simulates the dumping of a reactor. The transmitter, proportional controller, and control valve are simulated as discussed in Section 13.4.

Figure 7-3.4 shows the responses of the level and the outlet flow. In the case shown the first reactor starts dumping a $t = 1$ min, the second at $t = 4$ min, and the third at $t = 7$ min. So, the total inlet flow is zero, 70 ft^3/min, and 140 ft^3/min, at different periods of time.

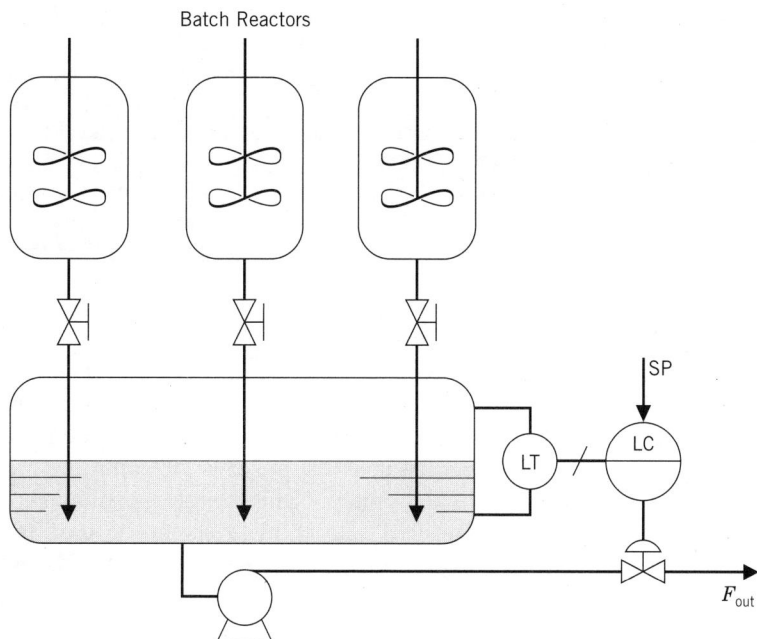

Figure 7-3.3 Level control on tank receiving discharge from batch reactors.

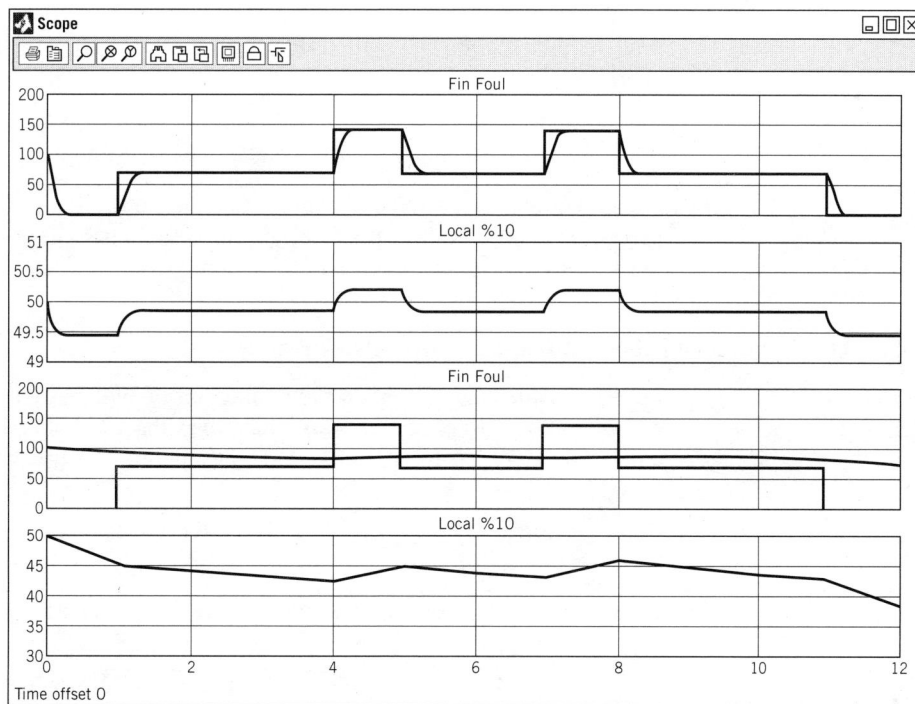

Figure 7-3.4 Response of outlet flow and tank level to discharge of batch reactors. Top two graphs are for tight level control $K_c = 100$ %CO/%TO. Bottom two graphs are for averaging level control. $K_c = 1$ %CO/%TO.

The top two plots in Fig. 7-3.4 are for tight level control, with a controller gain of 100 %CO/%TO. As you can see the level is maintained very close to the set point of 50 %TO, but the outlet flow changes almost as quickly as the inlet flow. These sudden variations would cause upsets to the downstream process and nullify the presence of the tank. The bottom two plots of Fig. 7-3.4 show the responses of the outlet flow and the level for averaging level control, with a controller gain of 1.0 %CO/%TO. This time the outlet flow variations are very smooth while the level is allowed to change from 10 to 20 %TO as the reactor contents are dumped into the tank. Averaging level control makes full use of the ability of the tank to accumulate the liquid.

7-3.4 Summary of Tuning for Integrating Processes

In this section we have presented the interesting problem of tuning proportional controllers for integrating processes. We have developed a formula, Eq. 7-3.12, that provides us the maximum controller gain resulting in a nonoscillatory response. We also found that the minimum gain that prevents the controlled variable from exceeding the transmitter range limits is unity. Gains near the maximum are to be used for tight level controller tuning, while a gain of unity should be used for averaging level control. A problem at the end of this chapter explores the tuning of proportional-integral (PI) level controllers.

7-4 SYNTHESIS OF FEEDBACK CONTROLLERS

In the preceding sections we have taken the approach of tuning a feedback controller by adjusting the parameters of the proportional-integral-derivative (PID) control structure. In this section we will take a different approach to controller design, that of controller synthesis,

Given the transfer functions of the components of a feedback loop, synthesize the controller required to produce a specific closed-loop response.

Although we get no assurances that the controller resulting from our synthesis procedure can be built in practice, we stand to gain some insight into the selection of the various controller modes and their tunings.

7-4.1 Development of the Controller Synthesis Formula

Let us consider the simplified block diagram of Fig. 7-4.1 in which the transfer functions of all the loop components other than the controller have been lumped into a single block, $G(s)$. From block diagram algebra, the transfer function of the closed loop is

$$\frac{C(s)}{R(s)} = \frac{G_c(s)G(s)}{1 + G_c(s)G(s)} \qquad \text{(7-4.1)}$$

Next, we use the above expression to solve for the controller transfer function.

$$G_c(s) = \frac{1}{G(s)} \cdot \frac{C(s)/R(s)}{1 - C(s)/R(s)} \qquad \text{(7-4.2)}$$

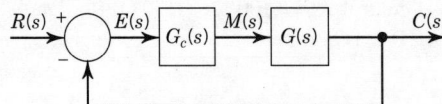

Figure 7-4.1 Simplified block diagram for controller synthesis.

This is the *controller synthesis formula*. It gives us the controller transfer function $G_c(s)$ from the process transfer function $G(s)$ and the specified closed-loop response $C(s)/R(s)$. In order to illustrate how this formula is used, consider the specification of perfect control, that is, $C(s) = R(s)$ or $C(s)/R(s) = 1$. The resulting controller is

$$G_c(s) = \frac{1}{G(s)} \cdot \frac{1}{1-1} = \frac{1}{G(s)} \cdot \frac{1}{0} \tag{7-4.3}$$

This says that in order to force the output equal to the set point at all times, the controller gain must be infinite. In other words, perfect control cannot be achieved with feedback control. This is because any feedback corrective action must be based on an error.

The controller synthesis formula, Eq. 7-4.2, results in different controllers for different combinations of closed-loop response specifications and process transfer functions. Let us look at each of these elements in turn.

7-4.2 Specification of the Closed-Loop Response

The simplest achievable closed-loop response is a first-order lag response. In the absence of process dead time, this response is the one shown in Fig. 7-4.2 and results from the closed-loop transfer function

$$\frac{C(s)}{R(s)} = \frac{1}{\tau_c s + 1} \tag{7-4.4}$$

where τ_c is the time constant of the closed-loop response and, being adjustable, becomes the single *tuning parameter* for the synthesized controller; the shorter τ_c, the tighter the controller tuning. *Note:* This response was originally proposed by Dahlin (1968), who defined the tuning parameter as the reciprocal of the closed-loop time constant, $\lambda = 1/\tau_c$. In this book we will use τ_c.

Substituting Eq. 7-4.4 into Eq. 7-4.2, we obtain

$$G_c(s) = \frac{1}{G(s)} \cdot \frac{\dfrac{1}{\tau_c s + 1}}{1 - \dfrac{1}{\tau_c s + 1}} = \frac{1}{G(s)} \cdot \frac{1}{\tau_c s + 1 - 1}$$

or

$$G_c(s) = \frac{1}{G(s)} \cdot \frac{1}{\tau_c s} \tag{7-4.5}$$

We can see that this controller has integral mode, which results from the specification of unity gain in the closed-loop transfer function, Eq. 7-4.4. This assures the absence of offset.

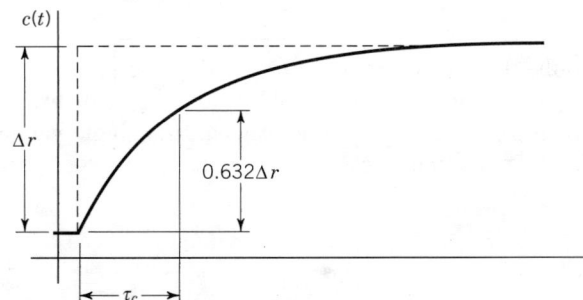

Figure 7-4.2 First-order closed-loop response specification for synthesized controller.

Although second- and higher-order closed-loop responses could be specified, it is seldom necessary to do so. However, when the process contains dead time, the closed-loop response must also contain a dead-time term with the dead time equal to the process dead time. We will look at this shortly, but first let us see how controller synthesis can guide us in the selection of controller modes for various process transfer functions.

7-4.3 Controller Modes and Tuning Parameters

Controller synthesis allows us to establish a relationship between the process transfer function and the modes of a PID controller. This is so because, for simple transfer function without dead time, the synthesized controller can be expressed in terms of the proportional, integral, and derivative modes. Controller synthesis also provides us with relationships for the controller tuning parameters in terms of the closed-loop time constant, τ_c, and the parameters of the process transfer function. In what follows we will derive these relationships by substituting process transfer functions of increasing complexity into Eq. 7-4.5.

Instantaneous Process Response

$$G(s) = K$$

From Eq. 7-4.5,

$$G_c(s) = \frac{1}{K\tau_c} \cdot \frac{1}{s} \tag{7-4.6}$$

where K is the process gain.

This is a *pure integral* controller, which is indicated for very fast processes such as flow controllers, steam turbine governors, and the control of outlet temperatures from reformer furnaces.

First-order process

$$G(s) = \frac{K}{\tau s + 1}$$

From Eq. 7-4.5,

$$G_c(s) = \frac{\tau s + 1}{K} \cdot \frac{1}{\tau_c s} = \frac{\tau}{K\tau_c}\left(1 + \frac{1}{\tau s}\right) \tag{7-4.7}$$

where τ is the process time constant and K is the process gain.

This is a proportional-integral (PI) controller with tuning parameters

$$K_c = \frac{\tau}{K\tau_c} \qquad \tau_I = \tau \tag{7-4.8}$$

or, in words, the integral time is set equal to the process time constant and the proportional gain is adjustable or tunable. Notice that if the process time constant τ is known, tuning is reduced to the adjustment of a single parameter, the controller gain. This is because the tuning parameter τ_c affects only the controller gain.

Second-order process

$$G(s) = \frac{K}{(\tau_1 s + 1)(\tau_2 s + 1)}$$

From Eq. 7-4.5,

$$G_c(s) = \frac{(\tau_1 s + 1)(\tau_2 s + 1)}{K} \cdot \frac{1}{\tau_c s} = \frac{\tau_1}{K \tau_c}\left(1 + \frac{1}{\tau_1 s}\right)(\tau_2 s + 1) \qquad \textbf{(7-4.9)}$$

where

τ_1 = the longer or dominant process time constant

τ_2 = the shorter process time constant

Equation 7-4.9 matches the transfer function of the actual or series PID controller discussed in Section 5.3, ignoring the noise filter term $(\alpha \tau_D' s + 1)$.

$$G_c(s) = K_c'\left(1 + \frac{1}{\tau_I' s}\right)\left(\frac{\tau_D' s + 1}{\alpha \tau_D' s + 1}\right) \qquad \textbf{(7-4.10)}$$

The tuning parameters are then

$$\boxed{K_c' = \frac{\tau_1}{K \tau_c} \qquad \tau_I' = \tau_1 \qquad \tau_D' = \tau_2} \qquad \textbf{(7-4.11)}$$

Again the tuning procedure is reduced to the adjustment of the process gain with the integral time set equal to the longer time constant and the derivative time set to the shorter time constant. This is dictated by experience that indicates that the derivative time should always be smaller than the integral time. In industrial practice, PID controllers are commonly used for temperature control loops so that the derivative mode can compensate for the sensor lag. Here we have arrived at this same result by controller synthesis.

We can easily see that a third-order process would demand a second derivative term in series with the first and with its time constant set to the third longest process time constant, and so on. A reason why this idea has not caught on in practice is that the controller would be very complex and expensive. Besides, the values of the third and subsequent process time constants are very difficult to determine in practice. The common practice has been to approximate high-order processes with low-order-plus-dead-time models. Let us next synthesize the controller for such an approximation of the process transfer function.

First-order-plus-dead-time process

$$G(s) = \frac{Ke^{-t_0 s}}{\tau s + 1}$$

From Eq. 7-4.5,

$$G_c(s) = \frac{\tau s + 1}{Ke^{-t_0 s}} \cdot \frac{1}{\tau_c s} = \frac{\tau}{K \tau_c}\left(1 + \frac{1}{\tau s}\right)e^{t_0 s} \qquad \textbf{(7-4.12)}$$

where t_0 is the process dead time.

We note immediately that this is an *unrealizable* controller because it requires knowledge of the future, that is, a negative dead time. This is even more obvious when the specified and the best possible closed-loop responses are graphically compared, as in Fig. 7-4.3. It is evident from this comparison that the specified response must be delayed by one process dead time.

$$\frac{C(s)}{R(s)} = \frac{e^{-t_0 s}}{\tau_c s + 1} \qquad \textbf{(7-4.13)}$$

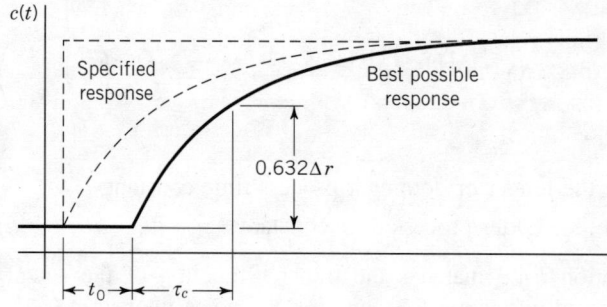

Figure 7-4.3 Specification for closed-loop response of system with dead-time t_0.

This results in the following synthesized controller transfer function:

$$G_c(s) = \frac{\tau s + 1}{K e^{-t_0 s}} \cdot \frac{e^{-t_0 s}}{\tau_c s + 1 - e^{-t_0 s}} = \frac{\tau s + 1}{K} \cdot \frac{1}{\tau_c s + 1 - e^{-t_0 s}} \quad (7\text{-}4.14)$$

Although this controller is now realizable in principle, its implementation is far from common practice. This is mostly because the original PID controllers were implemented with analog components and the term $e^{-t_0 s}$ cannot be implemented in practice with analog devices. Modern implementation of PID controllers on microprocessors and digital computers makes it possible to implement the dead-time term. When this is done the term is called a *predictor* or *dead-time compensation* term.

In order to convert the algorithm of Eq. 7-4.14 to the standard PI form, expand the exponential term by a first-order Padé approximation presented earlier, Eq. 6-2.6:

$$e^{-t_0 s} \approx \frac{1 - \dfrac{t_0}{2} s}{1 + \dfrac{t_0}{2} s} \quad (7\text{-}4.15)$$

Substituting Eq. 7-4.15 into Eq. 7-4.14 and simplifying, we obtain the following synthesized controller:

$$G_c(s) = \frac{\tau}{K(\tau_c + t_0)} \left(1 + \frac{1}{\tau s} \right) \left(\frac{1 + \dfrac{t_0}{2} s}{1 + \tau' s} \right) \quad (7\text{-}4.16)$$

where $\tau' = \dfrac{\tau_c t_0}{2(\tau_c + t_0)}$. This is equivalent to an actual PID controller, Eq. 7-4.10, tuned as follows:

$$K_c' = \frac{\tau}{K(\tau_c + t_0)} \qquad \tau_I' = \tau \qquad \tau_D' = \frac{t_0}{2} \quad (7\text{-}4.17)$$

Although a lag term is present in the transfer function of the actual controller to prevent high-frequency noise amplification, the time constant τ' is usually fixed and about one-tenth of τ_D'. Such value will be in agreement with Eq. 7-4.16 for tight control, since, as $\tau_c \to 0$, $\tau' \to 0$. Notice that if the process dead time is small, as $t_0 \to 0$, $\tau' \to t_0/2$, and the controller of Eq. 7-4.16 becomes a PI controller as the numerator and denominator terms in the last parenthesis cancel each other.

It is interesting to note that the derivative time of Eq. 7-4.17 is exactly the same as the value from the Ziegler–Nichols quarter decay ratio formulas (see Table 7-2.1). However, the proportional gain for quarter decay ratio is 20 % higher than the maximum synthesis gain (obtained by setting $\tau_c = 0$), and the integral time of the synthesis formula is based on the model time constant, whereas that of the quarter decay ratio formula is related to the model dead time.

Notice that the tuning relationship of Eq. 7-4.17 indicates that an increase in dead time results in a reduction of the controller gain for a given closed-loop time constant specification. Comparison of Eqs. 7-4.8 and 7-4.17 shows that the presence of dead time imposes a limit on the controller gain. In other words, for the first-order process without dead time, Eq. 7-4.8, the gain can be increased without limit to obtain faster and faster responses ($\tau_c \to 0$). However, for the process with an effective dead time, from Eq. 7-4.17, we have the following limit on the controller gain:

$$K_{c,\max} = \lim_{\tau_c \to 0} \frac{\tau}{K(\tau_c + t_0)} = \frac{\tau}{K t_0} \tag{7-4.18}$$

The closed-loop response will deviate from the specified first-order response as the controller gain is increased. That is, increasing the gain will eventually result in over-shoot and even instability of the closed-loop response. This is because the error of the first-order Padé expansion approximation increases with the speed of response, as s increases with speed. (Recall that s, the Laplace transform variable, has units of recip-rocal time or frequency. Thus higher speeds of response, or frequencies, correspond to a high magnitude of s.)

Integrating process

$$G(s) = \frac{K}{s}$$

From Eq. 7-4.5,

$$G_c(s) = \frac{s}{K} \cdot \frac{1}{\tau_c s} = \frac{1}{K \tau_c} \tag{7-4.19}$$

This is a proportional controller with an adjustable gain, which agrees with the discussion of Section 7-3 on tuning controllers for integrating processes.

We have now synthesized controllers for the most common process transfer functions. The synthesis of a controller for processes with inverse response is left as an exercise at the end of this chapter.

The same results obtained here by controller synthesis have been obtained by the technique of internal model control (IMC) by Rivera *et al.* (1986). In some literature articles the tuning formulas we have developed by synthesis are called IMC tuning rules.

7-4.4   Summary of Controller Synthesis Results

Table 7-4.1 summarizes the selection of controller modes and tuning parameters that results from the synthesis procedure for Dahlin's response. The fact that the controller gain is a function of the tuning parameter τ_c is both an advantage and a disadvantage of the tuning formulas derived by the synthesis procedure. It is an advantage in that it allows the engineer to achieve a specified response by adjusting a single parameter, the gain, regardless of the controller modes involved. The tunable gain is a disadvantage, however, because the formulas do not provide a ballpark value for it. The following guidelines are given in order to remedy this situation.

Minimum IAE

For disturbance inputs, $\tau_c = 0$ approximately minimizes the IAE when t_0/τ is in the range 0.1 to 0.5 for PI controllers ($\tau_D = 0$) and 0.1 to 1.5 for PID controllers. For

Table 7-4.1 Controller modes and tuning for Dahlin synthesis

Process	Controller	Tuning parameters
$G(s) = K$	I	$K_c = \dfrac{1}{K\tau_c}$
$G(s) = \dfrac{K}{\tau s + 1}$	PI	$K_c = \dfrac{\tau}{K\tau_c}$
		$\tau_I = \tau$
$G(s) = \dfrac{K}{(\tau_1 s + 1)(\tau_2 s + 1)}$ $\tau_1 > \tau_2$	PID	$K_c = \dfrac{\tau_1}{K\tau_c}$
		$\tau_I = \tau_1$
		$\tau_D = \tau_2$
$G(c) = \dfrac{K e^{-t_0 s}}{\tau s + 1}$	PID[a]	$K_c' = \dfrac{\tau}{K(t_0 + \tau_c)}$
		$\tau_I' = \tau$
		$\tau_D' = \dfrac{t_0}{2}$
$G(s) = \dfrac{K}{s}$	P	$K_c = \dfrac{1}{K\tau_c}$

[a] The PID formulas apply to both PID and PI ($\tau_D = 0$) controllers. PID is recommended when $t_0 > \tau/4$. The PID formula is for the actual or series PID controller, Eq. 5-3.19. To convert the tuning parameters for the ideal or parallel PID controller, see footnote to Table 7-2.2.

set point changes, the following formulas result in approximately minimum IAE when t_0/τ is in the range 0.1 to 1.5:

$$\text{PI controller } (\tau_D = 0): \qquad \tau_c = \frac{2}{3}t_0 \qquad \textbf{(7-4.20)}$$

$$\text{PID controller:} \qquad \tau_c = \frac{1}{5}t_0 \qquad \textbf{(7-4.21)}$$

These formulas are to be used with the next to last entry of Table 7-4.1.

5 % Overshoot

For set point inputs a response having an overshoot of 5 % of the change in set point is highly desirable. For this type of response Martin *et al.* (1976) recommend that τ_c be set equal to the effective dead time of the FOPDT model. This results in the following formula for the controller gain that produces 5 % overshoot on *set point changes*:

$$K_c = \frac{0.5}{K}\left(\frac{t_0}{\tau}\right)^{-1} \qquad \textbf{(7-4.22)}$$

Comparison of this formula with the one in Table 7-2.1 shows that this is about 40 % of the PID gain required for quarter decay ratio (50 % overshoot).

One interesting point about the controller synthesis method is that if controllers had been designed this way from the start, the evolution of controller modes would have followed the pattern I, PI, PID. This pattern follows from a consideration of the simplest process model to the more complex. Contrast this to the actual evolution of industrial controllers: P, PI, PID—that is, from the simplest controller to the more complex.

An important insight that we can gain from the controller synthesis procedure is that the main effect of the proportional mode, when added to the basic integral mode, is to compensate for the longest or dominant process lag, while that of the derivative mode is to compensate for the second-longest lag or for the effective process dead time. The entire synthesis procedure is based on the assumption that the primary closed-loop response specification is the elimination of offset or steady-state error. This is what makes the integral mode the basic controller mode.

EXAMPLE 7-4.1

Synthesis Tuning for Heat Exchanger

SOLUTION

Determine the tuning parameters of a PID controller for the heat exchanger of Example 6-2.1 using the formulas derived by the controller synthesis method. Compare these results with those obtained from the minimum IAE tuning formulas for disturbance inputs.

The FOPDT parameters obtained from the heat exchanger by fit 3 in Example 7-2.1 are

$$K = 0.80 \text{ \%TO/\%CO} \qquad \tau = 33.8 \text{ s} \qquad t_0 = 11.2 \text{ s}$$

Because the dead time in this case is greater than one-fourth the time constant, a PID controller is indicated. The proportional gain for minimum IAE on disturbance input is obtained with $\tau_c = 0$. From the next to last entry in Table 7-4.1, the tuning parameters are

$$K'_c = \frac{33.8}{(0.80)(11.2)} = 3.8 \frac{\% \text{ CO}}{\% \text{ TO}}$$

$$\tau'_I = \tau = 33.8 \text{ s } (0.56 \text{ min})$$

$$\tau'_D = \frac{t_0}{2} = 5.6 \text{ s } (0.093 \text{ min})$$

These parameters are for the actual PID controller. To compare them with those for minimum IAE we must convert them to the parameters for the ideal controller (see Table 7-2.1).

$$K_c = K'_c \left(1 + \frac{\tau'_D}{\tau'_I} \right) = 3.8 \left(1 + \frac{0.093}{0.56} \right) = 4.4 \frac{\% \text{ CO}}{\% \text{ TO}}$$

$$\tau_I = \tau'_I + \tau'_D = 0.56 + 0.093 = 0.65 \text{ min}$$

$$\tau_D = \frac{\tau'_I \tau'_D}{\tau'_I + \tau'_D} = \frac{0.56 \cdot 0.093}{0.56 + 0.093} = 0.080 \text{ min}$$

The tuning parameters for minimum IAE on disturbance inputs were determined in Example 7-2.4: $K_c = 5$ %CO/TO, $\tau_I = 0.28$ min, $\tau_D = 0.077$ min. We notice that the gain and derivative time are very close, but the integral time from the synthesis formula is much longer than the one for minimum IAE. Figure 7-4.4 compares the responses of the transmitter and controller outputs using these two sets of parameters for the heat exchanger. The heat exchanger is simulated by the transfer functions of Example 6-2.1 and the actual PID controller is modeled as described in Section 13-4. Figure 7-4.4 shows that the minimum IAE tuning results in a faster return to the set point after the change in process flow than the synthesis tuning. However, the synthesis tuning provides a smoother response to the change in set point with much less overshoot and less variation of the controller output than the minimum IAE tuning.

7-4.5 Tuning Rules by Internal Model Control (IMC)

A popular method to tune feedback controllers is commonly known as the IMC tuning rules. Rivera *et al.* (1986) show how tuning rules for feedback controllers can be

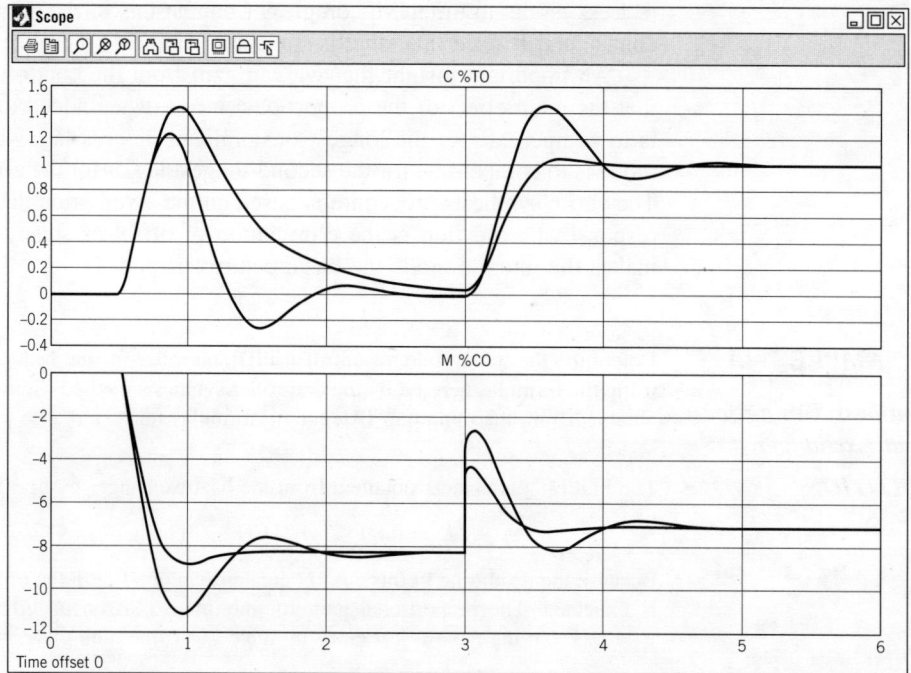

Figure 7-4.4 Comparison of PID controller responses tuned for minimum IAE for disturbance inputs versus synthesis tuning with $\tau_c = 0$ for the heat exchanger of Example 7-4.1. Input is a 2-kg/s increase in process flow followed by a 1°C change in set point.

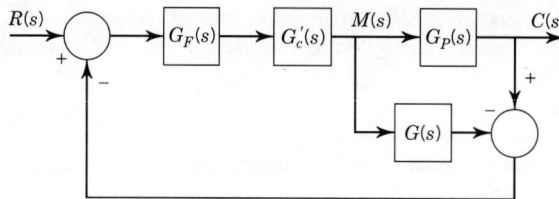

Figure 7-4.5 Block diagram for internal model control (IMC).

developed by internal model control (IMC). In this section we will look at the IMC method and see that the controller tuning rules are basically those we have developed in this section by the closed-loop synthesis method.

Consider the IMC feedback loop of Fig. 7-4.5. In this block diagram, $G(s)$ is the transfer function of an *internal model* of the true process transfer function $G_P(s)$. The feedback controller consists of three blocks: the internal model $G(s)$, an adjustable filter G_F, and a dynamic compensator $G_c'(s)$. The filter is usually a first-order filter with an adjustable time constant, while the internal model and the compensator are designed for each type of process.

From the block diagram of Fig. 7-4.5, the controller output is

$$M(s) = G_c'(s)G_F(s)[R(s) - C(s) + G(s)M(s)]$$

Solving for $M(s)$ and rearranging, we obtain the controller transfer function.

$$G_c(s) = \frac{M(s)}{R(s) - C(s)} = \frac{G_c'(s)G_F(s)}{1 - G_c'(s)G(s)G_F(s)} \qquad \textbf{(7-4.23)}$$

This is essentially the IMC controller synthesis formula. The filter in this formula is usually a first-order filter with unity gain:

$$G_F(s) = \frac{1}{\tau_c s + 1} \tag{7-4.24}$$

where τ_c is the adjustable filter time constant. The internal model, $G(s)$, is selected to match the true process transfer function $G_P(s)$, and the dynamic compensator is selected as the reciprocal of the process model, excluding any terms that are unrealizable such as dead time, or that would create instability, such as a positive zero. An important requirement is that the gain of $G'_c(s)$ be exactly the reciprocal of the gain of the process model $G(s)$. We can see from Eq. 7-4.23 that if this is the case, since the filter gain is also unity, the controller steady-state gain will be infinite, meaning that it will not produce an offset.

To show that the IMC synthesis formula is equivalent to the closed-loop response synthesis formula, let us assume for the moment that it is possible to implement $G'_c(s) = 1/G(s)$. Substitute, along with Eq. 7-4.31, into Eq. 7-4.30, and simplify to obtain

$$G_c(s) = \frac{\dfrac{1}{G(s)} \cdot \dfrac{1}{\tau_c s + 1}}{1 - \dfrac{1}{G(s)} \cdot G(s) \cdot \dfrac{1}{\tau_c s + 1}} = \frac{1}{G(s)} \cdot \frac{1}{\tau_c s} \tag{7-4.25}$$

This formula is identical to the synthesis formula of Eq. 7-4.5. It follows that the feedback controller modes and tuning formulas of Table 7-4.1, which were derived using Eq. 7-4.5, would have also resulted from Eq. 7-4.25, except for the last set of formulas which are for a process with dead time. For this case, the IMC dynamic compensator excludes the dead time.

$$G(s) = \frac{K \, e^{-t_0 s}}{\tau s + 1} \qquad G'_c(s) = \frac{\tau s + 1}{K}$$

Substitute these expressions and Eq. 7-4.24 into Eq. 7-4.23 and simplify to obtain

$$G_c(s) = \frac{\tau s + 1}{K} \cdot \frac{1}{\tau_c s + 1 - e^{-t_0 s}} \tag{7-4.26}$$

This equation is identical to Eq. 7-4.14 which was used to derive the next to last set of tuning formulas in Table 7-4.1.

We have shown that the controller modes and tuning formulas of Table 7-4.1 are also the popular IMC tuning rules.

7-5 TIPS FOR FEEDBACK CONTROLLER TUNING

Each of the tuning methods presented in the preceding sections requires some form of process dynamic testing to be carried out, either to find the ultimate gain and period of the loop, or the parameters of a simple process model. Unfortunately, for many processes, formal tests cannot be carried out because of safety, product quality, or other considerations. The tips presented in this section are intended to facilitate the tuning when no formal testing can be performed. We will also look at an important consideration which is often overlooked when tuning feedback controllers: the compromise between tight control of the controlled variable versus excessive movement of the manipulated variable.

The first tip is to realize that the performance of a feedback controller is relatively insensitive to the values of the tuning parameters, a point that was made earlier in this chapter. In modern control lingo we say that feedback control is a *robust* technique. Approximate values of the tuning parameters can produce a "good enough," if not "optimal," response. In fact, as discussed earlier, when tuning linear controllers for nonlinear processes there is no such thing as an optimal set of tuning parameters. Furthermore, it is possible to be off by 40 or 50 % on the reset and rate times, and still obtain a good response by adjusting the controller gain.

Based on this tip, our procedure for tuning a controller becomes the following:

1. Obtain approximate values for the reset and rate times.

2. Adjust the proportional gain to obtain an acceptable response.

7-5.1   Estimating the Reset and Rate Times

We can use the tuning formulas presented in this chapter as a guide in estimating the reset and rate times. It is usually possible to estimate the ultimate period or the time constant of the process in the loop without formally testing the process. On the other hand, the process dead time is harder to estimate. From the estimated process parameters, the reset and rate times can be estimated as follows:

- The formulas of Table 7-1.1 can be used to estimate the reset and rate times from the period of oscillation of the loop. The period of oscillation of the loop can be estimated by observing its closed-loop response when the controller is proportional only, achieved by detuning or turning off the reset and rate modes. Even if the oscillations are not sustained, the period between two peaks is usually about 40 to 60 % higher than the ultimate period, so the ultimate period can be estimated as two-thirds of the period of the decaying oscillations.

- If the process dominant time constant can be estimated from physical considerations, set the reset time equal to it (Table 7-4.1), and, based on the quarter decay ratio formulas (Table 7-1.1 or 7-2.1), set the rate time equal to one-fourth the reset time. However, if we suspect that the dead time is much smaller than the time constant, use a PI controller (zero rate time).

- If the process dead time can be estimated, the rate time can be set at one-half the dead time (Table 7-2.1 or 7-4.1). This method automatically results in a PI controller if the dead time is negligible.

Today's control systems provide convenient time trends of any measured process variable. The time scale on these trends can be adjusted from a few minutes to several hours, and the process variable scale can also be blown up as needed for greater precision. These trends are invaluable for estimating the necessary process parameters.

Figure 7-5.1 shows a response plot of the outlet temperature from a furnace coil being controlled by manipulating the flow through the coil, which is also shown in the plot. This plot is from an actual process furnace. If we assume the controller is proportional, we see that the period of oscillation is about 15 min. Since the oscillations are sustained, we can assume this is the ultimate period and estimate the reset time as 7.5 min, and the rate time as 1.9 min, using the formulas from Table 7-1.1.

The time constant of the process can sometimes be estimated from simple models based on basic principles, such as those presented in Chapters 3 and 4. For example, from the models of mixed tanks presented in Sections 3-3 and 3-7, we can estimate that the time constant of a tank is approximately equal to its residence time (volume/product

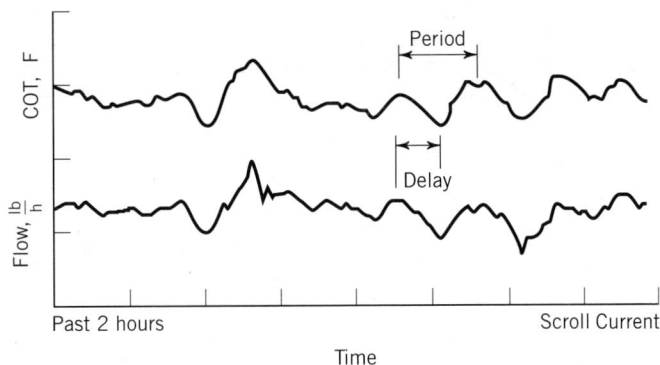

Figure 7-5.1 Trend of furnace coil outlet temperature (COT) and the manipulated flow through the coil.

flow). Similar formulas are derived in Chapter 3 for the time constants of gas tanks, reactors, etc.

The process time constant and dead time can also be estimated from the response of the closed loop, but this requires careful analysis. It can be shown, using the concepts of Section 2-5, that the response of several lags in series to a ramp or slow sine wave is eventually a ramp or sine wave that lags the input ramp by the sum of the time constants plus the dead time. Let us look, for example, at the response of the coil outlet temperature and flow in Fig. 7-5.1. At first the response may look wrong because the flow and the temperature go up and down together. From basic principles we know that when the flow through the coil goes up, the outlet temperature must come down, because there is more fluid to heat up with the same amount of energy. The reason they go up and down together is that the immediate action of the proportional controller causes the flow to increase and decrease with the temperature (direct-acting controller). If we watch carefully, we notice that the temperature does go down when the process fluid goes up, and vice versa, but not right away! Because of the process dynamics, there is a delay between the change in the flow and the corresponding change in temperature. This delay is marked in the figure and seems to be about 8 min. The question is: how are we going to model the coil?

- If we model the coil as a perfectly mixed system, the delay would be equal to its time constant, as in Section 3-3.
- If we model the coil as a plug flow system, the delay will be all caused by pure dead time, as in Section 3-4.

The first of these models results, from Table 7-4.1, in a PI controller with a reset time of 8 min. This value agrees with our earlier estimate based on the period of oscillation. The second model results, from Table 7-2.1, in a PID controller with a reset time of $2 \times 8 = 16$ min, and a rate time of $8/2 = 4$ min. These estimates are about twice those obtained from the period of oscillation.

Like any real system, the coil is neither perfectly mixed nor plug flow but rather a combination of the two, probably closer to the latter. This means that the 8-min delay is just the sum of the time constant and the dead time. From the two extremes we get estimates of the reset time ranging from 8 to 16 min, and estimates of the rate time ranging from 0 to 4 min. We could use a reset time of 12 min and a rate time of 3 min and be confident that these values are good enough to proceed with the adjustment of the proportional gain to obtain the desired response.

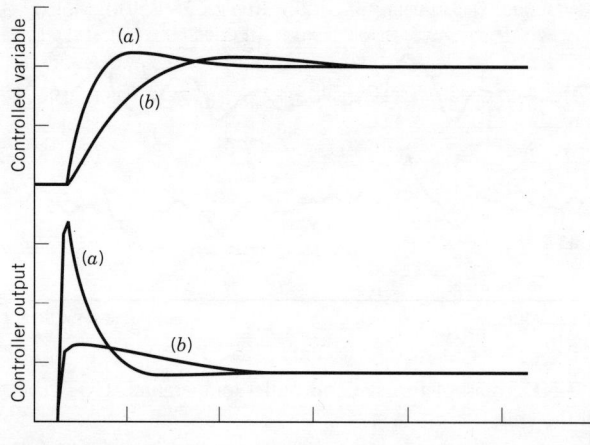

Figure 7-5.2 Comparison of tight (*a*) and reasonable (*b*) adjustment of the controller gain for set point change.

7-5.2 Adjusting the Proportional Gain

One of the problems with many tuning methods has been their rigidity; they provide a set of tuning formulas that leave no room for adjustment of any of the parameters. In contrast, the controller synthesis method and IMC tuning rules of Section 7-4 provide formulas with an adjustable gain.

When adjusting the proportional gain we must keep in mind that very tight control of the process variable usually requires large changes in the manipulated variable. This is undesirable, because changes in the manipulated variable cause upsets to the process and disturb other control loops. For example, a large decrease in the fuel flow to a furnace could cause the flame to go out, and a large increase in the reflux flow to a distillation column may cause the column to flood. Figure 7-5.2 shows the controlled and manipulated variable responses for two values of the controller gain. The higher gain results in tighter control, but also in a much larger upset of the process, because of the larger initial change in the controller output. The point is: when adjusting the gain consider both the tightness of control and the variability of the manipulated variable.

After selecting the reset and rate times, tuning is reduced to the adjustment of a single parameter, the gain. Thus no special procedure is required for adjusting the gain.

7-6 SUMMARY

This chapter has presented two methods for characterizing the process dynamic response and for tuning feedback controllers. We also looked at the tuning of feedback controllers for integrating processes. The controller synthesis method gave us insight into the functions of the proportional, integral, and derivative modes, plus a set of simple tuning formulas. Finally, the chapter closed with some practical tips for tuning feedback controllers.

REFERENCES

1. CHIU, K. C., A. B. CORRIPIO, and C. L. SMITH, "Digital Control Algorithms. Part III: Tuning PI and PID Controllers." *Instruments and Control Systems*, Vol. 46, December 1973, pp. 41–43.

2. CORRIPIO, A. B., *Tuning of Industrial Control Systems*. Research Triangle Park, NC: Instrument Society of America, 1990, Unit 6.

3. DAHLIN, E. B., "Designing and Tuning Digital Controllers." *Instruments and Control Systems*, Vol. 41, No. 6, June 1968, p. 77.

4. LOPEZ, A. M., P. W. MURRILL, and C. L. SMITH, "Controller Tuning Relationships Based on Integral Performance Criteria." *Instrumentation Technology*, Vol. 14, No. 11, November 1967, p. 57.

5. MARTIN, JACOB, Jr., Ph.D. dissertation, Department of Chemical Engineering, Louisiana State University, Baton Rouge, 1975.

6. MARTIN, JACOB, Jr., A. B. CORRIPIO, and C. L. SMITH, "How to Select Controller Modes and Tuning Parameters from Simple Process Models." *ISA Transactions*, Vol. 15, No. 4, 1976, pp. 314–319.

7. MOORE, C. F., C. L. SMITH, and P. W. MURRILL, "Simplifying Digital Control Dynamics for Controller Tuning and Hardware Lag Effects." *Instrument Practice*, Vol. 23, No. 1, January 1969, p. 45.

8. MURRILL, PAUL W., *Fundamentals of Process Control Theory.* 3rd ed., Research Triangle Park, NC: Instrument Society of America, 2000.

9. RIVERA, D. E., M. MORARI, and S. SKOGESTAD, "Internal Model Control, 4. PID Controller Design." *I&EC Process Design and Development*, Vol. 25, 1986, p. 252.

10. ROVIRA, A. A., Ph.D. Dissertation, Department of Chemical Engineering, Louisiana State University, Baton Rouge, 1981.

11. SMITH, CECIL, L., *Digital Computer Process Control.* Scranton, PA: Intext Educational Publishers, 1972.

12. ZIEGLER, J. G., and N. B. NICHOLS, "Optimum Setting for Automatic Controllers." *Transactions ASME*, Vol. 64, November 1942, p. 759.

PROBLEMS

7-1. A feedback control loop is represented by the block diagram of Fig. 7-2.2. The process transfer function is given by

$$G_1(s) = \frac{K}{(\tau_1 s + 1)(\tau_2 s + 1)(\tau_3 s + 1)}$$

where the process gain is $K = 2.5$ %TO/%CO and the time constants are

$$\tau_1 = 5 \text{ min} \qquad \tau_2 = 0.8 \text{ min} \qquad \tau_3 = 0.2 \text{ min}$$

Determine the controller tuning parameters for quarter decay ratio response by the ultimate gain method for

(a) A proportional (P) controller.

(b) A proportional-integral (PI) controller.

(c) A proportional-integral-derivative (PID) controller.

7-2. Using the tuning parameters determined for the loop of Problem 7-1,

(a) Find the roots of the characteristic equation, identify the dominant pair of roots, and calculate the damping ratio and the decay ratio of the response.

(b) Simulate the control loop and plot the responses of the controlled and manipulated variables to a step change in set point. (See Section 13.2 for the simulation of the loop.)

7-3. Given the feedback control loop of Fig. 7-2.2 and the following process transfer function

$$G_1(s) = \frac{K\, e^{-t_0 s}}{(\tau_1 s + 1)(\tau_2 s + 1)}$$

where the process gain, time constants, and dead time are: $K = 1.25$ %TO/%CO, $\tau_1 = 1$ min, $\tau_2 = 0.6$ min, $t_o = 0.20$ min, use Fig. 7-2.8 to estimate the first-order-plus-deadtime (FOPDT) model parameters. Then use these parameters to compare the tuning parameters for a proportional-integral (PI) controller using the following formulas:

(a) Quarter decay ratio response.

(b) Minimum IAE for disturbance inputs.

(c) Minimum IAE for set point inputs.

(d) Controller synthesis for 5 % overshoot on a set point change.

7-4. Do Problem 7-3 for a proportional-integral-derivative (PID) controller.

7-5. Do Problem 7-3 for a sampled-data (computer) controller with a sample time $T = 0.10$ min.

7-6. For the control loop of Problem 7-3, derive the tuning formulas for an actual PID controller using the controller synthesis procedure. Consider two cases:

(a) No dead time, $t_0 = 0$.

(b) The dead time given in the problem.

Check your answers with the entries in Table 7-4.1.

7-7. Prepare a simulation of the control loop in Problem 7-3 to obtain the responses to a step change in set point. Use the controller tuning parameters determined there. Can you improve on the control performance by trial-and-error adjustment of the tuning parameters? Have the program print the integral of the absolute error, IAE, and use it for the measure of control performance. *Note*: for an example of the simulation of a feedback control loop using transfer functions, see Section 13-2.

7-8. From the results of Problem 6-11, calculate the quarter decay ratio tuning parameters of a PI controller for the composition in the blending tank. *Note*: The simulation of this control loop is the subject of Problem 13-11.

7-9. From the results of Problem 6-12, calculate the quarter decay ratio tuning parameters of a PID controller for the reactor temperature controller. *Note*: The simulation of this control loop is the subject of Problem 13-21.

7-10. From the results of Problem 6-14, calculate the quarter decay ratio tuning parameters of a PI controller for the composition from the third tank.

7-11. For the composition controller for the three isothermal reactors in series of Problem 6-17, calculate the quarter decay ratio tuning parameters for a PI controller. Using these parameters, find the roots of the characteristic equation, identify the dominant root, and estimate the damping ratio and the actual decay ratio of the response.

Note: The simulation of this process is the subject of Problem 13-23.

7-12. Do Problem 7-11 for the compressor suction pressure controller of Problem 6-18.

7-13. Do Problem 7-11 for the temperature controller of the stirred tank cooler of Problem 6-19.

7-14. Do Problem 7-11 for the composition control of the reactors in series of Problem 6-22.

7-15. Consider the vacuum filter shown in Fig. P7-1. This process is part of a waste treatment plant. The sludge enters the filter at about 5 % solids. In the vacuum filter the sludge is dewatered to about 25 % solids. The filterability of the sludge in the rotating filter depends on the pH of the sludge entering the filter. One way to control the moisture of the sludge to the incinerator is by the addition of chemicals (ferric chloride) to the sludge feed to maintain the necessary pH. Figure P7-1 shows a proposed control scheme. The moisture transmitter has a range 55 to 95 %.

The following data have been obtained from a step test on the output of the controller (MIC70) of +12.5 %CO:

Time, min	Moisture, %	Time, min	Moisture, %
0	75.0	10.5	70.9
1	75.0	11.5	70.3
1.5	75.0	13.5	69.3
2.5	75.0	15.5	68.6
3.5	74.9	17.5	68.0
4.5	74.6	19.5	67.6
5.5	74.3	21.5	67.4
6.5	73.6	25.5	67.1
7.5	73.0	29.5	67.0
8.5	72.3	33.5	67.0
9.5	71.6		

When the input moisture to the filter was changed by 0.5 % the following data were obtained:

Time, min	Moisture, %	Time, min	Moisture, %
0	75	11	75.9
1	75	12	76.1
2	75	13	76.2
3	75	14	76.3
4	75.0	15	76.4
5	75.0	17	76.6
6	75.1	19	76.7
7	75.3	21	76.8
8	75.4	25	76.9
9	75.6	29	77.0
10	75.7	33	77.0

(a) Draw a block diagram for the moisture control loop. Include the possible disturbances.

(b) Use fit 3 to estimate the parameters of first-order-plus-dead-time models of the two transfer functions. Redraw the block diagram showing the transfer function for each block.

(c) Give an idea of the controllability of the output moisture. What is the correct controller action?

(d) Obtain the gain of a proportional controller for minimum IAE response. Calculate the offset for a 1 % change in inlet moisture.

(e) Tune a PI controller for quarter decay ratio response.

7-16. Consider the absorber shown in Fig. P7-2. The gas entering the absorber has a composition of 90 mole % air and 10 mole % ammonia (NH_3). Before this gas is vented to the atmosphere, it is necessary to remove most of the NH_3 from it. This will be done by absorbing it with water. The NH_3 concentration in the exit gas steam cannot be above 200 ppm. The absorber has been designed so that the outlet NH_3 concentration in the vapor is 50 ppm. From dynamic simulations of the absorber, the following data were obtained:

Response to a Step Change in Water Flow to the Absorber

Time, s	Water flow, gpm	Outlet NH_3 concentration
0	250	50.00
0	200	50.00
20	200	50.00
30	200	50.12
40	200	50.30
50	200	50.60
60	200	50.77
70	200	50.90
80	200	51.05
90	200	51.20
100	200	51.26
110	200	51.35
120	200	51.48
130	200	51.55
140	200	51.63
160	200	51.76
180	200	51.77
250	200	51.77

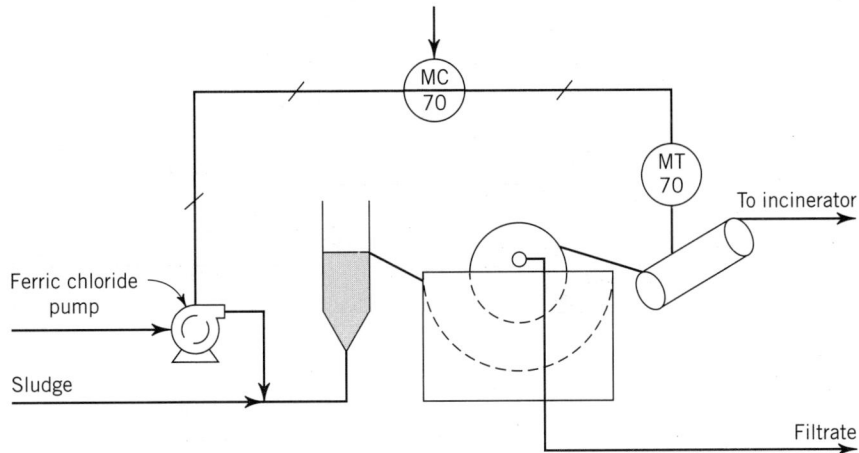

Figure P7-1 Vacuum filter for Problem 7-15.

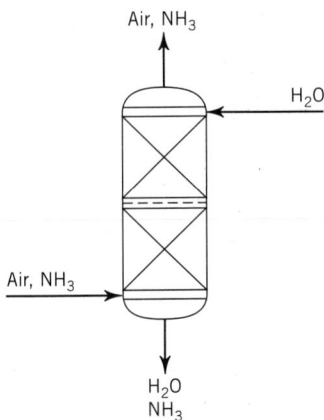

Figure P7-2 Absorber for Problem 7-16.

(a) Design a control loop for maintaining the outlet NH_3 concentration at a set point of 50 ppm. Draw the instrument diagram for the loop. There are some instruments in the stockroom that you can use for this purpose. There is an analyzer transmitter calibrated for 0–200 ppm. This instrument has a negligible time lag. There is also a control valve that, at full opening and for the 10-psi pressure drop that is available, will pass 500 gpm. The time constant of the valve actuator is negligible. You may need more instrumentation to complete the design, so go ahead and use anything you need. Specify the fail position of the control valve and the action of the controller.

(b) Draw a block diagram for the closed loop and obtain the transfer function for each block. Approximate the response of the absorber with a first-order-plus-dead-time model using fit 3.

(c) Tune a proportional-only controller for quarter decay ratio response and calculate the offset when the set point is changed to 60 ppm.

(d) Repeat part (c) using a PID controller.

7-17. Consider the furnace shown in Fig. P7-3, which is used to heat the supply air to a catalyst regenerator. The temperature transmitter is calibrated for 300–500°F. The following response data were obtained for a step change of +5 % in the output of the controller:

Time, min	$T(t)$, °F	Time, min	$T(t)$, °F
0	425	5.5	436.6
0.5	425	6.0	437.6
1.0	425	7.0	439.4
2.0	425	8.0	440.7
2.5	426.4	9.0	441.7
3.0	428.5	10.0	442.5
3.5	430.6	11.0	443.0
4.0	432.4	12.0	443.5
4.5	434.0	14.0	444.1
5.0	435.3	12.0	445.0

(a) Draw the complete block diagram specifying the units of each signal to/from each block. Identify each block and specify the fail-safe position of the valve and the correct action of the controller.

(b) Fit the process data by a first-order-plus-dead-time model using fit 3. Redraw the block diagram showing the transfer function for each block.

(c) Tune a PID controller for quarter decay ratio response.

(d) Tune a PID controller by the controller synthesis method for a 5 % overshoot.

7-18. Calculate the tuning parameters for a PID temperature controller tuned for quarter decay ratio response for the oil heater of Problem 6-24. *Note*: The simulation of this process is the subject of Problem 13-24.

Stack gases

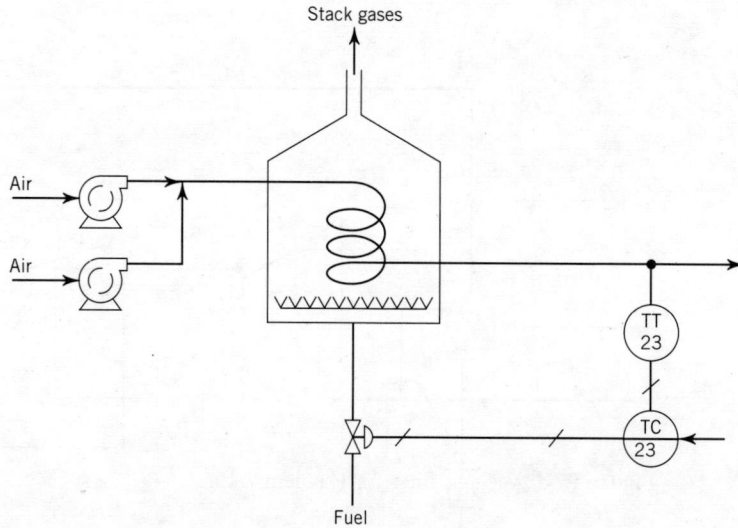

Figure P7-3 Furnace for Problem 7-17.

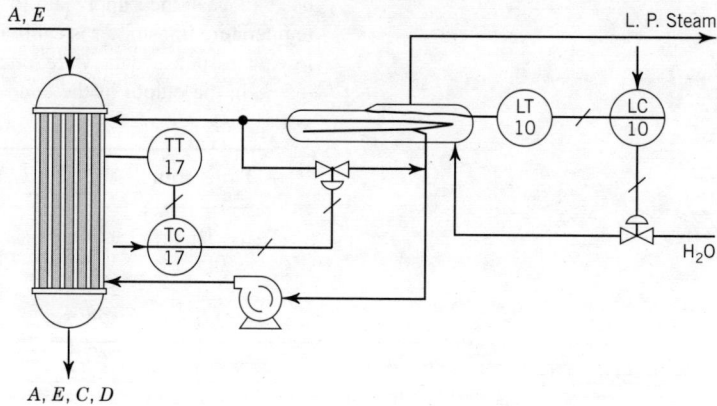

A, E, C, D

Figure P7-4 Catalytic reactor for Problem 7-19.

7-19. Consider the chemical reactor system shown in Fig. P7-4. An exothermic catalytic reaction takes place inside the reactor tubes. The reactor is cooled by oil flowing through the shell of the reactor. As the oil flows out of the reactor, it goes to a boiler where it is cooled by producing low-pressure steam. The temperature in the reactor is controlled by manipulating the bypass flow around the boiler. The following process conditions are known:

Reactor design temperature at point of measurement: 275°F

Oil flow the pump can deliver: 400 gpm (constant)

Control valve pressure drop at design flow: 10 psi

Flow at design conditions: 200 gpm

Range of the temperature transmitter: 150–350°F

Density of the oil: 55 lb/ft^3

Open loop test: A 5 % change in the valve position results in a temperature change of 4.4°F after a very long time.

Closed loop test: At a controller gain of 16 %CO/%TO, the temperature oscillates with constant amplitude and a period of 24 min.

(a) Size the temperature control valve for 100 % overcapacity. Specify the fail-safe position of the control valve and the required controller action.

(b) If the pressure drop across the boiler tubes varies with the square of the flow and the valve is equal percentage with a rangeability parameter of 50, what is the valve position at design conditions? What is the flow through the valve when fully open?

(c) Draw a block diagram for the temperature loop. What are your recommended valve fail-safe position and controller action?

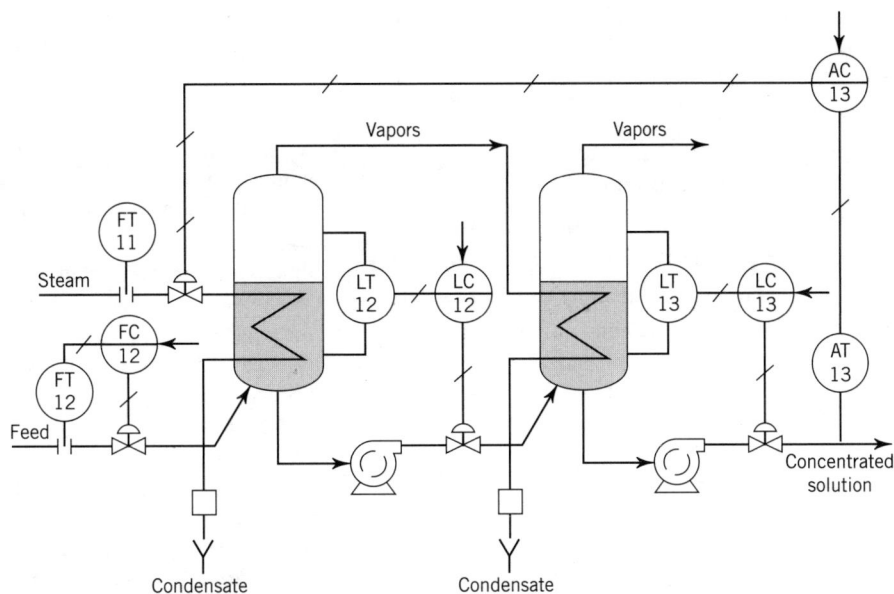

Figure P7-5 Double-effect evaporator for Problem 7-20.

(d)   Calculate the process gain, at design conditions, including the control valve and the temperature transmitter.

(e)   Calculate the tuning parameters for a PID controller for quarter decay ratio response. Report them as proportional band, repeats/minute, and minutes.

(f)   Tune a proportional controller for quarter decay ratio response and calculate the offset for a step change in set point of $-10°F$.

7-20.   Consider the typical control system for the double-effect evaporator shown in Fig. P7-5. Evaporators are characterized by slow dynamics. Manipulating the steam to the first effect controls the composition of the product out of the last effect. The design feed rate and composition are 50,000 lb/h and 5.0 weight %, respectively. Figure P7-6 shows the open loop step response of the product composition for a change of

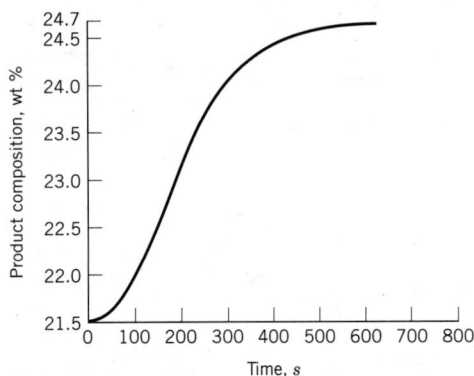

Figure P7-6 Response to step change in inlet composition for Problem 7-20.

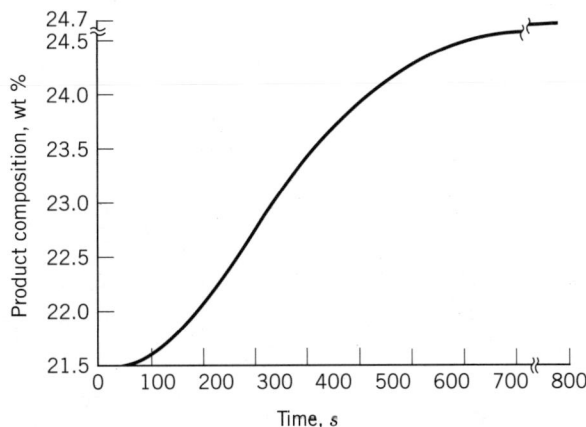

Figure P7-7 Response to step change in controller output for Problem 7-20.

0.75 % by weight in the composition of the solution entering the first effect. Figure P7-7 shows the response of the product composition for a change of 2.5 % in controller output. The composition sensor transmitter has a range of 10–35 weight %.

(a)   Draw a complete block diagram with the transfer function of each block. What should be the fail-safe position of the control valve? What is the correct controller action?

(b)   Tune a proportional-integral controller for quarter decay ratio response.

(c)   Tune a PI controller for 5 % overshoot using the controller synthesis method.

7-21. The temperature in a continuous stirred tank exothermic chemical reactor is controlled by manipulating the cooling water rate through a coil, as shown in Fig. P7-8. The following are the process design conditions:

Reactor temperature: 210°F

Cooling water flow: 350 gal/min

Pressure drop across the coil at design flow: 10 psi

Range of temperature transmitter: 190 to 230°F

Control valve trim: Equal percentage with $\alpha = 50$.

Open-loop test: A 10 gal/min increase in water rate results in a temperature change of 5.2°F after a long time.

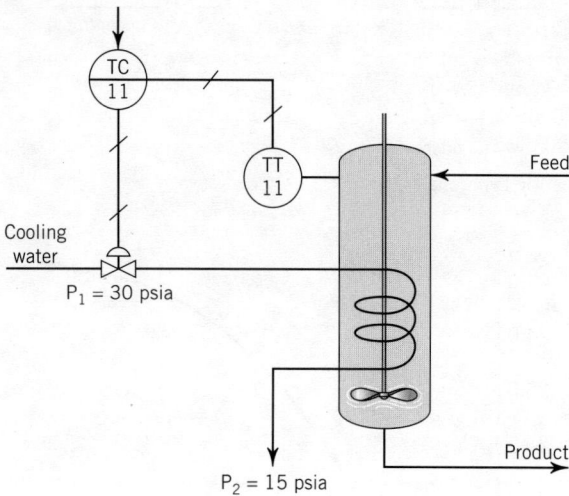

Figure P7-8 Reactor for Problem 7-21.

Closed-loop test: At a controller gain of 8.0 %CO/%TO, the temperature oscillates with constant amplitude and a period of 14 min.

(a) Size the valve for 100 % overcapacity, calculate the gain of the valve at design flow, and specify the fail-safe position of the valve.

(b) Draw the block diagram for the control loop and determine the total process gain, including the transmitter and the control valve.

(c) Calculate the PID controller tuning parameters for quarter decay ratio response. Report them as proportional band, repeats/minute, and derivative minutes. What is the required controller action?

7-22. Consider the process shown in Fig. P7-9 for drying phosphate pebbles. A table feeder transports the pebble–water slurry into the bed of the drier. In this bed the pebbles are dried by the direct contact with hot combustion gases. From the drier the pebbles are conveyed to a silo for storage. It is *most* important to control the moisture of the pebbles leaving the drier. If the pebbles are too dry, they may fracture into fine dust resulting in possible loss of material. If too wet, they may form large chunks, or clinkers, in the silo.

It is proposed to control the moisture of the exiting pebbles by the speed of the table feeder, as shown in Fig. P7-9. The speed of the feeder is directly proportional to its input signal. The moisture of the inlet pebbles is usually about 15 % and is reduced to 3 % in the drier. The transmitter has a range of 1 to 5 % moisture. An important disturbance to this process is the moisture of the inlet pebbles.

(a) Draw a complete block diagram of the control loop showing all units. Include the disturbances.

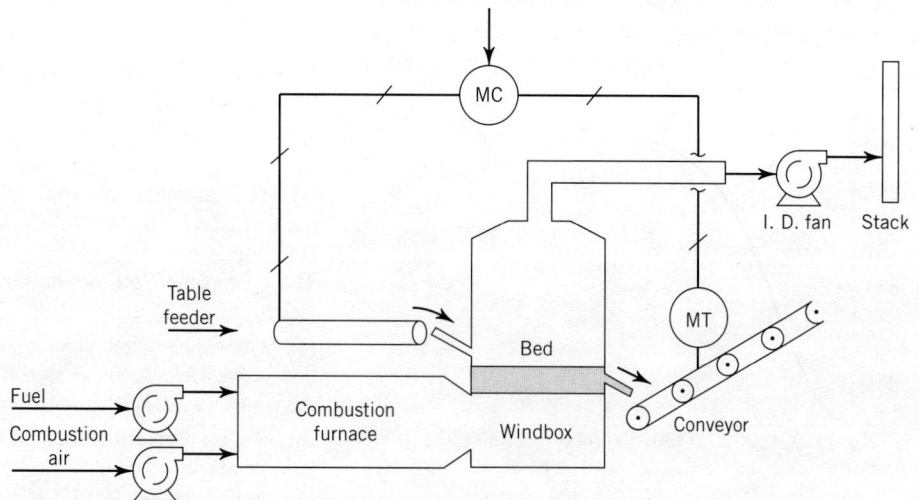

Figure P7-9 Drier of phosphate pebbles for Problem 7-22.

(b) Figure P7-10 shows the response of the outlet moisture to an increase of 8 % in controller output, while Fig. P7-11 shows the response of the outlet moisture to an increase of 3 % in inlet moisture. Approximate each process curve by a first-order-plus-dead-time model. Use fit 2. Redraw the block diagram showing the transfer functions of these approximate models.

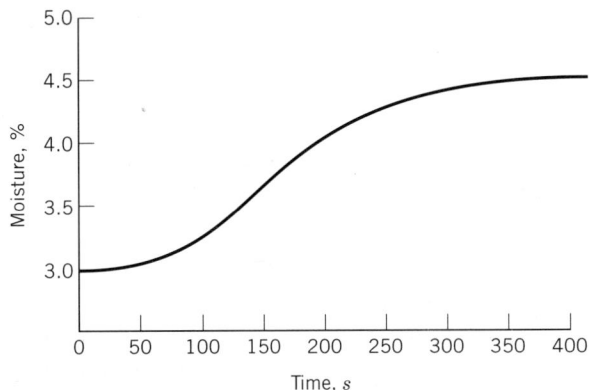

Figure P7-10 Response to step change in controller output for Problem 7-22.

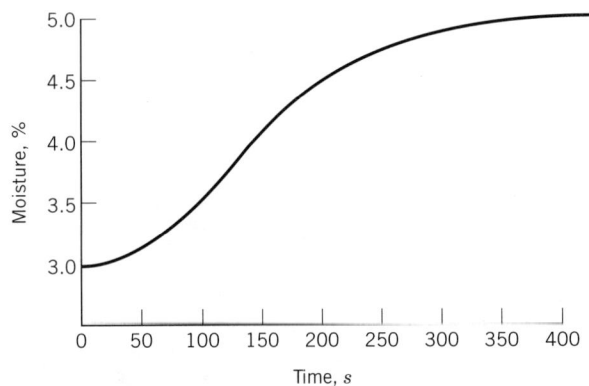

Figure P7-11 Response to step change in inlet moisture for Problem 7-22.

(c) Determine the tuning of a PID controller for minimum IAE response on disturbance inputs. Report the controller gain as proportional band. What is the correct controller action?

(d) If the inlet moisture of the pebbles drops by 2 %, what is the new steady-state value of the outlet moisture? Assume that the controller is proportional-only tuned for quarter decay ratio response from the information determined in part (b).

(e) What is the controller output required to avoid offset for the disturbance of part (d)?

7-23. In the liquid level control system of Fig. 7-3.1, assume that the level control valve is installed on the inlet line to manipulate the inlet flow, $f_i(t)$, and that the outlet flow is the disturbance. Modify the block diagrams of the loop (Fig. 7-3.2), and specify the correct action of the level controller assuming the control valve fails closed. Rewrite the closed-loop transfer function for the level. Do any of the formulas developed in Section 7-3.2 for the proportional level controller change?

7-24. Consider the first of the two tanks in series of Fig. 4-1.1. A model for this tank is derived in Section 4-1.1 that assumes the outlet flow depends on the level in the tank. This model results in a first-order lag transfer function for the level in the tank. A level control system is to be installed on that tank. The controller will manipulate the valve on the outlet line. Let the diameter of the tank be 3.0 m, the design level be 2.0 m, the range of the level controller be 1 to 3 m above the bottom of the tank, and the outlet valve be linear and sized for a maximum flow of twice the design flow of 0.003 m³/s. The pressure drop in the line is negligible, and the valve can be represented by a first-order lag with a time constant of 5 s.

(a) Draw the block diagram for the level control loop using the model developed for the first tank in Section 4-1.1. Specify the action of the controller and calculate the parameters of all the transfer functions. Calculate the maximum gain of a proportional level controller that will result in a nonoscillatory response. Calculate also the effective time constants of the loop at that gain, and the offset caused by a 0.5 m³/s change in inlet flow.

(b) Now model the tank as if the outlet flow is a function only of the valve position, as in Section 7-3.1 (integrating process), and carry out the calculations of part (a) using this model. Compare the answers from both models.

7-25. Although proportional control is ideal for both tight and averaging level control, many operators prefer to have a proportional-integral (PI) controller because they are not used to controllers with offset.

(a) Obtain the closed-loop transfer function and the characteristic equation of the loop for the block diagram of Fig. 7-3.2 using a PI controller. Show that there is no offset for either a change in set point or a change in flow.

(b) Determine the roots of the characteristic equation when the reset time is equal to the valve time constant. What is the level response? Is it dependent on the controller gain?

(c) Assuming a negligible valve time constant, show that the response of the PI level controller is oscillatory at low controller gains and nonoscillatory at high gains! Show that the minimum gain for which the response is not oscillatory is $4/\tau_I$. Show also that at very high gains the dominant time constant of the closed loop is equal to the reset time of the controller.

7-26. A level controller on a calandria-type evaporator needs to be tuned very tightly because its operation is very sensitive to the level. A high level results in a rise in the boiling temperature due to the hydrostatic pressure of the fluid, while a low level results in the formation of a scale of dried-up solids at the top of the hot tubes. Because of this, most evaporator levels are controlled by manipulating the feed flow to the evaporator which is the largest flow. Draw the schematic for the level control loop to an evaporator. The steam and product flows are the disturbances. Draw also the block diagram of the loop. Calculate the maximum proportional controller gain for nonoscillatory response, the effective time constants of the level loop at that gain, and the offset for a 10 % change in flow for the following design parameters: feed flow of 800 lb/min, density of concentrated solution of 98 lb/ft^3, evaporator cross-sectional area of 10 ft^2, valve time constant of 2.0 s, and level transmitter span of 4 ft. The valve is linear and sized for 100 % overcapacity.

7-27. The transfer function for a process with inverse response (see Section 4-4.3) is

$$G(s) = \frac{K(1 - \tau_3 s)}{(\tau_1 s + 1)(\tau_2 s + 1)}$$

Show that if we tried to synthesize a feedback controller for the standard first-order response specification, Eq. 7-4.4, the controller denominator would contain a positive (unstable) root. Then synthesize a feedback controller for the following closed-loop response specification.

$$\frac{C(s)}{R(s)} = \frac{1 - \tau_3 s}{\tau_c s + 1}$$

Identify the type of controller and derive the tuning formulas for it.

7-28. Simulate the nonisothermal chemical reactor described in Section 4-2.3. Using the methods described in Section 13-4, simulate the temperature control loop for the reactor. The control valve on the coolant to the jacket is equal percentage with a rangeability parameter $\alpha = 50$, constant pressure drop, and an actuator time constant of 0.1 min. It is sized for 100 % overcapacity. A temperature transmitter with a time constant of 1 min and a range of 640 to 700°R is to be used. Use the simulation to determine the open-loop step response of the transmitter output to a 2 % increase in the signal to the valve and then use fit 3 to estimate the gain, time constant, and dead time of the process. These parameters are then used to determine the controller tuning parameters. Obtain the closed-loop responses of the transmitter and controller outputs to a step change of −0.2 ft^3/min in reactants flow with a PID controller tuned for (a) quarter decay ratio response and (b) synthesis tuning with $\tau_c = 0$.

Chapter 8

Root Locus and Frequency Response Techniques

In Chapter 6 we began the study of the stability of control systems by presenting two techniques: Routh test and direct substitution. This chapter continues with this study by presenting the root locus technique and the frequency response technique. This presentation is done from a practical point of view, stressing what the techniques indicate about stability of the processes and how the different loop terms affect the stability. As the presentation of frequency response is independent of that of root locus, the reader may choose to skip the root locus section and go directly to Section 8-4.

8-1 SOME DEFINITIONS

Before the root locus and frequency response techniques are presented, some new terms must be defined. Consider the general closed-loop block diagram shown in Fig. 8-1.1. As we saw in Chapter 6, the closed-loop transfer functions are

$$\frac{C(s)}{R(s)} = \frac{G_c(s)G_v(s)G_{p_1}(s)}{1 + H(s)G_c(s)G_v(s)G_{p_1}(s)} \tag{8-1.1}$$

and

$$\frac{C(s)}{L(s)} = \frac{G_{p2}(s)}{1 + H(s)G_c(s)G_v(s)G_{p_1}(s)} \tag{8-1.2}$$

with the characteristic equation

$$1 + H(s)G_c(s)G_v(s)G_{p_1}(s) = 0 \tag{8-1.3}$$

The *open-loop transfer function (OLTF)* is defined as the product of all the transfer functions in the control loop; that is

$$\text{OLTF} = H(s)G_c(s)G_v(s)G_{p_1}(s) \tag{8-1.4}$$

Therefore, the characteristic equation can also be written as

$$1 + \text{OLTF} = 0 \tag{8-1.5}$$

Now suppose that the individual transfer functions are known and that the OLTF looks like this:

$$\text{OLTF} = \frac{K_c K_v K_{p_1} K_T (1 + \tau_D s)}{(\tau_T s + 1)(\tau_1 s + 1)(\tau_2 s + 1)}$$

Principles and Practice of Automatic Process Control/Third Edition, by C. A. Smith and A. B. Corripio
ISBN 0-471-66141-4   Copyright © 2006 John Wiley & Sons (Asia) Pte. Ltd.

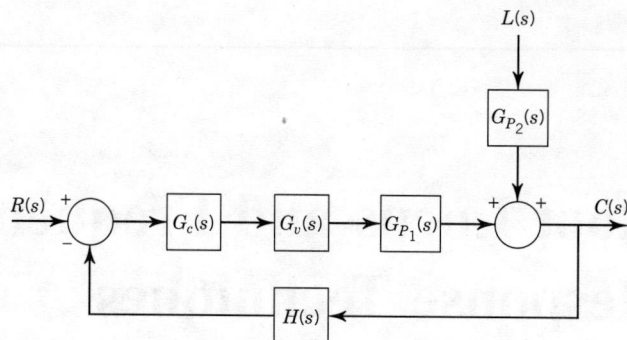

Figure 8-1.1 General closed-loop block diagram.

or

$$\text{OLTF} = \frac{K(1 + \tau_D s)}{(\tau_T s + 1)(\tau_1 s + 1)(\tau_2 s + 1)}$$

where $K = K_c K_v K_{p_1} K_T$

The *poles* are defined as the roots of the denominator of the OLTF. For the foregoing OLTF the poles are $-1/\tau_T, -1/\tau_1, -1/\tau_2$. The *zeros* are defined as the roots of the numerator of the OLTF, or $-1/\tau_D$ for the foregoing OLTF.

These definitions are generalized by writing the OLTF as

$$\text{OLTF} = \frac{K \prod_{i=1}^{m} (\tau_i s + 1)}{s \prod_{j=1}^{n} (\tau_j s + 1)}, \qquad n > m$$

or

$$\text{OLTF} = \frac{K' \prod_{i=1}^{m} \left(s + \frac{1}{\tau_i}\right)}{s \prod_{j=1}^{n} \left(s + \frac{1}{\tau_j}\right)}, \qquad n > m \tag{8-1.6}$$

where

$$K' = \frac{K \prod_{i=1}^{m} \tau_i}{\prod_{j=1}^{n} \tau_j}$$

From Eq. 8-1.6 the poles are recognized as equal to $-1/\tau_j$ for $j = 1$ *to* n, and as equal to 0 for the single s term. Similarly, the zeros are given by $-1/\tau_i$ from $i = 1$ *to* m. These definitions of poles and zeros will be frequently used in this chapter.

8-2 ANALYSIS OF FEEDBACK CONTROL SYSTEMS BY ROOT LOCUS

Root locus is a graphical technique that consists of graphing the roots of the characteristic equation, also referred to as eigenvalues, as a gain or any other control loop parameter changes. The resulting graph allows seeing at a glance whether a root crosses the imaginary axis to the right-hand side of the s plane. This crossing would indicate the possibility of instability of the control loop.

Several examples of how to draw the root locus are now presented. These examples also illustrate the effects of the different parameters of the control loop on its stability. These effects were presented in Chapter 6; so, the following should also serve as a review.

EXAMPLE 8-2.1

Consider the block diagram shown in Fig. 8-2.1. The characteristic equation for this system is

$$1 + \frac{K_c}{(3s+1)(s+1)} = 0 \tag{8-2.1}$$

and

$$\text{OLTF} = \frac{K_c}{(3s+1)(s+1)}$$

Note that this OLTF contains two poles, at $-1/3$ and -1, and no zeros. From Eq. 8-2.1, the polynomial

$$3s^2 + 4s + (1 + K_c) = 0$$

is obtained. This polynomial, being of second order, has two roots. Using the quadratic equation to solve for the roots, we develop the following expression

$$r_1, r_2 = -\frac{2}{3} \pm \frac{1}{3}\sqrt{1 - 3 K_c} \tag{8-2.2}$$

Equation 8-2.2 shows that the roots of the characteristic equation depend on the value of K_c. This is the same thing as saying that the closed-loop response of the control loop depends on the tuning of the feedback controller! This was also shown in Chapter 6. By giving values to K_c, we can determine the loci of the roots. The graph of the roots, or root locus, is shown in Fig. 8-2.2. Several things can be learned by examining this diagram.

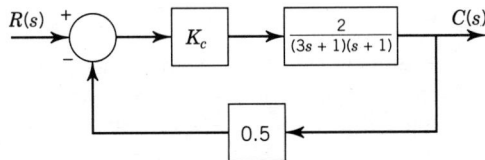

Figure 8-2.1 Block diagram of control loop for Example 8-2.1.

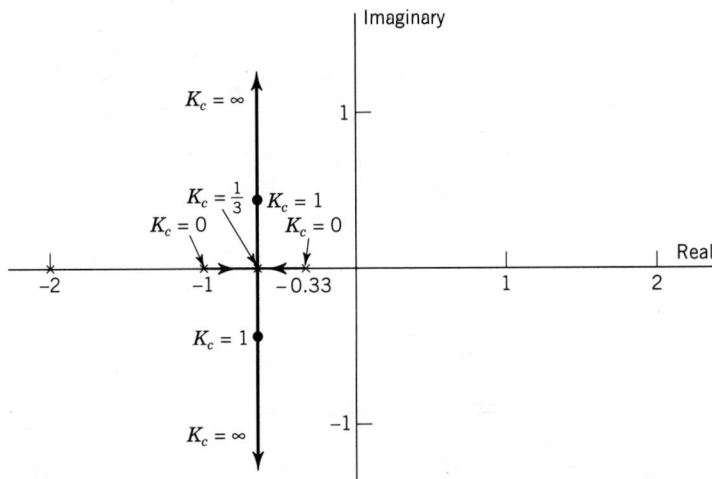

Figure 8-2.2 Root locus diagram for system in Fig. 8-2.1.

1. The most important point is that this particular control loop will never go unstable, no matter how high the value of K_c is set. As the value of K_c increases, the loop response becomes more oscillatory, or underdamped, but never unstable. The underdamped response is recognized because the roots of the characteristic equation move away from the real axis into the complex region as K_c increases. The fact that a control loop with a second-order (or first-order), and no dead time, characteristic equation does not go unstable, was also shown in Chapter 6 using the Routh test and direct substitution methods.

2. When $K_c = 0$, the root loci originate from the OLTF poles: $-1/3$ and -1.

3. The number of root loci, or branches, is equal to the number of OLTF poles, $n = 2$.

4. As K_c increases, the root loci approach infinity.

EXAMPLE 8-2.2

Suppose now that the sensor-transmitter combination of the previous example has a time constant of 0.5 time units. The block diagram is shown in Fig. 8-2.3. The new characteristic equation and open-loop transfer function are

$$1 + \frac{K_c}{(3s + 1)(s + 1)(0.5s + 1)} = 0$$

or

$$1.5s^3 + 5s^2 + 4.5s + (1 + K_c) = 0$$

$$\text{OLTF} = \frac{K_c}{(3s + 1)(s + 1)(0.5s + 1)}$$

with poles: $-1/3, -1, -2; \quad n = 3$
zeros: none; $\qquad m = 0$

In this case the characteristic equation is a third-order polynomial, and thus the calculation of its roots is not so straightforward. Figure 8-2.4 shows the root locus diagram. Again, several things can be learned by a simple glance at this diagram.

1. The most important thing is that this control system can go unstable. At some value of K_c, in this case $K_c = 14$, the root loci cross the imaginary axis. For values of K_c greater than 14, some roots of the characteristic equation will be on the right-hand side of the s-plane. The value of K_c at which the root locus crosses the imaginary axis, yielding a conditionally stable system, is called the ultimate gain, K_{c_u}, as we noted in Chapter 6. The ultimate frequency, ω_u, is given by the coordinate where the branches cross the imaginary axis. Any loop with a characteristic equation of third or higher order can go unstable; first- or second-order systems, with no dead time, will not go unstable. Any system with dead time can go unstable, as will be shown in this chapter.

2. The root loci again originate, when $K_c = 0$, at the OLTF poles: $-1/3, -1, -2$.

3. The number of root loci is again equal to the number of poles of the OLTF, $n = 3$.

4. The root loci again approach infinity as K_c increases.

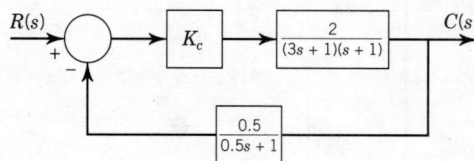

Figure 8-2.4 Block diagram of control loop for Example 8-2.2.

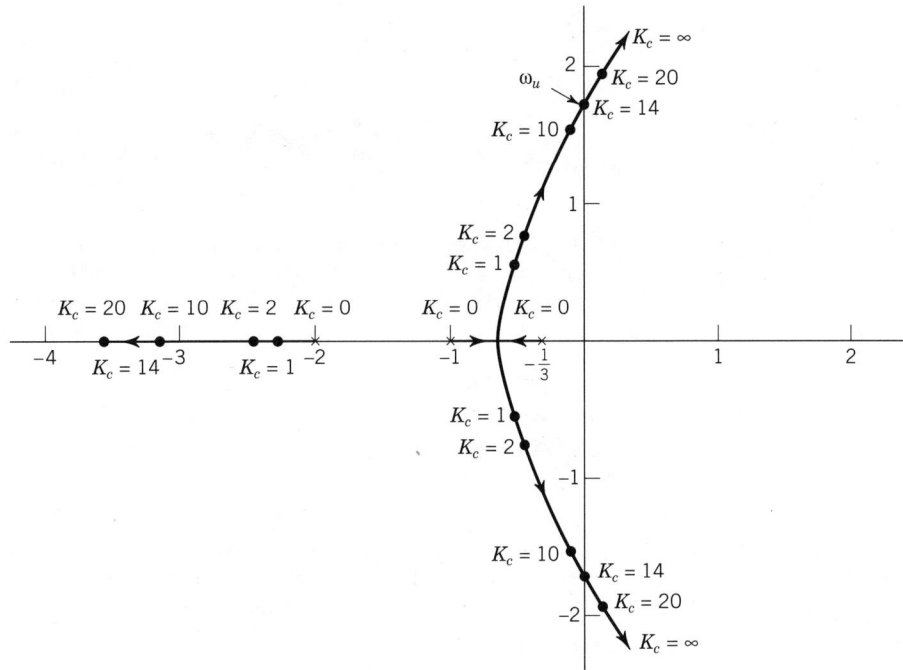

Figure 8-2.3 Root locus diagram for system in Fig. 8-2.3.

8-3 PLOTTING ROOT LOCUS DIAGRAMS

Example 8-2.1 shows that it is fairly simple to sketch the root locus diagram when the characteristic equation is of second order; first order is also simple. However, when the characteristic equation is of third order or higher, as in Example 8-2.2, the sketching of the diagram is not as simple. There are rules available that permit sketching the root locus without finding any root; the reader is referred to the previous edition of this book (Reference 1) for these rules. Obviously, and more importantly and convenient, there are several software packages, such as MATLAB (Reference 2), that draw the root locus directly from the transfer functions. There are also several software packages, such as TK Solver (Reference 3) and MATHCAD (Reference 4), that make the computation and drawing of the roots very convenient. The next example presents very briefly the use of MATLAB to draw the root locus.

EXAMPLE 8-3.1

Suppose it is desired to use a PI controller to control the process of Example 8-2.2. A reset time of 0.1 min (6 s) is used, so the transfer function for this controller is $G_c(s) = K_c(1 + (1/6s))$. Plot the root locus diagram for this new control system. What is the effect of adding reset action to the controller?

The characteristic equation and open-loop transfer function are

$$1 + \frac{K_c\left(1 + \dfrac{1}{6s}\right)}{(3s+1)(s+1)(0.5s+1)} = 0 \qquad \textbf{(8-3.1)}$$

$$\text{OLTF} = \frac{K_c\left(1 + \dfrac{1}{6s}\right)}{(3s+1)(s+1)(0.5s+1)} \qquad \textbf{(8-3.2)}$$

or

$$\text{OLTF} = \frac{K_c(6s + 1)}{6s(3s + 1)(s + 1)(0.5s + 1)} \tag{8-3.3}$$

with poles: $0, -\frac{1}{3}, -1, -\frac{1}{0.5}, \quad n = 4$

zeros: $-\frac{1}{6}, \qquad m = 1$ $\tag{8-3.4}$

The use of MATLAB to draw the root locus requires inputting the OLTF without any constants, K_c in this case. The program is shown in Fig. 8-3.1, and the root locus in Fig. 8-3.2. The ultimate gain in this case is 11.72, and the ultimate frequency is 1.58 rad/s.

Comparing Figs. 8-2.4 and 8-3.2 we see that the addition of the reset action to the proportional-only controller does not significantly change the shape of the root locus. The most significant effect, however, is the decrease in the ultimate gain (11.72 versus 14) and ultimate frequency. As we learned in Chapters 5 and 6, the reset action removes the offset, but it adds instability to the control system (decreases the ultimate gain).

num=[6 1]; den1=[6 0]; den2=[3 1]; den3=[1 1]; den4=[4 1];
den5=conv(den1,den2); den6=conv(den5,den3); den7=conv(den6,den4);
rlocus(num,den7)

Figure 8-3.1 MATLAB program to draw root locus diagram for Example 8-3.1.

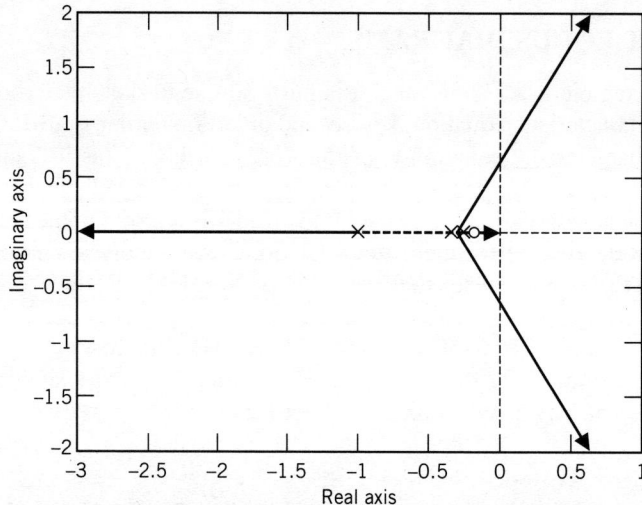

Figure 8-3.2 Root locus diagram for Example 8-3.1.

8-4 ANALYSIS OF CONTROL SYSTEMS BY FREQUENCY RESPONSE

Frequency response techniques are some of the most popular techniques for the analysis and design of control of linear systems. The following sections present what is meant by frequency response, and how to use these techniques to analyze and synthesize control systems.

Consider the general block diagram shown in Fig. 8-4.1. The control loop has been opened before the valve and after the transmitter. A variable-frequency generator provides the input signal to the valve, $x(t) = X_0 \sin \omega t$, and a recorder records the output signal from the transmitter and the input signal to the valve. Figure 8-4.2 shows the two recordings. After the transients have died out, the transmitter output reaches a sinusoidal response, $y(t) = Y_0 \sin(\omega t + \theta)$. This experiment is referred to as sinusoidal testing.

Let us now perform the same "experiment" using the transfer function that describes the process. Assume the following simple transfer function:

$$G(s) = \frac{Y(s)}{X(s)} = \frac{K}{\tau s + 1} \qquad \textbf{(8-4.1)}$$

This transfer function describes the valve, process, and sensor-transmitter combination. The input signal to the valve is

$$x(t) = X_0 \sin \omega t$$

and its Laplace transform, from Chapter 2, is

$$X(s) = \frac{X_0 \omega}{s^2 + \omega^2} \qquad \textbf{(8-4.2)}$$

Therefore,

$$Y(s) = \frac{K X_0 \omega}{(\tau s + 1)(s^2 + \omega^2)}$$

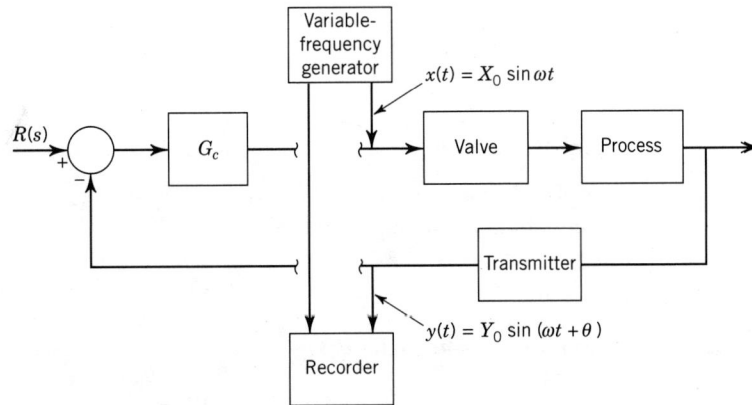

Figure 8-4.1 Block diagram showing variable frequency generator and recorder.

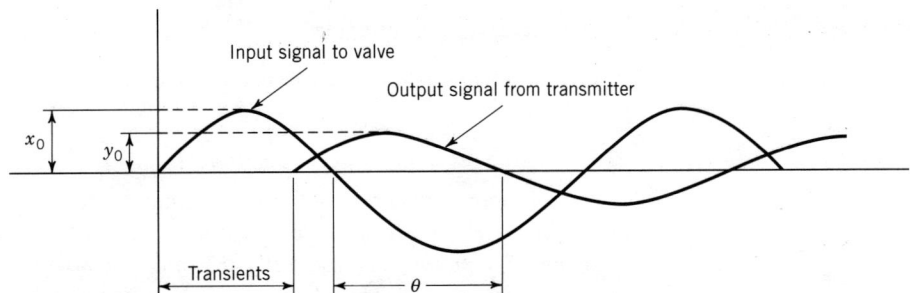

Figure 8-4.2 Recording from sinusoidal testing.

The time domain expression for $Y(t)$ can be obtained using the techniques learned in Chapter 2.

$$Y(t) = \frac{KX_0\omega\tau}{1 + \omega^2\tau^2}e^{-t/\tau} + \frac{KX_0}{\sqrt{1 + \omega^2\tau^2}}\sin(\omega t + \theta) \qquad \text{(8-4.3)}$$

with

$$\theta = \tan^{-1}(-\omega\tau) = -\tan^{-1}(\omega\tau) \qquad \text{(8-4.4)}$$

As time increases, the exponential term of Eq. 8-4.3 goes to zero; that is, the transient term dies out. When this happens, the output expression becomes

$$Y(t)|_{t \text{ very large}} = \frac{KX_0}{\sqrt{1 + \omega^2\tau^2}}\sin(\omega t + \theta) \qquad \text{(8-4.5)}$$

which constitutes the sinusoidal behavior of the output signal, and the one shown in Fig. 8-4.2. The amplitude of this output signal is

$$Y_0 = \frac{KX_0}{\sqrt{1 + \omega^2\tau^2}}$$

The minus sign in Eq. 8-4.4 indicates that the output signal "lags" the input signal by the amount θ calculated from the equation.

A word of advice is necessary here. Care should be taken when calculating the sine term in Eq. 8-4.5. The term ω is in radians/time, and the term ωt is in radians. Thus, for the operation $(\omega t + \theta)$ to be in the correct units, θ must be in radians. If degrees are to be used, then the term must be written as

$$\left(\frac{180}{\pi}\omega t + \theta\right)$$

In short, be careful with the units.

Some terms often used in frequency response studies are now defined.

Amplitude ratio (AR) is the ratio of the amplitude of the output signal to the amplitude of the input signal. That is,

$$AR = \frac{Y_0}{X_0} \qquad \text{(8-4.6)}$$

Magnitude ratio (MR) is the amplitude ratio divided by the steady-state gain.

$$MR = \frac{AR}{K} \qquad \text{(8-4.7)}$$

Phase angle(θ) is the amount, in radians or degrees, by which the output signal lags or leads the input signal. When θ is positive, it is a lead angle; when θ is negative, it is a lag angle.

For the foregoing first-order transfer function,

$$AR = \frac{K}{\sqrt{1 + \omega^2\tau^2}}; \qquad MR = \frac{1}{\sqrt{1 + \omega^2\tau^2}}; \qquad \theta = \tan^{-1}(-\omega\tau)$$

Note that all three terms are functions of the input frequency. Different processes have different AR (MR) and θ dependence on ω.

Frequency response is essentially the study of how the AR (MR) and θ of different components or systems behave as the input frequency changes. The following paragraphs show that frequency response is a powerful technique for analyzing and synthesizing control systems. We will discuss the development of the frequency response of process systems first and then its use for analysis and synthesis.

There are in general two different methods for generating the frequency response.

1. *Experimental method.* This method consists essentially of the experiment with the variable frequency generator and the recorder. The idea is to run the experiment at different frequencies, in order to obtain a table of AR versus ω and θ versus ω.

2. *Transforming the open-loop transfer function after a sinusoidal input.* This method consists of using the open-loop transfer function to obtain the expression that describes the response of the system to a sinusoidal input. From the expression the amplitude and phase angle of the output can then be determined. This method is the mathematical manipulations previously shown that resulted in Eqs. 8-4.4 and 8-4.5.

Fortunately, operational mathematics provides a very simple method to determine AR (MR) and θ without having to obtain the inverse Laplace transforms. The necessary mathematics have already been presented in Chapter 2. Consider

$$\frac{Y(s)}{X(s)} = G(s)$$

for $x(t) = X_0 \sin \omega t$, from Chapter 2

$$X(s) = \frac{X_0 \omega}{s^2 + \omega^2}$$

Then

$$Y(s) = G(s)\frac{X_0 \omega}{s^2 + \omega^2}$$

Expansion by partial fractions yields

$$Y(s) = \frac{A}{s + i\omega} + \frac{B}{s - i\omega} + [\text{terms for the poles of } G(s)] \qquad (8\text{-}4.8)$$

To obtain A, we use

$$A = \lim_{s \to 0} \frac{[(s + i\omega)X_0\omega G(s)]}{(s^2 + \omega^2)} = \frac{G(-i\omega)X_0\omega}{-2i\omega}$$

As shown in Chapter 2, any complex number can be represented by a magnitude and argument; then,

$$A = \frac{X_0|G(i\omega)|e^{-i\theta}}{-2i}$$

where $\theta = \underline{/G(i\omega)}$. To obtain B, we use

$$B = \lim_{s \to i\omega}\left[\frac{(s - i\omega)X_0\omega G(s)}{(s^2 + \omega^2)}\right] = \frac{X_0|G(i\omega)|e^{i\theta}}{2i}$$

Then, substituting the expressions for A and B into Eq. 8-4.8 yields

$$Y(s) = \frac{X_0|G(i\omega)|}{2i}\left[\frac{-e^{-i\theta}}{s + i\omega} + \frac{e^{i\theta}}{s - i\omega}\right] + [\text{terms of } G(s)]$$

Inverting back into the time domain, we get

$$Y(t) = \frac{X_0|G(i\omega)|}{2i}[-e^{-i\theta}e^{-i\omega t} + e^{i\theta}e^{i\omega t}]$$
$$+ \text{[transient terms resulting from the terms of } G(s)]$$

After the transient terms die out

$$Y(t)|_{t \text{ very large}} = X_0|G(i\omega)|\frac{e^{i(\omega t + \theta)} - e^{-i(\omega t + \theta)}}{2i}$$

or

$$Y(t)|_{t \text{ very large}} = X_0|G(i\omega)|\sin(\omega t + \theta)$$

The amplitude ratio is then

$$\boxed{AR = \frac{Y_0}{X_0} = \frac{X_0|G(i\omega)|}{X_0} = |G(i\omega)|} \tag{8-4.9}$$

and the phase angle is

$$\boxed{\theta = \underline{/G(i\omega)}} \tag{8-4.10}$$

Thus, in order to obtain the AR and θ, one simply substitutes iω for s in the transfer function and then calculates the magnitude and argument of the resulting complex number expression. The magnitude is equal to the amplitude ratio (AR), and the argument equals the phase angle (θ).

This manipulation simplifies the required calculations.

Let us apply these results to the first-order system used earlier

$$G(s) = \frac{K}{\tau s + 1}$$

Now substitute $i\omega$ for s

$$G(i\omega) = \frac{K}{i\omega\tau + 1}$$

This results in a complex number expression composed of the ratio of two terms: the numerator, a real number, and the denominator, a complex number. The equation can also be written as follows:

$$G(i\omega) = \frac{G_1}{G_2} = \frac{K}{i\omega\tau + 1} \tag{8-4.11}$$

As shown by Eq. 8-4.9, the AR is equal to the magnitude of this complex number expression

$$\boxed{AR = |G(i\omega)| = \frac{|G_1|}{|G_2|} = \frac{K}{\sqrt{\omega^2\tau^2 + 1}}} \tag{8-4.12}$$

which is the same AR as previously obtained.

The phase angle is equal to the angle of the complex number expression

$$\boxed{\theta = \underline{/G(i\omega)} = \underline{/G_1} - \underline{/G_2} = 0 - \tan^{-1}(\omega\tau) = -\tan^{-1}(\omega\tau)} \tag{8-4.13}$$

which is also the same θ as previously obtained.

Let us now look at several other examples.

EXAMPLE 8-4.1

Consider the following second-order system

$$G(s) = \frac{K}{\tau^2 s^2 + 2\tau\zeta s + 1}$$

Determine the expressions for AR and θ.

The first step is to substitute $i\omega$ for s.

$$G(i\omega) = \frac{K}{-\omega^2\tau^2 + i2\tau\zeta\omega + 1} = \frac{K}{(1 - \omega^2\tau^2) + i2\tau\zeta\omega}$$

Again, a complex number expression results that is a ratio of two other numbers

$$G(i\omega) = \frac{G_1}{G_2} = \frac{K}{(1 - \omega^2\tau^2) + i2\tau\zeta\omega}$$

The amplitude ratio is

$$\boxed{\text{AR} = |G(i\omega)| = \frac{|G_1|}{|G_2|} = \frac{K}{\sqrt{(1 - \omega^2\tau^2)^2 + (2\tau\zeta\omega)^2}}} \qquad \textbf{(8-4.14)}$$

and the phase angle is

$$\theta = \underline{/G(i\omega)} = \underline{/G_1} - \underline{/G_2}$$

$$\theta = 0 - \tan^{-1}\left(\frac{2\tau\zeta\omega}{1 - \omega^2\tau^2}\right)$$

$$\boxed{\theta = -\tan^{-1}\left(\frac{2\tau\zeta\omega}{1 - \omega^2\tau^2}\right)} \qquad \textbf{(8-4.15)}$$

This result was also presented in Chapter 2.

EXAMPLE 8-4.2

Consider the following first-order lead transfer function

$$G(s) = K(1 + \tau s) \qquad \textbf{(8-4.16)}$$

This is a transfer function composed of a gain times a first-order lead. Determine the expressions for AR and θ.

Substituting $i\omega$ for s results in the following complex number expression

$$G(i\omega) = K(1 + i\ \omega\tau)$$

which can also be thought of as being formed by two other numbers

$$G(i\omega) = G_1 G_2 = K(1 + i\omega t)$$

The amplitude ratio is

$$\boxed{\text{AR} = |G(i\omega)| = |G_1||G_2| = K\sqrt{1 + \omega^2\tau^2}} \qquad \textbf{(8-4.17)}$$

and the phase angle is

$$\theta = \underline{/G_1} + \underline{/G_2} = 0 + \tan^{-1}(\omega\tau)$$

$$\boxed{\theta = \tan^{-1}(\omega\tau)} \qquad \textbf{(8-4.18)}$$

The phase angles of the systems described by Eqs. 8-4.1 and 8-4.16 can be compared. Systems described by Eq. 8-4.1, which are referred to in Chapters 2 and 3 as first-order lags, provide negative phase angles, as shown by Eq. 8-4.13. Systems described by Eq. 8-4.16, which are referred to as a first-order leads, provide positive phase angles, as shown by Eq. 8-4.18. This fact is important in the study of process control stability by frequency response techniques.

Equation 8-4.17 helps explaining why real systems cannot be pure leads. The equation shows that the amplitude ratio increases with frequency, which means that high-frequency noise, which is always present in natural signals, would be infinitely amplified.

EXAMPLE 8-4.3

Determine the expressions for AR and θ for a pure dead time

$$G(s) = e^{-t_0 s}$$

Substituting $i\omega$ for s yields

$$G(i\omega) = e^{-i\, t_0 \omega}$$

Since this expression is already in polar form, using the principles learned in Chapter 2, we obtain

$$G(i\omega) = |G(i\omega)|e^{iG(i\omega)} = e^{-i\, t_0\omega}$$

which means that

$$\boxed{AR = |G(i\omega)| = 1} \tag{8-4.19}$$

and

$$\boxed{\theta = \underline{/G(i\omega)} = -t_0\omega} \tag{8-4.20a}$$

Recall the discussion on the units of θ. As written in Eq. 8-4.20a, the unit of θ is radians. If it is desired to obtain θ in degrees, then

$$\theta = \left(\frac{180°}{\pi}\right)(-t_0\omega) \tag{8-4.20b}$$

It is important to notice that θ becomes increasingly negative as ω increases. The rate at which θ drops depends on t_0; the larger t_0 is, the faster θ drops. This fact will become important in the analysis of control systems. The amplitude ratio and magnitude ratios are independent of frequency when the transfer function is a pure dead time.

EXAMPLE 8-4.4

Determine the expressions for AR and θ for an integrator

$$G(s) = \frac{1}{s}$$

Substituting $i\omega$ for s yields

$$G(i\omega) = \frac{1}{i\omega} = -\frac{1}{\omega}i$$

This is a pure imaginary number with amplitude ratio

$$\boxed{AR = |G(iw)| = \frac{1}{\omega}} \tag{8-4.21}$$

and the phase angle, as it lies on the negative imaginary axis, is

$$\boxed{\theta = \underline{/G(i\omega)} = -\frac{\pi}{2}; \text{or} - 90°} \tag{8-4.22}$$

For an integrator, then, the amplitude ratio is inversely proportional to the frequency, whereas the phase angle remains constant at $-90°$. That is, the integrator provides a constant phase lag.

At this point, we can generalize the expressions for AR and θ. Consider the following general OLTF:

$$\text{OLTF}(s) = \frac{K \prod_{i=1}^{m}(\tau_i s + 1)e^{-t_0 s}}{s^k \prod_{j=1}^{n}(\tau_j s + 1)}, \qquad (n+k) > m \qquad \textbf{(8-4.23)}$$

Substituting $i\omega$ for s yields

$$\text{OLTF}(i\omega) = \frac{K \prod_{i=1}^{m}(i\tau_i\omega + 1)e^{-i\omega t_0}}{(i\omega)^k \prod(i\tau_j\omega + 1)}$$

and finally we arrive at

$$AR = \frac{K \prod_{i=1}^{m}\sqrt{(\tau_i\omega)^2 + 1}}{\omega^k \prod_{j=i}^{n}\sqrt{(\tau_j\omega)^2 + 1}} \qquad \textbf{(8-4.24)}$$

and

$$\theta = \sum_{i=1}^{m}\tan^{-1}(\tau_i\omega) - t_0\omega - \sum_{j=1}^{n}\tan^{-1}(\tau_j\omega) - k\left(\frac{\pi}{2}\right) \qquad \textbf{(8-4.25)}$$

So far, expressions for AR and θ as a function of ω have been developed. There are several ways to represent these expressions graphically. The three most common ways are the Bode plots, the Nyquist plot, and the Nichols chart. Bode plots are presented in the next section.

8-4.1 Bode Plots

Bode plots are common graphical representations of AR (MR) and θ functions. A Bode plot consists of two graphs: (1) log AR (or log MR) versus log ω, and (2) θ versus log ω. Frequently the term 20 log AR, referred to as decibels, is plotted instead of log AR. This term is used extensively in the electrical engineering field and sometimes also in the process control field; this book plots log AR. Let us look at the Bode plots of some of the most common process transfer functions.

Gain Element   A pure gain element has the transfer function

$$G(s) = K$$

Substituting $i\omega$ for s gives

$$G(i\omega) = K$$

and using the mathematics previously presented yields

$$AR = |G(i\omega)| = K, \quad \text{or} \quad MR = 1$$

and

$$\theta = \tan^{-1} G(i\omega) = 0$$

(a) Gain element.

(b) First-order lag.

(c) Dead time.

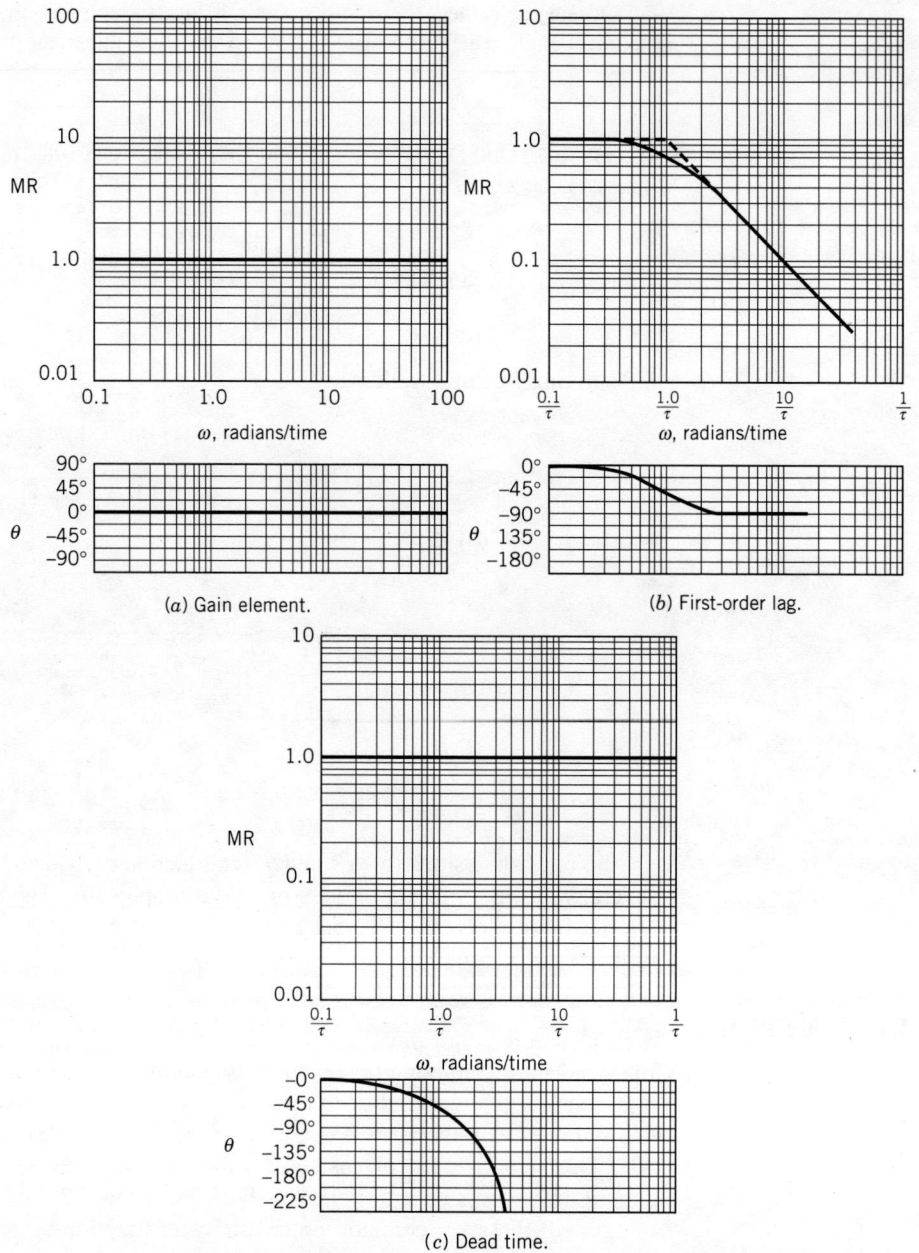

Figure 8-4.3 Bode plots. (a) Gain element. (b) First-order lag. (c) Dead time.

Figure 8-4.3a shows the Bode plot for this element; log-log and semi-log graph papers have been used.

First-Order Lag For a first-order lag, the AR and θ are given by Eqs. 8-4.12 and 8-4.13, respectively. From the AR equation the MR expression is obtained

$$\text{MR} = \frac{1}{\sqrt{\omega^2 \tau^2 + 1}}$$

(d) Second-order lag.

(e) First-order lead.

(f) Integrator.

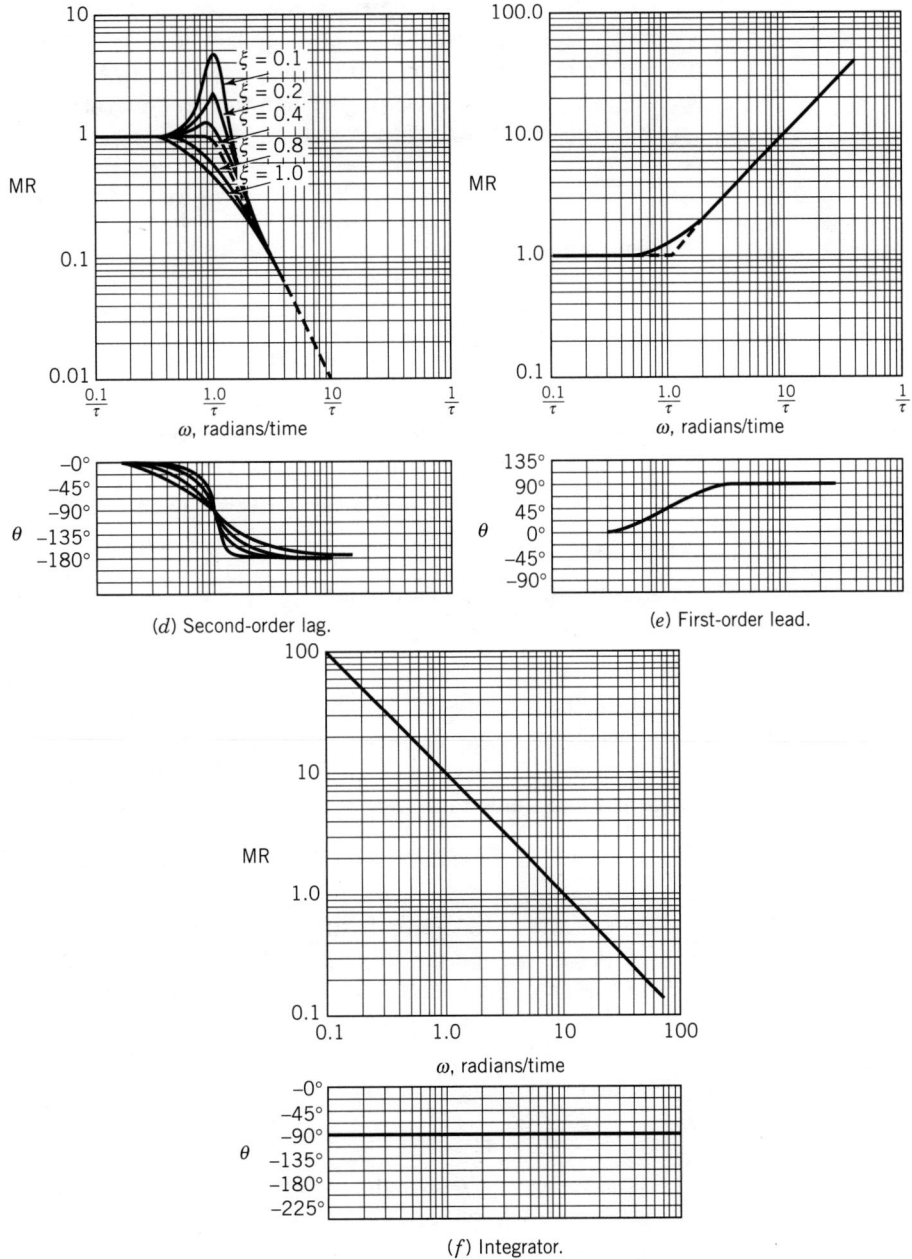

Figure 8-4.3(*Continued*) (d) Second-order lag. (e) First-order lead. (f) Integrator.

and

$$\theta = -\tan^{-1}(\omega\tau) \tag{8-4.13}$$

The Bode plot for this system is shown in Fig. 8-4.3b. The magnitude ratio plot also shows two dashed lines. These lines are asymptotes to the frequency response at low and high frequencies. The figure shows that these asymptotes do not deviate much from the actual frequency response. Therefore, frequency response analysis is quite often done with the asymptotes; they are easier to draw, and not much error is involved in their use. Let us now see how these asymptotes are developed.

From the magnitude ratio equation as $\omega \to 0$, MR $\to 1$, which results in the low-frequency asymptote. Before the high-frequency asymptote is developed, the magnitude ratio equation is written in log form:

$$\log \text{MR} = -\frac{1}{2} \log(\tau^2 \omega^2 + 1)$$

Now, as $\omega \to \infty$

$$\log \text{MR} \to -\frac{1}{2} \log \tau^2 \omega^2 = -\log \tau \omega$$

$$\log \text{MR} \to -\log \tau - \log \omega \tag{8-4.26}$$

which is the expression of a straight line in a log-log graph of MR versus ω; this straight line has a slope of -1. The location of the line in the graph must now be determined. The simplest way to do this is to find where the high-frequency asymptote meets the low-frequency asymptote. It is known that as $\omega \to 0$, MR $\to 1$ so,

$$\log \text{MR} \to 0$$

Equating this equation to Eq. 8-4.26 yields

$$\omega = \frac{1}{\tau} \tag{8-4.27}$$

This is the frequency, referred to as the *corner frequency* (ω_c) or *breakpoint frequency*, at which the asymptotes meet, as shown in Fig. 8-4.3b. It is also at this frequency that the maximum error between the frequency response and the asymptotes exists. The actual magnitude ratio is

$$\text{MR} = \frac{1}{\sqrt{\omega^2 \tau^2 + 1}} = \frac{1}{\sqrt{2}} = 0.707$$

and not MR $= 1$ as given by the asymptotes.

Before leaving the Bode plot of this system, it is necessary to go over the θ at low and high frequencies. At low frequencies, $\omega \to 0$,

$$\theta \to -\tan^{-1}(\omega \tau) = -\tan^{-1}(0) = 0$$

At high frequencies, $\omega \to \infty$,

$$\theta \to -\tan^{-1}(\infty) = -90°$$

These values of phase angle, $0°$ and $-90°$, are the asymptotes for the phase angle plot. At the corner frequency,

$$\omega_c = \frac{1}{\tau}, \qquad \theta = \tan^{-1}(1) = -45°$$

To summarize, the important characteristics of the Bode plot of a first-order lag are the following:

1. *AR (MR) graph.* The low-frequency asymptote has a slope of 0, whereas the high-frequency asymptote has a slope of -1. The corner frequency, where these two asymptotes meet, occurs at $1/\tau$.

2. *Phase angle graph.* At low frequencies the phase angle approaches $0°$, whereas at high frequencies it approaches $-90°$. At the corner frequency, the phase angle is $-45°$.

Second-Order Lag   As shown in Example 8-4.1, the AR and θ expressions for a second-order lag are given by Eqs. 8-4.14 and 8-4.15, respectively. From the AR equation, we obtain

$$\text{MR} = \frac{1}{\sqrt{(1 - \omega^2 \tau^2)^2 + (2\tau\zeta\omega)^2}}$$

and

$$\theta = -\tan^{-1}\left(\frac{2\tau\zeta\omega}{1 - \omega^2\tau^2}\right) \tag{8-4.15}$$

Giving values to ω for a given τ and ζ, the frequency response is determined as shown in Fig. 8-4.3d.

The asymptotes are obtained in a similar manner as for the first-order lag. At low frequencies, $\omega \to 0$,

$$\text{MR} \to 1$$

and

$$\theta \to 0°$$

At high frequencies, $\omega \to \infty$,

$$\log \text{MR} = -\frac{1}{2}\log[(1 - \omega^2\tau^2)^2 + (2\tau\zeta\omega)^2]$$

$$\log \text{MR} = -\frac{1}{2}\log(\omega^2\tau^2)^2 = -2\log\omega - 2\log\tau$$

which is the expression of a straight line with a slope of -2. At these high frequencies,

$$\theta \to -\pi \quad (-180°)$$

To find the corner frequency, ω_c, the same procedure as for the first-order lag is followed. It yields

$$\omega_c - \frac{1}{\tau}$$

Note from Fig. 8-4.3d that the transition of the frequency response from low to high frequencies depends on ζ.

At the corner frequency,

$$\theta = -\tan^{-1}(\infty) = -\frac{\pi}{2} \quad (-90°)$$

To summarize, the important characteristics of the Bode plot of a second-order lag are the following:

1. ***AR (MR) graph.*** The slope of the low-frequency asymptote is 0, and that of the high-frequency asymptote is -2. The corner frequency, ω_c, occurs at $1/\tau$. The transition of the AR from low to high frequency depends on the value of ζ.

2. ***Phase angle graph.*** At low frequencies the phase angle approaches $0°$, whereas at high frequencies it approaches $-180°$. At the corner frequency, the phase angle is $-90°$.

Chapter 8   Root Locus and Frequency Response Techniques

Dead Time   As shown in Example 8-4.3, the AR and θ expressions for a pure dead time are given by Eqs. 8-4.19 and 8-4.20, respectively.

$$AR = MR = 1 \tag{8-4.19}$$

and

$$\theta = -\omega t_0 \tag{8-4.20a}$$

The Bode plot is shown in Fig. 8-4.3c. Notice that as the frequency increases, the phase angle becomes more negative. The larger the value of the dead time, the faster the phase angle drops (becomes increasingly negative) without limit.

First-Order Lead   As shown in Example 8-4.2, the AR and θ expressions for a first-order lead are given by Eqs. 8-4.17 and 8-4.18, respectively. From the AR equation, we obtain

$$MR = \sqrt{1 + \omega^2 \tau^2}$$

and

$$\theta = \tan^{-1}(\omega \tau) \tag{8-4.18}$$

The Bode plot is shown in Fig. 8-4.3e. Note that the low-frequency asymptote has a slope of 0, and the high-frequency asymptote has a slope of $+1$. At low frequencies the phase angle approaches $0°$, whereas at high frequencies it approaches $+90°$. At the corner frequency, the phase angle is $+45°$. Thus a first-order lead provides a "phase lead."

Integrator   As shown in Example 8-4.4, the AR and θ expressions for an integrator are given by Eqs. 8-4.21 and 8-4.22, respectively.

$$AR = MR = \frac{1}{\omega} \tag{8-4.21}$$

and

$$\theta = -90° \tag{8-4.22}$$

The Bode plot is shown in Fig. 8-4.3f. Note that the MR graph consists of a straight line with slope of -1. This is easily shown by taking the log of Eq. 8-4.21,

$$\log MR = -\log \omega$$

This equation also shows that $MR = 1$ at $\omega = 1$ radians/time.

Development of Bode Plot of Complex Systems   Most complex transfer functions of process systems are formed by the product of simpler components. Let us now look at the Bode plot of these complex transfer functions; consider Eq. 8-4.23 as an example. From the AR expression, Eq. 8-4.24, we obtain

$$\log AR = \log K + \frac{1}{2}\sum_{i=1}^{m} \log[(\tau_i \omega)^2 + 1] - k\log \omega - \frac{1}{2}\sum_{j=1}^{n} \log[(\tau_j \omega)^2 + 1] \tag{8-4.28}$$

or

$$\log MR = \frac{1}{2}\sum_{i=1}^{m} \log[(\tau_i \omega)^2 + 1] - k\log \omega - \frac{1}{2}\sum_{j=1}^{n} \log[(\tau_j \omega)^2 + 1] \tag{8-4.29}$$

These equations together with Eq. 8-4.25 show that the Bode plot of complex systems consists of the sum of the individual components. To obtain the composite asymptote, we add the individual asymptotes.

EXAMPLE 8-4.5

Consider the following transfer function:

$$G(s) = \frac{K(s+1)e^{-s}}{s(2s+1)(3s+1)}$$

using the principles learned

$$MR = \frac{\sqrt{\omega^2 + 1}}{\omega\sqrt{4\omega^2 + 1}\sqrt{9\omega^2 + 1}}$$

or

$$\log MR = \frac{1}{2}\log(\omega^2 + 1) - \log(\omega) - \frac{1}{2}\log(4\omega^2 + 1) - \frac{1}{2}\log(9\omega^2 + 1)$$

and

$$\theta = \tan^{-1}(\omega) - \omega t_0 - \frac{\pi}{2} - \tan^{-1}(2\omega) - \tan^{-1}(3\omega)$$

From these last two equations, the Bode plot is developed as shown in Fig. 8-4.4. At low frequencies, $\omega < 0.33$, the slope is -1, because of the integrator term. At $\omega = 0.33$, one of the first-order lags starts to contribute to the graph and thus the slope changes to -2 at this frequency. At $\omega = 0.5$, the other first-order lag starts to contribute, changing the slope of the asymptote to -3. Finally, at $\omega = 1$ first-order lead enters with a slope of $+1$, and the slope of the asymptote changes back to -2. Similarly, the composite phase angle plot is obtained by algebraically adding the individual angles.

One final comment can be made about the slopes of the low-frequency and high-frequency asymptotes (initial and final slopes) and the angles of Bode plots. Consider a general transfer function such as

$$G(s) = \frac{K(a_m s^m + a_{m-1}s^{m-1} + \ldots + 1)}{s^k(b_n s^n + b_{n-1}s^{n-1} + \ldots + 1)}; \quad (n+k) > m$$

The slope of the low-frequency asymptote is given by

$$\text{slope of AR (MR)}|_{\omega \to 0} \to (-1)k$$

and the angle

$$\theta|_{\omega \to 0} \to (-90^\circ)k$$

The slope of the high-frequency asymptote is given by

$$\text{slope of AR (MR)}|_{\omega \to \infty} \to (n+k-m)(-1)$$

and the angle

$$\theta|_{\omega \to \infty} \to (n+k-m)(-90^\circ)$$

Most systems follow these slopes and angles; these systems are called *minimal phase systems*. There are three exceptions, however, which are called *nonminimal phase systems*. These exceptions are

1. Systems with dead time: $G(s) = e^{-t_0 s}$ (phase angle decreases without limit).
2. Systems that show inverse response (positive zeros): $G(s) = (1 - \tau_1 s)/(1 + \tau_2 s)$. A process example of this type of response is given in Chapter 4.

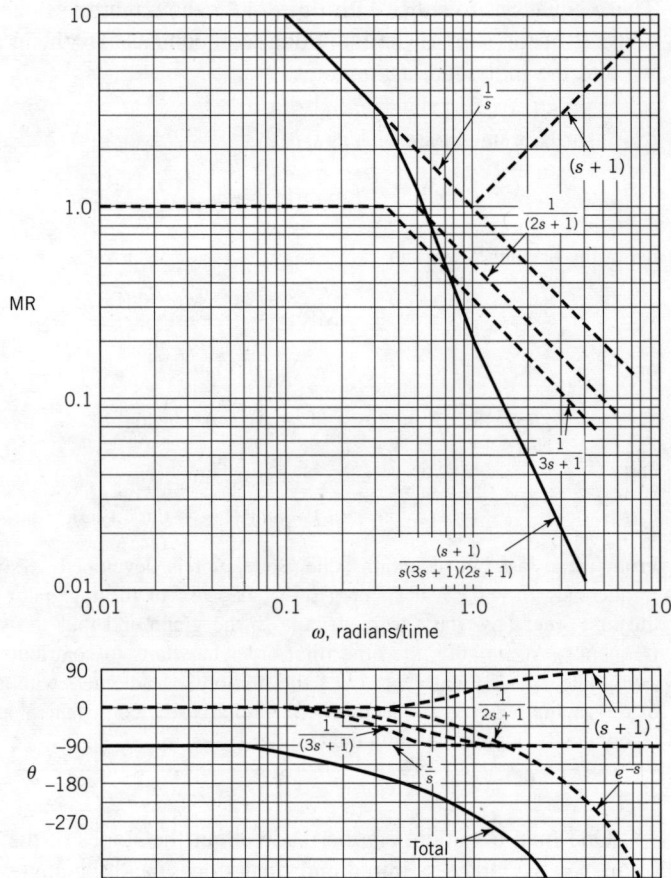

Figure 8-4.4 Bode plot of $G(s) = \dfrac{K(s+1)e^{-s}}{s(2s+1)(3s+1)}$.

3. Systems that are open-loop unstable (positive poles): $G(s) = 1/(\tau s - 1)$. A process example of this type of response is also given in Chapter 4.

In each of these cases, the magnitude ratio plot is not changed, but the phase angle plot is. The dead time term was presented earlier; the Bode plot of the other two systems is the subject of one of the problems at the end of the chapter.

The expression for the slope of the high-frequency asymptote also serves to show why transfer functions of real systems must have at least as many lags as leads. If $(n + k - m) < 0$, then the final slope is positive and noise of high frequency is amplified with infinite gain.

Example 8-4.5 has shown how to obtain the Bode plot using mainly the asymptotes. For a more precise graph, the expressions for AR, or MR, and θ are used. There are several software programs, such as MATLAB (Reference 2), that provide the Bode plot given the transfer function.

8-4.2 Frequency Response Stability Criterion

Chapter 6 presented two methods to determine the limits of stability of a feedback control loop, direct substitution and Routh's test. However, as we saw there, neither of these methods can handle the presence of dead time in the loop except through approximation. Frequency response helps us to determine the stability limits for feedback loops even when there is dead time in the loop. The frequency response stability criterion consists of determining the frequency at which the phase angle of the open-loop transfer function (OLTF) is $-180°$ ($-\pi$ radians), and the amplitude ratio of the OLTF at that frequency.

Consider the heat exchanger temperature control loop first presented in Example 6-2.1. For convenience, the exchanger is shown again in Fig. 8-4.5 and its block diagram in Fig. 8-4.6. The open-loop transfer function is

$$\text{OLTF} = \frac{0.8K_c}{(10s + 1)(30s + 1)(3s + 1)}$$

The MR and θ expressions are

$$\text{MR} = \frac{\text{AR}}{0.8K_c} = \frac{1}{\sqrt{(10\omega)^2 + 1}\sqrt{(30\omega)^2 + 1}\sqrt{(3\omega)^2 + 1}} \qquad \textbf{(8-4.30)}$$

$$\theta = -\tan^{-1}(10\omega) - \tan^{-1}(30\omega) - \tan^{-1}(3\omega) \qquad \textbf{(8-4.31)}$$

The Bode plot is shown in Fig. 8-4.7. From this figure, or from Eq. 8-4.31, the frequency at which $\theta = -180°$ (or $-\pi$ radians) is 0.219 rad/s. At this frequency, from the MR plot or from Eq. 8-4.30

$$\frac{\text{AR}}{0.8K_c} = 0.0524$$

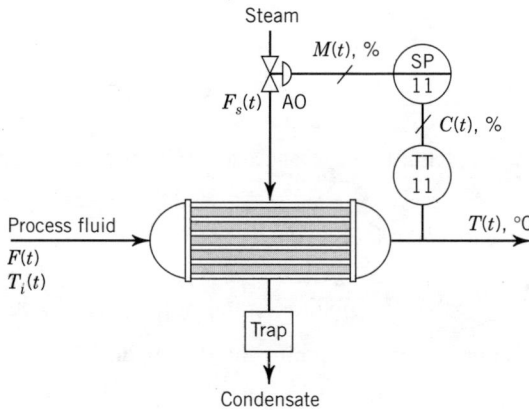

Figure 8-4.5 Feedback control loop for temperature control of heat exchanger.

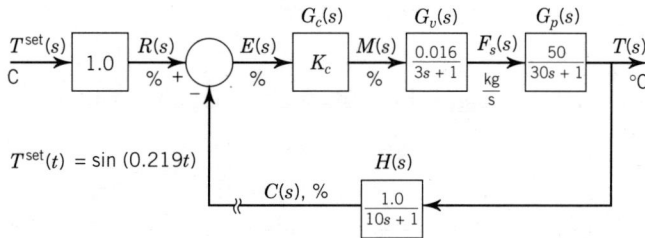

Figure 8-4.6 Block diagram of heat exchanger temperature control loop—P controller.

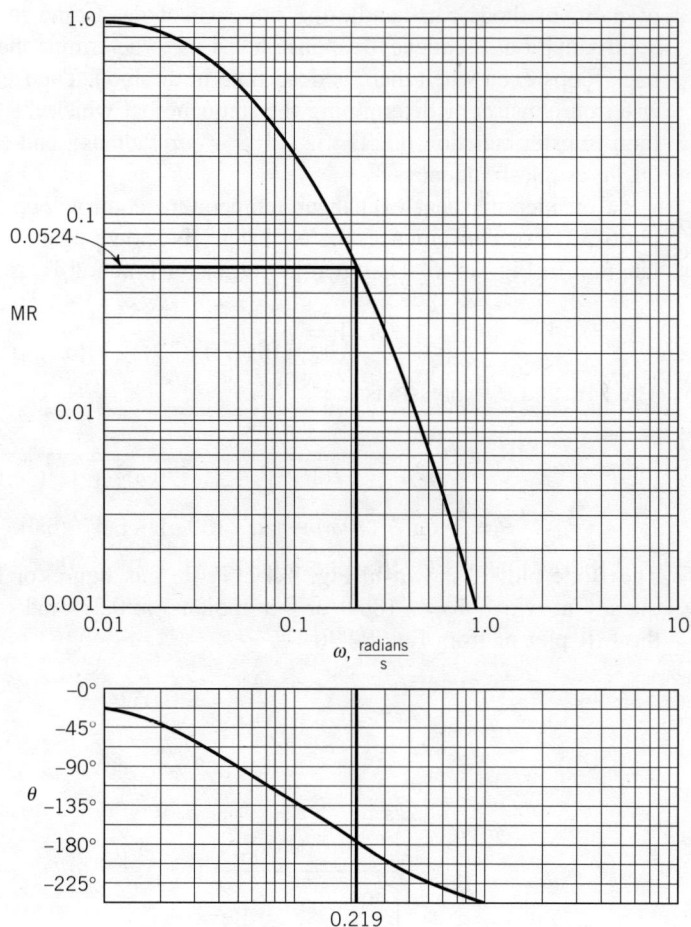

Figure 8-4.7 Bode plot of heat exchanger temperature control loop—P controller.

The controller gain that yields AR = 1 is

$$K_c = \frac{AR}{0.8(0.0524)} = \frac{1}{0.8(0.0524)} = 23.8\frac{\%CO}{\%TO}$$

These calculations are highly significant. The value $K_c = 23.8$ %CO/%TO is the gain of the controller that yields AR = 1 when the phase angle is $-180°$. Remember that AR is defined as the ratio of the amplitude of the output signal to the amplitude of the input signal, Y_o/X_o. This means that if the input set point to the temperature controller is varied as

$$T^{set}(t) = \sin(0.219t)$$

then the output signal from the transmitter, after the transients disappear, will vary as

$$TO(t) = \sin(0.219t - \pi) = -\sin(0.219t)$$

Note that the feedback signal is disconnected from the controller, as shown in the block diagram of Fig. 8-4.6, and that the frequency of the set point oscillation is 0.219 rad/s. This is the frequency at which $\theta = -180° = -\pi$ radians and, when the controller gain is 23.8, AR = 1. Under these conditions, the amplitude of the output signal is equal to that of the input signal.

Suppose now that at some time, $t = 0$, the set point oscillations are stopped, $T^{set}(t) = 0$, and at the same time the transmitter signal is connected to the controller. The error signal, $E(s)$, inside the controller remains unchanged, and the oscillations are sustained. If nothing changes in the control loop, the oscillations remain indefinitely. If at some time the controller gain is slightly increased to 25, the amplitude ratio becomes 1.04

$$AR = 0.0524(0.8)K_c$$
$$AR = 0.0524(0.8)(25) = 1.04$$

This means that as the signal goes through the control loop, it is amplified. After the first time, the output signal from the transmitter is $-1.04\sin(0.219t)$. After the second time, it is $-(1.04)^2\sin(0.219t)$ and so on. If this is not stopped, the outlet temperature will increase continuously, yielding an unstable control loop.

On the other hand, if the controller gain is slightly decreased to 23, the amplitude ratio becomes 0.96

$$AR = 0.0524(0.8)23 = 0.96$$

This means that as the signal goes through the control loop, it decreases in amplitude. After the first time, the output signal from the transmitter is $-0.96\sin(0.219t)$. After the second, time it is $-(0.96)^2\sin(0.219t)$, and so on. This results in a stable control loop.

In summary, the stability criterion based on frequency response can be stated as follows:

For a control system to be stable, the amplitude ratio must be less than unity when the phase angle is $-180°$ ($-\pi$ radians). If $AR < 1$ at $\theta = -180°$ the system is stable; if $AR > 1$ at $\theta = -180°$, the system is unstable.

The controller gain that provides the condition of $AR = 1$ at $\theta = -180°$ is the ultimate gain, K_{cu}. The frequency at which this condition happens is the ultimate frequency, ω_u. From this frequency the ultimate period can be calculated as $T_u = 2\pi/\omega_u$. Note that the values of ultimate frequency and gain obtained in this example are the same as those obtained in Example 6-2.1 by direct substitution.

Before proceeding with more examples, it is important to stress that the ultimate frequency and ultimate gain can be obtained directly from the MR and θ equations, Eqs. 8-4.29 and 8-4.30 for this example, without the need for the Bode plot. The Bode plot was developed from these equations. The determination of ω_u requires a small amount of trial and error using the θ equation—that is, finding which ω yields $\theta = -180°$. This ω is ω_u. Once ω_u has been determined, the equation for AR is used to calculate K_{cu}. This complete procedure is usually faster and yields more accurate results than drawing and using the Bode plot. The use of the Bode plot is still useful because it shows at a glance how AR and θ vary as the frequency varies.

The following examples will provide more practice with this powerful technique.

EXAMPLE 8-4.6

Consider the same heat exchanger (Fig. 8-4.5) previously used to present the frequency response stability criterion. Suppose now that for some reason, the outlet temperature cannot be measured at the exit of the exchanger but must be measured farther down the pipe. The effect of this new sensor location is the addition of some dead time, say 2 s, to the control loop.

The new OLTF is

$$OLTF = \frac{0.8K_c e^{-2s}}{(10s + 1)(30s + 1)(3s + 1)}$$

with

$$MR = \frac{AR}{0.8K_c} = \frac{1}{\sqrt{(10\omega)^2 + 1}\sqrt{(30\omega)^2 + 1}\sqrt{(3\omega)^2 + 1}}$$

and

$$\theta = -2\omega - \tan^{-1}(10\omega) - \tan^{-1}(30\omega) - \tan^{-1}(3\omega)$$

The last two equations can be used to determine the ultimate frequency and ultimate gain. Performing these calculations yields, for $\theta = -\pi$ radians,

$$\omega_u = 0.160 \text{ rad/s}$$

and

$$K_{cu} = 12.8$$

The Bode plot is shown in Fig. 8-4.8.

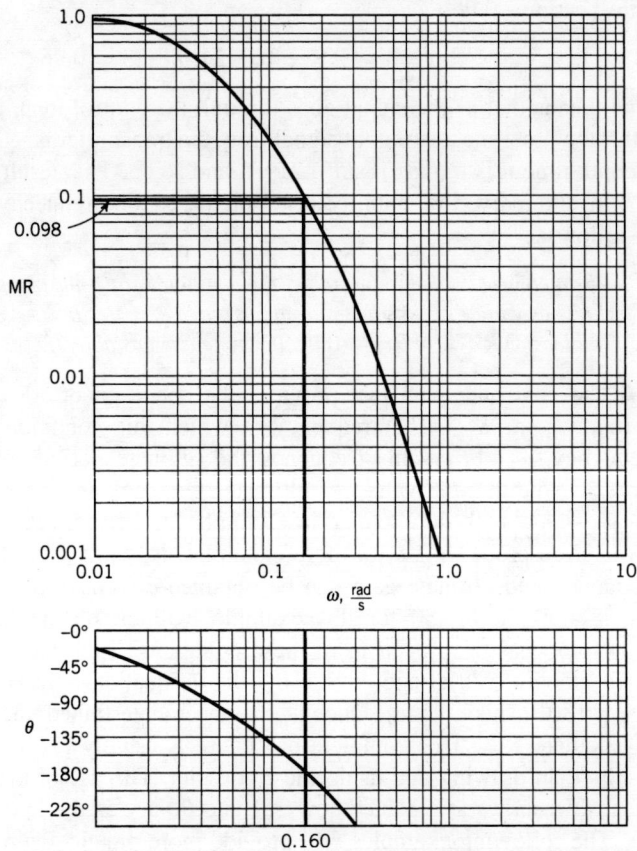

Figure 8-4.8 Bode plot of heat exchanger temperature control with dead time—P controller.

The results of Example 8-4.6 show the effect of dead time on the stability (and consequently also the controllability) of the control loop. The ultimate gain and ultimate period for the heat exchanger without dead time were previously found to be

$$K_{cu} = 23.8 \frac{\%CO}{\%TO} \qquad \text{and} \qquad \omega_u = 0.219 \text{ rad/s}$$

When the dead time of 2 s was added in Example 8-4.6, the results were

$$K_{cu} = 12.8 \frac{\%CO}{\%TO} \quad \text{and} \quad \omega_u = 0.160 \text{ rad/s}$$

Thus it is easier for the process with dead time to go unstable. The difference in ω_u also indicates that the closed-loop response of a process with dead time is slower than that of a process without dead time.

The preceding example demonstrates that the frequency response stability criterion can exactly analyze the effect of dead time in the control loop. As mentioned in previous chapters, dead time is the worst thing that can happen to any control loop; this example proves this point. The dead time term "adds phase lag" to the control loop and consequently the phase angle crosses the $-180°$ value at a lower frequency. The longer the dead time is, the lower the ultimate frequency and ultimate gain.

Example 8-3.1 showed that the addition of integral mode to a proportional controller decreases the ultimate frequency and ultimate gain. This can be explained, from a frequency response point of view, by saying that the addition of integral mode "adds phase lag" to the control loop. A proportional-only controller has a phase angle of $0°$, as shown in Fig. 8-4.3a. Consider now a proportional-integral controller

$$G(s) = K_c \left(1 + \frac{1}{\tau_I s} \right) = K_c \left(\frac{\tau_I s + 1}{\tau_I s} \right)$$

This transfer function is composed of a lead term, $\tau_I s + 1$, and an integrator term, $1/\tau_I s$. At low frequencies, $\omega \ll 1/\tau_I$, the lead term does not affect the phase angle, but the integrator term contributes $-90°$, thus adding phase lag. At higher frequencies, $\omega \gg 1/\tau_I$, the lead term cancels the integrator term with a resulting $0°$ phase angle. However, unless the reset time is very long, this canceling effect occurs at a frequency higher than that at which the phase angle crosses the $-180°$. Thus the loop with PI controller will have a lower ultimate frequency and gain than that with the P controller.

Remember that the integral mode in a controller is the one that removes the offset. However, as explained in the previous paragraph, its use provides phase lag to the loop.

The following example demonstrates the effect of derivative mode on the stability of a control loop.

EXAMPLE 8-4.7   Consider the same heat exchanger control loop, no dead time, with a proportional derivative controller. Suppose the derivative time is 0.25 min (15 s). The equation for a "real" PD controller as shown in Chapter 5 is

$$G_c(s) = K_c \left(\frac{1 + \tau_D s}{1 + \alpha \tau_D s} \right)$$

or, for this example, using $\alpha = 0.1$

$$G_c(s) = K_c \left(\frac{1 + 15s}{1 + 1.5s} \right)$$

The OLTF is then

$$\text{OLTF} = \frac{0.8 K_c (1 + 15s)}{(10s + 1)(30s + 1)(3s + 1)(1 + 1.5s)}$$

with

$$\text{MR} = \frac{\text{AR}}{0.8 K_c} = \frac{\sqrt{(15\omega)^2 + 1}}{\sqrt{(10\omega)^2 + 1}\sqrt{(30\omega)^2 + 1}\sqrt{(3\omega)^2 + 1}\sqrt{(1.5\omega)^2 + 1}}$$

and

$$\theta = \tan^{-1}(15\omega) - \tan^{-1}(10\omega) - \tan^{-1}(30\omega) - \tan^{-1}(3\omega) - \tan^{-1}(1.5\omega)$$

The Bode plot for this system is shown in Fig. 8-4.9. Comparing this Bode plot to the one shown in Fig. 8-4.7, we see that the phase angle plot has been "moved up"; the derivative action "adds phase lead." In this system the ultimate gain and period are found to be

$$K_{cu} = 33 \frac{\%CO}{\%TO} \qquad \text{and} \qquad \omega_u = 0.53 \text{ rad/s}$$

Thus these results show that the derivative mode makes the control loop more stable and faster.

In the preceding example and discussion, we used the terms *ultimate gain* and *ultimate frequency* for controllers other than proportional controllers. However, the ultimate gain and period used for tuning is still defined only for proportional controllers.

The examples presented in this section have demonstrated the use of frequency response for analysis of control loops. These examples have also shown the effect of different terms, dead time, and derivative mode, on the stability of control loops.

The frequency response stability criterion confirms that control loops with a pure (no dead time) first- or second-order open-loop transfer function will never go unstable;

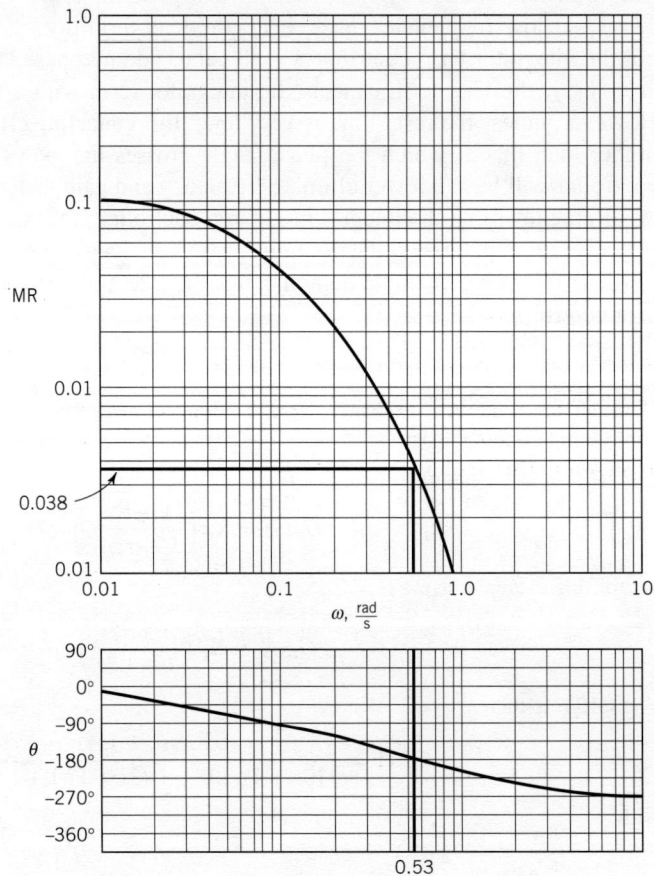

Figure 8-4.9 Bode plot of heat exchanger temperature control loop—PD controller.

their phase angles will never go below $-180°$. Once a dead time is added, no matter how small, the system can go unstable because the phase angle will always cross the $-180°$ value.

Controller Performance Specifications

Chapter 7 presented several ways to tune controllers to obtain a desired loop performance. The methods presented were the Ziegler–Nichols (quarter decay ratio), the error-integral criteria (IAE, ISE, and ITAE), and the controller synthesis. Frequency response provides a procedure by which to obtain the ultimate gain and ultimate frequency of a control loop. Once these terms have been determined, the Ziegler–Nichols equations can be used to tune the controller. Frequency response techniques provide still other performance specifications to tune controllers: gain margin, and phase margin. These are based on the frequency response of the open-loop transfer function (OLTF).

Gain Margin   The *gain margin* (GM) is a typical performance specification associated with the frequency response technique. The gain margin represents the factor by which the total loop gain must increase to just make the system unstable. The controller gain that yields a desired gain margin is calculated as follows

$$K_c = \frac{K_{cu}}{\text{GM}} = \frac{1}{K\,(\text{GM})\text{MR}|_{\theta=-180°}} \tag{8-4.32}$$

where K is the product of the gains of all other elements in the loop. Typical specification is for $\text{GM} \geq 2$. Note that the tuning of a proportional controller with the $\text{GM} = 2$ is the same as the Ziegler–Nichols quarter decay ratio tuning presented in Chapter 7.

Phase Margin   *Phase margin* (PM) is another specification commonly associated with the frequency response technique. The phase margin is the difference between $-180°$ and the phase angle at the frequency for which the amplitude ratio (AR) is unity. That is

$$\text{PM} = 180° + \theta|_{\text{AR}=1} \tag{8-4.33}$$

PM represents the additional amount of phase lag required to make the system unstable. Typical specification is for $\text{PM} > 45°$.

EXAMPLE 8-4.8   Consider the heat exchanger of Example 8-4.6. Tune a proportional controller for (**a**) $\text{GM} = 2$, and (**b**) $\text{PM} = 45°$.

(**a**) In Example 8-4.6 the ultimate gain of the controller was determined to be $K_c = 12.8$. To obtain GM specification of 2 the controller gain is then set to

$$K_c|_{\text{GM}=2} = \frac{K_{cu}}{2} = 6.4\frac{\%\text{CO}}{\%\text{TO}}$$

(**b**) In Example 8-4.6 expressions for MR and θ were determined to be

$$\text{MR} = \frac{\text{AR}}{0.8K_c} = \frac{1}{\sqrt{(10\omega)^2+1}\sqrt{(30\omega)^2+1}\sqrt{(3\omega)^2+1}}$$

and

$$\theta = -(2\omega t_0) - \tan^{-1}(10\omega) - \tan^{-1}(30\omega) - \tan^{-1}(3\omega)$$

On the basis of the definition of phase margin, for a PM $= 45°$, $\theta = -135°$. Using the equation for θ, or the Bode plot of Fig. 8-4.8, the frequency for this phase angle is

$$\omega|_{\text{PM}=45°} = 0.087 \text{ rad/s}$$

Then, substituting into the equation for the magnitude ratio, we get

$$\frac{\text{AR}}{0.8K_c} = 0.261$$

$$K_c|_{\text{PM}=45°} = \frac{\text{AR}}{0.8(0.261)} = \frac{1}{0.8(0.261)} = 4.8 \frac{\%\text{CO}}{\%\text{TO}}$$

Example 8-4.8 has shown how to obtain the tuning of the feedback controller for a certain GM, and PM. In part (**a**) the controller was tuned to yield a control loop with a GM of 2. This means that the overall loop gain must increase (because of process nonlinearities or for any other reason) by a factor of 2 before instability is reached. In choosing the value of GM, the engineer must understand the process to decide how much the process gain can change over the operating range. On the basis of this understanding, the engineer can choose a realistic GM value. The larger the GM value chosen, the greater the "safety factor" designed into the control system. However, the larger this safety factor is, the smaller the controller's gain that results and, therefore, the less sensitive the controller is to errors.

In part (**b**) of the example, the controller was tuned to yield a PM of $45°$. This means that $45°$ of phase lag must be added to the control loop before it goes unstable. Changes in phase angle of the control loop are mainly due to changes in its dynamic terms, time constants and dead time, because of process nonlinearities.

Gain margin and phase margin are two different performance criteria. The choice of one of them as the criterion for a particular loop depends on the process being controlled. If, because of process nonlinearities and characteristics, the gain is expected to change more than the dynamic terms, then the GM may be the indicated criterion. If, on the other hand, the dynamic terms are expected to change more than the gain, then the PM may be the indicated criterion.

In this section the meaning of gain margin and phase margin has been presented, as well as how to tune feedback controllers based on these performance specifications. In the process industries, however, the performance specifications of Chapter 7 are almost exclusively preferred.

8-5 SUMMARY

This chapter presented the root locus and frequency response techniques for the analysis and design of feedback control systems. Frequency response can handle the presence of dead time directly, without approximation.

Having studied the design and analysis of feedback control systems, we will next look into other important control strategies that are commonly used in industry.

REFERENCES

1. SMITH, C. A., and A. B. CORRIPIO, *Principles and Practice of Automatic Process Control.* 2nd ed., John Wiley & Sons, 1997, Chapter 8.

2. MATLAB, The MathWorks, Inc., Cochituate Place, 24 Prime Park Way, Natick, MA 01760–1520.

3. TK Solver, Universal Technical Systems, 1220 Rock Street, Rockford, IL 61101.

4. MATHCAD, MathSoft, Inc., 101 Main Street, Cambridge, MA 02142.

PROBLEMS

8-1. Draw the root locus diagram for each of the following open-loop transfer functions

(a) $G(s) = \dfrac{K}{s(s+1)(4s+1)}$

(b) $G(s) = \dfrac{K(3s+1)}{(2s+1)}$

(c) $G(s) = K\left(1 + \dfrac{1}{2s}\right)\left(\dfrac{0.5s+1}{0.05s+1}\right)$

8-2. Sketch the root locus diagram for the following two open-loop transfer functions.

(a) System with inverse response

$$G(s) = \frac{K(1 - 0.25s)}{(2s+1)(s+1)}$$

(b) Open-loop unstable system

$$G(s) = \frac{K}{(\tau_1 s + 1)(1 - \tau_2 s)}$$

for two cases: $\tau_1 = 2, \tau_2 = 1$, and $\tau_1 = 1, \tau_2 = 1$

8-3. Consider the pressure control system shown in Fig. P8-1. The pressure in the tank can be described by

$$\frac{P(s)}{F(s)} = \frac{0.4}{(0.15s+1)(0.8s+1)}, \text{ psi/scfm}$$

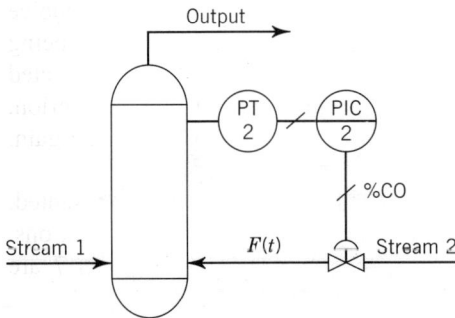

Figure P8-1 Pressure control system for Problem 8-3.

The valve can be described by the following transfer function:

$$\frac{F(s)}{M(s)} = \frac{0.6}{0.1s+1}, \text{ scfm/\%CO}$$

The pressure transmitter has a range of 0 to 185 psig. The dynamics of the transmitter are negligible.

(a) Draw the block diagram for this system including all the transfer functions.

(b) Determine the ultimate gain and ultimate period.

(c) Explain graphically how the addition of reset action to the controller affects the stability of the control loop. Use a $\tau_I = 0.8$ min.

(d) Explain graphically how the addition of rate action to the PI controller of part (e) affects the stability of the control loop.

8-4. The open-loop transfer function for Problem 7-25 on the control of the level in a tank using a PI controller is

$$\text{OLTF} = K_c\left(1 + \frac{1}{\tau_I s}\right)\left(\frac{K}{s(\tau s + 1)}\right)$$

Sketch the root locus for each of the following cases:
(a) $\tau_I > \tau$; (b) $\tau_I = \tau$; (c) $\tau_I < \tau$

Discuss, based on your sketches, how the stability of the loop is affected by the relationship between the controller integral time and the process time constant.

8-5. Consider the following transfer function of a certain process:

$$G(s)H(s) = \frac{1.5}{(s+1)(5s+1)(10s+1)}$$

Tune a proportional-only controller using as the gain half the ultimate gain.

8-6. Example 8-2.2 discusses the stability of a system with a P controller. Suppose now that a PD controller with $\tau_D = 0.2$ is used. Discuss the effects of increasing the value used for τ_D, for example using $\tau_D = 0.5$.

8-7. Draw the asymptotes of the Bode plot magnitude ratio (or amplitude ratio) and roughly sketch the phase angle plot for the following open-loop transfer functions.

(a) $G(s) = \dfrac{K}{(s+1)(2s+1)(10s+1)}$

(b) $G(s) = \dfrac{K(3s+1)}{(s+1)(2s+1)(10s+1)}$

(c) $G(s) = \dfrac{Ke^{-s}}{(s+1)(2s+1)(10s+1)}$

(d) $G(s) = \dfrac{K(3s+1)e^{-s}}{(s+1)(2s+1)(10s+1)}$

8-8. Draw the asymptotes of the Bode plot magnitude ratio (or amplitude ratio) and roughly sketch the phase angle plot for the transfer functions given in Problem 8-1.

8-9. Given the following transfer functions:

$$G(s) = \frac{4s+1}{s+1} \quad \text{and} \quad G(s) = \frac{1+2s}{(2s+1)(4s+1)}$$

(a) Sketch the asymptotes of the magnitude ratio part of the Bode plot, marking the breakpoint frequencies.

(b) Indicate the phase lag (or lead) at high frequencies $(\omega \to \infty)$.

8-10. Sketch the Bode plot of the transfer functions given in Problem 8-4.

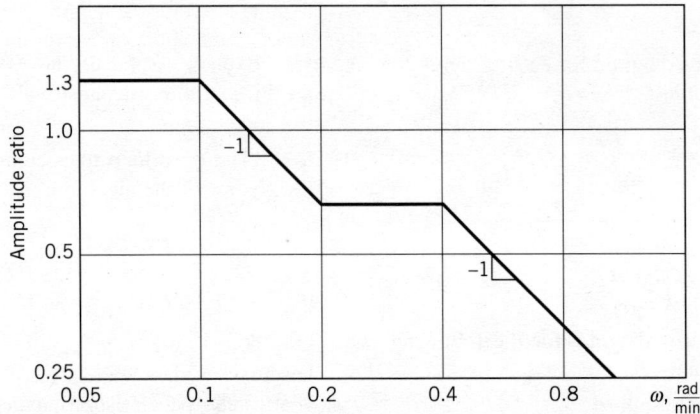

Figure P8-2 Amplitude ratio versus frequency plot for Problem 8-11.

8-11. The asymptotes of the amplitude ratio versus frequency plot for a process result in the sketch given in Fig. P8-2. The phase angle plot does not reach a high-frequency asymptote but becomes more negative as the frequency increases. At a frequency of 1.0 rad/min, the phase angle is $-246°$. Postulate a transfer function for the process and estimate the gain, time constants, and the dead time (if any).

8-12. Consider the vacuum filter process shown in Fig. P7-1. For this process using the data given in Problem 7-15 and applying frequency response techniques:

(a) Sketch the asymptotes of the Bode plot and the phase angle plot.

(b) Obtain the ultimate gain K_{cu} and ultimate period T_u.

(c) Tune the reset time of a PI controller by the controller synthesis method and determine the controller gain that would provide a gain margin of 2.

8-13. Consider the absorber presented in Problem 7-16. In part (a) of the problem, a feedback control loop was designed to control the exit concentration of ammonia. For this control loop:

(a) Sketch the asymptotes of the Bode plot and the phase angle plot.

(b) Obtain the ultimate gain, K_{cu} and ultimate period T_u.

(c) Tune a P controller for a phase margin of $45°$.

8-14. Consider a process with the following transfer function

$$G(s)H(s) = \frac{4.0}{(1s + 1)(0.8s + 1)(0.2s + 1)}$$

If a PI controller is used with Ziegler–Nichols setting, what gain margin, phase margin, are obtained?

8-15. Repeat Problem 8-14 for the following process transfer function

$$G(s)H(s) = \frac{6(1 - s)}{(s + 1)(0.5s + 1)}$$

8-16. Consider the block diagram shown in Fig. P8-3a. The input $N(s)$ represents noise that corrupts the output signal. If this process noise is significant, the control of the process may be difficult. To improve the control of noisy processes, filtering the feedback signal is usually done. A typical way to filter signals is by a filter device with a first-order transfer function. This device—pneumatic, electronic, or digital—is installed between the transmitter and the controller as shown in Fig. P8-3b. The gain of the filter is one (1) and its time constant, called the filter time constant, is τ_F. Using frequency response techniques explain how τ_F affects the filtering of the noisy signal and the performance of the control loop. Specifically, plot the ultimate gain as a function of τ_F.

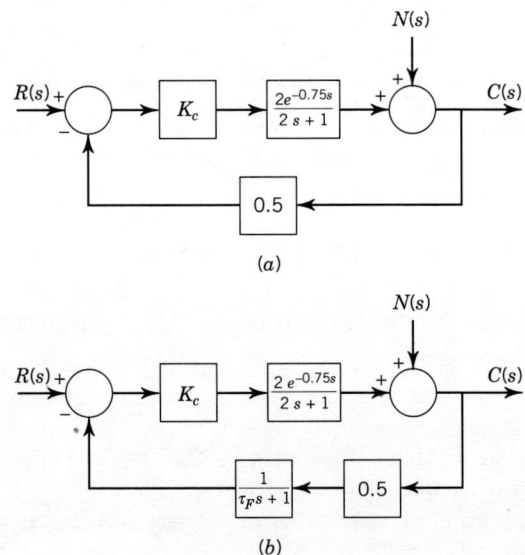

(a)

(b)

Figure P8-3 Block diagrams for Problem 8-16.

8-17. Consider a thermal process with the following transfer function for the process output versus controller output signal:

$$\frac{C(s)}{M(s)} = \frac{0.65e^{-0.35s}}{(5.1s + 1)(1.2s + 1)}$$

A sine wave of unity amplitude and a frequency of 0.80 rad/min is applied to the process (time constants and dead time are in minutes). Calculate the amplitude and phase lag of the sine wave out of the process (after the transient response dies out).

8-18. In Example 6-2.3 a feedback control loop with a first-order-plus-dead-time process was considered. The direct substitution method, coupled with an approximation of the dead-time term, resulted in approximate formulas for the ultimate gain and frequency of the loop. Compute the ultimate gain and frequency for that system using the frequency

response stability criterion, and compare your results with those of the approximate formulas. Use $t_0/\tau = 0.1, 0.2, 0.5, 1.0, 2.0$.

8-19. Using frequency response, verify the ultimate gains the and periods for the loops of Problem 6-8.

8-20. Using frequency response, verify the range of controller gains for which the closed loop is stable for the open-loop unstable process of Problem 6-10.

8-21. Using frequency response, verify the ultimate gain and ultimate period of the concentration loop of the three isothermal reactors in series of Problem 6-17.

8-22. Using frequency response, verify the ultimate gain and ultimate period of the compressor suction pressure control loop of Problem 6-18.

Chapter 9

Fundamentals and Benefits of Cascade Control

Chapters 5, 6, and 7 presented the design of feedback control. Feedback control is the simplest strategy of automatic process control that compensates for process upsets. However, the disadvantage of feedback control is that it only reacts after the process has been upset. That is, if a disturbance enters the process, it has to propagate through the process, make the controlled variable deviate from set point, and it is then that feedback takes corrective action. Thus, a deviation in the controlled variable is needed to initiate corrective action. Even with this disadvantage, probably 80 % of all control strategies used in industrial practice are simple feedback control. In these cases, the control performance provided by feedback is satisfactory for safety, product quality, and production rate.

As the processes requirements tighten, or in processes with slow dynamics, or in processes with too many, or frequently occurring upsets, the control performance provided by feedback control may become unacceptable. Thus, it is necessary to use other strategies to provide the required performance. These additional strategies are the subject of this, and some of the subsequent chapters. The strategies presented complement feedback control, not replace it. The reader must remember that it is always necessary to provide some feedback from the controlled variable.

Cascade control is a strategy that in some applications significantly improves the performance provided by feedback control. This strategy has been well known and used for a long time. This chapter explains in detail, the fundamentals and benefits of cascade control.

9-1 A PROCESS EXAMPLE

Consider the furnace/preheater and reactor process shown in Fig. 9-1.1. In this process the reaction A $\rightarrow$ B occurs in the reactor. Reactant A is usually available at a low temperature; therefore, it must be heated before being fed to the reactor. The reaction is exothermic, and to remove the heat of reaction, a cooling jacket surrounds the reactor.

The important controlled variable is the temperature in the reactor, T_R. The original control strategy called for controlling this temperature by manipulating the flow of coolant to the jacket. Manipulating the fuel valve controlled the inlet reactant temperature to the reactor. It was noticed during the start-up of this process that the cooling jacket could not provide the cooling capacity required. Thus, it was decided to open the cooling valve completely and control the reactor temperature by manipulating the fuel to the preheater, as shown in Fig. 9-1.1. This strategy worked well enough, providing automatic control during start-up.

Principles and Practice of Automatic Process Control/Third Edition, by C. A. Smith and A. B. Corripio
ISBN 0-471-66141-4   Copyright © 2006 John Wiley & Sons (Asia) Pte. Ltd.

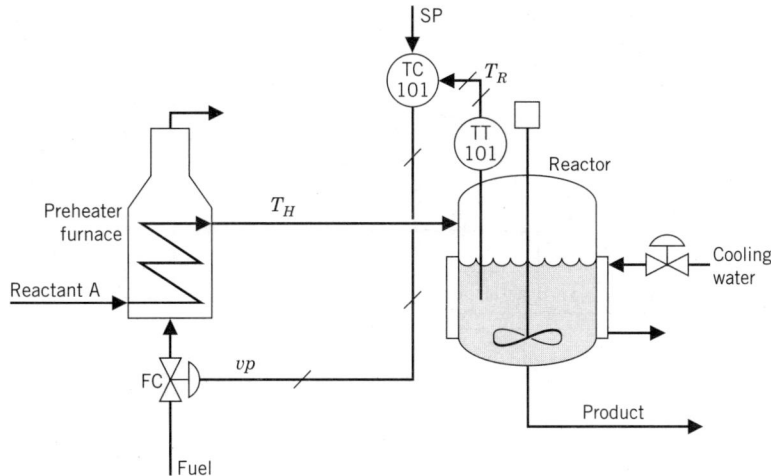

Figure 9-1.1 Preheater/reactor process—feedback control.

Once the process was "lined-out," the process engineer noticed that every so often the reactor temperature would move from set point enough to make off-spec product. After checking the feedback controller tuning to be sure that the performance obtained was the best possible, the engineer started to look for possible process disturbances. Several upsets were found around the reactor itself—cooling fluid temperature and flow variations—and others around the preheater—variations in inlet temperature of reactant A, in heating value of fuel, in inlet temperature of combustion air, etc. Furthermore, the engineer noticed that every once in a while the inlet reactant temperature to the heater would vary by as much as 25°C, certainly a major upset.

It is fairly simple to realize that the effect of an upset in the preheater results first in a change of the reactant exit temperature from the preheater, T_H, and then this affects the reactor temperature, T_R. Once the controller senses the error in T_R it manipulates the signal to the fuel valve. However, with so many lags in the process, preheater plus reactor, it may take a considerable amount of time to bring the reactor temperature back to set point. Due to these lags, the simple feedback control shown in the figure results in cycling, and in general sluggish control.

A superior control strategy can be designed by making use of the fact that the upsets in the preheater first affect T_H. Thus, it is logical to start manipulating the fuel valve as soon as a variation in T_H is sensed, before T_R starts to change. That is, the idea is not to wait for an error in T_R to start changing the manipulated variable. This corrective action uses an intermediate variable, T_H in this case, to reduce the effect of some dynamics in the process. This is the idea behind cascade control and it is shown in Fig. 9-1.2.

This strategy consists of two sensors, two transmitters, two controllers, and one control valve. One sensor measures the intermediate, or secondary, variable, T_H in this case, and the other sensor measures the primary controlled variable, T_R. Thus, this strategy results in two control loops: one loop controlling T_R, and the other loop controlling T_H. To repeat, the preheater exit temperature is used only as an intermediate variable to improve the control of the reactor temperature, which is the important controlled variable.

The strategy works as follows: controller TC-101 looks at the reactor temperature and decides how to manipulate the preheater outlet temperature to satisfy its set point. This decision is passed on to TC-102 in the form of a set point. TC-102 in turn

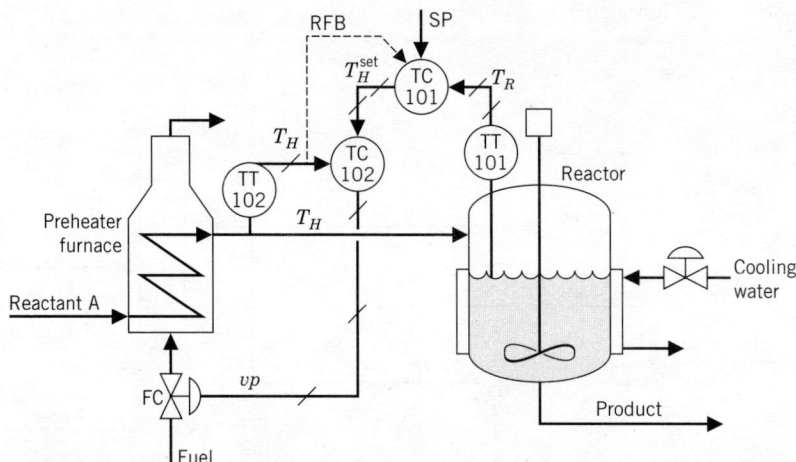

Figure 9-1.2 Preheater/reactor process—cascade control.

manipulates the signal to the fuel valve to maintain T_H at the set point given by TC-101. If one of the upsets mentioned earlier enters the preheater, T_H deviates from set point and TC-102 takes corrective action right away, before T_R changes. Thus, the dynamic elements of the process have been separated to compensate for upsets in the heater before they affect the primary controlled variable.

In general, the controller that keeps the primary variable at set point is referred to as the master controller, outer controller, or primary controller. The controller used to maintain the secondary variable at the set point provided by the master controller is usually referred to as the slave controller, inner controller, or secondary controller. The terminology primary/secondary is commonly preferred because for systems with more than two cascaded loops it extends naturally.

Note that the secondary controller receives a signal from the primary controller and this signal is used as the set point. To "listen" to this signal, the controller must be set in what is called *remote set point* or *cascade*. If one desires to set the set point manually, the controller must then be set in *local set point* or *auto*.

Figure 9-1.3 shows the response of the process to a $-25°C$ change in inlet reactant temperature, under simple feedback control, and under cascade control. The improvement is very significant and in all probability in this case pays for the added expenses in no time.

The following must be stressed: *in designing cascade control strategies, the most important consideration is that the secondary variable must respond faster to changes in the disturbance, and in the manipulated, than the primary variable does, the faster the better.* This requirement makes sense, and it is extended to any number of cascade loops. In a system with three cascaded loops, as shown in Section 9-3.2, the tertiary variable must be faster than the secondary variable and this variable in turn must be faster than the primary variable. Note that the most inner controller is the one that sends its output to the valve. The outputs of all other controllers are used as set points to other controllers; for these controllers their final control element is the set point of another controller.

As noted from this example, we are starting to develop more complex control schemes than simple feedback. It is helpful in developing these schemes, and others shown in the following chapters, to remember that every signal must have a physical

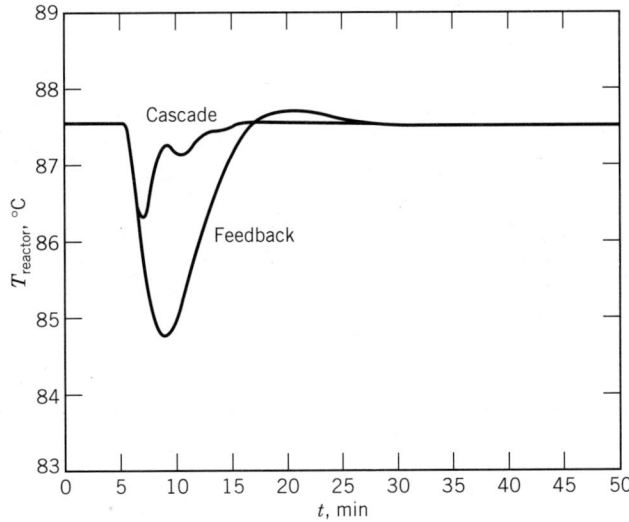

Figure 9-1.3 Response of feedback and cascade control to a $-25°C$ change in feed temperature to heater.

significance. In Figs. 9-1.1 and 9-1.2 we have labeled each signal with its significance. For example, in Fig. 9-1.2 the output signal from TT-101 indicates the temperature in the reactor, T_R, the output signal from TT-102 indicates the outlet temperature from the heater, T_H, and the output signal from TC-101 indicates the required temperature from the heater, T_H^{set}. Even though indicating the significance of the signals in control diagrams is not standard practice, we will continue to do so. This practice helps in understanding control schemes, and we again recommend the reader to do the same.

9-2 STABILITY CONSIDERATIONS

This section looks at how the implementation of cascade control affects the stability of the control system. Figure 9-2.1 shows the block diagram of the simple feedback control strategy shown in Fig. 9-1.1, and Fig. 9-2.2 shows the block diagram of the cascade strategy shown in Fig. 9-1.2. Simple transfer functions have been selected to represent the process.

The block diagram of Fig. 9-2.2 clearly shows why the secondary loop starts to compensate for any disturbance that affects the secondary controlled variable before

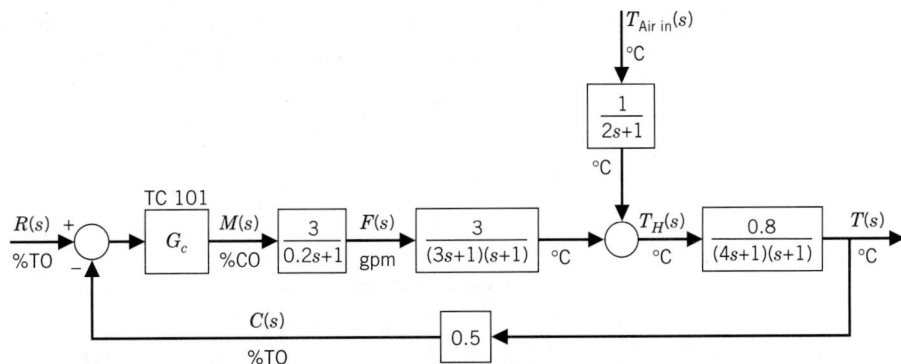

Figure 9-2.1 Block diagram of the process shown in 9-1.1.

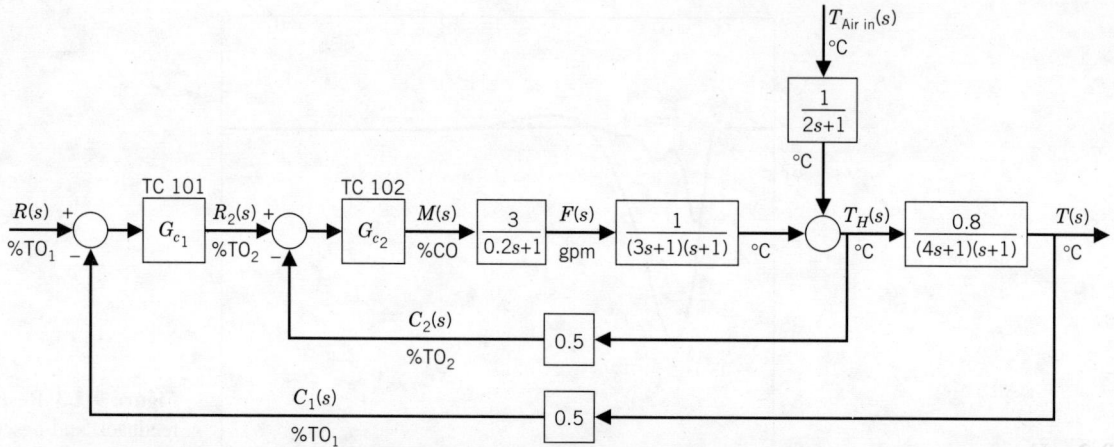

Figure 9-2.2 Block diagram of the process shown in 9-1.2.

its effect is felt by the primary controlled variable. The diagram also shows why the secondary loop is sometimes referred to as the "inner loop." This loop is imbedded inside the primary loop, or "outer loop."

The characteristic equation for the simple feedback control system, Fig. 9-2.1, is

$$1 + \frac{1.2\, G_{c_1}}{(0.2s+1)(3s+1)(s+1)(4s+1)(s+1)} = 0 \qquad (9\text{-}2.1)$$

Using the block diagram algebra techniques learned in Chapter 3 we obtain the characteristic equation for the cascade control strategy, Fig. 9-2.2, as

$$1 + \frac{\dfrac{1.2\, G_{c_1} G_{c_2}}{(0.2s+1)(3s+1)(s+1)(4s+1)(s+1)}}{1 + \dfrac{1.5\, G_{c_2}}{(0.2s+1)(3s+1)(s+1)}} = 0 \qquad (9\text{-}2.2)$$

Applying the direct substitution method, Chapter 6, or the frequency response technique, Chapter 8, to the feedback control system, Eq. 9-2.1, we can calculate the ultimate gain and ultimate frequency.

$$K_{cu} = 4.33\, \frac{\%CO}{\%TO} \qquad \text{and} \qquad \omega_u = 0.507\, \frac{\text{rad}}{\text{min}}$$

To determine the ultimate gain and frequency of the primary controller of the cascade strategy, the tuning of the secondary controller must first be obtained. This tuning can be obtained by determining the ultimate gain of the secondary loop in Fig. 9-2.2. The characteristic equation for this secondary process is

$$1 + \frac{1.5\, G_{c_2}}{(0.2s+1)(3s+1)(s+1)} = 0$$

and yields the following

$$K_{cu_2} = 17.06\, \frac{\%CO_2}{\%TO_2}$$

Assuming a proportional-only controller and using Ziegler–Nichols' suggestion, the tuning for the secondary controller becomes

$$K_{c_2} = 0.5 K_{cu_2} = 8.53\, \frac{\%CO_2}{\%TO_2}$$

When we use this tuning value for the secondary controller, the characteristic equation for the cascade strategy, Eq. 9-2.2, yields for the primary controller

$$K_{cu_1} = 7.2\,\frac{\%CO_1}{\%TO_1} \quad \text{and} \quad \omega_{u1} = 1.54\,\frac{\text{rad}}{\text{min}}$$

Comparing the results, we note that the cascade strategy yields a greater ultimate gain, or limit of stability, than the simple feedback control loop, 7.2 versus 4.33 %CO/%TO. The value of the ultimate frequency is also greater for the cascade strategy, 1.54 versus 0.57 rad/min, indicating faster process response.

The use of cascade control makes the overall control faster, and most times increases the ultimate gain of the primary controller. The methods of analysis are the same as for simple feedback loops.

9-3   IMPLEMENTATION AND TUNING OF CONTROLLERS

Two important questions still remain concerning how to put the cascade strategy into full automatic operation, and how to tune the controllers. The answer to both questions is the same: from inside out. That is, the most inner controller is first tuned and set into remote set point mode while the other loops are in manual. As this controller is set in remote set point, it is good practice to check how it performs, before proceeding to the next controller. This procedure continues outwardly until all controllers are in operation. For the process shown in Fig. 9-1.2, TC-102 is first tuned and set in automatic while TC-101 is in manual. The control performance of TC-102 is then checked before proceeding to TC-101. Very simply, varying the set point to TC-102 can usually do this checking. Remember, it is desired to make TC-102 as fast as possible, even if it oscillates a bit, to minimize the effect of the upsets. Once this is done, TC-102 is set in remote set point and TC-101 is tuned and set in automatic.

Tuning cascade control systems is more complex than simple feedback systems, if nothing else by the mere fact that there is more than one controller to tune. However, this does not mean it is difficult either. We first present the methods available to tune two-level cascade systems, and then continue by discussing the tuning methods available to tune three-level cascade systems.

9-3.1   Two-Level Cascade Systems

The control system shown in Fig. 9-1.2 is referred to as a two-level cascade system. Realize that the inner loop by itself is a simple feedback loop. Therefore, TC-102 can be tuned by any of the techniques discussed in Chapter 7. The recommendation, as previously mentioned, is to tune this controller as fast as possible—avoiding instability, of course. The objective is to make the inner loop fast and responsive to minimize the effect of upsets on the primary controlled variable. Tuning this system then reduces to tuning the primary controller.

There are several methods to obtain a first guess to the tuning of the primary controller. Trial and error if often used by experienced personnel. The other methods available follow a "recipe" to obtain the first tuning values. The first method available is the Ziegler–Nichols oscillatory technique presented in Chapter 7. That is, with the secondary controller in remote set point, after removing any integral or derivative action present in the primary controller, its gain is increased cautiously until the controlled variable oscillates with sustained oscillations. The controller gain that provides these oscillations is called the ultimate gain, K_{cu}, and the period of the oscillations is the ultimate period, T_u. The Ziegler–Nichols equations presented in Chapter 7 are then used.

The second method available is the one presented by Pressler (Reference 1). Pressler's method was developed assuming that the secondary controller is proportional only and that the primary controller is proportional integral; this P/PI combination is usually quite convenient. The method works well; however, it assumes that the inner loop does not contain dead time which limits its application to cascade systems with flow or liquid pressure loops as the inner loop. For processes with dead time in the inner loop, such as the one shown in Fig. 9-2.2, the application of Pressler's method would yield an unstable response if the master controller is ever set in manual.

The third method available is to extend the off-line methods presented in Chapter 7 to both the primary and secondary controllers. That is, with the secondary controller in manual, a step change in its output is introduced and the response of the temperature out of the heater (secondary variable) is recorded. From these data a gain, time constant, and dead time for the secondary loop are obtained and the controller tuned by whatever method presented in Chapter 7 the engineer desires. Once this is done, the secondary controller is set in the remote set point mode. With the primary controller in manual, a step change in its output is then introduced and the response of the reactor's temperature (primary variable) is recorded. From these data a gain, time constant, and dead time for the primary loop are obtained and the controller tuned by also whatever method presented in Chapter 7 the engineer desires.

The fourth method available to tune cascade systems is the one developed by Austin (1986), and extended by Lopcz (2003). The method provides a way to tune both the primary and secondary with only one step test. Tuning equations are provided for the primary controller, PI or PID, when the secondary controller is either P or PI. The method consists in generating a step change in signal to the control valve as explained in Chapter 7, and recording the response of the secondary and primary variables. The response of the secondary variable is used to calculate the gain, K_2 in %TT-102/%CO, time constant, τ_2, and dead time, t_{o_2}, of the inner loop. The response of the primary variable is used to calculate the gain, K_1 in %TT-101/%CO, time constant, τ_1, and dead time, t_{o_1}, of the primary loop. With the value of K_2, τ_2, and t_{o_2} the secondary controller is tuned using one of the methods of Chapter 7. Table 9-3.1 presents the equations to tune the primary controller. This method provides a simple procedure to obtain near-optimum tunings for the primary controller.

Figure 9-4.2a, discussed in Section 9-4, shows a temperature controller cascaded to a flow controller. Cascade systems with flow controllers in the inner loop are very common and thus, worthy of discussion. Following the previous presentation about the Lopez's method, after a change in the flow controller output is introduced, and the flow and temperature are recorded, the respective gains, time constants, and dead times can be obtained. Since flow loops are quite fast, the time constant will be in the order of seconds and the dead time very close to zero minutes, $t_{o_2} \approx 0$ min. Flow controllers are usually tuned with low gain, $K_c \approx 0.2$, and short reset time, $\tau_1 \approx 0.05$ min. However, in the process shown in Fig. 9-4.2a, the flow controller is the inner controller in a cascade system, and because it is desired to have a fast responding inner loop, the recommendation in this case is to increase the controller gain close to one, $K_c \approx 1.0$; to maintain stability the reset time may also have to be increased (Corripio, 1990). Once the flow controller has been tuned to provide fast and stable response, the temperature controller can be tuned following Lopez's guidelines. It is important to realize that t_{o_2} is not a factor in the equations and therefore, it will not have an effect in the tuning of the master controller.

Table 9-3.1   Tuning equations for two-level cascade system

Secondary controller	Primary controller	
	PI	PID
P	$K_{c_1} = \dfrac{\tau_1}{8.2048K_1t_{0_1}} \left(\dfrac{\tau_2}{\tau_1}\right)^{-1.3965} \left(\dfrac{t_{0_2}}{t_{0_1}}\right)^{0.2767}$	$K_{c_1} = \dfrac{\tau_1}{5.2416K_1t_{0_1}} \left(\dfrac{\tau_2}{\tau_1}\right)^{-1.1312} \left(\dfrac{t_{0_2}}{t_{0_1}}\right)^{-0.0966}$
	$\tau_{I_1} = \tau_1 \left(\dfrac{\tau_2}{\tau_1}\right)^{-0.0018} \left(\dfrac{t_{0_2}}{t_{0_1}}\right)^{0.2097}$	$\tau_{I_1} = \tau_1 \left(\dfrac{\tau_2}{\tau_1}\right)^{-0.0655} \left(\dfrac{t_{0_1}}{\tau_1}\right)^{0.5342}$
		$\tau_{D_1} = t_{0_1} \left(\dfrac{\tau_2}{\tau_1}\right)^{1.4397} \left(\dfrac{t_{0_1}}{\tau_1}\right)^{-1.6551}$
	Ranges: $0.1 \le \dfrac{\tau_2}{\tau_1} \le 0.7; 0.1 \le \dfrac{t_{0_2}}{t_{0_1}} \le 0.7$	Ranges: $\begin{array}{c} 0.2 \le \dfrac{t_{0_1}}{\tau_1} \le 1.0; 0.1 \le \dfrac{\tau_2}{\tau_1} \le 0.7; \\ 0.1 \le \dfrac{t_{0_2}}{t_{0_1}} \le 0.7 \end{array}$
PI	$K_{c_1} = \dfrac{1}{2.4468K_1} \left(\dfrac{t_{0_1}}{\tau_1}\right)^{-0.4485} \left(\dfrac{\tau_2}{\tau_1}\right)^{-0.3857} \left(\dfrac{t_{0_2}}{t_{0_1}}\right)^{-0.0995}$	$K_{c_1} = \dfrac{\tau_1}{5.2416K_1t_{0_1}} \left(\dfrac{\tau_2}{\tau_1}\right)^{-1.1312} \left(\dfrac{t_{0_2}}{t_{0_1}}\right)^{-0.0966}$
	$\tau_{I_1} = 0.8693\tau_1 \left(\dfrac{t_{0_1}}{\tau_1}\right)^{0.4195} \left(\dfrac{\tau_2}{\tau_1}\right)^{-0.3022} \left(\dfrac{t_{0_2}}{t_{0_1}}\right)^{-0.1334}$	$\tau_{I_1} = 1.1581\tau_1 \left(\dfrac{\tau_2}{\tau_1}\right)^{-0.0398} \left(\dfrac{t_{0_2}}{t_{0_1}}\right)^{0.1538}$
		$\tau_{D_1} = 0.6722t_{0_1} \left(\dfrac{\tau_2}{\tau_1}\right)^{-0.0905} \left(\dfrac{t_{0_2}}{t_{0_1}}\right)^{0.2750}$
	Ranges: $\begin{array}{c} 0.2 \le \dfrac{t_{0_1}}{\tau_1} \le 1.0; 0.1 \le \dfrac{\tau_2}{\tau_1} \le 0.7; \\ 0.1 \le \dfrac{t_{0_2}}{t_{0_1}} \le 0.7 \end{array}$	Ranges: $0.1 \le \dfrac{\tau_2}{\tau_1} \le 0.7; 0.1 \le \dfrac{t_{0_2}}{t_{0_1}} \le 0.7$

Another method to tune the cascade loop of a temperature controller cascaded to a flow controller, as in the heat exchanger, is to reduce the two-level cascade system to a simple feedback loop by realizing that the flow loop is very fast and thus just considering it as part of the valve, and therefore part of the process. This is done by first tuning the flow controller as previously explained and setting it in remote set point. Once this is done, the flow controller is receiving its set point from the temperature controller. Then introduce a step change from the temperature controller and record the temperature. From the recording calculate the gain, time constant, and dead time. With this information, tune the temperature controller by any of the methods presented in Chapter 7.

9-3.2   Three-Level Cascade Systems

Controller TC-102 in the cascade system shown in Fig. 9-1.2 manipulates the valve position to maintain the preheater outlet temperature at set point. The controller manipulates the valve position, not the fuel flow. The fuel flow depends on the valve position and on the pressure drop across the valve. A change in this pressure drop, a common

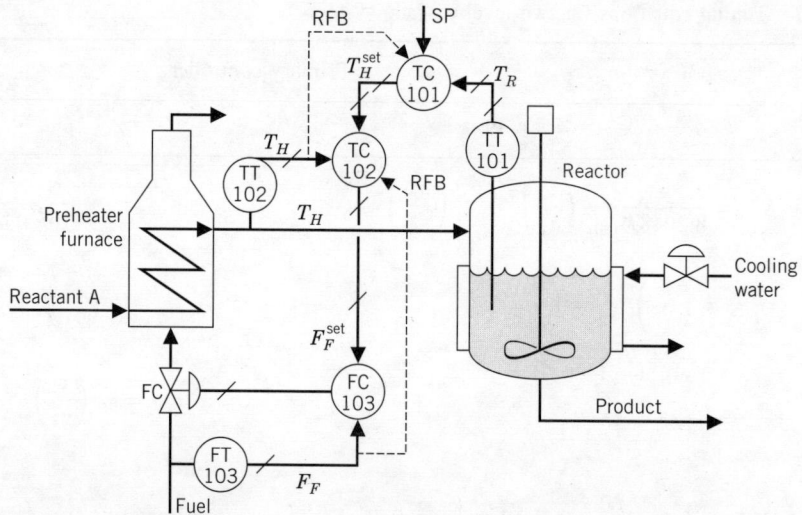

Figure 9-3.1 Preheater/reactor process—three-level cascade control.

upset, results in a change in fuel flow. The control system, as is, will react to this upset once the outlet preheater temperature deviates from set point. If it is important to minimize the effect of this upset, tighter control can be obtained by adding one extra level of cascade, as shown in Fig. 9-3.1. The fuel flow is then manipulated by TC-102, and a change in flow, due to pressure drop changes, would then be corrected immediately by FC-103. The effect of the upset on the outlet preheater temperature would be minimal.

In this new three-level cascade system, the most inner loop, the flow loop, is the fastest. Thus, the necessary requirement of decreasing loop speed from "inside out" is maintained.

To tune this three-level cascade system, first note that controllers FC-103 and TC-102 constitute a two-level cascade "subsystem," in which the inner controller is very fast. Furthermore, this is exactly the case just described at the end of the previous subsection. Following this discussion, the flow controller is first tuned and set in remote set point. Thus, the tuning of this three-level cascade system then reduces to tuning a two-level cascade system. Lopez's method is very easily applied. With TC-101 and TC-102 in manual and FC-103 in remote set point, introduce a step change in the signal from TC-102 to FC-103, and record the furnace and reactor temperatures responses. From the furnace temperature response, obtain the gain, K_2 in %TT-102/%CO; the time constant, τ_2; and the dead time, t_{o_2}. Using the reactor temperature response, obtain the gain, K_1 in %TT-101/%CO; the time constant, τ_1; and the dead time, t_{o_1}. With K_2, τ_2, and t_{o_2} tune the secondary controller using the equations presented in Chapter 7; then use Table 9-3.1.

9-4 OTHER PROCESS EXAMPLES

Consider the heat exchanger control system shown in Fig. 9-4.1 in which the outlet process fluid temperature is controlled by manipulating the steam valve position. The previous section discussed that the flow through any valve depends on the valve position *and* on the pressure drop across the valve. If a pressure surge in the steam pipe occurs,

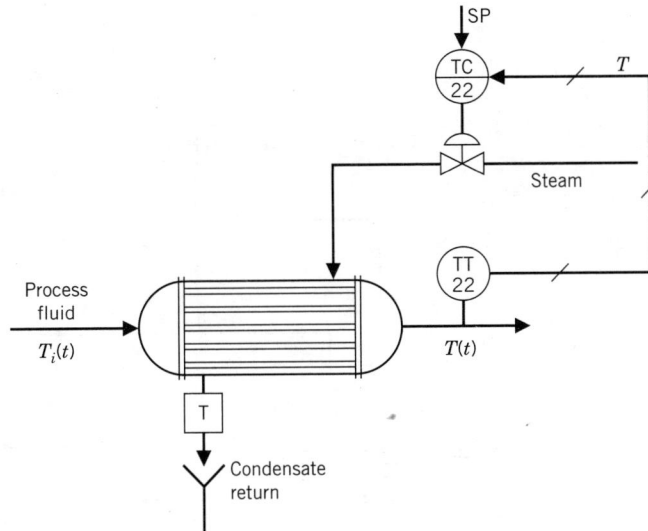

Figure 9-4.1 Heat exchanger temperature control loop.

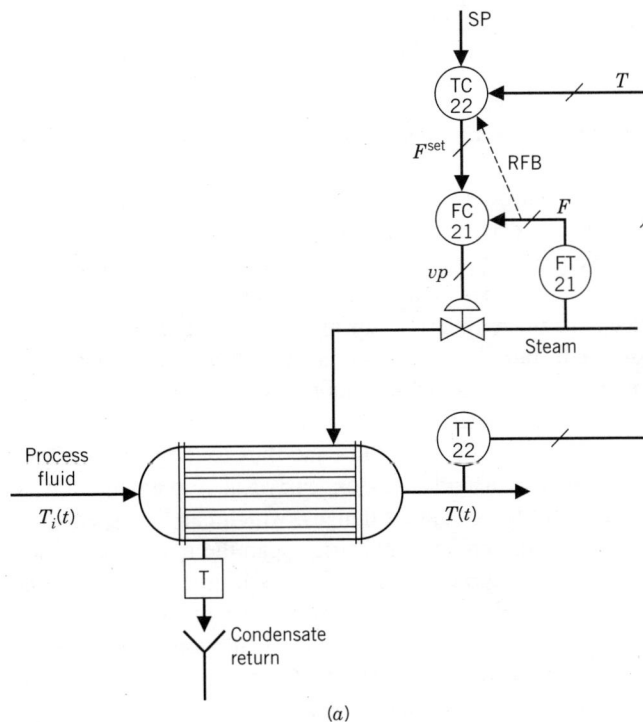

(a)

Figure 9-4.2 Cascade control schemes in heat exchanger temperature control.

the steam flow will change. The temperature control loop shown can compensate for this disturbance only after the process temperature has deviated away from set point.

Two cascade schemes that improve this temperature control, when steam pressure surges are important disturbances, are shown in Fig. 9-4.2. Figure 9-4.2a shows a cascade scheme in which a flow loop has been added; the temperature controller resets

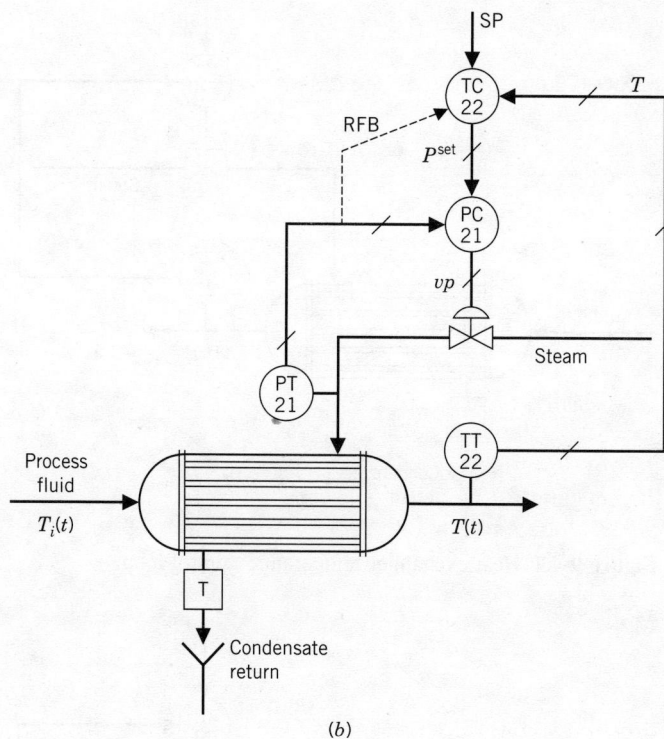

(b)

Figure 9-4.2 *(continued)*.

the flow controller set point. Any flow changes are now compensated by the flow loop. The cascade scheme shown in Fig. 9-4.2b accomplishes the same control but now the secondary variable is the steam pressure in the exchanger shell side. Actually, this steam pressure is usually measured in the line entering the shell; this is a safer and less expensive place than measuring the actual pressure in the shell. Any change in steam flow quite rapidly affects the shell side pressure, and it is then compensated by the pressure loop. This pressure loop also compensates for disturbances in the heat content (superheat and latent heat) of the steam, since the pressure in the shell side is related to the condensing temperature and thus to the heat transfer rate in the exchanger. This last scheme is usually less expensive in implementation since it does not require an orifice with its associated flanges, which can be expensive. Both cascade schemes are common in the process industries. Can the reader say which of the two schemes gives better initial response to disturbances in inlet process temperature $T_i(t)$?

The cascade control systems shown in Fig. 9-4.2 are very common in industrial practice. A typical application is in distillation columns where temperature is controlled to maintain the desired split. The temperature controller is often cascaded to the steam flow to the reboiler or the coolant flow to the condenser.

Finally, another very simple example of a cascade control system is that of a positioner on a control valve. The positioner acts as the inner controller of the cascade scheme.

9-5 FINAL COMMENTS

So far no comments have been made regarding the action of the controllers in a cascade strategy. This is important because, as learned in Chapter 5, if the actions are not correctly chosen, the controllers will not control. The procedure to choose the action is the same as explained in Chapter 5. That is, the action is decided by process requirements and the fail-safe action of the final control element. As previously mentioned, for some of the controllers in the cascade strategy, the final control element is the set point of another controller.

Consider the three-level cascade strategy shown in Fig. 9-3.1. The action of FC-103 is reverse (increase/decrease) because if the flow measurement increases above set point, indicating that more flow than required is being delivered by the valve, the valve opening must be reduced, and for a fail-closed valve this is accomplished by reducing the signal to it. The action of TC-102 is also reverse because if its measurement increases above set point, indicating a higher outlet preheater temperature than required, the fuel flow must be reduced, and this is accomplished by reducing the set point to FC-102. Finally, the action of TC-101 is also reverse because if its measurement increases above set point, indicating a higher reactor temperature than required, the way to reduce it is by lowering the inlet reactants temperature, which is accomplished by reducing the set point to TC-102. The decision regarding the controller action is simple and easy as long as we understand the significance of what each controller is doing.

Considering Fig. 9-1.2, the output from TC-101 is a signal, meaning 4–20 mA or 3–15 psig or, in general, 0–100 %. Then for a given output signal from TC-101, say 40 %, what is the temperature, in degrees, required from TC-102? This question is easy to answer by remembering that the job of the controller is to make its measurement equal to set point. Therefore, TC-102 will be satisfied when the signal from TT-102 is 40 %. Thus, the required temperature is 40 % of the range of TT-102.

Considering Fig. 9-1.2 again, it is important to realize what were to happen if TC-102 were taken off remote set point operation while leaving TC-101 in automatic. If this is done, and if TC-101 senses an error it would send a new signal (set point) to TC-102, however, TC-102 would be unable to respond to requests from TC-101. If TC-101 has reset action, it would windup since its output would have no effect in its input. That is, the effect of taking the secondary controller off remote set point is to "open" the feedback loop of the primary controller.

Computers, with their inherit flexibility, offer the necessary capabilities to avoid this windup possibility and thus provide for a safer cascade strategy. The computer can be programmed, or configured, so that at any time the secondary controller is taken off remote set point operation, the primary controller "automatically" goes into the manual mode if it is in automatic. The primary controller remains in manual as long as the secondary controller remains off remote set point. When the secondary controller is returned to remote set point, the primary controller could then return "automatically" to the automatic mode if the designer desires it. However, if while the secondary controller is off remote set point its set point changes, then at the moment it is returned to remote set point mode its present set point may not be equal to the output of the primary controller. If this occurs, the set point of the secondary controller will immediately jump to equal the output of the primary controller, thus generating a "bump" in the process operation. If a "bumpless" transfer is desired, most computer-based controllers can also be programmed so that while the secondary controller is off remote set point, the output from the primary controller is forced to equal either the process variable or the set point of the secondary controller. That is, the output from the primary controller "tracks" either variable of the secondary controller. Thus, when the secondary controller is returned to remote set point operation a smooth transfer is obtained.

The tracking option just explained, often referred to as *output tracking*, or *reset feedback*, (RFB), or "external reset feedback," is very important for the smooth and safe operation of cascade control systems. We represent this option by the dashed lines in Fig. 9-1.2.

9-6 SUMMARY

This chapter has presented in detail the fundamentals and benefits of cascade control, which is a simple strategy in concept and implementation that provides improved control performance. The reader must remember that the secondary variable must respond faster to changes in the manipulated variable than the primary variable! Typical two-level cascaded loops are temperature-to-flow, concentration-to-flow, pressure-to-flow, level-to-flow, and temperature-to-pressure.

REFERENCES

1. PRESSLER, GERHARD, Regelungs-Technik, Hochschultashenbucher, Band 63, Bibliographischer Institut, Mannheim, West Germany.

2. AUSTIN, VANESSA D., Development of Tuning Relations for Cascade Control Systems. Ph.D. Dissertation, Department of Chemical Engineering, University of South Florida, Tampa, FL, 1986.

3. LOPEZ, ROYMAN J., Desarrollo de Ecuaciones Robustas de Sintonizacion de Controladores PID Trabajando en Cascada Para Procesos Industriales Autorregulados. M.S. Thesis, Universidad del Norte, Barranquilla, Colombia, 2003.

4. CORRIPIO, A. B., *Tuning of Industrial Control Systems*. North Carolina: Instrument Society of America, 1990.

PROBLEMS

9-1. For the paper-drying process of Fig. P9-1, the following information is available: The flow control loop (FIC-47) can be represented by a first-order lag with a gain of 4 gpm/%CO and a time constant of 0.1 min. The transfer function of the air heater outlet temperature to the fuel flow is a second-order lag with time constants of 2 min and 0.8 min.

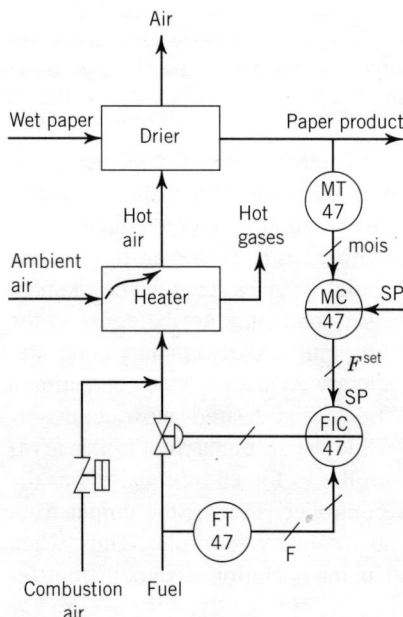

Figure P9-1 Drier for Problem 9-1.

A change in fuel flow of 1 gpm causes a change of $2°F$ in the outlet air temperature. The drier can be represented by a first-order lag with a time constant of 5 min. A change of $1°F$ in inlet air temperature causes a change in outlet moisture of 0.5 mass percent. The moisture transmitter (MT-47) has a range of 0 to 6 mass percent and a negligible time constant.

(a) For the control scheme shown in Fig. P9-1, draw the block diagram of the moisture control loop showing the transfer functions. Decide on the fail position of the control valve and the controller action, and make sure that the signs in your block diagram correspond to your decisions. Determine the ultimate gain and period of oscillation of the moisture control loop, then use these values to tune a PID controller for the moisture controller (MIC-47) for quarter decay response.

(b) Consider a cascade control scheme using an outlet temperature sensor with a range of 250 to $300°F$ installed on the air line from the heater to the drier and a temperature controller to manipulate the fuel flow set point; the output of the moisture controller (MIC-47) sets the set point of the outlet air temperature controller. Draw the instrumentation diagram and the block diagram of the cascade control scheme. Show the defined transfer functions on the block diagram and specify the action of each controller.

(c) Determine the ultimate gain and period of oscillation of the slave temperature control loop and use these values to tune a P controller for quarter decay response; then calculate the ultimate gain and period for the master moisture control loop, and the quarter decay response tuning parameters for a PID moisture controller. Comparing these parameters with those obtained in part (a), briefly comment on the advantages of cascade control for this application.

Note: The simulation of this process is the subject of Problem 13-25.

9-2. This is a real process with real data. Consider the heater shown in Fig. P9-2. In this process, the heater is actually a reactor where the catalytic reaction of C_3H_8 and steam takes place to form H_2 and CO_2. The temperature of the product gas is controlled by manipulating the flow of fuel. It was suspected that the controllers were not correctly tuned. This suspicion was confirmed looking at the unstable response, shown in Fig. P9-3, of the reactor after an upset, a change in feed flow, entered the reactor. Based on this response it was decided to retune the controllers. Both controllers were set in manual, and after the process reached a steady condition, shown in Fig. P9-3, the output from the

Table P9-1   Table for Problem 9-2

Time, s	Flow, mscfh	Time, s	Flow, mscfh
0	19.46	20	17.01
5	17.51	25	16.98
10	17.32	⋮	⋮
15	17.08	45	16.49

Time, min	T, °F	Time, min	T, °F	Time, min	T, °F
0	1463	24	1393	46	1293
2	1463	26	1382	48	1287
4	1463	28	1372	50	1281
6	1462	30	1361	52	1275
8	1461	32	1351	54	1275
10	1457	34	1341	56	1263
12	1452	36	1332	58	1258
14	1444	38	1324	⋮	⋮
16	1435	40	1316		
18	1426	42	1308		
20	1415	44	1301	94	1235
22	1405				

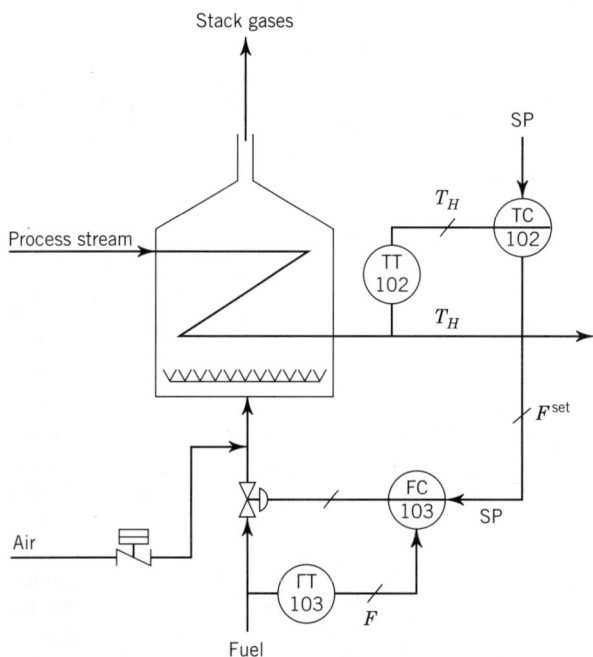

Figure P9-2 Heater for Problem 9-2.

flow controller to the valve was changed by -5 %. The flow and temperature responses are given in Table P9-1. Interestingly, the plant personnel were expecting a 20°F change in temperature, however, the actual change was about 230°F. FigureP9-3 also shows this temperature response. After the temperature reached bottom, the flow controller output was changed by $+5$ % to bring the temperature back to its desired operating condition. Based on the process response both controllers were tuned and set in automatic. Figure P9-3 shows the response under automatic control, with the new settings and for the same upset. Obtain the settings for the controllers using the method outlined in this chapter.

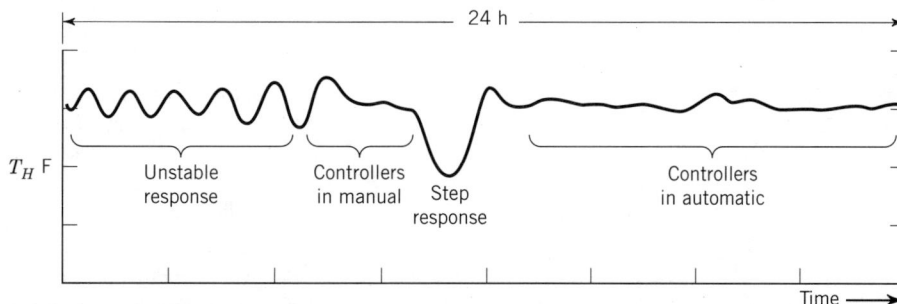

Figure P9-3 Temperature response for Problem 9-2.

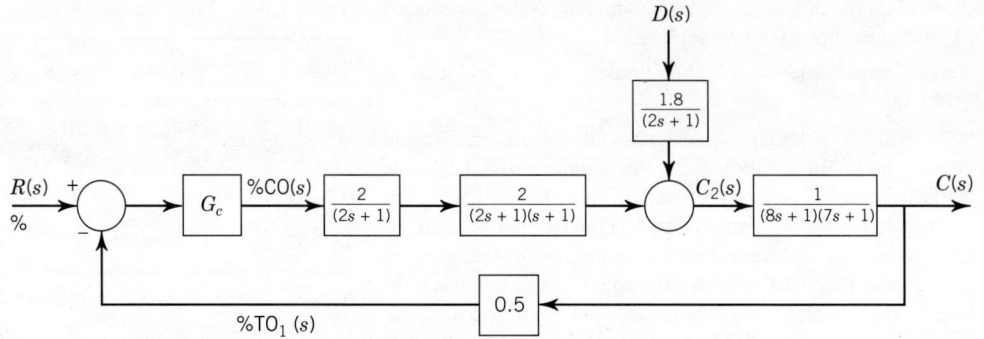

Figure P9-4 Block diagram for Problem 9-3.

The temperature transmitter range is 0 to 2000°F, and that of the flow transmitter is 0 to 24 mscfh.

9-3. Figure P9-4 shows the block diagram of a feedback control system. The control engineer in charge of the process decided that a cascade control system could improve the control performance. The proposed cascade scheme consists in measuring $C_2(s)$ with a transmitter with a gain of 0.5 and sending the signal to a controller (the slave controller). The controller shown in Fig. P9-4 becomes the master controller. Compare the stability of both systems, that is, obtain the ultimate gain and ultimate period for both systems. Assume the slave controller is proportional only, and tune it by the Ziegler–Nichols method.

Note: The simulation of this process is the subject of Problem 13-26.

9-4. Consider the jacketed continuous stirred tank reactor (CSTR) sketched in Fig. P9-5. The following information is obtained from testing the reactor and its control system: The transfer function of the reactor temperature to the jacket temperature is a first-order lag with a gain of 0.6°C/°C and a

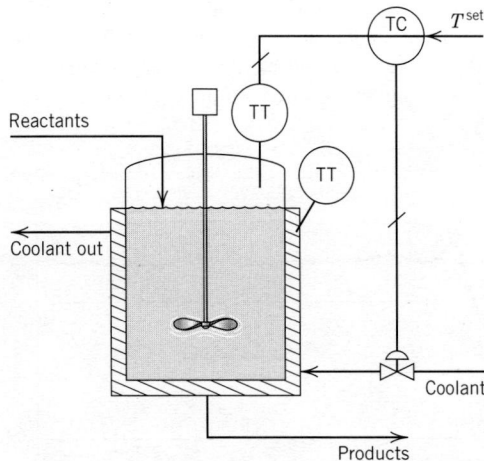

Figure P9-5 Jacketed reactor for Problem 9-4.

time constant of 13 min. The transfer function of the jacket temperature to the coolant flow is a first-order lag with a gain of −2.0°C/(kg/s) and a time constant of 2.5 min. The control valve is linear with constant pressure drop and is sized to pass 12 kg/s when fully opened. Its time constant is negligible. The reactor temperature transmitter is calibrated for a range of 50 to 100°C and has a time constant of 1 min. The jacket temperature transmitter is calibrated for a range of 0 to 100°C, and its time constant is negligible.

(a) Decide on the proper fail position of the control valve and the action of the controller for a simple feedback control loop with the reactor temperature controller manipulating the position of the coolant valve. Draw the block diagram showing all transfer functions and write the closed-loop transfer function of the reactor temperature to its set point. Pay particular attention to the signs which must correspond to the fail position of the valve and the controller action.

(b) Write the characteristic equation for the single feedback loop and calculate its ultimate gain and period by direct substitution.

(c) Design a cascade control system for the reactor temperature with the jacket temperature as the intermediate process variable, specifying the action of both controllers. Draw the complete block diagram for the cascade control system showing all transfer functions and their signs.

(d) Assuming a proportional slave controller with a gain of 2 %/%, write the transfer function for the jacket temperature loop and redraw the block diagram with the jacket temperature loop as a single block.

(e) Using the simplified block diagram from the previous part, write the characteristic equation of the reactor temperature loop in the cascade control system and calculate the ultimate gain and period of the loop by direct substitution.
Note: The simulation of this process is the subject of Problem 13-27.

9-5. The diagram for a reactor temperature controller cascaded to a coolant flow controller is shown in Fig. P9-6. The control valve is linear with constant pressure drop and

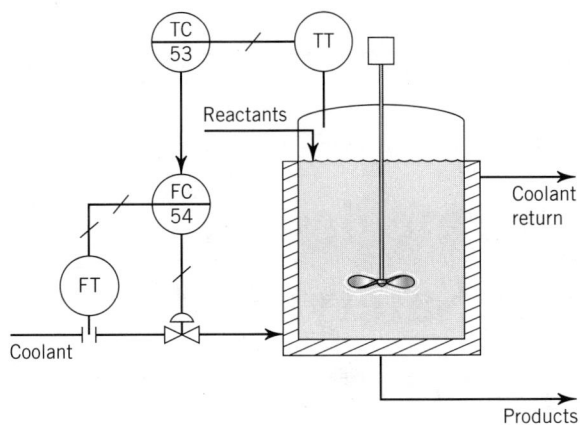

Figure P9-6 Jacketed reactor for Problem 9-5.

sized for a maximum flow of 500 gpm (gallons per minute); its time constant is 0.2 min. The flow transmitter (FT) has a range of 0 to 500 gpm and negligible time constant. The

flow controller (FC) is proportional integral (PI) with a gain of 1.0 %CO/%TO and the integral time set equal to the valve time constant. The transfer function of the reactor temperature to the coolant flow is first-order with a gain of $-2°F/gpm$ and a time constant of 5.0 min. The temperature transmitter (TT) has a range of 160 to 200°F and a time constant of 0.5 min.

(a) Draw the block diagram for the cascade control loop showing the transfer function of each device. Pay particular attention to the valve fail position and the action (direct or reverse) of the flow and temperature controllers, and show the appropriate signs for each of their transfer functions.

(b) Calculate the ultimate gain and period of the temperature control loop.

(c) Repeat parts (a) and (b) assuming the flow controller is removed and the temperature controller directly manipulates the valve. *Caution*: the controller action may change.

Chapter 10

Some Control Strategies for Productivity and Safety: Ratio, Selective, and Override Control

In Chapter 9, we began our presentation of control strategies that enhance simple feedback control to provide improved control performance. Specifically, Chapter 9 presented cascade control. The present chapter continues this presentation with three other strategies: *ratio, override*, and *selective control*; override control is also sometimes referred to as *constraint control*. Ratio control is commonly used to maintain two or more streams in a prescribed ratio. Override and selective control are usually implemented for safety and optimization considerations. These two strategies often deal with multiple control objectives (controlled variables) and a single manipulated variable. Up to now we have only dealt with processes with one control objective. The chapter begins with a presentation of the computer-based software and algorithms available for the implementation of these and other controls strategies.

10-1 SIGNALS, SOFTWARE, AND COMPUTING ALGORITHMS

Many of the control strategies presented in this and subsequent chapters require some amount of computing power. That is, many of these strategies require the multiplication, division, addition, subtraction, etc. of different signals. Several years ago these calculations were implemented with analog instrumentation. Computers allow for a simpler, more flexible, more accurate, more reliable, and less expensive implementation of these functions.

The computer systems used for process control are microprocessor-based. These systems are referred to as Distributed Control Systems (DCSs) because they use more than one microprocessor in the implementation of the control strategies. That is, instead of having a single processing unit (CPU) implementing all the strategies, these are distributed among several microprocessors; this distribution provides increased reliability.

10-1.1 Signals

There are two different ways field signals are handled once they enter a DCS. The first way is to convert the signal received into a number with engineering units. For example, if a signal is read from a temperature transmitter, the number kept in memory by the computer is the temperature in degrees. The computer is given the low value of the range and the span of the transmitter, and with this information it converts the

Principles and Practice of Automatic Process Control/Third Edition, by C. A. Smith and A. B. Corripio
ISBN 0-471-66141-4   Copyright © 2006 John Wiley & Sons (Asia) Pte. Ltd.

raw signal from the field into a number in engineering units. A possible command in the DCS to read a certain input is

Variable = AIN (input channel #, low value of range, span of transmitter)

or

T = AIN (3,50,100)

This command instructs the DCS to read an analog input signal (AIN) in channel 3, it tells the DCS that the signal comes from a transmitter with a low value of 50 and a span of 100, and to assign the name T to the variable read (possibly a temperature from a transmitter with a range of 50 to 150°C). If the signal read had been 60 %, or 0.6, then T = 110°C.

The second way of handling signals, and fortunately the least common, is not by converting them to engineering units but rather, by keeping them as a percentage, or fraction, of the span. In this case the input command is something as

Variable = AIN (input channel)

or

T = AIN(3)

and the result, for the same previous example, is T = 0.6.

During the presentation in earlier chapters we mentioned that instruments and controllers use signals to convey information and that, therefore, every signal has a "physical significance." That is, every signal used in the control scheme has some "meaning," or in other words, it is related to a meaning that makes sense from an engineering point of view. Signals are in percent but percent of "what" (pressure, temperature, flow, etc.)? This "what" is the meaning of the signal. It is now important to stress again this fact as we embark in the design of complex strategies to improve the control performance.

As already mentioned in this chapter, the new strategies frequently require the manipulation of signals in order to calculate controlled variables, set points, or to decide on control actions. To correctly perform these calculations it is important to understand the significance of the signals.

To help keep all of the information in order, and to understand the calculations, we will indicate next to each signal its significance and direction of information flow (see Fig. 9-1.2). This practice is not common in industry, but it helps in learning and understanding the subject.

10-1.2 Programming

There are two ways to program the mathematical manipulations in DCS systems: "block-oriented" programming and "software-oriented" programming.

Block-Oriented Programming

Block-oriented programming is software in a "subroutine-type" form, which are referred to as computing algorithms, or computing blocks. Each block performs a specified mathematical manipulation. Thus, to develop a control strategy, these computing blocks are linked together, the output of one block being the input to another block. This "linking," or programming, procedure is often referred to as *configuring* the control system.

Some typical calculations performed by computing blocks are

1. *Addition/subtraction.* Adding and/or subtracting the input signals obtain the output signal.

2. *Multiplication/division.* Multiplying and/or dividing the input signals obtain the output signal.

3. *Square root.* Extracting the square root of the input signal obtains the output signal.

4. *High/low selector.* The output signal is the highest/lowest of two or more input signals.

5. *High/low limiter.* The output signal is the input signal limited to some preset high/low limit value.

6. *Function generator, or signal characterization.* The output signal is a function of the input signal. The function is defined by configuring x and y coordinates.

7. *Integrator.* The output signal is the time integral of the input signal. The industrial term for integrator is "totalizer."

8. *Lead/lag.* The output signal is the response of the transfer function

$$\text{Output} = \left[\frac{\tau_{ld}s + 1}{\tau_{lg}s + 1} \right] \text{Input}$$

This calculation is often used in control schemes, such as feedforward, where dynamic compensation is required.

9. *Filter.* The output signal is the response of the transfer function given below. This calculation is frequently used to filter noisy signals.

$$\text{Output} = \left[\frac{1}{\tau s + 1} \right] \text{Input}$$

10. *Dead time.* The output signal is equal to a delayed input signal. This calculation is very easily done with computers; however, it is extremely difficult to do with analog instrumentation.

Table 10-1.1 shows the notation and algorithms we will use in this book for the mathematical calculations. Often, these blocks are graphically linked together using standard "drag-and-drop" technology.

Software-Oriented Programming

Manufacturers have developed their own programming language; however, all these languages are similar and resemble Fortran, Basic, or C languages. Table 10-1.2 presents the programming language we will use in this book; this language is similar to those used by different manufacturers.

Of particular interest is the PID controller. In DCSs that work in engineering units the range of the transmitter providing the controlled variable must be supplied to the PID controller (there are different ways to do so). With this information, the controller converts both the variable and the set point to percent values before applying the PID algorithm. This is done because the error is calculated in %TO. Remember, the K_C units are %CO/%TO. Thus, the controller output is then %CO. A possible way to "call" a PID controller is

OUT = PID (controlled variable, set point, low value of range, span of transmitter)

or,

$$\text{OUT} = \text{PID (T, 75, 50, 100)}$$

Table 10-1.1   Computing blocks

OUT = output from block

I_1, I_2, I_3 = input to blocks

K_0, K_1, K_2, K_3 = constants that are used to multiply each input

B_0, B_1, B_2, B_3 = constants

Summer

$$OUT = K_1 I_1 + K_2 I_2 + K_3 I_3 + B_0$$

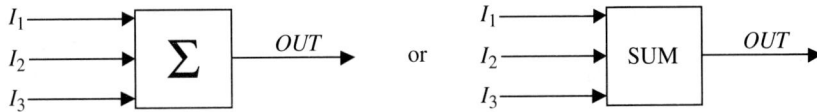

Multiplier/Divider

$$OUT = \frac{K_0(K_1 I_1 + B_1)(K_2 I_2 + B_2)}{K_3 I_3 + B_3} + B_0$$

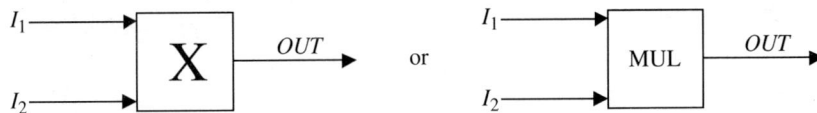

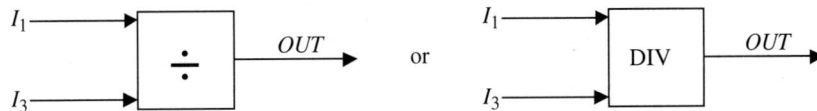

Square Root

$$OUT = K_1 \sqrt{I_1}$$

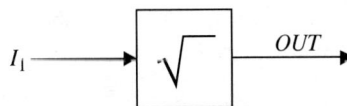

Lead/Lag

$$OUT = \frac{K_0(\tau_{ld}s + 1)}{(\tau_{lg}s_1 + 1)} I_1$$

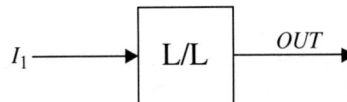

Selector

    OUT = maximum of inputs I_1, I_2, I_3

    OUT = minimum of inputs I_1, I_2, I_3

(continued overleaf)

Table 10-1.1 (*continued*)

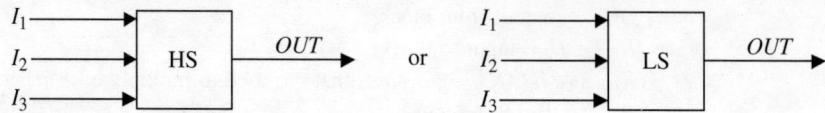

I₁ ─→ [HS] ─OUT→   or   I₁ ─→ [LS] ─OUT→
I₂ ─→
I₃ ─→

I₁ ─→ [HS], OUT →   or   I₁ ─→ [LS], OUT →
I₂ ─→   I₂ ─→
I₃ ─→   I₃ ─→

Dead Time

OUT = input I_1 delayed by t_0

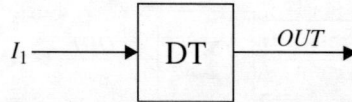

I_1 ─────→ [DT] ──OUT──→

Integrator

OUT = integrated, or totalized, value of I_1

I_1 ─────→ [TOT] ──OUT──→

Table 10-1.2 Programming language

Input/Output
AIN = analog in;   AOUT = analog out
Format:
In variable = AIN(input channel #, low value of range, span of transmitter)
"In variable" will be returned in engineering units.

AOUT(output channel #, out variable)
"out variable" must be in %.

Mathematical Symbols
$+, -, *, \wedge, /, <, >, =$

Statements
GOTO
IF/THEN/ELSE

Controller
Output = PID(variable, set point, low value of range of variable, span of variable).
"Output" will be return in %.
Every term in the PID argument must be in engineering units.

Lead/Lag

$$\text{Output} = K \left[\frac{\tau_{ld}s + 1}{\tau_{lg}s + 1} \right] \text{Input}$$

Table 10-1.2   (*continued*)

K = gain

τ_{ld} = lead time constant, in minutes

τ_{lg} = lag time constant, in minutes

Output = LdLg(input variable, lead time constant, lag time constant)

"Output" is in the same units as the input variable.

Lead time constant, and lag time constant in minutes.

Filter

Use same command as the lead/lag but set lead time constant = 0 min.

Integrator, or Totalizer

Output = TOT(input variable)

"Output" is in the integrated units of the input variable.

Dead Time

Output = DT(input variable, dead time)

"Output" is in the same units as the input variable.

Dead time is in minutes.

Comments

To insert a comment in any line use a semicolon followed by the comment.

This command instructs the DCS to control a variable T at 75 (degrees) that a transmitter supplies with a range from 50 to 150 (degrees). The controller output (OUT) is in percent (%CO). Manufacturers offer different ways to input the tuning parameters and other controller information, such as the action, tracking information, and the reset windup protection limits. We will not address these ways here, but the reader should remember that this information must be provided to the DCS to complete the programming of the controller.

It is instructive to complete this introduction to software programming by showing two different control loops; we will assume in both examples that the DCS works in engineering units. Consider first the process shown in Fig. 10-1.1 where two components, pure A and water, are mixed. It is desired to control the concentration of component A in the exiting mixture at a value of 0.2 lbmole/gal. There is an analyzer transmitter with a range of 0.05 to 0.5 lbmole/gal. The program to implement this loop is shown in Fig. 10-1.2.

As a second example consider the cascade strategy shown in Fig. 9-1.2. Assume that the range for TT101 is 320–360°C, and that the range for TT102 is 150–200°C. The program is shown in Fig. 10-1.3.

10-1.3   Scaling Computing Algorithms

The second way to handle field signals necessitates additional calculations before the required mathematical manipulations are performed. We will first explain the necessity and meaning of these additional calculations.

Consider the tank shown in Fig. 10-1.4 where temperature transmitters with different ranges measure temperatures at three different locations in the tank. The figure

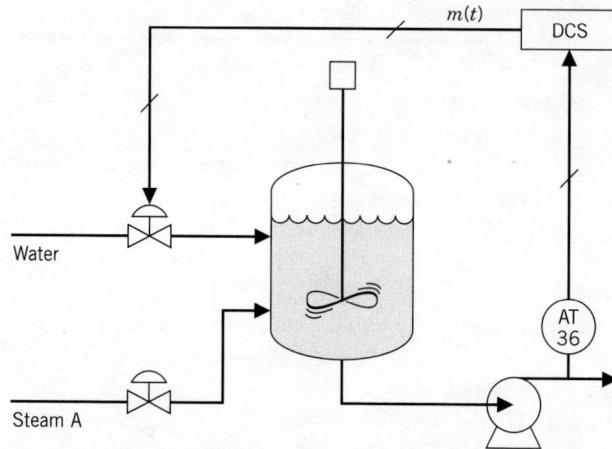

Figure 10-1.1 Mixing process.

```
CA = AIN (4,0.05,0.45)          ; reads, in input channel 4, in the
                                ; concentration of component A
CASET = 0.02                    ; set point for the concentration of A
m = PID (CA,CASET,0.05,0.45)    ; concentration controller
AOUT (1,m)                      ; outputs, in output channel 1, signal to valve
```

Figure 10-1.2 Software program to implement the control loop of Fig 10-1.1.

```
TR = AIN (12,320,40)            ; reads, in input channel 12, the temperature in
                                ; the reactor
TH = AIN (13,150,50)            ; reads, in input channel 13, the temperature out
                                ; the heater
TRSET = 341                     ; set point for temperature in reactor
M1 = PID (TR,TRSET,320,40)      ; output of master controller, %, TC101
THSET = 150 + M1*50/100         ; converts output of master controller in % to set
                                ; point of slave controller in °C
m = PID (TH,THSET,150,50)       ; output of slave controller, %, TC102
AOUT = (5,m)                    ; outputs, in output channel 5, signal to valve
```

Figure 10-1.3 Software program to implement the cascade control strategy of Fig 9-1.2.

shows the transmitter ranges and the steady-state values of each temperature, which are at mid-value of each range. It is desired to compute the average temperature in the tank. This computation is straightforward for the control system that reads each signal and converts it to engineering units: the three values are added together and divided by three; the program in Fig. 10-1.5 does just this. The first three lines read in the temperature, called T101, T102, and T103, and the fourth statement calculates the average temperature, TAVG.

For control systems that treat each signal as a percent of span this simple computation would result in an answer without much significance; Fig. 10-1.6 shows this program. That is, because each signal is 50 % of its range, the computation result is also 50 %. However, 50 % of what range? How do we translate this answer into a temperature? Furthermore, note that even though every input signal is 50 %, their

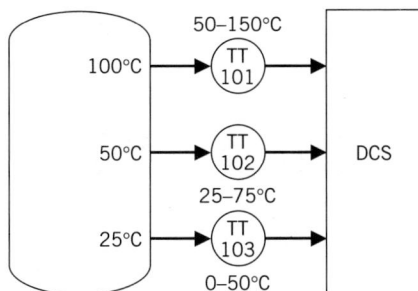

Figure 10-1.4 Tank with three temperature transmitters.

```
1   T101=AIN(1,50,100)  ; reads in T101
2   T102=AIN(2,25,50)   ; reads in T102
3   T103=AIN(3,0,50)    ; reads in T103
4   TAVG=(T101+T102+T103)/3 ; calculates average
```

Figure 10-1.5 Program to read in temperatures, in engineering units, and calculate average temperature.

```
1   T101=AIN(1)   ; reads in T101
2   T102=AIN(2)   ; reads in T102
3   T103=AIN(3)   ; reads in T103
4   TAVG=(T101+T102+T103)/3 ; calculates average
```

Figure 10-1.6 Program to read in temperatures, in percent of span, and calculate average temperature.

measured temperatures are different because the ranges are different. All of this indicates that for the computation to "make sense" the range of each input signal, and a chosen range for the output variable, must be considered. Considering each range ensures compatibility between input and output signals. The procedure to consider the range of each signal is referred to as "scaling." Fortunately not many DCSs handle signals in percent of span; however, for those that do, the previous edition of this book, Reference 1, presents the method to scale the computations.

10-2   RATIO CONTROL

A commonly used process control technique is ratio control. *Ratio control* is a strategy wherein one variable is manipulated to keep it as a ratio or proportion of another. This section presents two industrial examples to illustrate its meaning and implementation.

EXAMPLE 10-2.1

This first example is realistic and simple, and clearly demonstrates the need for ratio control. Assume that it is required to blend two liquid streams, A and B, in some proportion, or ratio, R; the process is shown in Fig. 10-2.1. That is,

$$R = F_B/F_A$$

where F_A and F_B are the flow rates of streams A and B, respectively.

An easy way of accomplishing this task is shown in the figure. Each stream is controlled by a flow loop in which the set points to the controllers are set such that the liquids are

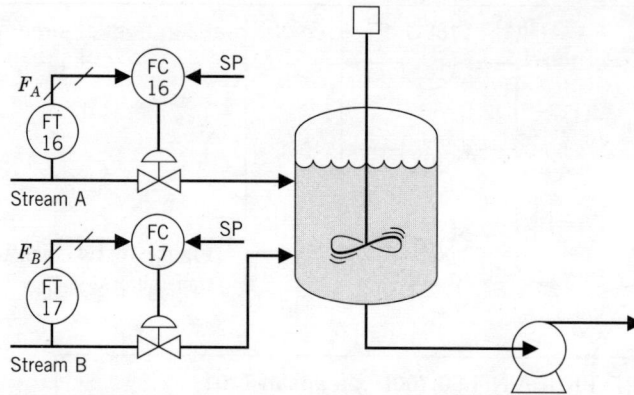

Figure 10-2.1 Control of blending of two liquid streams.

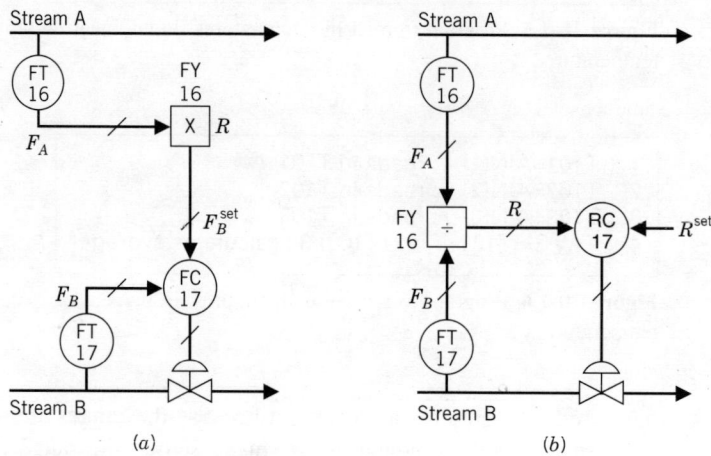

Figure 10-2.2 Ratio control of blending system.

blended in the correct ratio. However, suppose now that stream A cannot be controlled but only measured. The flow rate of this stream, often referred to as "wild flow," is usually manipulated to control something else, such as level or temperature, upstream. The controlling task is now more difficult. Somehow the flow rate of stream B must vary, as the flow rate of stream A varies, to maintain the blending in the correct ratio. Two possible ratio control schemes are shown in Fig. 10-2.2.

The first scheme, shown in Fig. 10-2.2a, consists of measuring the wild flow and multiplying it by the desired ratio, in FY16, to obtain the required flow rate of stream B; that is, $F_B^{set} = R \times F_A$. The output of FY16 is the required flow rate of stream B, and it is used as the set point to the flow controller of stream B, FC17. Thus as the flow rate of stream A varies, the set point to the flow controller of stream B will vary accordingly to maintain both streams at the required ratio. If a new ratio between the two streams is required, the new ratio is set in the multiplier. The set point to the flow controller of stream B is set from a computation, not locally. Figure 10-2.3a shows the software equivalent to Fig. 10-2.2a, and assumes that the control system works in engineering units. FT16LO, FT16SPAN, FT17LO, and FT17SPAN are the low value and span of FT16 and FT17.

The second ratio control scheme, shown in Fig. 10-2.2b, consists of measuring both streams and dividing them, in FY16, to obtain the actual ratio flowing through the system. The calculated ratio is then sent to a controller, RC17, which manipulates the flow of stream B to maintain set

```
1   FA=AIN(1, FT16LO, FT16SPAN) ; reads in flow of stream A
2   FB=AIN(2, FT17LO, FT17SPAN) ; reads in flow of stream B
3   FBSET=R*FA
4   CO17=PID(FB, FBSET, FT17LO, FT17SPAN) ; FC17
5   AOUT(1, CO17) ; outputs signal to valve
```

(a)

```
1   FA=AIN(1, FT16LO, FT16SPAN) ; reads in flow of stream A
2   FB=AIN(2, FT17LO, FT17SPAN) ; reads in flow of stream B
3   RCALC=FB/FA ; calculates ratio
4   CO17=PID(RCALC, R, RLO, RSPAN) ; RC17
5   AOUT(1, CO17) ; outputs signal to valve
```

(b)

Figure 10-2.3 Software equivalent of Fig. 10-2.2.

point. The set point to this controller is the required ratio, and it is set locally. Figure 10-2.3b shows the equivalent scheme using software. Note that in the controller it is necessary to specify RLO and RSPAN, which are the low value and span you expect the ratio to change. This is the same as selecting a "ratio transmitter range."

Both control schemes shown in Fig. 10-2.2 are used; however, the scheme shown in Fig. 10-2.2a is preferred because it results in a more linear system than the one shown in Fig. 10-2.2b. This is demonstrated by analyzing the mathematical manipulations in both schemes. In the first scheme FY16 solves the following equation $F_B^{set} = R \times F_A$. The gain of this device—that is, how much its output changes per change in flow rate of stream A—is given by

$$\frac{\partial F_B^{set}}{\partial F_A} = R$$

which, as long as the required ratio is constant, is a constant value. In the second scheme FY16 solves the following equation

$$R = \frac{F_B}{F_A}$$

Its gain is given by

$$\frac{\partial R}{\partial F_A} = \frac{F_B}{F_A^2} = \frac{R}{F_A}$$

so as the flow rate of stream A changes, this gain also changes, yielding a nonlinearity.

From a practical point of view, even if both streams can be controlled, the implementation of ratio control may still be more convenient than the control system shown in Fig. 10-2.1. Figure 10-2.4 shows a ratio control scheme for this case. If the total flow must be changed, the operator needs to change only one flow, the set point to FC16, then the set point to FC17 changes automatically once the flow rate of stream A changes. In the control system of Fig. 10-2.1 the operator needs to change two flows, the set points to both FC16 and FC17.

The schemes shown in Figs. 10-2.2a, and 10-2.4 are quite common in the process industries. Most DCSs offer a controller, referred to as PID-RATIO that accepts a signal, applies the same algorithm as the ratio unit, FY16, in Fig. 10-2.2a, and uses the internal result as its set point. Thus, if the PID-RATIO is used, the calculations done by FY16 and FC17 in Fig. 10-2.4 are performed in only one block.

As noted several times earlier, it is helpful in developing control schemes to remember that every signal must have a physical significance. In Figs. 10-2.2, and 10-2.4 we have labeled each signal with its significance. For example, in Fig. 10-2.2a the output signal from FT16 is related to the flow rate of stream A, and has the label F_A. If this signal is then multiplied by the ratio F_B/F_A, the output signal from FY16 is the required flow rate of stream B, F_B^{set}. Even

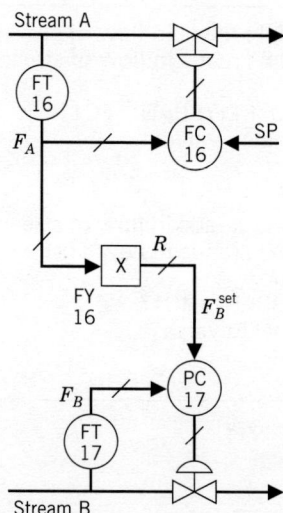

Figure 10-2.4 Ratio control of blending system.

though this use of labels is not standard practice, we will continue to label signals with their significance throughout the chapter for pedagogical reasons. We recommend that the reader do the same.

EXAMPLE 10-2.2

Another common example of ratio control used in the process industries is the control of the air/fuel ratio to a boiler or furnace. Air is introduced in a set excess of that stoichiometrically required for combustion of the fuel; this is done to ensure complete combustion. Incomplete combustion does not only result in the inefficient use of the fuel but may also result in smoke and the production of pollutants. The amount of excess air introduced is dependent on the type of fuel, fuel composition, and the equipment used. However, the greater the amount of excess air introduced, the greater the energy losses through the stack gases. Therefore, the control of the air and fuel flows is most important for proper safe, and economical operation.

The flow of combustibles is usually used as the manipulated variable to maintain the pressure of the steam produced in the boiler at some desired value. Figure 10-2.5 shows one way to control the steam pressure as well as the air/fuel ratio control scheme. This scheme is called parallel positioning control (O'Meara, 1979; Scheib and Russell, 1981; Congdon, 1981) with manually adjusted fuel/air ratio. The steam pressure is transmitted by PT22 to the pressure controller PC22, and this controller manipulates a signal, often referred to as *the boiler master signal*, to the fuel valve. Simultaneously, the controller also manipulates the air valve or damper through the ratio unit FY24. This ratio station sets the air/fuel ratio required.

The control scheme shown in Fig. 10-2.5 does not actually maintain a ratio of fuel flow to air flow; rather, it only maintains a ratio of signals to the final control elements, the actual flows are not measured and used. The flow through these elements depends on the signals and on the pressure drop across them. Consequently, any pressure fluctuation across the fuel valve or air damper changes the flow, even though the opening has not changed, and this in turn affects the combustion process and steam pressure. A better control scheme to avoid this type of disturbances is shown in Fig. 10-2.6, and it is referred to as full metering control (O'Meara, 1979). (Figure 10-2.6 is referred to as a "top-down" instrumentation diagram, and it is commonly used to present control schemes.) In this scheme the pressure controller sets the flow of fuel, and the air flow is ratioed from the fuel flow. The flow loops correct for any flow disturbances. The fuel/air ratio is still manually adjusted.

Note the differences between the two figures. In Fig. 10-2.6a the signal from FT23 is multiplied by the ratio F_F/F_A before it is used as the set point to FC24; note that the significance of all signals make sense. Figure 10-2.6b is the one that seems somewhat strange. The figure

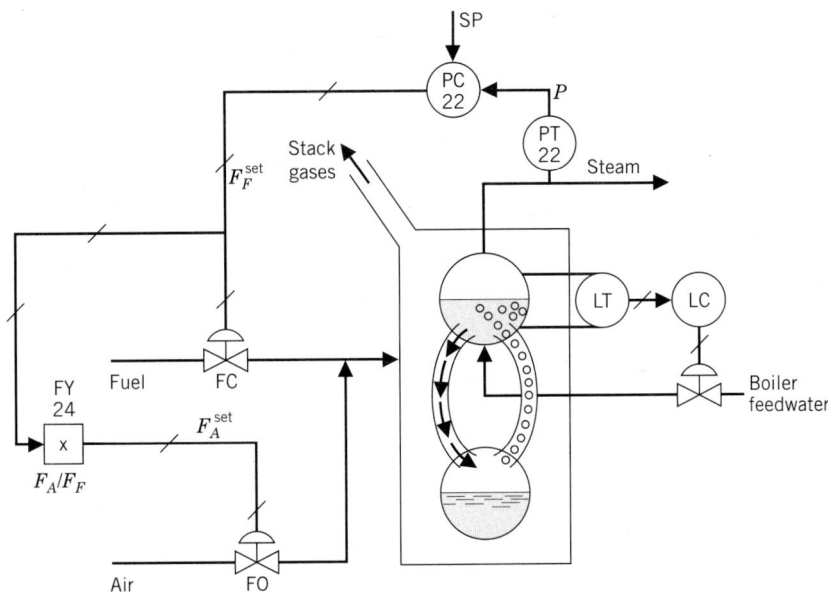

Figure 10-2.5 Parallel positioning control with manually adjusted air to fuel ratio.

shows that the signal setting the set point to FC24 comes from FT23 and, therefore, it is related to F_F; FC24 is the controller that moves the air flow! Note, however, that the signal from FT24, which is related to the air flow, is multiplied by F_F/F_A before it is used as the measurement to FC24. Thus, both the measurement and the set point to FC24 have the same meaning. It seems that Fig. 10-2.6*b* is somewhat more difficult to understand, however, its use in the following schemes results in less blocks (or computations) to use.

Let us analyze the control scheme shown in Fig. 10-2.6 in more detail. When the steam header pressure increases, probably due to a decrease in steam demand, the pressure controller reduces the demand for fuel. As the set point to the fuel flow controller reduces, the controller closes the valve to satisfy the set point. As the fuel flow decreases, the set point to the air flow controller also reduces. Thus, the air flow follows the fuel flow, and during a transient period the entering combustible mixture is richer than usual in air. Note that in the figure we have indicated the significance of each signal.

Now consider the case wherein the header pressure decreases, probably due to an increase in steam demand, and the pressure controller increases the demand for fuel. As the set point to the fuel flow controller increases, the controller opens the valve to satisfy the set point. As the fuel flow increases, the set point to the air flow is increased; the air flow again follows the fuel flow. In this last case, the entering combustible mixture is not richer in air during a transient period, and if we are not careful it may even be lean in air. This situation is certainly not desirable for two important reasons. First, a lean air mixture may result in pockets of pure fuel inside the combustion chamber: not a very safe (in fact, an explosive!) condition. Second, a lean air mixture results in unburned fuel in the stack gases, which constitutes an environmental hazard and a waste of energy and money. Therefore, a control scheme must be designed to avoid these situations. The control scheme must be such that when more combustibles are required to maintain pressure, it increases the air first and then the fuel. When less combustibles are required, it decreases the fuel first and then the air. This pattern ensures that the combustible mixture is air-rich during transient periods. Figure 10-2.7 shows a scheme, referred to as *cross-limiting control*, that provides the required control. Only two selectors, LS23 and HS24, are added to the control scheme of Fig. 10-2.6*b*. The selectors provide a way to decide which device sets the set point to the controller. The reader is encouraged to "go through" the scheme to understand how it works. As a way to do so, assume that the required air/fuel ratio is 2, and that at steady state

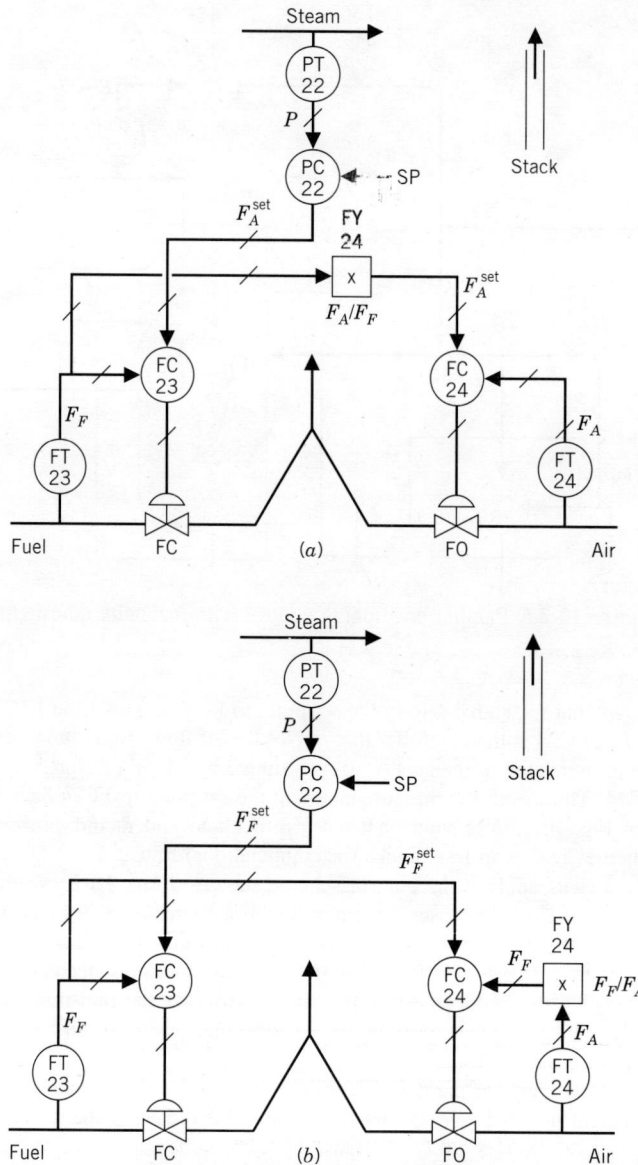

Figure 10-2.6 Full metering control with manually adjusted fuel/air ratio.

the required fuel is 10 units of flow. Consider next what happens if the header pressure increases and the pressure controller asks for only 8 units of fuel flow. Finally, consider what happens if the header pressure decreases and the pressure controller asks for 12 units of fuel flow.

Because the amount of excess air is so important for the economical and environmental operation of boilers, it has been proposed that some feedback based on an analysis of the stack gases be provided; the analysis is usually percent O_2, or percent CO. It is then proposed that the fuel/air ratio be adjusted on the basis of this analysis. This new scheme, which is shown in Fig. 10-2.8, consists of an analyzer transmitter, AT25, and a controller, AC25. The controller maintains the required percent O_2, for example, in the stack gases by setting the required fuel/air ratio. The figure shows the use of high and low limiters, HL26 and LL27. These two units are used mainly for safety reasons. They ensure that the air/fuel ratio will always be between some preset high and low values.

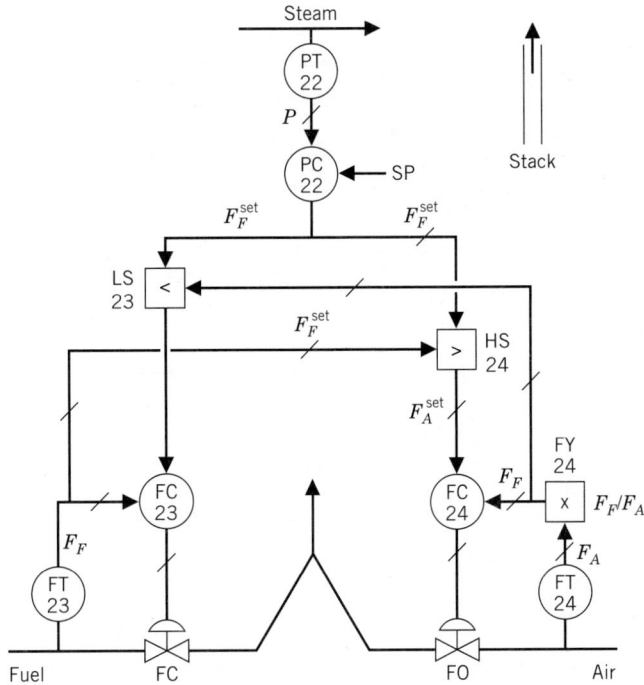

Figure 10-2.7 Cross-limiting control.

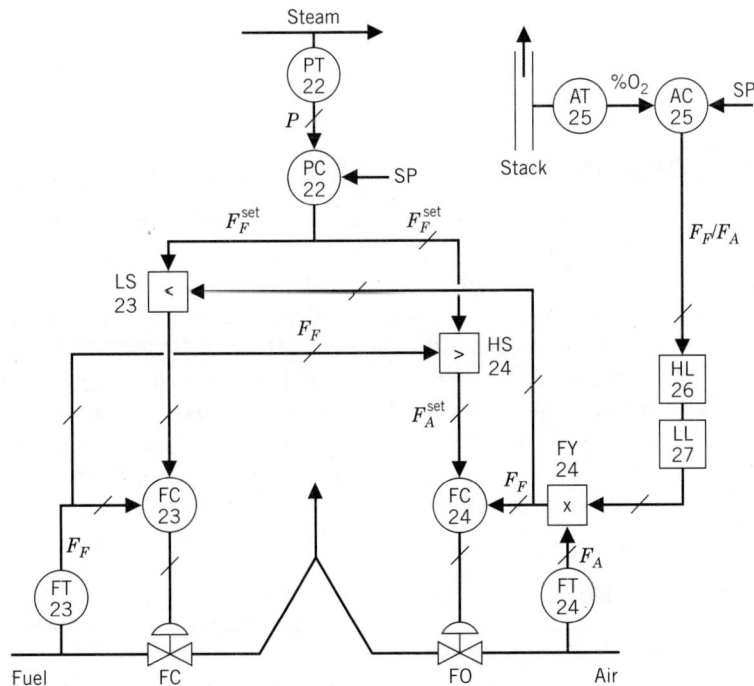

Figure 10-2.8 Cross-limiting with O_2 trim control.

```
1     P = AIN(1, P_low, P_span) ; reads in pressure
2     FA = AIN(2, FA_low, FA_span) ; reads in air flow
3     FF = AIN(3, FF_low, FF_span) ; reads in fuel flow
4     %O2 = AIN(4, %O2_low, %O2_span) ; reads in %O2
5     FOUT = PID(P, P^set, P_low, P_span) ; PC22
6     ROUT = PID(%O2, %O2^set, %O2_low, %O2_span) ; AC25
7     PFF^set = (FF_span/100)*FOUT + FF_low ; converts output of PC22 to
                                  ; fuel flow set point in engineering units
8     RATIO = (RATIO_span/100)*ROUT + RATIO_low ; converts output
                                  ; of AC25 to FA/FF ratio in engineering units
9     IF RATIO>RATIOMAX THEN RATIO=RATIOMAX ; HL26
10    IF RATIO<RATIOMIN THEN RATIO = RATIOMIN ; LL27
11    RFF = FA*RATIO ; FY24
12    IF PFF^set<RFF THEN FF^set = PFF^set ELSE FF^set = RFF ; LS23
13    COFUEL = PID(FF, FF^set, FF_low, FF_span) ; FC23
14    IF PFF^set>FF THEN FF^set = PFF^set ELSE FF^set = FF ; HS24
15    COAIR = PID(RFF, FF^set, FF_low, FF_span) ; FC24
16    AOUT(1, COFUEL) ; outputs signal to fuel valve
17    AOUT(2, COAIR) ; outputs signal to air valve
```

Figure 10-2.9 Software program equivalent to Fig. 10-2.8.

Figure 10-2.9 shows the software program equivalent to the control strategy of Fig. 10-2.8; the comments associated with each statement help to relate the program to Fig. 10-2.8.

This section has shown two applications of ratio control. As mentioned at the beginning of the section, ratio control is a common technique used in the process industries; it is simple and easy to use.

10-3   OVERRIDE, OR CONSTRAINT, CONTROL

Override, or constraint, control is a powerful yet simple control strategy usually used as (1) a protective strategy to maintain process variables within limits that must be enforced to ensure the safety of personnel and equipment, and product quality and (2) an optimization strategy that permits smooth transition between control schemes to obtain maximum benefit. As a protective strategy, override control is not as drastic as interlock control. Interlock controls are used primarily to protect against equipment malfunction. When a malfunction is detected, the interlock system usually shuts down the process. Interlock systems are not presented, but References 5 and 6 are provided for their study. Two examples of constraint control are now presented to demonstrate the concept and implementation of the strategy.

EXAMPLE 10-3.1

Consider the process shown in Fig. 10-3.1. A hot saturated liquid enters a tank and from there it is pumped under flow control back to the process. Under normal operation the level in the tank is at height h_1. If under any circumstances the liquid level drops below the height h_2, the liquid will not have enough net positive suction head (NPSH), and cavitation at the pump will result. It is therefore necessary to design a control scheme that avoids this condition. This new control scheme is shown in Fig. 10-3.2.

The level in the tank is now measured and controlled. The set point to LC50 is somewhat above h_2 as shown in the figure. It is important to notice the action of the controllers and final

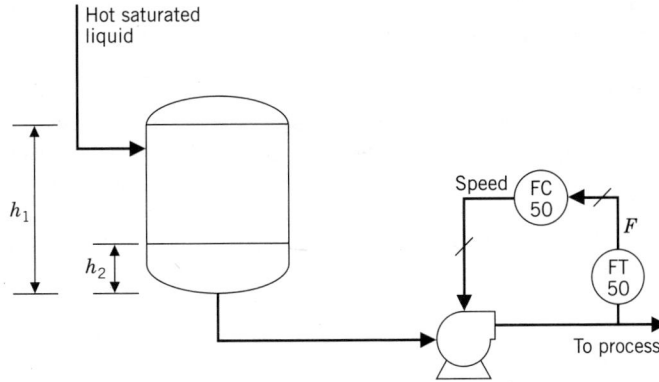

Figure 10-3.1 Tank and flow control loop.

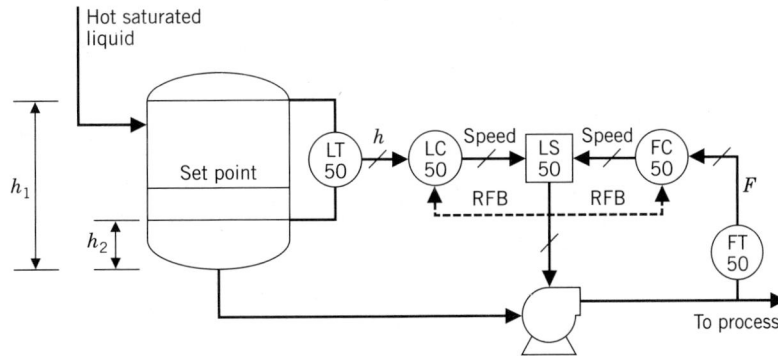

Figure 10-3.2 Override control scheme.

control element. The variable-speed pump is such that as the input energy (current in this case) to it increases, it pumps more liquid. Therefore, the FC50 is a reverse-acting controller while the LC50 is a direct-acting controller. The output of each controller is connected to a low selector, LS50, and the signal from this selector goes to the pump.

Under normal operating condition the level is at h_1, which is above the set point to the level controller. Consequently, the controller will try to speed up the pump as much as possible, increasing its output to 100 %. The output of the flow controller, under normal condition, may be 75 %, and the low selector selects this signal to manipulate the pump speed. Thus, under normal condition, the flow controller is manipulating the pump. The level controller is not connected to the pump because the level is not at an undesirable state. This is the desired operating condition.

Let us now suppose that the flow of hot saturated liquid into the tank slows down and the level starts to drop. As soon as the level drops below the set point on the level controller, this controller will slow down the pump by reducing its output. When the level controller's output drops below the output of the flow controller, the low selector selects the output of the level controller to manipulate the pump. It can be said that the level controller "overrides" the flow controller.

When the flow of hot liquid returns to its normal value, and the level increases above the set point, the level controller increases its output to speed up the pump. Once the output from the level controller increases above the output from the flow controller, the low selector selects the flow controller, and the operation returns to its normal condition.

An important consideration in designing an override control system is that of reset windup protection on any controller that has integration. The output of the controller not selected must stop at 100 %, not at a higher value, or at 0 %, not at a lower value. An even more desirable operation is the one that, if the selected controller output is 75 %, forces the nonselected output to be close to 75 %, not 100 %. That is, it forces the nonselected controller output to be close to the selected output. This desirable operation is easily accomplished using the *reset feedback (RFB)* technique presented in Chapter 5. Figure 10-3.2 shows the reset feedback connections (dotted lines) to both controllers. In this case the reset feedback signal to the controller(s) comes from an external computation, a low selector, not from the controller itself as shown in Fig. 5-3.16; sometimes we refer to this signal as external reset feedback. The output from the low selector is the one that goes to the pump; it is also used as the reset feedback signal to the controllers.

Figure 10-3.3 shows a schematic of the two controllers with the external reset feedback. To further explain how this system works, consider that at steady state the flow controller outputs a 75 % signal to maintain its set point, and the level in the tank is above its set point. In this case the level controller will increase its output, to speed up the pump, and control at its set point. Thus, the low selector selects the 75 % signal from the flow controller; this is the normal operating condition. The output signal from the selector is the RFB signal to both controllers, and the corresponding M_I signal will also be 75 %. At this steady state the error in the flow controller is zero, and the proportional calculation of this controller is also zero. Because the level in the tank is above the set point, the error in the level controller is positive (direct-acting) and the proportional calculation will have a certain output, depending on the error and controller gain, say 10 %. Since the M_I signal is 75 %, the output from the level controller to the low selector is 85 %. Now suppose that the input flow to the tank decreases and the level in the tank starts to drop. As this happens, the proportional calculation in the level controller also starts to decrease from 10 % down, and the controller output from 85 % down. As the level in the tank drops below the set point, the error in the controller, and the proportional calculation, become negative resulting in an output less than 75 %, say, 74 %. At this moment the selector selects this signal, and the level controller overrides the flow controller, and sends it to the pump to slow it down; this new value is also the RFB signal to the controllers. As the pump slows down to avoid a low level in the tank, the error in the flow controller becomes positive (reverse-acting), and the proportional calculation increases to increase the flow and correct for the error. However, the low selector will not allow this particular corrective action; it is more important to avoid a low level in the tank. Note that the output from the flow controller will then be equal to the output from the low selector plus its own proportional calculation.

Most controllers offer this external reset feedback capability. To summarize, this capability, also referred sometimes as output tracking, allows the controller not selected to override the

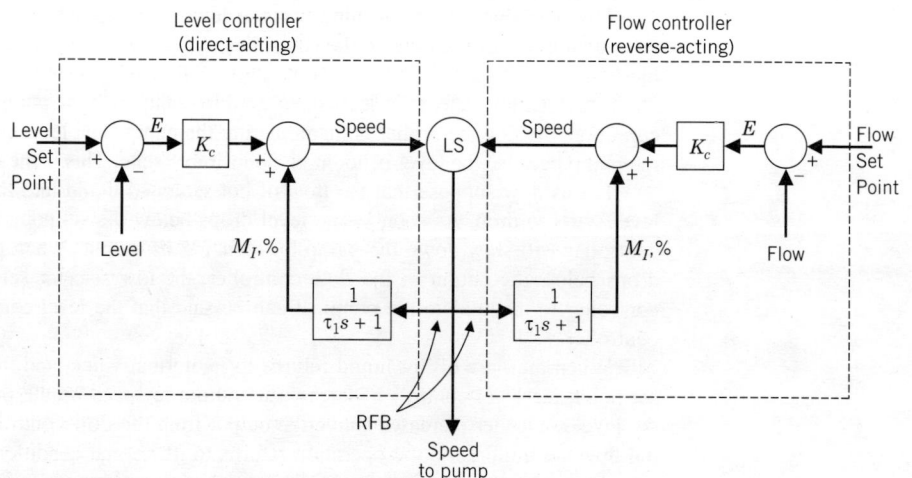

Figure 10-3.3 Controllers with reset feedback (RFB).

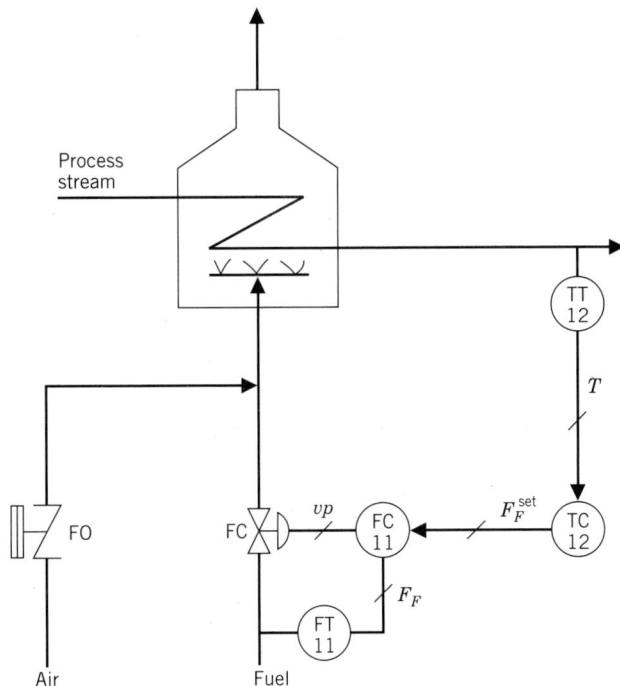

Figure 10-3.4 Heater temperature control.

controller selected as soon as its error changes sign. More than two controllers can provide signals to a selector and have RFB signals; this is shown in the following example.

EXAMPLE 10-3.2

A fired heater, or furnace, is another common process that requires the implementation of constraint control. Figure 10-3.4 shows a heater with temperature control manipulating the gas fuel flow. The manipulation of the combustion air has been omitted to simplify the diagram; however, it is the same that was discussed in detail in the ratio control section. There are several conditions in this heater that can prove quite hazardous. These conditions include (1) higher combustible pressure that can sustain a stable flame and (2) higher stack, or tube, temperature that the equipment can safely handle. If either of these two conditions exists, the combustible flow must decrease to avoid the unsafe condition; at this moment, the temperature control is certainly not so important as the safety of the operation. Only when the unsafe conditions disappear is it permissible to return to straight temperature control.

Figure 10-3.5 shows a constraint control strategy to guard against the unsafe conditions we have described. The gas fuel pressure is usually below the set point to PC14, so the controller will try to raise the set point to the fuel flow controller. Usually, the stack temperature is also below the set point to TC13, so the controller will also try to raise the set point to the fuel flow controller. Thus, under normal conditions, the exit heater temperature controller is the controller selected by the low selector, because its output will be the lowest of the three controllers. Only when one of the unsafe conditions exists is TC12 "overridden" by one of the other controllers.

As explained in Example 10-3.1, it is important to prevent windup of the controllers that are not selected. Thus the control system must be configured to provide reset feedback. This is shown by the dashed lines in the figure.

The constraint control scheme shown in Fig. 10-3.5 contains a possible safety difficulty. If at any time the operating personnel were to set the flow controller FC11 in local set point or in the manual mode (off remote set point), the safety provided by TC13 and PC14 would not be in effect. This would result in an unsafe and unacceptable operating condition. You may want to think how to design a new constraint control strategy to permit the operating personnel to

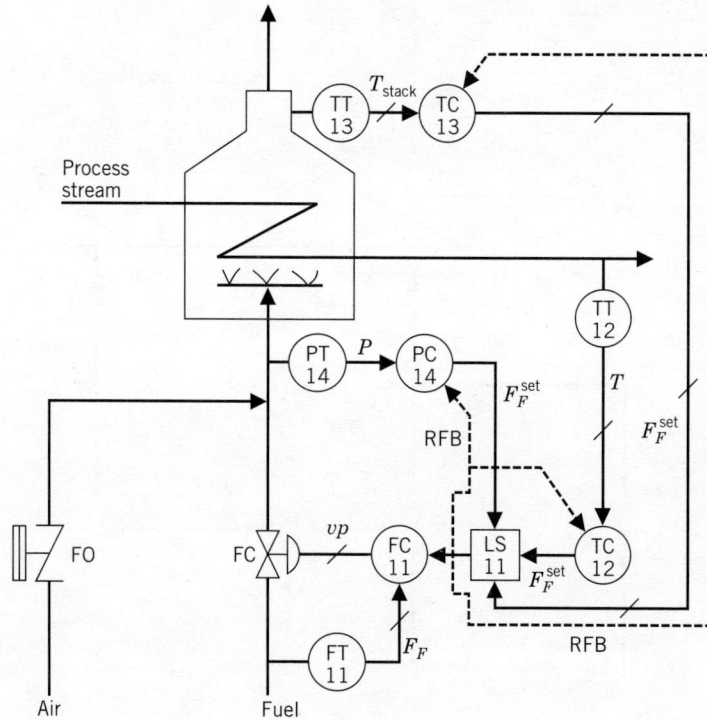

Figure 10-3.5 Heater temperature control—constraint control.

set the flow controller in automatic or manual and still have the safety provided by TC13 and PC14 in effect. This is a problem at the end of the chapter.

The introduction to this section mentioned that override control is commonly used as a protective scheme, and Examples 10-3.1 and 10-3.2 presented two such applications. As soon as the process returns to normal operating conditions, the override scheme returns automatically to its normal operating status. The two examples presented show multiple control objectives (controlled variables) with a single manipulated variable; however, only one objective is enforced at a time.

10-4   SELECTIVE CONTROL

Selective control is another interesting control strategy used for safety considerations and process optimization. Two examples are presented to show its principles and implementation.

EXAMPLE 10-4.1   Consider the plug flow reactor shown in Fig. 10-4.1 where an exothermic catalytic reaction takes place. The figure shows the reactor temperature control. The sensor providing the temperature measurement should be located at the hot spot. As the catalyst in the reactor ages, or as conditions change, the hot spot moves. It is desired to design a control scheme so that its measured variable "moves" as the hot spot moves. A control strategy that accomplishes the desired specification is shown in Fig. 10-4.2. The high selector in this scheme selects the transmitter

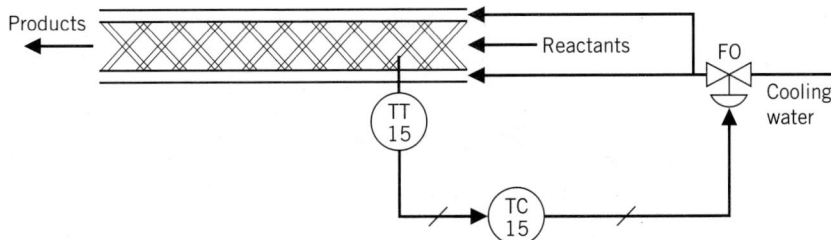

Figure 10-4.1 Temperature control of a plug flow reactor.

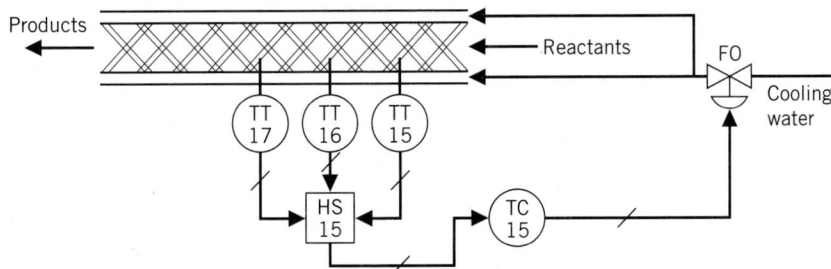

Figure 10-4.2 Selective control for a plug flow reactor.

with the highest output, and thus the controlled variable is always the highest, or closest to the highest, temperature.

In implementing this control strategy, an important consideration is that all temperature transmitters must have the same range so that their output signals can be compared on the same basis. Another consideration that may be important is to install an indication as to which transmitter is giving the highest signal. If the hot spot moves past the last transmitter, TT17, this may be an indication that it is time to either regenerate or change the catalyst. The length of reactor left for the reaction is probably not enough to obtain the desired conversion.

EXAMPLE 10-4.2

An instructive and realistic process wherein selective control can improve the operation is shown in Fig. 10-4.3. A furnace heats a heat transfer oil to provide an energy source to several process units. Each individual unit manipulates the flow of oil required to maintain its controlled variable at set point. In addition, the outlet oil temperature from the furnace is controlled by manipulating the fuel flow. A bypass control loop, DPC16, is provided.

Suppose it is noticed that the control valve in each unit is not open very much—that is, the output of TC13 is 20 %, that of TC14 is 15 %, and that of TC15 is 30 %. This indicates that the hot oil temperature provided by the furnace may be higher than required by the users. Consequently, not much oil flow is necessary and much of it will bypass the users. This situation is energy-inefficient because a large quantity of fuel must be burned to obtain a high oil temperature. Also, a significant amount of the energy provided by the fuel is lost to the surroundings in the piping system and through the stack gases.

A more efficient operation is the one that maintains the oil leaving the furnace at a temperature just hot enough to provide the necessary energy to the users with hardly any flow through the bypass valve. In this case the temperature control valves would be mostly open. Figure 10-4.4 shows a selective control strategy that provides this type of operation. The strategy first selects the most open valve using a high selector, HS16. The valve position controller, VPC16, controls the selected valve position—at, say, at 90 % open—by manipulating the set point of the furnace temperature controller. Thus, this strategy ensures that the oil temperature from the furnace is just "hot enough."

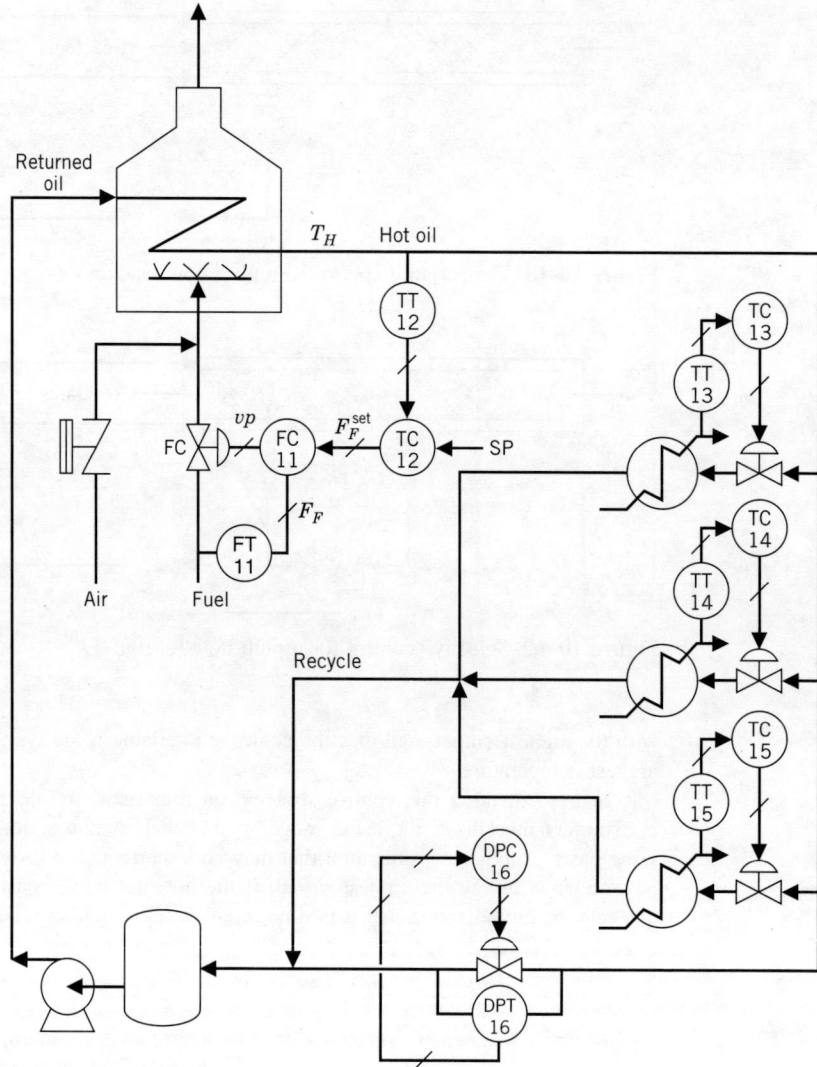

Figure 10-4.3 Hot oil system.

Note that because the most open valve is selected by comparing the signals to each valve, all of the valves should have the same characteristics.

The selective control strategy shows again that with a bit of logic, a process operation can be significantly improved.

10-5 DESIGNING CONTROL SYSTEMS

This section presents three examples to provide some hints on how to go about designing control schemes. To obtain maximum benefit from this section we recommend that you first read the example statement, and try to solve the problem yourself. Then check with the solution presented.

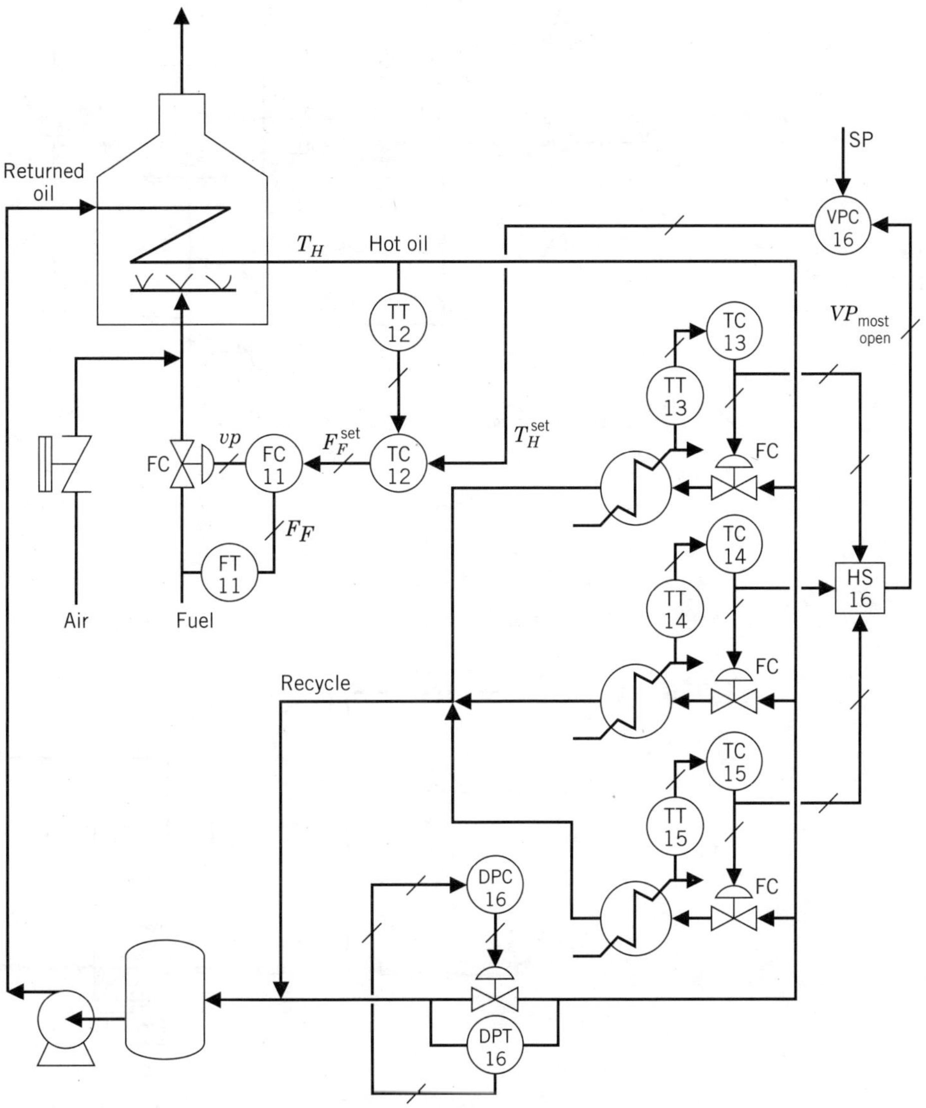

Figure 10-4.4 Selective control for hot oil system.

EXAMPLE 10-5.1

Consider a reactor, Fig. 10-5.1, where the exothermic reaction A + B → C takes place. The diagram shows the control of the temperature in the reactor by manipulating the cooling water valve.

(a) Design a control scheme to control the flow of reactants to the reactor. The flows of reactants A and B should enter the reactor at a certain ratio, R, that is, $R = F_B/F_A$. Both flows can be measured and controlled.

(b) Operating experience has shown that the inlet cooling water temperature varies somewhat. Because of the lags in the system this disturbance usually results in cycling of the temperature in the reactor. The engineer in charge of this unit has been wondering whether some other control scheme can help in improving the temperature control. Design a control scheme to help him.

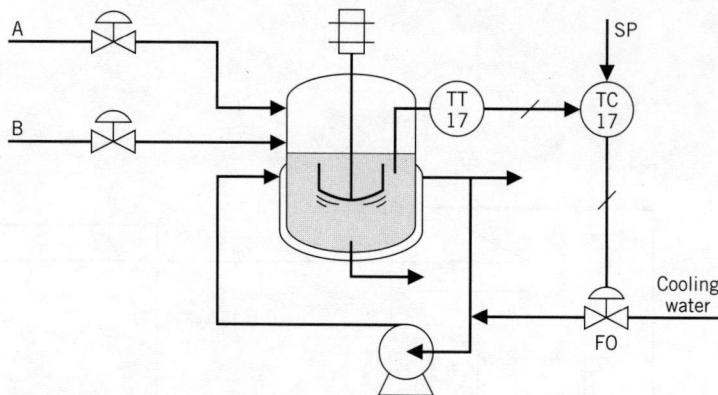

Figure 10-5.1 Reactor for Example 10-5.1.

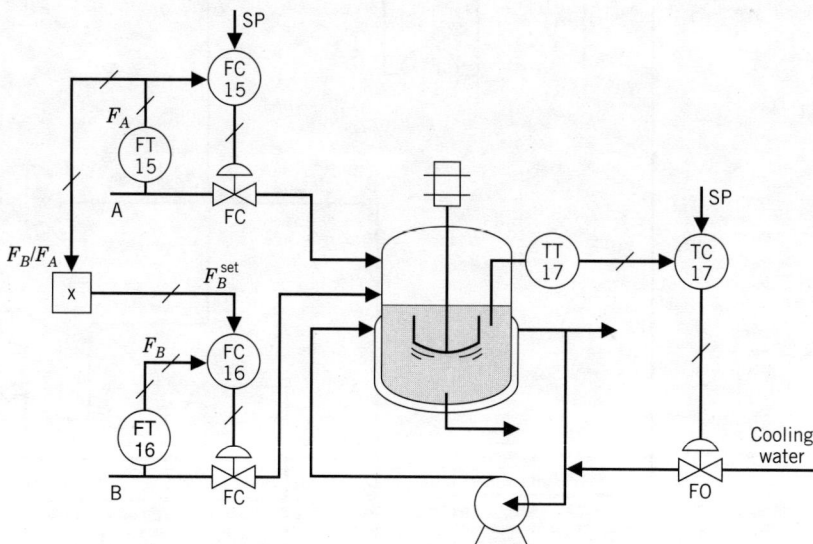

Figure 10-5.2 Ratio control scheme for Example 10-5.1 part (a).

(c) Operating experience has also shown that under some infrequent conditions the cooling system does not provide enough cooling. In this case the only way to control the temperature is by reducing the flow of reactants. Design a control scheme to do this automatically. The scheme must be such that when the cooling capacity returns to normal the scheme of part (b) is reestablished.

(a) Figure 10-2.2a provides a scheme that can be used to satisfy the ratio control objective; Fig. 10-5.2 shows the application of the scheme to the present process. The operator sets the flow of stream A, set point to FC15, and the flow of stream B is set accordingly.

(b) A common procedure we follow to design control schemes is to first think what we would do to control the process manually. In the case at hand, after some thinking you may decide that it would be nice if somehow you could find out as soon as possible a change in cooling water temperature. If this change is known then you could do something to negate its effect. For example, if the cooling water temperature increases then you could open the valve to feed in more fresh water; Fig. 10-5.3 shows this idea. But, you now think, I am not considering the temperature controller TC17 at all. Well, why not use the output of TC17 as my set point,

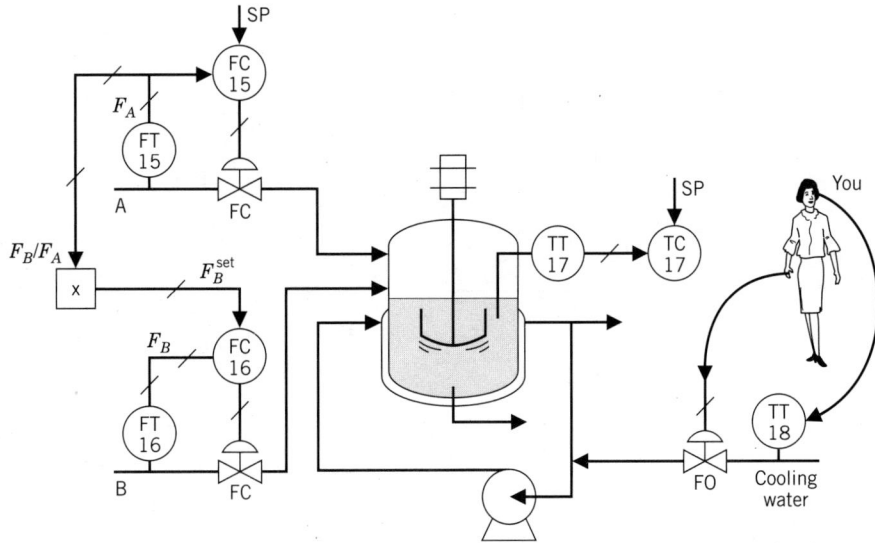

Figure 10-5.3 Proposed *manual* control scheme (first draft) to compensate for changes in inlet cooling water temperature.

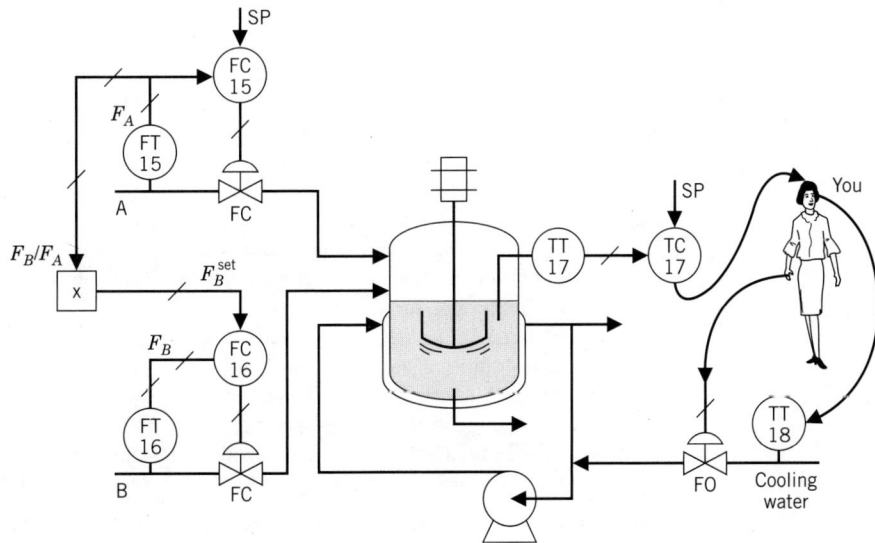

Figure 10-5.4 Proposed *manual* control scheme (second draft) to compensate for changes in inlet cooling water temperature.

as a cascade control scheme; Fig. 10-5.4 shows this proposed scheme. Next you decide to automate your idea and for that you sketch Fig. 10-5.5. You have replaced yourself with another intelligence: a controller.

Now that you have sketched your idea you need to analyze it further. The figure shows that the master controller, TC17, looks at the temperature in the reactor, compares it to its set point, and decides on the set point to the slave controller. That is, the master controller decides on the required inlet water temperature, T_{CW}^{set}. Now suppose that the inlet water temperature is not equal to the set point, for example, $T_{CW} > T_{CW}^{set}$? What would the slave controller do? Open the valve

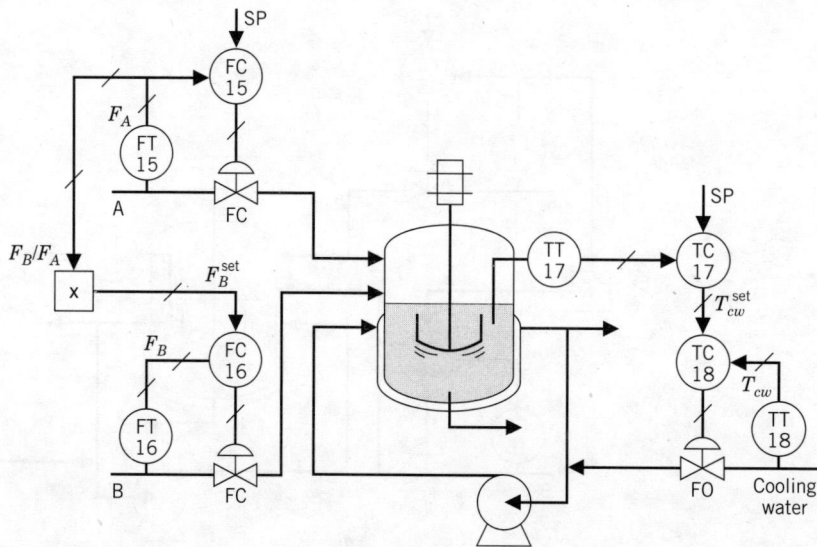

Figure 10-5.5 Proposed *automatic* control scheme (first draft) to compensate for changes in inlet cooling water temperature.

to add more water? Would this action make $T_{CW} = T_{CW}^{set}$? The answer is of course, no. The controller would open the valve but T_{CW} would not change. Opening or closing the valve does not have any effect on T_{CW}! The controller would keep opening the valve until it winds up. This is a perfect example where the action taken by the controller does not affect its measurement. Remember M-D-A in Chapter 1? Remember we said that these three operations—measurement, decision, and action—must be in a loop? That is, the action (A) taken by the controller must affect its measurement (M). The scheme showed in Fig. 10-5.5 does not provide a "closed-loop" but rather, we have an "open-loop."

Well, so this scheme does not work. But the idea is still valid, that is, learn as soon as possible that the cooling water temperature has changed. What about the scheme shown in Fig. 10-5.6? Go through the same analysis as previously and you will reach the same conclusion. That is, this last scheme still provides an "open-loop." Opening or closing the valve does not affect the temperature where it is measured.

The earliest you can detect a change in cooling water and have a closed-loop is any place in the recycle line or in the cooling jacket; Figure 10-5.7 shows the transmitter installed in the recycle line, and Fig. 10-5.8 shows the transmitter installed in the jacket. Go through the previous analysis until you convince yourself that both of these schemes indeed provide a closed-loop.

(c) For this part you again think of yourself as the controller. You know that as soon as the cooling system does not provide enough cooling you must reduce the flow of reactants to the reactor. But, how do you notice that you are short of cooling capacity? Certainly, if the temperature in the reactor, or in the jacket, reaches a high value the cooling system may not be providing the required cooling. But, what is this value? Further analysis (thinking) indicates that the best indication of the cooling capacity is the opening of the cooling valve. When this valve is fully open no more cooling is possible. At that time the temperature controller cannot do any more, and the process is out of control. Figure 10-5.9 shows what you may do as the controller.

The idea seems good but it is manual control so, how do I automate it? Figure 10-5.10 shows an override control scheme to do so. Let us explain. The output signal from TC18 that goes to the cooling valve is also used as the measurement to a controller which we call VPC20. This controller compares the measurement, which indicates the valve position, with the set point and sends a signal to the flow controller FC15. VPC20 is doing exactly what you were doing

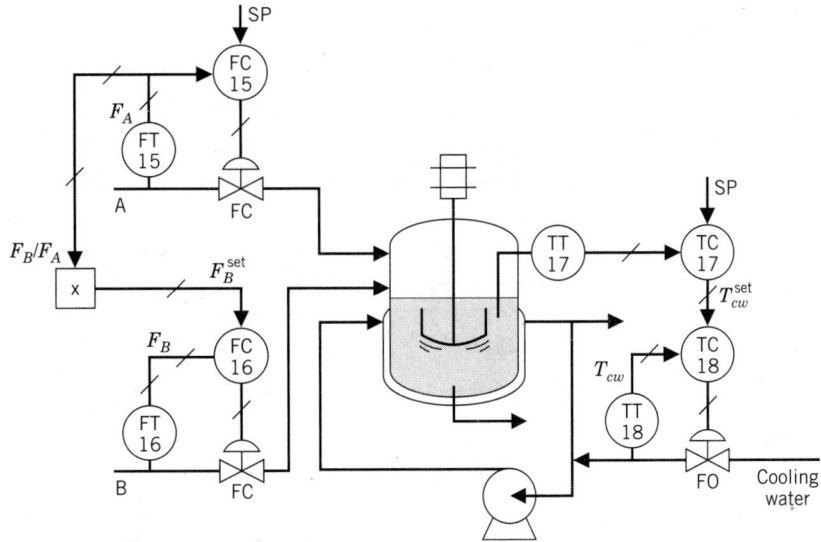

Figure 10-5.6 Proposed *automatic* control scheme (second draft) to compensate for changes in inlet cooling water temperature.

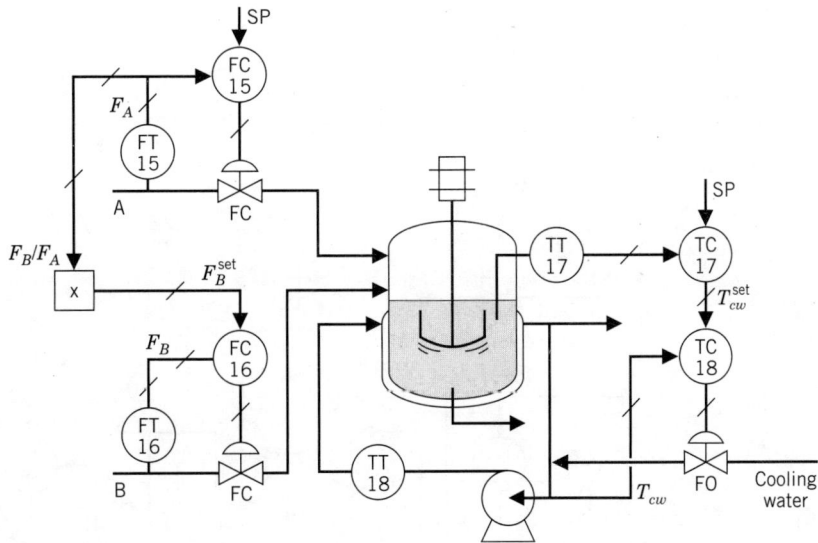

Figure 10-5.7 Cascade control scheme.

in the previous figure. Note that before the signal from VPC20 gets to FC15 it goes through a selector. The selector is used to select which signal, the operator set or the one from VPC20, really goes to FC15. At this moment we do not know what type of selector (high or low) to use, so lets decide that now.

Under normal condition the cooling valve is less that 90 % open, say 65 % open. As VPC20 "sees" that this valve is 65 % open it decides that the only way to make it open up to 90 % is by asking for a lot of reactants and for it to increase the output signal—say up to 100 %—to increase the set point to FC15. Obviously, under this condition there is plenty of cooling capacity left and there is no need to change the reactants flow required by the operator.

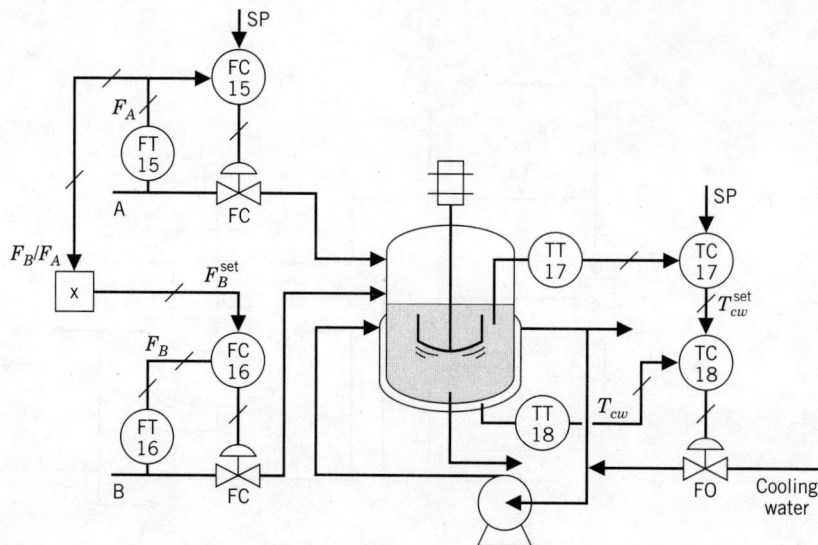

Figure 10-5.8 Cascade control scheme.

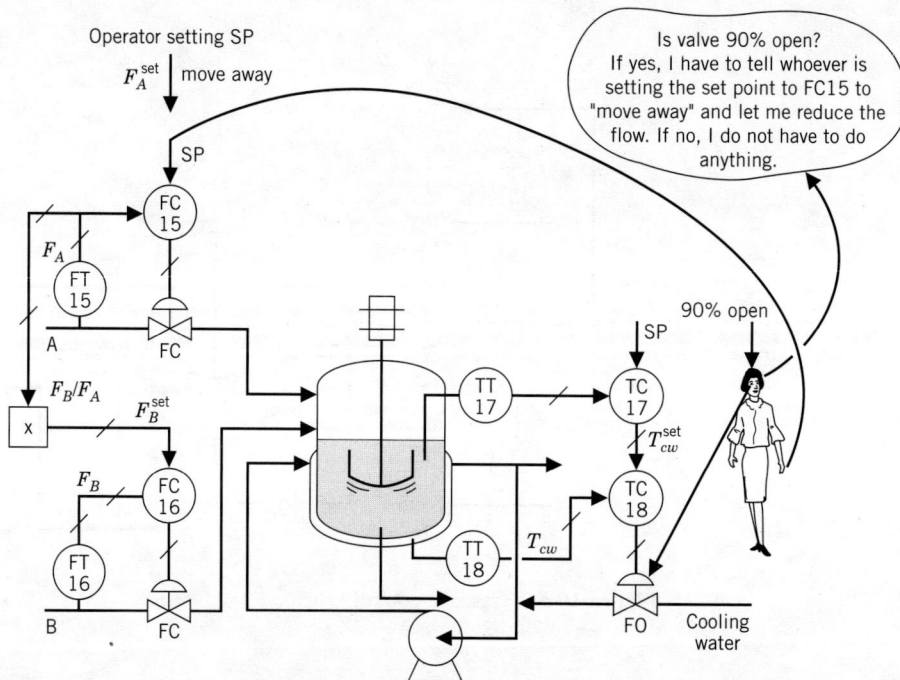

Figure 10-5.9 Avoiding that the cooling valve go over 90 % open, manual control.

The selector must be such that it selects the signal from the operator and not from VPC20. Because the signal from VPC20 is probably 100 %, the selector must be a low selector; this is shown in Fig. 10-5.11. *The selector is essentially used by VPC20 to tell the operator set point "move away" and let me set the set point.* Note the reset feedback indication from the selector to VPC20. Please analyze this scheme until you fully understand the use of the selector.

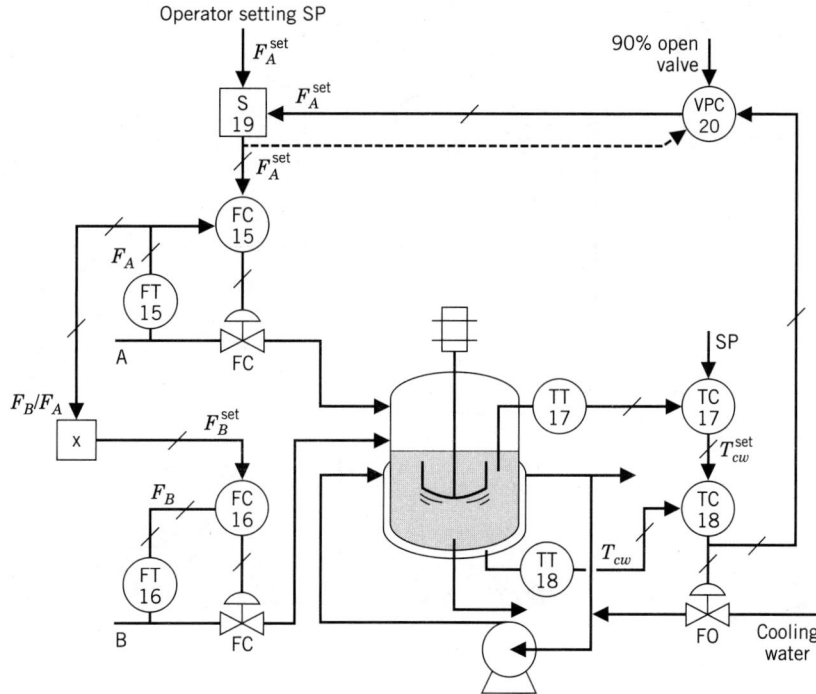

Figure 10-5.10 Override control scheme to compensate for loss of cooling capacity.

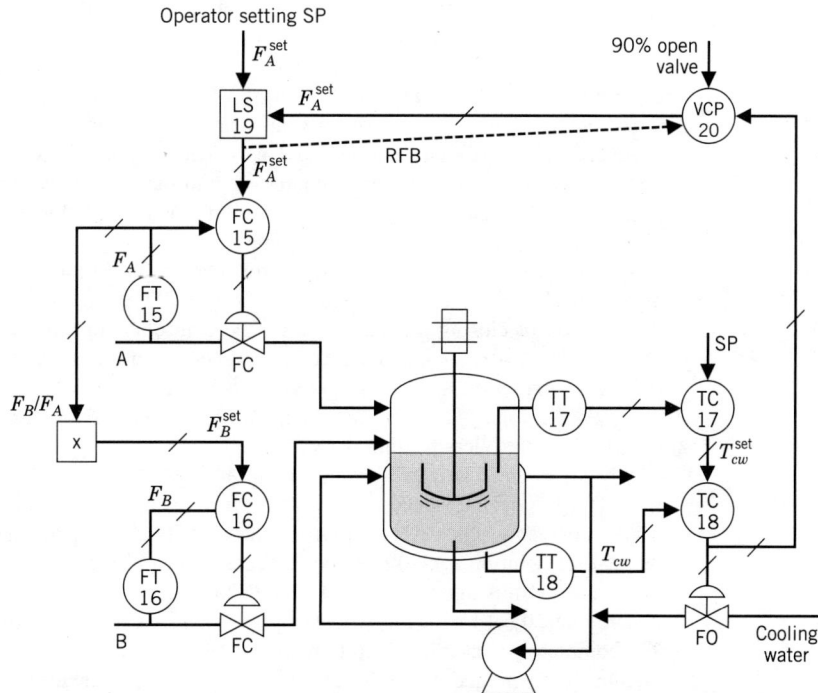

Figure 10-5.11 Override control scheme to compensate for loss of cooling capacity.

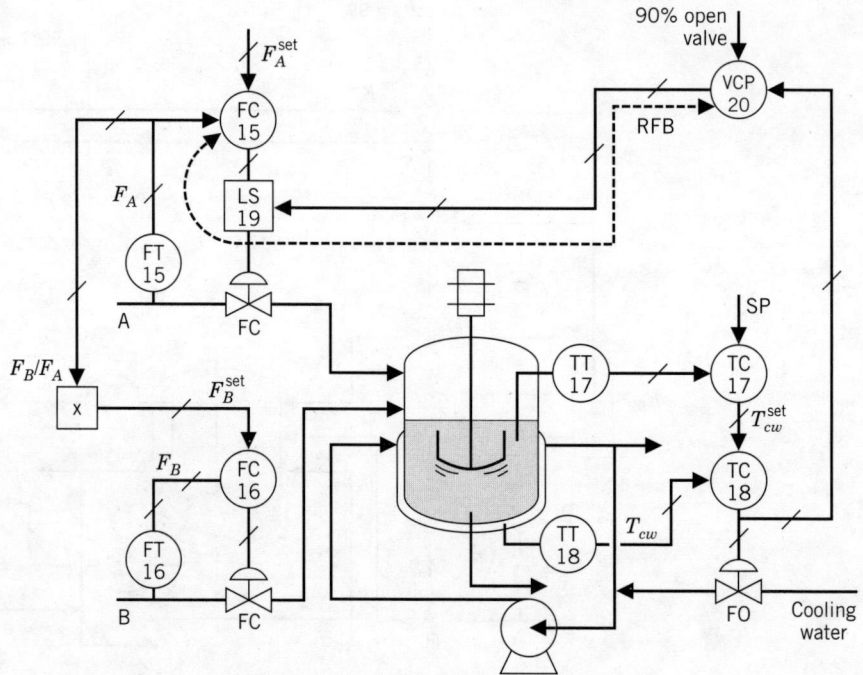

Figure 10-5.12 Another override control scheme.

Before leaving this example, there are a couple of additional things we need to discuss. First, as you may recall from the discussion on cascade control in Chapter 9, FC15 must be set to remote set point. However, what would happen if the operator sets FC15 to local set point, or manual? That is, what would happen if VPC20 asks for a lower set point but FC15 is not in remote set point? The answer is simply nothing, and the safety provided by VPC20 would not work, it is essentially not active. This is certainly not a safe operation! So, what can we do? That is, how can we design a control system that allows the operator set FC15 in local set point or manual, and at the same time at any moment VPC20 would be able to reduce the flow of reactants? Think about it before reading further . . . .

Figure 10-5.12 shows the new control scheme. In this scheme the output of VPC20 goes directly to the valve, and not to the set point of FC15. So no matter what the mode of FC15 is, the decision of closing the valve is after the controller. In this case the selector is also a low selector. The reader should convince himself/herself of this selection. *If for safety consideration it is necessary to manipulate a flow, it is always good practice to go directly to the valve and not to the set point of a controller manipulating the valve!* Note that the reset feedback as usual goes to the controllers feeding the selector.

Next, we need to address the set point of VPC20. The controller receives the output signal from TC18 to decide whether the valve is 90 % open or not. Realize, however, that the cooling water valve is fail-open (FO); therefore, the valve is 90 % open when the signal is 10 %! Thus, the real set point in VPC20 must be 10 %, as shown in Fig. 10-5.13. As an exercise, the reader may want to think about the action of VPC20.

Figure 10-5.14 shows a software program using the language presented in Section 10-1. The comments in each line help to relate the program to Fig. 10-5.13. For individuals with programming experience it may be easier to design control schemes thinking first in programming terms.

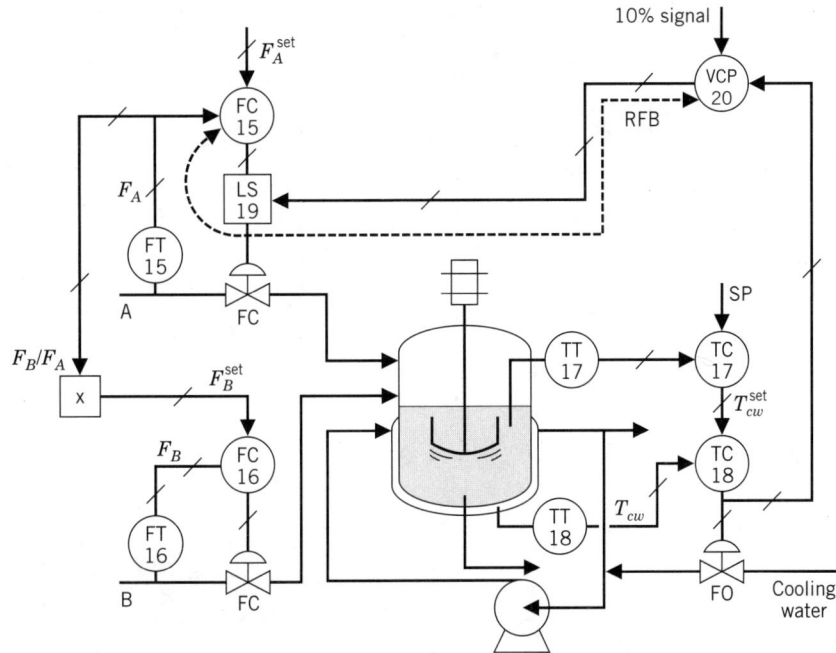

Figure 10-5.13 Override control scheme.

```
1    FA=AIN(1,FALO,FASPAN) ; reads in flow of stream A, FT15
2    FB=AIN(2,FBLO,FBSPAN) ; reads in flow of stream B, FT16
3    TR=AIN(3,TRLO,TRSPAN) ; reads in reactor's temperature, TT17
4    TJ=AIN(4,TJLO,TJSPAN) ; reads in jacket's temperature, TT18
5    FBSP=FA*RATIO ; calculates the SP for stream B, the RATIO is set by operator
6    VPB=PID(FB,FBSP,FBLO,FBSPAN) ; FC16
7    AOUT(1,VPB) ; outputs signal to valve in stream B
8    OUTTJSP=PID(TR,TRSP,TRLO,TRSPAN) ; master controller, TC17.
                              ; set point TRSP is set by operator
9    TJSP=TJLO+OUTTJSP*TJSPAN/100 ; converts controller output of TC17 to
                              ; engineering units of jacket temperature
10   VPCW=PID(TJ,TJSP,TJLO,TJSPAN) ; slave controller, TC18
11   AOUT(2,VPCW) ; outputs signal to cooling water valve
12   FAOVVP=PID(VPCW,10,0,100) ; VPC20
13   FAOPVP=PID(FA,FAOPSP,FALO,FASPAN) ; FC15, set point FAOPSP is set by
                              ; operator
14   IF FAOPVP<FAOVVP THEN VPFA=FAOPVP ELSE VPFA=FAOVVP ; LS19
15   AOUT(3,VPFA) ; outputs signal to valve in stream A
```

Figure 10-5.14 Computer program of control scheme in Fig. 10-5.13.

EXAMPLE 10-5.2

Consider a reactor, Fig. 10-5.15, in which the irreversible and complete liquid reaction $A + B \rightarrow C$ occurs. Product C is the raw material for several process units downstream from the reactor. Reactant A is available from two sources. Because of a long-term contract, Source 1 is less expensive than Source 2. However, the contract is written with two limitations: a maximum instantaneous rate of 100 gpm, and a maximum monthly consumption of 3.744×10^6 gals. If

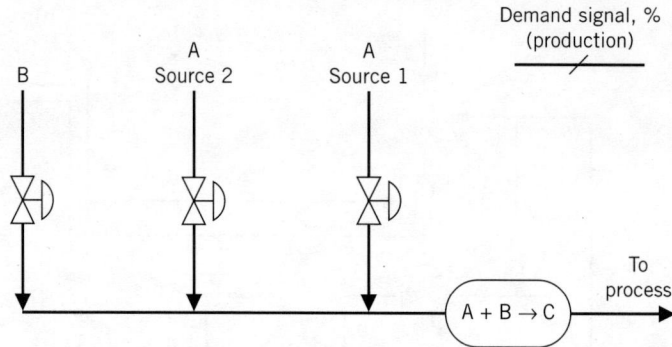

Figure 10-5.15 Reactor for Example 10-5.2.

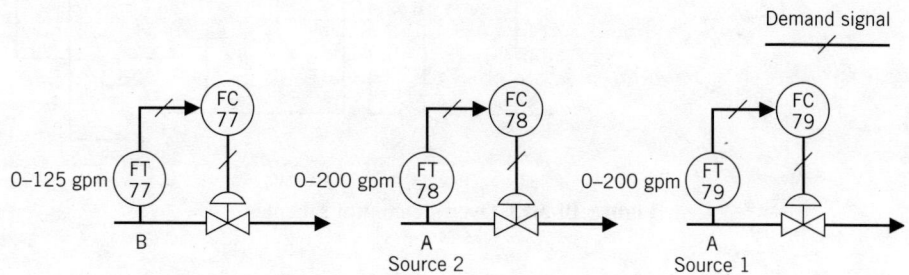

Figure 10-5.16 Flow loop installed in each stream.

either of these limitations is exceeded, a very high penalty must be paid, and thus it is less expensive to use the excess from Source 2. For example, if 120 gpm of reactant A are required, 100 gpm should come from Source 1 and the other 20 gpm from Source 2. Similarly, if on day 27 of the month 3.744×10^6 gals have been obtained from Source 1, from then on and until the end of the month, all of reactant A should come from Source 2. You may assume that the densities of each reactant, A and B and of product C do not vary much and, therefore, they can be assumed constant.

(a) Design a control system that will preferentially use reactant A from Source 1 and will not allow exceeding any contractual limitations. The feed ratio of A to B is 2:1 in gpm units.

(b) A few weeks after the control strategy designed in (a) was put into operation, it was noticed that for some unknown reason, the supply pressure from Source 2 was cut by the supplier every once in a while. Thus, the flow controller manipulating the flow from Source 2 would have to open the valve, and in some instances the valve would go wide open. At this moment there would not be enough flow from Source 2 to satisfy the demand. It was decided that the correct action to take in this case, while the lawyers investigate—which may take a long time—is to obtain from Source 1 whatever Source 2 does not supply. Design a control strategy to accomplish this action. Be sure that your design is such that whenever Source 2 provides the required flow, the scheme designed in part (a) is in effect.

(a) The first thing we will do is to install a flow loop in each of the three streams; this is shown in Fig. 10-5.16 along with each transmitter's range. Let us work out this design by developing the software program first. Note that the demand signal is the one that manipulates the flows, therefore, we first give this signal a significance. The significance can be anything that makes sense, for example, the flow of total A, or the flow of B. Let us call it the flow of B; we then convert this signal from % to units of flow of B:

```
1  FBSP=(Demand signal)*125 /100  ; converts the demand
                                   ; signal from percent
                                   ; to units of flow
                                   ; of stream B
```

We then send this set point to FC77:

```
2  FB=AIN(1,0,125) ; reads in the flow of B
3  VPB=PID(FB,FBSP,0,125) ; controls the flow of B (FC77)
4  AOUT(1,VPB) ; outputs signal to valve
```

Now that we know the flow of B, ratio the total flow of A, FA, to this flow:

```
5  FA=Ratio*FB ; the operator enters Ratio (FY75
               ; in Fig. 10-5.17)
```

Before we set the set point for the flow of A from Source 1 we must be sure that it does not exceed 100 gpm, and that the totalized flow A from Source 1 does not exceed 3.744×10^6 gals:

```
6  FA1=AIN(2,0,200) ; reads in flow A from Source 1
7  FA2=AIN(3,0,200) ; reads in flow A from Source 2
8  IF FA<100 THEN FA1SP=FA ELSE FA1SP=100 ; FASP1 is
                                          ; set point
                                          ; to the
                                     ; controller in A
                                     ; Source 1 (LS74)
9  FA1TOT=TOT(FA1) ; TOT( ) is a function that
                   ; totalizes its argument
                   ; (TOT71)
10 FA1LEFT=3.744 x 10⁶-FA1TOT ; calculates the flow
                              ; from Source 1
                              ; that is left to obtain
                              ; during the month (SUM72)
11 IF FA1LEFT<=0 THEN FA1LEFT=0 ; checks to make sure
                                ; FA1LEFT < 0
12 IF FA1LEFT<FA1SP THEN FA1SP=FA1LEFT ; (LS74)
13 VPA1=PID(FA1,FA1SP,0,200) ; FC79
14 AOUT(2,VPA1) ; outputs signal to valve in Source 1
15 FA2SP=FA-FA1SP ; calculates SP for A Source 2 (SUM73)
16 VPA2=PID(FA2,FA2SP,0,200) ; FC78
17 AOUT(3,VPA2) ; outputs signal to valve in Source 2
```

Lines 1 through 17 is the software program to accomplish the control objective. Figure 10-5.17 is the equivalent control scheme; it is wise to spend sometime analyzing both ways.

(b) On the basis of the decision, once the valve providing the A component from Source 2 is wide open, the flow that Source 2 is not providing must come from Source 1. Two lines of coding are enough to do this, and are the two new 13 and 14 lines shown below. Line 13 calculates the flow that is not given by Source 2 (NOTFA2); usually this quantity is zero. Line 14 checks if the controller output is 98 % or higher (valve is open at least 98 %), if this is the case then it adds NOTFA2 to the set point of Source 1 (FA1SP), if this is not the case then it does not add anything to FA1SP.

```
12 IF FA1LEFT<FA1SP THEN FA1SP=FA1LEFT ; (LS74)
13 NOTFA2 = FA2SP-FA2 ; Flow not provided by Source 2
14 IF VPA2>98 THEN FA1SP = FA1SP + NOTFA2 ELSE FA1SP =
      FA1SP
15 VPA1=PID(FA1,FA1SP,0,200) ; FC79
16 AOUT(2,VPA1) ; outputs signal to valve in Source 1
17 FA2SP=FA-FA1SP ; calculates SP for A Source 2 (SUM73)
18 VPA2=PID(FA2,FA2SP,0,200) ; FC78
19 AOUT(3,VPA2) ; outputs signal to valve in Source 2
```

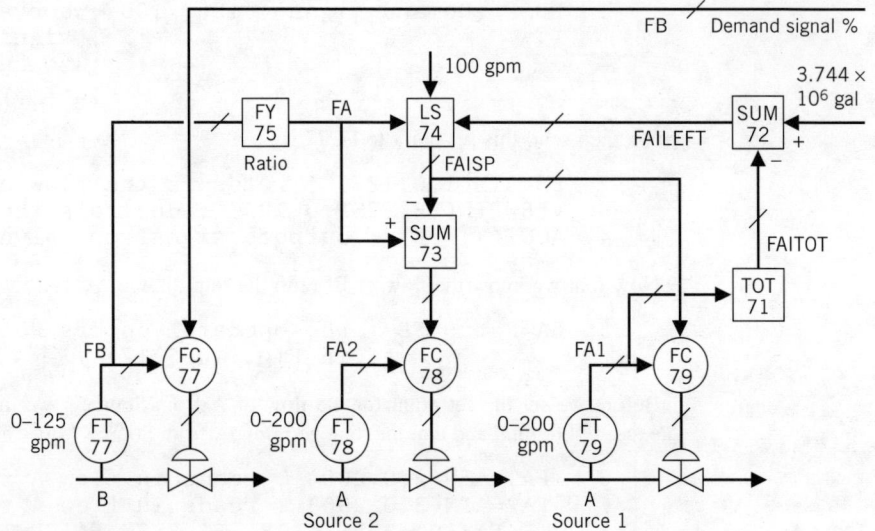

Figure 10-5.17 Control scheme for Example 10-5-2.

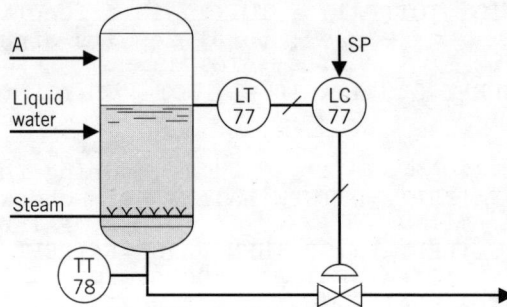

Figure 10-5.18 Reactor for Example 10-5.3.

EXAMPLE 10-5.3

Consider the reactor, shown in Fig. 10-5.18, where stream A reacts with water. Stream A can be measured but not manipulated. This stream is the by-product of another unit. The water enters the reactor in two different forms, as liquid and as steam. The steam is used to heat the reactor contents. It is necessary to maintain a certain ratio, R, between the total water and stream A into the reactor. It is also necessary to control the temperature in the reactor. It is important to maintain the ratio of total flow of water to flow of stream A below a value Y: otherwise, a very thick polymer may be produced plugging the reactor.

A situation has occurred several times during extended periods of time in which the flow of stream A reduces significantly. In this case the control scheme totally cuts the liquid water flow to the reactor to maintain the ratio. However, the steam flow to the reactor, to maintain temperature, still provides more water than required, and thus the actual ratio of water to stream A entering the reactor dangerously approaches Y. Design a control scheme to control the temperature in the reactor, and another scheme to maintain the ratio of total water to stream A, while avoiding reaching the value of Y even if it means that the temperature deviates from set point.

Figure 10-5.19 shows the temperature control and ratio control required. The temperature is controlled manipulating the steam flow using a cascade control scheme. The flow of stream A is measured and multiplied by R, in MUL 76, to obtain the total water flow required, F_{TW}. The steam flow is then subtracted from the total water to calculate the flow of liquid water required, which is then used as the set point to the liquid water controller.

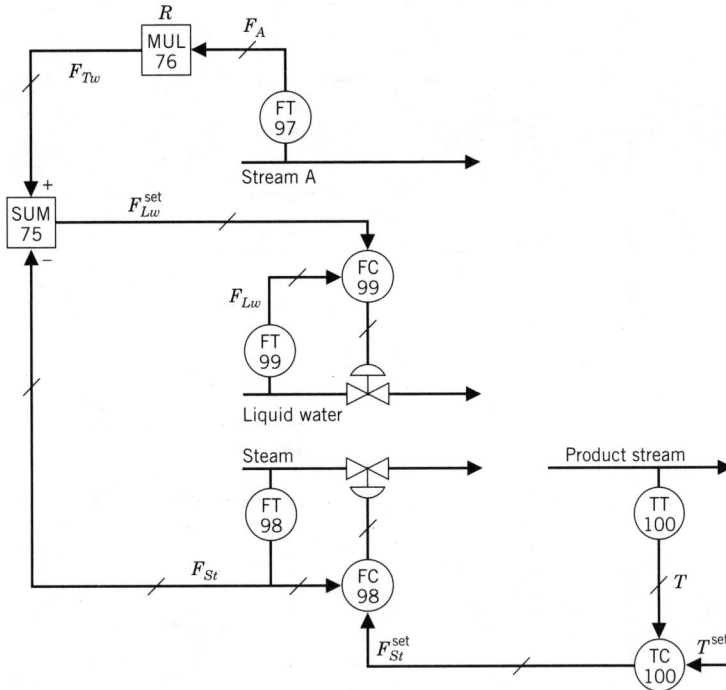

Figure 10-5.19 Temperature and ratio control for reactor of Example 10-5.3.

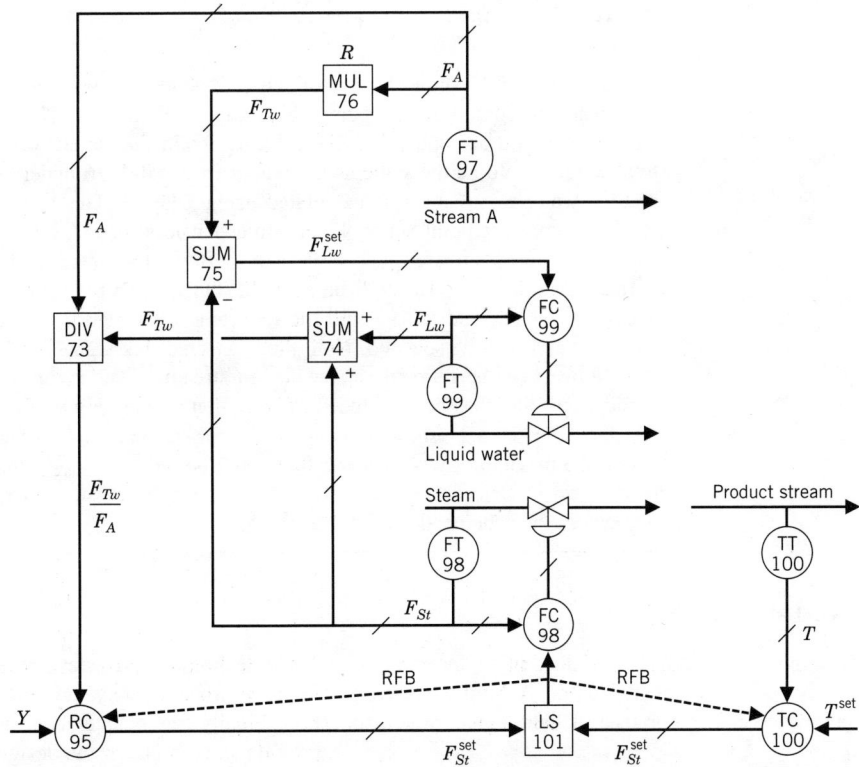

Figure 10-5.20 Complete control scheme for reactor of Example 10-5.3.

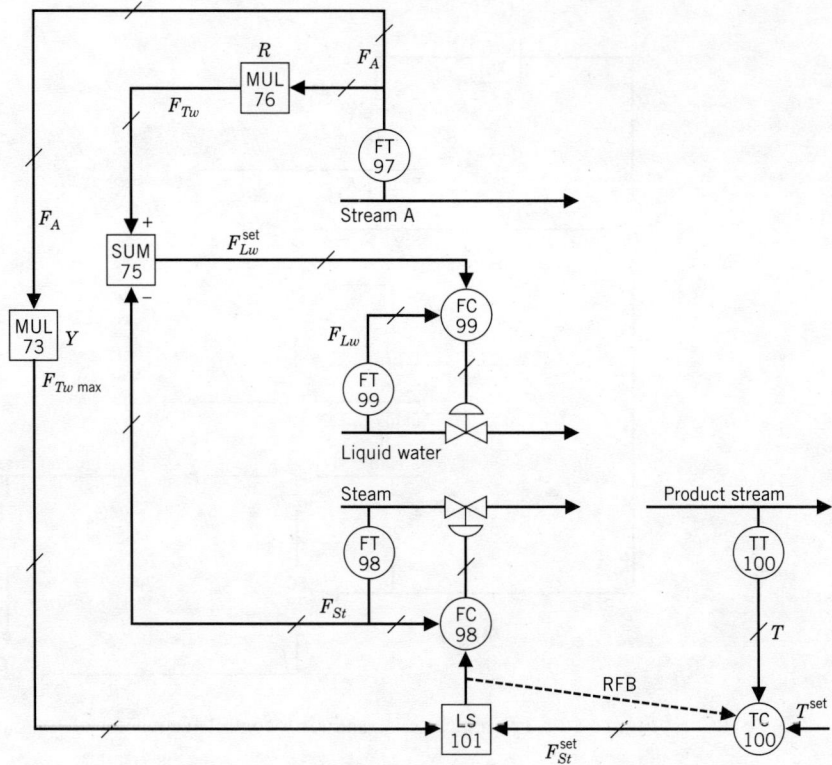

Figure 10-5.21 Another complete control scheme for reactor of Example 10-5.3.

Figure 10-5.20 shows a control scheme, added to the previous one, to avoid the ratio of total water to stream A exceeding the value of Y. In this scheme the actual flows of liquid water and steam are added, in SUM 74, to obtain the actual water flow into the reactor (this total water should be the same as the output from MUL 76 under normal conditions). The ratio of total water to stream A is calculated using DIV 73. This ratio is then sent to a controller, RC 95, with a set point set to Y, or somewhat less than Y for safety. The outputs of RC 95 and TC 100 are compared in LS 101 and the lowest selected as the set point to FC 98. Under normal conditions TC 100 will be selected. Only when the ratio of total water to stream A is above the set point to RC 95 will the ratio controller reduce its output enough, in an effort to cut the steam, and be selected. Note the reset feedback signals to RC 95 and TC 100.

Notice that the ratio of total water to stream A will be at, or close to, Y only after the liquid water flow has been reduced to zero, that is, the only water entering is the steam. Using this fact, Fig. 10-5.21 shows a simpler control scheme. In this case F_A is multiplied by Y to obtain the maximum flow of water that could be fed, $F_{TW\,max}$. This scheme is simpler because there is no need to tune a controller. The reader may want to write the software program to implement the scheme shown in Fig. 10-5.21.

10-6 SUMMARY

This chapter has presented an introduction to the computation tools provided by manufacturers. A brief discussion of the significance and importance of field signals was also presented.

The chapter also presented the concepts and applications of ratio control, override control, and selective control.

These techniques provide a realistic and simple method to improve process safety, product quality, and process operation. Finally, the chapter concluded with three examples to provide some hints on the design of control schemes.

REFERENCES

1. SMITH, C. A., and A. B. CORRIPIO, *Principles and Practice of Automatic Process Control*, 2nd Ed., John Wiley & Sons, 1997.

2. O'MEARA, J. E., "Oxygen Trim for Combustion Control." *Instrumentation Technology*, March 1979.

3. SCHEIB, T. J., and T. D. RUSSELL, "Instrumentation Cuts Boiler Fuel Costs." *Instrumentation and Control Systems*, November 1981.

4. CONGDON, P., "Control Alternatives for Industrial Boilers." *InTech*, December, 1981.

5. BECKER, J. V., "Designing Safe Interlock Systems." *Chemical Engineering Magazine*, October 15, 1979.

6. BECKER, J. V., "Designing Safe Interlock Systems." *Chemical Engineering Magazine*, October 15, 1979.

PROBLEMS

10-1. Consider a system, shown in Fig. P10-1, to dilute a 50 % by mass NaOH solution to a 30 % by mass solution. The NaOH valve is manipulated by a controller not shown in the diagram. Because the flow of the 50 % NaOH solution can vary frequently, it is desired to design a ratio control scheme to manipulate the flow of H_2O to maintain the required dilution. The nominal flow of the 50 % NaOH solution is 200 lbm/h. The flow element used for both streams is such that the output signal from the transmitters is linearly related to the mass flow. The transmitter in the 50 % NaOH stream has a range of 0 to 400 lbm/h, and the transmitter in the water stream has a range of 0 to 200 lbm/h. Specify the computing blocks required to implement the ratio control scheme, and write the complete DCS program that implements the ratio strategy.

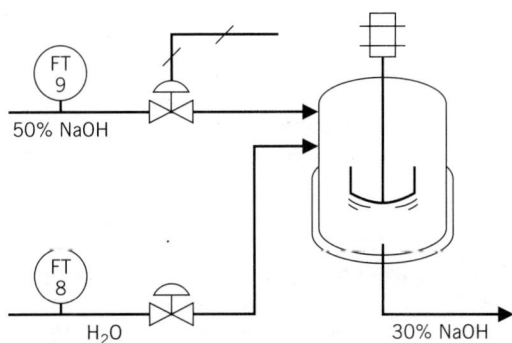

Figure P10-1 Mixing process for Problem 10-1.

10-2. A standard mass flow computer calculates the mass flow of a gas from the orifice equation

$$w(t) = K_0\sqrt{Mp(t)h(t)/RT(t)}$$

where $w(t)$ is the mass flow, K_0 is the orifice coefficient, M is the molecular weight of the gas, $p(t)$ is the absolute pressure, R is the ideal gas constant, $T(t)$ is the absolute temperature of the gas, and $h(t)$ is the differential pressure across the orifice. For this application

$$K_0 = 200(\text{lb/h})/(\text{in. } H_2O\text{-lb/ft}^3)^{0.5}$$

$$M = 40 \text{ lb/lbmole}; \qquad R = 10.73 \text{ psia-ft}^3/\text{lbmole-}°R$$

with transmitter ranges

$$p(t) = 0 \text{ to } 50 \text{ psig}; \qquad T(t) = 100 \text{ to } 200°F$$

$$h(t) = 0 \text{ to } 100 \text{ in. } H_2O; \qquad w(t) = 0 \text{ to } 1500 \text{ lb/h}$$

Develop the complete DCS program to implement the mass flow calculation.

10-3. The heat exchanger shown in Fig. 1-1.2 heats a process fluid by condensing steam. A control scheme calls for controlling the heat transferred to the fluid. This heat transfer is calculated using the following equation

$$Q = F\rho C_p(T - T_1)$$

The following information is known:

Variable	Range	Steady State
F	0–50 gpm	30 gpm
T	50–120°F	80°F
T_1	25–60°F	50°F
Q	0–?	?

The density ($\rho = 5.62$ lb/gal) and heat capacity ($C_p = 0.60$ Btu/lb-°F) are assumed to be constant. Using the blocks shown in Fig. 10-1.1, draw the control strategy to control the heat transfer rate to the process fluid. Then write the complete DCS program to implement this control.

10-4. Figure P10-2 shows the reflux to the top of a distillation column. The internal reflux, L_I, controls the separation in the column. This internal reflux cannot be directly manipulated; however, the external reflux L_E can be manipulated. The "internal reflux computer" computes the set point, L_E^{set}, of the external reflux flow controller so as to maintain the internal reflux at some desired value, L_I^{set}. The internal reflux is greater than the external reflux because of the condensation of vapors on the top tray, which is required to bring the subcooled reflux at T_L up to its bubble point, T_V. An energy balance on the top tray yields the following working equation

$$(L_I - L_E)\lambda = L_E C_{PL}(T_V - T_L)$$

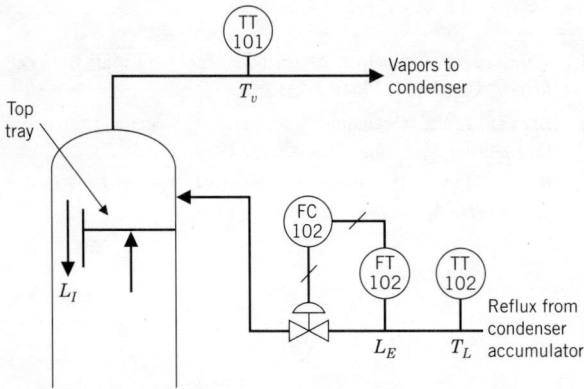

Figure P10-2 Distillation column for Problem 10-4.

For this process the heat capacity of the liquid and the latent heat can be assumed constants at values of $C_{PL} = 0.76$ Btu/lb-$°$F and $\lambda = 285$ Btu/lb. Other design specifications are as follows:

Transmitter	Range	Normal Value
FT102 (L_E)	0–5000 lb/h	3000 lb/h
TT102 (T_L)	100–300$°$F	195$°$F
TT101 (T_V)	150–250$°$F	205$°$F

Develop the complete DCS program to implement this control.

10-5. Consider the reactor shown in Fig. P10-3. This reactor is similar to a furnace in that the energy required for the reaction is provided by the combustion of a fuel with air (to simplify the diagram the temperature control is not completely shown). Methane and steam are reacted to produce hydrogen by the reaction

$$CH_4 + 2H_2O \longrightarrow CO_2 + 4H_2$$

The reaction occurs in tubes inside the furnace. The tubes are filled with a catalyst needed for the reaction. It is important that the reactant mixture always be steam-rich to avoid coking the catalyst. That is, if enough carbon deposits over the catalyst, it poisons the catalyst. This situation can be

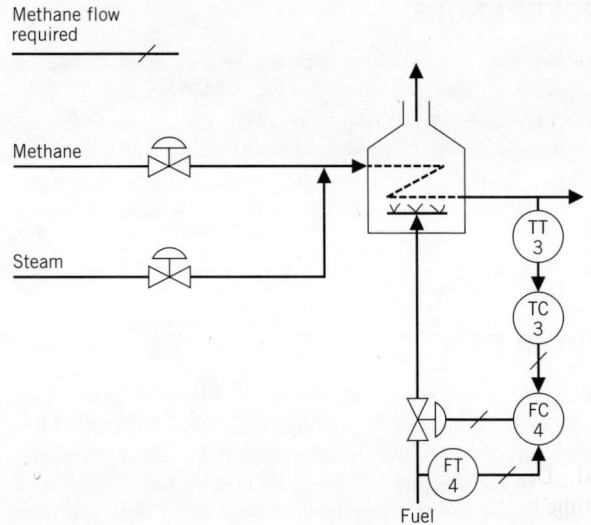

Figure P10-3 Reactor for Problem P10-5.

avoided by ensuring that the entering mixture is always rich in steam. However, too much steam is also costly in that it requires more energy (fuel and air) consumption. The engineering department has estimated that the optimum ratio R_1 (methane to steam) must be maintained. Design a control scheme that ensures the required ratio is maintained, and that during production rate changes, when it increases or decreases, the reactant mixture be steam-rich. Please note that there is a signal that sets the methane flow required.

10-6. Chlorination is used for disinfecting the final effluent of a wastewater treatment plant. The EPA requires that certain chlorine residual be maintained. To meet this requirement, the free chlorine residual is measured at the beginning of the chlorine contact basin, as shown in Fig. P10-4. An aqueous solution of sodium hypochlorite is added to the filter effluent to maintain the free chlorine residual at the contact basin. The amount of sodium hypochlorite required is directly proportional to the flow rate of the effluent. The wastewater plant has two parallel filter effluent streams, which are combined in the chlorine contact basin. Sodium hypochlorite is added to each stream based on free chlorine residual in the basin.

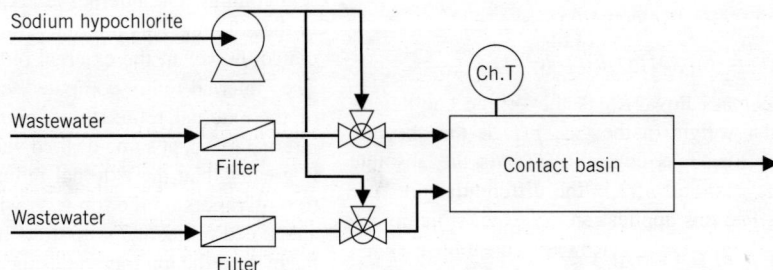

Figure P10-4 Chlorination process for Problem 10-6.

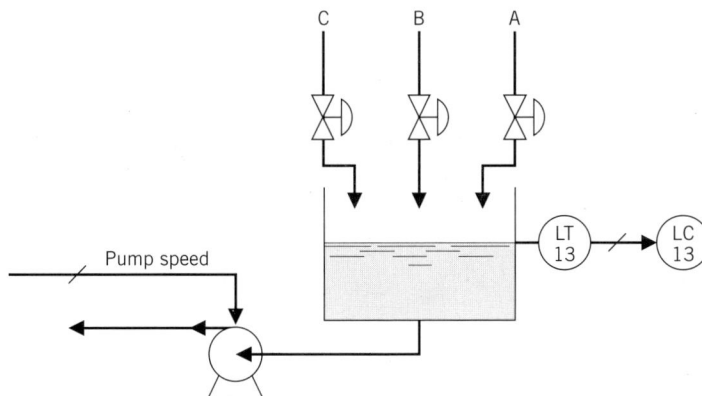

Figure P10-5 Process for Problem 10-7.

(a) Design a control scheme to control the chlorine residual at the beginning of the basin.

(b) Due to a number of reactions occurring in the contact basin, the chlorine residual exiting the basin is not equal to the chlorine residual entering the basin (the one being measured). It happens that EPA is interested in the exiting chlorine residual. Thus, a second analyzer is added at the effluent of the contact basin. Design a control scheme to control the effluent chlorine residual.

10-7. Consider the tank shown in Fig. P10-5. In this tank three components are mixed in a given proportion so as to form a stock that is supplied to another process. For a particular formulation the final mixture contains 50 mass % of A, 30 mass % of B, and 20 mass % of C. Another process provides the signal to the pump depending on its demand.

(a) Design a control system to control the level in the tank and at the same time maintain the correct formulation.

(b) Suppose now that because of operating problems in one or more of the units supplying A, B, or C the supply pressure in one or more of these lines decreases. Obviously, the controller(s) manipulating the corresponding valve(s) will start opening the valve(s). Once a valve goes to 100 % open there is no control any longer. Actually, once a valve opens to about 90 % there is not much control action. So, what would you do to avoid any valve to open more than 90 %? Remember, you must still maintain level control. How would you automate your proposed strategy?

10-8. Fuel cells are used in spacecraft and proposed extraterrestrial bases for generating power and heat. The cell produces electric power by the reaction between liquid hydrogen and liquid oxygen

$$2H_2 + O_2 \longrightarrow 2H_2O$$

Design a ratio controller to maintain the flows of liquid hydrogen and oxygen into the cell in the exact stoichiometric ratio (both hydrogen and oxygen are valuable in space, so we cannot supply either in excess). Calculate the design flows of hydrogen and oxygen required to produce 0.5 kg/h

of water, and the design ratio of oxygen to hydrogen flow. Sketch a ratio control scheme that will manipulate the flow of oxygen to maintain the exact stoichiometric ratio between the two flows. You may assume that the signals from the flow transmitters are linear with the mass flow rates. Calculate reasonable ranges for the flow transmitters and the ratio in terms of the transmitter signals.

10-9. Consider the process shown in Fig. P10-6. Mud is brought into a storage tank, T-3, from which it is pumped to two filters. Manipulating the exit flow controls the level in the tank. This flow must be split between the two filters in the following known ratio:

$$R = \frac{\text{flow to filter 1}}{\text{total flow}}$$

The two flow transmitters and control valves shown in the figure cannot be moved from their present locations, and no other transmitters or valves can be added. Design a control system that controls the level in T-3, while maintaining the desired flow split between the two filters.

10-10. Consider the furnace shown in Fig. P10-7. This furnace is used to partially vaporize water. Liquid water enters the furnace, and a mixture of saturated liquid and vapor exits. The mixture then goes into a tank where residence time is provided to separate the saturated vapor from the saturated liquid. The process engineers talk about the "efficiency" of this process. They define efficiency as the fraction of the liquid that is vaporized; the desired efficiency is 80 %. Design a control scheme to maintain the efficiency at this desired value.

10-11. Consider the bottom of the distillation column shown in Fig. P10-8. There are two streams leaving the column. One of the streams is manipulated by User 1 to satisfy its own needs. The other stream is manipulated to maintain liquid level in the bottom of the column. Under some upset conditions the level drops enough so that the level controller closes the valve; when this happens level control is lost. If the level ever drops below 1.5 ft it would be very difficult

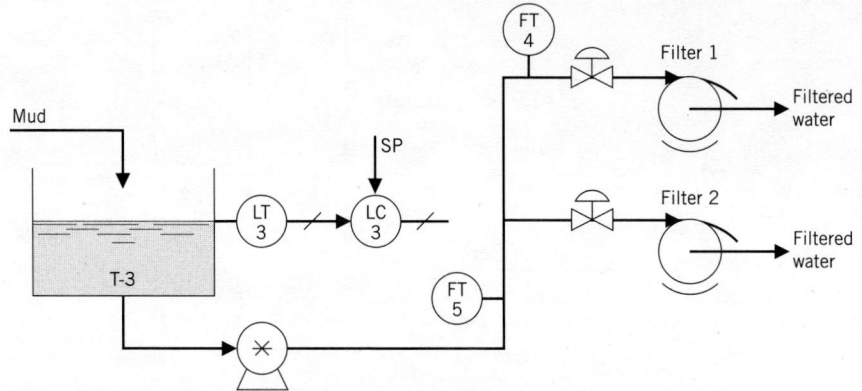

Figure P10-6 Process for Problem 10-9.

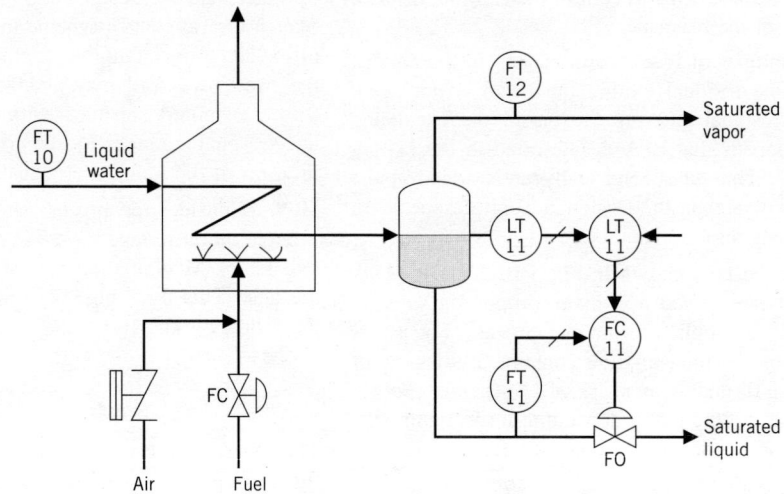

Figure P10-7 Process for Problem 10-10.

to have enough flow through the reboiler to provide boil-up to the column; this condition would be catastrophic to the operation of the column. The level transmitter is calibrated for 1–5 ft. What would you propose to avoid this condition? Design the control scheme to implement your proposal.

10-12. Consider the process shown in Fig. P10-9. The feed to the reactor is a gas, and the reaction produces a polymer. The outlet flow from the reactor is manipulated to control the pressure in the reactor. Exiting the reactor is polymer with entrained gas. This outlet flow goes to a separator that provides enough residence time to separate the gas from the polymer. The polymer product is manipulated to control the level in the separator; the gases flow out of the separator freely. These gases contain the unreacted reactants and an amount of wax components that have been produced. The gases are compressed before being returned to the reactor. A portion of the gases are cooled and mixed with the reactor effluent to control the temperature in the separator, as shown in the figure. If the temperature in the separator is too high

the wax components will exit with the gases. This wax will damage the compressor, and it is why cyclones are installed in the recycle line. If the temperature in the separator is too low the polymer will not flow out of the separator. Thus, the separator temperature must be controlled.

When the separator temperature increases, the temperature controller opens the recycle valve to increase the flow of cool gas. Under some significant upsets, as when a new polymer product is being produced, the recycle valve may go wide open in an effort to control the temperature. At this time the operator manually opens the chilled-water valve to the gas coolers. This action reduces the gas temperature providing more cooling capacity to the separator, and thus the recycle valve can close. Design a control scheme that provides this operation automatically, that is, avoids the recycle valve opening more than 90 %.

10-13. Figure P10-10 shows a system to filter an oil before it is processed. The oil enters a header in which the pressure is controlled, for safe operation, by manipulating the inlet

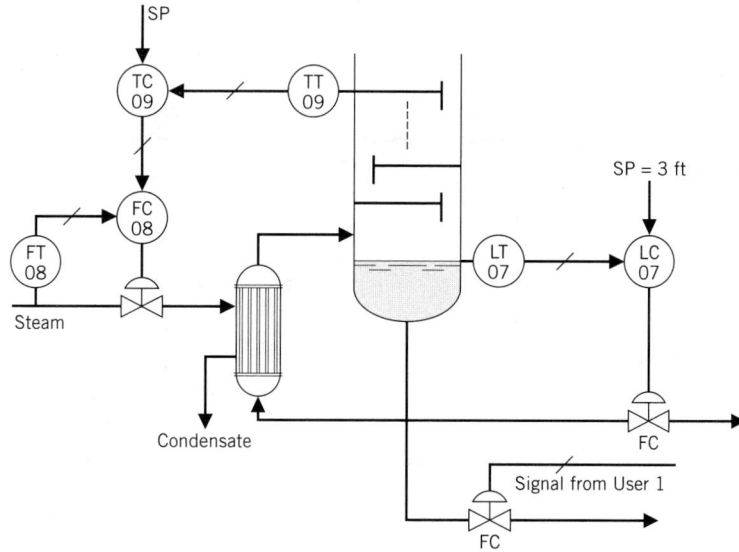

Figure P10-8 Process for Problem 10-11.

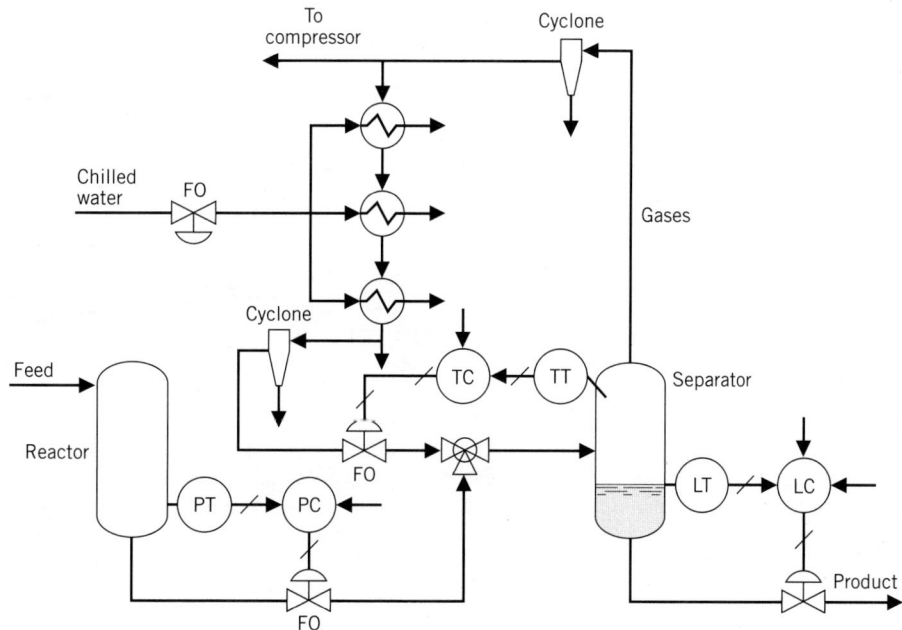

Figure P10-9 Process for Problem 10-12.

valve. From the header, the oil is distributed to four filters. The filters consist of a shell with tubes inside similar to heat exchangers. The tube wall is the filter medium through which the oil must flow to be filtered. The oil enters the shell and flows through the medium into the tubes. As time passes, the filter starts to build up a cake, and consequently, the oil pressure required for flow increases. If the pressure increases too much, the walls may collapse. Thus at some point, the

filter must be taken out of service and cleaned. Under normal conditions, three filters can handle the total oil flow.

(a) Design a control system to maintain the total oil flow through the system. The pressure loop must not be used for this purpose; it is there to maintain a safe supply pressure to the filters.

(b) Design a control system so that as the oil pressure in the shell side of each filter increases above some predetermined

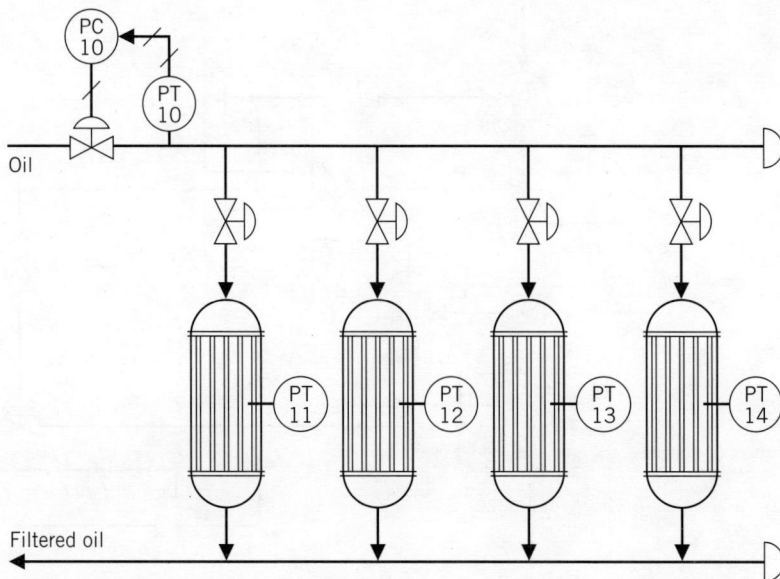

Figure P10-10 Filters for Problem 10-13.

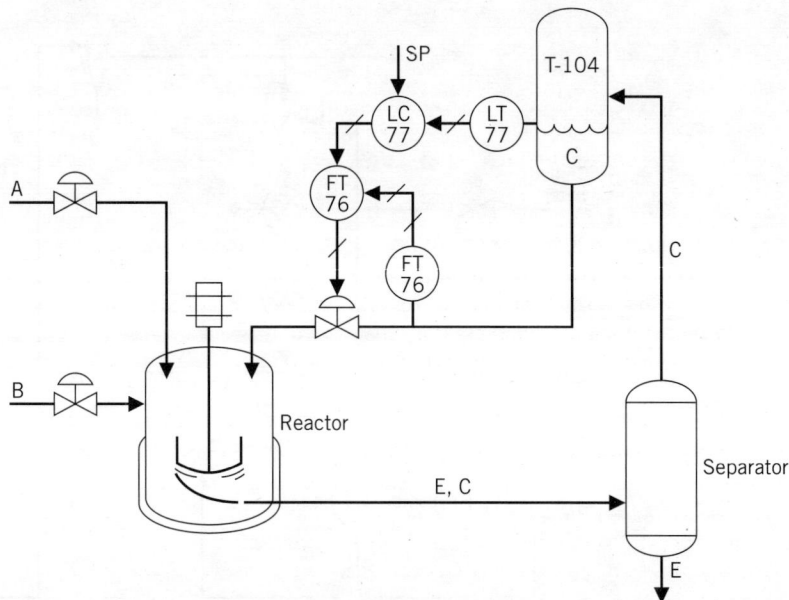

Figure P10-11 Process for Problem 10-14.

value, the oil flow to that filter starts to decrease. Once the feed valve to the filter is 10 % open, an interlock system will shut down the filter for cleaning (this system is not shown). The total oil flow through the system must still be maintained.

10-14. Consider the process shown in Fig. P10-11. This process is used to manufacture product E from the reaction of A and B. The output from the reactor is product E and some unreactants, mainly A, referred to as liquid C. E and

C are separated, and liquid C is recycled back to tank T-104 to be fed back to the reactor as shown in the figure. The amount of B fed to the reactor depends on the amount of A and on the amount of C fed to the reactor. That is, there must be some B to react with the A fed, given by the ratio $R_1 = B/A$, and some B to react with the C fed, given by the ratio $R_2 = B/C$. You may assume that all the flowmeters provide a signal related to mass flow. Design a control scheme to control the total flow, T in lb/min, into the reactor.

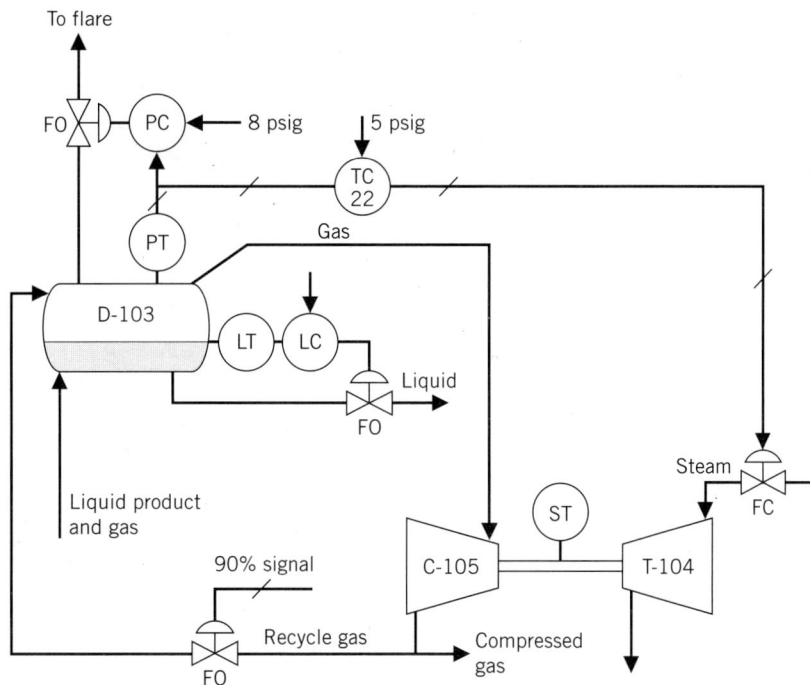

Figure P10-12 Process for Problem 10-15.

10-15. Consider the process shown in Fig. P10-12. In this process a liquid product is separated from a gas; the gas is then compressed. Drum D-103 provides the necessary residence time for the separation. The pressure in the drum is controlled at 5 psig as shown in the figure. Another pressure controller opens the valve to the flare if the drum pressure reaches 8 psig. There is always a small amount of recycle gas to the drum. The turbine driving the compressor is rather old, and for safety considerations its speed must not exceed 5600 rpm nor drop below 3100 rpm. Design a control scheme that provides this limitation.

10-16. Figure P10-13 shows a process often found in chemical plants. R-101 is a reactor where a high-pressure gas is generated. It is necessary to transfer the gas to a low-pressure vessel, V-102, at about 50 psig. As an energy-saving measure, the gas pressure is dropped across a power recovery turbine, T-102. The work produced in T̂-102 is used to drive a compressor, C-102. However, the work produced in T-102 is usually not enough to run C-102, so a steam turbine, T-103, is connected in series with T-102 to provide the necessary work. The figure shows the control systems to control the pressure in R-101 and the pressure of the gas leaving the compressor. ST-16 is a speed transmitter connected to the turbine's shaft; SC-16 is a controller controlling the shaft rotational speed. Out of R-101 there is a line that bypasses T-102 and goes directly to V-102. This line is used in case T-102 is down or in case any emergency develops and it is necessary to relief the pressure in R-101 quickly. The set

point to PC-14 is 500 psig, whereas that to PC-15 is 510 psig. A condition occurs when the gas produced in R-101 increases significantly, thus increasing the pressure. In this case PC-14 opens the valve to T-102 more to relieve the pressure. When this occurs, the steam valve to T-103 backs off to control the pressure of the compressed gas. It has been determined that if the steam valve is less than 10 % open, then severe mechanical damage can occur in T-103. Design a control scheme to avoid this condition while still controlling the pressure of the compressed gas.

10-17. In Example 10-3.2 the constraint control scheme shown in Fig. 10-3.5 was presented and discussed. In the discussion of the scheme, it was mentioned that "if at any time the operating personnel were to set the flow controller FC11 in local set point or in the manual mode, that is, off remote set point, the safety provided by TC13 and PC14 would not be in effect. This would result in an unsafe and unacceptable operating condition." Modify the control scheme shown such that even when FC11 is off remote set point, controllers TC13 and PC14 still provide the necessary safety override.

10-18. Consider the distillation column shown in Fig. P10-14. The diagram shows the controls associated with the tops of the column. Specifically, the column pressure control, reflux flow control, accumulator level control, and cooling water flow control are shown.

Consider the column pressure control in more detail. The pressure controller manipulates the outlet valve from the condenser to maintain the column pressure at its set point. If the

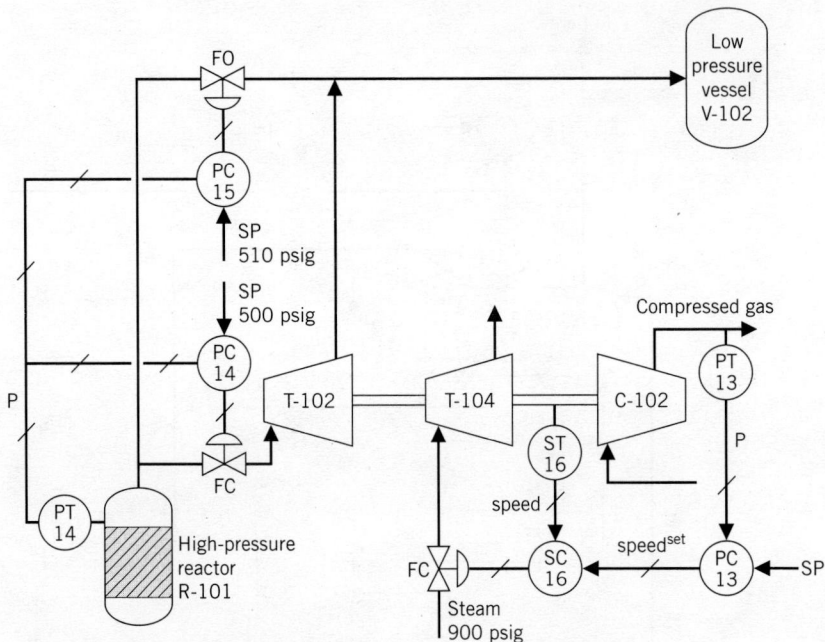

Figure P10-13 Process for Problem 10-16.

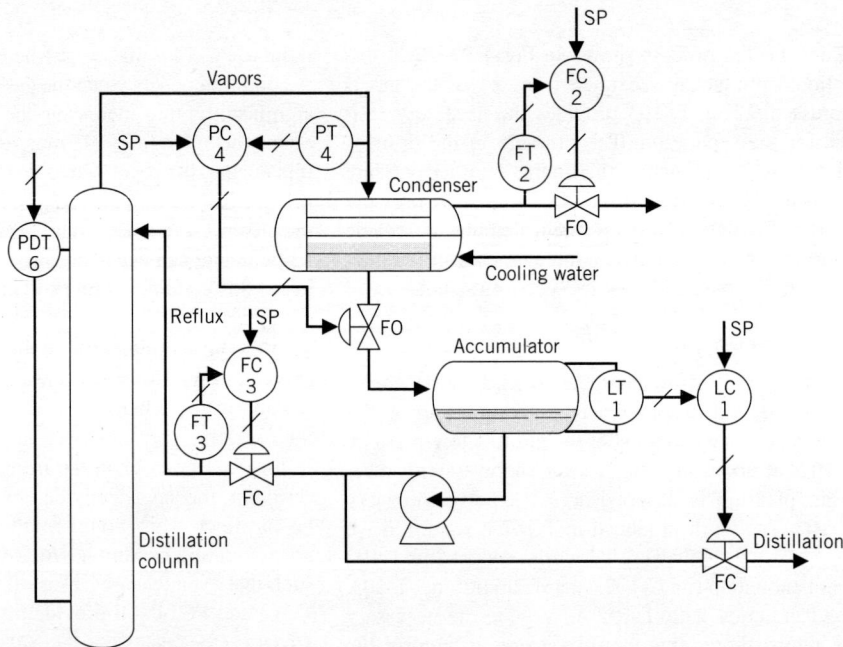

Figure P10-14 Process for Problem 10-18.

pressure in the column rises, the controller opens the valve to discharge liquid from the condenser, exposing more heat transfer area to condense more vapors and lower the pressure. Thus, essentially, the pressure controller varies the exposed heat transfer area to control the column pressure. For most mixtures the separation is easier at lower pressures. However, there are limits on how low the pressure can be:

(a) Column flooding—for a given molar flow, the pressure drop across the column increases as the pressure decreases. PDT-6 (pressure differential transmitter) indicates

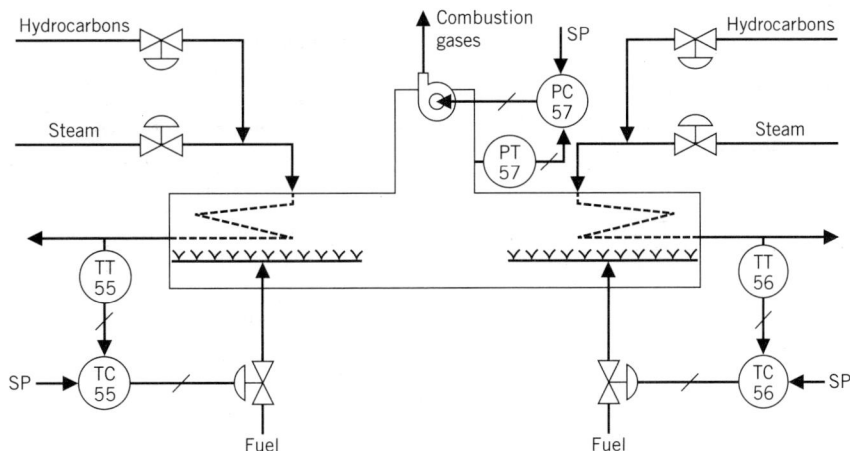

Figure P10-15 Furnace for Problem 10-19.

the pressure drop. A high pressure drop indicates column flooding.

(b) Condenser capacity—at lower pressures the condensation temperature is lower and, therefore, the temperature difference between the condensing vapors and the cooling water in the condenser is less. Because this condensing driving force (the temperature difference) is less, the condenser capacity is also less. To recover condenser capacity the outlet valve from the condenser must open to expose transfer area. Once the valve is completely opened there is no pressure control.

(c) Distillate valve capacity—at lower pressure the pressure drop across the distillate valve decreases. The level controller in the accumulator must then open the valve to control the level in the accumulator at its set point. Once the valve is completely opened there is no level control.

Design a control system to run the column at the lowest possible pressure while avoiding the three limit conditions explained in the previous paragraph.

10-19. Consider the furnace shown in Fig. P10-15, consisting of two sections with one common stack. In each section the cracking reaction of hydrocarbons with steam takes place. Manipulating the fuel to the particular section controls the temperature of the products in each section. Manipulating the speed of a fan installed in the stack controls the pressure in the stack. This fan induces the flow of flue gases out of the stack. As the pressure in the stack increases, the pressure controller speeds up the fan to lower the pressure.

(a) Design a control scheme to ratio the steam flow to the hydrocarbons flow in each section. The operating personnel set the hydrocarbons flow.

(b) During the last few weeks the production personnel have noticed that the pressure controller's output is consistently reaching 100 %. This indicates that the controller is doing the most possible to maintain pressure control.

However, this is not a desirable condition because it means that the pressure is really out of control—not a safe condition. A control strategy must be designed such that when the speed of the fan is greater than 90 %, the flow of hydrocarbons starts to reduce to maintain the fan speed at 90 %. As the flow of hydrocarbons decreases, less fuel is required to maintain exit temperature. This in turn reduces the pressure in the stack, and the pressure controller will slow down the fan. Whenever the fan speed is less than 90 %, the feed of hydrocarbons can be whatever the operating personnel require. It is known that the left section of the furnace is less efficient than the right section. Therefore, the correct strategy to reduce the flow of hydrocarbons calls for reducing the flow to the left section first, up to 35 % of the flow set by the operating personnel. If further reduction is necessary, the flow of hydrocarbon to the right section is then reduced, also up to 35 % of the flow set by the operating personnel. (If even further reduction is necessary, an interlock system would then drop off line the furnace). Design the control strategy to maintain the pressure controller's output below 90 %.

10-20. Consider the furnace of Fig. P10-16, wherein two different fuels—a waste gas and fuel oil—are manipulated to control the outlet temperature of a process fluid. The waste gas is free to the operation, so it must be used to full capacity. However, environmental regulations dictate that the maximum waste gas flow be limited to one-quarter of the fuel oil flow. The heating value of the waste gas is HV_{wg}, and that of the fuel oil is HV_{oil}. The air to waste gas ratio is R_{wg}, and the air to fuel oil ratio is R_{oil}.

(a) Design a cross-limiting control scheme to control the furnace product temperature.

(b) Assume now that the heating value of the waste gas varies significantly as its composition varies. It is difficult to measure on-line the heating value of this gas. However, laboratory analysis has shown that there is definitely a correlation between the density of the gas and its heating value.

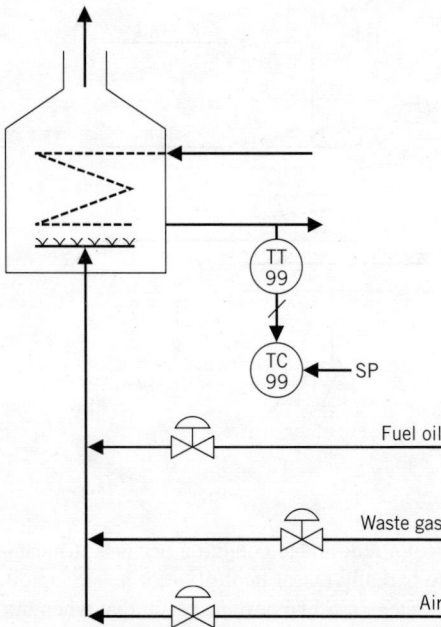

Figure P10-16 Furnace for Problem 10-20.

There is a densitometer available to measure the density of the waste gas, so the heating value is known. Adjust the control scheme design in (a) to consider variations in HV_{wg}.

(c) For safety reasons, it is necessary to design a control scheme such that in case of loss of burner flame, the waste gas and fuel oil flows cease; the air valve must open wide. Available for this job is a burner switch whose output is 20 mA as long as the flame is present, and whose output drops to 4 mA as soon as the flame stops. Design this control scheme into the previous one.

10-21. Consider the turbine/compressor process shown in Fig. P10-17. The motive force for the turbine, T-30, is a high-pressure gas, and the compressor, C-30, compresses a refrigerant gas. The operator sets the valve position of the high-pressure gas valve, which in turn results in a certain compressor speed. A lag unit is used to avoid sudden changes in the high-pressure valve position. Under normal operating conditions, the valve should respond to the operator's set value. However, there are some special conditions that a control system must guard against.

- Under normal conditions, the pressure in the refrigerant gas line is about 15 psig. However, during start-up and other circumstances, the pressure in the line tends to drop below 8 psig, which is a dangerous condition to the compressor. In this case the compressor velocity must be reduced to pull in less gas, thus increasing the pressure in the line. The lowest safe pressure in the refrigerant line is 8 psig.

- Because of mechanical difficulties, the compressor velocity must not exceed 95 % of its maximum rated velocity. Also, it must not drop below 50 % of its maximum velocity.

Design a control scheme that avoids violating the foregoing constraints.

10-22. Consider the compressor shown in Fig. P10-18. This two-stage compressor has two different suction points. In each suction line there is a volumetric flowmeter calibrated at $0°C$ and 1 atm, a pressure transmitter, and a temperature transmitter. An important consideration in the control of the compressor is to avoid the surge condition. Figure P10-18 also shows a curve indicating the minimum inlet flow, in acfm (actual ft^3/min), required, for a given inlet pressure, to avoid surge. Each stage can go into surge independently. Under normal operating conditions, the operator sets the position to each suction valve. However, for safety reasons, the operator must not close the valves below the surge limit. Design a control scheme to avoid closing the valves below the surge limit. Also, for safety considerations, it is permissible to open the valves very fast; closing the valves must be done slowly. Design this constraint into your previous design.

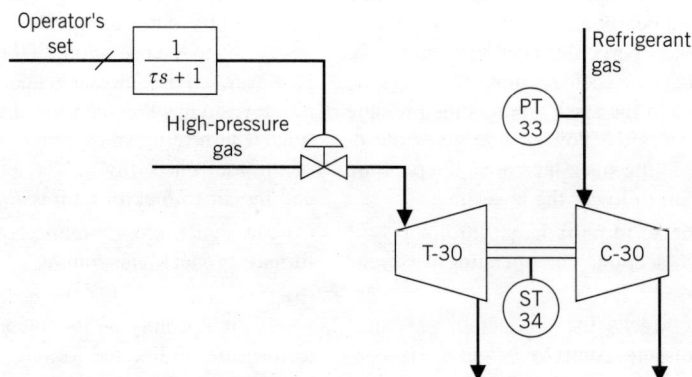

Figure P10-17 Turbine/compressor process for Problem 10-21.

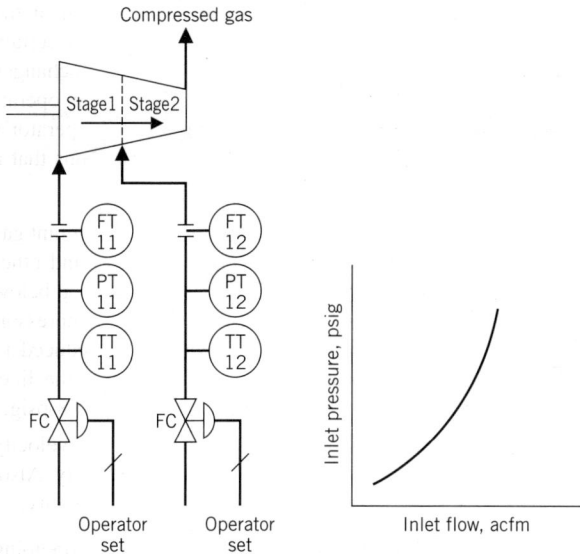

Figure P10-18 Compressor for Problem 10-22.

pressure drop required to flow through it increases. Once this pressure drop reaches a critical value the filter must be removed from service and cleaned. The cleaning, which consists of back flushing with a solvent, is a relatively fast operation. From the filters the water flows to a scrubber, T-46, where the dissolved oxygen is stripped. The scrubber consists of a top section in which the water flows countercurrent to natural gas, which is the stripping agent. The internals of this top section are similar to that of a plate distillation column. From the top section the water flows to the bottom section which looks like a surge tank. The level in the tank must be controlled. The dissolved O_2 in the water out of this tank must also be controlled. From the scrubber the water is then pumped to three oil wells.

(a)   Design a control scheme to control the level of water in the bottom section of the tank.

(b)   Design an override control scheme that reduces the water flow through any filter to avoid the pressure drop reaching the critical value. Once the water valve in the filter reaches a minimum prescribed opening, sequential logic takes over to stop the operation and start the cleaning cycle. You are not required to show this.

(c)   Design a control scheme to control the dissolved O_2 in the water leaving the scrubber. Most of the control action can be obtained by manipulating the natural gas; however, sometimes this gas by itself is not enough to reach set point. In such cases chemicals are used to reduce the O_2 to its desired value. The natural gas is less expensive than the chemicals. It is known that the most important disturbance to this control is the water flow to the scrubber.

10-23.   High-pressure water is often used to force petroleum crude out of oil wells. Figure P10-19 shows the process to prepare the water and its injection into the wells. The water must be filtered, from suspended solids, and deoxygenated before entering the wells. The figure shows three filters—F-43, F-44, and F-45—where the suspended solids are removed. Two filters are enough to filter the water. As a filter plugs, due to the solids removed from the water, the

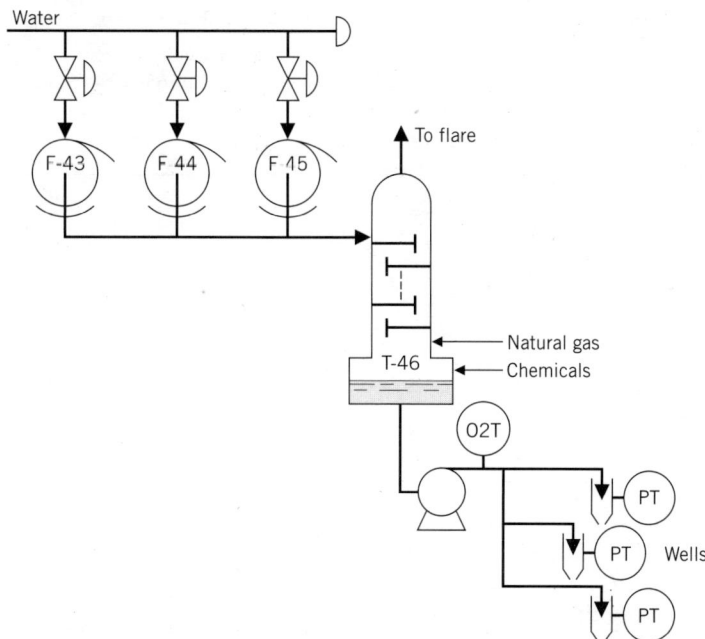

Figure P10-19 Process for Problem 10-23.

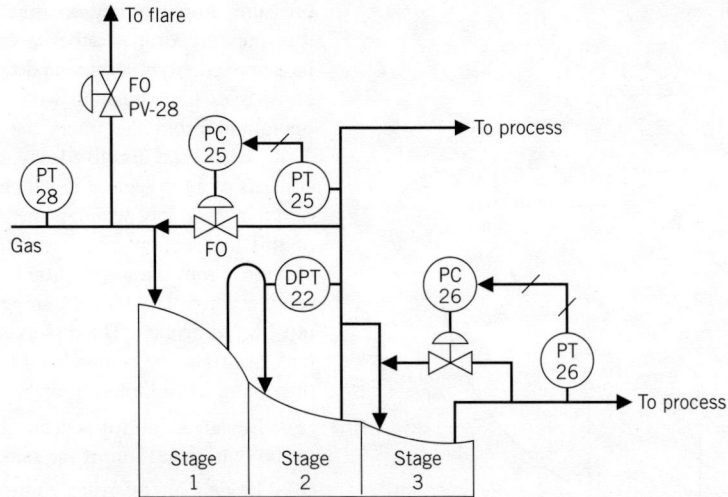

Figure P10-20 Process for Problem 10-24.

(d) Design a control scheme to control the flow of water to each well. An important consideration is the water pressure in each well. As the well ages, or internal disturbances occur, the pressure in the well increases. If the pressure reaches a certain value it may crack the well. Thus, the pressure in each well must be considered in this control scheme.

10-24. Consider the three-stage compressor shown in Fig. P10-20. The figure shows the control schemes associated with the compressor. PC-25 controls the discharge pressure from stage 2 (set point = 65 barg), and PC-26 controls the discharge pressure from stage 3 (set point = 110 barg). PV-28 opens to discharge the gas to the flare stack if the suction pressure rises above 21 barg.

There are several interlock systems that drop off-line the compressor because of different causes. These causes are (a) low suction pressure to stage 1 (<14 barg), (b) high pressure rise across stage 2 (>40 bars), and (c) high pressure rise across stage 3 (>65 bars). The compressor drops off-line often, and anytime it drops it takes approximately 3 h before returning to normal operation. This amount of time represents a significant loss to the operation. Obviously, sending gas to the flare is another loss.

Design the necessary override control scheme to minimize these losses. Specify the necessary set points.

Steady-state information:

P suction	P discharge
Stage 1 = 18 barg	Stage 1 = 42 barg
P discharge	P discharge
Stage 2 = 65 barg	Stage 3 = 110 barg
P rise across	P rise across
Stage 2 = 23 bars	Stage 3 = 45 bars

Chapter 11

Feedforward Control: Principles and Application

This chapter presents the principles and application of feedforward control, quite often a most profitable control strategy. Feedforward is not a new strategy; the first reports date back to the early 1960s (Dobson, 1960; Shinskey, 1963). However, the use of computers has significantly simplified and expands its implementation which has resulted in an increased application of the technique. Feedforward requires a thorough knowledge of the steady-state and dynamics characteristics of the process. Thus good process engineering knowledge is basic to its application.

11-1 THE FEEDFORWARD CONCEPT

To help us understand the concept of feedforward control, let us briefly recall feedback control; Fig. 11-1.1 depicts the feedback concept. As different disturbances, $d_1(t), d_2(t), \ldots, d_n(t)$, enter the process, the controlled variable, $c(t)$, deviates from set point, and feedback compensates by manipulating another input to the process, the manipulated variable, $m(t)$. The advantage of feedback control is its simplicity. Its disadvantage is that in order to compensate for disturbances, the controlled variable must first deviate from set point. Feedback acts on an error between the set point and the controlled variable. It may be thought of as a *reactive* control strategy: it waits until the process has been upset before it begins to take corrective action.

By its very nature, feedback control results in a temporary deviation in the controlled variable. Many processes can permit some amount of deviation. In many other processes, however, this deviation must be minimized to such an extent that feedback control may not provide the required performance. For these cases, feedforward control may prove helpful.

The idea of feedforward control is to compensate for disturbances *before* they affect the controlled variable. Specifically, feedforward calls for measuring the disturbances before they enter the process, and, on the basis of these measurements, calculating the required manipulated variable to maintain the controlled variable at set point. If the calculation and action are done correctly, the controlled variable should remain undisturbed. Thus feedforward control may be thought of as being a *proactive* control strategy; Fig. 11-1.2 depicts this concept.

To further explain, consider a disturbance, $d(t)$, as shown in Fig. 11-1.3, entering the process. As soon as the feedforward controller (FFC) realizes that a change has occurred, it calculates a new value of $m_{FF}(t)$ and sends it to the process (valve). This is done such that path G_M negates the effect of path G_D. G_M is the transfer function that describes how the manipulated variable, $m_{FF}(t)$, affects the controlled variable, and G_D is the transfer function that describes how the disturbance, $d(t)$, affects the controlled variable.

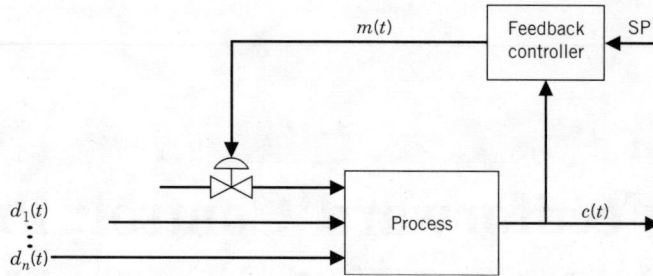

Figure 11-1.1 Feedback control.

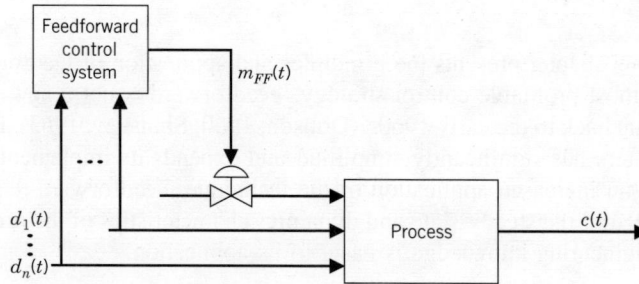

Figure 11-1.2 Feedforward control.

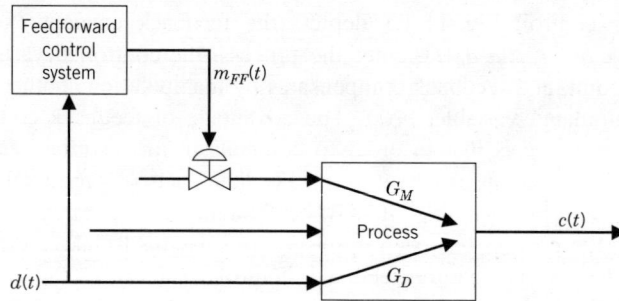

Figure 11-1.3 Feedforward control.

To attain perfect negation, the feedforward controller must be designed considering the steady-state characteristics of the process. That is, assume that a change of plus one unit in $d(t)$ affects $c(t)$ by plus ten units and that a change of plus one unit in $m_{FF}(t)$ affects $c(t)$ by plus five units. Thus if $d(t)$ changes by plus one unit, affecting $c(t)$ by plus ten units, the feedforward controller must change $m_{FF}(t)$ by minus two units, affecting $c(t)$ by minus ten units, and, therefore, negating the effect of $d(t)$.

The preceding paragraph explains how the feedforward control strategy compensates through consideration of the steady-state characteristics of the process. However, to avoid any change in the controlled variable, the dynamic characteristics of the process must also be considered. It is desired that the effects of $m_{FF}(t)$ and $d(t)$ affect $c(t)$ at the same time. That is, consider that when $d(t)$ changes, the feedforward controller changes $m_{FF}(t)$ almost at the same time. If as a result of the process dynamics the effect of $m_{FF}(t)$ on $c(t)$ is faster than the effect of $d(t)$ on $c(t)$, then $c(t)$ will deviate from its desired value due to $m_{FF}(t)$, not due to $d(t)$! In this case perfect compensation requires "slowing-down" the feedforward controller. That is, the feedforward controller should not take immediate corrective action; rather, it should wait a certain

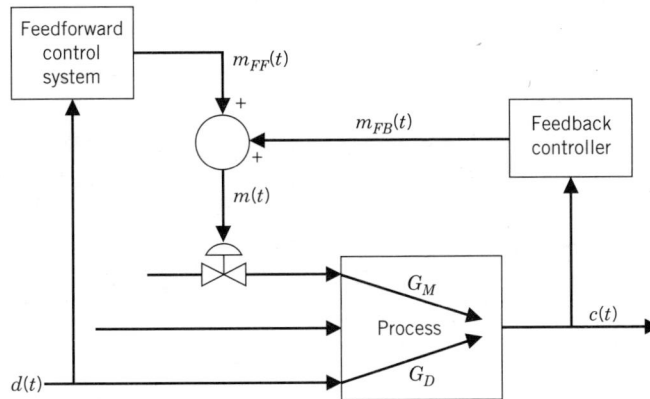

Figure 11-1.4 Feedforward/feedback control.

time before taking action so that the negating effects reach $c(t)$ at the same time. In other processes, the effect of $d(t)$ on $c(t)$ may be faster than the effect of $m_{FF}(t)$ on $c(t)$. In these cases perfect compensation requires "speeding up" the feedforward controller. Thus *the feedforward controller must be designed to provide both the required steady-state and dynamic compensations.*

Figure 11-1.2 shows feedforward compensation for all the disturbances entering the process. However, very often it may be difficult, if not impossible, to measure some disturbances. In addition, some of the possible measurable disturbances may occur so infrequently that the need for feedforward compensation is questionable. Therefore, feedforward control is used to compensate for the major measurable disturbances. It is up to the operating personnel to define "major" disturbances (those that occur often and/or cause significant deviations in the controlled variable). Feedback control is then used to compensate for those disturbances that are not compensated by feedforward. Figure 11-1.4 shows a possible implementation of this feedforward/feedback control.

The above paragraphs have explained the objective and some design considerations of feedforward control. The things to keep in mind are that the feedforward controller must contain, most times, steady-state and dynamic compensations, and that feedback compensation must always be present.

11-2   BLOCK DIAGRAM DESIGN OF LINEAR FEEDFORWARD CONTROLLERS

This section and the following three show the design of feedforward controllers. The mixing system shown in Fig. 11-2.1 is used to illustrate this design; Table 11-2.1 gives the steady-state conditions and other process information. In this process, three different streams are mixed and diluted with water to a final desired composition of component A, $x_6(t)$. Process considerations dictate that the mixing must be done in three constant-volume tanks, as shown in the figure. All the input streams represent possible disturbances to the process; that is, the flows and compositions of streams 5, 2, and 7 may vary. However, the major disturbances usually come from stream 2. Commonly, the stream flow, $f_2(t)$, may double, whereas the mass fraction, $x_2(t)$, may drop as much as 20 % of its steady-state value. Figure 11-2.2 shows the control, provided by feedback control, when $f_2(t)$ changes from 1000 to 2000 gpm. The composition changes from its steady-state value of 0.472 mass fraction (mf) to about

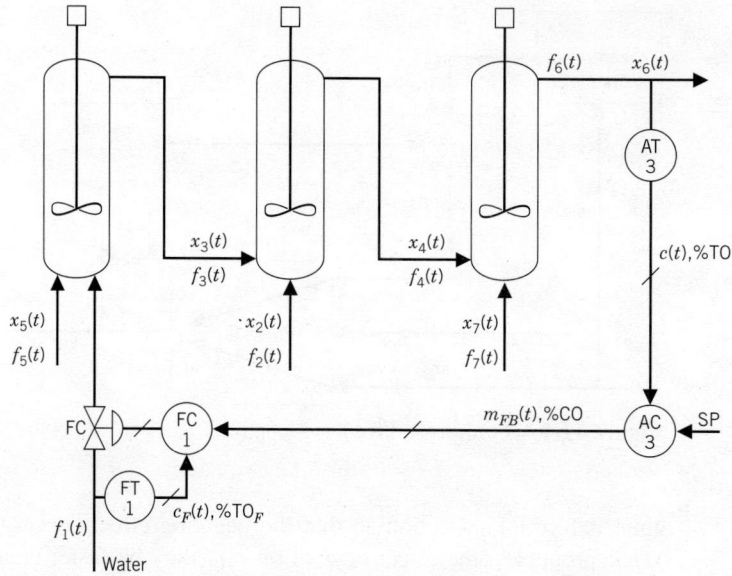

Figure 11-2.1 Mixing process.

Table 11-2.1 Process information and steady-state values for mixing process

Information

Concentration transmitter range: 0.3–0.7 mass fraction. Its dynamic can be described by a time constant of 0.1 min.

The pressure drop across the valve can be considered constant, and the maximum flow provided by the valve is 3800 gpm. The valve dynamics can be described by a time constant of 0.1 min.

The density of all streams are also considered similar and constant.

Steady-state values		
Stream	Flow, gpm	Mass fraction
1	1900	0.000
2	1000	0.990
3	2400	0.167
4	3400	0.049
5	500	0.800
6	3900	0.472
7	500	0.900

0.525 mf, a 11.23 % change from set point. However, process specifications dictate that the maximum deviation permitted for this process is ±1.5 % from set point. That is, the maximum value of composition permitted is 0.479 mf, and the minimum value permitted is 0.465 mf. Thus it does not appear that simple feedback can provide the required performance; feedforward control may be justified.

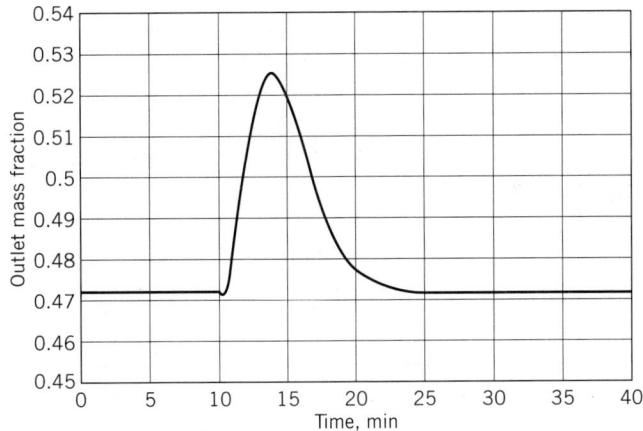

Figure 11-2.2 Feedback control when $f_2(t)$ changes from 1000 to 2000 gpm.

An index often used to evaluate control performances is the integral of the absolute value of the error (IAE), which is the total area under the curve, or error. The IAE value for the response shown in Fig. 11-2.2 is 73.06 mf-min.

Assuming for the moment that $f_2(t)$ is the major disturbance, the application of feedforward to this process calls for measuring this flow and, on a change, take corrective action. Let us examine the design of this feedforward controller.

The block diagram for the process is shown in Fig. 11-2.3a. The diagram shows that $f_2(t)$ is the disturbance of concern. We next note the significance of each transfer function.

G_c = transfer function of the composition controller.

G_{cF} = transfer function of the flow controller.

G_v = transfer function of the valve. It describes how the water flow is affected by the flow controller output.

G_{T_1} = transfer function of the mixing process. It describes how $x_6(t)$ is affected by the water flow.

G_{T_2} = transfer function of the mixing process. It describes how $x_6(t)$ is affected by $f_2(t)$.

H_F = transfer function of the flow sensor and transmitter.

H = transfer function of the concentration sensor and transmitter.

Because the flow loop, once tuned, is fast and stable, Fig. 11-2.3a can be simplified as shown in Fig. 11-2.3b. The new transfer function is

G_F = transfer function that describes how the water flow is affected by the composition controller, $G_F = G_{cF}G_v/1 + G_{cF}G_vH_F$

A more condensed block diagram, shown in Fig. 11-2.4, can be drawn and compared to that of Fig. 11-2.3b. The significance of each transfer function follows.

G_M = transfer function that describes how the manipulated variable, $M(s)$, affects the controlled variable, $C(s)$. In this case, $G_M = G_FG_{T_1}H$.

G_D = transfer function that describes how the disturbance, $F_2(s)$, affects the controlled variable. In this case, $G_D = G_{T_2}H$.

(a)

(b)

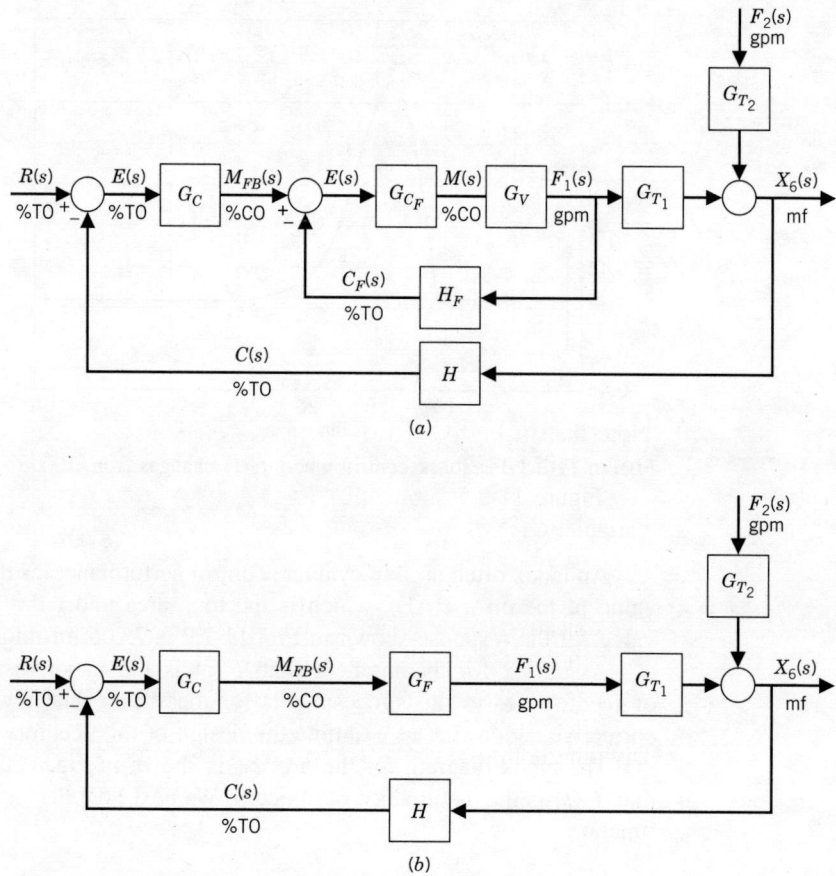

Figure 11-2.3 Block diagrams of mixing process.

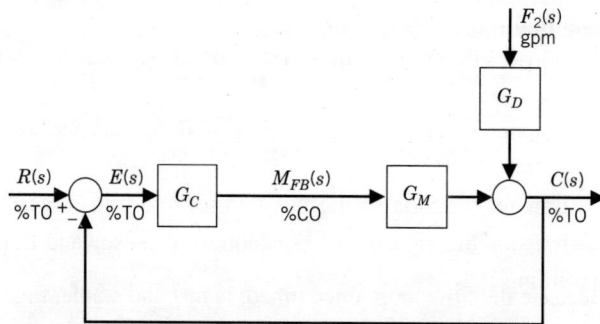

Figure 11-2.4 Condensed block diagram.

To review, the objective of feedforward control is to measure the input(s) and, if a disturbance is detected, to adjust the manipulated variable to maintain the controlled variable at set point. This control operation is shown in Fig. 11-2.5. The significance of each new transfer function follows.

H_D = transfer function that describes the sensor and transmitter that measures the disturbance.

FFC = transfer function of feedforward controller.

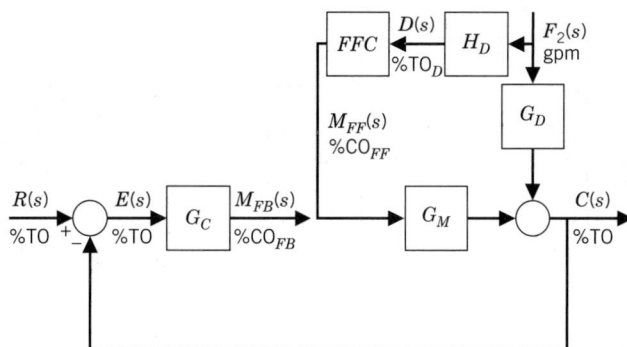

Figure 11-2.5 Block diagram of feedforward control system.

Note that in Fig. 11-2.5 the feedback controller has been "disconnected." This controller will be "connected" again later.

Figure 11-2.5 shows that the way the disturbance, $F_2(s)$, affects the controlled variable, $C(s)$, is given by the following:

$$C(s) = G_D F_2(s) + H_D(\text{FFC})G_M F_2(s)$$

The objective is to design FFC such that a change in $F_2(s)$ does not affect $C(s)$—that is, such that $C(s) = 0$. Thus

$$0 = G_D F_2(s) + H_D(\text{FFC})G_M F_2(s)$$

Dividing both sides by $F_2(s)$ and solving for FFC, we get

$$\boxed{\text{FFC} = -\frac{G_D}{H_D G_M}} \tag{11-2.1}$$

Equation 11-2.1 is the design formula for the feedforward controller.

As learned in previous chapters, first-order-plus-dead-time transfer functions are commonly used as an approximation to describe processes; Chapter 7 showed how to evaluate this transfer function from step inputs. Using this type of approximation for this process:

$$G_D = \frac{K_D e^{-t_{0_D} s}}{\tau_D s + 1}, \ \frac{\%\text{TO}}{\text{gpm}} \tag{11-2.2}$$

$$G_M = \frac{K_M e^{-t_{0_M} s}}{\tau_M s + 1}, \ \frac{\%\text{TO}}{\%\text{CO}} \tag{11-2.3}$$

and assuming that H_D is only a gain

$$H_D = K_{T_D}, \ \frac{\%\text{TO}_D}{\text{gpm}} \tag{11-2.4}$$

Substituting Eqs. 11-2.2, 11-2.3, and 11-2.4 into Eq. 11-2.1 yields

$$\text{FFC} = \frac{M_{FF}(s)}{\text{TO}_D(s)} = -\frac{K_D}{K_{T_D} K_M}\left(\frac{\tau_M s + 1}{\tau_D s + 1}\right) e^{-(t_{0_D} - t_{0_M})s} \tag{11-2.5}$$

Equation 11-2.5 contains several elements that will be explained in detail next; the implementation of this equation is shown in Fig. 11-2.6.

The first element of the feedforward controller, $-K_D/K_{T_D} K_M$, contains only gain terms. This element is the part of the feedforward controller that compensates for the

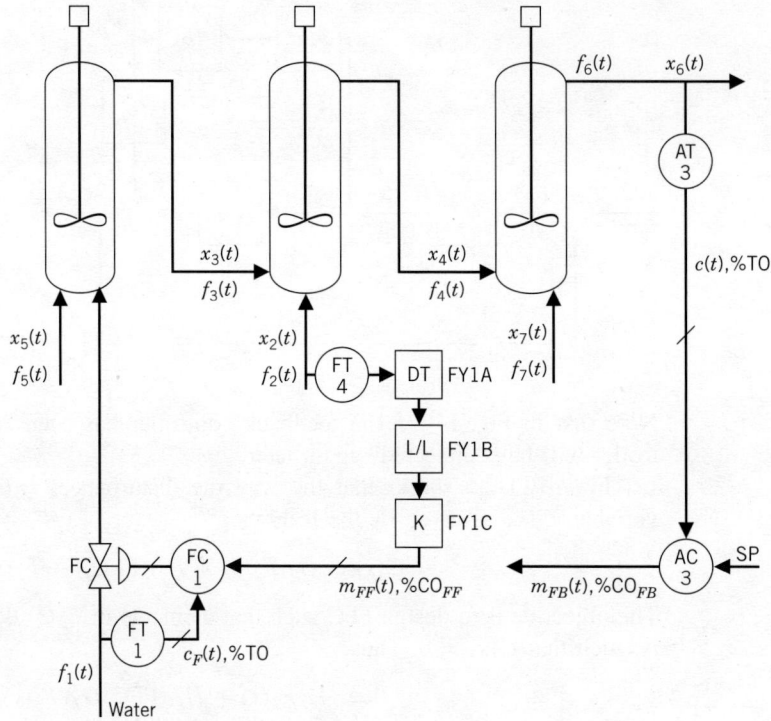

Figure 11-2.6 Feedforward control.

steady-state differences between the "G_M and G_D paths." The units of this element help in understanding its significance:

$$\frac{K_D}{K_{T_D} K_M} \Rightarrow \frac{\dfrac{\%\text{TO}}{\text{gpm}}}{\dfrac{\%\text{TO}_D}{\text{gpm}} \dfrac{\%\text{TO}}{\%\text{CO}_{FF}}} = \frac{\%\text{CO}_{FF}}{\%\text{TO}_D}$$

Thus the units show that the term indicates how much the feedforward controller output, $m_{FF}(t)$, changes per unit change in transmitter's output, $\%\text{TO}_D$.

Note the minus sign in front of the gain element in Eq. 11-2.5. This sign helps to decide the "action" of the controller. In the process at hand, K_D is positive; as $f_2(t)$ increases, the outlet concentration, $x_6(t)$, also increases because stream 2 is more concentrated than the outlet stream. K_M is negative; as the signal to the water flow controller increases, the valve opens and the outlet concentration decreases. Finally, K_{T_D} is positive; as $f_2(t)$ increases, the signal from the flow transmitter also increases. Thus, the sign of the gain element is negative:

$$\frac{K_D}{K_{T_D} K_M} \Rightarrow \frac{+}{+-} = -$$

A negative sign means that as the signal from the flow transmitter increases, indicating an increase in $f_2(t)$, the feedforward controller output, $m_{FF}(t)$, should decrease, closing the valve and reducing the water flow. This action does not make sense. As $f_2(t)$ increases, tending to increase the concentration of the output stream, the water flow should also increase to dilute the outlet concentration thus, negating the effect of $f_2(t)$. Therefore, the sign of the gain element should be positive. Note that when the negative sign in front of the gain element is multiplied by the sign of this element it results in the correct feedforward action.

The second element of the feedforward controller includes only the time constants of the "G_D and G_M paths." This element, referred to as a lead-lag (L/L), compensates for the differences in time constants between the two paths. Section 11-3 discusses this element in detail.

The last element of the feedforward controller contains only the dead-time terms of the "G_D and G_M paths." This element, referred to as a dead-time compensator, compensates for the differences in dead time between the two paths. Sometimes the term $(t_{0_D} - t_{0_M})$ may be negative, yielding a positive exponent. As we noted in Chapter 2, the Laplace representation of dead time includes a negative sign in the exponent. When the sign is positive, it is definitely not a dead time and cannot be implemented. A negative sign in the exponential is interpreted as "delaying" an input; a positive sign may indicate a "forecasting." That is, the controller requires taking action before the disturbance happens. This is not possible! When this occurs, quite often there is a physical explanation, as the present example will show.

Thus the first element of the feedforward controller is a steady-state compensator, whereas the last two elements are dynamic compensators. All these elements are easily implemented using computer control software. Years ago, however, when analog instrumentation was solely used, the dead-time compensator was either extremely difficult or impossible to implement. At that time, the state-of-the-art was to implement only the steady-state and lead-lag compensators. Figure 11-2.6 shows a component for each calculation needed for the feedforward controller—that is, one component for the dead time, one for the lead-lag, and one for the gain. Very often however, lead-lags have adjustable gains and in this case we can combine the components FY4B and FY4C into only one component. This is more efficient since it uses fewer components. From now on we will show only one component, the lead-lag, and the gain will be included.

Returning to the mixing system, under open-loop conditions, a step change of 5 % in the signal from the feedback controller provides a process response from which the following first-order-plus-dead-time approximation is obtained:

$$G_M = \frac{-1.095 \; e^{-0.93s}}{3.82s + 1}, \qquad \frac{\%\text{TO}}{\%\text{CO}} \tag{11-2.6}$$

Also, under open-loop conditions, $f_2(t)$ was allowed to change by 10 gpm in a step fashion, and from the process response the following approximation is obtained:

$$G_D = \frac{0.0325 \; e^{-0.75s}}{2.75s + 1}, \qquad \frac{\%\text{TO}}{\text{gpm}} \tag{11-2.7}$$

Finally, assuming that the flow transmitter in stream 2 is calibrated from 0 to 3000 gpm, its transfer function is given by

$$H_D = K_{T_D} = \frac{100 \; \%\text{TO}_D}{3000 \; \text{gpm}} = 0.0333 \; \frac{\%\text{TO}_D}{\text{gpm}}$$

Substituting the previous three transfer functions into Eq. 11-2.1 yields

$$\text{FFC} = \frac{M_{FF}(s)}{D(s)} = 0.891 \left(\frac{3.82s + 1}{2.75s + 1} \right) e^{-(0.75 - 0.93)s} \tag{11-2.8}$$

The dead time indicated, 0.75−0.93, is negative, so the dead time compensator cannot be implemented. Thus the implementable, or realizable, feedforward controller is

$$\text{FFC} = \frac{M_{FF}(s)}{\text{TO}_D(s)} = 0.891 \left(\frac{3.82s + 1}{2.75s + 1} \right) \tag{11-2.9}$$

Figure 11-2.7 shows the implementation of this controller. The figure shows that the feedback compensation has also been implemented. This implementation has been accomplished by adding the output of both feedforward and feedback controllers in FY3. Section 11-4 discusses how to implement this addition. Figure 11-2.8 shows the block diagram for this combined control scheme.

Figure 11-2.9 shows the response of the composition, when $f_2(t)$ doubles from 1000 to 2000 gpm. The figure compares the control provided by feedback control (FBC), steady-state feedforward control (FFCSS), and dynamic feedforward control (FFCDYN). In steady-state feedforward control no dynamic compensation is implemented; that is, in this case FFC = 0.891. Dynamic feedforward control includes the complete controller, Eq. 11-2.9. Under steady-state feedforward, the mass fraction

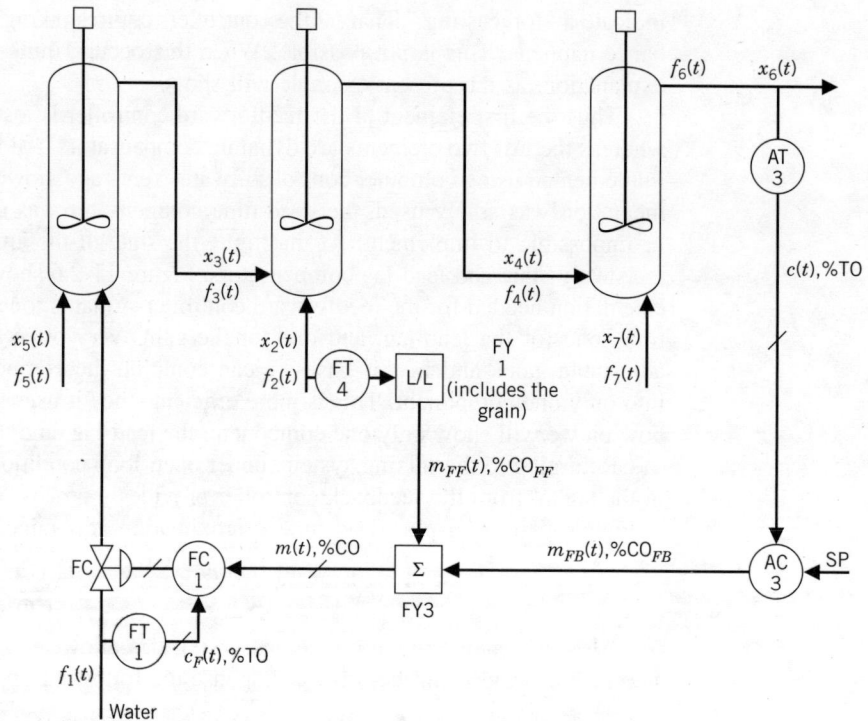

Figure 11-2.7 Implementation of feedforward/feedback controller.

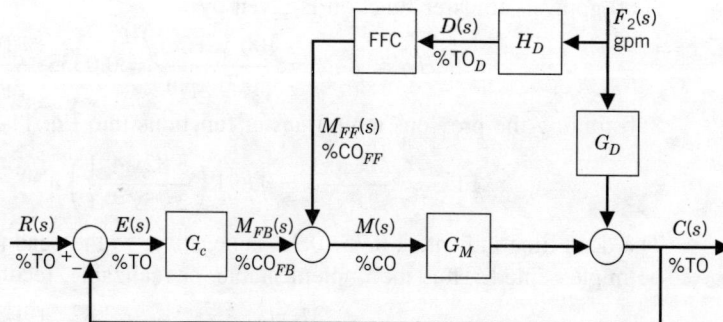

Figure 11-2.8 Block diagram of feedforward/feedback controller.

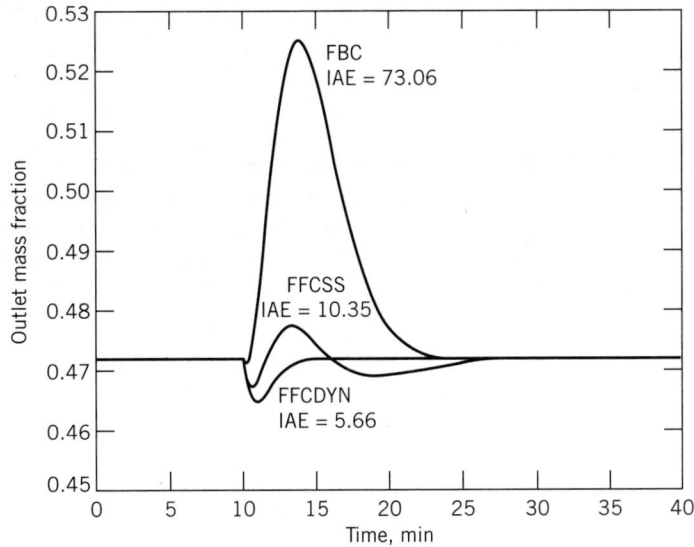

Figure 11-2.9 Feedforward and feedback responses when $f_2(t)$ changes from 1000 to 2000 gpm.

increased up to 0.477 mf, or a 1.05 % change from set point. Under dynamic feedforward the mass fraction increased up to 0.473 mf, or a 0.21 % change. The improvement provided by feedforward control is quite impressive. Figure 11-2.9 also shows that the process response tends to decrease first and then increase; we will discuss this response later.

The previous paragraphs and figures have shown the development of a linear feedforward controller and the responses obtained. At this stage, since we have not offered an explanation of the lead-lag element, the reader may be wondering about it. Thus we will explain this element before further discussing feedforward control.

11-3   LEAD-LAG ELEMENT

As indicated in Eqs. 11-2.5 and 11-2.9, the lead-lag element is composed of a ratio of two $(\tau s + 1)$ terms. More specifically, its transfer function is

$$\frac{O(s)}{I(s)} = \frac{\tau_{ld}s + 1}{\tau_{lg}s + 1} \tag{11-3.1}$$

where

$O(s)$ = Laplace transform of output variable

$I(s)$ = Laplace transform of input variable

τ_{ld} = lead time constant

τ_{lg} = lag time constant

This lead-lag element was previously discussed in Chapter 2; however, a brief review is warranted here. Let us suppose that the input changes in a step fashion of A units of amplitude; that is,

$$I(s) = \frac{A}{s}$$

Substituting this expression for $I(s)$ in Eq. 11-3.1 and inverting the equation back to the time domain, we get

$$O(t) = A \left[1 + \frac{\tau_{ld} - \tau_{lg}}{\tau_{lg}} e^{-t/\tau_{lg}} \right] \qquad \textbf{(11-3.2)}$$

Figure 11-3.1a shows the response for different values of the ratio τ_{ld}/τ_{lg} while keeping $\tau_{lg} = 1$. The figure shows that as the ratio τ_{ld}/τ_{lg} increases, the initial response also increases; as time increases then the response approaches exponentially its final value of A. For values of $\tau_{ld}/\tau_{lg} > 1$ the initial response is greater than its final value, whereas for values of $\tau_{ld}/\tau_{lg} < 1$ the initial response is less than its final value. Therefore, the initial response depends on the ratio of the lead time constant to the lag time constant, τ_{ld}/τ_{lg}. The approach to the final value depends only on the lag time constant. Thus, in tuning a lead-lag, both τ_{ld} and τ_{lg} must be provided.

Figures 11-3.1b through 11-3.1d help to further understand the effects of τ_{ld} and τ_{lg} on the response of the lead-lag to a step change of A units of magnitude. Figure 11-3.1b shows how τ_{lg} affects the response while keeping τ_{ld} constant. The figure shows that as τ_{lg} decreases, the ratio τ_{ld}/τ_{lg} increases, the magnitude of the initial output response increases, and the rate at which the response approaches its final value increases. Figure 11-3.1c shows how τ_{ld} affects the response while keeping τ_{lg} constant. The figure shows that as τ_{ld} increases, the ratio τ_{ld}/τ_{lg} also increases, and the magnitude of the initial output response increases. The figure also shows that all curves reach the final value at the same time because τ_{lg} is the same in all cases. Finally, Fig. 11-3.1d shows two response curves with identical values of the ratio τ_{ld}/τ_{lg} but different individual values of τ_{ld} and τ_{lg}. The figure shows that the magnitude of the initial output response is the same, because the ratio is the same, but the response with the larger τ_{lg} takes longer to reach the final value.

Equation 11-2.5 reflects the use of a lead-lag element in the feedforward controller. The equation indicates that τ_{ld} should be set equal to τ_M, and that τ_{lg} should be set equal to τ_D.

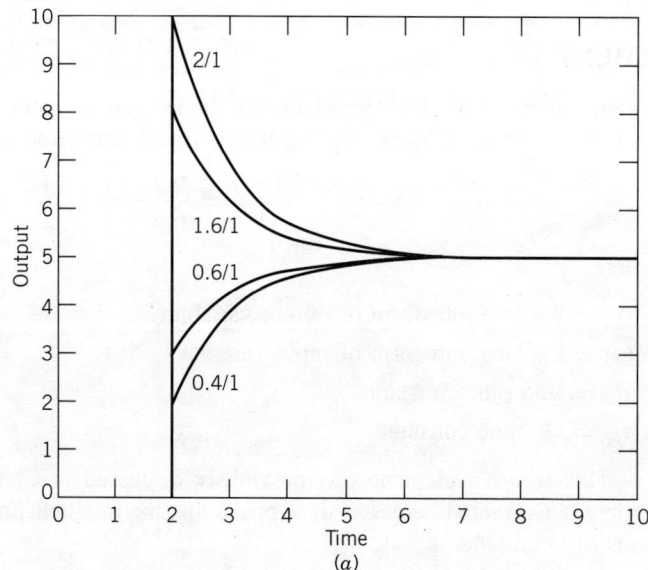

Figure 11-3.1a Response of lead-lag to an input change of 5 units.

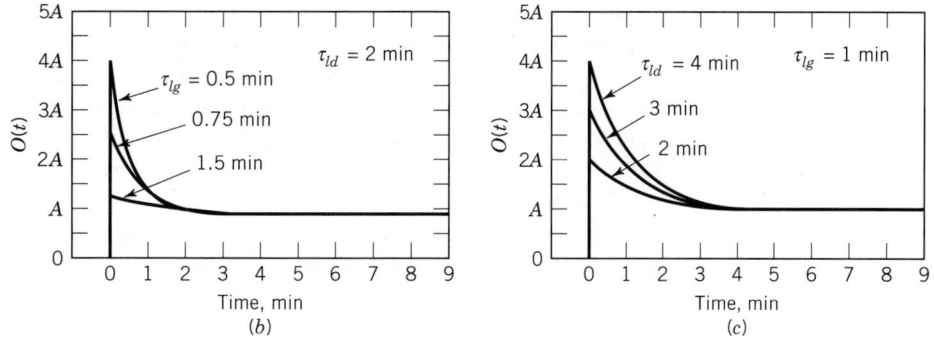

Figure 11-3.1b,c Response of lead-lag to an input of A units of magnitude.

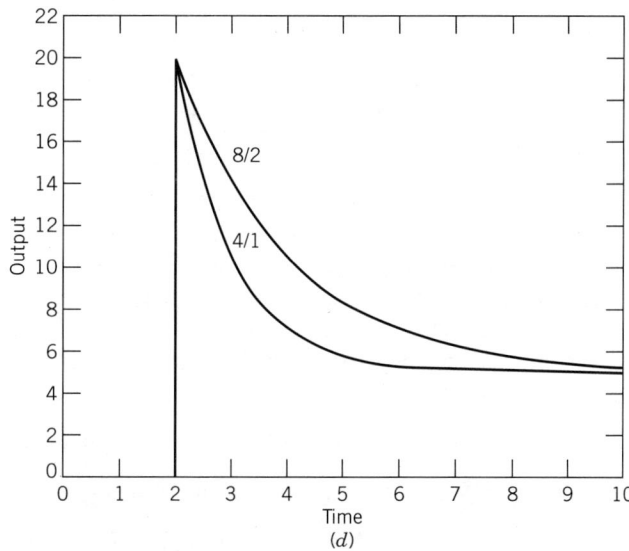

Figure 11-3.1d Response of lead-lag to an input change of 5 units.

11-4   BACK TO THE PREVIOUS EXAMPLE

With an understanding of the lead-lag element we can now return to the example of Section 11-2 and specifically to a discussion of the dynamic compensation of the feedforward controller.

Comparing the transfer functions given by Eqs. 11-2.6 and 11-2.7 it is easy to realize that the controlled variable, $c(t)$, responds slower to a change in the manipulated variable, $m(t)$, than to a change in the disturbance, $f_2(t)$. Recall that one design consideration for a feedforward controller is to compensate for the dynamic differences between the manipulated and the disturbance paths, the G_D and G_M paths. The feedforward controller for this process should be designed to "speed up" the response of the controlled variable to a change in the manipulated variable. That is, the feedforward controller should "speed up" the G_M path; the resulting controller, Eq. 11-2.9, does exactly this. First, realize that the resulting lead-lag term has a τ_{ld}/τ_{lg} ratio greater than one, $\tau_{ld}/\tau_{lg} = 3.82/2.75 = 1.39$. This means that at the moment the signal from the flow transmitter changes by 1 %, indicating a certain change in $f_2(t)$, the lead-lag output changes by 1.39 %, resulting in an initial output change from the feedforward controller of $(0.891)(1.39) = 1.238$ %. Eventually, the lead-lag output approaches 1 %, and the feedforward controller output approaches 0.891 %. This type of action results in an initial increase in $f_1(t)$ greater than the one really needed for the specific increase in $f_2(t)$. This initial greater increase provides a "kick" to the G_M path to move faster, resulting in a tighter control as compared to the control provided by steady-state feedforward control; this is shown in Fig. 11-2.9. Second, note that the feedforward controller equation does not contain a dead-time element. There is no need to delay the feedforward action. On the contrary, the present process makes it necessary to "speed up" the feedforward action. Thus, the absence of a dead-time term makes sense [it would be great if we could "forecast" a change in $f_2(t)$!].

It is important to realize that this feedforward controller, Eq. 11-2.9, compensates only for changes in $f_2(t)$. Any other disturbance will not be compensated by the controller, and in the absence of a feedback controller it would result in a deviation of the controlled variable. The implementation of feedforward control requires the presence of feedback control. Feedforward control compensates for the major measurable disturbances, while feedback control takes care of all other disturbances. In addition, the feedback controller compensates for any inexactness in the feedforward controller. Thus *feedforward control must be implemented with feedback compensation*! Feedback from the controlled variable must be present.

Figure 11-2.7 shows a summer where the signals from the feedforward controller, $m_{FF}(t)$, and from the feedback controller, $m_{FB}(t)$, are combined. A way to think about the significance of the signals entering the summer is to note that the signal from the feedforward calculation, $m_{FF}(t)$, is related to the flow, $f_1(t)$, required to compensate for the major disturbances. The signal from the feedback controller, $m_{FB}(t)$, is $\Delta f_1(t)$. That is, this feedback signal biases the feedforward calculation to correct for unmeasured disturbances or for errors in the feedforward calculation. The summer solves the following equation

$$OUT = \text{feedback signal} + \text{feedforward signal} + \text{constant}$$

To be more specific, consider the use of the summer shown in Table 10-1.1

$$OUT = K_x X + K_y Y + K_z Z + B_o$$

Let the feedback signal be the X input, the feedforward signal the Y input, and the Z input is not used in this case. Therefore,

$$OUT = K_x[m_{FB}(t)] + K_y[m_{FF}(t)] + B_o$$

As previously discussed, the sign of the steady-state part of the feedforward controller, $-K_D/K_{T_D}K_m$, is positive for this process, so the value of K_y is set to $+1$ (if the sign had been negative then K_y would have been set to -1). The value of K_x is also set to $+1$. Note that setting K_y to 0 or to 1 provides an easy way to "turn off" or "turn on" the feedforward controller.

Let us suppose that the process is at steady-state under feedback control only ($K_y = 0$, $B_o = 0$) and it is now desired to "turn on" the feedforward controller. Furthermore, because the process is at steady state, it is desired to turn on the feedforward controller without upsetting the signal to the valve. That is, a "bumpless-transfer" from simple feedback control to feedforward/feedback control is desired. To accomplish this transfer, the summer is first set to "manual" which freezes its output; K_y is set to $+1$, the output of the feedforward controller, $m_{FF}(t)$, is read from FY1; the constant term, B_o, is set equal to the negative of the value read in FY1; and finally, the summer is set to automatic again. This procedure results in the constant term canceling the feedforward controller output. To be a bit more specific, suppose that the process is running under feedback control only, with a signal to the valve equal to $m_{FB}(t)$. It is then desired to "turn on" the feedforward controller, and at this time the process is at steady state with $f_2(t) = 1500$ gpm. Under this condition the output of the flow transmitter is at 50 %, and $m_{FF}(t)$ is at (50 %)(0.891), or 44.55 %. Then, the procedure just explained is followed, yielding

$$\text{OUT} = (1)m_{FB}(t) + (1)(44.5) - 44.55 = m_{FB}(t)$$

Now suppose $f_2(t)$ changes from 1500 to 1800 gpm, making the output from the flow transmitter equal to 60 %. After the transients through the lead-lag have died out, the output from the feedforward controller becomes equal to (60 %)(0.891) = 53.46 %. Thus, the feedforward controller asks for 8.91 % more signal to the valve to compensate for the disturbance. At this moment, the summer output signal becomes

$$\text{OUT} = (1)m_{FB}(t) + (1)(53.46) - 44.55 = m_{FB}(t) + 8.91 \%$$

which changes the signal to the valve by the required amount.

The procedure just described to implement the summer is easy; however, it requires the manual intervention of the operating personnel. Most control systems can be easily configured to perform the procedure automatically. For instance, consider the use of an on–off switch and two constant terms, B_{FB} and B_{FF}. The switch is to indicate feedback control (switch is OFF) or feedforward control (switch is ON). B_{FB} is used when only feedback control is used ($K_y = 0$), and B_{FF} is used when feedforward is used ($K_y = 1$). Thus, the summer is

$$m(t) = K_x m_{FB}(t) + K_y m_{FF}(t) + (B_{FB} \text{ if switch is OFF or } B_{FF} \text{ if switch is ON})$$

Originally, $B_{FB} = B_{FF} = 0$. While only feedback is used, the following is being calculated:

$$B_{FF} = -m_{FF}(t) + B_{FB}$$

As soon as the switch goes ON this calculation stops, K_y is set to 1, and B_{FF} remains constant at the last value calculated. While feedforward is being used, the following is being calculated:

$$B_{FB} = m_{FF}(t) + B_{FF}$$

As soon as the switch goes OFF this calculation stops, K_y is set to 0, and B_{FB} remains constant at the last value calculated. This procedure guarantees automatic bumpless transfer. The reader is encouraged to test this algorithm.

So far in our example, feedforward control has been implemented to compensate for $f_2(t)$ only. But what if it is necessary to compensate for another disturbance such as $x_2(t)$? The technique for designing this new feedforward controller is the same as before; Fig. 11-4.1 shows a block diagram including the new disturbance with the new feedforward controller FFC$_2$. Following the previous algebraic development, the new controller equation is

$$\text{FFC}_2 = -\frac{G_{D_2}}{H_{D_2} G_M} \tag{11-4.1}$$

Step testing yields the following transfer function:

$$G_{D_2} = \frac{64.1\, e^{-0.85s}}{3.15s + 1}, \quad \frac{\%\text{TO}}{\text{mf}} \tag{11-4.2}$$

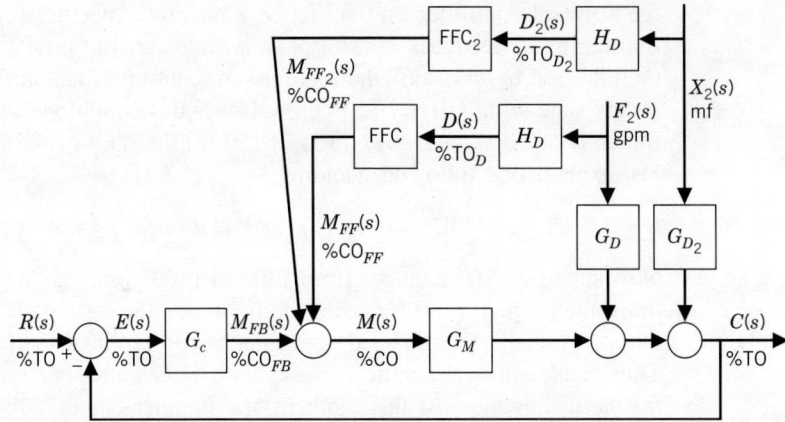

Figure 11-4.1 Block diagram of feedforward control for two disturbances.

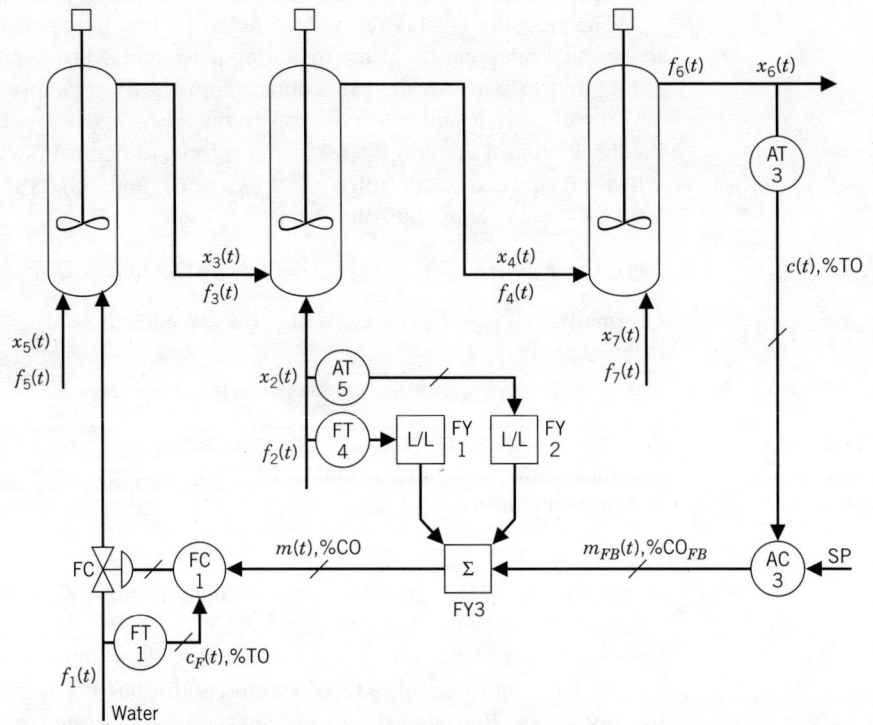

Figure 11-4.2 Implementation of feedforward/feedback control for two disturbances.

Assuming that the concentration transmitter in stream 2 has a negligible lag and that it has been calibrated from 0.5 to 1.0 mf, its transfer function is given by

$$H_{D_2} = \frac{100 \ \%\text{TO}_{D_2}}{0.5 \ \text{mf}} = 200 \ \frac{\%\text{TO}_{D_2}}{\text{mf}} \cdot \tag{11-4.3}$$

Finally, substituting Eqs. 11-2.6, 11-4.2, and 11-4.3 into Eq. 11-4.1 yields

$$\text{FFC}_2 = 0.293 \left(\frac{3.82s + 1}{3.15s + 1} \right) e^{-(0.85 - 0.95)s} \tag{11-4.4}$$

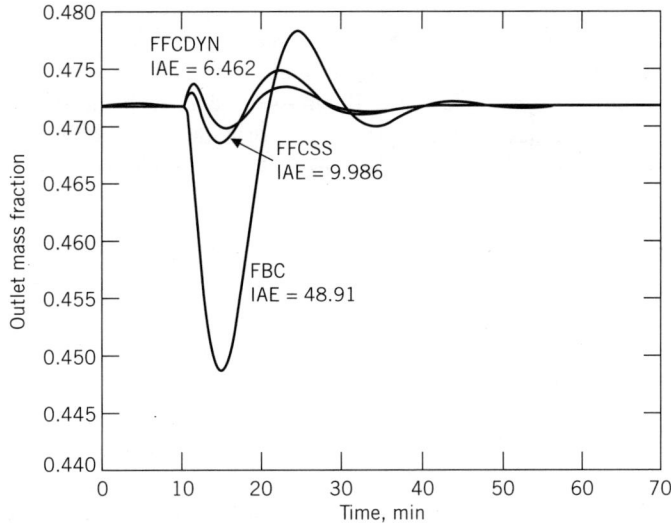

Figure 11-4.3 Control performance of $x_6(t)$ to a change of -0.2 mf in $x_2(t)$.

Because the dead time is again negative,

$$\text{FFC}_2 = 0.293 \left(\frac{3.82s + 1}{3.15s + 1} \right) \tag{11-4.5}$$

Figure 11-4.2 shows the implementation of this new feedforward controller added to the previous one and to the feedback controller. Figure 11-4.3 shows the response $x_6(t)$ to a change of -0.2 mass fraction in $x_2(t)$ under feedback, steady-state feedforward, and dynamic feedforward control. The improvement provided by feedforward control is certainly significant. Most of the improvement in this case is provided by the steady-state element; arguably, the addition of the lead-lag provides an improvement. It is a judgment call in this case whether to implement the lead-lag. Note that the ratio of the lead time constant to the lag time constant is 1.25 which is close to 1.0. On the basis of our discussion on the lead-lag element, the closer the ratio is to 1.0, the less the need for lead-lag compensation. Here is a rule of thumb that can be used to decide whether to use the lead-lag: if $0.65 < \tau_{ld}/\tau_{lg} < 1.3$ do not use lead-lag. Outside these limits, the use of lead-lag may significantly improve the control performance.

11-5   DESIGN OF NONLINEAR FEEDFORWARD CONTROLLERS FROM BASIC PROCESS PRINCIPLES

The feedforward controllers developed thus far, Eqs. 11-2.9 and 11-4.5, are linear controllers. They were developed from linear models of the process that are valid only for small deviations around the operating point where the step tests were performed. These controllers are then used with the same constant parameters without consideration of the operating conditions. As we noted in Chapters 3 and 4, processes have most often nonlinear characteristics, consequently, as operating conditions change the control performance provided by linear controllers degrades.

As shown in Section 11-2, the feedforward controllers are composed of steady-state and dynamic compensators. Very often the steady-state compensator, represented by the $-K_D/K_{T_D}K_M$ element, can be obtained by other means that yield a nonlinear compensator. This nonlinear compensator provides an improved control performance over a wide range of operating conditions.

A method for obtaining a nonlinear steady-state compensator consists of starting from first principles, usually mass or energy balances. Using first principles, it is desired to develop an equation that provides the manipulated variable as a function of the disturbances and of the set point of the controlled variable. That is,

$$m(t) = f[d_1, d_2(t), \ldots, d_n(t), \text{set point}]$$

For the process at hand,

$$f_1(t) = f[f_5(t), x_5(t), f_2(t), x_2(t), f_7(t), x_7(t), x_6^{\text{set}}(t)]$$

where $x_6^{\text{set}}(t)$ is the desired value of $x_6(t)$.

In the previous section we decided that for this process, the major disturbances are $f_2(t)$ and $x_2(t)$ and that the other inlet flows and compositions are minor disturbances. Thus we need to develop an equation, the steady-state feedforward controller, that relates the manipulated variable $f_1(t)$ in terms of the disturbances $f_2(t)$ and $x_2(t)$. In this equation, we consider all other inlet flows and compositions at their steady-state values. That is,

$$f_1(t) = f[\overline{f}_5, \overline{x}_5, f_2(t), x_2(t), \overline{f}_7, \overline{x}_7, x_6^{\text{set}}(t)]$$

where the overbar indicates the steady-state values of the variables.

Because we are dealing with compositions and flows, mass balances are appropriate to use. There are two components, A and water, so we can write two independent mass balances. We start with a total mass balance around the three tanks

$$\rho \overline{f}_5 + \rho f_1(t) + \rho f_2(t) + \rho \overline{f}_7 - \rho f_6(t) = 0 \qquad \textbf{(11-5.1)}$$

$$\text{1 eq., 2 unk. } [f_1(t), f_6(t)]$$

Note that $f_2(t)$ is not considered an unknown because it will be measured, and thus its value will be known. A mass balance on component A provides the other equation

$$\rho \overline{f}_5 \overline{x}_5 + \rho f_2(t) x_2(t) + \rho \overline{f}_7 \overline{x}_7 - \rho f_6(t) x_6^{\text{set}}(t) = 0 \qquad \textbf{(11-5.2)}$$

$$\text{2 eq., 2 unk.}$$

Because $x_2(t)$ will also be measured, it is not considered an unknown. Solving for $f_6(t)$ from Eq. 11-5.1, substituting into Eq. 11-5.2, and rearranging yield

$$f_1(t) = \frac{1}{x_6^{\text{set}}(t)} \left[\overline{f}_5 \overline{x}_5 + \overline{f}_7 \overline{x}_7 \right] - \overline{f}_5 - \overline{f}_7 + \frac{1}{x_6^{\text{set}}(t)} [x_2(t) - x_6^{\text{set}}(t)] f_2(t) \quad \textbf{(11-5.3)}$$

Substituting the steady-state values into Eq. 11-5.3 yields

$$f_1(t) = \frac{1}{x_6^{\text{set}}(t)} [850 + f_2(t) x_2(t)] - f_2(t) - 1000 \qquad \textbf{(11-5.4)}$$

Equation 11-5.4 is the desired steady-state feedforward controller.

The implementation of Eq. 11-5.4 depends on how the feedback correction, the output of the feedback controller, is implemented. This implementation depends on the physical significance given to the feedback signal; there are several ways to decide this. One way is to decide that the significance of the feedback signal is $\Delta f_1(t)$ and use a summer similar to that in Fig. 11-4.2. In this case, we first substitute $x_6^{\text{set}}(t) = 0.472$ into Eq. 11-5.4 and obtain

$$f_1(t) = 800.85 + f_2(t) \left[\frac{x_2(t)}{0.472} - 1 \right] \qquad \textbf{(11-5.5)}$$

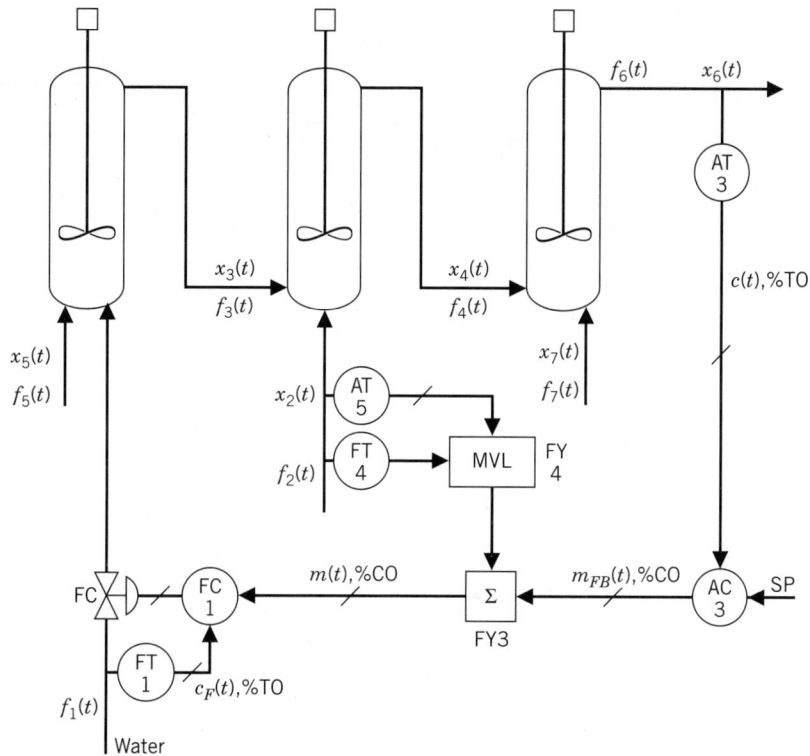

Figure 11-5.1 Nonlinear feedforward control.

This equation is written in engineering units; Fig. 11-5.1 shows the implementation of this controller—only a multiplier is needed, with no dynamic compensation. The second way to implement the feedback compensation is by deciding that the significance of the feedback signal is $1/x_6^{\text{set}}(t)$. This signal is then input into Eq. 11-5.4 to calculate $f_1(t)$. Thus in this case, the feedback signal is used directly in the feedforward calculation, and not to add to it; Fig. 11-5.2 shows the implementation of this controller. For simplicity, only one block, referred to as CALC, is shown. The implementation of Eq. 11-5.4, whether using computing blocks or software, is included in this block. Figure 11-5.3 shows the response of the process under feedback controller, linear steady-state feedforward, and the two nonlinear steady-state feedforward controllers, to disturbances of 1000 gpm increase in $f_2(t)$ and -0.2 mf change in $x_2(t)$. The improvement in control performance obtained with the nonlinear controllers is obvious. The performance obtained with the second nonlinear controller is quite impressive. This controller maintains the nonlinearity characteristics of the process, and can provide a better control.

Instead of calling the output of the feedback controller $1/x_6^{\text{set}}(t)$, we could call it $x_6^{\text{set}}(t)$; the control performance would be the same. But what about the action of the feedback controller in both cases? Think about it.

The previous paragraphs have shown two different ways to implement the nonlinear steady-state feedforward controller depending on the significance given to the feedback signal, $\Delta f_1(t)$ or $1/x_6^{\text{set}}(t)$. The designer has complete freedom to make this decision. In the first case, the feedback controller added to the feedforward calculation. This is a simple and valid choice, and the one usually used when the steady-state part of the

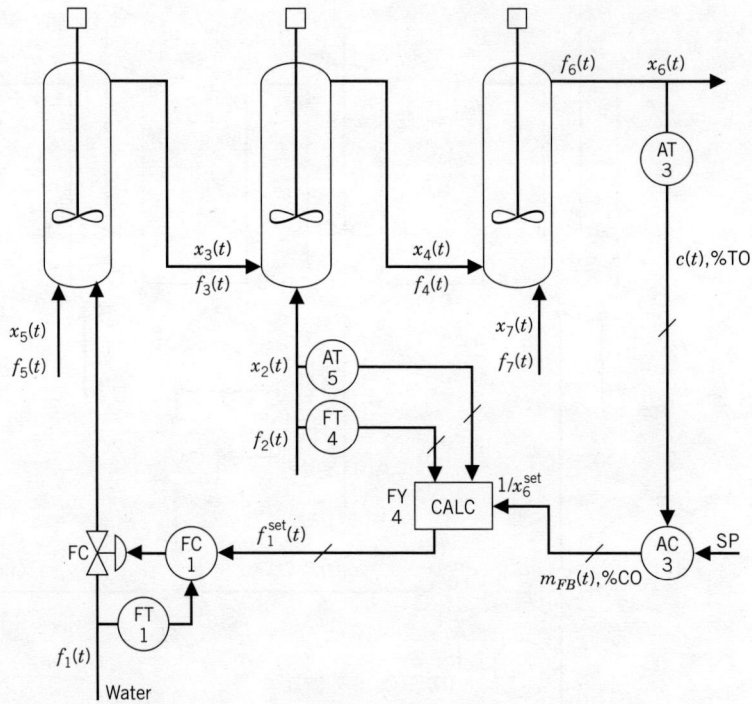

Figure 11-5.2 Nonlinear feedforward control.

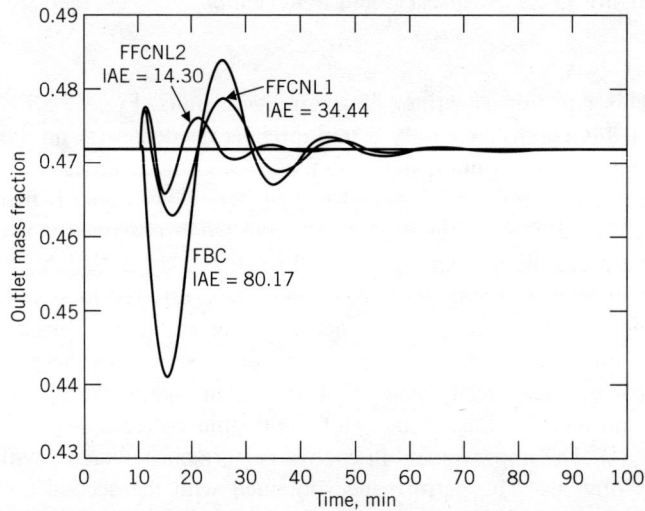

Figure 11-5.3 Control performance of $x_6(t)$ to a decrease of 500 gpm in $f_2(t)$ and a decrease of 0.2 in $x_2(t)$.

controller is obtained as in Section 11-2. The second choice, $1/x_6^{set}(t)$, or $x_6^{set}(t)$, is also a simple choice that helps maintain the nonlinearity of the controller. Please note that the actual desired value of $x_6(t)$ is the set point to the feedback controller. The controller changes the term $1/x_6^{set}(t)$, or $x_6^{set}(t)$, in the feedforward equation to keep its own set point. This way of providing the feedback compensation usually results in better performance.

Sometimes, developing a nonlinear steady-state compensator from first principles may be difficult. Fortunately, process engineering tools provide yet another way to develop the controller. Processes are usually designed by either steady-state flowsheet simulators or any other steady-state simulation. These simulators, along with regression analysis tools, provide another means of designing the steady-state compensator. The simulation can be run at different conditions—that is, different $f_2(t)$, $x_2(t)$, and $x_6^{set}(t)$—and the required manipulated variable, $f_1(t)$, can be calculated to keep the controlled variable at set point. This information can then be fed to a multiple regression program to develop an equation relating the manipulated variable to the disturbances and set point.

11-6   SOME CLOSING COMMENTS AND OUTLINE OF FEEDFORWARD CONTROLLER DESIGN

Before proceeding with more examples, we want to make the following comments about the process and example presented in this chapter and about feedforward control in general.

The first comment refers to the process itself. Figure 11-2.9 shows the response of the control system when $f_2(t)$ changes from 1000 to 2000 gpm. The composition of this stream is quite high (0.99 mf), so this change tends to increase the composition of the outlet stream 6. However, the response shown in Fig. 11-2.9 shows that the composition $x_6(t)$ first tends to decrease and then increases. This behavior is an inverse response of the type presented in Section 4-4, and, of course, there is an explanation for this behavior. Because the tanks are constant volume tanks, an increase in $f_2(t)$ results in an immediate increase in $f_4(t)$. The composition of stream 4, which enters the third tank, is less than the composition of stream 6 which exits the third tank. Thus this increase in $f_4(t)$ tends to initially dilute—decrease—the composition $x_6(t)$. Eventually, the increase in $f_2(t)$ results in an increase in the composition entering the third tank, and a corresponding increase in $x_6(t)$. The transfer function relating $f_2(t)$ and $x_6(t)$ should show a negative zero (see Problem 4-3). Figure 11-2.9 shows that the response under feedforward control exhibits a more pronounced inverse response. What happens is that when $f_2(t)$ increases, $f_1(t)$ is also increased by the feedforward controller. Thus the total flow to the third tank increases more, and the dilution effect in that tank is more pronounced. Can you explain why the inverse response is more pronounced under dynamic feedforward than under steady-state feedforward?

The second comment refers to the lead-lag element. The lead-lag element is a simple unit used to implement the dynamic compensation for the linear and nonlinear feedforward controllers. We showed how to tune the lead-lag, obtain τ_{ld} and τ_{lg}, on the basis of step testing the process. This method gives an initial tuning for the unit. But what if the step testing cannot be done? How do we go about tuning the unit? Obviously, a good dynamic simulation can provide the required tunings. However, when this simulation is not available, what we learned in Section 11-3, and in Section 2-4.5 suggest that we can provide some guidelines to answer these questions. Figure 2-4.6 shows the response of a lead-lag unit to a ramp input. Note in Fig. 2-4.6 that the amount of time the output lags the input depends on the *net lag*, defined as $\tau_{lg} - \tau_{ld}$. The amount of time the output leads the input depends on the *net lead*, defined as $\tau_{ld} - \tau_{lg}$. The response of the lead-lag unit to a ramp input is important because most often disturbances look more like ramps than like steps. Some tuning guidelines follow.

- If you need to lag the input signal, set τ_{ld} to zero and select a τ_{lg}. The lead will not make much difference; it is the net lag that matters.

- If you need to lead the input signal, concentrate on the net lead term; however, you must also choose a τ_{lg}.

- From the response of the lead-lag unit to a step change in input, it is clear that if $\tau_{ld} > \tau_{lg}$, it amplifies the input signal. For noisy signals, such as flow, do not use ratios greater than 2.

- Because dead time just adds to the lag, a negative dead time would effectively decrease the net lag if it could be implemented. Thus we could decrease τ_{lg} in the lead-lag element by the positive dead time; that is,

$$\tau_{lg} \text{ to be used} = \tau_{lg} \text{ calculated} + (t_{0_D} - t_{0_M})$$

Alternatively, we could increase the lead in the lead-lag element by the negative of the dead time, that is,

$$\tau_{ld} \text{ to be used} = \tau_{ld} \text{ calculated} - (t_{0_D} - t_{0_M})$$

- If significant dead time is needed, then use a τ_{lg}, with no τ_{ld}, and a dead time. It would not make sense to delay the signal and then lead it, even if the transfer functions called for it.

The third comment also refers to the lead-lag element, specifically, to the location of the unit when multiple disturbances are measured and used in the feedforward controller. If linear compensators are implemented, all that is needed is a single lead-lag unit with adjustable gain for each input. The outputs from the units are then added in the summer, as shown in Fig. 11-4.2. When dynamic compensation is required with nonlinear steady-state compensators, the individual lead-lag elements should be installed just after each transmitter—that is, on the inputs to the steady-state compensator. This permits the dynamic compensation for each disturbance to be implemented individually. It would be impossible to provide different dynamic compensations after the measurements were combined in the steady-state compensator.

The fourth comment refers to the steady-state portion of the feedforward controller. This chapter has shown the development of a linear and of a nonlinear compensator. The nonlinear compensator has shown better performance. Often it is easy to develop this nonlinear compensator using first principles, or a steady-state simulation. If the development of a nonlinear compensator is possible, this is the preferred method. However, if this development is not possible, a linear compensator can be set up with a lead-lag element, with adjustable gain, for each input, and a summer. Which method to use depends on the process.

The fifth comment refers to the comparison of feedforward control to cascade and ratio control; all three of these techniques take corrective action before the controlled variable deviates from set point. Feedforward control takes corrective action *before, or at the same time as*, the disturbance enters the process. Cascade control takes corrective action before the primary controlled variable is affected, but after the disturbance has entered the process. The block diagrams of Figs. 11-2.8 and 9-2.2 graphically show these differences. Figure 11-2.6 shows the implementation of feedforward control only, that is, with no feedback compensation. Interestingly, this scheme is similar to the ratio control scheme shown in Fig. 10-2.2a. The ratio control scheme does not have dynamic compensation, but the ratio station in Fig. 10-2.2a provides the same function as the gain element shown in Fig. 11-2.6. Thus we can say that ratio control is the simplest form of feedforward control.

Finally, an outline of the different steps in designing a feedforward control strategy should be useful. The eight steps outlined next can serve as a design procedure. (Corripio, 1990).

Step 1. State the control objective; that is, define which variable needs to be controlled, and what its set point is. The set point should be adjustable by the operator, not a constant.

Step 2. Enumerate the possible measured disturbances. Which disturbances can be easily measured? How much and how fast should each be expected to vary? Have an idea of the cost and maintenance of each sensor. Knowing the answers to these questions should help designers decide which disturbance(s) will be considered major and will thus be compensated by the feedforward controller.

Step 3. State which variable is going to be manipulated by the feedforward controller. In a cascade arrangement, wherein the feedforward controller is cascaded to a slave controller, the manipulated variable is the set point to the slave controller. This makes sense because the feedforward controller manipulates the set point to the slave controller. Such was the case presented in this chapter.

Step 4. Now you are ready to design the feedforward controller. The feedforward controller consists of two parts: steady-state and dynamic compensators. Develop the steady-state compensator first, and specifically, a nonlinear compensator using first principles, or an existing steady-state simulation. The compensator should be an equation such that the manipulated variable, identified in step 3, can be calculated from the measured disturbances, identified in step 2, and the control objective (set point), identified in step 1. Keep the equation as simple as possible. If the steady-state nonlinear compensator cannot be developed by any of the methods mentioned, then use the step testing procedure developed in Section 11-2.

Step 5. At this point, we can reevaluate the list of disturbances. If a nonlinear compensator has been developed, it may help in the reevaluation. The effect of a disturbance on the controlled variable can be calculated from the equation. A disturbance that was not in the original list may appear in the equation and may be important. The final decision as to which disturbance to compensate using feedforward depends on its effect on the controlled variable, on the frequency and magnitude of variation, and on the capital cost and maintenance of the sensor. Unmeasured disturbances can be treated as constants, at their steady-state or expected values.

Step 6. Introduce the feedback compensation. The way feedback is introduced depends on the physical significance assigned to the feedback signal.

Step 7. Decide whether dynamic compensation, lead-lag and/or dead time, is required, and decide how it is to be introduced into the design.

Step 8. Draw the instrumentation diagram for the feedforward strategy. The diagram detail largely depends on the control system being used. A good design should be able to continue to operate safely when some of its input measurement fails. This characteristic of the design is known as "graceful degradation."

11-7   THREE OTHER EXAMPLES

EXAMPLE 11-7.1   This example presents the control of the process shown in Fig. 11-7.1. The process is similar to the one presented in Section 11-2; however, it is dissimilar enough to require a different feedforward controller, principally in its dynamic compensation; Table 11-7.1 presents the steady-state and other information. It is desired to maintain the outlet composition $x_6(t)$ at 0.472 mass fraction of component A. Any flow or composition entering the process is a possible disturbance. However, operating experience has shown that the flow of stream 2, $f_2(t)$, is the major upset; the stream can double its flow from 1000 to 2000 gpm almost instantaneously. This upset occurs when another process upstream comes on line. The flow of water, $f_1(t)$, is the manipulated variable.

Figure 11-7.2 shows the process response, to the previously mentioned upset, when feedback control is used. This response shows that $x_6(t)$ deviates from its set point of 0.472 mf to 0.487 mf, a 3.18 % change; however, this process requires a tighter quality control. Feedforward can be used to minimize the effect of the disturbance.

We have already completed the first three steps of the design procedure outlined in the previous section. That is, we have stated the control objective (step 1), we have enumerated the disturbances and chosen the major one (step 2), and we have stated the manipulated variable (step 3). We now proceed to design the feedforward controller (step 4).

The block diagram for this process is identical to the one shown in Fig. 11-2.8. As we saw in previous sections, it is necessary to determine the transfer functions G_D and G_M to design the feedforward controller. G_D is determined by step changing $f_2(t)$ and recording $x_6(t)$. Using this procedure we obtain the following transfer function:

$$G_D = \frac{0.014\,e^{-2.2s}}{3.6s + 1}, \qquad \frac{\%\text{TO}}{\text{gpm}} \qquad \textbf{(11-7.1)}$$

G_M is determined by step changing the controller's output and recording $x_6(t)$. The following transfer function is obtained:

$$G_M = \frac{-1.065\,e^{-1.2s}}{3.15s + 1}, \qquad \frac{\%\text{TO}}{\%\text{CO}} \qquad \textbf{(11-7.2)}$$

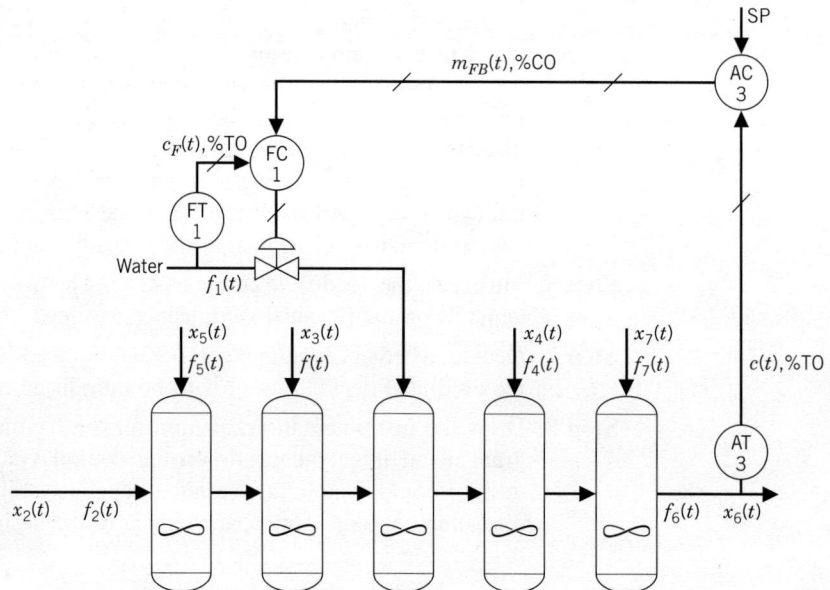

Figure 11-7.1 Mixing process for Example 11-7.1.

Table 11-7.1 Process information and steady-state values for Example 11-7.1

Information

Concentration Transmitter Range: 0.3–0.7 mass fraction. Its dynamic can be described by a time constant of 0.1 min.

The pressure drop across the valve can be considered constant, and the maximum flow provided by the valve is 8000 gpm. The valve dynamics can be described by a time constant of 0.1 min.

The density of all streams is also considered similar and constant.

Steady-state values		
Stream	Flow, gpm	Mass fraction
1	3983	0.000
2	1000	0.990
3	1100	0.850
4	1500	0.875
5	500	0.800
6	8483	0.472
7	500	0.900

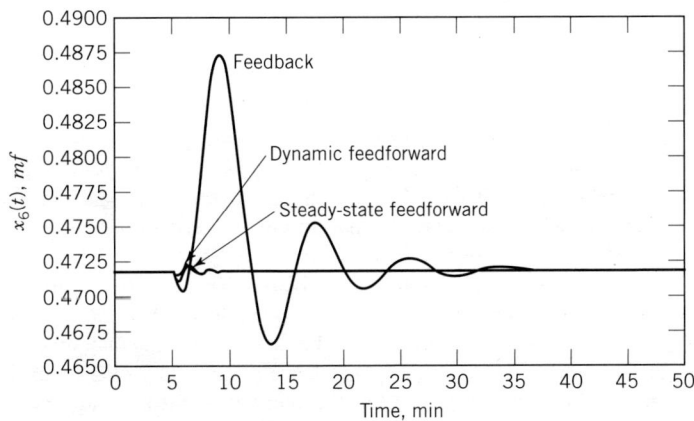

Figure 11-7.2 Process response under feedback, nonlinear steady-state feedforward, and dynamic (dead time only) feedforward control— $f_2(t)$ changed from 1000 to 2000 gpm.

Because the flow transmitter for $f_2(t)$ has a range of 0 to 3000 gpm and has negligible dynamics, its transfer function is given by

$$H_D = 0.0333 \frac{\%\text{TO}_D}{\text{gpm}} \tag{11-7.3}$$

Using Eqs. 11-7.1, 11-7.2, and 11-7.3 the following feedforward controller is designed

$$\text{FFC} = 0.395 \left(\frac{3.15s + 1}{3.60s + 1} \right) e^{-1.0s} \tag{11-7.4}$$

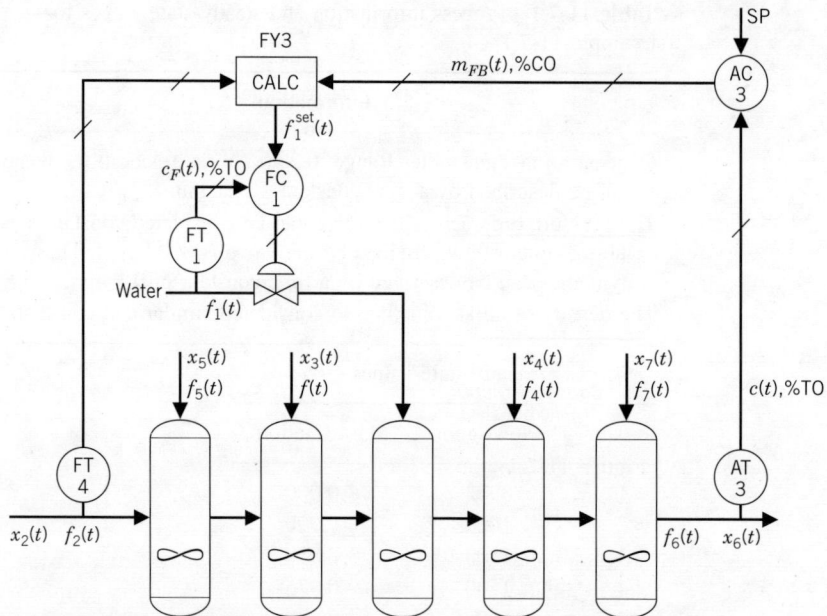

Figure 11-7.3 Implementation of feedforward control.

Even though a steady-state compensator is available (0.395) from the process testing, let us develop a nonlinear steady-state compensator starting from first principles. This compensator, in engineering units, is

$$f_1(t) = \frac{1}{x_6^{set}(t)}[3012.5 + 0.99\,f_2(t)] - 3500 - f_2(t) \qquad \textbf{(11-7.5)}$$

The feedback compensation will be used to adjust the $1/x_6^{set}(t)$ term in Eq. 11-7.5. Choosing how the feedback is to be introduced is step 6 in the procedure.

Step 7 calls for deciding on the dynamic compensation. In this controller, the dynamic compensation elements, lead-lag and dead time, indicate that the action of the manipulated variable should be delayed. The ratio of the lead to the lag time constants is less than 1, and the dead time is positive. However, the lead-lag term in Eq. 11-1.24 indicates that the $\tau_{ld}/\tau_{lg} = 3.15/3.6 = 0.875$. The ratio is relatively close to 1, and applying the rule of thumb presented at the end of Section 11-4, we can neglect it. Furthermore, the dead time itself does not seem very large when compared to the time constants. This may give an indication that it can also be neglected. As an exercise, we will leave it in the controller to test its contribution. Figure 11-7.3 shows the implementation of this feedforward controller (step 8). The block referred to as CALC implements Eq. 11-7.5.

Figure 11-7.2 compares the control performance provided by feedback, nonlinear steady-state feedforward, and dynamic (nonlinear steady-state and dead time) feedforward control. Obviously the addition of a dead time to the feedforward controller does not improve the control performance in this case, so it can be neglected.

EXAMPLE 11-7.2

An interesting and challenging process is the control of the liquid level in a boiler drum. Figure 11-7.4 shows a schematic of a boiler drum. The control of the level in the drum is very important. A high level may result in carrying over of liquid water, and perhaps impurities, into the steam system; a low level may result in tube failure due to overheating for lack of liquid water in the boiling surfaces.

Figure 11-7.4 shows steam bubbles flowing upward through riser tubes into the water; this is an important phenomenon. The specific volume (volume/mass) of the bubbles is very large,

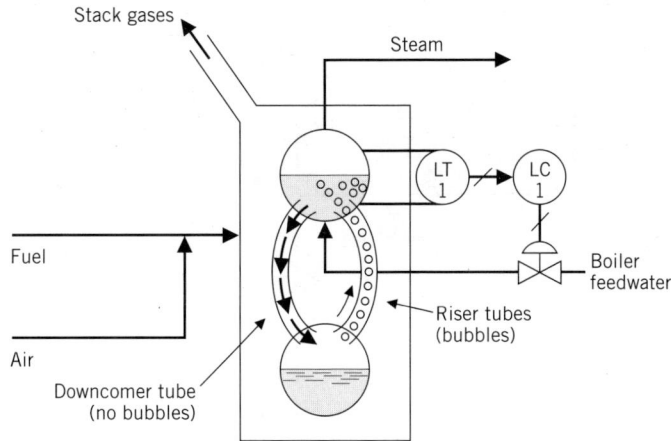

Figure 11-7.4 "Single-element" control in a boiler drum.

so these bubbles displace the liquid water. This results in a higher apparent level than the level due to liquid water only. The presence of these bubbles also presents a problem under transient conditions. Consider the situation when the pressure in the steam header drops because of an increased demand for steam by the users. This drop in pressure results in a certain quantity of water flashed into steam bubbles. These new bubbles tend to increase the apparent level in the drum. The drop in pressure also causes the volume of the existing bubbles to expand, further increasing the apparent level. Such surge in level resulting from a decrease in pressure is called swell. An increase in steam header pressure, brought about by a decreased demand for steam by the users, has the opposite effect on the apparent level and is called shrink.

The swell/shrink phenomena, combined with the importance of maintaining a good level, as explained earlier, makes the level control even more critical. The following paragraphs develop some of the level control schemes presently used in industry.

The drum level control is accomplished by manipulating the flow of feedwater. Figure 11-7.4 shows the simplest type of level control, referred to as *single-element control*. A standard differential pressure sensor-transmitter is usually used. Because this control scheme relies only on the drum level measurement, that measurement must be reliable. Under frequent transients the swell/shrink phenomena do not render a reliable measurement, so a control scheme that compensates for these phenomena is required. Single element is good for boilers that operate at a constant load.

Two-element control shown in Fig. 11-7.5, is essentially a feedforward/feedback control system. The idea behind this scheme is that the major reason for level changes is changes

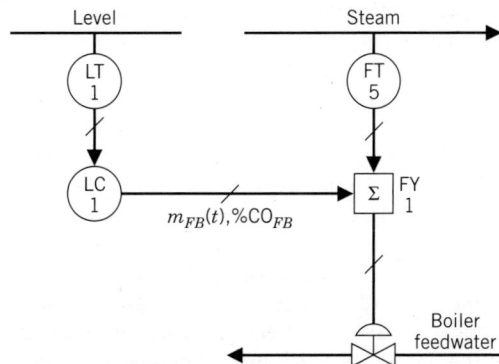

Figure 11-7.5 "Two-element" control.

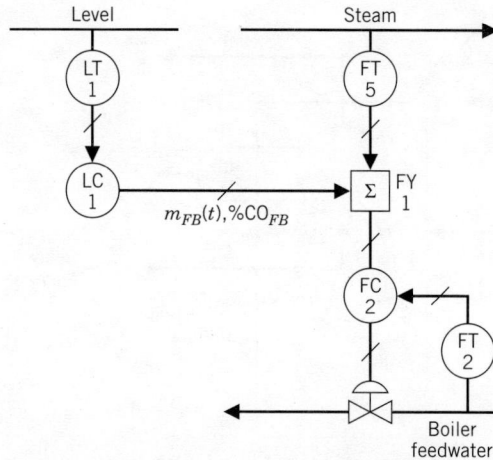

Figure 11-7.6 "Three-element" control.

in steam demand and that for every pound of steam produced, a pound of feedwater should enter the drum; that is, there should be a mass balance. The signal output from FT5 provides the feedforward part of the scheme, while LC1 provides the feedback compensation for any unmeasured flows such as blowdown. The feedback controller also helps to compensate for errors in the flowmeters.

The two-element control scheme works quite well in many industrial boiler drums. However, there are some systems that exhibit variable pressure drop across the feedwater valve. The two-element control scheme does not directly compensate for this disturbance, and consequently, it upsets drum level control by momentarily upsetting the mass balance. The *three-element control* scheme, shown in Fig. 11-7.6, provides the required compensation. This scheme provides a tight mass balance control during transients. It is interesting to note that all that has been added to the two-element control scheme is a cascade control system.

The boiler drum level provides a realistic example wherein the cascade and feedforward control schemes are used to improve the performance provided by feedback control. In this particular example, the use of these schemes is almost mandatory to avoid mechanical and process failures. Every step taken to improve the control was justified. Otherwise, there is no need to complicate matters.

EXAMPLE 11-7.3

We now present another industrial example that has proved to be a successful application of feedforward control. The example is concerned with the temperature control in the rectifying section of a distillation column. Figure 11-7.7 shows the bottom of the column and the control scheme originally proposed and implemented. This column uses two reboilers. One of the reboilers, R-10B, uses a condensing process stream as the heating medium, and the other reboiler, R-10A, uses condensing steam. For efficient energy operation, the operating procedure calls for using as much of the process stream as possible. This stream must be condensed anyway, and thus serves as a "free" energy source. The steam flow is used to control the temperature in the column.

After start-up of this column, it was noticed that the process stream serving as heating medium experienced changes in flow and in pressure. These changes acted as disturbances to the column and, consequently, the temperature controller needed to continually compensate for these disturbances. The time constants and dead time in the column and reboilers complicated the temperature control. After the problem was studied, it was decided to use feedforward control. A pressure transmitter and a differential pressure transmitter had been installed in the process stream, and from them the amount of energy given off by the stream in condensing could be calculated. Using this information the amount of steam required to maintain the temperature

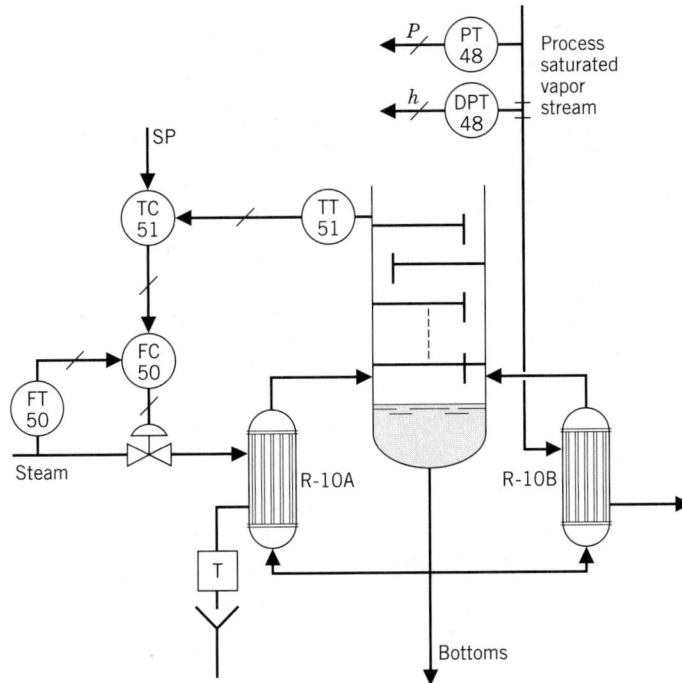

Figure 11-7.7 Temperature control in a distillation column.

at set point could also be calculated, and, thus, corrective action could be taken before the temperature deviated from set point. This is a perfect application of feedforward control.

Specifically, the procedure implemented was as follows. Since the process stream is pure and saturated, the density, ρ, is a function of pressure only. Therefore, using a thermodynamic correlation, the density of the stream can be obtained.

$$\rho = f_1(P) \tag{11-7.6}$$

Using this density and the differential pressure, h, obtained from the transmitter DPT48, the mass flow of the stream can be calculated from the orifice equation

$$\dot{m} = K_o\sqrt{h\rho}, \quad \text{lbm/hr} \tag{11-7.7}$$

where K_o is the orifice coefficient.

Also, knowing the stream pressure and using another thermodynamic relation, the latent heat of condensation, λ, can be obtained

$$\lambda = f_2(P) \tag{11-7.8}$$

Finally, multiplying the mass flow rate times the latent heat, the energy, q_1, given off by the process stream in condensing is obtained.

$$q_1 = \dot{m}\lambda \tag{11-7.9}$$

Figure 11-7.8 shows the implementation of Eqs. 11-7.6 through 11-7.9 and the rest of the feedforward scheme. Block PY48A performs Eq. 11-7.6, block PY48B performs Eq. 11-7.7, block PY48C performs Eq. 11-7.8, and block PY48D performs Eq. 11-7.9. Therefore, the output of relay PY48D is q_1, the energy given off by the condensing process stream.

To complete the control scheme, the output of the temperature controller is considered to be the total energy required, q_t, to maintain the temperature at its set point. Subtracting q_1 from q_t the energy required from the steam, q_s, is determined.

$$q_s = q_t - q_1 \tag{11-7.10}$$

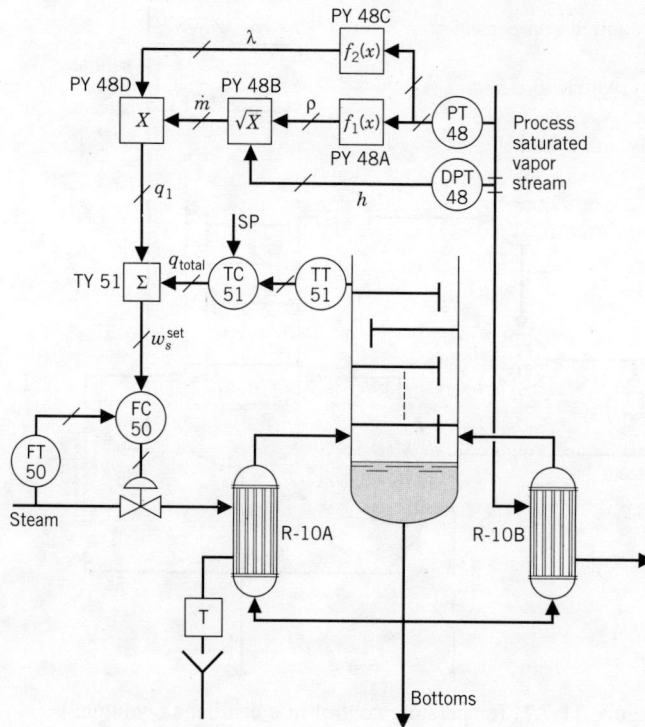

Figure 11-7.8 Implementation of feedforward control.

Finally, dividing q_s by the latent heat of condensation of the steam, h_{fg}, the required steam flow, w_s, is obtained

$$w_s = \frac{q_s}{h_{fg}}, \quad \text{lbm/hr} \tag{11-7.11}$$

Block TY51 performs Eqs. 11-7.10 and 11-7.11, and its output is the set point to the flow controller FC50. In Eq. 11-7.11 h_{fg} is assumed constant.

Several things must be noted in this feedforward scheme. First, the model of the process is not one equation but, rather, several equations. This model was obtained using several process engineering principles. This makes process control interesting, and challenging. Second, the feedback compensation is an integral part of the control strategy. This compensation is q_t or total energy required to maintaining set point in TC51. Finally, the control scheme shown in Fig. 11-7.8 does not show dynamic compensation. This compensation may be installed later if needed.

11-8 SUMMARY

This chapter has presented in detail the concept, design, and implementation of feedforward control. The technique has been shown to provide significant improvement over the control performance provided by feedback control. However, undoubtedly the reader has noted that the design, implementation, and operation of feedforward control require a significant amount of engineering, extra instrumentation, and training of the operating personnel to understand the technique.

All of this means that feedforward control is more costly than feedback control and thus, it must be justified. The reader must also understand that feedforward is not the solution to all the control problems. It is another good "tool" to aid feedback control in some cases.

It was shown that feedforward control is generally composed of (a) steady-state compensation and (b) dynamic compensation. Not in every case are both compensations needed. The amount of each required compensation depends on each particular process.

There are three means by which to design the steady-state compensator. The best way is based on process engineering principles, usually mass and energy balances. Steady-state simulations—flowsheet simulators—provide still another realistic avenue. Both of these methods were discussed in Section 11-5. The process testing method presented in Section 11-2 provides a way to design a linear steady-state compensator when neither of the other two methods can be achieved. The design of the dynamic compensator, lead-lag and/or dead time, requires dynamic information, usually obtained by process testing.

Finally, the reader must remember that feedforward control must be designed with some amount of feedback compensation.

REFERENCES

1. DOBSON, J. G., "Optimization of Distillation Columns Using Feed-Forward Computer Control." Presented at Interkama 1960, Duesseldorf, Germany, October 1960.

2. SHINSKEY, F. G., "Feedforward Control Applied." *ISA Journal*, November 1963.

3. CORRIPIO, A. B., Tuning of Industrial Control Systems. North Carolina: Instrument Society of America, 1990.

PROBLEMS

11-1. Consider the process of Problem 10-1. A few weeks after the ratio control system was put into operation, the operating personnel noted that the concentration of the exit solution varied somewhat every once in a while; it is suspected that the concentration of the 50 % solution is the culprit. Because it is important to control the concentration of the exit solution at 30 %, it is necessary to provide a tighter control. Design a feedforward controller, based on process engineering principles to compensate for the flow and composition of the 50 % NaOH solution. There is available an analyzer with a range of 55 to 60 % that could be used in that stream.

11-2. Consider the kiln drier sketched in Fig. P11-1. A slurry is fed to the drier and directly contacted with gases from the combustion of fuel and air. The contact of the solid with the hot gases vaporizes the water, the heat of vaporization being provided by the hot gases. It is desired to manipulate the flow of the fuel into the drier to maintain the outlet moisture content of the solid at its set point. The major disturbances are the slurry feed flow, $w_i(t)$; its moisture content, $x_i(t)$; and the heating value of the fuel gas. Of these, only the feed flow and its moisture content can be measured, as shown in the figure. The outlet moisture content of the solid, $x_o(t)$, can also be measured. The design conditions and process parameters are

Slurry feed flow:	100 lb/h
Slurry feed composition:	0.60 mass fraction of water
Desired outlet moisture content:	0.05 mass fraction of water
Design fuel gas flow:	80 scfh

The flow transmitter signals are linear with flow and compensated for temperature and pressure. Design a feedforward controller with feedback trim for the drier; use the software of Table 10-1.2 for the implementation. Specify which disturbances are measured, what the controlled objective is, and which variable is manipulated. Derive the following equation by combining the total and water mass balances on the drier

$$w_i(t)x_i(t) = [w_i(t) - k_v f_g(t)]x_o(t) + k_v f_g(t)$$

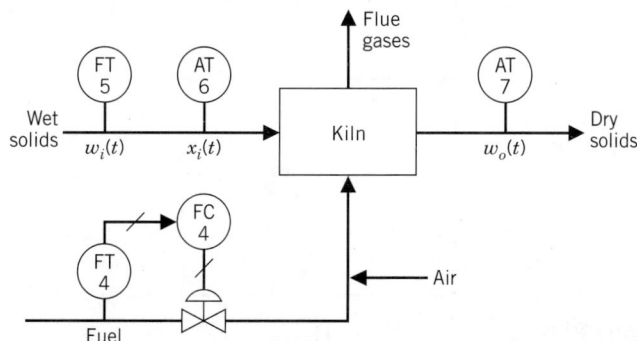

Figure P11-1 Kiln for Problem 11-2.

where $f_g(t)$ is the fuel gas flow in scfh, and k_v represents the pounds of water vaporized per scf of fuel gas. Specify how you plan to introduce dynamic compensation and feedback trim if needed.

11-3. It is desired to design a simple linear feedforward controller to compensate for the effect of changes in feed rate to a distillation column on the overhead product composition. The reflux flow is to be manipulated to control the overhead composition. Two steps tests are applied to the column, one on feed rate and one on reflux flow set point; in each case, the response of the overhead composition is recorded and analyzed. Results of the tests are summarized in the following table.

Step Test in:	Process Gain, %TO/(klb/h)	Time Constant, min	Dead Time, min
Feed rate, klb/h	0.8	18.0	8.0
Reflux flow, klb/h	1.2	4.0	1.0

The feed flow transmitter has a calibrated range of 0 to 50 klb/h, and the reflux flow transmitter has a range of 0 to 100 klb/h. Both transmitters generate signals that are linear with flow.

Draw the block diagram that shows the effect of feed rate and reflux flow on the overhead composition. Approximate the transfer functions with first-order-plus-dead-time models, and show the numerical values of all parameters. Using the block diagram, design a linear feedforward controller to compensate for the feed disturbance on the control objective. Draw the instrumentation diagram for your feedforward controller using (a) the block of Table 10-1.1, and (b) the software of Table 10-1.2. Include feedback trim, assuming an on-line analyzer is installed on the overhead product stream.

11-4. Design a feedforward controller to compensate for the feed flow and temperature of the continuous stirred tank heater of Example 6-1.1. Assume that the feed flow transmitter has a range of 0 to 25 ft^3/min and a negligible time constant, and that the inlet temperature transmitter has a range of 60 to 120°F and a time constant of 0.6 min. Identify the control objective, the disturbances, and the manipulated variable. Determine the corrections needed in steam flow for the expected changes in disturbances: 10 ft^3/min change in feed flow, and 20°F change in feed temperature. Discuss the need for feedback trim and how to incorporate it into your design. Also discuss the need for and form of dynamic compensation. Draw the instrumentation diagram for your design.

11-5. Consider the control scheme for the solid drying system shown in Fig. P11-2. The major disturbance to this process is the moisture content of the incoming solids. For this disturbance, the control system responds quite slowly. It is desired to implement a feedforward system to improve this control. After some initial work, the following data have been obtained:

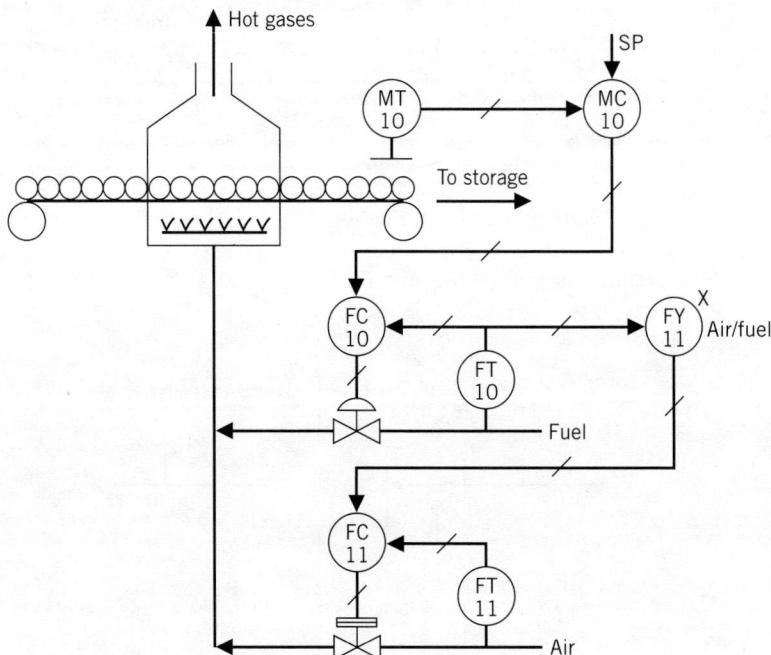

Figure P11-2 Drier for Problem 11-5.

Step change in inlet moisture = +2 %

Time, min	Exit Moisture, %	Time, min	Exit Moisture, %
0	5.0		
0.5	5.0	5.0	6.6
1.0	5.1	5.5	6.7
1.5	5.2	6.0	6.8
2.0	5.4	6.5	6.9
2.5	5.7	7.0	7.0
3.0	5.9	7.5	7.0
3.5	6.1	8.0	6.9
4.0	6.3	8.5	7.0

Step change in output signal from moisture controller, MC-10 = +25 %CO

Time, min	Exit Moisture, %	Time, min	Exit Moisture, %
0	5.0	5.0	3.81
0.5	5.0	5.5	3.70
1.0	4.95	6.0	3.55
1.5	4.93	6.5	3.45
2.0	4.85	7.0	3.35
2.5	4.70	7.5	3.25
3.0	4.60	8.5	3.10
3.5	4.40	9.5	3.03
4.0	4.20	11.5	3.00

The feedback moisture analyzer has a range of 1 to 7 % moisture. There is another analyzer with a range of 10 to 15 % moisture that can be used to measure the inlet moisture. These analyzers have been used before in this particular process and have proved to be reliable. Identify the control objective, the disturbances, and the manipulated variable.

(a)   On the basis of process engineering principles (mass balances, energy balances, etc.), develop a feedforward control scheme. The statement of the problem has not provided all of the necessary information. Assume that you can obtain this information from the plant's files. Show the implementation of this scheme using the block of Table 10-1.1, and the software of Table 10-1.2.

(b)   Draw a complete block diagram for this process that shows the effect of the inlet moisture on the controlled variable. Include all known transfer functions.

(c)   Develop a feedforward control scheme using the block diagram approach. Show the implementation of this scheme, including the feedback trim, using the block of Table 10-1.1, and the software of Table 10-1.2.

11-6.   Consider the vacuum filter shown in Problem 7-15. Using the information given in that problem, design a feedforward/feedback control scheme to compensate for changes in inlet moisture. You may assume that there is a moisture transmitter with a range of 60 to 95 % moisture. Show the complete implementation using the computing blocks of Table 10-1.1. Identify the eight steps outlined in Section 11-6.

11-7.   Consider the evaporator system of Problem 7-20. Design a feedforward/feedback control scheme to compensate for composition changes in the solution entering the first effect. The sensor to measure to inlet composition has a range of 0 to 20 % sugar. The data given in the problem is enough to design a linear feedforward controller. It is also known that the total evaporation in both effects is approximately 1.8 lb of water vaporized per lb of steam. Using this information also design a nonlinear feedforward controller. Use the computing blocks of Table 10-1.1 to implement the scheme. Identify the eight steps outlined in Section 11-6.

11-8.   Consider the process shown in Problem 7-22 to dry phosphate pebbles. As mentioned in the problem, an important disturbance to the process is the moisture of the inlet pebbles. Using the information provided, design a feedforward/feedback control scheme to compensate for its disturbance. There is a moisture transmitter available to measure the inlet moisture. This transmitter has a range of 12 to 16 % moisture. Use the computing blocks of Table 10-1.1. Identify the eight steps outlined in Section 11-6.

11-9.   Design a feedforward controller for controlling the outlet temperature of the furnace sketched in Fig. P11-3. The controller must compensate for variations in feed rate and feed temperature. Use the software of Table 10-1.2. The following information is known: gas specific heat: $C_p = 0.26$ Btu/lbm-$°$F; fuel heating value = 980 Btu/scf; inlet gas temperature: $T_i = 90°$F; outlet gas temperature: $T_0 = 850°$F; furnace efficiency $= 0.75$; range of FT42: 0 to 20,000 lb/h; range of TT42: 50 to 120$°$F.

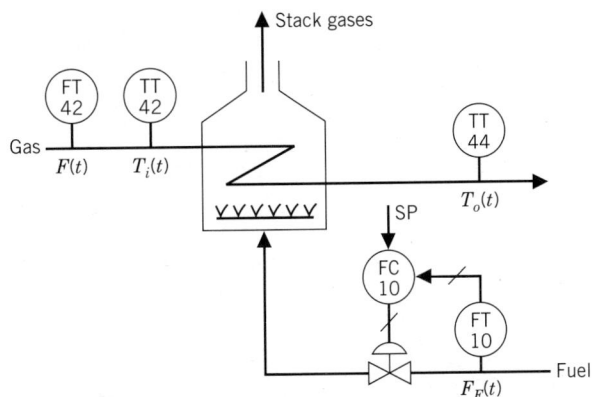

Figure P11-3 Furnace for Problem 11-9.

11-10. Let us propose that some of the process data, of the furnace of Problem 11-9, such as the fuel heat of combustion or the gas specific heat, are not known. Thus it is necessary to design the feedforward controller using the block diagram approach. The following data are obtained from the step tests on the furnace.

Variable	Step Change	Change in T_0, °F	Time Constant, min	Dead Time, min
F	750 lbm/h	38	0.30	0.10
T_i	10°F	7.5	0.10	0.40
F_F^{set}	5 % of range	20	0.95	0.15

(a) Draw the complete block diagram including all transfer functions.

(b) Design the feedforward controller to compensate for variations in feed rate and feed temperature. Include the dynamic compensator.

(c) If it is decided to compensate for only one of the disturbances, which one would it be and why? How would your design be modified?

11-11. For the stripping section of a distillation column shown in Fig. P11-4, the objective is to maintain the bottom's purity at a desired value. This objective is commonly attained by controlling the temperature in one of the trays (the column pressure is assumed constant) by using the steam flow to the reboiler as the manipulated variable. A usual "major" disturbance is the feed flow to the column.

(a) Sketch a feedforward/feedback control scheme to compensate for this disturbance; describe it briefly.

(b) Briefly describe the dynamic tests that you would perform on the column in order to tune the feedback controller and the feedforward controller. Would you expect

the dynamic compensation on the feedforward controller to be a net lead or a net lag?

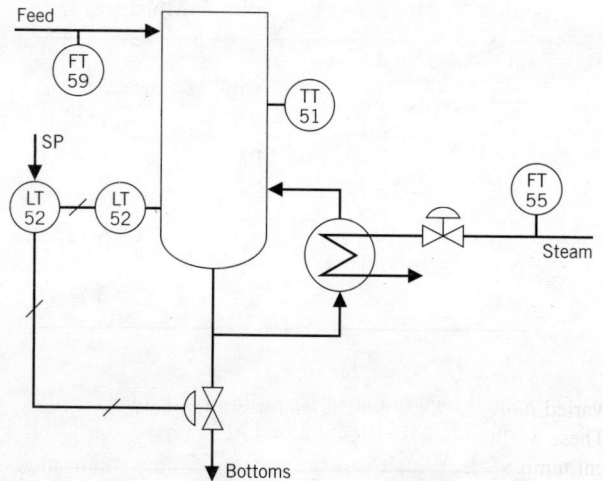

Figure P11-4 Distillation column for Problem 11-11.

11-12. Consider the drying process shown in Fig. P11-5. In this process, wet paper stock is being dried to produce the final paper product. The drying is done using hot air; this air is heated in a heater in which fuel is burned to provide the energy. The controlled variable is the moisture of the paper leaving the drier. Figure P11-5 shows the original control scheme proposed and installed.

(a) A few weeks after start-up, the process engineer noticed that even though the moisture controller was keeping the moisture within certain limits from set point, the oscillations were more than desired. After searching for possible causes and making sure that the moisture controller was well tuned, they found that the hot air temperature leaving the heater

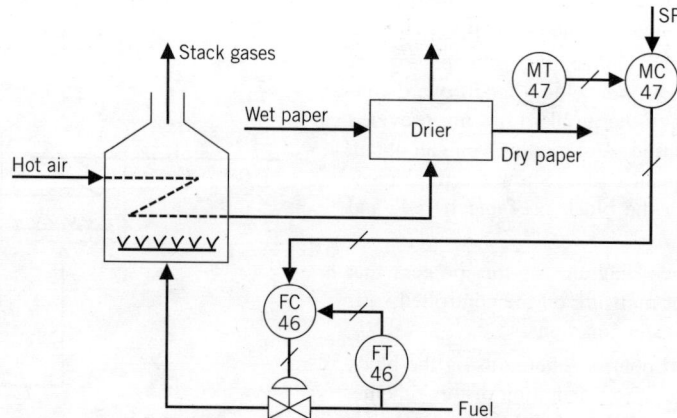

Figure P11-5 Wet paper drier for Problem 11-12.

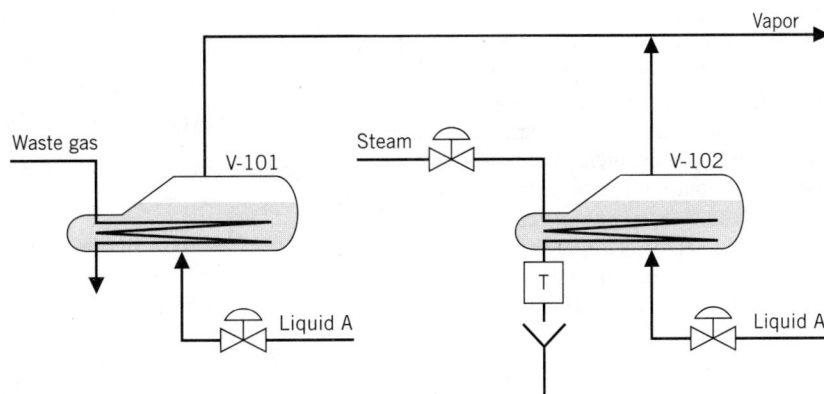

Figure P11-6 Vaporizers for Problem 11-13.

varied more than had been assumed during the design stage. These variations were attributed to daily changes in ambient temperature and possible disturbances in the combustion chamber of the heater. Design a control scheme to maintain the hot air temperature at the desired value to help maintain moisture set point.

(b) The control scheme just described significantly improved the moisture control. A few weeks later, however, the operators complained that every once in a while, the moisture would go out of set point considerably, though the control scheme would eventually bring it back to set point. This disturbance would require that the paper produced during this period be reworked, so it represented a production loss. After searching through the production logs, the process engineers discovered that changes in inlet moisture were the cause of this disturbance. Design a feedforward control scheme that compensates for these disturbances. There is a moisture transmitter, with a range of 5 to 20 moisture %, available to measure the inlet moisture.

11-13. The process shown in Fig. P11-6 is used to vaporize a certain liquid A. Two vaporizers, V-101 and V-102, are used for this purpose. The heating fluid in V-101 is a waste gas, which can be measured but not controlled; steam is used in V-102 as the heating fluid. The pressure in the vapor header needs to be controlled. It is also necessary to maintain a level of liquid A, above the heating surface, in each vaporizer.

(a) Design a system to control the vapor header pressure and the level in the vaporizers.

(b) Operating experience has shown that the waste gas flow, and sometimes its temperature, vary often enough to swing (upset) the vapor header pressure significantly. The production engineer wonders whether something can be done to minimize the pressure swings. Propose a control system to accomplish this control.

11-14. The "well-known" reaction A + B → C takes place in the reactor shown in Fig. P11-7. This reaction does not

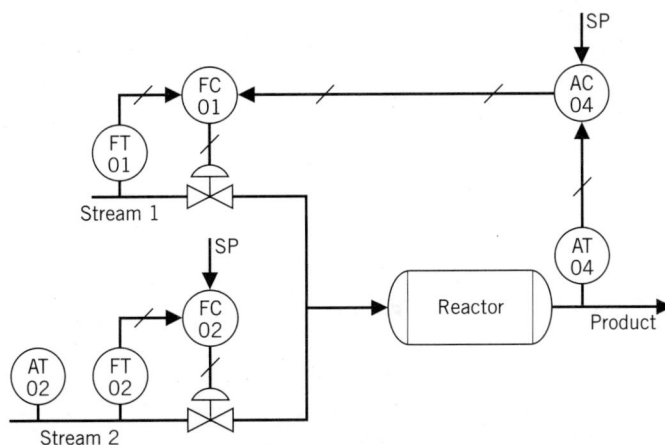

Figure P11-7 Reactor for Problem 11-14.

release, or require any energy. Two streams mix and enter the reactor. Stream 1 is pure A, and stream 2 is mainly B with varying amounts of A. The original control scheme calls for controlling the rate from this reactor by setting the flow of stream 2. The analysis of the product, the concentration of C in the product stream, manipulates the flow of stream 1; the figure shows this control scheme. After the start-up of this unit, it was noticed that the product concentration deviated significantly from set point once or twice every shift. An analyzer, AT-03, of component A in stream 2 indicated that, a few seconds before this deviation in product concentration occurred, a change in stream 2 concentration had also occurred. Furthermore, the research department indicated that if too much reactant A is present during the reaction, it may start reacting with product C, reducing the amount of C in the product stream. Thus the production department has contacted you, the "well-known" control engineer, for possible ideas on how to maintain tighter control of the product concentration. The transmitter ranges are as follows:

FT-01:0–800 lb/hr; FT-02:0–300 lb/hr;

AT-03:0–0.3 mass fraction

Design a control scheme that provides the control performance required by the production department.

11-15. Problem 10-9 describes a furnace with two sections and a single stack. Referring to that process, if the flow of hydrocarbons changes, the outlet temperature will deviate from set point and the feedback controller will have to react to bring the temperature back to set point. This seems a natural for feedforward control. Design this strategy for each section.

11-16. Comment on how the addition of a linear feedforward affects the stability of the control system.

Chapter 12

Multiple-Input/Multiple-Output Control Systems

Up to this point in our study of automatic process control, we have only considered processes with a single control objective or controlled variable. Often, however, we encounter processes in which more than one variable must be controlled, that is, multiple control objectives. In such processes, we can still consider each control objective separate from the others, as long as they do not interact with each other. In this chapter we will study and design control systems for processes in which the various control objectives interact with each other. We refer to these systems as *multivariable control systems* or *multiple-input, multiple-output (MIMO)* control systems. The problem we will be addressing is that of *loop interaction*. We will find that the response and stability of the multivariable system can be quite different from that of its constituent loops taken separately. We will learn how to pair the controlled and manipulated variables to minimize the effect of interaction, and how to design decouplers that reduce or eliminate the effect of interaction.

12-1 LOOP INTERACTION

Figure 12-1.1 shows several examples of multivariable control systems. For the blending tank of Fig. 12-1.1*a* it is necessary to control both the flow and composition of the outlet stream. To accomplish this objective we manipulate the flow of each of the two inlet streams. Figure 12-1.1*b* shows a chemical reactor for which it is necessary to control the outlet temperature and composition. The manipulated variables in this process are the cooling water flow and the process flow. In the evaporator of Fig. 12-1.1*c* the level, the process flow, and the outlet composition are controlled by manipulating the steam, inlet, and outlet flows. Figure 12-1.1*d* shows a paper-drying machine in which the controlled variables are the moisture and dry basis weight (fibers per unit area) of the final paper product. The two manipulated variables are the stock flow to the machine and the steam flow to the last set of heated drums. Finally, Figure 12-1.1*e* depicts a typical distillation column with the necessary controlled variables: column pressure, product compositions, and the levels in the accumulator and column base. The five manipulated variables are the coolant flow to the condenser, distillate flow, reflux flow, bottoms flow, and steam flow to the reboiler.

All of the foregoing are examples of processes with multiple interacting control objectives, the control of which can be quite complex and challenging to the process engineer. There are usually three questions the engineer must ask when faced with a control problem of this type:

1. What is the effect of interaction on the response of the feedback loops?

Principles and Practice of Automatic Process Control/Third Edition, by C. A. Smith and A. B. Corripio
ISBN 0-471-66141-4   Copyright © 2006 John Wiley & Sons (Asia) Pte. Ltd.

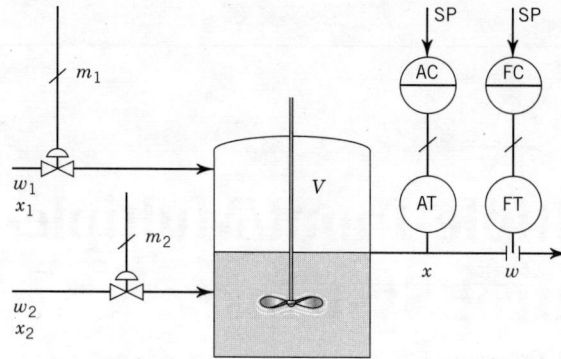

Figure 12-1.1 Examples of multivariable control systems. (*a*) Blending tank.

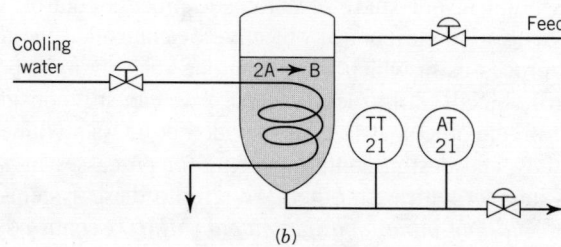

Figure 12-1.1(*Continued*) (*b*) Chemical reactor.

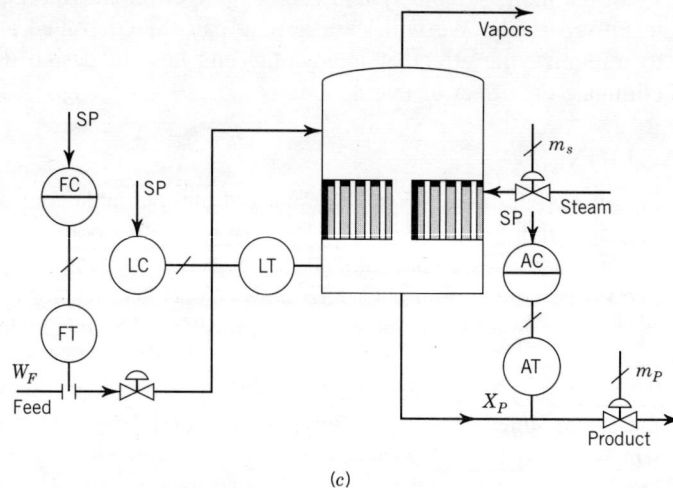

Figure 12-1.1(*Continued*) (*c*) Evaporator.

2. How much interaction exists between the loops and which is the best way to pair the controlled and manipulated variables to reduce the effect of interaction?

3. Can the interaction between loops be reduced or eliminated through the design of an appropriate control system?

We will now address the first of these questions. The other two questions are addressed in other sections of this chapter.

To understand the effect of interaction, let us consider the blending tank of Fig. 12-1.1*a*. Let us say that, at design conditions, the tank blends a solution containing 10 weight % salt with a concentrated solution containing 35 weight % salt, to produce 100 lb/h of a solution containing 20 weight % salt. Steady-state balances on

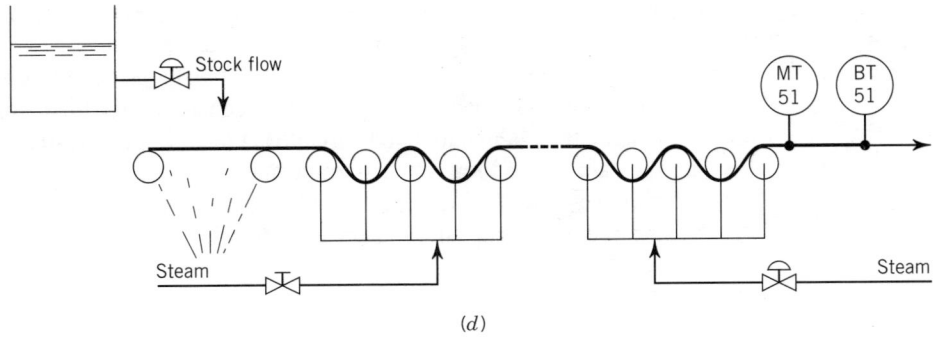

Figure 12-1.1(*Continued*) (*d*) Paper-drying machine.

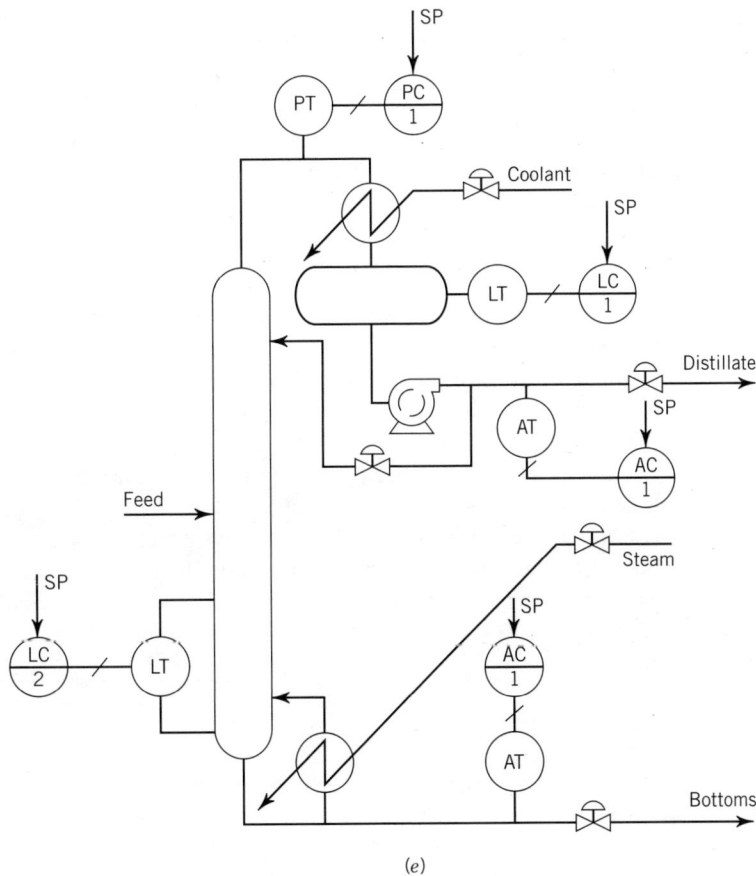

(*e*)

Figure 12-1.1(*Continued*) (*e*) Distillation column.

total mass and mass of salt around the tank result in the following two equations:

$$w = w_1 + w_2$$
$$wx = w_1 x_1 + w_2 x_2$$
$$(12\text{-}1.1)$$

where w is the stream flow in lb/h, x is the mass fraction of salt in each stream, and the subscripts mark the two inlet streams. At the design conditions of $\overline{w} = 100$ lb/h,

$\bar{x} = 0.20$, $\bar{x}_1 = 0.1$, and $\bar{x}_2 = 0.35$, solution of Eq. 12-1.1 results in the required inlet flows of $\bar{w}_1 = 60.0$ lb/h and $\bar{w}_2 = 40.0$ lb/h.

To show how interaction between the loops affects the response parameters of each loop, specifically the gain, let us consider only the composition control loop and assume that its manipulated variable is the concentrated stream, say stream 2. To obtain the gain of the loop for this arrangement, we apply a small change in the manipulated variable—say an increase of 2.0 lb/h in w_2 to 42.0 lb/h. From Eq. 12-1.1 we find that the product composition increases to 20.3 weight % salt, resulting in the following steady-state gain:

$$K_{x2} = \frac{20.3 - 20.0}{42.0 - 40.0} = 0.15 \frac{\% \text{ salt}}{\text{lb/h}}$$

where K_{x2} is the gain of the flow of stream 2 on the weight % salt in the product stream. At the same time, the total flow increases to $60.0 + 42.0 = 102.0$ lb/h.

Let us now consider the control system of Fig. 12-1.2 in which the product flow is controlled by manipulating the flow of the dilute inlet stream, w_1. With this scheme, when the flow of stream 2 is increased from 40.0 to 42.0 lb/h, the product flow controller decreases the flow of the dilute stream to 58.0 lb/h, to keep the product flow at 100.0 lb/h. Solving Eq. 12-1.1 for these conditions, we find that the product composition now increases to 20.5 weight % salt. The gain of the product composition loop becomes

$$K'_{x2} = \frac{20.5 - 20.0}{42.0 - 40.0} = 0.25 \frac{\% \text{ salt}}{\text{lb/h}}$$

This represents an increase of 67 % in the gain of the product composition loop, from 0.15 to 0.25 weight % salt/(lb/h), caused by interaction with the flow controller. To distinguish the two gains, we call K'_{x2} the *closed-loop gain*, because it is the gain of the composition loop when the flow loop is closed, while K_{x2} is the *open-loop gain*, that is, the composition loop gain when the flow loop is open.

Interaction occurs because when the composition controller changes the concentrated stream flow it causes a change in the product flow; this in turn causes the flow controller to change the dilute stream flow which causes a change in the product composition. This additional change in the product composition would not be there if the product flow were not controlled. It is easy to show that installation of the composition controller causes a similar change in the gain of the flow controller.

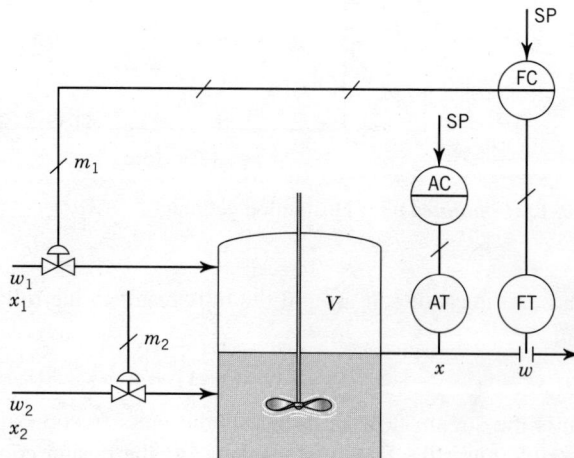

Figure 12-1.2 Blending tank control with product flow controller manipulating stream 1.

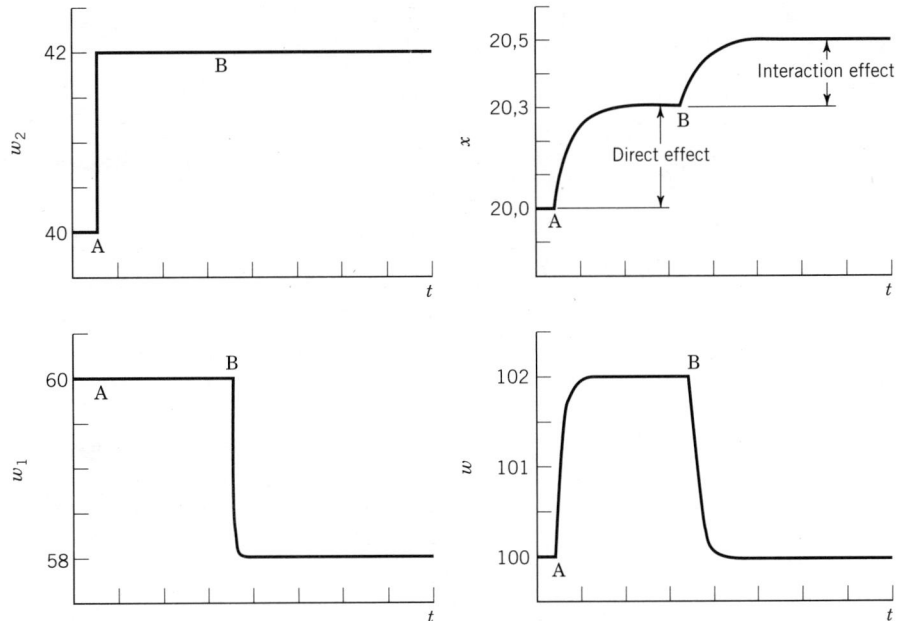

Figure 12-1.3 Response of blender to step change in concentrated stream flow with flow controller manipulating the dilute stream flow.

Figure 12-1.3 shows the response of the controlled and manipulated variables for the change in concentrated stream flow. The change in flow takes place at point A with the flow controller in *manual*. At point B the flow controller is switched to *automatic* and brings the product flow back to its set point. The additional increase in product composition is the result of interaction.

Positive and Negative Interaction

In the blending tank example, the interaction is such that the two loops help each other, that is, closing the flow loop causes a change in the product composition that is in the same direction as the original change. This case, known as *positive interaction*, results in an increase in the loop gain when the other loop is closed. When the two loops fight each other, the interaction is said to be *negative* and the gain of a loop decreases or changes sign when the other loop is closed. This is because the interaction causes a change in the controlled variable that is in the opposite direction to the original change. The change in the sign of the gain occurs when the interaction change is greater than the original change. Because of this possible change in the sign of the gain, negative interaction can be a more severe problem than positive interaction.

The preceding presentation of interaction has been limited to its effect on the steady-state gains. A more complete analysis based on block diagrams will be presented in a latter section of this chapter. We will next look at a quantitative measure of interaction and its application to the proper pairing of controlled and manipulated variables in multivariable control systems.

12-2 PAIRING CONTROLLED AND MANIPULATED VARIABLES

The second question we posed earlier about interaction was about a quantitative measure of interaction and on how to pair controlled and manipulated variables. Oftentimes

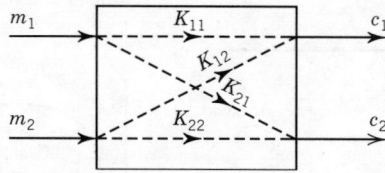

Figure 12-2.1 Schematic of general 2×2 process.

it is simple to decide on the pairing, but many times, such as the systems of Fig. 12-1.1, it is more difficult. In this section we will learn to calculate a quantitative measure of the interaction between control loops. Then we will use the interaction measure to select the pairing of controlled and manipulated variables that minimizes the effect of interaction.

To fix ideas, let us consider the system of Fig. 12-2.1 with two controlled variables c_1 and c_2, and two manipulated variables, m_1 and m_2. We call this system a 2×2 system. It makes sense to pair each controlled variable with the manipulated variable which has the greatest "influence" on it. In this context "influence" and "gain" have the same meaning; consequently, to make a decision we must find the gains of each manipulated variable on each controlled variable.

Open-Loop Gains

The four open-loop steady-state gains for the 2×2 system are

$$K_{11} = \frac{\Delta c_1}{\Delta m_1}\bigg|_{m_2} \qquad K_{12} = \frac{\Delta c_1}{\Delta m_2}\bigg|_{m_1}$$

$$K_{21} = \frac{\Delta c_2}{\Delta m_1}\bigg|_{m_2} \qquad K_{22} = \frac{\Delta c_2}{\Delta m_2}\bigg|_{m_1}$$

(12-2.1)

where K_{ij} is the gain relating the ith controlled variable to the jth manipulated variable. The vertical bars indicate that the gains are determined with the loops opened, that is, a change is made on each manipulated variable keeping the other manipulated variables constant.

It may appear that the pairing of the controlled and manipulated variables can be done by simply comparing the open-loop gains. For example, if K_{12} is larger in magnitude than K_{11}, then m_2 is chosen to control c_1. However, this is not quite correct because the gains usually have different units and cannot be compared to each other. Even if the gains are expressed in %TO/%CO, factors such as the transmitter ranges and valve sizes, which have nothing to do with the process interaction, would affect the pairing.

Bristol (1966) proposed a measure of interaction that is independent of units of measure, transmitter ranges, and valve sizes. The measure of interaction is based on what we found in the previous section: in an interacting control system the gain of a loop changes when the other loop or loops are closed. This means that for each of the open-loop gains of Eq. 12-2.1 there is a corresponding closed-loop gain, to be defined next.

Closed-Loop Gains

For each pair of controlled and manipulated variables, the *closed-loop gain* is the change in the controlled variable divided by the change in the manipulated variable when all the other *controlled variables are held constant*. Notice that this requires that the other manipulated variables be adjusted to bring the other controlled variables back to their base values.

For the 2×2 system of Fig. 12-2.1, the four steady-state closed-loop gains are

$$K'_{11} = \frac{\Delta c_1}{\Delta m_1}\bigg|_{c_2} \quad K'_{12} = \frac{\Delta c_1}{\Delta m_2}\bigg|_{c_2}$$

$$K'_{21} = \frac{\Delta c_2}{\Delta m_1}\bigg|_{c_1} \quad K'_{22} = \frac{\Delta c_2}{\Delta m_2}\bigg|_{c_1} \tag{12-2.2}$$

It is important to realize that the closed-loop gains, defined by Eq. 12-2.2, are not the same as the gains of the closed loops that we learned to compute in Section 6-1.3.

Interaction Measure or Relative Gains

For a particular pair, the measure of interaction proposed by Bristol (1966) is simply the ratio of the open-loop gain to the closed-loop gain. For the general case

$$\mu_{ij} = \frac{K_{ij}}{K'_{ij}} \tag{12-2.3}$$

where μ_{ij} is the interaction measure or *relative gain* for the pair $c_i - m_j$. For the 2×2 system of Fig. 12-2.1, the relative gains are

$$\mu_{11} = \frac{K_{11}}{K'_{11}} \quad \mu_{12} = \frac{K_{12}}{K'_{12}}$$

$$\mu_{21} = \frac{K_{21}}{K'_{21}} \quad \mu_{22} = \frac{K_{22}}{K'_{22}} \tag{12-2.4}$$

It is evident that each relative gain is dimensionless, since it is defined as the ratio of two gains relating the same controlled and manipulated variables. This may be made clearer by substituting the gains from Eqs. 12-2.1 and 12-2.2, into Eq. 12-2.4. For example, for μ_{12},

$$\mu_{12} = \frac{K_{12}}{K'_{12}} = \frac{\dfrac{\Delta c_1}{\Delta m_2}\bigg|_{m_1}}{\dfrac{\Delta c_1}{\Delta m_2}\bigg|_{c_2}}$$

The relative gain is also independent of such things as transmitter ranges and valve sizes, because these parameters affect the open-loop and closed-loop gains in exactly the same way.

The usefulness of the relative gains as measures of interaction derives directly from their definition. If the gain for a given controlled–manipulated variable pair were not affected by interaction with other loops, the relative gain for that pair would be unity, because the closed-loop gain would be exactly the same as the open-loop gain. The more a pair is affected by interaction with the other loops, the further from unity its relative gain. When the interaction is positive, the relative gain is positive and less than unity because the closed-loop gain is of the same sign and greater than the open-loop gain. When the interaction is negative, the relative gain is either greater than unity or negative, because the closed-loop gain is either less than the open-loop gain, or has the opposite sign.

Pairing Rule

The rule for using the relative gain to pair controlled and manipulated variables is simply stated as follows:

To minimize the effect of interaction in a multivariable control system, the controlled and manipulated variables must be paired so that the relative gain for each pair is closest to unity.

It is important to realize that this rule considers only the steady-state effect of interaction. In some systems, particularly those exhibiting negative interaction, dynamic effects may negate the validity of the preceding rule.

Pairings with negative relative gains must be avoided at all costs. A negative relative gain means that the action of the process changes when the other loops are opened and closed, or when their manipulated variables reach their limits (e.g., fully opened or closed valves). In such cases the loop with the negative relative gain will become unstable unless the action of the controller is changed, a difficult thing to do automatically.

It is possible, although not recommendable, to select a pair with a relative gain of zero. In this case the manipulated variable has no direct effect on the controlled variable and depends on interaction with the other loops to control it (zero open-loop gain). A pairing with a large relative gain will not work well, because the interaction with the other loops cancels most of the direct effect of the manipulated variable on the controlled variable (closed-loop gain near zero).

To further understand the meaning of the relative gain, let us look at the two cases of the following examples.

EXAMPLE 12-2.1

Consider the following matrix of relative gains:

	m_1	m_2
c_1	0.20	0.80
c_2	0.80	0.20

As we shall soon see, the relative gains for each row and column always add up to unity. Also, for a 2×2 system, there are only two possible pairing options with two pairs each, c_1-m_1, c_2-m_2, and c_1-m_2, c_2-m_1; the relative gains for the two pairs in each option are exactly the same.

The relative gain for one pairing, $\mu_{11} = \mu_{22} = 0.2 = 1/5$, indicates that for this pairing the gain of each loop increases by a factor of 5 when the other loop is closed. The relative gain for the other pairing, $\mu_{12} = \mu_{21} = 0.8 = 4/5$, indicates that for this pairing the gain increases only by a factor of 1.25 when the other loop is closed. Obviously, the pairing c_1-m_2, c_2-m_1 results in less sensitivity to interaction than the other one.

EXAMPLE 12-2.2

Consider the following matrix of relative gains:

	m_1	m_2
c_1	2.0	-1.0
c_2	-1.0	2.0

In this case the relative gain $\mu_{11} = \mu_{22} = 2.0 = 1/0.50$ indicates that the gain of each loop is cut in half when the other loop is closed, while the relative gain $\mu_{12} = \mu_{21} = -1.0 = 1/-1$ indicates that the gain of each loop changes sign when the other loop is closed. Certainly, this last case is undesirable because it means that the action of the controller depends on whether the other loop is closed or open. The correct pairing is obviously c_1-m_1, c_2-m_2.

It is convenient to arrange the relative gains in a matrix, as demonstrated in the previous examples, because it gives a graphical association of the values with the controlled and manipulated variables. This can also be done with the open-loop and the closed-loop gains. The first of the two examples involved a process with positive interaction, and the second had negative interaction.

We will next look at the calculation of the relative gains from the open-loop gains. First, we will derive a simple formula for the relative gains of 2×2 systems, and then for a general $n \times n$ system, that is, for a system with n interacting control objectives.

12-2.1 Calculating the Relative Gains for a 2 × 2 System

The major advantage of the relative gain analysis presented here is that it requires only steady-state process parameters, specifically, the steady-state gains. In many cases these gains can be calculated from simple steady-state material and energy balances on the process. More complex processes such as distillation columns and reactors may require a simulation of the process, but steady-state flowsheet simulations are readily available and commonly used to design the process and to analyze its performance. In this section we show how to compute the relative gains from just the open-loop gains for a 2×2 system.

Relative gain analysis is based on a linear approximation of the process around some operating conditions. As we have discussed numerous times in this text, most processes are nonlinear. This means that just as the steady-state gains vary with operating conditions, so will the relative gains.

For the 2×2 system of Fig. 12-2.1, the steady-state changes in the controlled variables caused by simultaneous changes in both manipulated variables are

$$\Delta c_1 = K_{11}\Delta m_1 + K_{12}\Delta m_2$$
$$\Delta c_2 = K_{21}\Delta m_1 + K_{22}\Delta m_2 \tag{12-2.5}$$

where we have used the principle of superposition and assumed a linear approximation for small changes in the manipulated variables. The gains are the open-loop steady-state gains.

To determine the closed-loop gain for the pairing c_1-m_1, we must connect a feedback controller connecting c_2 with m_2, as in Fig. 12-2.2. If this controller has integral mode, when we apply a change in m_1, it will adjust m_2 to bring c_2 back to its set point, making the steady-state change in c_2 equal to zero.

$$\Delta c_2 = K_{21}\Delta m_1 + K_{22}\Delta m_2 = 0$$

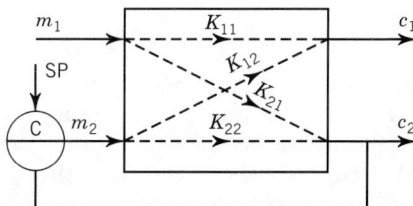

Figure 12-2.2 Schematic of general 2×2 process with one loop closed.

The change in m_2 required to compensate for the change in m_1 can be calculated from this formula.

$$\Delta m_2 = -\frac{K_{21}}{K_{22}}\Delta m_1$$

Substitute this change in m_2 into Eq. 12-2.5 to obtain the total change in c_1 caused by the change in m_1 when c_2 is kept constant.

$$\Delta c_1 = K_{11}\Delta m_1 - K_{12}\frac{K_{21}}{K_{22}}\Delta m_1$$

This is the value we need to compute the closed-loop gain; from its definition, Eq. 12-2.2,

$$K'_{11} = \frac{\Delta c_1}{\Delta m_1}\bigg|_{c_2} = K_{11} - K_{12}\frac{K_{21}}{K_{22}}$$

This formula shows that the closed-loop gain can be calculated from the four open-loop gains. The relative gains for each of the other three pairs of variables are obtained by appropriately rearranging the connections of the feedback controller of Fig. 12-2.2. The resulting formulas are, after minor rearrangement,

$$K'_{11} = \frac{K_{11}K_{22} - K_{12}K_{21}}{K_{22}} \quad K'_{12} = \frac{K_{11}K_{22} - K_{12}K_{21}}{-K_{21}}$$
$$K'_{21} = \frac{K_{11}K_{22} - K_{12}K_{21}}{-K_{12}} \quad K'_{22} = \frac{K_{11}K_{22} - K_{12}K_{21}}{K_{11}}$$

$$(12\text{-}2.6)$$

The relative gains can now be obtained from their definition, Eq. 12-2.3.

$$\mu_{11} = \frac{K_{11}K_{22}}{K_{11}K_{22} - K_{12}K_{21}} \quad \mu_{12} = \frac{-K_{12}K_{21}}{K_{11}K_{22} - K_{12}K_{21}}$$
$$\mu_{21} = \frac{-K_{12}K_{21}}{K_{11}K_{22} - K_{12}K_{21}} \quad \mu_{22} = \frac{K_{11}K_{22}}{K_{11}K_{22} - K_{12}K_{21}}$$

$$(12\text{-}2.7)$$

As mentioned earlier, the relative gains for the pair c_1-m_1 are equal to those for the pair c_2-m_2 ($\mu_{11} = \mu_{22}$) because they represent the same pairing option. Similarly, $\mu_{12} = \mu_{21}$. Also, it can be readily shown that $\mu_{11} + \mu_{12} = 1$, and that $\mu_{12} + \mu_{22} = 1$; that is, the sums of the relative gains in any row or column is equal to 1. This last property extends to $n \times n$ systems.

You can easily memorize the formulas for the relative gains by noticing that the denominator is the determinant of the matrix of open-loop gains, and the numerator is the product term of the denominator that contains the open-loop gains for the pair of interest.

Having shown that the relative gains can be calculated from the open-loop gains, let us next demonstrate the procedure for calculating the relative gains.

EXAMPLE 12-2.3

Relative Gains of Blending Process

The equations relating the variables of the blending process of Fig. 12-1.1a were presented earlier in this chapter (Eq. 12-1.1). These equations are so simple that the gains can be obtained analytically using differential calculus. Determine the general formulas for the open-loop and relative gains for the blending process, and derive a general strategy for pairing variables in any blending process.

SOLUTION

The first step is to solve for the controlled variables in terms of the manipulated variables. From Eq. 12-1.1,

$$w = w_1 + w_2$$
$$x = \frac{w_1 x_1 + w_2 x_2}{w_1 + w_2}$$

(12-2.8)

where w and x are the controlled variables and w_1 and w_2 are the manipulated variables. Notice that we use the manipulated flows instead of the control valve positions as manipulated variables. This is because the gains of the valves have no effect on the relative gains since they affect the open-loop and closed-loop gains in exactly the same way.

Differential calculus can be used to determine the open-loop gains: take limits of the formulas for the gains as the change in manipulated variable approaches zero.

$$K_{ij} = \lim_{\Delta m_j \to 0} \left. \frac{\Delta c_i}{\Delta m_j} \right|_{\substack{m_k \\ k \neq j}} = \frac{\partial c_i}{\partial m_j}$$

(12-2.9)

Applying this definition to Eq. 12-2.8, we obtain the following open-loop gains:

$$K_{w1} = 1 \qquad\qquad K_{w2} = 1$$
$$K_{x1} = \frac{-w_2(x_2 - x_1)}{(w_1 + w_2)^2} \quad K_{x2} = \frac{w_1(x_2 - x_1)}{(w_1 + w_2)^2}$$

(12-2.10)

The relative gains are then obtained using Eq. 12-2.7.

$$\mu_{w1} = \frac{w_1}{w_1 + w_2} \quad \mu_{w2} = \frac{w_2}{w_1 + w_2}$$
$$\mu_{x1} = \frac{w_2}{w_1 + w_2} \quad \mu_{x2} = \frac{w_1}{w_1 + w_2}$$

(12-2.11)

Both relative gains are positive and less than unity, indicating positive interaction. The correct pairing is the one resulting in the relative gain being closest to unity. Because the blending process is nonlinear, the relative gains and therefore the correct pairing is a function of the operating conditions. The correct pairing depends on which of the two inlet flows is larger.

- If $w_1 > w_2$, then the relative gains closer to 1 are μ_{w1} and μ_{x2} (equal to each other). The correct pairing is $w-w_1$, $x-w_2$.

- If $w_2 > w_1$, then the relative gains closer to 1 are μ_{w2} and μ_{x1} (equal to each other). The correct pairing is $w-w_2$, $x-w_1$.

Notice that this results in a general pairing strategy for blending processes:

Control the product flow with the inlet stream having the larger flow, and the composition with the inlet stream having the smaller flow.

The composition represents any intensive property of the product stream that is to be controlled. It could be, for example, temperature, as in the case of a household shower, where cold and hot water streams are mixed to obtain the desired shower temperature.

For processes in which the level in the tank is controlled, the stream with the largest flow should be manipulated to control the level. In this example, the stream with the largest flow is the product stream, but it may be desirable to have the operator directly set the product flow. In such case, the product stream is flow-controlled, the inlet stream with the larger flow is manipulated to control the level, and the inlet stream with the smaller flow controls the intensive product property (e.g., composition, temperature).

EXAMPLE 12-2.4

Control of Distillation Product Purities

The distillation column of Fig. 12-2.3 is designed to separate a mixture consisting of 60 mole % benzene and 40 mole % toluene into a distillate product with 95 mole % benzene and 5 mole % toluene, and a bottoms product with 5 mole % benzene and 95 mole % toluene. The feed rate is 1000 lbmoles/h and enters on the seventh tray from the top. The column has 12 sieve

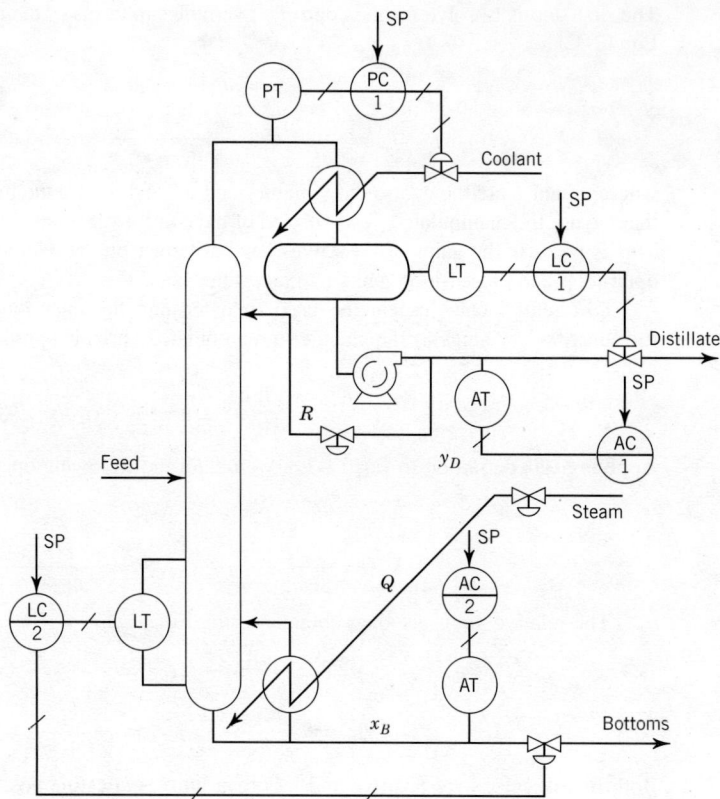

Figure 12-2.3 Distillation column for Example 12-2.4.

trays, a partial reboiler, total condenser, and the Murphree tray efficiency is 70 %. The distillate composition, measured as mole % toluene, and the bottoms composition, measured as mole % benzene, are to be controlled by manipulating the reflux flow and the heat rate to the reboiler. Determine the open-loop gains for the 2×2 system, the relative gains, and the pairing of the manipulated and controlled variables that minimizes the effect of interaction. Assume that the pressure controller maintains the column pressure constant, and the two level controllers maintain the material flows in balance at the top and bottom by manipulating the two product flows.

SOLUTION

A ChemSep (Taylor, Haket, and Kooijman, 2004) simulation of the column was used to determine the steady-state gains. Normally, three runs of the simulation are needed to estimate the steady-state gains: one at the design conditions, one in which the reflux flow is changed keeping the reboiler heat rate constant, and one in which the reboiler heat rate is changed at constant reflux. However, because distillation columns exhibit asymmetric responses when the variables are increased and decreased, it is necessary to make five runs so that the reflux flow and reboiler heat rate can each be increased and decreased from their design values. The results of these runs are shown in the accompanying table.

	Base Case	Test 1$^+$	Test 1$^-$	Test 2$^+$	Test 2$^-$
Reflux, klb/h	131.94	132.59	131.29	131.94	131.94
Heat rate, MBtu/h	30.10	30.10	30.10	30.23	29.97
Distillate mole %	5.00	4.43	5.69	5.87	4.34
Bottoms mole %	5.00	5.84	4.25	4.05	6.06

Notice the asymmetric response of the compositions: although the change of reflux flow in each direction is of the same magnitude, 0.65 klb/h, the composition changes are different for the increase and decrease in flow, and the same happens for the changes in heat rate. These differences are small in this case because the changes in flow and heat rate are small, as they should be. Nevertheless, these small differences are enough to throw the calculation of the gains off. This is why two changes, one up and one down, are required for each variable. The open-loop gains are calculated as follows:

$$K_{DR} = \frac{4.43 - 5.69}{132.59 - 131.29} = -0.97 \frac{\text{mole \%}}{\text{klb/h}} \qquad K_{DQ} = \frac{5.87 - 4.34}{30.23 - 29.97} = 5.88 \frac{\text{mole \%}}{\text{MBtu/h}}$$

$$K_{BR} = \frac{5.84 - 4.25}{132.59 - 131.29} = 1.22 \frac{\text{mole \%}}{\text{klb/h}} \qquad K_{BQ} = \frac{4.05 - 6.06}{30.23 - 29.97} = -7.73 \frac{\text{mole \%}}{\text{MBtu/h}}$$

where the first subscript is D for the distillate composition and B for the bottoms composition and the second subscript is R for the reflux flow and Q for the reboiler heat rate.

The relative gains are, from Eq. 12-2.7,

$$\mu_{DR} = \mu_{BQ} = \frac{(-0.97)(-7.73)}{(-0.97)(-7.73) - (1.22)(5.88)} = 23.1$$

$$\mu_{DQ} = \mu_{BR} = \frac{-(1.22)(5.88)}{(-0.97)(-7.73) - (1.22)(5.88)} = -22.1$$

or, in matrix form:

	R	Q
y_D	23.1	−22.1
x_B	−22.1	23.1

The relative gains are either greater than unity or negative, indicating negative interaction. To avoid negative relative gains, the pairing should be the obvious one: the reflux flow controls the distillate composition, and the reboiler heat rate controls the bottoms composition. However, even for the best pairing, the high value of the relative gain, 23.1, shows that the two loops fight each other to the point of essentially canceling each other's actions. For example, assume the distillate composition controller changes the reflux flow by 0.2 klb/h. If the bottoms composition loop is open, the change in distillate composition, using the open-loop gain, is

$$\Delta y_D = K_{DR} \Delta R = (-0.97)(0.20) = -0.194 \text{ mole \% toluene}$$

and the bottoms composition will change by

$$\Delta x_B = K_{BR} \Delta R = (1.22)(0.20) = 0.244 \text{ \% benzene}$$

If the bottoms composition controller is switched to automatic, it will bring the bottoms composition back to set point. The required change in reboiler heat rate is

$$\Delta Q = \frac{-\Delta x_B}{K_{BQ}} = \frac{-0.244}{-7.73} = 0.0316 \frac{\text{MBtu}}{\text{h}}$$

The total change in distillate composition will then be

$$\Delta y_D = K_{DR} \Delta R + K_{DQ} \Delta Q = (-0.97)(0.20) + (5.88)(0.0316) = -0.008 \text{ \% toluene}$$

Notice how the change caused by interaction between the two controllers, 0.186, essentially cancels the change the manipulated variable caused on its controlled variable, −0.194. The gain of each controller is reduced by a factor of 23.1, the relative gain.

12-2.2 Calculating the Relative Gains for an $n \times n$ System

When there are more than two interacting control loops, the system is called an $n \times n$ system, where n is the number of interacting control objectives. Such a system consists of n controlled variables and n manipulated variables. The definition of the relative gain is the same as for 2×2 systems, that is, the ratio of the open-loop gain to the closed-loop gain for each controlled–manipulated variable pair:

$$\mu_{ij} = \frac{\dfrac{\Delta c_i}{\Delta m_j}\bigg|_{\substack{m_k \\ k \neq j}}}{\dfrac{\Delta c_i}{\Delta m_j}\bigg|_{\substack{c_k \\ k \neq i}}} = \frac{K_{ij}}{K'_{ij}} \tag{12-2.12}$$

where K_{ij} is the open-loop gain, that is, the gain when all other manipulated variables are kept constant, and K'_{ij} is the closed-loop gain, when all other controlled variables are kept constant.

The formulas for the relative gain matrix for a 2×2 system, Eq. 12-2.7, are simple. For an $n \times n$ system, it is easier to develop a procedure based on matrix operations. The procedure, as proposed by Bristol (1966), is as follows:

Obtain the transpose of the inverse of the steady-state gain matrix and multiply each term of the resulting matrix by the corresponding term in the original matrix. The terms thus obtained are the relative gains.

Algebraically, let **B** be the inverse of the steady-state gain matrix **K**; that is, let $\mathbf{B} = \mathbf{K}^{-1}$. Then

$$\mu_{ij} = B_{ji}K_{ij} \tag{12-2.13}$$

where K_{ij} and B_{ji} are elements of matrices **K** and **B**, respectively. With today's personal calculators and programs such as MATLAB (2001) and MathCad (2001), the required matrix operations can be readily performed. For a 2×2 system, this procedure yields the same relative gains as Eq. 12-2.7.

The derivation of Eq. 12-2.13 is carried out as follows: in matrix notation, the change in the controlled variables caused by simultaneous changes in the manipulated variables is

$$\Delta \mathbf{c} = \mathbf{K} \, \Delta \mathbf{m} \tag{12-2.14}$$

To solve for the changes in the manipulated variables required to produce specific changes in the controlled variables, we multiply by the inverse of the steady-state gain matrix **K**.

$$\Delta \mathbf{m} = \mathbf{K}^{-1} \, \Delta \mathbf{c} = \mathbf{B} \, \Delta \mathbf{c} \tag{12-2.15}$$

We now write the jth element of this matrix equation.

$$\Delta m_j = B_{j1}\Delta c_1 + \cdots + B_{ji}\Delta c_i + \cdots + B_{jn}\Delta c_n \tag{12-2.16}$$

from where we notice that

$$B_{ji} = \frac{\Delta m_j}{\Delta c_i}\bigg|_{\substack{c_k \\ k \neq i}} = \frac{1}{K'_{ij}} \tag{12-2.17}$$

In other words, the elements of the inverse matrix **B** are the inverse of the closed-loop gains, transposed. Combining Eq. 12-2.17 with Eq. 12-2.12, we obtain Eq. 12-2.13. Let us demonstrate the use of these formulas with an example.

EXAMPLE 12-2.5

Control of Gasoline Blending System

Figure 12-2.4 shows an in-line blender to produce gasoline of a given formulation by mixing three refinery products: alkylate (stream 1), light straight run (stream 2), and reformate (stream 3). It is desired to control the research octane x, Reid vapor pressure y, and flow f of the gasoline product by manipulating the flow set points on the inlet streams. Calculate the steady-state gains, the relative gains, and decide which inlet stream should control each product variable so that the effect of interaction is minimized. The design values of the process variables are shown in the accompanying table.

		Octane, x	RVP, y	Flow, kbl/day
Alkylate	1	97.0	5.00	7.50
Straight run	2	80.0	11.0	28.12
Reformate	3	92.0	3.00	24.38
Gasoline		87.0	7.00	60.00

SOLUTION

Assuming the properties of the product gasoline are the average of the feed streams, weighted by the volume rate of each stream, and that the densities of all streams are approximately equal, we obtain the following equations relating the controlled variables to the manipulated flows.

$$x = \frac{f_1 x_1 + f_2 x_2 + f_3 x_3}{f_1 + f_2 + f_3}$$

$$y = \frac{f_1 y_1 + f_2 y_2 + f_3 y_3}{f_1 + f_2 + f_3}$$

$$f = f_1 + f_2 + f_3$$

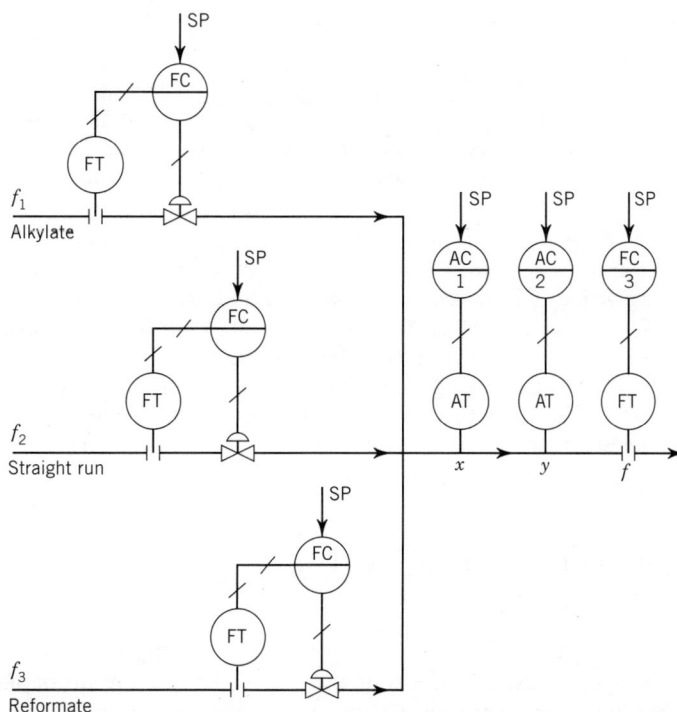

Figure 12-2.4 In-line gasoline blending process for Example 12-2.5.

Notice that the design values given in the statement of the problem satisfy these equations, as they must. This means that three of the design values are not independent of the rest.

As in Example 12-2.3, we obtain the steady-state gains by taking partial derivatives of the controlled variables with respect to the manipulated variables. The results are

$$K_{1j} = \frac{\partial x}{\partial f_j} = \frac{f_1(x_j - x_1) + f_2(x_j - x_2) + f_3(x_j - x_3)}{(f_1 + f_2 + f_3)^2}$$

$$K_{2j} = \frac{\partial y}{\partial f_j} = \frac{f_1(y_j - y_1) + f_2(y_j - y_2) + f_3(y_j - y_3)}{(f_1 + f_2 + f_3)^2}$$

$$K_{3j} = \frac{\partial f}{\partial f_j} = 1$$

Substitute the design values to obtain the open-loop gain matrix.

$$\mathbf{K} = \begin{bmatrix} 0.167 & -0.117 & 0.083 \\ -0.033 & 0.067 & -0.067 \\ 1 & 1 & 1 \end{bmatrix}$$

The inverse of this matrix is

$$\mathbf{B} = \mathbf{K}^{-1} = \begin{bmatrix} 7.50 & 11.250 & 0.125 \\ -1.875 & 4.688 & 0.469 \\ -5.625 & -15.938 & 0.406 \end{bmatrix}$$

where the inverse was calculated using MathCad (2001). From Eq. 12-2.13, the relative gains are

	f_1	f_2	f_3
x	1.250	0.219	−0.469
y	−0.375	0.312	1.063
f	0.125	0.469	0.406

The boxed values are the relative gains closest to 1 which minimize the effect of interaction. The proper pairing is then to control the octane (x) with the alkylate (stream 1), the Reid vapor pressure (y) with the reformate (stream 3), and the gasoline flow (f) with the light straight run (stream 2). Two of the relative gains are slightly greater than unity, which means the interaction is negative for each of these loops and thus their gains decrease when the other loops are closed. The other relative gain is 0.469, which means the loop gain is more than doubled when the other loops are closed. This loop is helped by the combination of the other two loops (positive interaction). Notice that the sum of the relative gains in each row and each column of the matrix is unity, as it should be. Since an in-line mixer has very fast response, the steady-state analysis of the interactions should be sufficient to decide on the best pairing.

In this example, as in the two-stream blender of Example 12-2.3, the correct pairing has the inlet stream with the largest flow controlling the product flow. This makes sense, since the largest stream has the greatest relative influence on the product flow.

12-3 DECOUPLING OF INTERACTING LOOPS

The third and final question to answer is, can the interaction between loops be reduced or eliminated through the design of an appropriate control system? The answer is of course yes, and the simplest way to do it is by *decoupling*. Decoupling can be a profitable, realistic possibility when applied carefully. The relative gain matrix provides an indication of when decoupling could be beneficial. If for the best pairing option, one or more of the relative gains is far from unity, decoupling may help. For existing systems, operating experience usually helps in making the decision.

This section presents the design of decouplers. The design of decouplers is very similar to the design of feedforward controllers presented in Chapter 11. Decouplers can be designed from block diagrams or from basic engineering principles. The basic difference is that, unlike feedforward controllers, decouplers form part of feedback control of loops. Because of this, they must be selected and designed with great care. One basic characteristics of decoupling that we must keep in mind is this: *in interacting systems, decoupling does to each loop what the other loops were going to do anyway.* The difference is that the performance of each loop becomes independent of the tuning and open/closed condition of the other loops, provided the action of the decoupler is not blocked.

12-3.1 Decoupler Design from Block Diagrams

Figure 12-3.1 presents the block diagram for a 2×2 interacting system. This block diagram shows graphically that the interaction between the two loops is caused by the process "cross" blocks with transfer functions $G_{12}(s)$ and $G_{21}(s)$. To circumvent this interaction, two decoupler blocks with transfer functions $D_{12}(s)$ and $D_{21}(s)$ are installed, as in Fig. 12-3.2. The purpose of the decouplers is to cancel the effects of the process cross blocks so that each controlled variable is not affected by changes in the manipulated variable of the other loop. In other words, decoupler $D_{21}(s)$ cancels the effect of manipulated variable $M_1(s)$ on controlled variable $C_2(s)$, and $D_{12}(s)$ cancels the effect of $M_2(s)$ on controlled variable $C_1(s)$. From block diagram algebra, these effects are

$$\frac{C_1(s)}{M_2(s)} = D_{12}(s)G_{v1}(s)G_{11}(s) + G_{v2}(s)G_{12}(s) = 0$$

$$\frac{C_2(s)}{M_1(s)} = D_{21}(s)G_{v2}(s)G_{22}(s) + G_{v1}(s)G_{21}(s) = 0$$

(12-3.1)

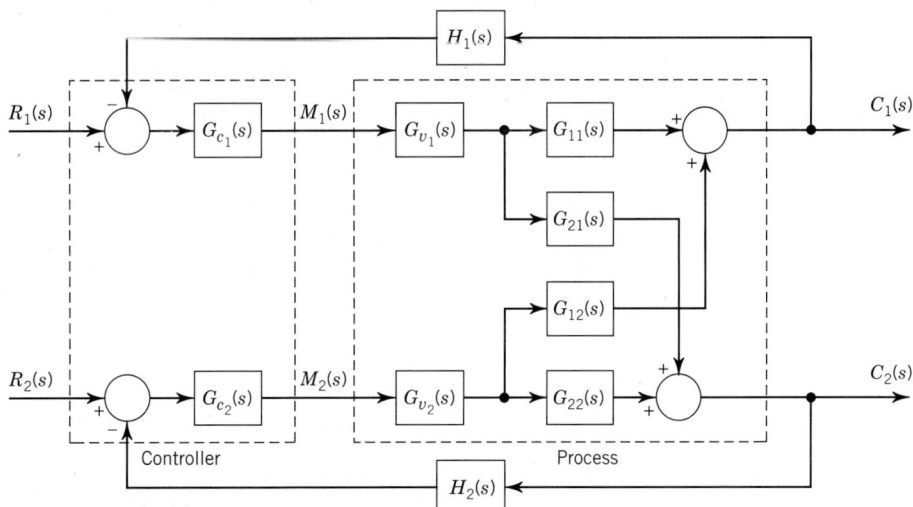

Figure 12-3.1 Block diagram of general 2×2 control system.

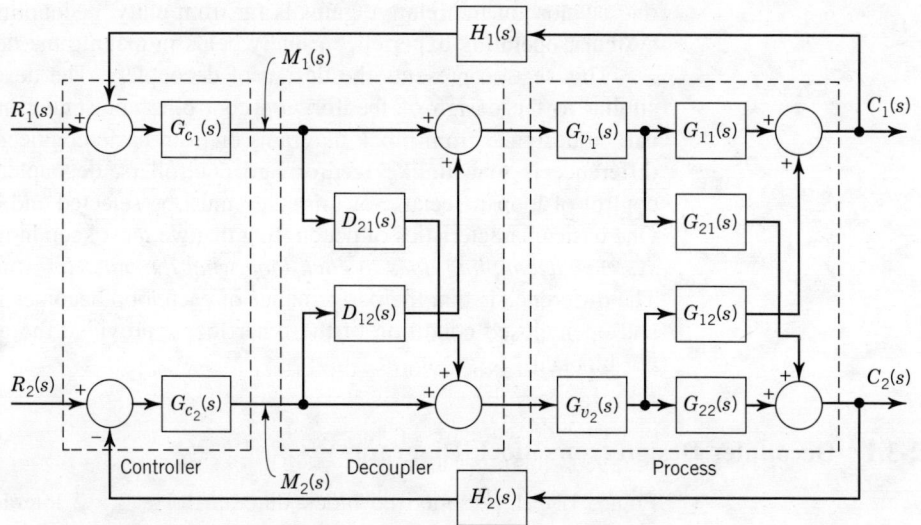

Figure 12-3.2 Block diagram of 2×2 control system with decouplers.

Notice that this is similar to designing feedforward controllers in which $M_2(s)$ is a disturbance to $C_1(s)$, $M_1(s)$ is a disturbance to $C_2(s)$. To obtain the design formulas for the decouplers, solve for the decoupler transfer functions from Eq. 12-3.1.

$$D_{12}(s) = -\frac{G_{v2}(s)G_{12}(s)}{G_{v1}(s)G_{11}(s)}$$

$$D_{21}(s) = -\frac{G_{v1}(s)G_{21}(s)}{G_{v2}(s)G_{22}(s)}$$

(12-3.2)

These formulas can be used to design linear decouplers for any 2×2 system as long as the transfer functions of the process and valves can be derived from basic principles, as we learned to do in Chapters 3, 4, 5, and 6, or approximated from the step responses of the controlled variables to the manipulated variables, as we learned in Chapter 7. In the latter approach the transfer functions of the valve, process, and transmitter are lumped into a single transfer function for each combination. It can easily be shown that the decoupler formulas then become

$$D_{12}(s) = -\frac{G_{P12}(s)}{G_{P11}(s)} \quad D_{21}(s) = -\frac{G_{P21}(s)}{G_{P22}(s)}$$

(12-3.3)

where $G_{Pij}(s) = G_{vj}(s)G_{ij}(s)H_i(s)$.

As pointed out earlier, the decouplers form part of the loops, as can be seen from the block diagram of Fig. 12-3.2. The relationship between each controlled variable and its manipulated variable in the decoupled diagram is obtained by block diagram algebra:

$$\frac{C_1(s)}{M_1(s)} = G_{v1}(s)G_{11}(s) + D_{21}(s)G_{v2}(s)G_{12}(s)$$

$$\frac{C_{2(s)}}{M_2(s)} = G_{v2}(s)G_{22}(s) + D_{12}(s)G_{v1}(s)G_{21}(s)$$

(12-3.4)

The two terms in each formula tell us what the block diagram of Fig. 12-3.2 tells us graphically—that in the decoupled system each manipulated variable affects its controlled variable through two parallel paths. As with the interacting system, these two paths may help each other if their effects are additive (positive interaction), or fight each other if the effects are of the opposite sign (negative interaction).

If the decouplers can be implemented exactly as designed by Eq. 12-3.2, substitution into Eq. 12-3.4 gives, after some simplification

$$\frac{C_1(s)}{M_1(s)} = G_{v1}(s)\left[G_{11}(s) - \frac{G_{21}(s)G_{12}(s)}{G_{22}(s)}\right]$$

$$\frac{C_2(s)}{M_2(s)} = G_{v2}(s)\left[G_{22}(s) - \frac{G_{12}(s)G_{21}(s)}{G_{11}(s)}\right]$$

$$(12\text{-}3.5)$$

It is easy to show that the steady-state gain of the terms in brackets, obtained by letting $s = 0$ in the transfer functions, is exactly the same as the closed-loop gains of Eq. 12-2.6. In other words, the effect of perfect decouplers on the gain of the loop is the same as the effect of closing both loops without decouplers. This makes sense, since each decoupler is keeping each controlled variable constant when the manipulated variable in the other loop changes. This is exactly what the controller on the other loop does, at steady state, if it has integral mode.

EXAMPLE 12-3.1

Decoupler Design for Blending Tank

SOLUTION

Design a linear decoupler for the blending tank of Fig. 12-1.1a. Neglect the time constants of the control valves as they are small relative to the time constant of the tank.

To obtain the transfer functions for the tank, we write total and salt balances. Assuming perfect mixing, constant volume and density, and constant inlet compositions,[1] the differential equations are

$$w_1(t) + w_2(t) - w(t) = \frac{d}{dt}(\rho V) = 0$$

$$x_1(t)w_2(t) + x_2(t)w_2(t) - x(t)w(t) = \frac{d}{dt}[\rho V x(t)]$$

$$(12\text{-}3.6)$$

where ρ is the density of the solution, lb/ft^3, and V is the volume of solution in the tank, ft^3. The flows $w_1(t)$, $w_2(t)$, and $w(t)$ are in lb/min, and the compositions are in mass fraction salt. We next linearize and take Laplace transforms, as in Chapter 3, to obtain the transfer functions.

$$W(s) = W_1(s) + W_2(s)$$

$$X(s) = \frac{K_{x1}}{\tau s + 1}W_1(s) + \frac{K_{x2}}{\tau s + 1}W_2(s)$$

$$(12\text{-}3.7)$$

where

$$\tau = \frac{\rho V}{\overline{w}} \text{ min}$$

$$K_{x1} = \frac{x_1 - \overline{x}}{\overline{w}} \quad \frac{\text{mf}}{\text{lb/min}}$$

$$K_{x2} = \frac{x_2 - \overline{x}}{\overline{w}} \quad \frac{\text{mf}}{\text{lb/min}}$$

These gains are the same as those of Eq. 12-2.10 because, as the base conditions are at steady state, it can be shown that

$$\frac{x_2 - x_1}{\overline{w}} = \frac{x_2 - \overline{x}}{\overline{w}_1} = -\frac{x_1 - \overline{x}}{\overline{w}_2}$$

and $\overline{w} = \overline{w}_1 + \overline{w}_2$.

---

[1] The inlet compositions are the disturbance inputs, but we will consider them constant to keep the analysis simple.

The transfer functions we have developed match the block diagram of Fig. 12-3.2 with $C_1(s) = W(s)$, and $C_2(s) = X(s)$. Neglecting the time constants of the control valves, from Eq. 12-3.7, the transfer functions shown in the block diagram are

$$G_{11}(s) = 1 \qquad\qquad G_{12}(s) = 1$$

$$G_{21}(s) = \frac{K_{x1}}{\tau s + 1} \qquad G_{22}(s) = \frac{K_{x2}}{\tau s + 1}$$

$$G_{v1}(s) = K_{v1} \qquad\qquad G_{v2}(s) = K_{v2}$$

where K_{v1} and K_{v2} are the valve gains in (lb/h)/%CO.

The decouplers are now obtained by substituting the above transfer functions into the decoupler design formulas, Eq. 12-3.2,

$$D_{12}(s) = -\frac{G_{v2}(s)G_{12}(s)}{G_{v1}(s)G_{11}(s)} = -\frac{K_{v2}}{K_{v1}} \frac{\%CO_1}{\%CO_2}$$

$$D_{21}(s) = -\frac{G_{v1}(s)K_{x1}}{G_{v2}(s)K_{x2}} = \frac{K_{v1}\overline{w}_2}{K_{v2}\overline{w}_1} \frac{\%CO_1}{\%CO_2}$$

$$(12\text{-}3.8)$$

where we have substituted the gains from Eq. 12-2.10. Figure 12-3.3 shows the instrumentation diagram for the tank with the linear decoupler. In the diagram, FY-1 implements $D_{12}(s)$, and AY-2 implements $D_{21}(s)$.

Both decouplers are simple gains. This makes sense, because both inlet flows have exactly the same dynamic effects on the outlet flow and composition. Decoupler $D_{12}(s)$ (FY-1) has a negative gain, because its purpose is to keep the outlet flow constant when the second inlet stream changes. This requires that the first flow change in the opposite direction by exactly the same amount. The decoupler gain corrects for the different capacities of the two valves.

Decoupler $D_{21}(s)$ (AY-2) is positive because its purpose is to keep the outlet composition constant when the first inlet flow changes. To keep the outlet composition constant, the concentrated stream flow must change in the same direction as the dilute stream flow, and the ratio of the two streams must remain constant. Notice that this is exactly what the decoupler does, since its gain is the ratio of the two streams corrected for the capacities of the two valves.

To illustrate, suppose that valve 2 has half the gain (capacity) of valve 1, $K_{v1} = 2\,K_{v2}$, and that, at the base conditions, flow 1 is 50 % greater than flow 2, $\overline{w}_1 = 1.5\,\overline{w}_2$. Then, if the composition controller changes its output by 1 %, decoupler $D_{12}(s)$, with a gain of $-1/2$, changes the signal to valve 1 by -0.5 %. Since valve 1 has twice the gain as valve 2, the stream 1 flow will decrease by exactly the same amount stream 2 increased, keeping the total flow constant. If now the flow controller changes its output by 1 %, decoupler $D_{21}(s)$, with

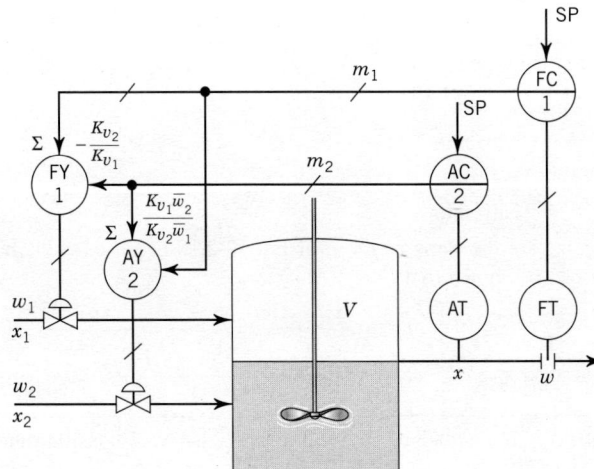

Figure 12-3.3 Instrumentation diagram for linear decoupler of blending tank.

a gain of $(2)(1/1.5) = 1.33$, changes the signal to valve 2 by 1.33 %. Since valve 2 has half the gain of valve 1, the change in the flow of stream 2 is $1.33/2 = 0.67$ of the change in the flow of stream 1. This maintains the flow ratio of stream 2 to stream 1 at 0.67, that is, the flow of stream 1 continues to be 50 % greater than the flow of stream 2, keeping the outlet composition constant.

Variable Pairing and Decoupling

We have shown that decoupling has the same effect on each loop as the interacting loops had before decoupling. Because of this, and because the decoupler action could be blocked if the valve it actuates is driven to saturation, it is important to properly pair the controlled and manipulated variables, even when a decoupler is used.

In the preceding example the variables are properly paired, with stream 1, the larger flow, controlling the product flow, and stream 2, the smaller flow, controlling the composition. Notice that this makes the ratio of the two streams less than unity, so that the action of the flow controller in changing stream 1 has a greater effect on the product flow than the action of the decoupler. Imagine what would happen if the ratio of stream 2 to stream 1 were 10 and the same pairing is used: the decoupler would have 10 times greater effect on the product flow than the direct action of the controller. It can be shown that the same would happen to the composition controller.

Partial Decoupling

It is not always necessary to decouple all the interactions. When the control of one of the variables is more important than the control of the others, better control of that variable would result if only the decoupler terms that keep that variable constant are implemented. Suppose that in the blending tank example it is more important to keep the product composition constant than it is to keep the product flow constant. We would then implement only decoupler $D_{21}(s)$ (AY-2 in Fig. 12-3.3). This keeps the composition from changing when the flow controller takes action. When the composition controller takes action, the flow is allowed to vary and the flow controller must take action to bring the flow back to set point. However, this action of the flow controller will not affect the composition as the decoupler compensates for it.

The following example shows how to design a decoupler from simple process models.

EXAMPLE 12-3.2

Decoupler Design for an Evaporator

In the evaporator of Fig. 12-3.4 the product composition, x_P, and feed flow, w_F, are to be controlled by manipulating the signals to the control valves on the steam and product lines, m_S and m_P. As shown in the diagram, although there is a control valve on the feed line, this valve must control the level in the evaporator (LC). This is because the feed is the largest of the three flows in the system and thus it has the largest influence on the level. The level in a calandria-type evaporator must be controlled very tightly, because it has great influence on the heat transfer rate.

Step tests are performed on the signals to the steam and product valves, m_S and m_P, one at a time, and the responses of the feed flow and product compositions are carefully recorded. The resulting first-order-plus-dead-time models are

$$G_{FS}(s) = \frac{0.84e^{-0.30s}}{1.05s + 1} \quad G_{FP}(s) = \frac{0.20e^{-0.03s}}{0.65s + 1}$$

$$G_{xS}(s) = \frac{1.68e^{-0.80s}}{2.70s + 1} \quad G_{xP}(s) = \frac{-1.60e^{-0.50}}{2.97s + 1}$$

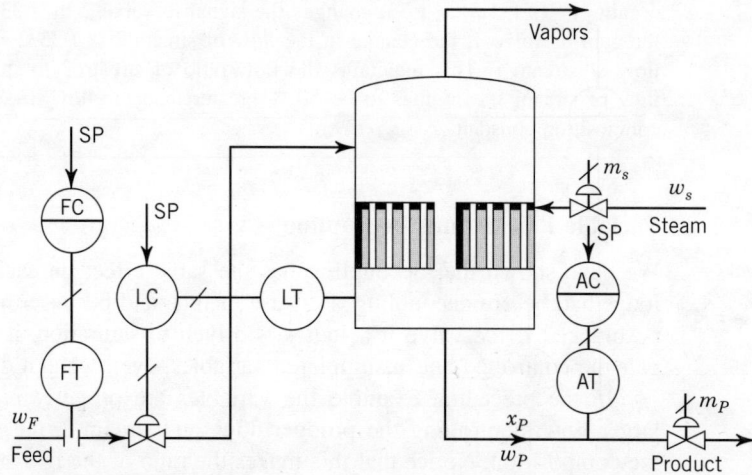

Figure 12-3.4 Evaporator control system for Example 12-3.2.

where the first subscript is F for feed rate or x for product mass percent, and the second subscript is S for steam or P for product. Each transfer function is in units of %TO/%CO; that is, they each combine the transfer functions of the control valve, the process, and the sensor-transmitter. Calculate the relative gains, pair the controlled and manipulated variables so as to minimize the effect of interaction, and design the decouplers.

SOLUTION

The open-loop steady-state gains are obtained by setting $s = 0$ in the model transfer functions, and the relative gains are calculated from Eq. 12-2.7. In matrix form they are

K	m_S	m_P
w_F	0.84	0.20
x_P	1.68	−1.60

μ	m_S	m_P
w_F	0.80	0.20
x_P	0.20	0.80

The pairing that minimizes interaction has the steam valve control the feed flow and the valve on the product stream control the product composition. The relative gain for this pairing is 0.8.

The decouplers are designed using Eq. 12-3.3, since we have the combined transfer functions for all the field devices. The transfer functions of the decouplers are

$$D_{12}(s) = -\frac{G_{FP}(s)}{G_{FS}(s)} = -\frac{0.20}{0.84}\left(\frac{1.05s + 1}{0.65s + 1}\right)e^{0.27s}$$

$$D_{21}(s) = -\frac{G_{xS}(s)}{G_{xP}(s)} = \frac{1.68}{1.60}\left(\frac{2.97s + 1}{2.70s + 1}\right)e^{-0.30s}$$

The first decoupler has a positive dead time which cannot be implemented. To provide this necessary lead in the action of the decoupler, we increase the lead to $1.05 + 0.27 = 1.32$ min. This reduces the decoupler to a simple lead-lag unit with a gain of $-0.20/0.84 = -0.24$, a lead of 1.32 min and a lag of 0.65 min. Although the dead-time compensation of the second decoupler is negative and could be implemented, it is not recommended to use the dead time in a decoupler. This is because, as we saw earlier, the decoupler forms part of the feedback loop and dead time decreases the controllability of feedback loops. To remove the dead-time term and keep the same total lag, we decrease the lead to $2.97 - 0.30 = 2.67$ min. This makes the lead almost the same as the lag, which means that the second decoupler can be a simple gain.

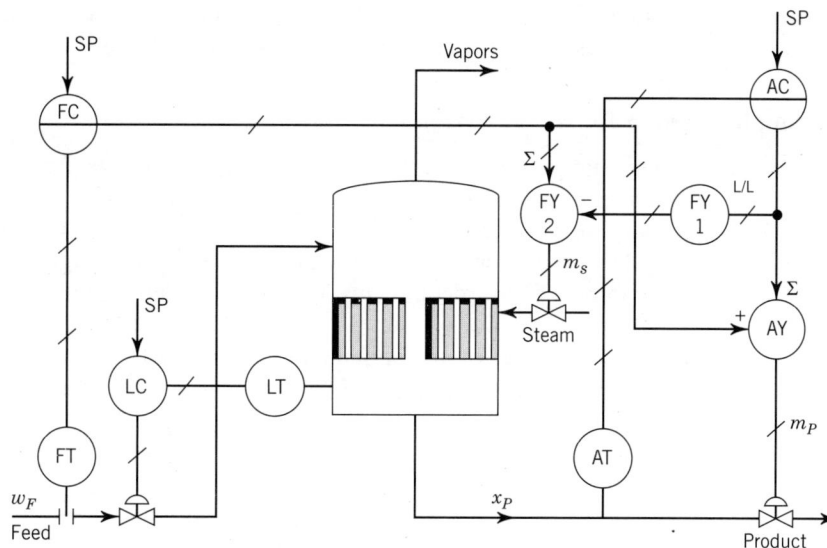

Figure 12-3.5 Instrumentation diagram for linear decoupler of evaporator.

The resulting decouplers are

$$D_{12}(s) = -0.24 \left(\frac{1.32s + 1}{0.65s + 1} \right) \frac{\%CO_S}{\%CO_P}$$

$$D_{21}(s) = \frac{1.68}{1.60} = 1.05 \frac{\%CO_P}{\%CO_S}$$

Figure 12-3.5 shows the instrumentation diagram for the evaporator decouplers. The purpose of the first decoupler, $D_{12}(s)$ (FY-2 in the diagram), is to keep the feed flow constant when the product flow is changed. Its gain is negative because an increase in product flow requires a decrease in steam flow to reduce the rate of vapor generation. If these two effects cancel each other, as they will if the decoupler gain is correct, the level in the evaporator, and consequently the feed flow, do not change. This decoupler requires a net lead because the vapor rate has a lag to the steam flow change whereas the product flow has an immediate effect on the level.

The purpose of the second decoupler, $D_{21}(s)$ (AY), is to maintain the product composition constant when the steam flow changes. Its gain is positive because when the steam flow increases, so does the vapor rate, causing the product concentration to increase. To keep this increase from happening, the decoupler will increase the product flow just enough to cause the level controller to increase the flow of dilute feed into the evaporator. This increased feed flow will decrease the product composition, balancing the effect of the increase in vapor rate. No dynamic compensation is required in this decoupler because the lag between the steam and vapor flows is approximately matched by the mixing lag of the feed flow on the composition.

In practice it is more important to maintain the product composition constant than the feed flow. So, partial decoupling is indicated, using only the term $D_{21}(s)$ (AY), and leaving out the other decoupler term (FYs).

Static Decoupling

As with feedforward control, decoupling can be accomplished statically and dynamically. Static decoupling is accomplished by leaving out the dynamic terms. The advantage is that only the steady-state gains of the transfer functions need to be used in designing a static decoupler. The steady-state gains can be obtained off-line by the methods discussed in the determination of the relative gains.

12-3.2   Decoupler Design for $n \times n$ Systems

The simple procedure for designing the decouplers for 2×2 systems, Eq. 12-3.1, cannot be extended to systems with more than two interacting control objectives. Once more we must turn to matrix notation to outline a design procedure for the general $n \times n$ system. However, even with matrix notation, the design of decouplers is not simple for these higher-order systems because of the large number of decouplers required. For example, for a 4×4 system, 12 decouplers must be designed and implemented.

By the rules of matrix multiplication, it can be shown that the relationship between the controlled and manipulated variables in a decoupled $n \times n$ system is

$$
\begin{bmatrix} C_1(s) \\ C_2(s) \\ \cdots \\ C_n(s) \end{bmatrix} = \begin{bmatrix} G_{P11}(s) & G_{P12}(s) & \cdots & G_{P1n}(s) \\ G_{P21}(s) & G_{P22}(s) & \cdots & G_{P2n}(s) \\ \cdots & \cdots & \cdots & \cdots \\ G_{Pn1}(s) & G_{Pn2}(s) & \cdots & G_{Pnn}(s) \end{bmatrix}
$$
$$
\times \begin{bmatrix} 1 & D_{12}(s) & \cdots & D_{1n}(s) \\ D_{21}(s) & 1 & \cdots & D_{2n}(s) \\ \cdots & \cdots & \cdots & \cdots \\ D_{n1}(s) & D_{n2}(s) & \cdots & 1 \end{bmatrix} \begin{bmatrix} M_1(s) \\ M_2(s) \\ \cdots \\ M_n(s) \end{bmatrix}
$$

or, equivalently,

$$
\mathbf{c}(s) = \mathbf{G}_P(s)\mathbf{D}(s)\mathbf{m}(s) \tag{12-3.9}
$$

where the terms of matrix $\mathbf{G}_P(s)$ combine the transfer functions of all the field elements. The objective of the decoupler matrix $\mathbf{D}(s)$ is to obtain the following decoupled system:

$$
\mathbf{c}(s) = \mathbf{G}'_P(s)\mathbf{m}(s) = \begin{bmatrix} G'_{P11}(s) & 0 & \cdots & 0 \\ 0 & G'_{P22}(s) & \cdots & 0 \\ \cdots & \cdots & \cdots & \cdots \\ 0 & 0 & \cdots & G'_{Pnn}(s) \end{bmatrix} \mathbf{m(s)}
$$

in other words, a diagonal matrix. Comparing this desired result with Eq. 12-3.9, we find that

$$
\mathbf{G}_P(s)\mathbf{D}(s) = \mathbf{G}'_P(s)
$$

Next we solve for the decoupler matrix by premultiplying by the inverse of the process transfer function matrix.

$$
\mathbf{D}(s) = \mathbf{G}_P^{-1}(s)\mathbf{G}'_P(s) \tag{12-3.10}
$$

This is the design formula for decoupling an $n \times n$ system. The matrix equation represents n^2 independent equations. The n^2 unknowns are the $(n^2 - n)$ nondiagonal elements of $\mathbf{D}(s)$ and the n elements of the diagonal matrix $\mathbf{G}'_p(s)$.

Equation 12-3.10 is not easy to solve because it requires the inversion of a matrix of transfer functions. Because of this, modern computer multivariable control systems are designed by different methods. One popular method is Dynamic Matrix Control, DMC (Cutler and Ramaker, 1979).

If the inverse $\mathbf{B}(s) = \mathbf{G}_P^{-1}(s)$ can be obtained, then the elements of the decoupler matrix are calculated as follows:

$$
D_{ij}(s) = \frac{B_{ij}(s)}{B_{jj}(s)} \tag{12-3.11}
$$

This calculation forces the diagonal terms of the decoupler matrix to be unity. The following example shows how Eq. 12-3.10 is used to design a static decoupler.

EXAMPLE 12-3.3

Static Decoupler for Gasoline Blending System

SOLUTION

Design a static decoupler for the gasoline blending control system of Example 12-2.5.

Since the decoupler is static, the steady-state open-loop gains can be substituted for the process transfer function in Eq. 12-3.10. However, before this equation is applied to design the decoupler, we must rearrange the gain matrix to conform to the pairing that minimizes interaction. Recall from Example 12-2.5 that the correct pairing has stream 2 controlling the third variable and stream 3 controlling the second variable. With this pairing, columns 2 and 3 of the open-loop gain matrix must be swapped:

	f_1	f_3	f_2
x	0.167	0.083	−0.117
y	−0.033	−0.067	0.067
f	1	1	1

When this matrix is inverted, the inverse is the same matrix as in Example 12-2.5, with the last two rows swapped.

$$\mathbf{B} = \mathbf{K}^{-1} = \begin{bmatrix} 7.500 & 11.250 & 0.125 \\ -5.625 & -15.938 & 0.406 \\ -1.875 & 4.688 & 0.469 \end{bmatrix}$$

Equation 12-3.11 gives us the decoupler matrix

$$\mathbf{D} = \begin{bmatrix} 1 & -0.71 & 0.27 \\ -0.75 & 1 & 0.87 \\ -0.25 & -0.29 & 1 \end{bmatrix}$$

where all the numbers are in (kbl/day)/(kbl/day). Assuming the control signals are in these units instead of in %CO, the equations required to carry out the decoupling are

$$f_1^{\text{set}} = m_1 - 0.71m_2 + 0.27m_3$$

$$f_3^{\text{set}} = -0.75m_1 + m_2 + 0.87m_3$$

$$f_2^{\text{set}} = -0.25m_1 - 0.29m_2 + m_3$$

where m_1, m_2, and m_3 are in kbl/day, and so are the flow set points. Figure 12-3.6 presents the instrumentation diagram required to implement these calculations. Notice that stream 3, the reformate, is directly manipulated by m_2, the output of the Reid vapor pressure controller, and stream 2, the light straight run, is directly manipulated by m_3, the output of the gasoline flow controller. This is the pairing we found in Example 12-2.5.

If each manipulated stream has the same dynamic effect on its controlled variable as the others, this static decoupler is all that is needed to decouple the three control objectives. The decoupled transfer function matrix is then the matrix product **KD**, which is a diagonal matrix with diagonal terms 0.133, −0.063, and 2.133. The transfer functions of the decoupled system are then

$$X(s) = 0.133\ M_1(s)$$

$$Y(s) = -0.063\ M_2(s)$$

$$F(s) = 2.133\ M_3(s)$$

These gains are dimensional, since they come from the process gains that are in engineering units. This means we cannot compare them with each other.

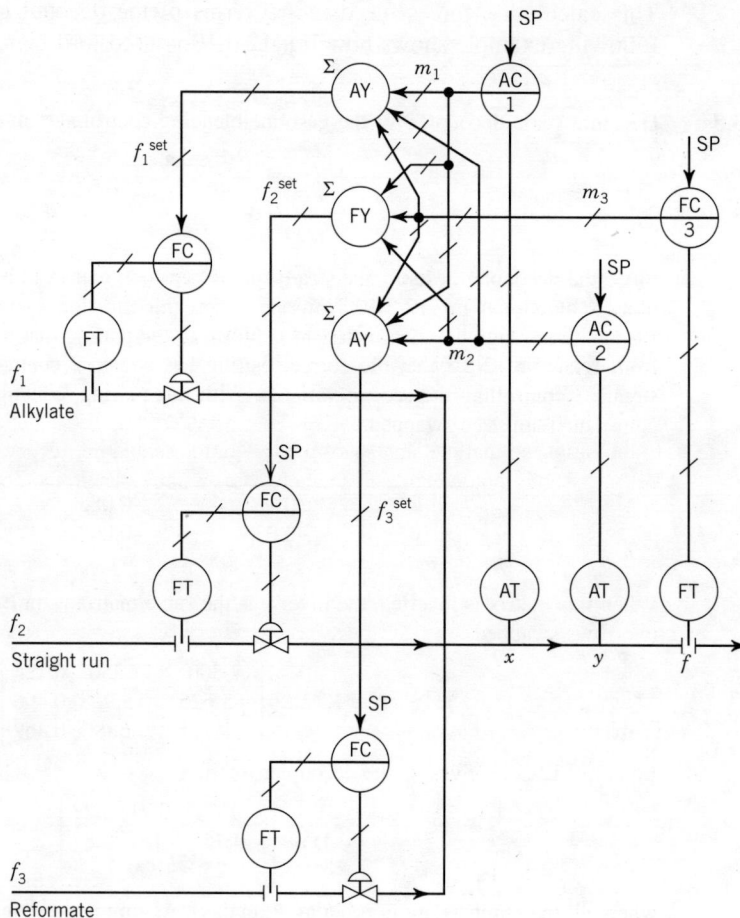

Figure 12-3.6 Instrumentation diagram for static linear decoupler on gasoline blending system.

12-3.3 Decoupler Design from Basic Principles

The decoupler design techniques discussed previously were derived from the block diagram of the multivariable control system. Since block diagrams show only linear relationships between the variables, only linear decouplers can be designed by that procedure. This section presents an example of the design of decouplers from process models that are obtained from applying basic engineering principles to the process. The procedure is basically the same as the one presented in Chapter 11 for designing feed-forward controllers. The only difference is that, in decoupler design, the disturbances to one loop are the manipulated variables for the other loops.

EXAMPLE 12-3.4

Nonlinear Decoupler for Blending Tank

Design a decoupler for the blending tank of Fig. 12-1.1*a* from the basic model equations that represent the tank.

SOLUTION

In designing the decoupler we ignore the disturbances, which for the tank are the inlet stream compositions. After identifying the controlled and manipulated variables, the simple mass and energy balances that relate them are written. For the blending tank, total mass and salt balances result in Eq. 12-1.1, which can be slightly rearranged into the following form:

$$w = w_1 + w_2$$

$$x = \left(\frac{w_1}{w_1 + w_2} \right) x_1 + \left(\frac{w_2}{w_1 + w_2} \right) x_2 \tag{12-3.12}$$

Shinskey (1981) proposes the idea of assigning to the controller output the physical significance of the process variable or combination of variables that has the most influence on the controlled variable. We see from the preceding formula that the product flow w is most affected by the *sum of the two inlet flows*, rather than by either one of them. Also, the product composition is most affected by the ratios of the input flows to the total flow. However, these two ratios are not independent of each other, since they must add up to unity. Therefore, we must pick one of them. According to the results of the relative gain analysis for the tank, the smaller of the two flows should control the composition of the product (see Example 12-2.3). This result extends to the ratios, so that the smaller of the two ratios should be the one to control the composition. Let us assume that stream 2 is smaller than stream 1. Then, the outputs of the two controllers must be made to have the following significance:

$$m_1 = w_1 + w_2$$

$$m_2 = \frac{w_2}{w_1 + w_2} = \frac{w_2}{m_1} \tag{12-3.13}$$

where m_1 and m_2 are, respectively, the outputs of the product flow and composition controllers, assumed to be in the appropriate engineering units.

The implementation of any nonlinear control scheme is more precise if flow controllers are installed on the manipulated streams so that the controller outputs are the set points of these flow controllers. The decoupler scheme then consists of calculating the set points to the two

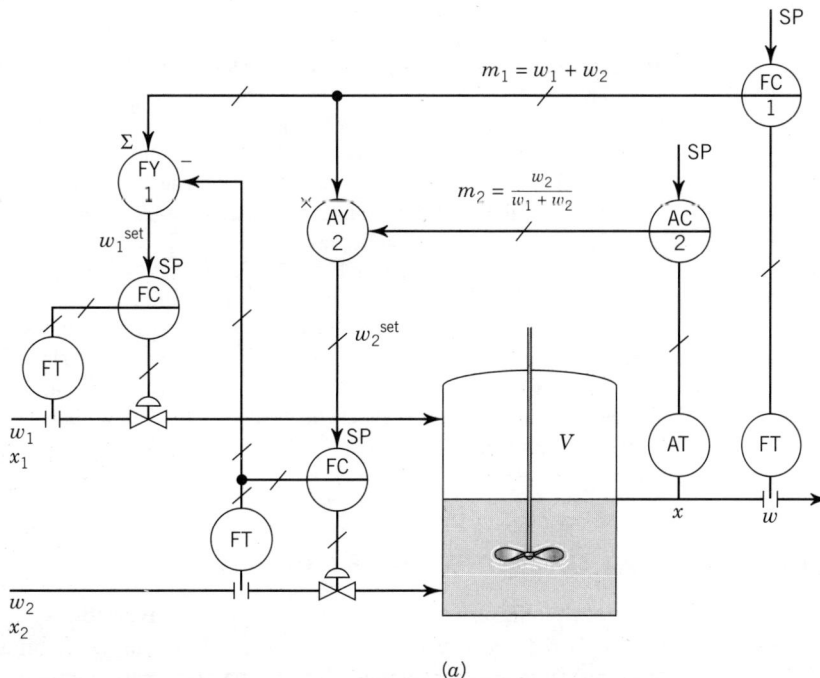

(a)

Figure 12-3.7 Nonlinear decoupler for blending tank. (a) Combining the controller outputs.

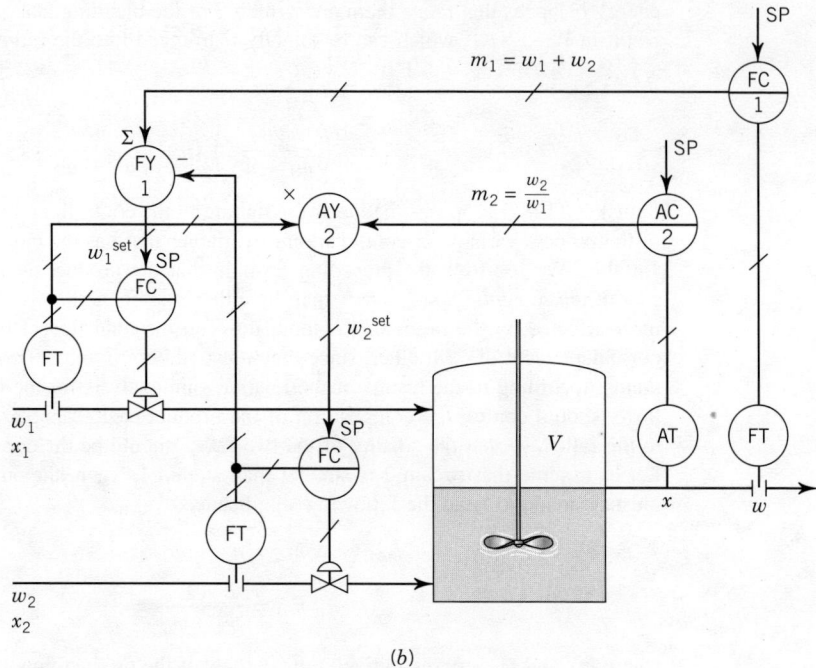

$$m_1 = w_1 + w_2$$

$$m_2 = \frac{w_2}{w_1}$$

(b)

Figure 12-3.7(*Continued*) (*b*) Using a ratio controller.

inlet streams, w_1^{set} and w_2^{set}. From Eq. 12-3.13, these set points are

$$w_1^{set} = m_1 - w_2$$
$$w_2^{set} = m_1 m_2$$

(12-3.14)

The implementation of these equations is shown in Fig. 12-3.7a. Notice that the actual flow w_2, from the flow transmitter signal, is used in the calculation of w_1^{set}. This ensures that the correct set point of stream 1 is computed even when the flow controller on stream 2 is unable to keep its set point. You might ask, why not also use the measured total flow w instead of m_w in the formulas? Shinskey (1996) points out that if the outlet flow is used in calculating the set points of the inlet flows, a positive feedback loop is created around the tank (an increase in the outlet flow increases the inlet flow, which increases the outlet flow further, and so on). Such positive feedback may result in instability and should not be done.

The decoupling scheme of Fig. 12-3.7a may be simplified to the scheme in Fig. 12-3.7b. The difference is that the output of the composition controller is the ratio w_2/w_1 between the two inlet streams. It is easy to show that this simpler ratio still decouples the two loops because it makes the product composition independent of total flow. This scheme also makes it possible to use only partial decoupling by leaving out the summer that calculates w_1^{set}. If this is done, the product composition is independent of the total flow, but the product flow is affected by changes in composition.

12-4 MULTIVARIABLE CONTROL VERSUS OPTIMIZATION

Often in the design of control systems for multivariable systems, a feedback controller is installed for every flow in the process that can be manipulated. This is not always good practice, especially when it is not necessary to maintain the controlled variable at a specific set point. An alternative to providing a controlled variable for each manipulated variable is to use the additional manipulated variables to optimize the operation of the process.

Consider the distillation column of Example 12-2.4. In that example we found that the scheme that tries to control both distillate and bottoms composition has little chance for success because the action of each controller is almost exactly cancelled by the interaction with the other. Suppose that it was not really important to maintain the bottoms composition at 5 mole % benzene, but that this controller had just been included to keep the losses of benzene in the bottom from becoming too high. If benzene is the valuable product from the column, excessive losses of benzene in the bottom represent lower profits because of lower distillate rates. However, the economic value of the benzene lost in the bottom stream must be balanced against the cost of the steam required to provide the heat to the reboiler. Lower benzene compositions in the bottom require higher reboiler heat rates and thus higher steam costs.

A simple optimization scheme can be applied to the column by removing the bottoms composition controller and setting the steam rate at the value that maximizes the profit rate. The profit rate is the difference between the values of the products and the costs of the feed, steam, and condenser coolant.

$$\text{Profit rate} = V_D w_D + V_B w_B - V_R Q_R - V_C Q_C - V_F w_F$$

where V are the economic values in \$/klb or \$/MBtu, w are the mass rates in klb/h, Q are the heat rates in MBtu/h, and the subscripts refer to the distillate, bottoms, reboiler, condenser, and feed.

Several methods have been proposed to determine the reboiler heat rate that maximizes the profit rate, but they are outside the scope of this text. See, for example, the paper by Moore and Corripio (1991).

12-5   DYNAMIC ANALYSIS OF MULTIVARIABLE SYSTEMS

To analyze the dynamic response of a multivariable control system we must determine the transfer functions of the system. In this section we show that the characteristic equation of a set of interacting control loops does not have the same roots as the characteristic equations of the individual loops taken separately. Because of the complexity of analyzing block diagrams with multiple loops, we will simply look at the transfer function for a 2×2 system to draw some conclusions, and then illustrate them using simulation examples.

12-5.1   Dynamic Analysis of a 2 × 2 System

Figure 12-5.1 shows the general block diagram for a 2×2 control system in which the field devices, valve, process, and transmitter, have been combined into the process transfer functions $G_{Pij}(s)$. For simplicity the disturbances have been omitted from the diagram.

To obtain the closed-loop transfer functions for the diagram, we first use block diagram algebra to write the transfer functions of the two outputs in terms of the two set point inputs.

$$\begin{aligned}
C_1(s) &= G_{P11}(s)M_1(s) + G_{P12}(s)M_2(s) \\
&= G_{P11}(s)G_{c1}(s)[R_1(s) - C_1(s)] + G_{P12}(s)G_{c2}(s)[R_2(s) - C_2(s)] \\
C_2(s) &= G_{P21}(s)M_1(s) + G_{P22}(s)M_2(s) \\
&= G_{P21}(s)G_{c1}(s)[R_1(s) - C_1(s)] + G_{P22}(s)G_{c2}(s)[R_2(s) - C_2(s)]
\end{aligned} \qquad \textbf{(12-5.1)}$$

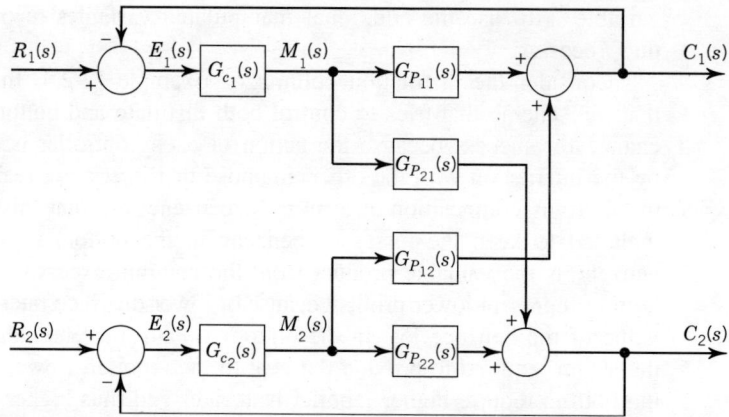

Figure 12-5.1 Block diagram of a general 2×2 system.

After considerable algebraic manipulation, the four individual transfer functions can be obtained as follows. Let

$$\Delta = 1 + G_{c1}(s)G_{P11}(s) + G_{c2}(s)G_{P22}(s) - G_{c1}(s)G_{P21}(s)G_{c2}(s)G_{P12}(s)$$
$$+ G_{c1}(s)G_{P11}(s)G_{c2}(s)G_{P22}(s)$$

(12-5.2)

Then

$$\frac{C_1(s)}{R_1(s)} = \frac{G_{c1}(s)G_{p11}(s)\lfloor 1 + G_{c2}(s)G_{P22}(s)\rfloor - G_{c1}(s)G_{P21}(s)G_{c2}(s)G_{P12}(s)}{\Delta}$$

$$\frac{C_2(s)}{R_1(s)} = \frac{G_{c1}(s)G_{P21}(s)}{\Delta}$$

$$\frac{C_1(s)}{R_2(s)} = \frac{G_{c2}(s)G_{P12}(s)}{\Delta}$$

$$\frac{C_2(s)}{R_2(s)} = \frac{G_{c2}(s)G_{P22}(s)[1 + G_{c1}(s)G_{P11}(s)] - G_{c2}(s)G_{P12}(s)G_{c1}(s)G_{P21}(s)}{\Delta}$$

(12-5.3)

As with any dynamic system, the response is determined by the location of the roots of the denominator polynomial or characteristic equation. To obtain the characteristic equation just set the denominator equal to zero: $\Delta = 0$. It is enlightening to rearrange it, Eq. 12-5.2, into the following form:

$$\Delta = [1 + G_{c1}(s)G_{P11}(s)][1 + G_{c12}(s)G_{P22}(s)] - G_{c1}(s)G_{P21}(s)G_{c2}(s)G_{P12}(s) = 0$$

(12-5.4)

The roots of this equation determine the stability and response of the interacting 2×2 system. As in any other system, the response will oscillate if there is at least one pair of complex conjugate roots, and will be unstable if any of the real roots is positive, or if a pair of complex conjugate roots has a positive real part (see Section 2-3). Furthermore, since the denominator is the same for all four transfer functions, the response characteristics are the same for the two controlled variables.

Equation 12-5.4 tells us the following:

- The tuning of each controller affects the response of both controlled variables, because it affects the roots of the common characteristic equation.

- The effect of interaction on one loop may be eliminated by interrupting the other loop. For example, if controller 2 is switched to manual, $G_{c2}(s) = 0$, and the characteristic equation for the system becomes

$$1 + G_{c1}(s)G_{P11}(s) = 0 \qquad \textbf{(12-5.5)}$$

This is the characteristic equation for the control loop 1 taken separately. The change in the characteristic equation means that if we tune controller 1 with controller 2 in manual, the response of control loop 1 will change when loop 2 is switched back to automatic. Similarly, if controller 1 is switched to manual, $G_{c1}(s) = 0$, and the characteristic equation of the system becomes

$$1 + G_{c2}(s)G_{P22}(s) = 0 \qquad \textbf{(12-5.6)}$$

This is the characteristic equation of control loop 2 taken separately.

- For interaction to affect the response of the loops, it must act both ways. That is, each manipulated variable must affect the controlled variable of the other loop. This is because, if either of the "cross" process transfer functions, $G_{P12}(s)$ or $G_{P21}(s)$, is equal to zero, the characteristic equation of the system becomes

$$[1 + G_{c1}(s)G_{P11}(s)][1 + G_{c2}(s)G_{P22}(s)] = 0$$

The roots of this characteristic equation are the same as the roots of the characteristic equations of each feedback loop taken separately, Eqs. 12-5.5 and 12-5.6. Thus, there would be no interaction. Notice, however, that one of the manipulated variables will still affect both controlled variables.

The reader is encouraged to study Eq. 12-5.4 to gain an understanding of how these important properties of interacting systems follow from the characteristic equation of the system.

The following example shows the effect of interaction on the closed-loop response of the system using simulation.

EXAMPLE 12-5.1 *Feedback Control of Blender*	Study by simulation the effect of interaction on the dynamic response of the blending tank of Fig. 12-1.1*a*. The tank holds 1000 lb of solution, and the inlet stream compositions are 10 and 30 mass % salt, respectively. It is desired to produce 100 lb/min of product solution with a concentration of 20 mass % salt. The control valves are linear with constant pressure drop and sized to pass 150 lb/min when fully opened. Each has a time constant of 0.1 min. The flow transmitter is linear with a range of 0 to 150 lb/min and a time constant of 0.1 min. The analyzer transmitter has a range of 5 to 35 mass % salt, and can be represented by a first-order lag with a time constant of 5.0 min. Assume the composition controller is PI tuned for quarter decay ratio response with the flow controller in manual. How does the decay ratio change when the flow controller, also PI, is switched to automatic?
SOLUTION	From the steady-state mass balances, Eq. 12-1.1, we determine that the design inlet flows are each equal to 50 lb/min. The relative gains, determined in Example 12-2.3, are, from Eq. 12-2.11, equal to 0.50, no matter how the loops are paired. We will assume the flow controller manipulates the dilute stream and the composition controller manipulates the concentrated stream. Assuming constant mass of solution in the tank and perfect mixing, the equations that describe the tank are obtained by total and solute balances on the tank.

$$\frac{dM}{dt} = w_1(t) + w_2(t) - w(t) = 0$$

$$\therefore w(t) = w_1(t) + w_2(t)$$

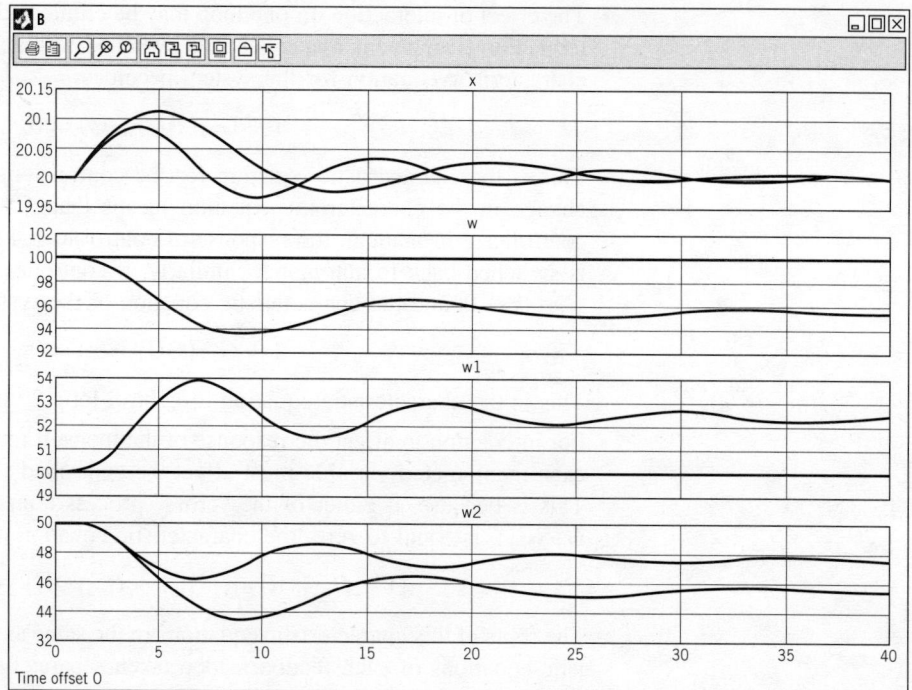

Figure 12-5.2 Responses of blender variables to a 1 mass % increase in concentrated solution composition. Product composition control versus product composition and flow control.

$$M\frac{dx(t)}{dt} = w_1(t)x_1(t) + w_2(t)x_2(t) - w(t)x(t)$$
$$= w_1(t)[x_1(t) - x(t)] + w_2(t)[x_2(t) - x(t)]$$

These equations are then solved in MATLAB to simulate the tank, and the transmitters, controllers, and control valves are simulated as described in Section 13-4. The initial conditions are $w(0) = 100$ lb/min, $x(0) = 20$ mass %, $w_1(0) = 50$ lb/min, $w_2(0) = 50$ lb/min.

With the flow controller opened, the composition controller is tuned for quarter decay ratio response. The resulting tuning parameters are $K_{c2} = 15$ %CO/%TO, and $\tau_{I2} = 12$ min. The flow controller is tuned with a gain of 0.9 %CO/%TO, and an integral time of 0.1 min, typical values for flow control loops.

Figure 12-5.2 shows the responses of the product composition and flow, and of the inlet flows, to a step change in the composition of the concentrated solution from 30 to 31 mass %. Responses are shown for the case in which only the product composition is controlled, and the case in which both the product flow and the composition are controlled. The first case shows a higher initial deviation of the composition from set point, and a drop in total flow while the flow of the dilute solution, w_1, remains constant. The case in which both the flow and composition are controlled shows a smaller initial deviation from set point at the cost of a more oscillatory response. This is because, as the two loops help each other, the gain of the composition loop is higher when the flow loop is closed. Notice how the product flow is maintained constant while both the inlet flows change in opposite directions.

In the preceding example the basic response of the flow controller was not affected by the interaction. This is because the flow controller is very fast and controllable so that its response is insensitive to the process gain. The composition controller is affected by the interaction in that it oscillates faster and the oscillations decay more slowly. This is because its gain increases by a factor of 2 when the flow control loop is closed.

12-5.2 Controller Tuning for Interacting Systems

It is evident from the discussion in the preceding section that interaction makes the tuning of multivariable control systems a difficult task. However, a proper procedure can overcome the difficulties. This section presents a brief outline of the tuning procedure. For a more detailed discussion the reader is referred to the book by Shinskey (1996).

The first step in the tuning procedure, after properly pairing the controlled and manipulated variables, is to determine the relative speed of the loops. If one loop is much faster than the other, say, having a dominant time constant five times smaller, then it is tuned first with the other loop opened. Then the slower loop is tuned with the faster loop closed. The assumption is that the closing of the slower loop has minor effect on the performance of the faster loop, because the faster loop can respond fast to the actions of the slow loop. When the slow loop is tuned with the faster loop closed, the tuning takes full consideration of the interaction. Flow and pressure loops are usually much faster than temperature and composition loops. For example, in the blending tank of Example 12-5.1, the flow loop had a closed-loop time constant of 0.1 min, whereas the analyzer loop had a dominant time constant of 10 min. In that example we saw that the response of the flow control loop was not affected when the composition loop was closed.

If the two loops are of about the same speed of response, then detuning one loop by setting a small gain and long reset time will have the same effect of making that loop slow. This will reduce its effect on the response of the other loop because the detuned loop will appear to be opened. Such an approach is followed when the control of one variable is more important than the control of the other one. The controller on the less important variable is the one that is detuned.

Finally, when both loops are of about the same speed of response and of equal importance, each controller should be tuned with the other loop opened. Then the tuning should be adjusted to account for interaction. If the interaction is positive (see Section 12-1), the two loops help each other. This means that the dynamic effects of interaction are usually insignificant. In such case, adjust each controller gain from its initial (open-loop) tuning by multiplying it by the relative gain for the pair. The principle is that such an adjustment will keep a constant overall gain for each loop. For the loop gain to remain the same,

$$K_{ci}K_{Pii} = K'_{ci}K'_{Pii}$$

where the primes denote the controller and process gains when the other loop is closed. Solving for the controller gain with the other loop closed yields

$$K'_{ci} = K_{ci}\frac{K_{Pii}}{K'_{Pii}} = K_{ci}\mu_{ii} \qquad (12\text{-}5.7)$$

where we have used the definition of the relative gain, Eq. 12-2.3.

For example, in the blending tank of Example 12-5.1, the relative gain for either pairing is 0.5. This means that the gain obtained for each controller when the other loop is opened should be reduced by half to account for the increase in the process gain when the other loop is closed. In that case, however, the flow control loop is so fast that it is not necessary to adjust its gain. The reader is invited to check that if the gain of the composition controller is reduced to (15) (0.5) = 7.5 %CO/%TO, the response of the product composition is the same as when the flow loop was opened and the composition controller had the original gain.

If the interaction is negative, its effect on the dynamic response is quite significant because the two loops fight each other. This means that inverse response is possible when the two loops are closed. Inverse response causes the controlled variable to move the wrong way first and is quite detrimental to the performance of the loop. Because of this, adjustment of the controller gains must be done by trial and error, after both loops are closed.

Tuning of $n \times n$ Systems

The tuning procedure we have just presented for 2×2 systems can be extended to any number of loops. What needs to be done is to first rank the loops in order of speed of response and tune those loops that are much faster than the others. In doing this the gains of the fast loops must be adjusted to account for interaction between them. Then the slower loops must be ranked in order of decreasing importance and the less important loops detuned if possible. Finally, the gains of the slow loops must be adjusted to compensate for the effect of interaction. Equation 12-5.7 is valid to do the adjustment for those loops for which the interaction is positive. Recall that the interaction is positive when the relative gain for a pair is between 0 and +1.

12-6 DESIGN OF PLANTWIDE CONTROL SYSTEMS

One of the important phases of the design of a new process is the selection of which process variables must be controlled and which streams are going to be manipulated to control them. Such a task is best carried out by a team consisting of the engineer who designed the process, the instrument or control engineer who will design the control system and specify the instrumentation, and the operating personnel who will run the process. We have so far considered control systems for individual units—heat exchangers, reactors, distillation columns, etc.—without looking at how these units may interact with one another.

In the past, control schemes for chemical processes were designed on a unit by unit basis without consideration for interaction between these units. It was assumed that sufficiently large surge tanks would be installed between units to smooth out any interactions. Today, because of safety, environmental, and economic considerations, it is industry's desire to reduce the amount of intermediate storage in the process. If a well-designed control system can deal with interaction between units and reduce the storage of intermediate chemicals, the inventory costs are reduced, and the environmental and safety liability are also reduced because there is less storage of flammable and hazardous chemicals.

To accomplish the reduction in inventory, control schemes must be designed in a way to account for the dynamic interactions introduced by streams leaving units coupled to each other when there is no intermediate storage. One way to do this is to design the control system by taking the entire plant into account. This is the goal of *plantwide control*.

In designing control schemes in this manner, there are usually more choices than can realistically be evaluated. There are many excellent references that treat the design of plantwide control. This section refers to some of these references, and the readers are strongly encouraged to read them; a particular book that we enjoy is the one by Luyben *et al.* (1998).

On the basis of extensive studies, some guidelines (Kight, 1998) are offered next. These guidelines attempt to ease the task of designing control systems on a plantwide basis.

The most critical decision is the identification of the *throughput manipulated variable* (TPM), that is, the stream flow that determines the *production rate* of the plant. This stream affects the operation of the entire process, not only because of the direct effect of the stream flow, but because the location of the TPM dictates the arrangement of many of the remaining control loops. When choosing a location for the TPM, there is often more than one choice that will satisfy the rules shown below. Each location can be rationalized on the basis of some perceived process benefit. Several rules that have been developed for placement of the TPM are as follows:

1. The TPM must be located in the primary process path, and on the feed stream to the most susceptible unit as, for example, the main reactor. The path can be traced by starting with the feeds and tracing the flow as it is converted to products on to the final product stream. The production rate must be set by manipulating one of the flows in this path. If more than one product is produced, a determination must be made for which is the primary one; usually this decision is based on economic considerations.

2. The stream manipulated should be a product-rich stream or a feed stream containing a high fraction of a reactant which is mostly converted to products. If the reactant or product is in low concentration the stream flow is no longer a good indication of production rate.

3. The feed to a unit that represents a major bottleneck in the process is a good candidate for the TPM. Controlling the feed rate to the limiting unit minimizes the disturbances seen by that unit, allowing maximum production.

4. Whenever possible a TPM internal to the process, rather than a feed or product stream, is preferred (Price, 1993). Using an internal stream minimizes the propagation of flow disturbances. The TPM serves as a buffer for the units on either side of it in regards to disturbances.

5. Although the most common scheme is to directly manipulate the flow of the TPM stream, it is not always necessary to do this. When the flow of the TPM stream is not directly manipulated, the manipulated flow must be one that has a substantial effect on the TPM stream flow. An example would be steam to a flash drum where the vapor exit stream is the primary process path (Lyman, 1992).

6. Avoid controlling the flows of product streams from a distillation column to set values. This point will be discussed in the example below.

Once the location for the TPM is determined, the inventory controls upstream and downstream of the TPM must be designed. In most cases, proportional-only controllers tuned for averaging level control (see Section 7-3.3) give the best overall performance, except when the pressures and levels must be maintained at their set points.

1. Inventories upstream of the TPM must be controlled by manipulating inlet flows, and those downstream must manipulate exit flows. The inventory control structure must be self-consistent; that is, the set of controls must be able to propagate a change in production rate back to the feed stream and forward to the product stream.

2. Proximity rules should be used when choosing manipulated variables. That is, manipulated variables which are physically close to the inventory being controlled are likely to offer the best performance.

It is recommended to first design a simple feedback control loop for each controlled variable. Once this has been done, the engineer should decide whether another control system (ratio, cascade, feedforward, or a decoupler) may be needed to achieve the required control performance; experience with similar processes provides a basis for the decision. In any case, the best recommendation is to *design the simplest strategy that provides the required performance.*

It is worthwhile to mention Luyben's (2002) laws of plumbing:

- *First Law of Plumbing.* Locate valves in liquid lines downstream of centrifugal pumps. If the valve is located at the pump suction, the NPSH requirement may not be met because of the pressure drop across the valve (see Example 5-2.3).

- *Second Law of Plumbing.* Use only one control valve in a line, except when the second valve controls the pressure between the two valves. This law refers to two *control* valves on the same line. If the valves are manipulated by different controllers they will fight each other, because there can only be one flow in the line.

- *Third Law of Plumbing.* Do not throttle the discharge of a compressor. Installing a valve on the outlet of the compressor affects only the work and the compression ratio, not the flow, and drives the compressor toward an unstable condition known as *surge*. Installing a valve at the suction of the compressor directly affects the mass flow and drives the compressor away from surge. The most energy-efficient method of controlling a compressor is by adjusting the compressor speed.

Let us consider an example to look at the applications of these guidelines.

EXAMPLE 12-6.1

Design a Control System for an Acetone Process

Figure 12-6.1 is a schematic of a process to produce acetone by catalytic dehydrogenation of isopropyl alcohol (IPA). It has been reported that this process can achieve 85 % conversion per pass through the reactor at a pressure of 200 kPa and a temperature of 350°C over a metallic copper catalyst (Austin, 1984). The highly endothermic gas-phase reaction is given by

$$(CH_3)_2CHOH \longrightarrow (CH_3)_2CO + H_2$$

The schematic of Fig. 12-6.1 is simplified in that it assumes no side reactions. This means that after separation of the noncondensable hydrogen, only one distillation column, T-100, is required to separate the product acetone from the unreacted IPA. The unreacted IPA is recycled back to the reactor as shown in the figure. A compressor K-100 and refrigerated flash drum V-101 are used to recover the acetone and IPA carried out with the hydrogen. The final hydrogen stream is 98 mole % pure and can be used as a chemical feedstock or combusted for its heating value.

To provide the energy for the reaction, a furnace is used. Energy integration is achieved by cross-exchanging the stream out of the reactor with the feed to the reactor in exchanger E-100. The heat duty is sufficient to vaporize the combined fresh feed and recycle streams. As the reaction is endothermic, there is no danger of a runaway.

The design of the control system for the process must consider control of production rate, the control of the distillation column, and the control of the temperatures and pressures in the reactor and the flash drums, and the inventories in the flash drums and in the distillation column.

SOLUTION

The first thing we must do is select the stream that adjusts the production rate, the TPM. There are two streams out of this process, the product P and the hydrogen H, and one inlet stream, the

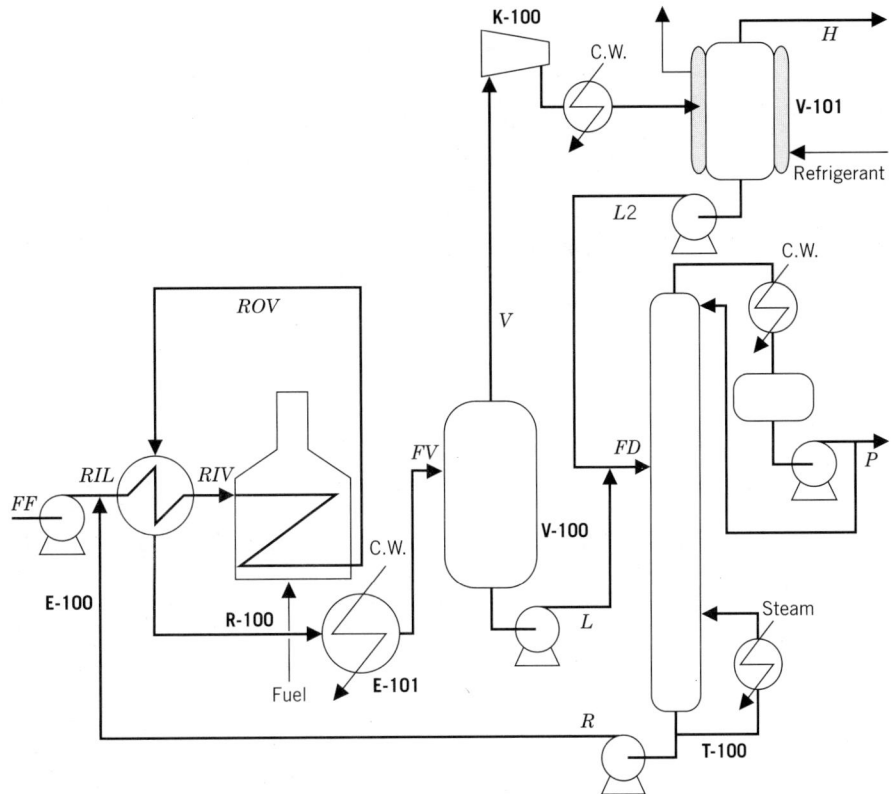

Figure 12-6.1 Schematic for process of acetone from isopropyl alcohol.

fresh IPA feed *FF*. There are also several internal streams around the reactor, the flash drums, and the distillation column, and the recycle stream *R*.

Although guideline 4 already tells us that an internal stream should be selected for the TPA, this example demonstrates why this is so.

Let us first consider the product stream *P*. It is the distillate product of the distillation column. According to guideline 6 it is seldom a good idea to control the product flow from a distillation column to a set value. Let us see why. Suppose that we kept the flow of distillate constant and that an upset causes the conversion in the reactor to drop. This would cause the flow of acetone into the column to decrease and become less than the flow of distillate. To maintain the distillate flow higher than the flow of acetone in the feed, more IPA will have to go to the product and the product purity will decrease. Notice that the same thing will happen if we maintained the bottoms product *R* from the column constant while allowing the distillate rate to vary. The drop in conversion will cause more IPA to enter the column than is removed in the bottoms. The additional IPA will have to go to the distillate, increasing its flow and decreasing the product purity.

Similar decreases in product purity will result from upsets that cause the losses of acetone with the hydrogen stream to increase. Although an increase in reactor conversion will not adversely affect the purity of the product stream, it will cause the extra acetone to flow to the bottom of the column. This would not cause the acetone to be lost, since it is recycled to the reactor, but it may upset the operation of the column.

As the product stream should not be the TPM, let us now consider the other two external streams. It is obvious from the schematic of Fig. 12-6.1 that the hydrogen stream *H* is not in the main path between the feed and the product and, according to guideline 1, should not be the TPM. This leaves the fresh IPA feed stream *FF*. The problem with using *FF* as the TPM is that it does not directly control the production of acetone in the reactor. When the conversion in the reactor varies, the recycle stream R, carrying the unreacted IPA, also varies, varying the flow into the reactor and on to the separation system. Although at steady state the conversion rate of IPA will match closely the fresh feed rate, variations in conversion will cause temporary upsets on all of the equipment in the recycle loop. So we have demonstrated that guideline 4 is applicable to this process.

Of the internal streams in the path between the feed and the product we have the streams in and out of the reactor, and the feed stream to the distillation column. The column feed FD could be selected if it is determined that the column is the bottleneck in this process, but this would present problems on how to control the level in flash drum V-100, because the feed to the drum FV is a two-phase stream, difficult to manipulate. So it would be preferable to select one of the streams in the path in and out of the reactor. Of these, the streams out of the reactor have a low mole fraction of acetone and IPA—about half of them is hydrogen—and, according to guideline 2, should not be selected. There is the additional disadvantage that they are either vapor or two-phase streams.

The final two streams to consider are the streams in the path into the reactor, RIL and RIV, one liquid and the other one vapor. The flow of the liquid stream RIL can be measured more accurately than that of the vapor, because the density of the liquid does not vary much. Thus we select the liquid stream RIL as our TPM stream. This stream is the combination of the fresh IPA feed FF, and the recycle stream, R. So the next question is how to control the flow of our TPM stream.

One approach is to feed both FF and R into a tank, directly control the flow of the stream out of the tank, the RIL stream, with a control valve on that line, and control the level in the tank with a valve on the fresh IPA feed line, FF. This is the simplest approach, probably preferred by the operating personnel, but we said we wanted to eliminate intermediate storage because of safety, environmental, and economic considerations. So, we must manipulate one of the two streams that make up the RIL stream to control its flow. We already saw that the distillation column must manipulate the recycle stream R so that it can return all of the unreacted IPA in its feed. So our final solution is

> Select stream RIL as the TPM stream and control its flow by manipulating the flow of fresh IPA feed FF.

By Luyben's first law of plumbing, the control valve must be installed on the discharge of the pump. Figure 12-6.2 shows the proposed control scheme (FC on FF). In the final design we are following guideline 5 that says that it is not necessary to directly manipulate the TPM stream.

Distillation column controls. As we saw in Example 12-2.4, a simple distillation column with two products involves five controlled variables and five control valves. Normally an interaction measure analysis must be made to decide how to pair these variables. Here we will just use a simple preliminary scheme to be modified later after further study. Figure 12-6.2 shows the control scheme, which is as follows:

- Control column pressure by manipulating the condenser heat rate (PC on cooling water).
- Control inventory in condenser accumulator by manipulating the distillate product flow P (LC on P).
- Control distillate product purity by manipulating the reflux flow (AC).
- Control bottoms column inventory by manipulating the bottoms product flow, the recycle R (LC on R).
- Control the heat rate to the reboiler constant by manipulating steam flow (FC on steam).

The decision here is to control the purity of product P constant and not attempt to control the acetone content of the recycle, as the acetone in the recycle is not lost but goes back to the reactor. By not controlling the bottoms purity we avoid interaction effects on the product composition control.

Control of pressures. Other than the column, the three pressures in the process are the pressures in the reactor and the two flash drums. For the reactor and flash drum V-101, being downstream

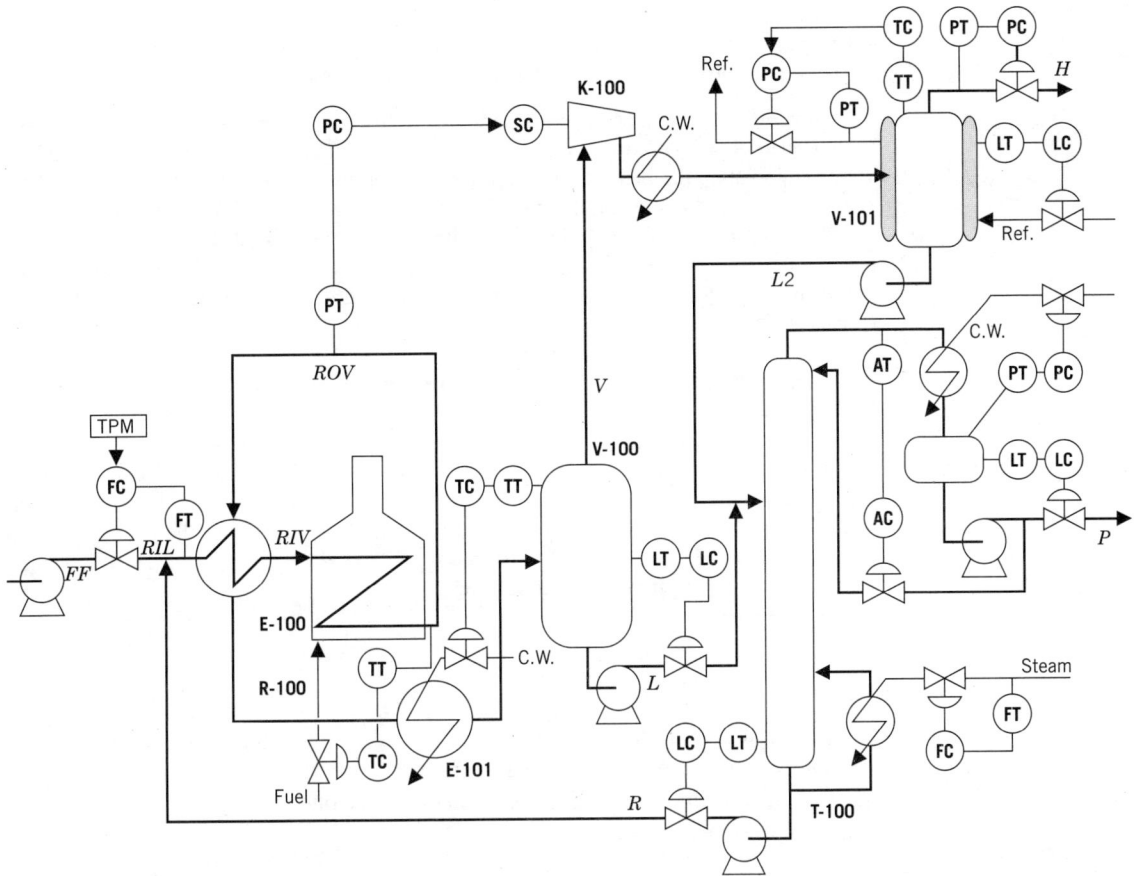

Figure 12-6.2 Control scheme for acetone process.

of the TPM stream, the pressure is controlled by manipulating control valves on the vapor streams leaving the vessels. The pressure controllers are then simple back-pressure regulators which are relatively simple to tune and operate. For flash drum V-100 the vapor flow is the flow to the compressor. Following Luyben's third law of plumbing, the control valve should be on the compressor suction, in which case this pressure controller becomes also a back-pressure regulator. If the compressor is driven by a steam turbine whose speed can be adjusted, it is more efficient to control the pressure in V-100 by manipulating the speed of the compressor. This results in lower overall compressor energy consumption because it does not require a pressure drop through a control valve.

It is important to ask if it is really necessary to control the pressure in both the reactor and in flash drum V-100. If only one of these pressures is controlled, the control valve between the two vessels is eliminated as is the pressure drop required by the control valve. Then the pressure in the flash drum would be higher and the composition of the acetone and IPA in the vapor stream from the drum would be reduced, reducing the flow and the power required by the compressor. Pressure is a fast-responding variable, so not much controllability is lost when the pressure in the reactor is controlled by manipulating the speed of the compressor. This is the scheme shown in Fig. 12-6.2 (PC on SC, compressor speed controller). The pressure at the compressor suction will be the pressure in the reactor minus the sum of the pressure drops in exchangers E-100, E-101, and vessel V-100.

Control of inventories. Other than the inventories in the column, the liquid levels in the two flash drums must be controlled. Being downstream of the TPM stream, these levels are controlled by manipulating the liquid streams out of the drums, with the valve on the discharges of the pumps (Luyben's first law of plumbing). Figure 12-6.2 shows the control loop for V-100 (LCs). The level control loop for V-101 is not shown as it essentially the same.

Control of temperatures. The temperatures in the reactor and in the two flash drums are to be controlled. As the reaction is carried out in a furnace, the temperature in the reactor is controlled by manipulating the fuel flow to the furnace firebox (TC on fuel). This simple feedback control can be enhanced by adding a fuel-to-feed flow ratio controller. Furnaces also require significant safety instrumentation and controls which will not be discussed here.

The temperature in flash drum V-100 can be controlled by adjusting the coolant to heat exchanger E-101 (TC on cooling water). This temperature must be kept as low as possible to keep low the composition of acetone and IPA in the vapor stream leaving the drum, thus reducing the flow to the compressor.

Flash drum V-101 is cooled by refrigeration. The refrigerant usually enters as a liquid and vaporizes in the jacket around the drum. So the temperature in the drum is controlled with a cascade control scheme: the temperature controller adjusts the set point of the refrigerant pressure controller which in turn manipulates the control valve in the refrigerant exit vapor stream (TC to PC on refrigerant). The level of refrigerant in the jacket is controlled by manipulating the inlet liquid flow of refrigerant (LC on refrigerant). This replenishes the refrigerant at the rate at which it vaporizes. Flash drum V-101 must be kept as cold as possible to reduce the losses of acetone and IPA in the hydrogen stream.

The preceding example shows how the complete control scheme is designed by applying sound process engineering principles. In arriving at the solution it is essential to understand the process and the operation of the various process units. The discussion also indicates that the scheme presented is only one of several correct solutions to the control design problem.

12-7 SUMMARY

This chapter has presented the design and tuning of multivariable feedback control systems. We first explained the effect of interaction and then introduced its measurement from the open-loop gains. The relative gains are used not only as quantitative measures of interaction, but also to pair the controlled and manipulated variables so as to minimize the effect of interaction. The proper pairing has each pair with a relative gain near unity. The design of decouplers was presented next. Decoupling can totally or partially eliminate the effect of interaction. Finally, the dynamic response of interacting loops was analyzed and a procedure for tuning them was presented.

REFERENCES

1. AUSTIN, G. T., *Shreve's Chemical Process Industries*, 5th ed. New York: McGraw-Hill, 1984, p. 764.

2. BRISTOL, EDGAR H., "On a New Measure of Interaction for Multivariable Process Control." *IEEE Transactions on Automatic Control AC-11*, January 1966, pp. 133–134.

3. CUTLER, C. R., and B. L. RAMAKER, "DMC-A Computer Control Algorithm." Paper read at AIChE 1979 Houston Meeting. Paper #516. New York: AIChE, 1979.

4. HULBERT, D. G., and E. T. WOODBURN, "Multivariable Control of a Wet Grinding Circuit." *AIChE Journal*, March 1983.

5. KIGHT, DARRYL, *Evaluation of Plant Wide Control Guidelines with a series of Two-Phase Reactors with Separation.* M.S. Thesis, University of South Florida, 1998.

6. LUYBEN, W. L., B. D. TYREUS, and M. L. LUYBEN, *Plantwide Process Control.* New York: McGraw-Hill, 1998.

7. LUYBEN, WILLIAM L., *Plantwide Dynamic Simulators in Chemical Processing and Control.* New York: Marcel Dekker, 2002.

8. LYMAN, PHILIP, *Plant-Wide Control Structures for the Tennessee Eastman Process.* M.S. Thesis, Lehigh University, 1992.

9. MATLAB, The MathWorks, Inc., Natick, MA, 2001.

10. MathCad, MathSoft, Inc., Cambridge, MA, 2001.

11. MOORE, J. D., and A. B. CORRIPIO, "On-Line Optimization of Distillation Columns in Series." *Chemical Engineering Communications* 106: 71–86, 1991.

12. PRICE, R. M., *Design of Plant Wide Regulatory Control Systems*, Ph.D. Dissertation, Lehigh University, 1993.

13. SHINSKEY, F. G., *Controlling Multivariable Processes*. Research Triangle Park, NC: ISA, 1981.

14. SHINSKEY, F. G., *Process Control Systems, Application, Design, and Tuning*, 4th ed. New York: McGraw-Hill, 1996.

15. TAYLOR, R., A. HAKET, and H. KOOIJMAN, *ChemSep*. Potsdam, NY: Chemical Engineering Department, Clarkson University, 2004.

PROBLEMS

12-1. For each of the processes of Fig. 12-1.1, using your knowledge of process mechanisms and basic principles, explain what causes interaction between the variables shown. Is the interaction negative or positive? List your assumptions.

12-2. *Control of Caustic Dilution Process.*

The tank of Fig. P12-1 is used to mix a 50 % solution of caustic with demineralized water to produce a dilute solution. The specified product composition is 30 mass % NaOH, and, at design conditions, the product flow is 40 klb/h. It is desired to control the product flow, w_P (FC-3), and composition, x_P (AC-3), by manipulating the set points on the flow controllers on the two inlet streams (FC-1 and FC-2). The tank is assumed perfectly mixed, and the total mass M of the solution in the tank is constant. Obtain the steady-state open-loop gains for the 2×2 system, and the relative gains at design conditions. Is the interaction positive or negative? Which pairing of the controlled and manipulated variables minimizes the effect of interaction? By how much does the gain of the product composition loop increase or decrease (specify which) when the product flow loop is closed? *Note:* The simulation of this process is the subject of Problem 13-34.

12-3. *Automatic Control of a Household Shower.*

The design of automated showers for Mr. Trump's mansion requires the control of the temperature and flow of the water to the showerhead. The design calls for a system that can deliver 3 gpm of water at $110°F$ by mixing hot water at $170°F$ with cold water at $80°F$. The density of water is 8.33 lb/gal, and its specific heat is 1.0 Btu/lb-$°F$. Control valves are used to manipulate the flows of cold and hot water. A flow transmitter and a temperature transmitter are also used in each shower.

(a) Show that steady-state mass and energy balances result in the following relationships, assuming constant density and specific heat of the water.

$$f_o = f_1 + f_2$$
$$f_o T_o = f_1 T_1 + f_2 T$$

(b) Calculate the required flows of hot and cold water at design conditions, and the steady-state open-loop gains. Specify the units.

(c) Calculate the relative gains for the system, and decide which of the two inlet flows is to control the flow and which is to control the temperature so that the effect of interaction is minimized.

12-4. *Control of an Evaporator.*

Consider the evaporator of Fig. 12-3.4. The level in the evaporator is controlled very tightly by manipulating a valve on the feed line because this is the largest of the inlet and outlet streams (LC). It is desired to control the feed flow, w_F (FC) and the product composition, x_P (AC), by manipulating the

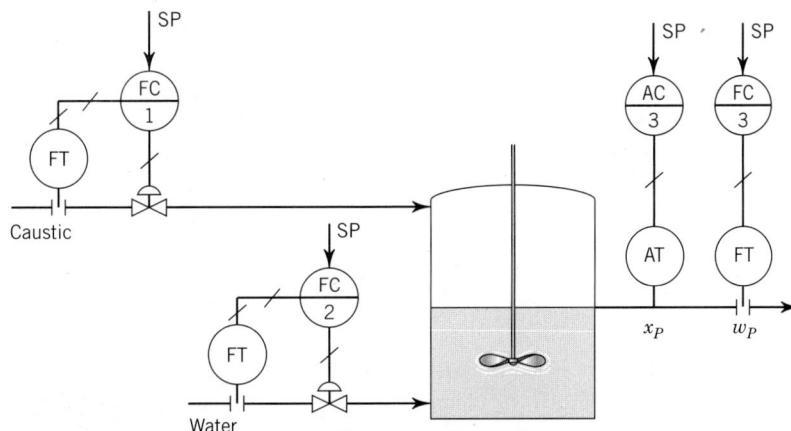

Figure P12-1 Caustic blending tank for Problem 12-2.

signals to the valves on the product, m_P, and steam, m_S, streams. The evaporator economy is approximately constant at E lb of vapor per lb of steam, and the inlet composition is x_F, weight % solute.

(a) Show, from steady-state balances on the evaporator, that the relationships between the variables are

$$w_F = w_P + E w_s$$

$$x_P = x_F \left(1 + \frac{E w_s}{w_P}\right)$$

(b) Derive the general formulas for the steady-state open-loop gains and for the relative gains. Is the interaction positive or negative?

(c) Develop a general pairing strategy that will minimize the effect of interaction for any evaporator depending on the design values of the process variables.

12-5. The evaporator of Problem 12-4 is designed to concentrate a 30 % solution of caustic to produce a product solution with 50 mass % NaOH, and, at design conditions, the feed flow is 80 klb/h. The economy is 0.9 klb of vapor per klb of steam. Solve the problem for these design conditions. By how much does the gain of the product composition controller increase or decrease (specify which) when the feed flow control loop is closed? *Note*: The simulation of this process is the subject of Problem 13-35.

12-6. *Distillation Product Composition Control.*

An example of highly interactive control loops is the control of top and bottom compositions in a distillation column. Two tests were performed on a simulation of a butane–pentane column: one was a change in steam flow keeping the reflux flow constant, and the other one was a change in reflux flow keeping the steam flow constant. The following table summarizes the flows and compositions for each of the tests, after steady state is established:

	Base	Test 1	Test 2
Steam flow, klb/h	24.0	25.0	24.0
Reflux flow, lbmole/h	70.0	70.0	75.0
Bottoms percent butane	6.22	3.08	8.77
Distillate percent butane	93.50	91.79	96.88

The controlled variables are the percent butane in the bottoms and distillate products, and the manipulated variables are the steam and reflux flow set points. Calculate the steady-state open-loop gains, and the relative gains for the two possible pairings. Is the interaction positive or negative? Which variable should be manipulated to control the distillate composition so that the interaction between the two loops is minimized? What is the steady-state gain for the distillate composition loop when the bottoms composition loop is closed?

12-7. For the gasoline blending control system of Example 12-2.5, calculate the required flows of the three feed streams if the specified gasoline octane number is 89.0 instead of 87.0. Then recalculate the steady-state open-loop gains and the relative gains. Does the pairing of controlled and manipulated variables that minimizes interaction change from the one obtained in the example? Design a static decoupler for this system and calculate the gains of the decoupled system, as in Example 12-3.3.

12-8. *Control of Wet Grinding Circuit.*

Hulbert and Woodburn (1983) present the multivariable control of a wet grinding circuit as the one shown in Fig. P12-2. The controlled variables are the torque required to turn the mill (TOR), the density of the cyclone feed (DCF), and the flow from the mill (FML). These variables were selected on the basis of their importance to the metallurgical process. The torque and flow from the mill describe the operation of the

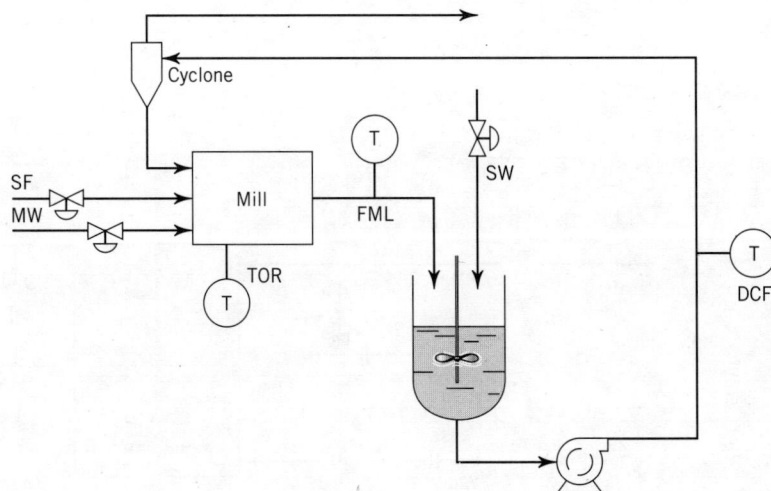

Figure P12-2 Wet grinding circuit for Problem 12-8.

mill, and the density of the cyclone feed describes the operation of the slurry tank. The flow from the mill is not measured directly, but from a combination of other measurements and a mass balance around the mill. However, for simplicity, a single transmitter is shown in the diagram.

The manipulated variables are the flows of the three feed streams: solids (SF) and water to the mill (MW), and water to the slurry tank (SW). The feed rate of solids to the mill is manipulated by adjusting the speed of the conveyor carrying the solids.

Open-loop step tests in each manipulated variable results in the following transfer functions:

	SF, kg/s	MW, kg/s	SW, kg/s
TOR, Nm	$\dfrac{119}{217s+1}$	$\dfrac{153}{337s+1}$	$\dfrac{-21}{10s+1}$
FML, m³/s	$\dfrac{0.00037}{500s+1}$	$\dfrac{0.000767}{33s+1}$	$\dfrac{-0.000050}{10s+1}$
DCF, kg/m³	$\dfrac{930}{500s+1}$	$\dfrac{-667e^{-320s}}{166s+1}$	$\dfrac{-1033}{47s+1}$

where all time parameters are in seconds.

(a) Calculate the relative gains and select the pairings of the controlled and manipulated variables that minimize the effect of interaction.

(b) Using the pairing of part (a), draw the block diagram and design a linear decoupler for the system. Show the instrumentation diagram for the complete control system.

12-9. *Decoupler Design for Distillation Column.*

In the distillation column of Fig. 12-2.3, the flows of reflux and steam to the reboiler are manipulated to control the distillate and bottoms product purities. Open-loop step tests on each of these manipulated variables result in the following transfer functions:

$$Y_D(s) = -3.0\left(\frac{1-0.11s}{1+0.35s}\right)M_R(s) + \left(\frac{0.75}{1+0.35s}\right)M_S(s)$$

$$X_B(s) = \left(\frac{2.5}{1+0.35s}\right)M_R(s) - 3.5\left(\frac{1+0.25s}{1+0.35s}\right)M_S(s)$$

where $Y_D(s)$ is the composition of the heavy key in the distillate, $X_B(s)$ is that of the light key in the bottoms, in

%TO, and $M_R(s)$ and $M_S(s)$ are the signals to the reflux and steam valves, respectively, in %CO. The time parameters are in hours.

(a) Calculate the relative gains for this system and determine the correct way to pair the controlled and manipulated variables. Do the loops help or fight each other?

(b) Draw the block diagram and design the decouplers for this system. Briefly discuss any implementation problems and suggest modifications to ensure the control system is stable.

12-10. *Decoupler Design for 2 × 2 Process.*

A process has two controlled variables that are affected by two manipulated variables and one disturbance, $U(s)$. The transfer functions are

$$C_1(s) = G_{P11}(s)M_1(s) + G_{P12}(s)M_2(s) + G_{P1U}(s)U(s)$$
$$C_2(s) = G_{P21}(s)M_1(s) + G_{P22}(s)M_2(s) + G_{P2U}(s)U(s)$$

where

$$G_{P11}(s) = \frac{0.81e^{-0.6s}}{1.4s+1} \quad G_{P12}(s) = \frac{1.2e^{-1.1s}}{2.4s+1}$$

$$G_{P1U}(s) = \frac{0.5}{2.2s+1}$$

$$G_{P21}(s) = \frac{1.1e^{-0.3s}}{1.5s+1} \quad G_{P22}(s) = \frac{0.9e^{-s}}{2.0s+1}$$

$$G_{P2U}(s) = \frac{-1.5}{1.8s+1}$$

(a) Calculate the relative gains and select the correct pairing for this system.

(b) Design a decoupler for this system and show its implementation on the block diagram.

12-11. Show that if the exact decoupler is implemented on the system represented by the block diagram of Fig. 12-3.2, the system is indeed decoupled.

12-12. Design a decoupler for the blending tank described in Example 12-5.1. Show the instrumentation diagram for implementing the decoupler.

12-13. Design, from basic principles, a nonlinear static decoupler for the evaporator of Problem 12-4.

Chapter 13

Control Systems and Dynamic Simulation

Dynamic simulation is a most convenient method for analyzing the performance of process control systems. This chapter presents the basics for learning how to develop and use dynamic simulations of processes and instruments—controllers, valves, sensors, etc. Most of the simulations presented here will be based on the dynamic modeling technique presented in Chapters 3 and 4, as well as the instrument models presented in Chapter 5.

Two methods of dynamic simulation are presented:

- Simulation of linear systems through transfer function representation.
- Simulation of control systems from basic mechanistic models.

The first of these methods has the advantage of not requiring concern about the initial conditions because, as indicated in Chapter 2, transfer functions assume the variables are deviations from steady-state initial conditions, that is, the initial conditions are zero. However, it requires the development of the transfer functions and, in the case of nonlinear systems, the linearization of the model equations.

The second method is more fundamental and is most suitable for the modeling of nonlinear systems. The major difficulty is that the modeler must be very concerned with using the correct initial conditions of all the variables. When simulating control systems for continuous processes it is also necessary to ensure that the initial conditions represent a steady state.

The bulk of the chapter assumes that the processes are lumped, that is, the variables are functions only of time. A brief discussion of the simulation of stiff systems is included at the end.

The examples presented use the program MATLAB with Simulink, because it is one of the most powerful block-oriented simulation programs that is commonly available to students and engineers. Simulations using VisSim are also included in the web page that accompanies this book.

13-1 USES AND TOOLS OF DYNAMIC SIMULATION

This section presents the uses of dynamic simulation and some of the simulation programs currently available in the market.

Principles and Practice of Automatic Process Control/Third Edition, by C. A. Smith and A. B. Corripio
ISBN 0-471-66141-4   Copyright © 2006 John Wiley & Sons (Asia) Pte. Ltd.

13-1.1   Uses of Dynamic Simulation

Some common uses of process dynamic simulation are

- Study the dynamic behavior of processes.
- Design, tune, and evaluate process control systems.
- Train operators on the operation of processes.
- Design batch processes and their operation.

Notice that the design of continuous processes is left out of this list. The reason is that such design is best carried out by steady-state simulators such as ASPEN PLUS, HySys, ProII, Chemcad, and similar programs.

The preceding chapters of this book have presented pencil-and-paper methods of studying process dynamic behavior, and of designing and tuning control systems. However, the methods presented have required considerable algebraic manipulation to obtain the results—linearization, Laplace transforms, combination of simultaneous equations, etc. Also, little has been required in terms of analyzing the final response of the process and the control system. In contrast, dynamic simulation cannot only produce the same results with little or no algebraic manipulations, but it can also produce the time responses of the control systems to various inputs. Nevertheless, the analytical procedures presented so far are most useful for understanding the concepts of stability, the root causes of instability, the need for dynamic compensators and how to design them, and the like.

As pointed out in the list above, dynamic simulation is an excellent tool for training operators in the operation of highly integrated processes and their control systems. This practice, modeled after the training of astronauts for space travel, has been practiced in industry for at least four decades. Such training is sometimes carried out with a simulator interfaced to a control console that looks and feels like the console the operators use to control the real process.

13-1.2   Tools for Dynamic Simulation

Before the personal computer revolution, dynamic simulation required considerable programming skills from the modeler. Initially simulation was carried out in analog computers. The differential equations were solved by wiring electrical components that were analogous to the physical systems being simulated. Not only was the required wiring of electrical components cumbersome, but the requirement of scaling the problem variables to the set ranges of the simulator was a skill difficult to master. However, analog simulation had the advantage of graphically connecting components in a way that matched the way that control system components are connected, as, for example, in block diagrams.

When analog simulation was replaced in the 1970s and 1980s by digital computer programs, the set of skills required from the modeler changed, but the degree of difficulty was not much reduced. Procedural languages such as FORTRAN, BASIC, Pascal, and C were required to put together and solve the differential equations representing the dynamics of the system. Fortunately, a number of numerical integration routines became available that eliminated the need for the modeler to develop his or her own.

Currently there are a good number of computer programs that make dynamic simulation extremely accessible to modelers with little or no programming skills. A list of the most commonly used programs is given in Table 13-1.1. The ones that are most useful have the following features:

Table 13-1.1 Dynamic simulation programs

Program	Company	Web page	Features
MATLAB/ Simulink	MathWorks	www.mathworks.com	Procedural language, block-oriented programming
VisSim	Visual Solutions, Inc.	www.vissim.com	Block-oriented programming, included with some versions of MathCad
TK Solver	UTS	www.uts.com	Equation-formatted notation, handles units conversion, symbolic manipulation
MathCad	MathSoft	www.mathsoft.com	Equation-formatted notation, handles units conversion, symbolic manipulation
Mathematica	Wolfram Research, Inc.	www.wolfram.com	Equation-formatted notation, handles units conversion, symbolic manipulation
LabVIEW	National Instruments	www.ni.com	Block-oriented data collection and control program that can be used for simulation
Maple	Waterloo Maple, Inc.	www.maplesoft.com	Command language includes integration and plotting subprograms
Aspen Plus Dynamic	AspenTech	www.aspentech.com	Provides unit operation modules and thermodynamic property packages
HySys	AspenTech	www.aspentech.com	Provides unit operation modules and thermodynamic property packages

- Graphical interface for block-oriented programming.
- Blocks to generate input signals (sources or signal generators) such as steps, ramps, and sine waves.
- Time plots of variables.
- Blocks to simulate the most common control devices—transportation lag, transfer functions, signal limiters, sample-and-hold, etc.
- Ability to develop custom simulation blocks using a procedural language.

The examples in this chapter use MATLAB with Simulink because it is one of the most powerful dynamic simulation programs that is readily available to students and engineers. The web page accompanying this book also has VisSim versions of most simulations. VisSim is currently available as part of MathCad.

As pointed out in the introduction to this chapter, there are two basic methods to carry out the dynamic simulation of a system. The next section presents the simulation of linear systems by simulating the transfer functions of their components.

The rest of the sections of this chapter present the simulation of models developed from basic principles.

13-2   SIMULATION OF LINEAR TRANSFER FUNCTIONS

In Chapters 2, 3, 4, and 5 we learned to develop transfer functions of processes, controllers, control valves, and sensor-transmitters; we did this by application of the Laplace transform to linear dynamic models. Processes represented by nonlinear differential equations required linearization, so that the transfer functions were valid only in a narrow range of operating conditions over which the linear approximations were valid. When such transfer function models are available, the transfer functions can be simulated to study the response of the control system to various inputs. The following example shows how to carry out such a simulation.

EXAMPLE 13-2.1

Simulation of Continuous Stirred Tank Heater

In Example 6-1.1 we developed the transfer functions for the feedback control loop to control the temperature of a continuous stirred tank heater. Simulate the loop using the transfer functions developed there. Then use the simulation to verify the ultimate gain and period of the loop determined in Example 6-2.2. Check also the response of a proportional controller tuned for quarter decay ratio response to step change in process flow.

SOLUTION

Figure 13-2.1 shows the block diagram for the simulation of this problem in MATLAB with Simulink. The blocks used are as follows:

- Step change blocks are used for the inputs: the set point *SP* and the process flow *F*.
- Gain blocks are used for the proportional controller K_c and the set point gain K_{sp}.

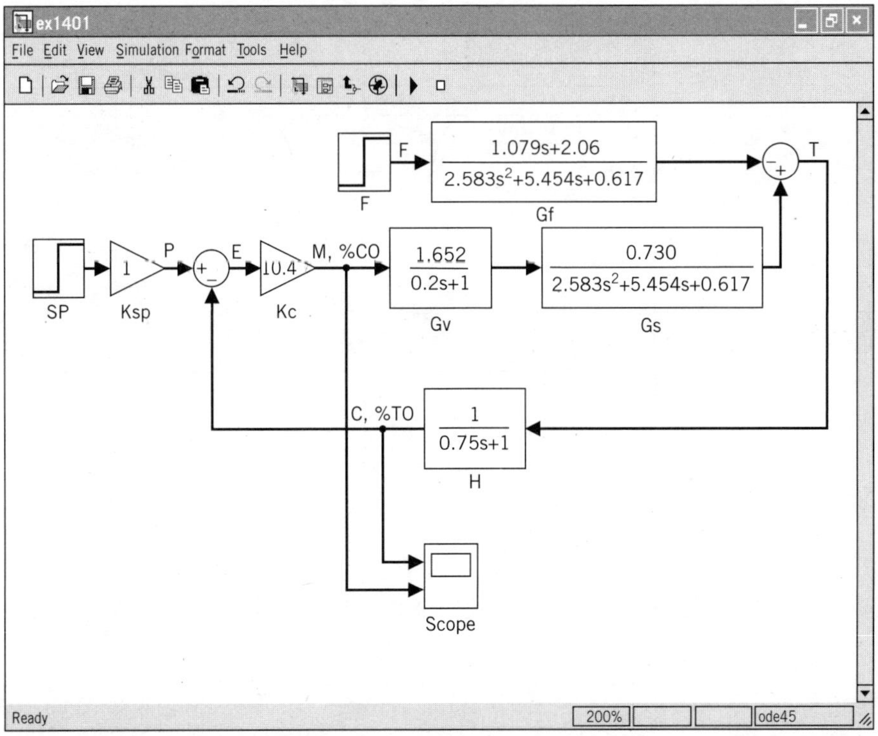

Figure 13-2.1 Simulink diagram for simulation of stirred tank heater using transfer functions.

- Summers are used for the error computation E and the sum of the process flow and steam flow effects on the outlet temperature T.
- Transfer function blocks are used for the valve G_v, the sensor-transmitter H, and the process transfer functions G_F and G_s.
- A scope block is used to plot the temperature response versus time.

The values of the transfer functions are obtained from the results of Example 6-1.1. Most of the numerical values are from Table 6-1.1, except for the transfer functions G_F and G_s that must be computed into the form required by the Simulink block. These calculations, using values from Table 6-1.1, are

$$G_F(s) = \frac{K_F(\tau_c s + 1)}{(\tau s + 1)(\tau_c s + 1) - K_s} = \frac{2.06(0.524s + 1)}{(4.93s + 1)(0.524s + 1) - 0.383}$$

$$= \frac{1.079s + 2.06}{2.583s^2 + 5.454s + 0.617}$$

$$G_s(s) = \frac{K_w K_s}{(\tau s + 1)(\tau_c s + 1) - K_s} = \frac{1.905 \times 0.383}{(4.93s + 1)(0.524s + 1) - 0.383}$$

$$= \frac{0.730}{2.583s^2 + 5.454s + 0.617}$$

To obtain the ultimate gain and period the controller gain is set and the simulation is run, checking the response in the scope after each run. The gain is then increased until the response is a sustained oscillation. When sustained oscillations are obtained, the controller gain is the ultimate gain and the ultimate period is the period of the response. Figure 13-2.2 shows the response of the transmitter and controller outputs when the controller gain is set at 10.4 %CO/%TO. From the figure we see that three complete oscillations occur in 13.9 min, for a period of

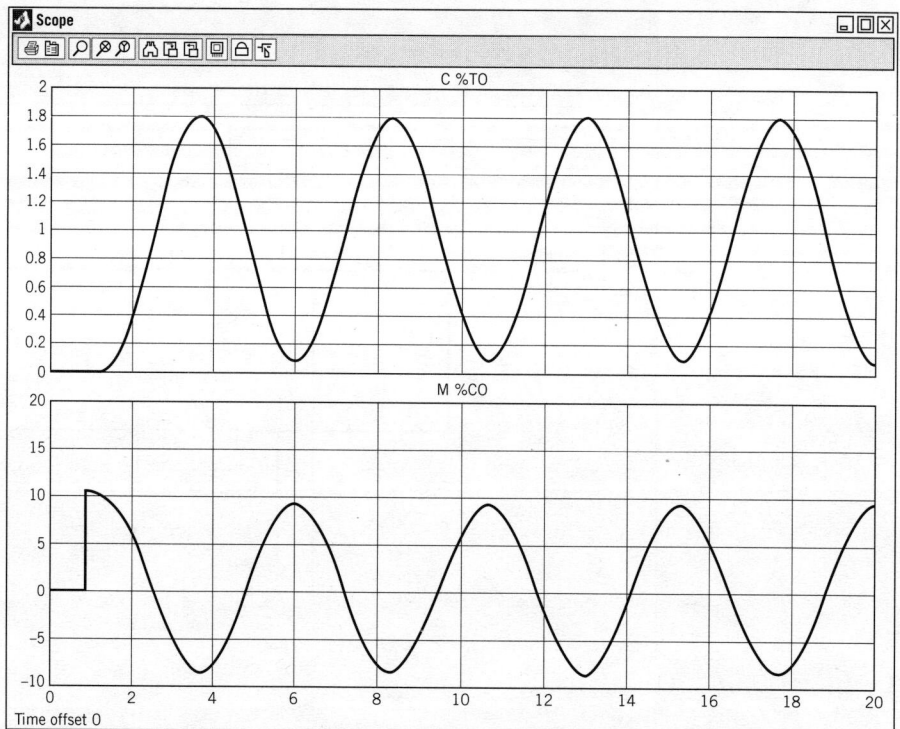

Figure 13-2.2 Response of transmitter and controller outputs for the ultimate controller gain.

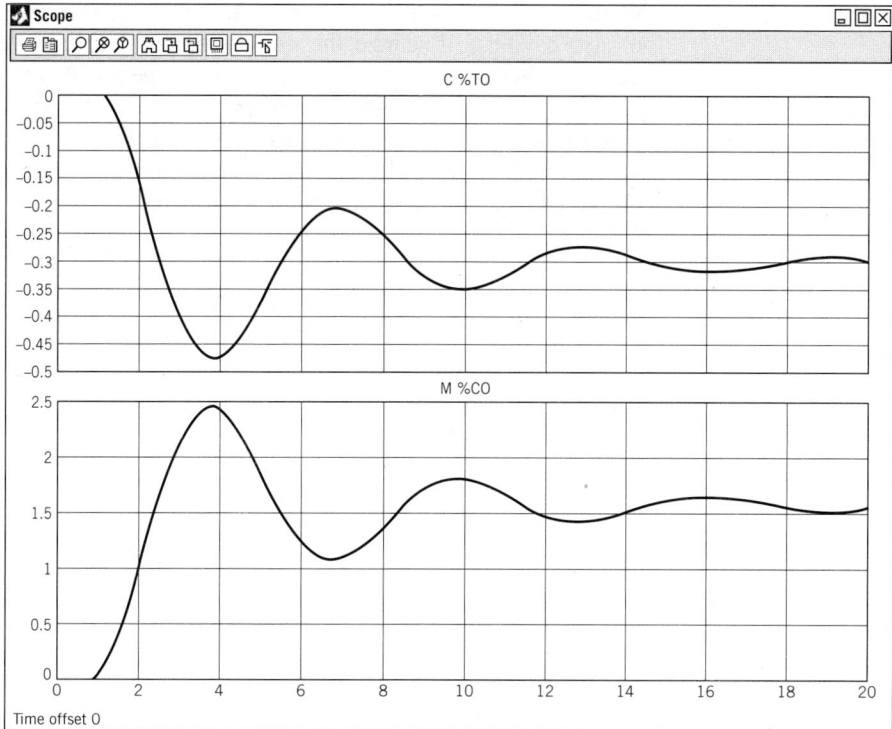

Figure 13-2.3 Response of transmitter and controller outputs to a step change in process flow.

13.9/3 = 4.6 min. These are the values obtained by direct substitution in Example 6-2.2. The responses to a step change in process flow of 1.0 lb/min are shown in Fig. 13-2.3. The controller gain is set to 10.4/2 = 5.2 %CO/%TO, as indicated by the quarter decay ratio tuning formulas of Table 7-1.1 for a proportional controller. Such responses, easy to obtain from the simulation, require significant algebraic manipulations to obtain by analytical means.

Comparison of the Simulink diagram of Fig. 13-2.1 to the block diagram of Fig. 6-1.7 shows a one-to-one correspondence of the blocks. This is one of the major advantages of block-oriented simulation for studying process control systems.

The initial conditions of all the variables in the diagram are zero because, being transfer functions, the variables are deviations from initial steady-state conditions (see Section 2-3.1 and Example 6-1.1).

Just as the simulation of Example 13-2.1 was used to tune the controller by the ultimate gain method, it could have been used to determine the unit step response of the process. To do this, the controller is simply replaced by a step input block.

Having demonstrated how to simulate linear transfer function models of processes, the next section presents a more fundamental method of simulating differential equation models of the process.

13-3 PROCESS SIMULATION

An alternative way to simulate the process to the one presented in the preceding section is to numerically solve the differential equations that represent the model. In this manner we not only save having to develop the transfer function of the process, but also having to linearize the process model if it happens to be nonlinear. In fact both linear and nonlinear process models can be simulated with equal ease.

Today's simulation software offers a variety of very powerful methods for the numerical solution of differential equations. Further, it is completely transparent to the user of the software which method is used. All that is needed is a model in which each differential equation is in first-order form, such as:

$$\frac{dx(t)}{dt} = f[x(t), u(t), t] \tag{13-3.1}$$

where

$x(t) =$ state variable

$u(t) =$ input variable

$t =$ time

and f is any linear or nonlinear function of the variables. Fortunately, process models developed from basic principles are exactly in the form of Eq. 13-3.1, as you can easily verify by examining all of the models developed in Chapters 3 and 4.

Block-diagram oriented simulation software provides an integrator block designed to numerically integrate equations in the form of Eq. 13-3.1. In addition to the input derivative, the integrator block requires the initial condition of the state variable x. The symbol $1/s$ is used in Simulink and VisSim to indicate an integrator block. This does not mean that the block uses Laplace transforms to carry out the integration. It is only a convenient label for the block.

The following examples show how to simulate processes using MATLAB with Simulink.

EXAMPLE 13-3.1

Simulation of Gas Process

This is the gas process modeled in Section 3-6 to obtain a response of the pressure of the gas in the tank to a step input in the signal to the fan pushing the gas into the tank. The model developed in Section 3-6 is Eq. 3-6.10:

$$\overline{\rho} f_i(t) - \overline{\rho} f_o(t) = \frac{V}{RT} \frac{dp(t)}{dt}$$

$$f_i(t) = 0.16 m_i(t)$$

$$f_o(t) = 0.00506 m_o(t)\sqrt{p(t)[p(t) - p_1(t)]}$$

where

$m_i(t) =$ signal to the input fan, %

$m_o(t) =$ signal to outlet valve, %

$p(t) =$ pressure in the tank, psia

$p_1(t) =$ pressure downstream of valve, psia

$\overline{\rho} =$ gas density, 0.00263 lbmole/scf

$V =$ tank volume, 20 ft^3

$R =$ ideal gas law constant, 10.73 psia-ft^3/lbmole-°R

$T =$ gas temperature, 520°R

and the constants are in units of scf/min-%. The tank model consists of one differential equation and two algebraic equations. The initial condition of the pressure, $p(0)$, is needed to define the state of the tank, so the pressure is said to be the *state variable*. There are three inputs or forcing functions, the two signals $m_i(t)$ and $m_o(t)$, and the downstream pressure $p_1(t)$.

For a simulation to properly represent the response of the process to the input signals it is necessary that the initial conditions be at steady state. This steady state is usually obtained from the process design conditions. However, we must realize that the steady-state requirement imposes a restriction on the values the design conditions can take. Of the design conditions given in Section 3-6, the minimum required specifications are

$$m_i(0) = \overline{m}_i = 50\ \% \qquad m_o(0) = \overline{m}_o = 50\ \% \qquad p_1(0) = \overline{p}_1 = 14.7\ \text{psia}$$

Using the three model equations, with $dp(0)/dt = 0$ at steady state, the other three initial conditions can be determined.

$$f_i(0) = 0.16\,m_i(0) = (0.16)(50) = 8.0\ \text{scf/min}$$
$$f_o(0) = f_i(0) = 8.0\ \text{scf/min}$$
$$f_o(0) = 0.00506\,m_o(0)\sqrt{p(0)[p(0) - p_1(0)]}$$
$$\therefore p(0) = 39.8\ \text{psia}$$

Having the initial conditions, the differential equation is solved for the derivative as follows:

$$\frac{dp(t)}{dt} = \frac{\overline{\rho}RT}{V}[f_i(t) - f_o(t)]$$
$$= \frac{(0.00263)(10.73)(520)}{20}[f_i(t) - f_o(t)]$$
$$= 0.734[f_i(t) - f_o(t)]$$

Figure 13-3.1 shows the Simulink block diagram used to simulate the gas process. Step input blocks are used to generate the three input signals, and multiplication, addition, and a square-root function block are used to carry out the calculation of the outlet flow. Gain blocks are used to introduce the constant coefficients in the model, and an integrator block integrates

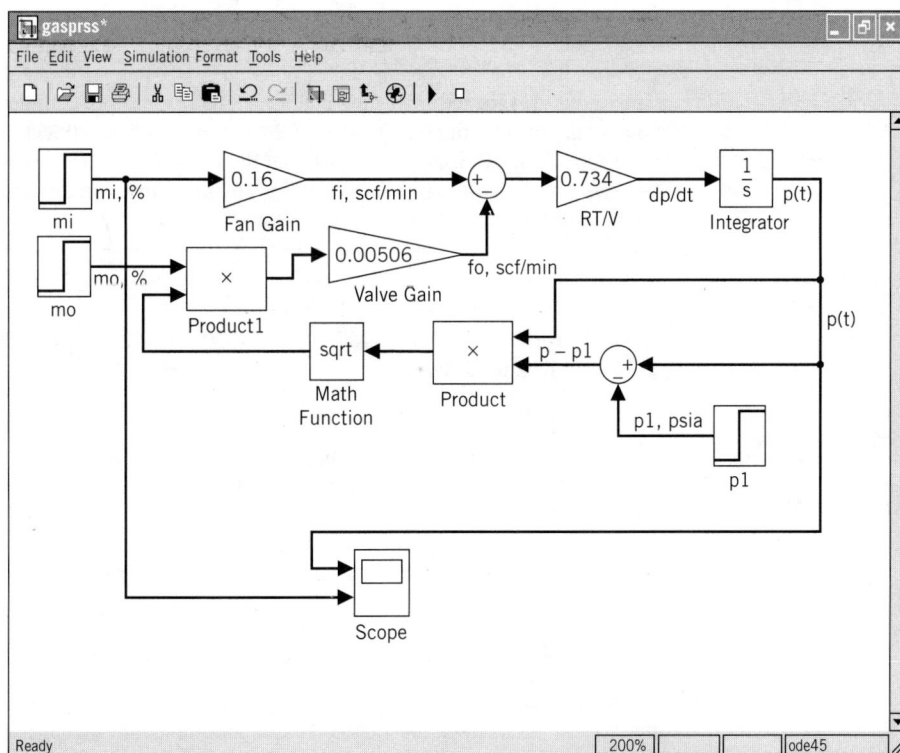

Figure 13-3.1 Simulink diagram for the simulation of the gas process.

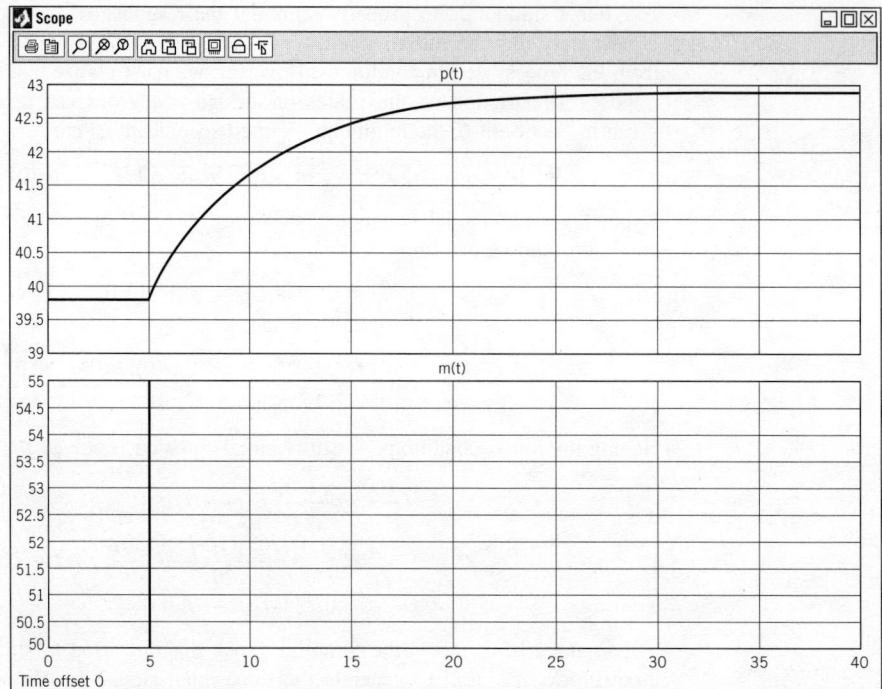

Figure 13-3.2 Response of gas pressure to step change in the signal to the fan (inlet flow).

the differential equation to obtain the pressure in the tank. The initial condition on the integrator is set to 39.8 psia.

Figure 13-3.2 shows the response of the pressure to a step change of 5 % in the signal to the inlet fan. The step change takes place at $t = 5$ min, to show that the pressure is indeed at steady state before the inlet flow changes. The response is complete in about 25 min, which is consistent with the time constant of 5.24 min determined in Section 3-6.

You might ask what determines that the units of time are minutes. The fact that the units of $dp(t)/dt$ are psia/min, is what determines that the time variable is in minutes, as the integration with respect to time gives the tank pressure in psia.

The following example shows how to simulate a process represented by multiple differential equations.

EXAMPLE 13-3.2

Simulation of Stirred-Tank Heater

This is the steam heater of Example 6-1.1, the same heater that was simulated from the linear transfer functions in Example 13-2.1. Here we will simulate it from the basic differential equations derived in Example 6-1.1.

$$\frac{dT(t)}{dt} = \frac{1}{V}f(t)[T_i(t) - T(t)] + \frac{UA}{V\rho c_v}[T_s(t) - T(t)]$$

$$\frac{dT_s(t)}{dt} = \frac{1}{C_M}\{\lambda w(t) - UA[T_s(t) - T(t)]\}$$

where we have solved for the derivatives. The explanation of the variables and the values of the system parameters and of the design conditions are given in Example 6-1.1. There it was determined that the initial steady-state conditions are $T(0) = 150°$F and $T_s(0) = 230°$F. Also the initial design conditions are $f(0) = 15$ ft³/min, $T_i(0) = 100°$F, and $w(0) = 42.2$ lb/min.

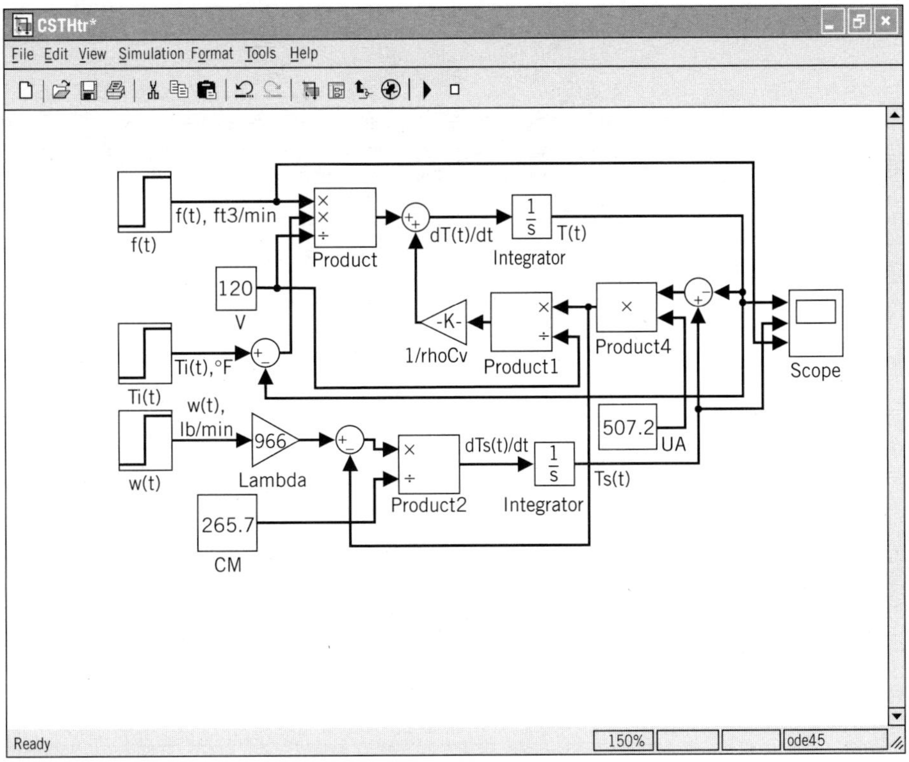

Figure 13-3.3 Simulink diagram for stirred tank heater.

Figure 13-3.3 shows the Simulink diagram for the simulation of the heater. Step input signal generators are used for the three inputs: the process and steam flows, and the inlet temperature. Two integrators ($1/s$) are used to integrate the two differential equations and arithmetic blocks—summers, multipliers, gains, and constant blocks are used to calculate the derivative functions from the equations given above. A scope is used to plot the response of the input signal and the two temperatures.

Figure 13-3.4 shows the responses of the temperatures to a step change in process flow. The process gain can be determined from its definition $(136.9 - 150)°\text{F}/(16 - 15)(\text{ft}^3/\text{min}) = -3.1°\text{F}/(\text{ft}^3/\text{min})$. This compares with the analytical result obtained in Example 6-1.1, $K_F/(1 - K_s) = [-2.06°\text{F}/(\text{ft}^3/\text{min})]/(1 - 0.383) = -3.3°\text{F}/(\text{ft}^3/\text{min})$. The slight difference is due to the linear approximation.

EXAMPLE 13-3.3

Simulation of a Batch Bioreactor

Many important specialty chemical products are produced in bioreactors by processes such as fermentation. Most of these processes are carried out in batch mode by filling a tank with a substrate solution and inoculating it with a small amount of biomass. The biomass, feeding on the substrate, reproduces to produce the desired product, until the substrate is consumed. This example is presented here to show some of the special characteristics of biochemical processes.

A dynamic model of the growth of the biomass concentration $x(t)$ and of the consumption of the substrate concentration, $s(t)$, is given on a per unit volume basis as follows:

$$\frac{dx(t)}{dt} = \mu(t)x(t)$$

$$\frac{ds(t)}{dt} = -\frac{1}{y}\mu(t)x(t)$$

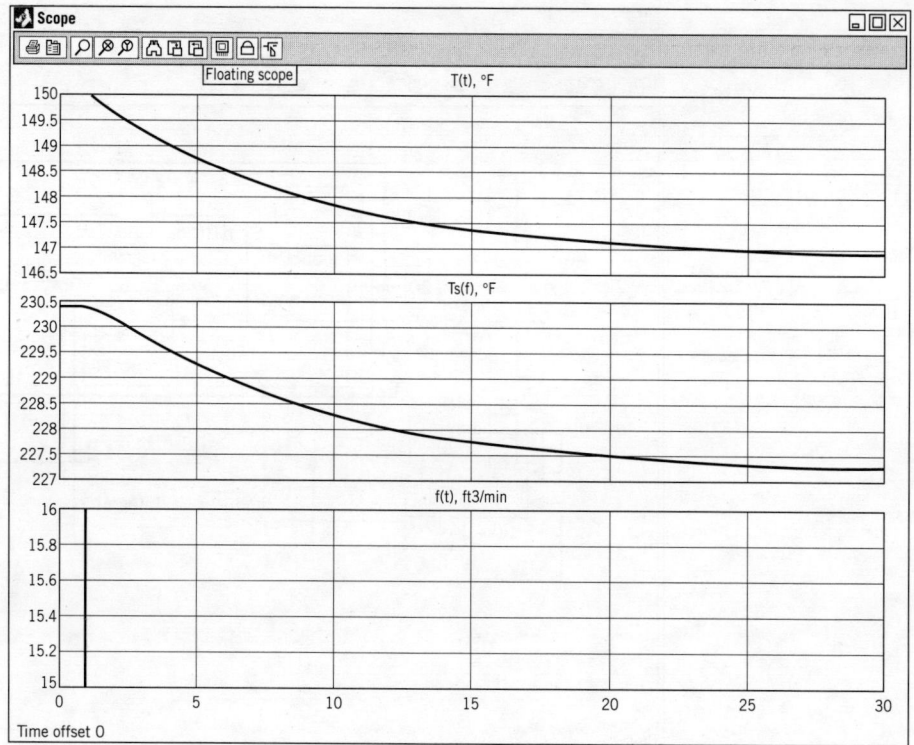

Figure 13-3.4 Response of heater outlet temperature and steam chest temperature to a step change in process flow.

where y is the yield in biomass per unit mass of substrate and $\mu(t)$ is the biomass growth rate function, usually in h^{-1}. This growth rate function is analogous to the kinetic models used to model chemical reactors. It is designed to match experimental reactor data. Here we will use the Monod model with adaptability which has the following form:

$$\frac{d\mu(t)}{dt} = \alpha \left[\mu_m \frac{s(t)}{k + s(t)} - \mu(t) \right]$$

where α is the adaptability parameter, and k and μ_m are the parameters of the model. These three differential equations are solved in the Simulink diagram of Fig. 13-3.5 with $\alpha = 15$ h^{-1}, $k = 0.5$ g/liter, $s(0) = 2.5$ g/liter, $\mu(0) = \mu_m = 1.2$ h^{-1} and $x(0) = 0.001$ g/liter.

Figure 13-3.6 shows a plot of the biomass and substrate concentrations with time in hours. It takes about 3 h for the growth rate of the biomass to become noticeable, but it then takes off exponentially while the substrate concentration plot is a mirror image of the biomass plot. Eventually, about 10 h from the start, the substrate has been converted into biomass, the growth stops, and the batch is complete.

Biomass reactors have been the subject of numerous control studies that have been reported in the literature.

The preceding examples showed how to simulate dynamic models consisting of multiple differential equations. Each derivative function is calculated from the state and input variables and fed to a separate integrator block.

A convenient feature of Simulink and other block-oriented simulation programs is the ability to combine the blocks for the simulation of a system into a *subsystem*. In

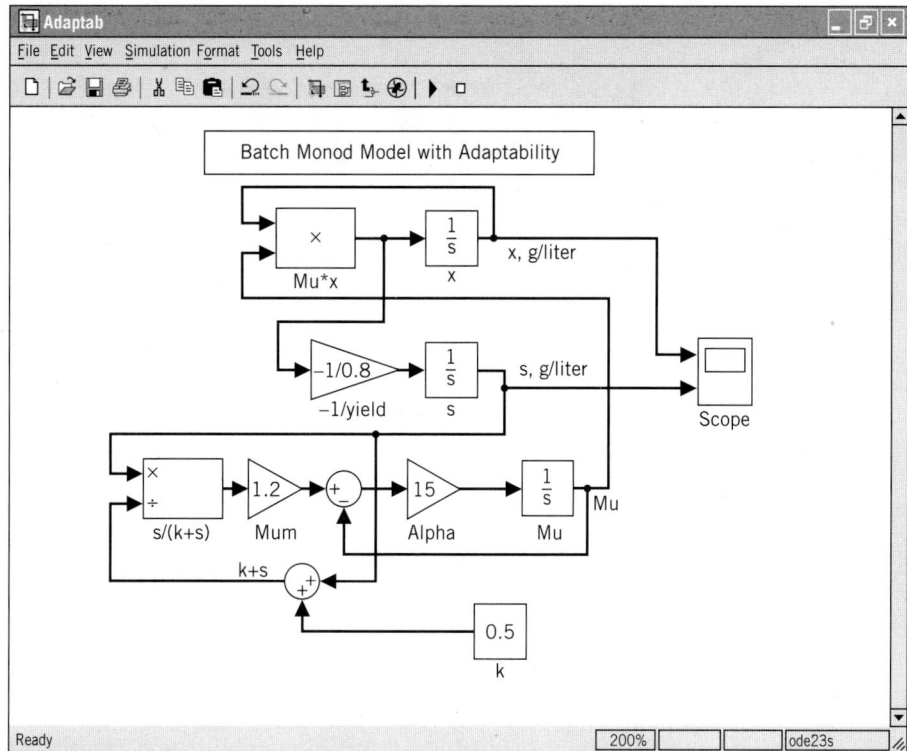

Figure 13-3.5 Simulink diagram for batch bioreactor.

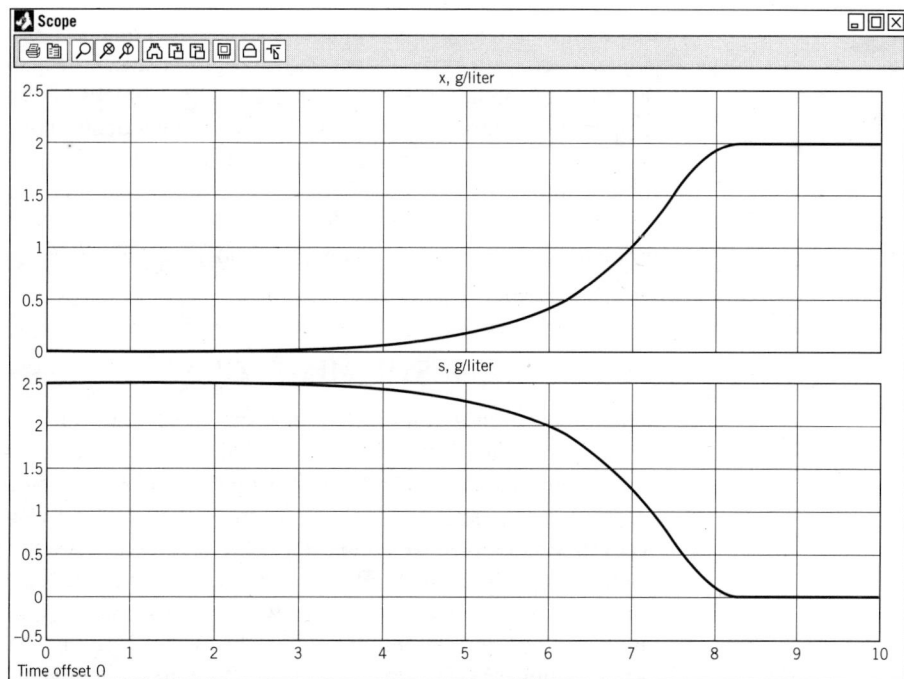

Figure 13-3.6 Plot of biomass and substrate concentrations versus time in hours.

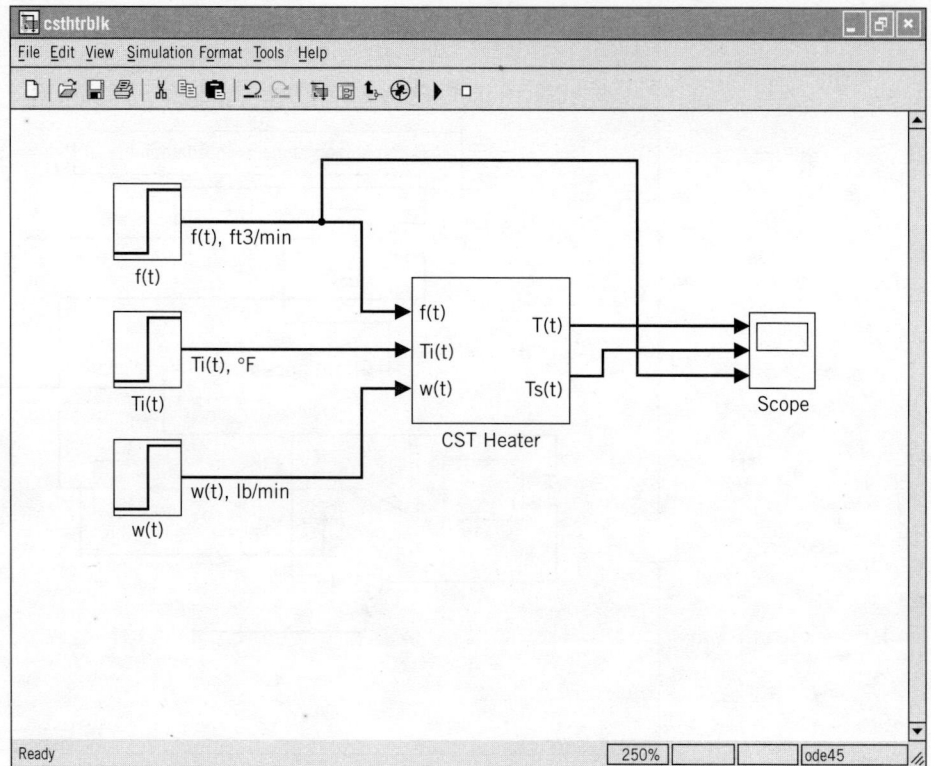

Figure 13-3.7 Subsystem block for the continuous stirred tank heater.

this manner the simulation of the heater shown in Fig. 13-3.3 can be represented as a single block with three inputs and two outputs, as in Fig. 13-3.7. If we now replace the block of the heater for the transfer function blocks of Fig. 13-2.1, we can study the temperature control system with the more fundamental nonlinear model of the heater. However, this would present a problem: the transfer function models of the valve and the transmitter in Fig. 13-2.1 have zero initial conditions, but we need nonzero initial conditions of the steam flow for the fundamental model of the heater. To resolve this problem we need more fundamental models of the instrumentation. This is the subject of the next section.

13-4   SIMULATION OF CONTROL INSTRUMENTATION

When simulating a control system using linear transfer functions, as in Section 13-2, the instrumentation can also be simulated using linear transfer functions. Chapters 5 and 11 presented the linear transfer functions for sensor-transmitters, control valves, feedback controllers, and lead-lag units. However, if we would limit ourselves to these types of simulations, we would not be able to study by simulation the effect of process nonlinearities. Such study requires simulating the process from the original differential equations, as in Section 13-3. As we saw there, an important characteristic of these simulations is that the initial conditions are not zero and must be carefully taken into consideration. As simple transfer function models do not allow for nonzero initial conditions, we must develop special simulation models of the various instruments. This is the subject of this section.

Most of the instrument models to be developed here will require the simulation of a first-order lag with a nonzero initial condition. The transfer function of a first-order lag with unity gain is

$$\frac{Y(s)}{X(s)} = \frac{1}{\tau s + 1} \tag{13-4.1}$$

Where $Y(s)$ is the output signal. $X(s)$ is the input signal, and τ is the time constant. To model the response of the output, the transfer function is rearranged thus:

$$(\tau s + 1)Y(s) = X(s)$$
$$\tau s Y(s) = X(s) - Y(s)$$
$$Y(s) = \frac{1}{s}\left\{\frac{1}{\tau}[X(s) - Y(s)]\right\} \tag{13-4.2}$$

Equation 13-4.2 is simulated by feeding an integrator block ($1/s$) the difference between the input and the output signals divided by the time constant. To start at steady state, the initial condition of the integrator must be set to the initial value of the input. This model of a first-order lag will appear in every instrument model presented in this section.

13-4.1   Control Valve Simulation

Figure 13-4.1 shows the Simulink block diagram for the simulation of a control valve with constant pressure drop. This diagram assumes that the valve model is a subsystem

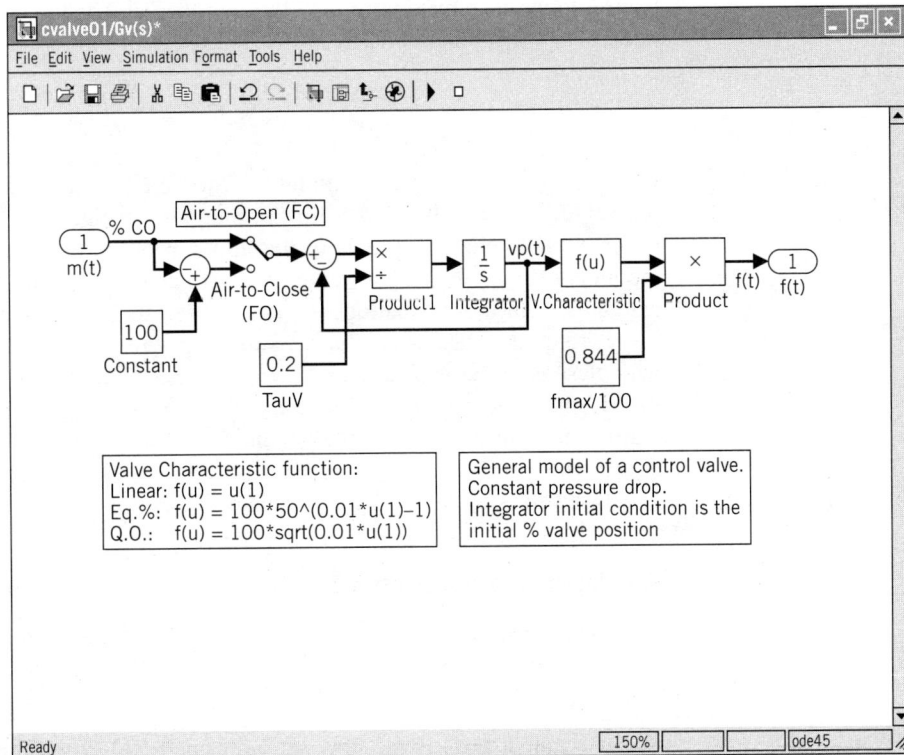

Figure 13-4.1 Simulink block diagram for a general control valve with constant pressure drop.

with an input $m(t)$, the controller output in % CO, and an output $f(t)$, the flow through the valve in the most convenient units.

The first item in the valve model is a switch to select the valve as air-to-open (fail closed), in which case the signal $m(t)$ is fed through, or air-to-close (fail opened), in which case the signal is changed to $[100 - m(t)]$% CO. Notice that the controller output signal $m(t)$ varies in the range 0 to 100 % CO and is not a deviation variable. Therefore, changing the sign of the valve gain is not sufficient to change its action, because that would only work on deviation variables.

The next part of the valve model is the first-order lag with the valve time constant τ_v. This is simply solving Eq. 13-4.2. The output of the first-order lag is the valve position $vp(t)$, in % VP. To start at steady state the initial condition of the integrator must be set to the initial value of the input. Notice that this could be $m(0)$ or $100 - m(0)$, depending on the action of the valve. If the valve time constant is negligible, the entire simulation of the first-order lag must be deleted or bypassed, because the time constant cannot be set to zero.

The valve dynamic term is followed by a general purpose "Math Function" block to enter the valve characteristics function. Such a block has, in this case, a single input $u(1)$, and allows any calculation using this input. To simulate a linear valve, enter simply $u(1)$, and for an equal percentage valve the statement $100 * \alpha^{\wedge} (0.01 * u(1) - 1)$ is required, where α is the rangeability parameter of the valve, for example, 50 (see Eq. 5-2.7). A quick-opening valve could be simulated by the square root function.

Finally, the capacity of the valve is entered as a constant multiplier, $f_{max}/100$, and the 100 because the valve position is in % VP. When the pressure drop across the valve is not affected by the simulation variables, the maximum flow through the valve is a constant. This is usually the case in the simulation of temperature or composition control loops. The maximum flow through the valve is the flow when the valve is fully opened. For a liquid valve, from Eq. 5-2.1,

$$f_{max} = C_{v,max}\sqrt{\frac{\Delta p_v}{G_f}}$$

(13-4.3)

where f_{max} is in gallons per minute. However, for the purposes of the simulation of the valve, f_{max} can be entered in the diagram of Fig. 13-4.1 in any convenient units (lb/hr, ft^3/min, etc.). The units of the flow out of the model, $f(t)$, will be the same as the units of f_{max}. For a gas valve, f_{max} are calculated from either Eq. 5-2.3 or 5-2.4, as required. In these equations, $C_{v,max}$ is used for C_v to obtain the maximum flow.

For liquid valves, when the pressure drop across the valve is affected by the process variables, as in the simulation of level or pressure control loops, the block diagram of Fig. 13-4.2 should be used. The only differences with the diagram of Fig. 13-4.1 are that the flow through the valve is calculated using Eq. 5-2.1, and an additional input is required for the pressure drop across the valve Δp_v. A similar diagram could be developed for simulating a gas or steam valve.

Simulation of Flow Control Loops

The diagram of Fig. 13-4.1 can be used to simulate a flow control loop when the pressure drop across the valve is not affected by the process variables. When this is done,

- the input $m(t)$ is the flow set point in % TO of the flow transmitter,
- the switch is in the air-to-open position,

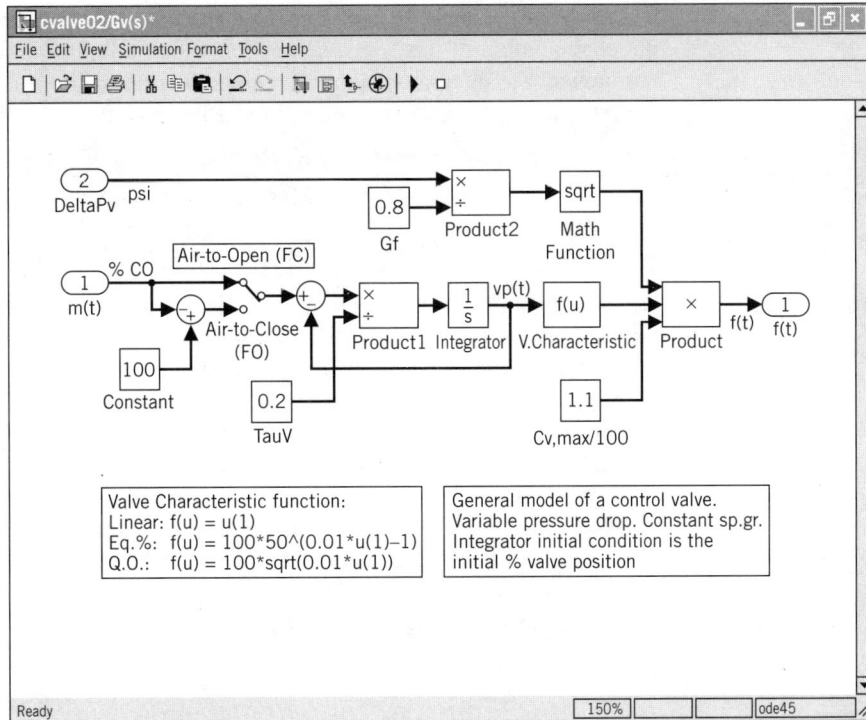

Figure 13-4.2 Simulink block diagram for a control valve with variable pressure drop and constant specific gravity.

- the time constant τ_v is the closed-loop time constant of the flow loop,
- the valve characteristic function is usually linear, and
- f_{max} is the upper limit of the flow transmitter in the loop.

This way it is not necessary to simulate and tune the flow controller. If, on the other hand, the pressure drop across the valve is a function of the process variables, the diagram of Fig. 13-4.2 must be used to simulate the valve in a loop that will include a flow controller and a flow transmitter.

13-4.2 Simulation of Feedback Controllers

This subsection develops a model of a proportional-integral-derivative (PID) controller that provides for nonzero initial conditions, direct limits of 0 and 100 % in the controller output, and a dynamic gain limit on the derivative mode. This model is highly recommended when there is a possibility that the controller output may saturate. The model will be presented in two parts, the proportional-integral (PI) and the proportional-derivative (PD). Then the two parts will be combined to simulate a series PID controller. A model of a parallel PID controller is also presented.

Proportional-Integral (PI) Controller

The transfer function of a PI controller was given by Eq. 5-3.14,

$$G_c(s) = \frac{M(s)}{E(s)} = K_c\left(1 + \frac{1}{\tau_I s}\right) = K_c\left(\frac{\tau_I s + 1}{\tau_I s}\right) \tag{5-3.14}$$

where K_c is the proportional gain and τ_I is the integral time. The problem with the direct simulation of this transfer function is that a limiter on the controller output does not prevent the integrator output from winding up, while it is not easy to directly limit the integrator. A more convenient model is developed as follows:

$$\tau_I s M(s) = K_c E(s)(\tau_I s + 1)$$

$$\tau_I s M(s) + M(s) = K_c E(s)(\tau_I s + 1) + M(s)$$

$$M(s) = K_c E(s) + \frac{1}{\tau_I s + 1} M(s) \tag{13-4.4}$$

Equation 13-4.4 is programmed in the Simulink block diagram of Fig. 13-4.3. The diagram shows that the controller output $m(t)$ is the sum of the terms $K_c E$ and the output lagged by a first-order time constant τ_I. A limiter with limits 0 and 100 % CO is inserted before the output is fed back to the calculation. Thus there is no way for any of the variables in the loop to go outside those limits. To start at steady state, the integrator initial condition must be the desired initial value of the controller output, $m(0)$, counting on the initial error to be zero. The only undesirable feature of the PI model given here is that it is not possible to combine it with a derivative term to simulate a parallel PID controller.

Proportional-Derivative (PD) Unit

We learned in Section 5-3 that the derivative mode requires a filter in practice:

$$\frac{U(s)}{X(s)} = \frac{\tau_D s}{\alpha \tau_D s + 1} \tag{13-4.5}$$

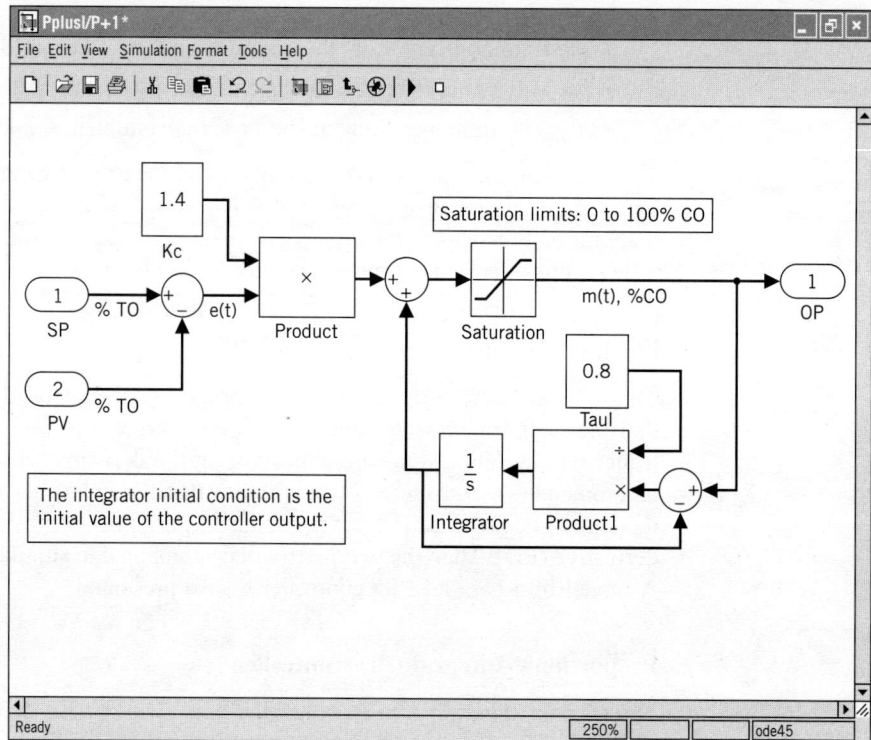

Figure 13-4.3 Simulink block diagram for proportional-integral (PI) controller with saturation limits.

where $U(s)$ is the output, $X(s)$ is the input, and τ_D is the derivative time. The filter consists of a first-order lag with a time constant $\alpha\tau_D$. The value of α is selected small enough—0.05 to 0.2—so that the filter does not affect the performance of the controller. The filter limits the output of the derivative mode on a sudden input, such as a step change, the limit being equal to $1/\alpha$ times the magnitude of the step. This is why the filter is also called the *dynamic gain limit*. The model of the filter derivative of Eq. 13-4.5 is developed thus:

$$\alpha\tau_D s U(s) + U(s) = \tau_D s X(s)$$

$$U(s) = \frac{1}{\alpha\tau_D s}(\tau_D s X(s) - U(s)) = \frac{1}{\alpha}\left(X(s) - \frac{1}{\tau_D s}U(s)\right) \tag{13-4.6}$$

This equation gives a filtered derivative mode. To obtain the PD unit we simply add the input variable.

$$Y(s) = X(s) + U(s) \tag{13-4.7}$$

Where $Y(s)$ is the output. Figure 13-4.4 shows the Simulink block diagram for the PD unit. The constant multiplier $1/\alpha$ is set equal to 8, which is then the dynamic gain limit. To prevent a bump at the start the integrator initial condition must be the initial value of the input $x(0)$.

Combining Eqs. 13-4.5 and 13-4.7, the transfer function for the PD unit is

$$\frac{Y(s)}{X(s)} = 1 + \frac{\tau_D s}{\alpha\tau_D s + 1} = \frac{(\alpha+1)\tau_D s + 1}{\alpha\tau_D s + 1} \tag{13-4.8}$$

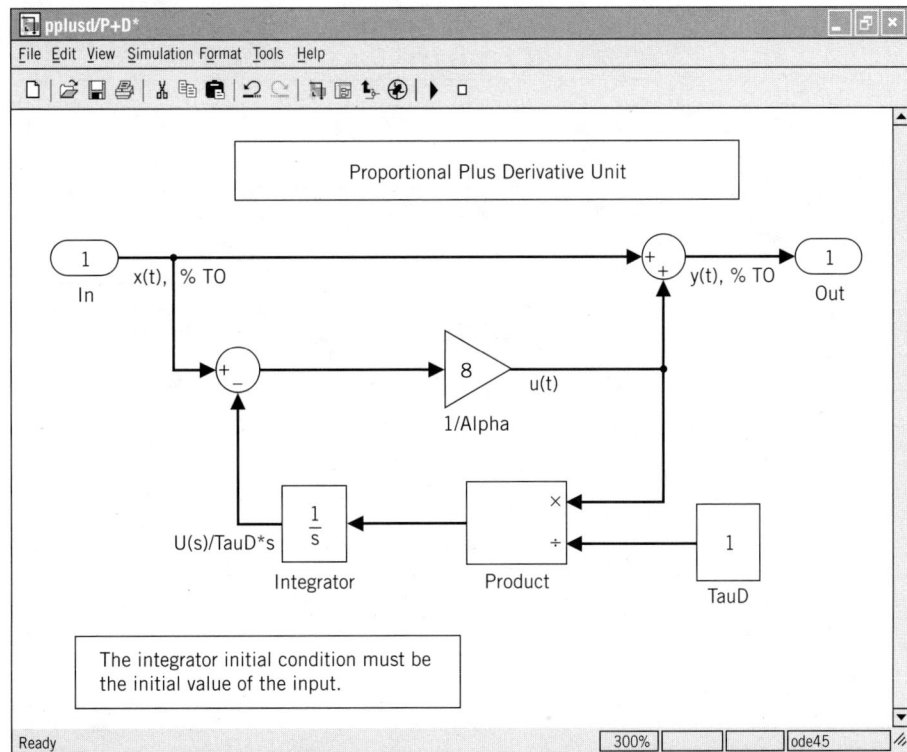

Figure 13-4.4 Simulink block diagram for proportional-plus-derivative (PD) unit. This also serves to introduce a net lead of TauD with an attenuation factor of Alpha.

We see that the PD unit is a lead-lag unit with a net lead of τ_D and a ratio of the lead to the lag of $(\alpha + 1)/\alpha$. This means that the diagram of Fig. 13-4.4 can also be used to insert a net lead with a negligible lag into other control schemes such as feedforward controllers and decouplers.

Simulation of a Series Proportional-Integral-Derivative (PID) Controller

The diagrams of Figs. 13-4.3 and 13-4.4 are Simulink subsystem blocks that can now be put together into a series PID controller as in Fig. 13-4.5. Notice that the PD unit is connected on the measurement input instead of on the error signal. This is the recommended arrangement. If derivative action on the error signal is desired, the PD block should be moved into the PI block of Fig. 13-4.3 and inserted just after the error calculation. The diagram of Fig. 13-4.5 is itself a Simulink subsystem that can be used to simulate the PID controllers in the simulation of a control system.

The diagram of Fig. 13-4.5 includes an auto–manual switch to introduce a step change into the controller output. This is where the step change must be applied to carry out the process step test procedure described in Section 7-2.1. To obtain the step response the switch is put in the bottom position and the step function $M(t)$ is set up to introduce a step change. Notice that the initial value of the step function must match the initial value of the controller output, and its final value is the initial value plus the magnitude of the step change.

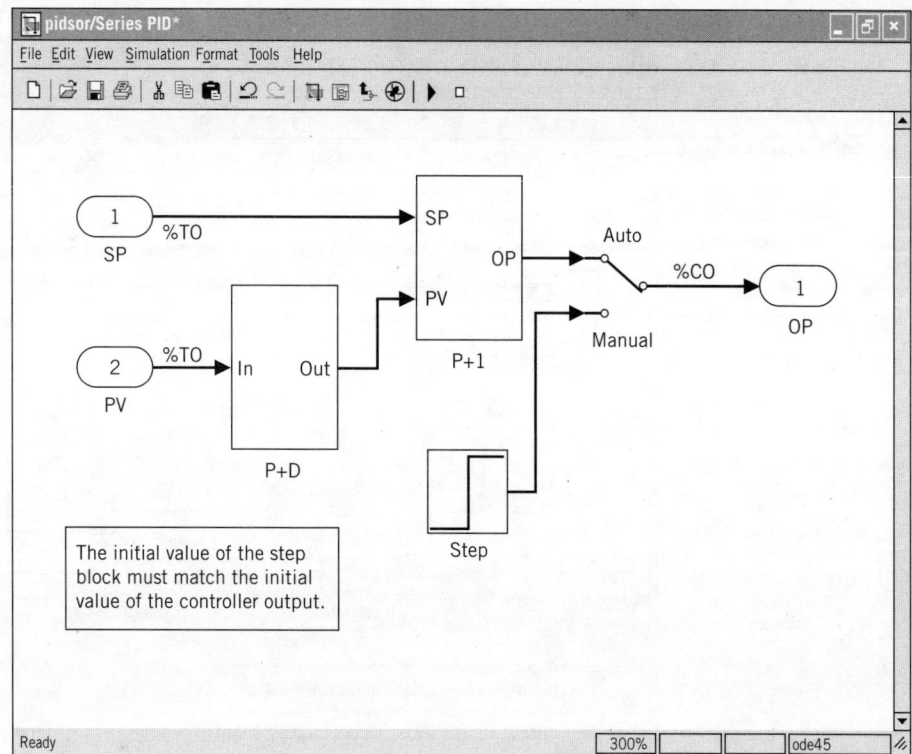

Figure 13-4.5 Block diagram for series PID controller with derivative on the measurement. The P + I and P + D blocks are subsystems containing the diagrams of Figs. 13-4.2 and 13-4.4, respectively. The auto–manual switch allows a step change in controller output.

Simulation of a Parallel Proportional plus Integral plus Derivative (P + I + D) Controller

Figure 13-4.6 shows a Simulink block diagram for a parallel proportional-integral-derivative (P + I + D) controller. This model was developed by adding the integral term to the PD block of Fig. 13-4.4, and it has the following features:

- The derivative term is filtered with a first-order lag with a time constant equal to $\tau_D/8$.
- The derivative term acts on the process variable and not on the error. This is the recommended arrangement.
- If the initial conditions are not zero, the diagram explains how to set the initial conditions for the two integrators.

The transfer function of the controller of Fig. 13-4.6 was presented in Chapter 5 as Eq. 5-3.17.

13-4.3 Simulation of Sensors-Transmitters

The Simulink block diagram of a sensor-transmitter model shown in Fig. 13-4.7 has two parts, a dynamic term consisting of first-order lag with time constant τ_T, and the change of scale from engineering units to %TO. Also shown in the model is the change of scale of the set point from engineering units to %TO, which is identical to that of

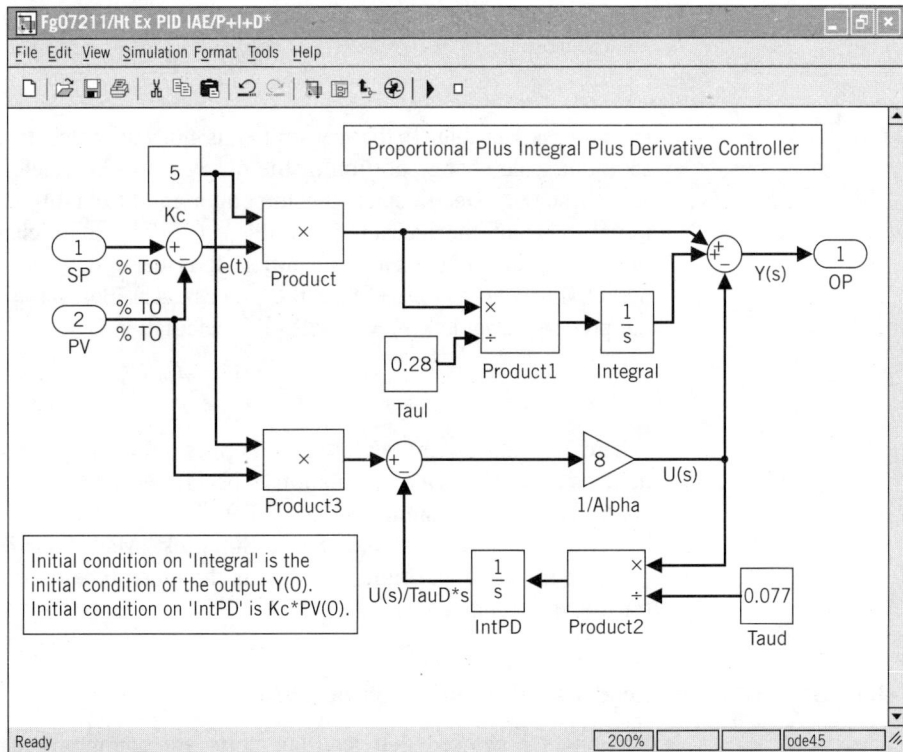

Figure 13-4.6 Simulation of parallel proportional-integral-derivative (P + I + D) controller with derivative on the measurement.

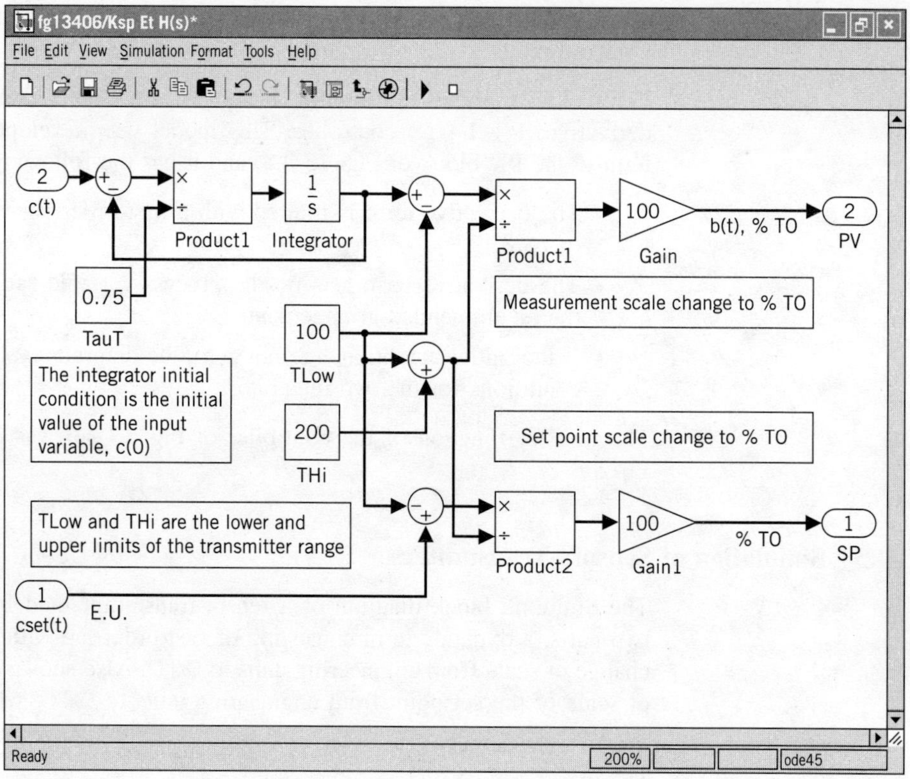

Figure 13-4.7 Simulink block diagram to simulate a sensor transmitter and change the set point scale.

the process variable. If the transmitter is not connected to a feedback controller, the set point scale change portion of the diagram can be deleted.

To start at steady state, the integrator initial condition must be the same as the initial value of the measured variable, $c(0)$. The scale change calculation shown in Fig. 13-4.7 is for a linear transmitter. It is derived by knowing that the transmitter output signal $b(t)$ is zero when the process variable $c(t) = T_{low}$, and 100 %TO when the process variable $c(t) = T_{hi}$. So the calculation is

$$b(t) = \frac{c(t) - T_{low}}{T_{hi} - T_{low}} 100 \text{ \%TO} \qquad \textbf{(13-4.9)}$$

Where T_{low} and T_{hi} are the lower and upper limits of the transmitter range, in the same units as $c(t)$. The same calculation is performed on the set point input to produce the controller reference signal $r(t)$ in %TO.

If the transmitter is nonlinear, a Simulink "Math Function" block must be used to program the nonlinear function. This is the block type used in Figs. 13-4.1 and 13-4.2 for the valve characteristics.

13-4.4 Simulation of Lead-Lag Dynamic Compensation

Section 11.3 showed that lead-lag units are commonly used to introduce dynamic compensation in feedforward control schemes. In many situations, the tuning of the lead-lag device can be simplified considerably by using either the required net lead without a lag or the required net lag without a lead. When this is the case, no special device is required to simulate the lead-lag unit. For example,

- A net lead with negligible lag can be simulated using the PD unit of Fig. 13-4.4 with $\tau_D = \tau_{lead} - \tau_{lag}$. If a nonnegligible lag is desired, then set $\alpha = \tau_{lag}/\tau_D$.
- A net lag without a lead can be simulated using a first-order lag with the time constant set to $\tau_{lag} - \tau_{lead}$.

The only situation not covered by the above two cases is when a net lag with a nonnegligible lead is desired.

Figure 13-4.8 shows the Simulink block diagram for a lead-lag unit. It is developed as follows:

$$\frac{Y(s)}{X(s)} = \frac{\tau_{lead}s + 1}{\tau_{lag}s + 1} \tag{13-4.10}$$

$$\tau_{lag}s\,Y(s) + Y(s) = \tau_{lead}s\,X(s) + X(s)$$

$$Y(s) = \frac{\tau_{lead}}{\tau_{lag}}X(s) + \frac{1}{\tau_{lag}s}[X(s) - Y(s)]$$

To prevent an initial sudden change in output, the integrator initial condition must be set to $(\tau_{lag} - \tau_{lead})/\tau_{lag}$ times the initial value of the input, $x(0)$.

Simulation of Dead Time

Simulink, VisSim, and most other block-oriented simulation programs provide a dead-time block with the ability to set nonzero initial values.

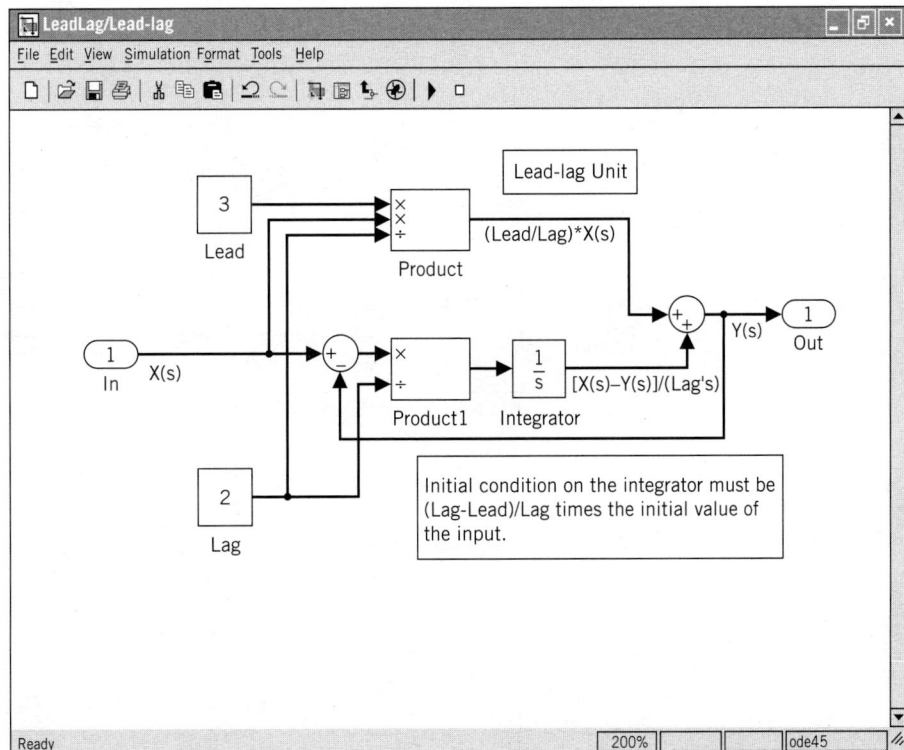

Figure 13-4.8 Simulink block diagram for a lead-lag unit.

The following examples will demonstrate how to use the models of the instruments presented here for simulating of feedback and feedforward control of the temperature in the stirred tank heater of Example 13-3.1.

EXAMPLE 13-4.1

Feedback Temperature Control of Stirred Tank Heater

SOLUTION

Simulate the feedback temperature control of the stirred tank heater of Example 6-1.1. Use a series PID controller with the tuning parameters determined in Example 7-1.1 and show the responses to step changes in process flow.

Figure 13-4.9 shows the Simulink block diagram of the control loop. Step input signal generators are used for the three inputs, the temperature set point $T^{SP}(t)$, the process flow $f(t)$, and the process inlet temperature $T_i(t)$. Subsystem blocks are used for the temperature sensor-transmitter, K_{sp} and $H(s)$, the series PID controller, the control valve $Gv(s)$, and the process (STHtr). The first three blocks are those detailed, respectively, in Figs. 13-4.7, 13-4.5, and 13-4.1, while the process block is detailed in Fig. 13-3.3. The parameters for the instrument blocks are those given in Examples 6-1.1 and 7-1.1, namely,

Sensor-transmitter:	$\tau_T = 0.75$ min	$T_{low} = 100°F$	$T_{hi} = 200°F$
PID controller:	$K_c = 6.1$ %CO/%TO	$\tau_I = 2.3$ min	$\tau_D = 0.58$ min
Control valve:	$\tau_v = 0.20$ min	$f_{max} = 84.4$ lb/min	

Here the maximum flow through the valve is obtained by doubling the steady-state flow of 42.2 lb/min determined in Example 6-1.1, as the valve is sized for 100 % overcapacity.

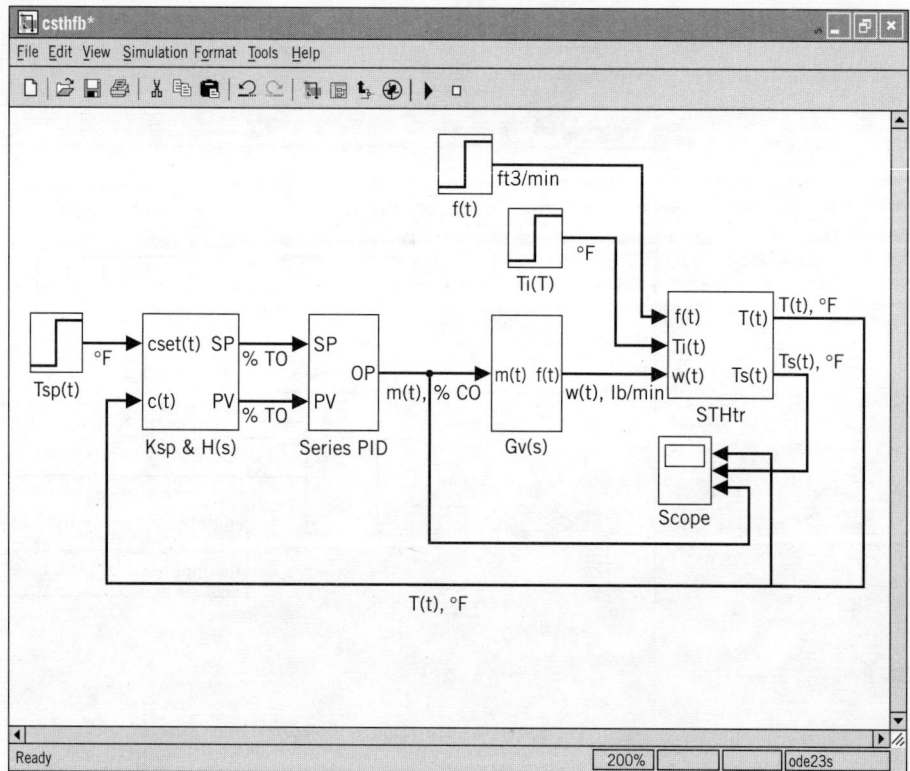

Figure 13-4.9 Simulink block diagram of the temperature control loop for the continuous stirred tank heater.

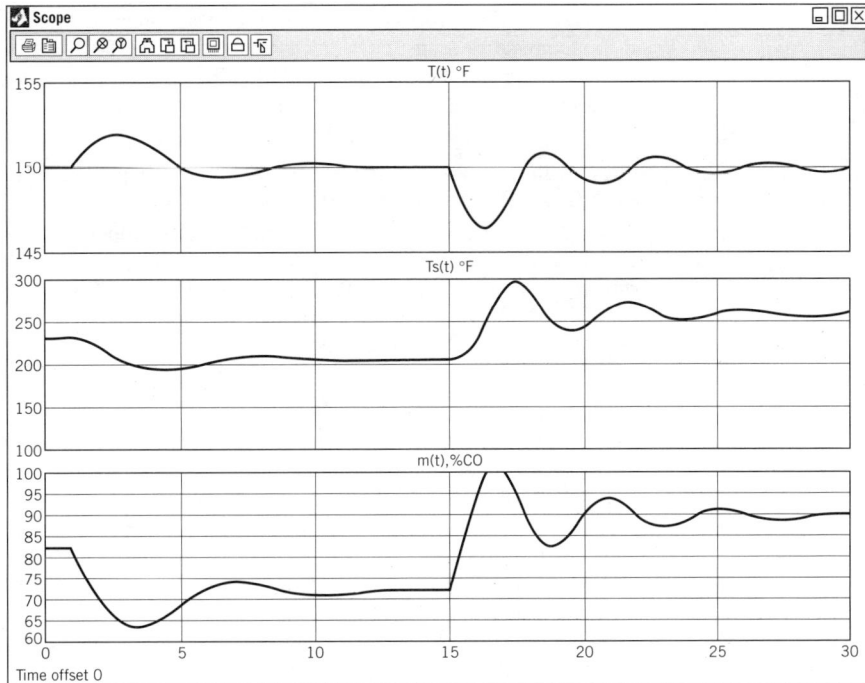

Figure 13-4.10 Responses of stirred tank heater to step changes in process flow.

The initial condition of the valve position, which is also that of the controller output, is calculated so that the initial flow is 42.2 lb/min, for an equal percentage valve with a rangeability parameter of 50,

$$\overline{w} = w_{max} \cdot 50^{\overline{vp} - 1}$$

$$\overline{vp} = 1 + \frac{\ln(\overline{w}/w_{max})}{\ln 50} = 1 + \frac{\ln(42.2/84.4)}{\ln 50} = 0.823$$

or 82.3 % VP. The other initial conditions are the design conditions from Example 6-1.1, that is $T(0) = 150°$F, $T_s(0) = 230°$F, $b(0) = 50$ % TO.

Figure 13-4.10 shows the responses of the outlet and steam temperatures, and of the controller output, to a step change in process flow from 15 to 10 ft³/min at $t = 1$ min, followed by another step change from 10 to 20 ft³/min at $t = 15$ min. Notice that the response to the increase in flow is much more oscillatory than the step in the decrease of the flow. This is probably due to the increase in the gain of the equal percentage valve which is not really needed here, given the assumptions of the problem. Notice also that the controller output saturates briefly at 100 % CO for the increase in process flow. These effects cannot be detected when the linear transfer function simulation of Fig. 13-2.1 is used. There is another effect that could require a more detailed simulation of the control valve. Notice how much the steam temperature rises for the increase in process flow. As the steam temperature increases, the pressure in the steam chest increases, because it is the vapor pressure of water at the condensing temperature. This could affect the pressure drop across the valve and reduce the capacity of the valve, depending on the supply pressure of the steam. This difficulty is addressed in one of the problems at the end of this chapter.

EXAMPLE 13-4.2

Feedforward Control of Stirred Tank Heater

Design a feedforward controller for the stirred tank heater of Example 13-4.1. The controller is to compensate for changes in process flow and inlet temperature. Simulate the feedforward control scheme with and without dynamic compensation. Assume a flow controller is installed to control the steam flow and the feedforward controller is to determine the set point of the flow controller. The process flow transmitter has a range of 0 to 25 ft^3/min and a negligible lag, and the temperature transmitter has a range of 50 to 150°F and a lag of 0.75 min. The steam flow transmitter has a range of 0 to 75 lb/min, and the time constant of the flow control loop is 0.2 min.

SOLUTION

To design the feedforward controller we follow the procedure outlined in Section 11-2. The control objective is to maintain the process temperature at the set point, $T(t) = T^{set}(t)$. The disturbances are the process flow $f(t)$ and the inlet temperature $T_i(t)$, and the manipulated variable is the steam flow set point, $w^{set}(t)$. A steady-state enthalpy balance on the tank establishes a simple relationship between these variables.

$$w(t)\lambda = f(t)\rho c_p (T(t) - T_i(t))$$
$$\therefore w^{set}(t) = \frac{\rho c_p}{\lambda} f(t)(T^{set}(t) - T_i(t))$$

where ρ is the fluid density in lb/ft^3, c_p is the specific heat in Btu/lb-°F, and λ is the steam latent heat of condensation in Btu/lb.

Figure 13-4.11 shows the feedforward controller based on the equation derived above. It consists of a summer to compute the change in temperature and a multiplier to obtain the product of the flow, the temperature change, and the constant parameter. For simplicity, the scale conversion to the fraction of the transmitter outputs has been omitted. In practice the process flow and inlet temperature transmitter ranges must be taken into consideration in the calculation. The inlet temperature transmitter $H_t(s)$ is simulated as a first-order lag with unity gain, and the steam flow controller as a linear valve.

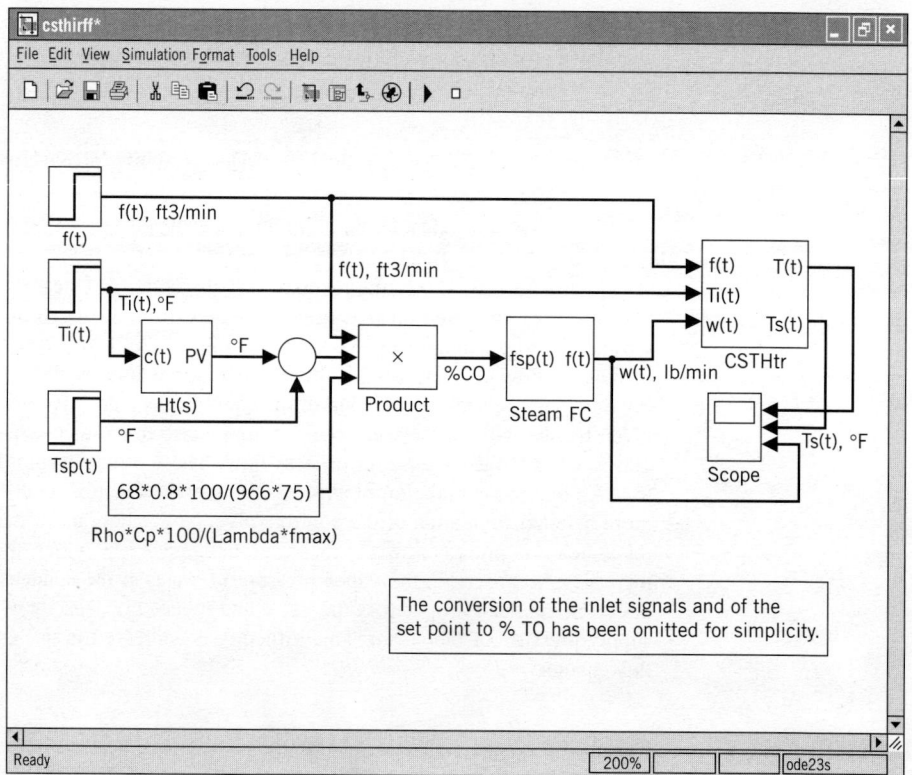

Figure 13-4.11 Simulink block diagram of feedforward control of stirred tank heater without dynamic compensation.

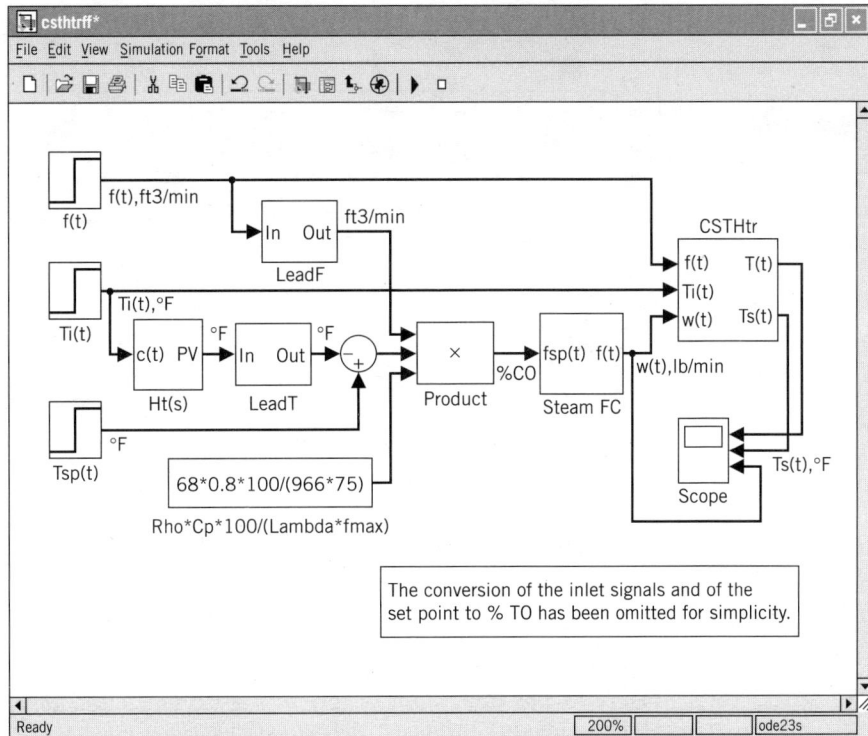

Figure 13-4.12 Block diagram of feedforward controller with dynamic compensation.

Figure 13-4.12 shows the feedforward controller with dynamic compensation. The lead blocks have been simulated as the PD unit of Fig. 13-4.4. After some tuning, the net lead for the flow disturbance was set at 1.0 min with $\alpha = 1/1.5$—this is equivalent to a lead of 1.67 min with a lag of 0.67 min. A net lead of 1.7 min with $\alpha = 1/4$ was used on the inlet temperature disturbance, equivalent to a lead of 2.12 min with a lag of 0.42 min. The requirement for a longer lead on the temperature is to compensate for the transmitter lag.

Figure 13-4.13 shows the responses of the heater variables to a step change in process flow from 15 to 10 ft^3/min at $t = 1$ min, followed by a step change in inlet temperature from 100 to 110°F at $t = 30$ min. The responses shown are for feedback control and feedforward control with and without dynamic compensation. Compared with the feedback response, the static feedforward controller reduces the initial deviation in temperature by about half and dynamic compensation further halves the initial deviation. This is because the static feedforward controller takes action as soon as the process flow changes, while the feedback controller must wait for the deviation in temperature to take action. The dynamic lead units give additional initial correction, as Fig. 13-4.13 shows.

Although slower than for the feedback controller, the feedforward controllers return the process temperature to its set point of 150°F after the initial transient. This is because in this ideal simulation situation, the values of the process parameters—density, specific heat, and latent heat—are known exactly. In practice a feedback controller must be installed in combination with the feedforward controller to compensate for errors in the feedforward model. When this is done the output of the feedback controller can be connected to the set point of the feedforward controller.

The preceding examples show how to put together the models presented in Sections 13-3 and 13-4 to simulate complete process control systems. One advantage

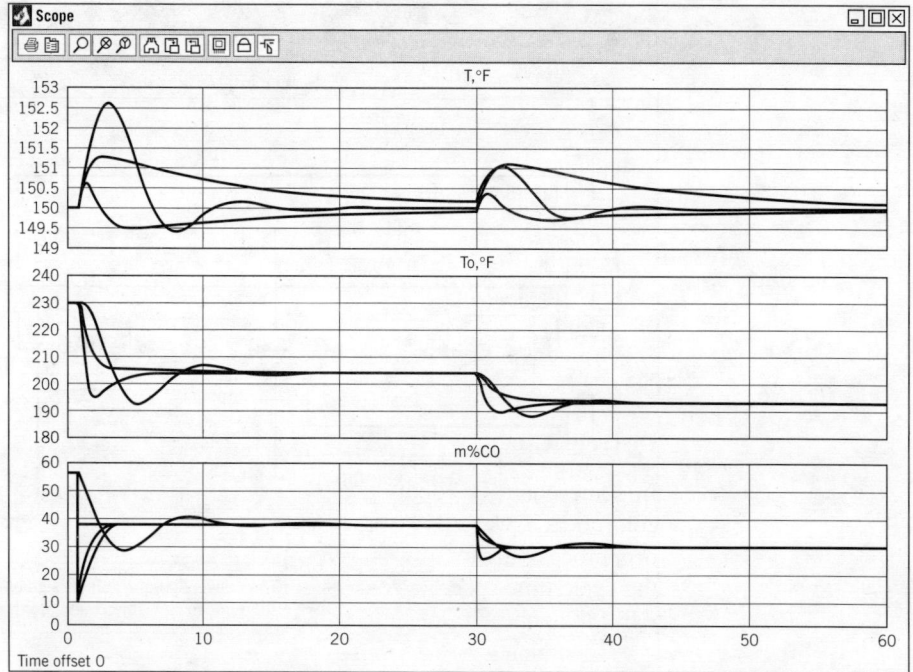

Figure 13-4.13 Comparison of responses of the stirred tank heater for feedback control and feedforward control with and without dynamic compensation.

of setting the models as subsystems is that they can be easily copied and pasted from one simulation to another.

13-5   OTHER SIMULATION ASPECTS

So far we have discussed just the basic aspects of dynamic simulation as it applies to continuous systems. Most process simulators provide a number of features and specialized blocks that are extremely useful to the engineer in modeling processes. Some of these will be briefly discussed in this section.

Stiffness

When numerically solving a system of differential equations, some equations may respond much faster than others. This characteristic is known as stiffness, and it requires special numerical integration methods. For example, in the simulations above, the PD and lead units introduce the time constant $\alpha \tau_D$ which is much smaller than the other time constants in the simulation. When a standard integration method is used to solve a stiff system, such as MATLAB's ODE45 or ODE23, the response of some variables will be noisy due to integration errors. When this happens, select a method designed to handle stiffness—such as ODE23s—the "s" at the end of the method's name in MATLAB designates the method as appropriate for stiff systems.

Complex Models

Some dynamic models are too complex to simulate using computation blocks. Examples are models of tray distillation columns and distributed parameter systems, that is, systems in which the variables are functions of both space and time. To program the many equations, procedure-oriented programs such as C++, Pascal, Basic, or MAT-LAB *m*-files, are much more convenient than programming with blocks. However, block programming is still desirable to simulate the control and instrumentation components and to put together the control system. It is therefore convenient to be able to create customized blocks that solve the equations in a procedural language while accepting inputs from and generating outputs to the control components. Many block-oriented programs provide such a feature. In MATLAB the S-function block provides that feature.

The S-function block accepts a vector of input variables and generates a vector of output variables. Procedural statements contained in a function stored in an *m*-file, which is associated with the block, perform the calculations.

Discrete Time Operations

In some control systems the process variables are not continuously available to the controllers. For example, when a gas chromatograph is used to measure the composition of a process stream, the analyzer takes a sample and analyzes it over a period of time, the cycle time. The controller does not see the composition until the end of the cycle. Also, the controller cannot tell if the composition changes during the cycle.

To simulate discrete time operations, simulation programs provide a set of blocks, known as *discrete blocks*. Two of the most useful are the *sample-and-hold* and the delay blocks. The sample-and-hold block samples its input variable and maintains it constant for one sample time, just like the analyzer. However, to fully simulate the analyzer a delay block of one sample must be connected after the sample-and-hold block to simulate the fact that the composition is not available until the end of the cycle.

Other Features

Modern simulation programs provide many other features, including the storage of the response data in files to use with other programs, the manipulation of vectors and matrices, blocks for the simulation of special effects such as hysteresis and dead bands, and so on. The user of such programs should carefully review the on-line help they provide for learning how to use these features when they are needed.

EXAMPLE 13-5.1

Simple Model of a Distillation Column

Although distillation columns are some of the most complex operations modeled, a simple model is presented here to demonstrate the use of the S-function in Simulink. Figure 13-5.1 is a schematic of the column.

A saturated liquid feed, consisting of 55 mole % benzene (x_F) and the balance toluene, flows at the rate F of 3500 lbmole/h into a distillation column. For the purpose of this model it may be assumed that the relative volatility α is constant and equal to 2.46, and that the latent heat of the mixture λ is constant and equal to 13,860 Btu/lbmole. The column can be represented by 6 ideal plates with the feed entering between plates 3 and 4. Each plate has a constant hold-up of 450 lbmole. A condenser may be assumed to condense all the vapors leaving the column to maintain the column pressure constant, and the condensate leaves the condenser at its bubble point. In the condenser accumulator drum, the hold-up varies from 50 to 550 lbmole as the level transmitter signal varies from 0 to 100 % TO. It may be assumed that the rate V_s of the vapors leaving the reboiler is proportional to the steam flow W_s. Saturated steam at 15 psig is used. At this pressure the latent heat of condensation λ_s is 952 Btu/lb. The partial reboiler behaves as an additional equilibrium stage. The column bottom has a hold-up of 100 lbmole when the

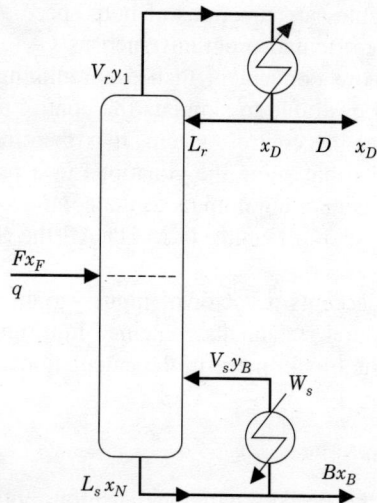

Figure 13-5.1 Sketch of the distillation column for Example 13-5.1.

level is at the bottom of the transmitter range and 1100 lbmole when the level is at the top of the range.

At the design conditions the distillate composition x_D is 92 mole % benzene and the bottoms composition x_B is 10 mole %. The reflux ratio L_r/D is approximately 2.5.

Model the column assuming equimolal overflow, that is, no change in liquid and vapor rates from stage to stage, and perfectly mixed trays, accumulator, and bottom.

SOLUTION

In modeling processes it is a good idea to first define the input variables to the process. For the columns these are the feed rate F, benzene mole fraction x_F, and enthalpy condition q, that is, the fraction that is liquid at the conditions of the feed tray. For saturated liquid feed $q = 1$. The variables to be manipulated by the control system are the reflux Lr, the steam flow to reboiler W_s, the distillate flow D, and the bottoms product flow B.

By the assumption of equimolal overflow, the liquid leaving each tray is equal to L_r in the rectifying section—above the feed tray—and L_s in the stripping section—below the feed tray. Also the vapor rates are V_r and V_s in the rectifying and stripping sections, respectively.

From an enthalpy balance on the reboiler, neglecting sensible heat effects, we obtain the proportionality between the steam rate and the stripping vapor rate:

$$V_s(t) = \frac{\lambda_s}{\lambda} W_s(t) = \frac{952 \text{ Btu/lb}}{13{,}860 \text{ Btu/lbmole}} W_s(t) = 0.0687 W_s(t)$$

From the definition of the feed enthalpy condition, total mass balances on the feed tray give

$$L_s(t) = L_r(t) + q F(t)$$
$$V_r(t) = V_s(t) + (1 - q) F(t)$$

A benzene balance on each stage, assuming constant hold-up and perfect mixing, gives the following:

$$M_i \frac{dx_1(t)}{dt} = L_r(t)[x_D(t) - x_1(t)] + V_r(t)[y_2(t) - y_1(t)]$$

$$M_i \frac{dx_2(t)}{dt} = L_r(t)[x_1(t) - x_2(t)] + V_r(t)[y_3(t) - y_2(t)]$$

$$M_i \frac{dx_3(t)}{dt} = L_r(t)[x_2(t) - x_3(t)] + V_r(t)[y_4(t) - y_3(t)]$$

$$M_i \frac{dx_4(t)}{dt} = L_r(t)x_3(t) + Fx_F(t) + V_s(t)y_5(t) - V_r(t)y_4(t) - L_s(t)x_4(t)$$

$$M_i \frac{dx_5(t)}{dt} = L_s(t)[x_4(t) - x_5(t)] + V_s(t)[y_6(t) - y_5(t)]$$

$$M_i \frac{dx_6(t)}{dt} = L_s(t)[x_5(t) - x_6(t)] + V_s(t)[y_B(t) - y_6(t)]$$

6 eq., 14 unk. ($6\,x_i, 6\,y_i, x_D, y_B$)

The equilibrium relationships, based on the relative volatility, give us the vapor compositions:

$$y_i(t) = \frac{\alpha x_i(t)}{1 + (\alpha - 1)x_i(t)} \qquad i = 1, \ldots, 6$$

12 eq., 14 unk.

The equilibrium relationship also applies to the reboiler, as it is assumed to be an equilibrium stage.

$$y_B(t) = \frac{\alpha x_B(t)}{1 + (\alpha - 1)x_B(t)}$$

13 eq., 15 unk. (x_B)

Total and benzene mole balances are then carried out on the condenser accumulator.

$$\frac{dM_D(t)}{dt} = V_r(t) - L_r(t) - D(t)$$

$$\frac{d[M_D(t)x_D(t)]}{dt} = V_r(t)y_1(t) - [L_r(t) + D(t)]x_D(t)$$

15 eq., 16 unk. (M_D)

Similarly, total and benzene mole balances are written around the column bottoms and reboiler, assumed perfectly mixed:

$$\frac{dM_B(t)}{dt} = L_s(t) - B(t) - V_s(t)$$

$$\frac{d[M_B(t)x_B(t)]}{dt} = L_s(t)x_6(t) - B(t)x_B(t) - V_s(t)y_B(t)$$

17 eq., 17 unk. (Solved!)

This completes the simple model of the column. It must be kept in mind that this model requires two level controllers to maintain the material balances around the accumulator and the column bottoms.

Because the hold-ups are not constant, the benzene balances on the accumulator and bottoms do not give the derivative of the compositions directly. This problem may be resolved with the aid of a little algebra. Expanding the derivative of the accumulator, and substituting the total mole balance, we obtain

$$\frac{d[M_D(t)x_D(t)]}{dt} = M_D(t)\frac{dx_D(t)}{dt} + x_D(t)\frac{dM_D(t)}{dt}$$

$$= M_D(t)\frac{dx_D(t)}{dt} + x_D(t)[V_r(t) - D(t) - L_r(t)]$$

$$= V_r(t)y_1(t) - [D(t) + L_r(t)]x_D(t)$$

$$M_D(t)\frac{dx_D(t)}{dt} = V_r(t)[y_1(t) - x_D(t)]$$

Similarly, the benzene balance around the column bottoms is converted to:

$$M_B(t)\frac{dx_B(t)}{dt} = L_s(t)[x_6(t) - x_B(t)] - V_s(t)[y_B(t) - x_B(t)]$$

The model of the column consists of 10 differential equations and the 7 equilibrium relationships, plus the 2 total mole balances around the feed stage. As in most control system simulations, it is desirable to start from the initial conditions at steady state. With modern computation tools such

Table 13-5.1 Distillation column conditions

	Design conditions		Initial conditions
$F = 3500$ lbmole/h	$x_F = 0.55$		$x_D = 0.9206$
$D = 1921$ lbmole/h	$q = 1.00$		$x_1 = 0.8274$
$B = 1579$ lbmole/h			$x_2 = 0.7037$
$L_r = 4802$ lbmole/h	$V_r = 6723$ lbmole/h		$x_3 = 0.5703$
$L_s = 8302$ lbmole/h	$V_s = 6723$ lbmole/h		$x_4 = 0.4524$
	$W_s = 97873$ lb/h		$x_5 = 0.3188$
			$x_6 = 0.1928$
			$x_B = 0.0993$

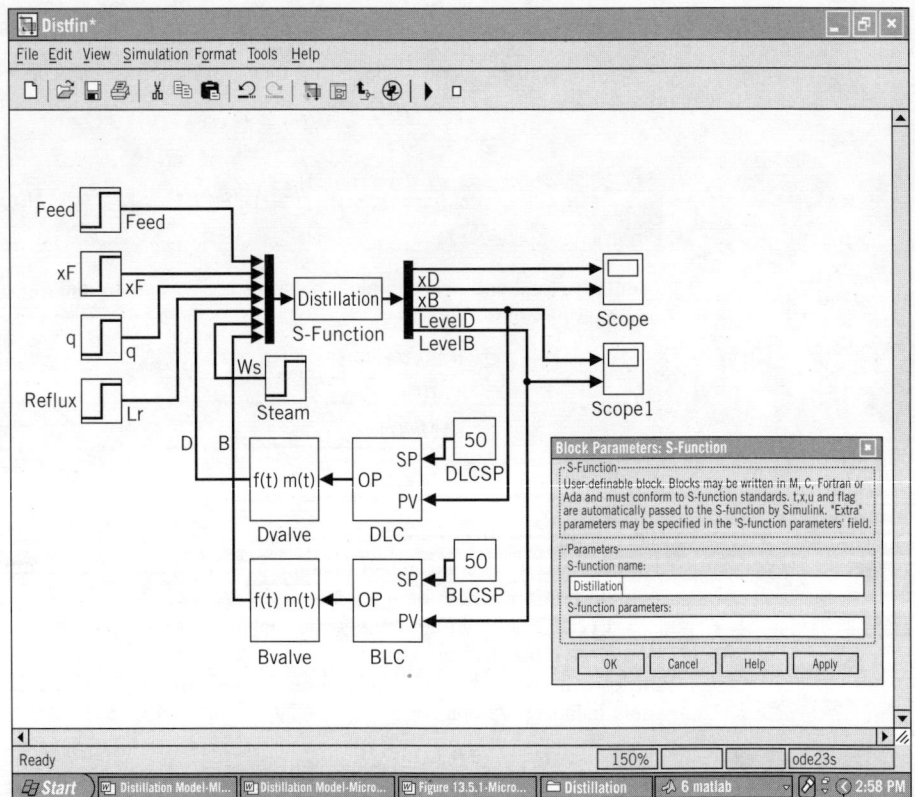

Figure 13-5.2 Simulink schematic for the distillation column using an S-function.

as MATLAB and MathCAD, the equations can be solved simultaneously for the steady-state conditions by setting the derivatives with respect to time equal to zero. We must also set the feed rate, composition, and enthalpy condition to their design values, as well as the bottoms composition and the reflux ratio L_r/D. The solution of the 19 equations with the remaining 19 unknowns produce the steady-state conditions listed in Table 13-5.1. Notice that the initial values of the hold-ups in the condenser accumulator and column bottoms cannot be calculated from the steady-state column equations, so we assume them to correspond to 50 % of the level transmitter outputs. The initial value of the steam rate is obtained from the enthalpy balance on the column bottoms presented above.

The simulation of the distillation column model using Simulink computing blocks would be very complex and difficult to manage. A more convenient method is the use of the Simulink *S-function* block, which is like building a custom block to simulate the column. Figure 13-5.2 shows the Simulink block diagram for the column showing the S-function block. The block refers to an *m*-file called Distillation.*m*, which contains the equations for the simulation of the column. Figure 13-5.3 shows the contents of the *m*-file for the distillation model. The following are the features of this program:

```
function [sys, xo] = Distillation (t, x, flag)

%  t = time (independent) variable; x = state variable vector
%  u = input variable vector
%  flag (See below. set by Simulink according to the stage of the problem)
%
%    Binary distillation column with 6 equilibrium stages, feed on stage 4
%    Equimolal overflow, constant relative volatility, constant pressure

xL = zeros (1, 6);
y  = zeros (1, 6);

          %        Set problem constants here

Alpha    = 2.46;          %  Relative volatility
LambdaV = 13860;          %  Latent heat of vapors, BTU/lbmole
LambdaS = 952;            %  Latent heat of steam, BTU/lb (sat. 15 psig steam)
Mi       = 450;           %  Tray hold–up, lbmole
MDmin    = 50;            %  minimum accumulator hold-up, lbmole
MDmax    = 550;           %  maximum accumulator hold-up, lbmole
MBmin    = 100;           %  minimum bottoms hold-up, lbmole
MBmax    = 1100;          %  maximum bottoms hold-up, lbmole

if flag == 1    %  Calculation of the derivatives of the state variables

          %        Input variables

Feed = u(1)/60;        %  Feed rate, lbmole/min
xF    = u(2);          %  Feed mole fraction
q     = u(3);          %  Feed fraction liquid
Lr    = u(4)/60;       %  Reflux flow, lbmole/min
D     = u(5)/60;       %  Distillate rate, lbmole/min
Ws    = u(6)/60;       %  Steam flow to reboiler, lb/min
B     – u(7)/60;       %  Bottoms flow, lbmole/min

          %        Initial calculations

Ls   = Lr + q*Feed;            %  stripping liquid flow, lbmole/min
Vs   = Ws*LambdaS/LambdaV;     %  stripping vapor rate, lbmole/min
Vr   = Vs + (1–q) *Feed;       %  rectifying vapor rate, lbmole/min

          %        State variables

for   i = 1:6
        xL(i) = x(i);
        y(i)  = Alpha*xL(i)/(1+(Alpha – 1) *xL(i));
end
xD  = x(7);
xB  = x(8);
MD = x(9);
MB = x(10);
yB  = Alpha*xB/(1+(Alpha – 1) *xB);
```

Figure 13-5.3 *m*-file for distillation model *S*-function.

```
%   Calculate the derivatives sys(i) = dx(i)/dt of the state variables

sys (1) = (Lr* (xD – xL(1)) + Vr* (y(2) – y(1)))/Mi;              % Stage 1
sys (2) = (Lr* (xL(1) – xL(2) + Vr*(y(3) – y(2))))/Mi;            % Stage 2
sys (3) = (Lr* (xL(2) – xL(3) + Vr*(y(4) – y(3)))/Mi;             % Stage 3
sys (4) = (Lr*xL(3) + Feed*xF – Ls*xL(4) + Vs*y(5) – Vr*y(4))/Mi;% Feed
sys (5) = (Ls* (xL(4) – xL(5) + Vs*(y(6) – y(5)))/mi;             % Stage 5
sys (6) = (Ls* (xL(5) – xL(6) + Vs*(yB – y(6)))/mi                % Stage 6

sys (7) = Vr* (y(1) – xD)/MD;                         % Distillate mole fraction
sys (8) = (Ls*(xL(6) – xB) – Vs*(yB – xB))/MB;        % Bottoms mole fraction
sys (9) = Vr – Lr – D;                                % Accumulator hold–up
sys (10) = Ls – Vs – B;                               % Bottoms hold–up

elseif flag = = 3   % Calculate outputs
  sys (1) = x(7);                                     % Distillate mole fraction
  sys (2) = x(8);                                     % Bottoms mole fraction
  Dlevel = (x(9) – MDmin)/(MDmax – MDmin) *100;       % To accmulator level
  Dlevel = min ( [Dlevel 100] );
  sys (3) = max ( [Dlevel 0] );
  Blevel = (x(10) – MBmin)/(MBmax – MBmin) *100;        % To bottoms level
  Blevel = min ( [Blevel 100] );
  sys (4) = max ( [Blevel 0] );

elseif flag = = 0   % Initialize: [# cont. states, # disc. states = 0,
  sys = [10, 0, 4, 7, 0, 0]; %   # outputs, # inputs]

MDo = MDmin + 0.5* (MDmax – MDmin); % Initial hold–up at 50%TO
MBo = MBmin + 0.5* (MDmax – MBmin); % Intitial compostns. and hold–ups
xo = [0.8274, 0.7037, 0.570, 0.452, 0.3188, 0.1928, 0.9206, 0.0993, MDo,MBo];
  else
  sys = [ ];
end
```

Figure 13-5.3 (*Continued*)

- Starts by defining the parameters of the problem.
- Stores the seven input variables to the S-function into variable names that are easier to recognize as model variables. Notice that the flows are assumed to be in lbmole/h and lb/h, but they are divided by 60 to convert them to lbmole/min and lb/min. The unit of the independent variable in the simulation is then minute.
- It moves the state variables, contained in vector **x**, into variable names that are easier to identify as model variables, as was done with the input variables.
- It calculates the derivatives of the state variables.
- It outputs the four output variables: x_D, x_B, and the levels in the accumulator and the column bottom.
- It sets the initial conditions for the 10 state variables.

The rest of the statements are requirements of the S-function that are described in the Simulink help files.

The diagram of Fig. 13-5.2 shows the level controllers for the accumulator drum (DLC) and the column bottoms (BLC). It is important to realize that without these level controllers the model would run away from the steady state, just as a real distillation column would.

Figure 13-5.4 shows the responses of the distillate and bottoms compositions to a step change of 2000 lb/h in steam flow followed by a step in 140 lbmole/h in reflux flow. Notice how the increase in reflux flow almost exactly cancels the effect of the steam flow on both

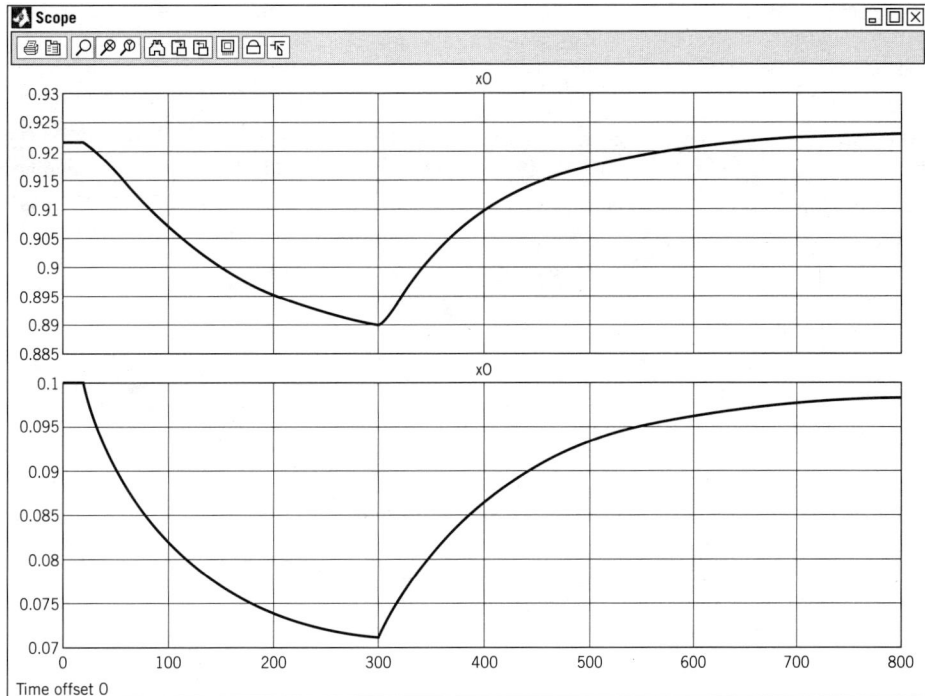

Figure 13-5.4 Responses of the distillate and bottoms composition to a step change of 2000 lb/h in steam flow followed by a step of 140 lbmole/h in reflux flow.

compositions. If the steam and reflux flows were to be manipulated by a control system to control both product compositions, the interaction between them would cause the controllers to cancel each other's actions. The concept of loop interaction is studied in Chapter 12. Figure 13-5.4 also shows that the response of the column is very slow. It takes about 400 min, over 6 h, for the column to reach a new steady state after a disturbance.

13-6 SUMMARY

Simulation is a very important tool for analyzing, designing, and tuning control systems. We have seen a few examples on how to simulate simple processes and their control systems using a block-oriented program. Although simulation cannot completely replace the understanding you have acquired through the mathematical analysis of dynamic responses, it

can really save time and effort when analyzing specific systems. If you are planning to make a career in control systems engineering, you should resolve to gain expertise in the use of some dynamic simulation program. We can tell you from experience that you will find a satisfying and exciting future in this field.

PROBLEMS

13-1. Simulate the response of the bird mobile of Problem 2-9 and plot the response. Assume an initial condition $y(0) = 0.1$ m, and no external force, $f(t) = 0$.

13-2. Model and simulate the response of a pendulum with mass $M = 0.5$ kg, and a length—distance from the fulcrum to the center of the mass—$L = 0.8$ m. For small swing angles and negligible air resistance, show that the dynamic

response is represented by

$$M\frac{d^2x(t)}{dt^2} = -\frac{Mg}{L}x(t)$$

where $g = 9.8$ m/s^2 is the acceleration of gravity and $x(t)$ is the horizontal deviation of the pendulum from the vertical position. Assume an initial condition $x(0) = 0.1$ m. *Note:*

Construct a pendulum with a string and a weight, or just use a yo-yo, and time the oscillations with a watch. See how it matches the response of the simulation when you match the length of the string.

13-3.   Simulate the punctured tank of Problem 2-23 and plot the response of the pressure in the tank.

13-4.   Simulate the response of the temperature of the turkey of Problem 2-24. Assume the oven temperature is constant at $800°R$, the turkey weighs 12 lb, is initially at $535°R$, and has an area of $3.5 ft^2$. Use a specific heat of 0.95 Btu/lb-$°R$, and an emissivity of 0.6.

13-5.   Simulate the chemical reactor of Section 4-2.3 and plot the responses of the reactor temperature to a step change of $0.25 ft^3/min$ in process flow, and of $0.1 ft^3/min$ in coolant flow. The reactor is initially at the design conditions. Compare your responses with the responses of the linear transfer functions given in Section 4-2.3 for those responses by simulating the transfer functions. *Note*: The differences in the responses will be due to the nonlinear nature of the reactor. You may decrease the size of the step changes to observe how the responses of the nonlinear reactor approach the responses of the transfer functions.

13-6.   Simulate the mixing process of Problem 3-1 and plot the response of the outlet concentration to a step change of 5 gpm in flow f_1. At the initial steady-state conditions the flow from the tank is 100 gpm, and its concentration is 0.025 moles/cm^3. The tank volume is 200 gallons, and the feed compositions are 0.010 and 0.05 moles/cm^3. Assume a tight level controller keeps the volume in the tank constant.

13-7.   Add a feedback control loop for the concentration in the tank of Problem 13-6. Use a transmitter with a time constant of 1.0 min and a range of 0 to 0.1 moles/cm^3, and a linear control valve with a time constant of 0.1 min and sized for a maximum flow of 100 gpm. The valve is installed in the more concentrated inlet stream. Tune a PI controller and obtain the plot of the outlet concentration to a 5-gpm step change in the flow of the other inlet stream.

13-8.   Simulate the isothermal rector of Problem 3-2 and plot the response of the concentration of reactants at point 3 for a unit step change of $5 ft^3/min$ in reactants flow. At the initial steady-state conditions the reactant flow is $50 ft^3/min$, the feed concentration is 2 lbmoles/ft^3, and the outlet concentration is 0.5 lbmoles/ft^3. The tank volume is $150 ft^3$ (constant), the pipe length between the tank exit and point 3 is 400 ft, and the inside diameter of the pipe is 5.500 in. *Note*: You may use a delay block with variable delay to simulate the transportation lag in the pipe.

13-9.   Simulate the flash drum of Problem 3-11 and plot the response of the liquid and vapor compositions for a step change of 0.10 mole fraction in the inlet composition $z(t)$. Assume that the flow rates are constant.

13-10.   Simulate the distillation tray of Problem 3-12 and obtain the response of the level $h(t)$ and the flow from the tray $f_o(t)$ for step changes of 10 and $20 ft^3/min$ in the flow into the tray $f_i(t)$. Determine the effect of nonlinearity by comparing the responses for the two step changes.

13-11.   Simulate the blending tank of Problem 3-18 using the additional parameters given for this process in Problem 6-11. Use a PID controller tuned for quarter decay ratio response and obtain the responses of the transmitter and controller outputs to a step change of $0.1 m^3/min$ in the flow of the concentrated solution.

13-12.   Simulate the tanks of Fig. 4-1.1 to obtain the responses of the levels in the tanks and of the flow out of the second tank, $f_2(t)$, to a step change in the flow to the first tank, $f_i(t)$, of $0.2 m^3/min$. At the initial steady state the inlet flow is $5 m^3/min$, the pump flow $f_o(t)$ is $2 m^3/min$, and the level in each tank is 2.5 m. Each tank has a cross-sectional area of $9 m^2$. Compare the gains and time constants measured from the simulation responses with those calculated from the results given in Section 4-1.1.

13-13.   Do Problem 13-12 for the interacting tanks of Fig. 4-2.1. In this case the initial levels are 5.0 m in the first tank and 2.5 m in the second tank.

13-14.   Simulate the thermal tanks in series of Fig. 4-1.5 to obtain the responses of the temperatures in the tanks for a unit step change of 10 K in the inlet temperature to the first tank, $T_1(t)$. Assume the flow into the first tank f_A is constant and equal to $1.0 m^3/min$, and ignore the flow f_B. Each tank has a constant volume of $5.0 m^3$ and the liquid is water. Compare the gains and time constants from the simulation responses with those calculated from the results given in Section 4-1.2.

13-15.   Do Problem 13-14 for the interacting thermal tanks of Fig. 4-2.4. Obtain the responses for the following values of the recycle flow f_R: 0, 1, 5, 20, and $100 m^3/min$. Discuss what happens to the temperature responses as the recycle flow increases.

13-16.   Simulate the reactors of Problem 4-9 to obtain the responses of the reactant concentrations in each tank to a step change of 0.5 lbmole/ft^3 in the feed concentration. Use the parameters given for these reactors in Problem 6-15. Obtain the responses for the following values of the recycle flow f_R: 0, 10.0, 50.0, and $500 ft^3/min$. Discuss what happens to the concentration responses as the recycle flow is increased.

13-17.   Simulate the extraction process of Problem 4-6 to obtain the response of the exit raffinate concentration $c_1(t)$ to a step change of $2 m^3/min$ in solvent rate $f_2(t)$. At the initial steady-state conditions the inlet flow f_1 is $5 m^3/min$ and the solute concentration is 0.4 kmole/m^3. It is desired to recover 90 % of the solute. The total volume of the extractor is $25 m^3$, the equilibrium slope is $m = 3.95$, and the mass transfer coefficient is $K_a = 3.646 min^{-1}$.

13-18. Simulate the temperature control loop for the stirred tank cooler of Problem 4-7. Use the parameters given for this process in Problem 6-19. Simulate the controller as a PID tuned for quarter decay ratio response and obtain the responses of the transmitter and controller outputs to a step change in the flow of cooling water of 0.02 m^3/min, and to a set-point step change of $2°C$.

13-19. Simulate the tank of Problem 4-5 and plot the responses of the outlet temperature to step changes of 5 gpm in the process flow and of 5 % in the signal to the steam control valve.

13-20. Modify the simulation of the control valve given in Section 13-4.1 to calculate the flow of a gas in which both the inlet pressure p_1 and the pressure drop across the valve Δp_v can vary. Using the data for the steam valve of Example 5-2.2, obtain a plot of the flow versus pressure drop as the pressure drop is ramped from 5 to 25 psi. Keep the inlet pressure and the valve position constant.

13-21. Simulate a simple temperature control loop for the nonisothermal reactor of Section 4-2.3 that was simulated in Problem 13-5. Use the data given for the control loop in Problem 6-12, but assume the temperature transmitter has a time constant of 1.0 min, and the control valve has a time constant of 0.1 min. Tune a PID controller for quarter decay ratio response and obtain the responses of the transmitter and controller outputs to a step change of -0.2 ft^3/min in reactants flow and to a step change of $1°R$ in set point.

13-22. In Example 13-4.1 the steam valve was simulated as a constant pressure drop valve. However, as the temperature of the condensing steam T_s varies, so does the pressure in the steam coil, so the pressure drop across the valve is lower the higher the steam temperature. To simulate this effect we must calculate the pressure in the coil as a function of the temperature of the steam using the Antoine equation for water. Then the model of the valve must be modified to calculate the flow of steam $w(t)$ using Eqs. 5-2.3 and 5-2.4. Assume the steam is supplied at 30 psia and $250°F$ (saturated). Size the valve for 100% overcapacity and use $C_f = 0.9$. Simulate the heater with the new valve and obtain the responses of the temperatures and the controller output to a step change in process flow of -5 ft^3/min followed by a step of $+5$ ft^3/min. Compare your responses to those of Fig. 13-4.10.

13-23. Simulate the composition control loop of Problem 6-17 and obtain the responses of the transmitter and controller outputs to a step change of 5 gpm in inlet flow. Use a PID controller tuned for quarter decay ratio response.

13-24. Simulate the level and temperature control loops of Problem 6-24 and obtain the responses of the transmitter and controller outputs for the temperature loop to a step change of 5 gpm in outlet oil flow. Use a tight level controller and a PID temperature controller tuned for quarter decay ratio response.

13-25. Simulate the drier of Problem 9-1 to compare the responses of the simple feedback control loop of part (a) and the cascade control loop of part (c). Obtain the responses of the transmitter outputs and of the signal to the control valve for a step change of $10°F$ in ambient air temperature, and to a step change of 0.3 mass % in set point.

13-26. Simulate the loop of Problem 9-3 to compare the responses of the simple feedback loop and the cascade loop proposed in the problem. Obtain the responses of the transmitter and controller outputs to unit step changes in disturbance and in set point. Use a PI controller for the master loop tuned for minimum IAE response.

13-27. Simulate the jacketed reactor of Problem 9-4 to compare the responses of the simple feedback loop and the cascade loop using the jacket temperature as the intermediate variable. Obtain the responses of the transmitter and controller outputs to a step change of $5°C$ in set point. Use a PID controller for the reactor temperature tuned for minimum IAE response.

13-28. Show that when the inlet temperature disturbance is neglected in Example 13-4.2, the feedforward controller becomes a simple ratio controller. Modify the simulation given in that example and compare the responses of the outlet temperature to those given in the example for step changes in both a process flow and inlet temperature.

13-29. In the simulation of the blending tank in Problem 13-11, add a ratio controller between the concentrated stream and the dilute stream and compare the responses to a step change in concentrated stream. Simulate a flow control loop on the dilute stream, instead of a control valve, as shown in Section 13-4.1.

13-30. The nonisothermal chemical reactor of Section 4-2.3 was simulated in Problem 13-21 with a simple feedback controller. Using the data from that problem, simulate and tune a cascade control system for the reactor with the jacket outlet temperature $T_c(t)$ as the intermediate variable. The jacket temperature transmitter has a time constant of 1.0 min and a range of 560 to $660°R$. Use a PI slave controller tuned by the synthesis formulas of Table 7-4.1 for 5 % overshoot on a set-point change, and a master PID controller tuned for quarter decay ratio response. Obtain the closed-loop responses of the reactor temperature and the signal to the valve for a step change of -0.2 ft^3/min in reactants, and for a step change of $1°R$ in set point.

13-31. The nonisothermal chemical reactor of Section 4-2.3 was simulated in Problem 13-21 with a simple feedback controller. Using the data from that problem, simulate and tune a feedforward control system for the reactor with the process flow $f_c(t)$ as the major disturbance. The process flow transmitter has a time constant of 0.1 min and a range of 0 to 2.0 ft^3/min. Instead of the control valve on the coolant, simulate a flow control loop (see Section 13-4.1). Obtain the responses of the reactor temperature and the signal to the valve for a step change of -0.2 ft^3/min in reactant flow, with and without feedback trim.

13-32. Do Problem 13-31 using a ratio controller with the feedback controller adjusting the ratio.

13-33. Design a cascade control scheme for the reactors in series of Problem 13-23 using the concentration out of the first reactor as the intermediate variable. Assume a transmitter range of 0 to 4 lb/gal of reactant with a time constant of 1.0 min. The slave controller is a PI controller tuned by the synthesis formulas of Table 7-4.1. Compare the responses to those of the single feedback control loop obtained in Problem 13-23.

13-34. Simulate the caustic blending tank of Problem 12-2 with PI controllers for both the product flow and composition. Assume the tank has a constant hold-up of 10,000 lb, the flow transmitter has a negligible time lag and a range of 0 to 60 klb/h, and the composition transmitter has a time constant of 1.0 min and a range of 0 to 50 mass % NaOH. The flow control loops have ranges of 0 to 60 klb/h and time constants of 0.1 min. Tune the controllers for quarter decay ratio response and obtain the responses to a 10 klb/h step change in product flow set point. Do the problem for

(a) the correct pairing of the controlled and manipulated variables,

(b) the other pairing,

(c) using a decoupler as the one in Example 12-3.1.

13-35. *Simulation of an Evaporator.*

The evaporator sketched in Fig. 12-3.4 is to concentrate a sugar solution by vaporizing the water using steam. Model the evaporator to find the response of the level and product composition to changes in feed rate, feed composition, steam flow, and product flow. Assume the contents of the evaporator are perfectly mixed and that the economy is constant. The economy is defined as the mass of water vaporized per unit mass of steam supplied. By assuming a constant economy you avoid having to write energy balances on the evaporator. The level is a linear function of the mass of solution in the evaporator. At design conditions the feed is 50,000 lb/h with a composition of 12 weight % sugars. The desired product composition is 70 weight % sugars. The economy is 0.95 lb of water evaporated per lb of steam. The evaporator is initially at steady state with the level at 50 % of the transmitter. When the level in the evaporator is at 0 % of the transmitter range the mass of solution in the evaporator is 394 lb, and when the level is at 50 %, the mass is 747 lb. *Control system*: Simulate a proportional level controller with a gain of 20 %CO/%TO—tight level control—and a PI composition controller with the feed and product flows as the manipulated variables. The control valves are to be linear with constant pressure drop and sized for twice the design flow. Assume negligible dynamics of the level transmitter, and a 0.1 min time constant on the control valve actuators. The composition transmitter has a time constant of 0.6 min and a range of 40 to 90 weight % sugars. Notice that there are two ways

to connect the controllers to the control valves. In Problem 12-4 a general pairing strategy is developed for the evaporator. You may want to pair the controller both ways and see which pairing results in better control. Tune the composition controller for each pairing and plot the responses to a step change of 1000 lb/h in steam flow.

13-36. *Distillation column control.*

Use the simulation of the distillation column of Example 13-5.1 to study composition control. Assume PI controllers tuned for minimum IAE. The composition transmitters have ranges of 0 to 1.0 mole fraction benzene, and time constants of 1.0 min, and the reflux and steam flow control loops have time constants of 0.1 min and have ranges of 0 to 200 % of the respective design flows. Obtain the responses of the transmitter and controller outputs to a step change of 200 lbmole/h in feed flow. Do the problem for the following cases:

(a) Control only the distillate composition.

(b) Control only the bottoms composition.

(c) Control both compositions.

13-37. *Gasoline blending.*

Simulate the gasoline blending system of Example 12-2.5 with PI controllers or the product flow, octane, and Reid vapor pressure (RVP). Assume constant densities and a constant mixer volume of 450 barrels, instead of an on-line mixer. The flow transmitter has a time constant of 0.1 min and a range of 0 to 100 kbl/day, and the analyzer transmitters have time constants of 1.0 min and ranges of 60 to 110 octane and 0 to 20 RVP. The flow control loops on the inlet streams have time constants of 0.1 min and ranges of 0 to 150 % of the design flows. Obtain the responses of the transmitter and controller outputs to a step change of 10 kbl/day in product flow set point. Do the following cases,

(a) Use the correct pairing.

(b) Use an alternate pairing.

(c) Use the decoupler of Example 12-3.3.

13-38. *Three pump and tank problem.*

This problem was created by chemical engineers at duPont in the early 1960s, the heyday of process simulation. The tank shown in Figure P13-1 is fed by three identical pumps and discharges through a very long pipe. Because of the inertia of the liquid in the pipe, the tank level response is underdamped. This causes a problem because, although the tank is tall enough at steady state, the overshoot in level can cause the tank to overflow when all three pumps are turned on. Your assignment is to simulate the tank and determine if there is a time sequence for turning the pumps on that prevents overflowing the tank. Each pump can be simulated as a step function with the step time being the time at which the pump is turned on. The step time for the first pump can be assumed to be 1 min, and the problem is to find the step

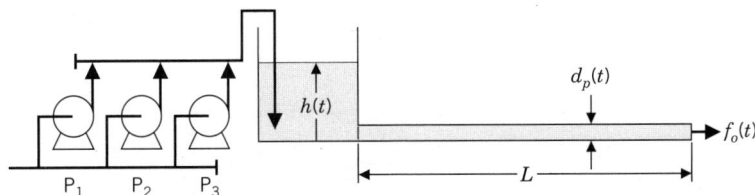

Figure P13-1 Schematic for Problem 13-38.

times for the other two pumps so that the tank level does not exceed the height of the tank. A force balance on the exit pipe results in the following equation:

$$\rho L \frac{df_o(t)}{dt} = \frac{\pi d_p^2}{4}\left(7.48\,\frac{\text{gal}}{\text{ft}^3}\right)\rho g[h(t) - 2.22 \times 10^{-3} f_o(t)$$
$$- 0.5184 \times 10^{-6} f_o^2(t)]$$

where $f_o(t)$ is the flow out of the tank in gal/min, $h(t)$ is the level in the tank in ft, L is the length of the pipe in ft, d_p is the inside diameter of the pipe in ft, ρ is the fluid density in lb/gal, and g is the acceleration of gravity in ft/min^2. The term in brackets contains a correlation for the friction losses in the pipe in ft of head. Each pump has a capacity of 750 gal/min, the tank is a cylindrical vertical tank 9.0 ft high by 3.41 ft inside diameter. The pipe is 3170 ft long by 1.19 ft inside diameter. Simulate the system, obtain responses of the level in the tank and the outlet flow, and determine the time sequence for starting the pumps that prevents the level in the tank from exceeding the height of the tank.

13-39. *Ecological interaction of host–parasite populations.*

A problem that illustrates the interaction between species in an ecological system, also from the heyday of dynamic simulation, considers only two species in the system, a host (e.g., rats) and parasite (e.g., fleas). Assume that the net growth rate of rats—birth rate minus death rate—is 5 rat/rat-year, in the absence of fleas. The fleas infect the rats and increase their death rate by 0.05 rat/flea-rat-year. The fleas need the rats to build their nests, and their growth rate is 0.2 flea/rat-flea-year, while the normal death rate of fleas is 20 fleas/flea-year. Set up the differential equations and determine the population cycle for rats and fleas as a function of time. Assume there are initially 100 rats and 20 fleas.

13-40. *Environmental impact of chlorine barge accident.*

In the summer of 1965 a barge loaded with liquid chlorine sank in the Mississippi River near downtown Baton Rouge, Louisiana. When they were preparing to bring it up, one of the authors was asked to estimate the rates of release of chlorine in the event one of the 4-in. nozzles on the barge were to be knocked off. The barge originally contained 136,000 kg of liquid chlorine at river ambient temperature, 25°C. The area of contact between the tank and the river is 212 m^2 and, for worse-case scenario, it is assumed that the nozzle is knocked off when the barge is still submerged in the water but the nozzle is just above the surface of the water. Heat of radiation from the sun is neglected. The overall heat transfer coefficient between the flowing river and the barge is estimated as 1000 kjoule/hr-m^2-°C. Assume the liquid chlorine has a constant latent heat of vaporization of 288 kjoule/kg, and a constant specific heat of 0.946 kjoule/kg-°C, and that the pressure inside the barge is the vapor pressure of chlorine at the temperature of the chlorine in the barge. The flow of chlorine gas through the nozzle opening is given by the following equation:

If subcritical flow: $w_c = \dfrac{\pi D^2}{4}\sqrt{\dfrac{2M_w}{RT}[P(P - P_o)]}$

If critical flow: $w_c = \dfrac{\pi D^2}{4}\sqrt{\dfrac{M_w}{RT}}\,P$

where w_c is the flow of chlorine gas through the opening, D is the diameter of the hole, M_w is the molecular weight of chlorine, R is the ideal gas constant, T is the absolute temperature, P is the pressure in the tank, and P_o is atmospheric pressure. The pressure P in the tank is the vapor pressure of chlorine that can be calculated from the temperature T using the Antoine equation. Simulate the tank by writing the mass and enthalpy balances on the barge and determine the initial and steady-state flows of chlorine out of the barge. How long will it take for the barge to empty? *Note*: A simple way to simulate the critical/subcritical flow is to calculate the flow using both formulas and select the smaller of the two flows using a *minmax* block or a *min* function.

13-41. *Control of semibatch reactor.*

A semibatch reactor is to produce butyl propionate from 2-butanol and propionic anhydride using a sulfuric acid catalyst. The reaction is given by

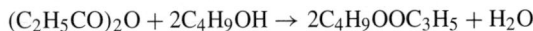

$$(C_2H_5CO)_2O + 2C_4H_9OH \rightarrow 2C_4H_9OOC_3H_5 + H_2O$$

Feliu *et al.* (2003) report the following kinetic model for this reaction:

$$r_B = (k_1 C_A + k_2 C_{Cat1} + k_3 C_{Cat2})C_B$$

where r_B is the rate of reaction of the anhydride, kmol/min-m^3, C_A, C_B, C_{Cat1}, and C_{Cat2} are the concentrations of the butanol, the anhydride, and two forms of the catalyst, in kmol/m^3. A reaction is proposed for the conversion of one form of the catalyst to the other, as follows

$$Cat_1 \rightarrow Cat_2$$

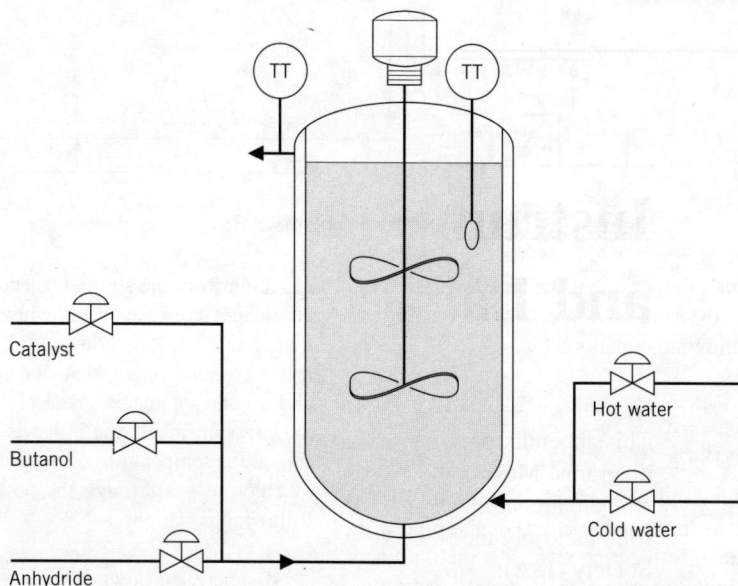

Figure P13-2 Semibatch reactor for Problem 13.41.

with kinetics

$$r_{Cat1} = k_4 10^{-H_r} C_{Cat2} C_B$$

$$H_r = -(p_1 C_{Cat1} + p_2 C_{Cat2}) \left(p_3 + \frac{p_4}{T} \right)$$

where T is the reactor temperature in K. The heat of reaction is $-80,000$ kJ/kmol and the reaction rate coefficients are given by the standard Arrhenius expression

$$k_i = A_i e^{-E_i/RT}$$

The parameters of the kinetic model are

Subscript i	P_i	A_i	E_i, kJ/kmol
1	0.2002	1.93×10^{11}	80,480
2	0.3205	1.01×10^{14}	79,160
3	-21.38	1.42×10^{14}	69,970
4	1271	5.05×10^{11}	76,620

Figure P13-2 shows a schematic of the jacketed reactor. Hot and cold water, at 90 and 15°C, respectively, are fed to the jacket to be able to heat and cool the reactor as necessary. Assume the reactor is initially filled with 600 kg of essentially pure 2-butanol at 20°C and the flows of cold and hot water are adjusted to produce an inlet jacket temperature of 30°C, which is also the initial temperature of the jacket. The catalyst, consisting of 70 weight% 2-butanol, 20 weight % sulfuric acid, and the rest water, at 20°C, is fed at the rate of 1200 kg/h for a period of 5 min. Then the anhydride, assumed pure and also at 20°C, is added at a rate that must not exceed 1200 kg/hr. An important operational constraint is that the reactor temperature must not exceed 60°C. Assume the heat transfer coefficient between the reactor and the jacket is constant at 20 kJ/min-m²-°C, and the heat transfer area is 15 m². The total volume of the reactor is 3 m³, and the jacket holds 415 kg of water. Get the molecular weights, specific gravities, and specific heats from Perry's Chemical Engineers' Handbook.

Simulate the reactor and design a control system and operating procedure that minimizes the duration of the batch cycle. Reference: Feliu, J. A., I. Grau, M. A. Alós, and J. J. Macías-Hernández, "Match Your Process Constraints Using Dynamic Simulation." *CEP*, New York: AIChE: December 2003, 42–48.

Instrumentation Symbols and Labels

This appendix presents the symbols and labels used in this book for the instrumentation diagrams. Most companies have their own symbols and labels, and even though most of them are similar, they are not all identical. The symbols and labels used in this book follow closely the standard published by the Instrumentation, Systems, and Automation Society (ISA); see the References. The appendix presents just the information needed for this book. For more information, see the ISA standard.

In general, the instrument identification, also referred to as tag number, is of the following form:

Typical Tag Number

LRC	101	Instrument identification or tag number
L	101	Loop identification
	101	Loop number
LRC		Functional identification
L		First letter
RC		Succeeding letters

Expanded Tag Number

20-TAH-6A	Tag number
20	Optional prefix
A	Optional suffix

Note: Hyphens are optional as separators.

The meanings of some identification letters are given in Table A-1.

Some symbols used in this book to designate the functions of computing blocks, or software, are presented in Table A-2. Table A-3 presents some instrument symbols, and Table A-4 presents some instrument line (signal) symbols.

Figure A-1 shows different ways to draw a control system, particularly a flow control loop. Figure A-1*a* shows a flow element, FE-10, which is an orifice plate with flange taps, connected to an electronic flow transmitter, FT-10. The output of the transmitter goes to a square root extractor, FY-10A, and from here the signal goes to a flow-indicating controller, FIC-10. The output from the controller goes to an I/P transducer, FY-10B, to convert the electrical signal to a pneumatic signal. The signal from the transducer then goes to a flow valve, FV-10. Often the labels for the flow

Principles and Practice of Automatic Process Control/Third Edition, by C. A. Smith and A. B. Corripio
ISBN 0-471-66141-4   Copyright © 2006 John Wiley & Sons (Asia) Pte. Ltd.

Table A-1 Meanings of identification letters (Courtesy of the Instrumentation, Systems, and Automation Society)

	First letter		Succeeding letters		
	Measured or initiating variable	Modifier	Readout or passive function	Output function	Modifier
A	Analysis		Alarm		
B	Burner, combustion		User's choice	User's choice	User's choice
C	User's choice			Control	
D	User's choice	Differential			
E	Voltage		Sensor (primary element)		
F	Flow rate	Ratio (fraction)			
G	User's choice		Glass, viewing device		
H	Hand				High
I	Current (electrical)		Indicate		
J	Power	Scan			
K	Time, time schedule	Time rate of change		Control station	
L	Level		Light		
M	User's choice	Momentary			Middle, intermediate
N	User's choice		User's choice	User's choice	User's choice
O	User's choice		Orifice, restriction		
P	Pressure, vacuum		Point (test) connection		
Q	Quantity	Integrate, totalize			
R	Radiation		Record		
S	Speed, frequency	Safety		Switch	
T	Temperature			Transmit	
U	Multivariable		Multifunction	Multifunction	Multifunction
V	Vibration, mechanical analysis			Valve, damper, louver	
W	Weight, force		Well		
X	Unclassified	X axis	Unclassified	Unclassified	Unclassified
Y	Event, state, or presence	Y axis		Relay, compute convert	
Z	Position, dimension	Z axis		Driver, actuator, unclassified final control element	

Table A-1 Meanings of identification letters (*Continued*)

First letters	Initiating or measured variable	Controllers — Recording	Controllers — Indicating	Controllers — Blind	Self-actuated control valves	Readout devices — Recording	Readout devices — Indicating	Switches and alarm devices — High	Switches and alarm devices — Low	Switches and alarm devices — Comb.	Transmitters — Recording	Transmitters — Indicating	Transmitters — Blind	Solenoids, relays, computing devices	Primary element	Test point	Well or probe	Viewing device, glass	Safety device	Final element
A	Analysis	ARC	AIC	AC		AR	AI	ASH	ASL	ASHL	ART	AIT	AT	AY	AE	AP	AW			AV
B	Burner combustion	BRC	BIC	BC		BR	BI	BSH	BSL	BSHL	BRT	BIT	BT	BY	BE		BW	BG		BZ
C	User's choice																			
D	User's choice																			
E	Voltage	ERC	EIC	EC		ER	EI	ESH	ESL	ESHL	ERT	EIT	ET	EY	EE					EZ
F	Flow rate	FRC	FIC	FC	FCV, FICV	FR	FI	FSH	FSL	FSHL	FRT	FIT	FT	FY	FE	FP		FG		FV
FQ	Flow quantity	FQRC	FQIC			FQR	FQI	FQSH	FQSL			FQIT	FQT	FQY	FQE					FQV
FF	Flow ratio	FFRC	FFIC	FFC		FFR	FFI	FFSH	FFSL						FE					FFV
G	User's choice																			
H	Hand		HIC	HC	HC					HS										HV
I	Current	IRC	IIC			IR	II	ISH	ISL	ISHL	IRT	IIT	IT	IY	IE					IZ
J	Power	JRC	JIC			JR	JI	JSH	JSL	JSHL	JRT	JIT	JT	JY	JE					JV
K	Time	KRC	KIC	KC	KCV	KR	KI	KSH	KSL	KSHL	KRT	KIT	KT	KY	KE					KV
L	Level	LRC	LIC	LC	LCV	LR	LI	LSH	LSL	LSHL	LRT	LIT	LT	LY	LE		LW	LG		LV
M	User's choice																			
N	User's choice																			
O	User's choice																			
P	Pressure vacuum	PRC	PIC	PC	PCV	PR	PI	PSH	PSL	PSHL	PRT	PIT	PT	PY	PE	PP			PSV, PSE	PV
PD	Pressure differential	PDRC	PDIC	PDC	PDCV	PDR	PDI	PDSH	PDSL	PDSL	PDRT	PDIT	PDT	PDY	PE	PP				PDV
Q	Quantity	QRC	QIC			QR	QI	QSH	QSL	QSHL	QRT	QIT	QT	QY	QE					QZ
R	Radiation	RRC	RIC	RC		RR	RI	RSH	RSL	RSHL	RRT	RIT	RT	RY	RE		RW			RZ

(*continued overleaf*)

Table A-1 Meanings of identification letters (*Continued*)

First letters	Initiating or measured variable	Controllers Record-ing	Controllers Indica-ting	Controllers Blind	Self-actuated control valves	Readout devices Record-ing	Readout devices Indica-ting	Switches and alarm devices High	Switches and alarm devices Low	Switches and alarm devices Comb.	Transmitters Record-ing	Transmitters Indica-ting	Transmitters Blind	Solenoids, relays, computing devices	Primary element	Test point	Well or probe	Viewing device, glass	Safety device	Final element
S	Speed frequency	SRC	SIC	SC	SCV	SR	SI	SSH	SSL	SSHL	SRT	SIT	ST	SY	SE					SV
T	Temperature	TRC	TIC	TC	TCV	TR	TI	TSH	TSL	TSHL	TRT	TIT	TT	TY	TE	TP	TW		TSE	TV
TD	Temperature differential	TDRC	TDIC	TDC	TDCV	TDR	TDI	TDSH	TDSL	TDSL	TDRT	TDIT	TDT	TDY	TE	TP	TW			TDV
U	Multivariable					UR	UI							UY						UV
V	Vibration machinery analysis					VR	VI	VSH	VSL	VSHL	VRT	VIT	VT	VY	VE					VZ
W	Weight force	WRC	WIC	WC	WCV	WR	WI	WSH	WSL	WSHL	WRT	WIT	WT	WY	WE					WZ
WD	Weight force, differential	WDRC	WDIC	WDC	WDCV	WDR	WDI	WDSH	WDSL	WSHL	WDRT	WDIT	WDT	WDY	WE					WDZ
X	Unclassified																			
Y	Event state presence		YIC	YC		YR	YI	YSH	YSL	YSHL			YT	YY	YE					YZ
Z	Position dimension	ZRC	ZIC	ZC	ZCV	ZR	ZI	ZSH	ZSL	ZSHL	ZRT	ZIT	ZT	ZY	ZE					ZV
ZD	Gauging deviation	ZDRC	ZDIC	ZDC	ZDCV	ZDR	ZDI	ZDSH	ZDSL	ZDSL	ZDRT	ZDIT	ZDT	ZDY	ZDE					ZDV

Table A-2   Function and symbols of computing blocks or software

Function	Symbol	Function	Symbol
Summation	$\sum$	Integral	$\int$
Multiplication	$\times$ or $*$	Division	$\div$
Square root	$\sqrt{}$	Function	$f(x)$
High selector	$>$ or HS	Low selector	$<$ or LS
High limiter	$>$ or HL	Low limiter	$<$ or LL
Bias	B_0	Lead-lag	L/L

Table A-3   General instrument symbols

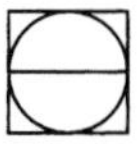

Computer-based algorithm

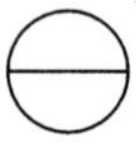

Analog instrument, accessible board-mounted

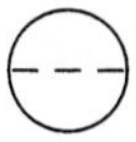

Analog instrument, mounted behind board

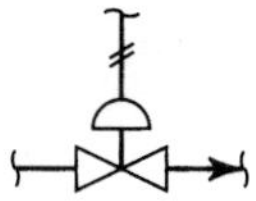

Pneumatic-operated globe valve

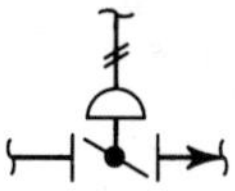

Pneumatic-operated butterfly valve, damper, or louver

Hand-actuated control valve

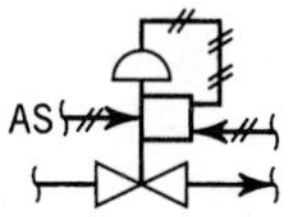

Control valve with positioner

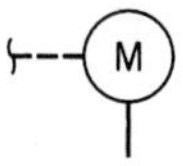

Motor

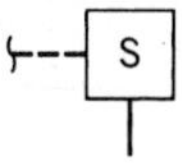

Solenoid

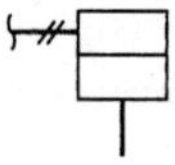

Single-acting cylinder

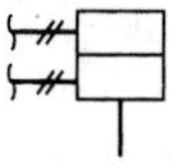

Double-acting cylinder

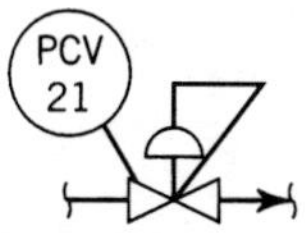

Pressure-reducing regulator, self-contained

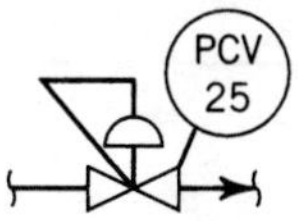

Back-pressure-reducing regulator, self-contained

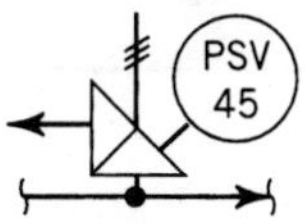

Pressure relief or safety valve, angle pattern

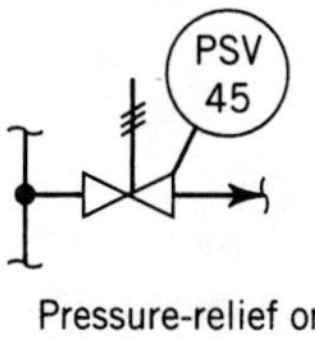

Pressure-relief or safety valve, straight through pattern

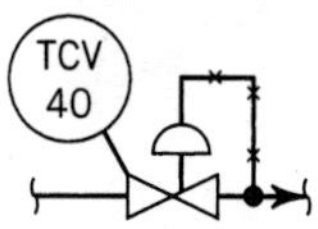

Temperature regulator, filled-system type

(continued overleaf)

Table A-3   (*Continued*)

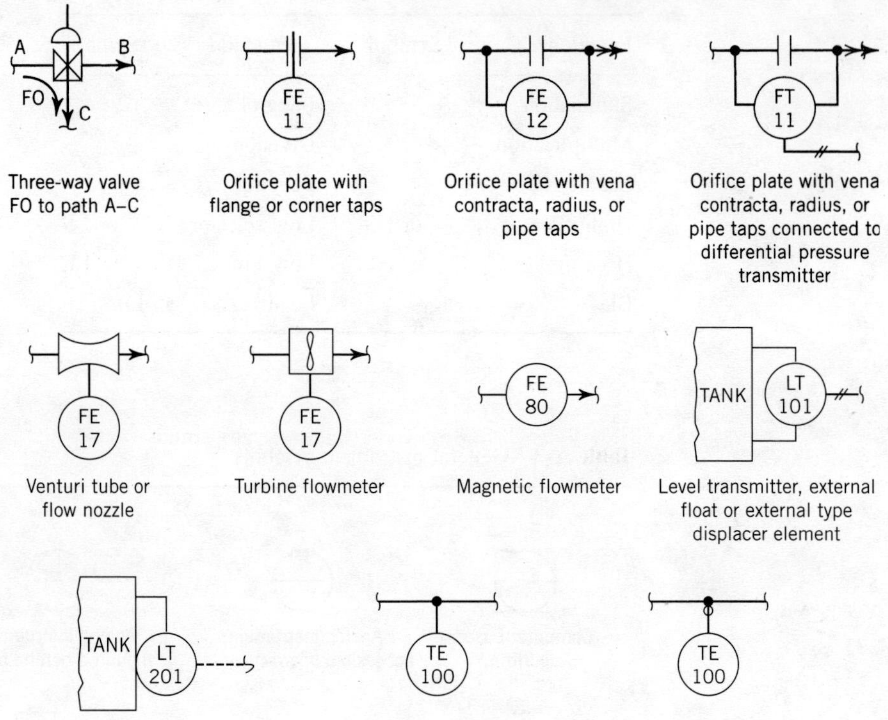

Three-way valve
FO to path A–C

Orifice plate with
flange or corner taps

Orifice plate with vena
contracta, radius, or
pipe taps

Orifice plate with vena
contracta, radius, or
pipe taps connected to
differential pressure
transmitter

Venturi tube or
flow nozzle

Turbine flowmeter

Magnetic flowmeter

Level transmitter, external
float or external type
displacer element

Level transmitter,
differential pressure
type element

Temperature element
without well

Temperature element with well

Table A-4   Instrument line (signal) symbols

Pneumatic Signal	Electrical Signal (or)	Generic Signal
Software or Data Link	Mechanical Data Link	Electrical Binary Signal (or)

element and valves are omitted for the sake of simplicity; the resulting diagram is shown in Fig. A-1*b*. The signals drawn in Fig. A-1*b* indicate that the control system used is electrical. Figure A-1*c* shows the control system when a computer control system is used; note the difference in signals. Figure A-1*d* shows the symbols used in this book. The figure shows the control concept without concern for specific hardware.

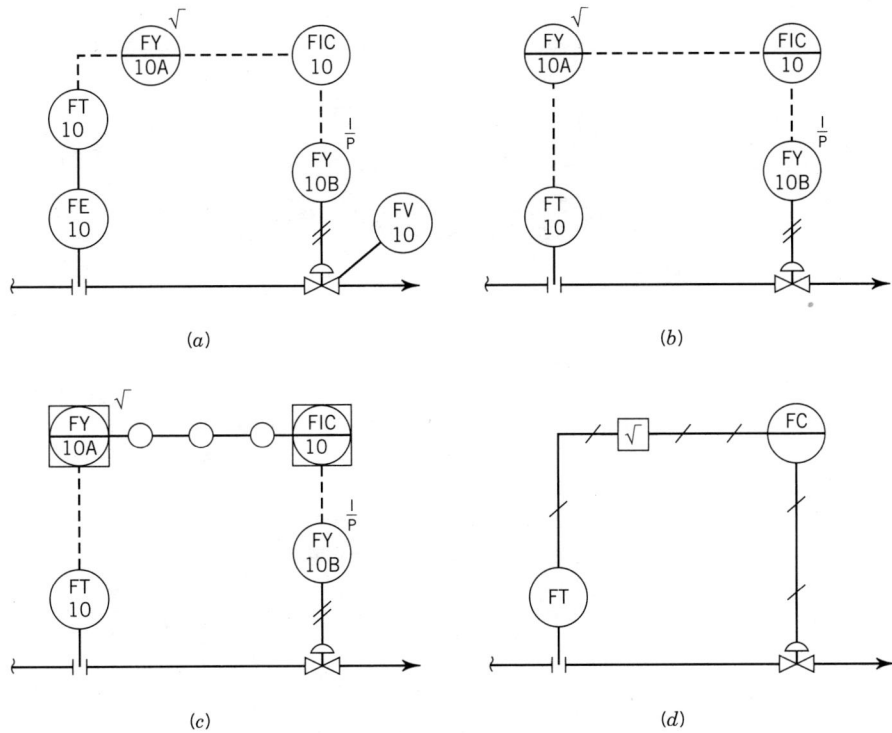

Figure A-1 Flow control system.

REFERENCES

1. "Instrumentation Symbols and Identification." Standard ISA-S5.1-1984. Research Triangle Park, NC: Instrumentation, Systems, and Automation Society.

Appendix B

Design Case Studies

This appendix presents a series of design case studies that provide the reader with an opportunity to design process control systems from scratch. The first step in designing control systems for a process plant is deciding which process variables must be controlled. This decision should be made jointly by the process engineer who designed the process, the instrument or control engineer who will design the control system and specify the instrumentation, and the operating personnel who will run the process. This is certainly very challenging and requires team effort. The second step is the actual design of the control system. It is the second step that is the subject of these case studies; the first step has been done. Please note that, like any design problem, these problems are open-ended. That is, there are multiple correct answers.

Section 12-6 discussed the topic of plantwide control, presented some guidelines to ease the task of designing control systems. This appendix presents another example, and then provides several case studies.

EXAMPLE B-1

Sodium hypochlorite (NaOCl) is formed by the following reaction:

$$2NaOH + Cl_2 \longrightarrow NaOCl + H_2O + NaCl$$

Figure B-1 shows the process for its manufacture.

Dilute caustic (NaOH) is continuously prepared, to a set concentration (30 % solution), by water dilution of a 50 % caustic solution and stored in an intermediate tank. From this tank, the solution is then pumped to the hypochlorite reactor. Chlorine gas is introduced into the reactor for the reaction.

A. How would you set the production rate from this unit?

B. Design the control system to accomplish the following:

1. Control the level in the dilution tank.

2. Control the level in the bleach liquor storage tank.

3. Control the dilution of the 50 % caustic solution. The concentration of this stream is to be measured by a conductivity cell. When the dilution of this stream decreases, the output from this cell increases.

4. Control the ratio of excess NaOH/available Cl_2 in the outlet stream from the hypochlorite reactor. This ratio is measured by an ORP (oxidation–reduction potential) technique. As the ratio increases, the ORP signal also increases.

Following the guidelines presented in Section 12-6, we must first decide where to locate the throughput manipulated variable, or production rate. Because there are three tanks in the process, providing enough surge capacity, there are several possible locations for this manipulated variable. The very first guideline indicates that the TPM must be located in the primary

Principles and Practice of Automatic Process Control/Third Edition, by C. A. Smith and A. B. Corripio
ISBN 0-471-66141-4   Copyright © 2006 John Wiley & Sons (Asia) Pte. Ltd.

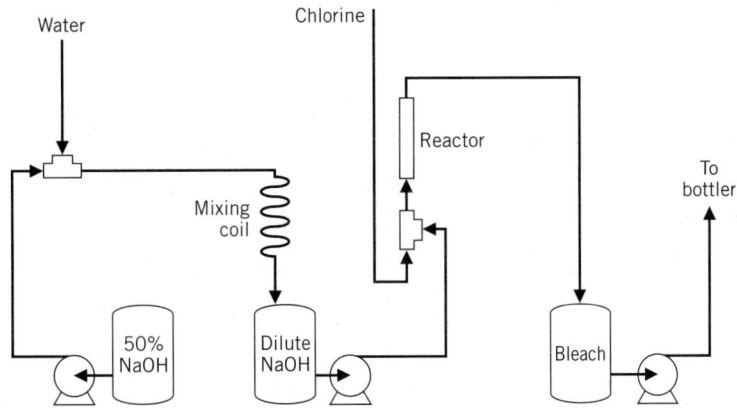

Figure B-1 Sodium hypochlorite process.

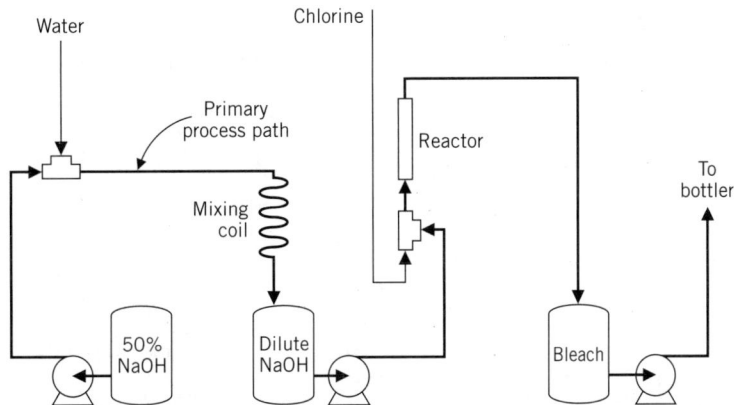

Figure B-2 Primary process path.

process path; Fig. B-2 shows this path. Figure B-3 shows three possible locations for controlling the production rate. The guidelines indicate that the preferred location is on the feed stream to the most sensitive unit, and whenever possible, internal to the process. If the engineer considers that the reactor is the most sensitive unit, then location B is the indicated location; actually, this location is also internal in the process. However, if the bottling process (not shown) is the most sensitive unit, then location C is the proper place for the production rate. Location A is another possible location, and there may be reasons for installing the production rate there; we will look at one of these reasons shortly.

Assuming that location B is the one the engineer selects, Fig. B-4 shows the locations of the inventory controls (levels in this process) and the two product quality controls (the dilution of the 50 % NaOH and the NaOH/available Cl_2 at the exit of the reactor). On the basis of Luyben's first law of plumbing (Luyben, 2002) the level control in the bleach storage tank is incorrect. Actually, it is the location of the valve that is incorrect; the valve should not be located in the inlet of the pump. This can easily be remedied by installing the valve downstream from the pump, as shown in Fig. B-5. Also, on the basis of Luyben's second law of plumbing (Luyben, 2002), the location of the level control in the dilute NaOH tank shown in Fig. B-4 is incorrect; here again, it is the location of the valve that is incorrect. Note that there are two control valves in the water path to the tank. One valve is manipulated by the level controller, and the other valve is manipulated by the conductivity controller. The possibility exists in this case that one controller wants to close its valve while the other controller wants to open its

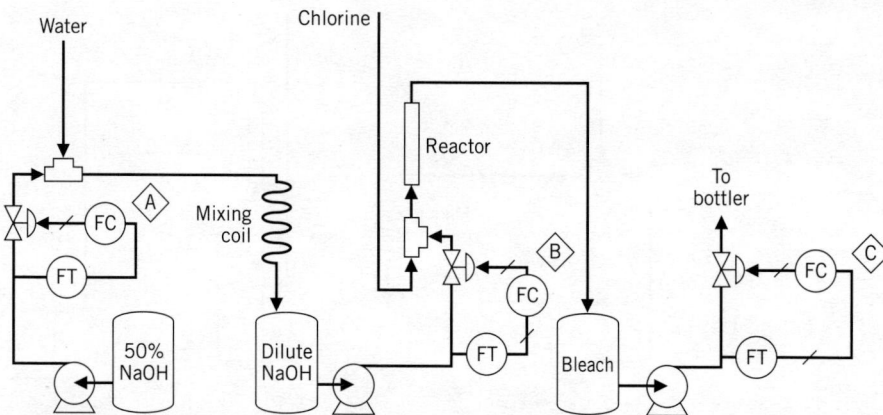

Figure B-3 Possible locations for control of production rate.

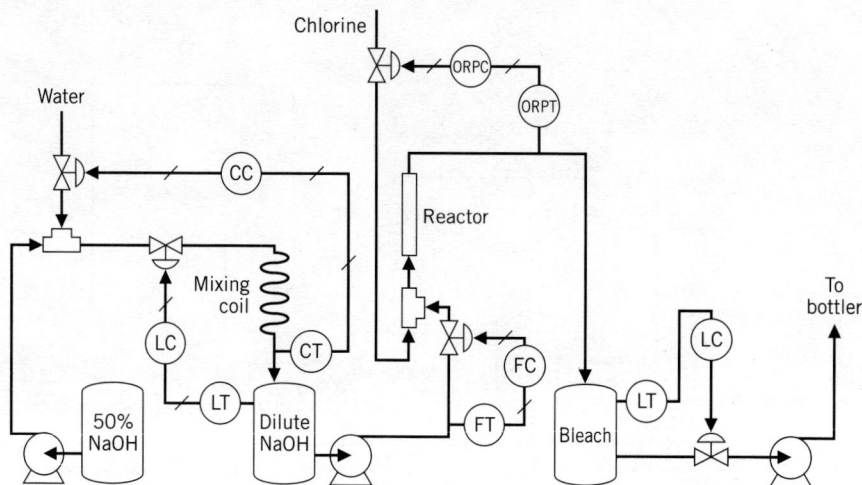

Figure B-4 First proposed schemes.

valve. Certainly, this is an undesired situation; it can be remedied very easily by installing the level valve upstream from the mixing tee, as shown in Fig B-5. Note that the inventory control upstream from the production control (the level in the NaOH dilution tank) is controlled manipulating the inlet flow, and the inventory control downstream from the production control (the level in the bleach tank) is controlled manipulating the exit flow. This inventory control is self-consistent.

Figure B-6 shows the control schemes if the engineer decides that the bottling process is the bottleneck—no pun intended—because it cannot handle many upsets. In this case all inventory controls manipulate the inlet flows.

Let us now focus our attention to the control of the level in the dilute NaOH tank, and to the control of the dilution of the concentrated NaOH; these controls are the same in Figs. B-5 and B-6. Note that any disturbance or set point change that affects any of these controlled variables eventually affects the other variable. For example, suppose that it is decided to decrease the dilution of the concentrated NaOH, and, consequently, the set point to the conductivity controller is increased. As the controller realizes that the NaOH is more diluted

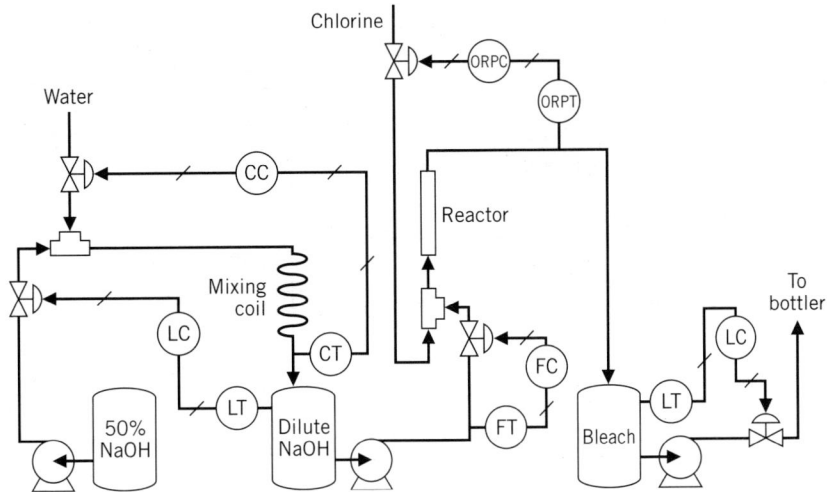

Figure B-5 Production control manipulating feed to the reactor.

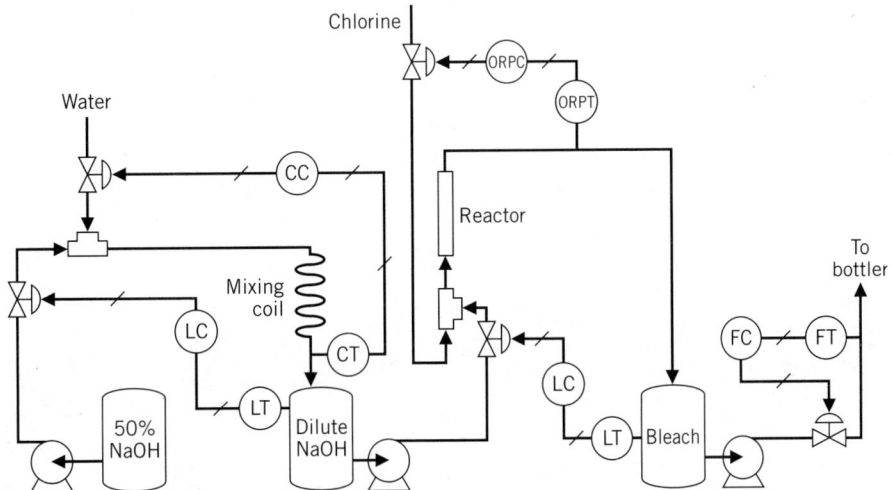

Figure B-6 Production control manipulating feed to the bottling process.

than required, it closes somewhat the water valve. As the water flow reduces, the level in the dilution tank decreases, and the level controller opens the concentrated NaOH valve. This action results in increasing the concentration of the diluted NaOH and the conductivity controller in turn changes the water valve again. Obviously, this is a classic example of a multivariable, fully coupled, control system; Chapter 12 discusses this important topic.

If the engineer decides that this interaction affects the control performance of the conductivity controller, and that this control is most important for the performance of the process, then controlling the production rate from location A in Figure B-3 is appropriate. Fig. B-7 shows the control scheme. In this case, the inventory controls manipulate the exit flows.

So far we have designed strictly feedback loops for each controlled variable. Suppose that previous experience indicates that the water supply pressure may vary, and that this disturbance affects the performance of the dilution control strategy. The cascade control shown in Fig. B-8

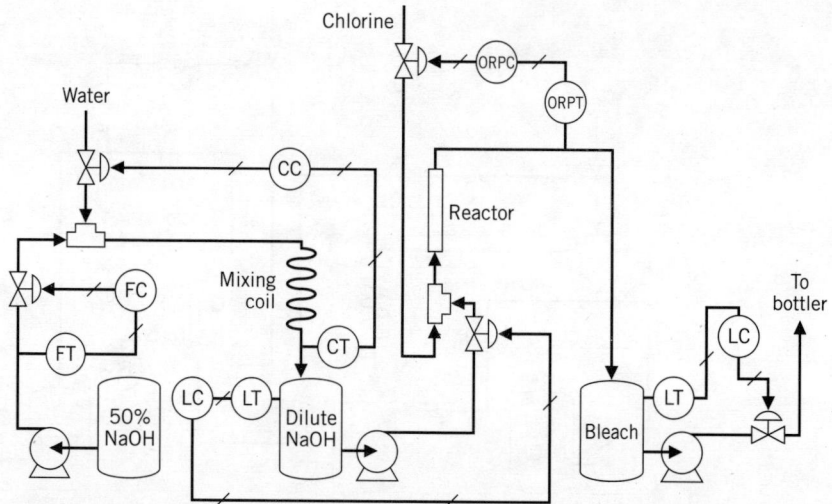

Figure B-7 Production control manipulating flow of concentrated NaOH.

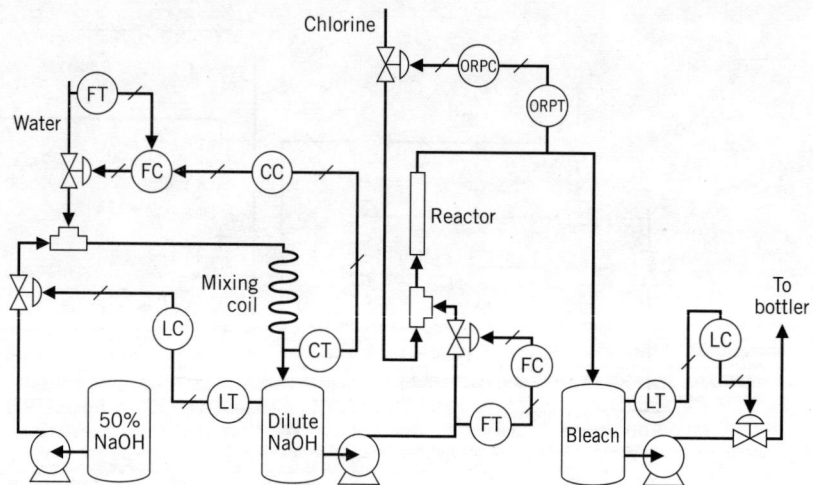

Figure B-8 Cascade control in dilution water stream.

provides a simple way to efficiently compensate for this disturbance. Obviously, similar cascade loops can be installed in the other loops as needed.

Suppose also that previous experience indicates that ratio control between the concentrated NaOH and the dilution water may prove profitable; Fig. B-9 shows this control. This ratio control works well, and will maintain the flow of water in correct proportion to the flow of concentrated NaOH. But, now the dilution control is out of the picture, and the dilution may not be at the desired value. Changes in the concentration of the entering NaOH affect the dilution but there is no control to compensate for these changes. If the control of the dilution is still required, we must somehow bring the conductivity controller into the strategy; this is shown in Fig. B-10. In this strategy the conductivity controller decides on the ratio of water to NaOH solution, R, required to maintain the dilution at set point. The reader may recognize this "ratio control with feedback compensation" as a feedforward control, and indeed, this is what it is. The feedforward is a simple steady-state compensator (the ratio),

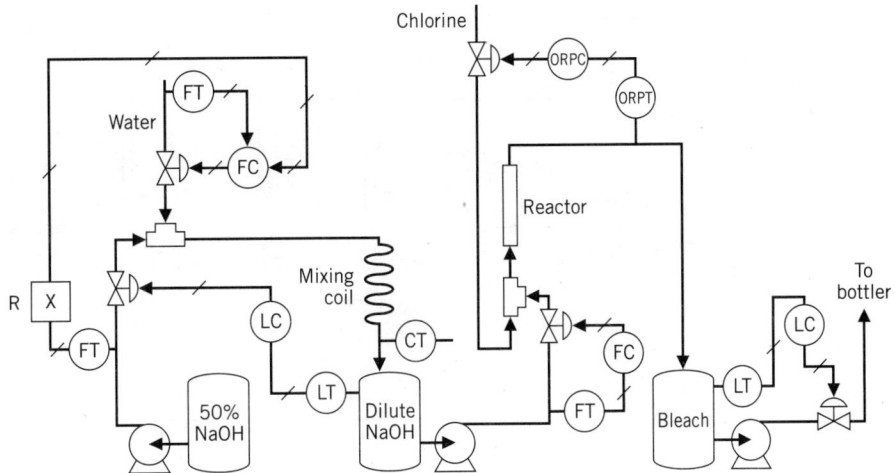

Figure B-9 Ratio control to improve performance of dilution control.

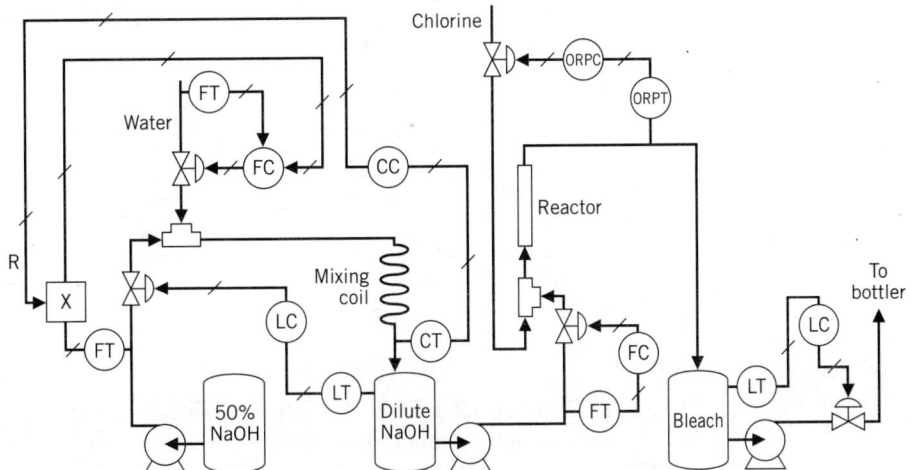

Figure B-10 Feedforward control to improve performance of dilution control.

with no dynamic compensation. A similar strategy can be implemented around the reactor if required for better control.

For simplicity, Fig. B-10 shows the level controllers and the ORP controller directly manipulating valves. As discussed in Chapter 9, often a flow loop manipulates the valve receiving the set point from these other controllers in a cascade arrangement. This strategy compensates for disturbances, usually pressure drop changes across the valve, affecting the flow. The flow transmitters are also useful for material balances and any other accounting issues.

CASE 1. METHANOL SYNTHESIS PROCESS

A petrochemical manufacturing company produces a hydrocarbon residue by-product. The residue was originally used for fuel within the plant; however, with *increased environmental regulations*, the company has chosen to incorporate an additional process that yields crude methanol by partial oxidation of the residue (Reference 2). The crude

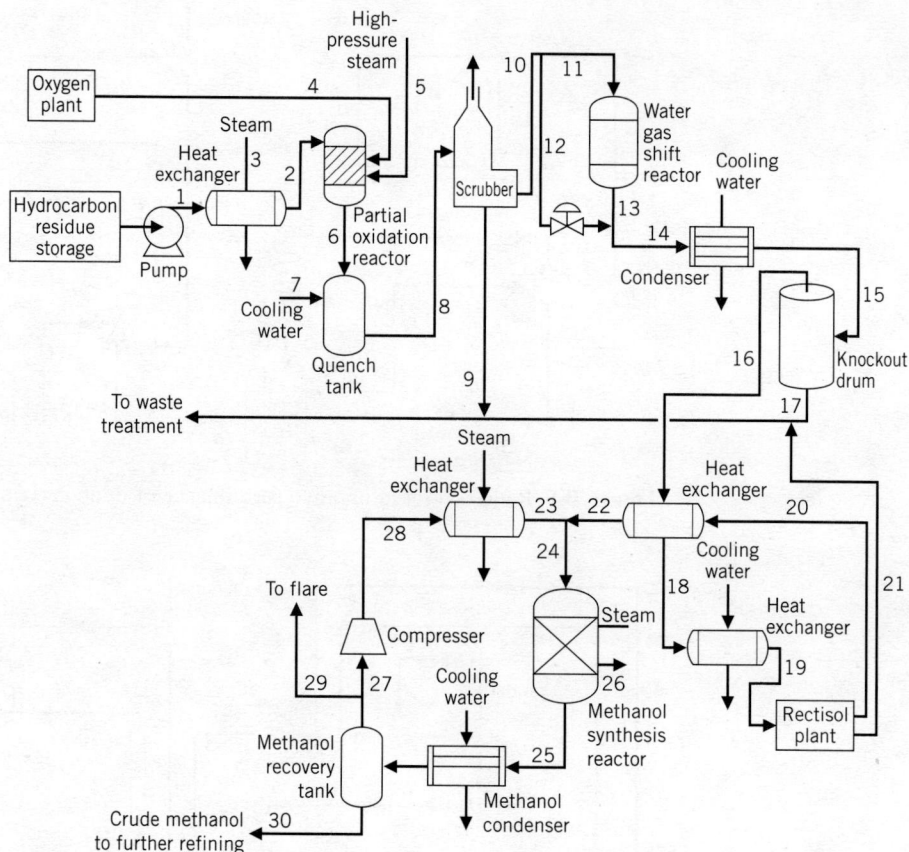

Figure B-11 Methanol process.

methanol is further refined by additional distillation and sent to storage for later use, and the unwanted remnants of the process are sent to an aqueous waste treatment unit: Fig. B-11 shows the process.

The residue is oxidized in a partial oxidation reactor to hydrogen and carbon monoxide gas in the presence of high-pressure oxygen that is supplied by an oxygen plant. High-pressure steam is added to the reactor to maintain a temperature of $1125°C$. At this temperature, some of the CO reacts with steam to form CO_2 and H_2 by the following reaction:

$$CO + H_2O \longleftrightarrow H_2 + CO_2$$

The exit gas from the reactor is quenched to $301°C$ by direct injection of cooling water. The soot and ash present in the gas is removed via a venturi scrubber and is sent to the wastewater treatment plant. The gas from the scrubber is split into two streams, one stream flows through a Co/Mo catalyzed packed bed reactor where 96 % of the CO is converted with H_2O to CO_2 and H_2 gas by the above reaction, and the other stream bypasses the reactor. The reactor effluent combines with the bypass stream resulting in a single stream with a H_2:CO ratio of 2.5:1; an analyzer is available to measure this ratio. The water in the resulting stream is removed from the gas by condensation in a condenser and removing it in a knockout drum. The gas exiting the drum flows through two heat exchangers that cool the gas to $50°C$ before entering a rectisol process. In the rectisol process the combustion by-products COS, H_2S, and CO_2 in the gas stream are removed, and sent to the aqueous waste treatment unit, to avoid poisoning the catalyst in the methanol synthesis reactor. The other product stream from the rectisol process contains CO, H_2, and trace amounts of Ar and N_2. This stream is heated by the exiting gas stream from the knockout drum and combined with a recycle stream before entering the methanol synthesis reactor. The combined feed to the reactor should enter at $200°C$. The synthesis reactor looks like a heat exchanger; the reaction occurring inside the tubes, which are filled with catalyst, and steam condensing in the shell. The H_2 and CO gas is converted into methanol and trace amounts of water and dimethyl ether; the conversion of CO in the reactor is 12.7 %. The gas out of the reactor goes to a condenser to condense the methanol, and to a tank for separation; the methanol liquid goes to a distillation train for further purification. The gas leaving the tank is mainly unconverted H_2 and CO with some Ar and N_2. Part of this stream goes to the flare, to relieve the nonreactive gases, and the other portion is recycled back to the reactor.

Design and draw the systems to control:

A. Production rate.

B. The temperature of the residue entering the partial oxidation reactor.

C. The flow of O_2 entering the partial oxidation reactor.

D. The temperature in the partial oxidation reactor.

E. The temperature in the quench tank.

F. The H_2:CO ratio of 2.5:1 in the combined stream exiting the water gas shift reactor.

G. The temperature in the stream entering the knockout tank.

H. The level in the knockout tank.

I. The temperature of the stream entering the rectisol process.

J. The temperature of the stream entering the methanol synthesis reactor.

K. The temperature of the stream exiting the methanol condenser.

L. The level in the methanol tank.

M. The pressure in the methanol tank.

Specify the fail-save action of every valve and the action of every controller.

CASE 2. HYDROCARBON PROCESS

A hydrocarbon processing plant purifies the incoming feed stream containing light hydrocarbons (two through six carbon chains) using a series of heat exchangers and distillation columns that capitalize on the relative volatilities of the three final products: propane, butane, and pentane (Reference 2). Figure B-12 shows the process.

The feed to the process is first heated in a heat exchanger (HEAT1), then it is pumped to a second heat exchanger (HEAT2), and flows through a valve, where the pressure is reduced to 253 psi, into a distillation column (COLUMN1). This column is responsible for the separation of propane from the feed stream as distillate. Both liquid streams leaving the column are at 253.1 psi, with the distillate at $119.9°F$ and the bottoms at $229.5°F$. The distillate flashes through a valve into a chamber operating at $100°F$ and 196.5 psi, to purge the ethane present in the stream; the purge stream is a vapor. The residual liquid stream is pumped to 450 psi, resulting in the product stream at the temperature of $104.7°F$.

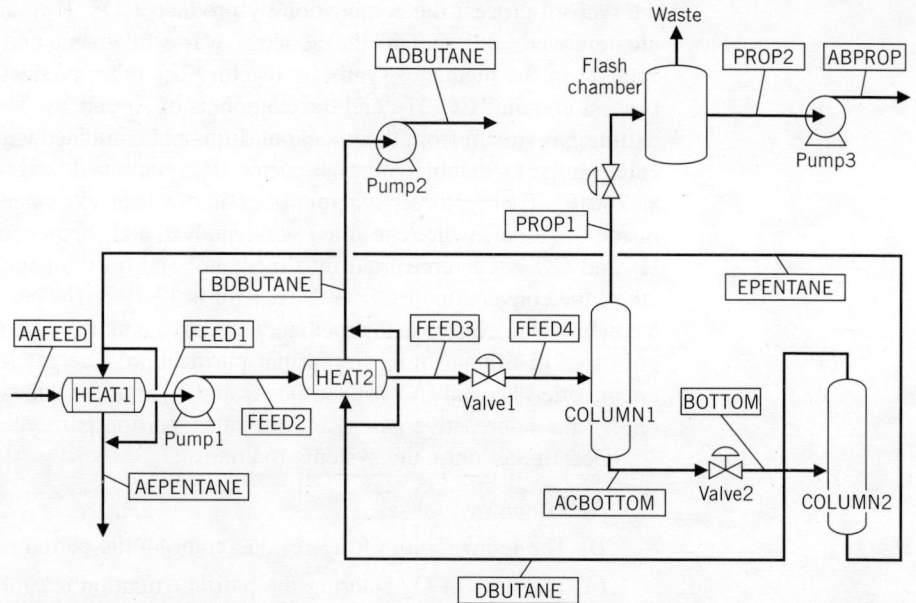

Figure B-12 Process for Case 2.

The bottoms stream from COLUMN1 flashes through a valve to a pressure of 89 psi and a temperature of 146.8°F; the stream is fed into another distillation column (COLUMN2). The resulting streams from this column are both desired products. Both exit at 89 psi, with the distillate at 120.7°F and the bottoms at 222.7°F. To meet delivery specifications these product streams must be cooled to 105°F. To cool the two product streams, each is run cross-current to the feed in the two previously mentioned heat exchangers (HEAT1 and HEAT2).

Design and draw the systems to control:

A. Production rate.
B. The pressure of stream FFED4.
C. The temperature of stream AEPENTANE.
D. The pressure of stream ADBUTANE.
E. The temperature of stream BDBUTANE.
F. The pressure in the flash chamber.
G. The level in the flash chamber.
H. The level in the base of COLUMN1.
I. The level in the base of COLUMN2.
J. The pressure of stream BOTTOM.

Specify the fail-save action of every valve and the action of every controller.

CASE 3. FATTY ACID PROCESS

Consider the process shown in Fig. B-13. The process hydrolyzes crude fats into crude fatty acids (CFA) and dilutes glycerine by using a continuous high-pressure fat splitter

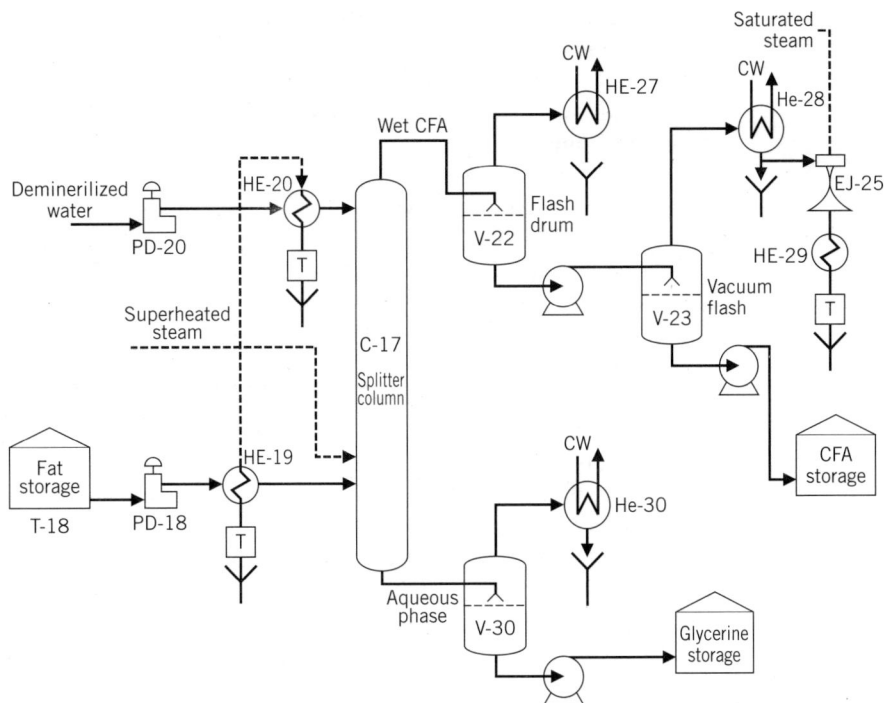

Figure B-13 Fatty acid process.

column (C-17). The main product is high-quality CFA. The CFA quality is primarily a function of the acid value. In the column the following reaction takes place:

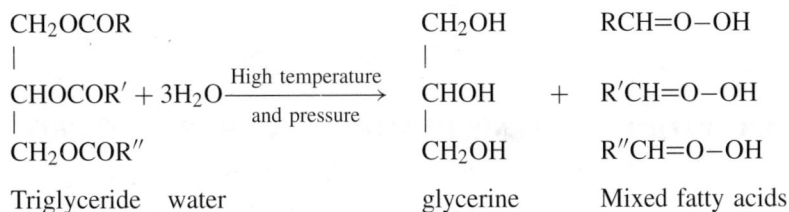

$$
\begin{array}{llll}
CH_2OCOR & & CH_2OH & RCH=O-OH \\
| & & | & \\
CHOCOR' + 3H_2O \xrightarrow[\text{and pressure}]{\text{High temperature}} & CHOH & + & R'CH=O-OH \\
| & & | & \\
CH_2OCOR'' & & CH_2OH & R''CH=O-OH \\
\end{array}
$$

Triglyceride   water   glycerine   Mixed fatty acids

Fat is stored (T-18) at 120°F and pumped into the column by means of a positive displacement pump (PD-18). The fat is preheated (HE-19) to 400°F with superheated steam before it enters the column. The column operates continuously at 700 psig and 500°F with a crude fat feed rate of 25,000 lb/hr.

Demineralized water is pumped into the column using a positive displacement pump (PD-20). The water is preheated (HE-21) to 500°F. Excess water is required to assure complete hydrolysis of the crude fat.

Superheated steam at 800 psig and 700°F is sparged directly into the column. The steam provides heat and mixing to break up the fat.

The splitter is basically a countercurrent contactor. Water feed at the top has a higher specific gravity than the CFA. Crude fat feed at the bottom is insoluble in water and rises as the water migrates down the column. The glycerine produced by the reaction is soluble in water and increases the specific gravity of the aqueous phase.

An interface forms in the column. Above the interface the material is mostly fat and CFA. Below the interface is mostly aqueous phase of water and glycerine. The best operation of the column is achieved when this interface is located near the steam sparger. If the interface level is low, then the amount of CFA in the aqueous phase increases. If the level is too high, fat dispersion into the water is lost and incomplete hydrolysis results. High temperature is required to produce the hydrolysis reaction, but boiling must be avoided, as this condition causes the aqueous phase to rise and upset the column.

The material removed overhead contains CFA and a small amount of water. This wet CFA is a light brown, milky material. The overhead product is dried by a two-step flash process. The sensible heat of the material is enough to dry the material without heat. The material is sprayed into the first vessel (V-22) and most of the water evaporates. The overhead water is condensed (HE-27). The resulting CFA is then sent to a vacuum flash (V-23) to fully dry the material. A steam jet ejector (EJ-25) is used to draw vacuum. The overhead water in the vacuum flash is condensed in the precondenser (HE-28) and sent to the sewer. The noncondensables from the precondenser are pulled through the steam jet ejector and the motive steam condensed in the barometric condenser (HE-29). The vacuum flash tank should be operated at 100 mm Hg. The ejector is significantly oversized for normal duty and consumes 2500 lb/hr of 150 psig saturated steam. Very low pressure will cause low-molecular-weight elements of the CFA to vaporize and foul the precondenser. Loss of vacuum will allow wet CFA to be stored in the tank, which will cause problems in downstream processes.

The aqueous phase is removed from the bottom of the column and should be 20 weight % glycerine dissolved in water. Like the CFA, the aqueous phase is flashed at atmospheric pressure (V-30). Any fatty material in the aqueous phase makes purification of the glycerine very difficult. Excess water in the aqueous phase requires additional energy in the glycerine purification. Glycerine is a clear, colorless liquid.

A. How would you set the production rate for this plant?

B. Prepare a detailed instrument diagram to control:

    1. The level in the splitter column.

    2. The level in all flash tanks.

CASE 4. CONTROL SYSTEMS IN THE SUGAR-REFINING PROCESS

The process units shown in Fig. B-14 form part of a process to refine sugar. Raw sugar is fed to the process through a screw conveyor. Water is sprayed over it to form a sugar syrup. The syrup is heated in the dilution tank. From the dilution tank the syrup flows to the preparation tank where more heating and mixing are accomplished. From the preparation tank the syrup flows to the blending tank. Phosphoric acid is added to the syrup as it flows to the blending tank. In the blending tank lime is added. This treatment with acid, lime, and heat serves two purposes. The first is that of clarification; that is, the treatment causes the coagulation and precipitation of the no-sugar organics. The second purpose is to eliminate the coloration of the raw sugar. From the blending tank the syrup continues to the process.

A. How would you control the production rate?

B. The following variables are thought to be important to control:

    1. Temperature in the dilution tank.

    2. Temperature in the preparation tank.

    3. Density of the syrup leaving the preparation tank.

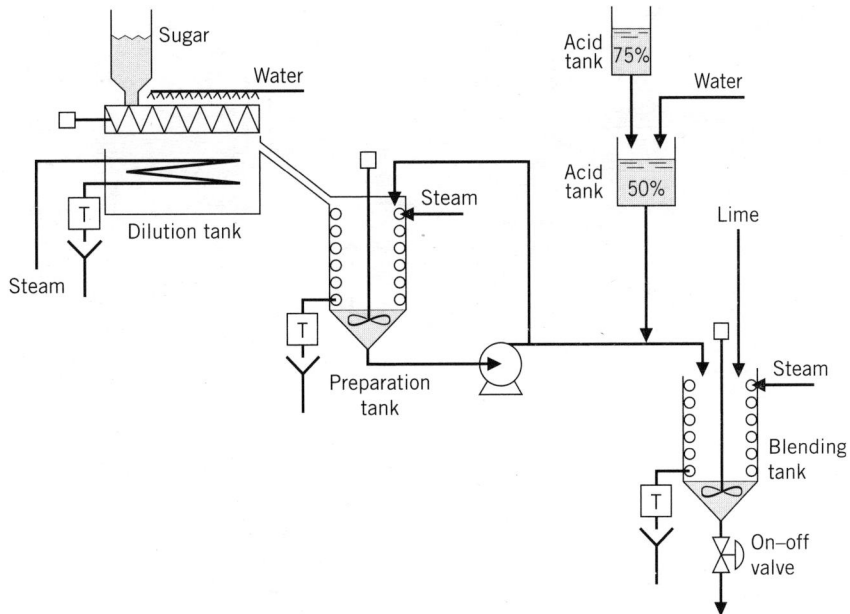

Figure B-14 Sugar-refining process.

4. Level in the preparation tank.

5. Level in the 50 % acid tank. The level in the 75 % acid tank can be assumed constant.

6. The strength of the 50 % acid. The strength of the 75 % acid can be assumed constant.

7. The flow of syrup and 50 % acid to the blending tank.

8. The pH of the solution in the blending tank.

9. Temperature in the blending tank.

10. The blending tank requires only a high-level alarm.

The flowmeters used in this process are magnetic flowmeters. The density unit used in the sugar industry is °Brix, which is roughly equivalent to the percentage of sugar solids in the solution by weight.

Design the control systems necessary to control all of the above variables. Show the action of control valves and controllers.

CASE 5. SULFURIC ACID PROCESS

Figure B-15 shows a simplified flow diagram for the manufacture of sulfuric acid (H_2SO_4).

Sulfur is loaded into a melting tank, where it is kept in the liquid state. From this tank the sulfur goes to a burner, where it is reacted with the oxygen in the air to produce SO_2 by the following reaction

$$S_{(l)} + O_{2(g)} \longrightarrow SO_{2(g)}$$

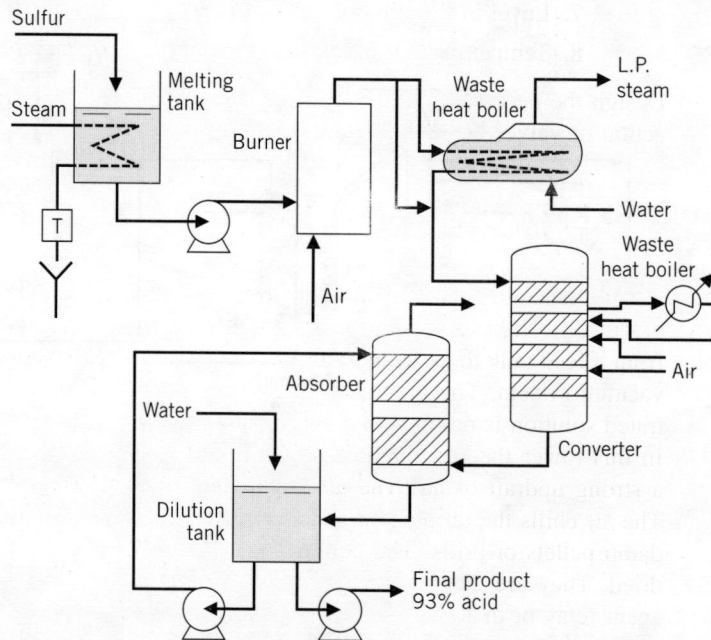

Figure B-15 Sulfuric acid process.

From the burner the gases are passed through a waste heat boiler where the heat of reaction of the above reaction is recovered by producing steam. From the boiler, the gases are then passed through a four-stage catalytic converter (reactor). In this converter the following reaction takes place:

$$SO_{2(g)} + \tfrac{1}{2}O_{2(g)} \rightleftharpoons SO_{3(g)}$$

From the converter, the gases are then sent to an absorber column, where the SO_3 gases are absorbed by dilute H_2SO_4 (93 %). The water in the dilute H_2SO_4 reacts with the SO_3 gas, producing H_2SO_4

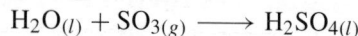

$$H_2O_{(l)} + SO_{3(g)} \longrightarrow H_2SO_{4(l)}$$

The liquid leaving the absorber, concentrated H_2SO_4 (98 %), goes to a circulation tank where it is diluted back to 93 % using H_2O. Part of the liquid from this tank is then used as the absorbing medium in the absorber.

A. How would you set the production rate for this plant?

B. The following variables are thought to be important to control.

   1. Level in the melting tank.

   2. Temperature of sulfur in the melting tank.

   3. Air to the burner.

   4. Level of water in the waste heat boiler.

   5. The concentration of SO_3 in the gas leaving the absorber.

   6. Concentration of H_2SO_4 in the dilution tank.

7. Level in the dilution tank.

8. Temperature of the gases entering the first stage of the converter.

Design the necessary control systems to accomplish the above. Be sure to specify the action of valves and controllers. Briefly discuss your design.

CASE 6. AMMONIUM NITRATE PRILLING PLANT CONTROL SYSTEM

Ammonium nitrate is a major fertilizer. Figure B-16 shows the process for its manufacture (Reference 3). A weak solution of ammonium nitrate (NH_4NO_3) is pumped from a feed tank to an evaporator. At the top of the evaporator there is a steam ejector vacuum system. The air fed to the system controls the vacuum drawn. The concentrated solution is pumped to a surge tank and then fed into the top of a prilling tower. In this tower the concentrated solution of NH_4NO_3 is dropped from the top against a strong updraft of air. The air is supplied by a blower at the bottom of the tower. The air chills the droplets in spherical form and removes part of the moisture, leaving damp pellets or prills. The pellets are then conveyed to a rotary drier where they are dried. They are then cooled, conveyed to a mixer for the addition of an antisticking agent (clay or diatomaceous earth), and bagged for shipping.

A. How would you control the production rate of this unit?

B. Design the system to implement the following:

    1. Control the level in the evaporator.

    2. Control the pressure in the evaporator. This can be accomplished by manipulating the flow of air to the exit pipe of the evaporator.

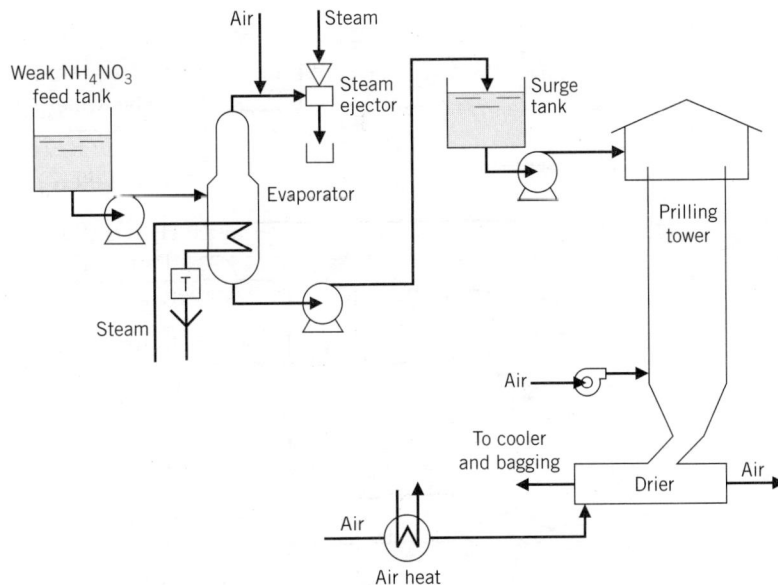

Figure B-16 Ammonium nitrate process.

   3. Control the level in the surge tank.

   4. Control the temperature of the dried pellets leaving the dryer.

   5. Control the density of the strong solution leaving the evaporator.

   Be sure to specify the action of valves and controllers.

   C. If the flow to the prilling tower varies often, it may also be desired to vary the air flow through the tower. How would you implement this?

CASE 7. NATURAL GAS DEHYDRATION CONTROL SYSTEM

Consider the process shown in Fig. B-17. The process is used to dehydrate the natural gas entering the absorber, using a liquid dehydrant (glycol). The glycol enters the top of the absorber and flows down the tower countercurrent to the gas, picking up the moisture in the gas. From the absorber, the glycol flows through a cross-heat-exchanger into the stripper. In the reboiler, at the base of the stripper, the glycol is stripped of its moisture, which is boiled off as steam. This steam leaves the top of the stripper and is condensed and used for the water reflux. This water reflux is used to condense the glycol vapors that might otherwise be exhausted along with the steam.

The process engineer who designed the process has decided that the following must be controlled:

1. The liquid level at the bottom of the absorber.

2. The water reflux into the stripper.

3. The pressure in the stripper.

4. The temperature in the top third of the stripper.

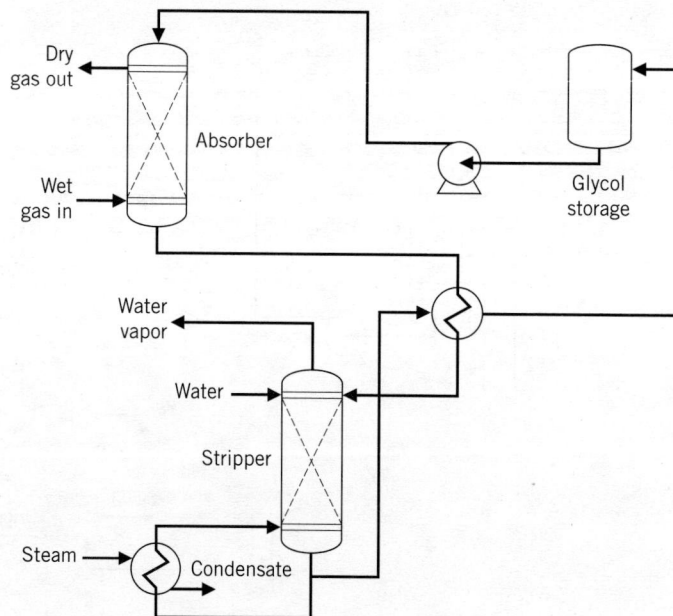

Figure B-17 Natural gas dehydration system.

5. The liquid level at the bottom of the stripper.

6. Efficient absorber operation at various throughputs.

Design the control system to accomplish the desired control.

REFERENCES

1. LUYBEN, WILLIAM L., *Plantwide Dynamic Simulators in Chemical Processing and Control*. New York: Marcel Dekker, 2002.

2. Provided by Dr. AYDIN K. SUNOL and Ms. RENEE DOCKENDORF of the University of South Florida, Tampa.

3. The Foxboro Co., Application Engineering Data, January 1972.

Appendix C

Sensors, Transmitters, and Control Valves

This appendix presents some of the hardware necessary to implement control systems and is closely related to Chapter 5. Some of the most common sensors—pressure, flow, level, and temperature—are presented as well as two different types of transmitters, one pneumatic and the other electronic. The appendix ends with a presentation of the different types of control valves and of additional considerations when sizing these valves.

C-1 PRESSURE SENSORS

There are many types of pressure sensors (see References 1–4). The most common pressure sensor is the *Bourdon tube*, developed by the French engineer Eugene Bourdon. The Bourdon tube, shown in Fig. C-1.1, is basically a piece of tubing in the form of a horseshoe with one end sealed and the other end connected to the pressure source. The cross section of the tube is elliptical or flat, so the tubing tends to straighten as pressure is applied, and when the pressure is released the tubing returns to its original form so long as the elastic limit of the material of the tubing was not exceeded. The amount of straightening that the tubing undergoes is proportional to the applied pressure. Thus if the open end of the tubing is fixed, then the closed end can be connected to a pointer to indicate pressure or to a transmitter to generate a signal.

The pressure range measured by the Bourdon tube depends on the wall thickness and the material of the tubing. An extended Bourdon tube in the form of a helical spiral, called the *helix*, permits additional motion of the sealed end. The helix can handle pressure ranges of about 10:1 with an accuracy of ± 1 % of the calibrated span · (Ryan, 1975). Another common type of Bourdon tube is the *spiral* element.

Another type of pressure sensor is the *bellows* which looks like a corrugated capsule made up of a somewhat elastic material such as stainless steel or brass. On increasing pressure the bellows expands, and on decreasing pressure it contracts. The amount of expansion or contraction is proportional to the applied pressure. Similar to the bellows is the *diaphragm* sensor. As the process pressure increases, the center of the diaphragm moves away from the pressure. The amount of motion is proportional to the applied pressure.

C-2 FLOW SENSORS

Flow is one of the two most commonly sensed process variables, the other being temperature; consequently, many different types of flow sensors have been developed

---

Principles and Practice of Automatic Process Control/Third Edition, by C. A. Smith and A. B. Corripio
ISBN 0-471-66141-4   Copyright © 2006 John Wiley & Sons (Asia) Pte. Ltd.

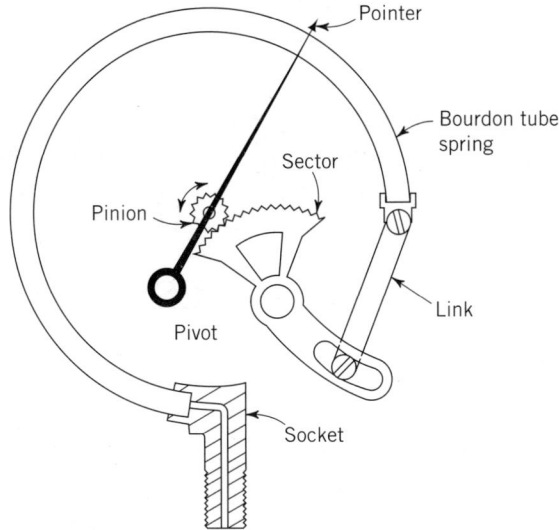

Figure C-1.1 Simple bourdon tube. (Courtesy of the Instrumentation, Systems, and Automation Society.)

(see References 2 and 4–7). This section describes the most used flow sensors, and mentions some others. Table C-2.1 (Zientara, 1972) shows several characteristics of some common sensors.

A common flow sensor is the *orifice meter*, which is a flat disk with a machined hole; see Fig. C-2.1. The disk is inserted in the process line perpendicular to the fluid motion with the intention of producing a pressure drop, Δp_o. This pressure drop across the orifice is a nonlinear function of the volumetric flow rate through the orifice. Accurate orifice meter flow equations are complex and are presented in many fine references (see References 8, 9, and 10); however, most installations probably use the following simple equation

$$f = C_o A_o \sqrt{\frac{\Delta p_o}{\rho(1 - \beta^4)}} \qquad \textbf{(C-2.1)}$$

where

f = volumetric flow rate

Δp_o = pressure drop across orifice

A_o = area of orifice

C_o = orifice coefficient

ρ = fluid density

β = dimensionless ratio of the diameter of the orifice, d, to the diameter of the pipe, D

Equation C-2.1 derives from the application of a mass balance and a mechanical energy balance (Bernoulli). The previously cited references also show how to size the required orifice diameter. Most orifice diameters vary between 10 and 75 % of the pipe diameter, $0.1 < \beta < 0.75$.

The pressure drop across the orifice is usually measured with taps:

1. Flange taps are the most common. Their technique consists of measuring the pressure drop across the flanges holding the orifice in the process line.

Table C-2.1 Characteristics of typical flow sensors

Primary element	Type of fluid	Pressure loss[a]	Flow rangeability	Error	Upstream piping[b]	Viscosity effect	Readout
Concentric orifice	Liquids, gases, and steam	50–90 %	3:1	3–4 %	10–30D	High	Square root
Segmental orifice	Liquid slurries	60–100 %	3:1	2.5 %	10–30D	High	Square root
Eccentric orifice	Liquid–gas combination	60–100 %	3:1	2 %	10–30D	High	Square root
Quadrant edged orifice	Viscous liquids	45–85 %	3:1	1 %	20–50D	Low	Square root
Segmental wedge	Slurries and viscous liquids	30–80 %	3:1	1 %	10–30D	Low	Square root
Venturi tube	Liquids and gases	10–20 %	3:1	1 %	5–10D	Very high	Square root
Dall tube	Liquids	5–10 %	3:1	1 %	5–10D	High	Square root
Flow nozzle	Liquids, gases, and steam	30–70 %	3:1	1.5 %	10–30D	High	Square root
Elbow meter	Liquid	None	3:1	1 %	30D	Negligible	Square root
Rotameter	All fluids	1–200″ WG	10:1	2 %	None	Medium	Linear
V-notch weir	Liquids	None	30:1	4 %	None	Negligible	5/2
Trapezoidal weir	Liquids	None	10:1	4 %	None	Negligible	3/2
Parshall flume	Liquid slurries	None	10:1	3 %	None	Negligible	3/2
Magnetic flowmeter	Liquid slurries	None	30:1	1 %	None	None	Linear
Turbine meter	Clean liquids	0–7 psi	14:1	0.5 %	5–10D	High	Linear
Pitot tube	Liquids	None	3:1	1 %	20–30D	Low	Square root
Pitot venturi	Liquids and gases	None	3:1	1 %	20–30D	High	Square root
Positive displacement	Liquids	0–15 psi	10:1	0.5–2 %	None	None	Linear totalization
Swirlmeter	Gases	0–2 psi	10:1–100:1	1 %	10D	None	Linear
Vortex shedding	Liquids and gases	0–6 psi 0–5″ WG	30:1–100:1	0.25 %	15–30D	Minimum Reynolds no. 10,000	Linear
Ultrasonic	Liquids	None			None	None	Linear

[a]Pressure loss percentages are stated as percentages of differential pressure produced.

[b]Upstream piping is stated in the number of straight pipe diameters required before the primary element.

C-3

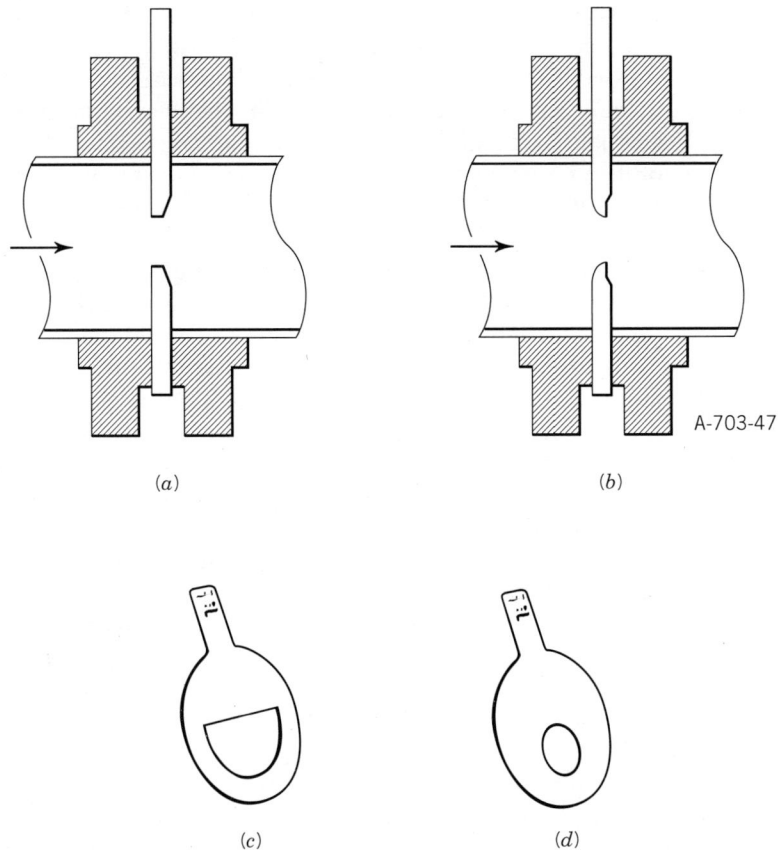

Figure C-2.1 Schematic of orifice meters. (*a*) Sharp edge. (*b*) Quadrant edge. (Courtesy of ABB Kent-Taylor.) (*c*) Segmental edge. (*d*) Eccentric edge. (Courtesy of Foxboro Co.)

2. Other types include vena contracta taps, radius taps, corner taps, and line taps. These are not as popular as flange taps.

The tap upstream from the orifice is called the high-pressure tap, and the one downstream from the orifice is called the low-pressure tap. Most tap diameters vary between 1/4 and 3/4 in. The pressure drop sensed will be a function of tap location as well as flow rate. A differential pressure sensor is used to measure the pressure drop across the orifice.

Several things must be stressed about the use of orifice meters to measure flows. The first is that the output signal from the orifice/transmitter combination is the pressure drop across the orifice, not the flow. Equation C-2.1 shows that this pressure drop is related to the square of the volumetric flow rate, or

$$\Delta p_o \propto \rho f^2 \qquad \textbf{(C-2.2)}$$

Consequently, if the flow is desired, then the square root of the pressure drop must be obtained; Chapter 10 presents square root extractors. Most manufacturers offer the option of installing a square root extraction unit within the transmitter. In this case, the output signal from the transmitter is linearly related to the volumetric flow. In distributed control systems (DCSs) and other microprocessor-based systems, a square root extractor is not needed because the square root is an input option. That is, the control system can be configured such that when it reads a signal it extracts the square root automatically and keeps that result in memory. The second thing that must be stressed is that not the entire pressure drop measured by the taps is lost by the process fluid. A certain amount is recovered by the fluid, in the next few pipe diameters, as it reestablishes its flow regime. Finally, the rangeability of the orifice meter—the ratio of the maximum measurable flow to the minimum measurable flow—is about 3:1, as indicated in Table C-2.1. This is important to know; it indicates the expected accuracy when running the process at low or high loads.

Several conditions may prevent the use of orifice sensors. Among such causes are not enough available pressure to provide pressure drop, as in the case of gravity flow; the flow of corrosive fluids; fluids with suspended solids that may plug the orifice; and fluids close to their saturated vapor pressure that may flash when subjected to a drop in pressure. These cases require the use of other sensors to measure flow.

Another common type of sensor is the *magnetic flowmeter*. The operating principle of this element is Faraday's law; that is, as a conductive material (a fluid) moves at right angles through a magnetic field, it induces a voltage. The voltage created is proportional to the intensity of the magnetic field and to the velocity of the fluid. If the intensity of the magnetic field is constant, then the voltage is proportional to the velocity of the fluid. Furthermore, the velocity measured is the average velocity, and thus this sensor can be used for both regimes, laminar and turbulent. During calibration of this flowmeter, the cross-sectional area of the pipe is taken into consideration so that the electronics associated with the meter can calculate the volumetric flow. Thus, the output is linearly related to the volumetric flow rate.

Because the magnetic flowmeter does not restrict flow, it is a zero pressure drop device suitable for measuring gravity flow, slurry flows, and flow of fluids close to their vapor pressure. However, the fluid must have a minimum required conductivity of about 10 μohm/cm^2, which makes the meter unsuitable for the measurement of both gases and hydrocarbon liquids.

Table C-2.1 shows that the rangeability of the magnetic flowmeter is 30:1, which is significantly greater than that of orifice meters; however, their cost is also greater. The cost differential increases as the size of the process pipe increases.

An important consideration in the application and maintenance of magnetic flow-meters is coating of the electrodes. This coating represents another electrical resistance resulting in erroneous readings. Manufacturers offer techniques such as ultrasonic cleaners for maintaining clean electrodes.

Another important flowmeter is the *turbine meter*. This meter is one of the most accurate of the commercially available flowmeters. Its working principle consists of a rotor that the fluid velocity causes to spin. The rotation of the blades is detected by a magnetic pickup coil that emits pulses the frequency of which is proportional to the volumetric flow rate; this pulse is equally converted to a 4- to 20-mA signal. The problems most commonly associated with turbine meters arise with the bearings, which require clean fluids with some lubricating properties.

The mass flow rate measurement of liquids using the *Coriolis effect* is common in the process industries. The measurement accuracy using this effect is unaffected by changes in the fluid's temperature, density, pressure, and viscosity, or by changes in velocity profile. Because the measurement principle is only based on mass flow, once the calibration of the flowmeter has been established with a conventional fluid such as water, it applies equally as well to other fluids. The meter has a maximum error of ± 0.15 % of reading over a dynamic range of 10:1, and is suitable for most fluids over a range of 100:1 with an accuracy of ± 1.5 % of reading.

We have briefly discussed four of the most common flowmeters in use in the process industries. There are many other types. They range from rotameters, flow nozzles, venturi tubes, pitot tubes, and annubars, which have been used for many years, to more recent developments such as vortex-shedding meters, ultrasonic meters, thermal conductivity mass meters, and swirlmeters. Limited space does not permit a discussion of these meters. The reader is directed to the many fine references given at the beginning of this section for discussion of these meters.

C-3   LEVEL SENSORS

The three most important level sensors are the differential pressure, float, and air bubbler sensors (see References 2, 5, 6, and 11). The *differential pressure* method consists of sensing the difference in pressure between the pressure at the bottom of a liquid and that above the liquid level, as shown in Fig. C-3.1. This differential pressure

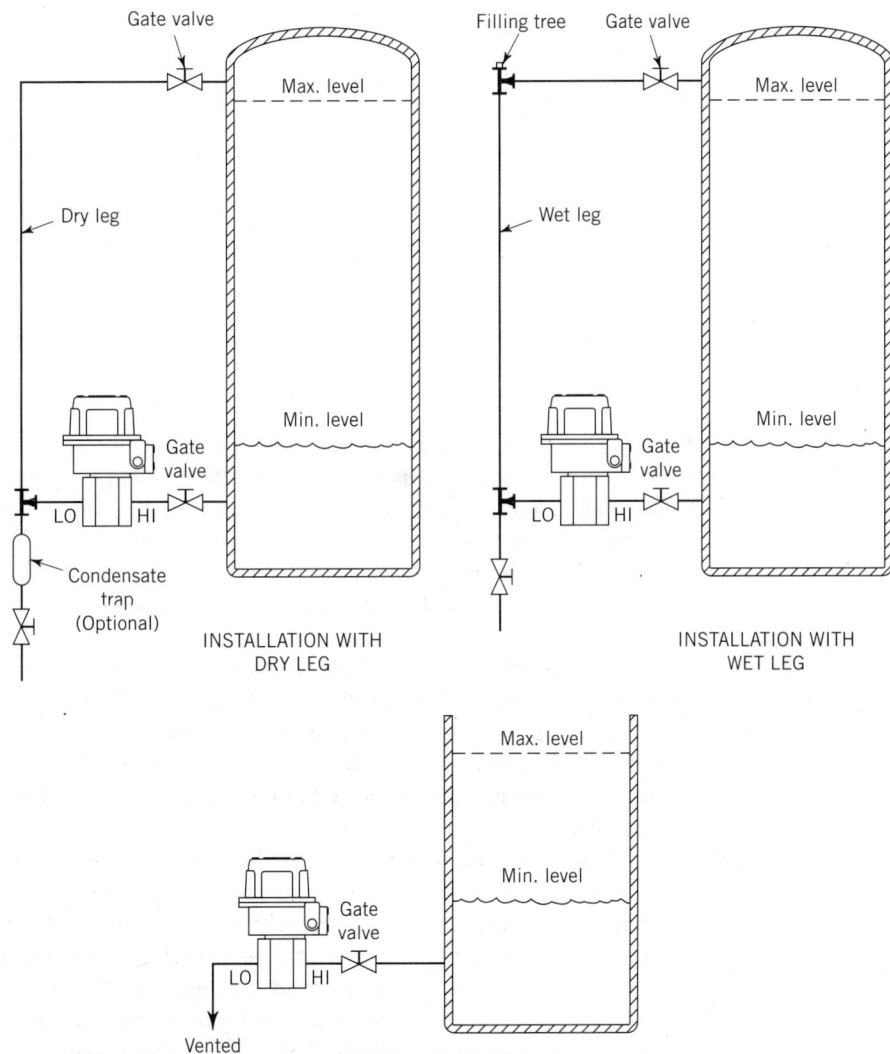

Figure C-3.1 Differential pressure transmitters installed in closed and in open vessels. (Courtesy of ABB Kent-Taylor.)

is caused by the hydrostatic head developed by the liquid level. The side that senses the pressure at the bottom of the liquid is referred to as the high-pressure side and the one that senses the pressure above the liquid level is referred to as the low-pressure side. Knowing the differential pressure and the density of the liquid makes it possible to obtain the level. Figure C-3.1 shows the installation of the differential pressure sensor in open and closed vessels. If the vapors above the liquid level are noncondensable, then the low-pressure piping, also known as the wet leg, can be empty. However, if the vapors are likely to condense, then the wet leg must be filled with a suitable seal liquid. If the density of the liquid varies, then some compensation technique must be employed.

The *float sensor* detects the change in buoyant force on a body immersed in the liquid. This sensor is generally installed in an assembly mounted externally to the vessel. The force required to keep the float in place, which is proportional to the liquid level, is then converted to a signal by the transmitter. This type of sensor is less expensive than most other level sensors; however, a major disadvantage lies in their inability to change their zero and span. To change the zero requires relocation of the complete housing.

The *bubbler sensor* is another type of hydrostatic pressure sensor. It consists of an air or inert gas pipe immersed in the liquid. The air or inert gas flow through the pipe is regulated to produce a continuous stream of bubbles. The pressure required to produce this continuous stream is a measure of the hydrostatic head or liquid level.

There are some other methods to measure level in tanks, such as capacitance gauges, ultrasonic systems, and nuclear radiation systems. The last two sensors are also used to measure the level of solid material. The references cited at the beginning of this section are recommended for further reading.

C-4 TEMPERATURE SENSORS

Along with flow, temperature is the most frequently measured variable in the process industries. A simple reason is that very few physical phenomena are not affected by it. Temperature is also often used to infer other process variables. Two of the most common examples are in distillation columns and in chemical reactors. In distillation columns, temperature is commonly used to infer the purity of one of the exit streams. In chemical reactors, temperature is used as an indication of the extent of reaction or conversion.

Because of the many effects produced by temperature, numerous devices have been developed to measure it (see References 2, 3, and 6). With a few exceptions, the devices fall into four general classifications as shown in Table C-4.1. Quartz thermometers, pyrometric cones, and specialized paints are some of the sensors that do not fit into the classifications shown in Table C-4.1. Table C-4.2 (Zientara, 1972) shows some characteristics of typical sensors.

Liquid-in-glass thermometers indicate temperature change caused by the difference between the temperature coefficient of expansion for glass and the liquid employed. Mercury and alcohol are the most widely used liquids. Mercury-in-glass thermometers made from ordinary glass are useful between -35 and $600°F$. The lower limit is due to the freezing point of mercury and the upper limit to its boiling point. By filling the space above the mercury with an inert gas (usually nitrogen) to prevent boiling, the useful range may be extended to $950°F$. Such thermometers usually bear the inscription "nitrogen filled." For temperatures below the freezing point of mercury ($-38°F$) another liquid must be employed. Alcohol is the most widely used fluid for temperatures down

Table C-4.1   Popular sensors for temperature measurement

I.   Expansion thermometer
    A.   Liquid-in-glass thermometer
    B.   Solid-expansion thermometers (bimetallic strip)
    C.   Filled-system thermometers (pressure thermometers)
        1. Gas-filled
        2. Liquid-filled
        3. Vapor-filled
II.   Resistance-sensitive devices
    A.   Resistance thermometers
    B.   Thermometers
III.   Thermocouples
IV.   Noncontact methods
    A.   Optical pyrometers
    B.   Radiation pyrometers
    C.   Infrared techniques

---

to $-80°$F, pentane for temperatures down to $-200°$F, and toluene for temperatures below $-230°$F.

The *bimetallic strip thermometer* works on the principle that metals expand with temperature and that the expansion coefficients are not the same for all metals. The temperature-sensitive element is a composite of two different metals fastened together into a strip. One metal has a high thermal expansion coefficient and the other metal has a low thermal expansion coefficient. A common combination is invar (64 % Fe, 36 % Ni), which has a low coefficient and another nickel–iron alloy that has a high coefficient. Usually the expansion with temperature is low, and this is the reason for having the bimetallic strip wound in the form of a spiral. As the temperature increases, the spiral will tend to bend toward the side of the metal with the low thermal coefficient.

Another common temperature sensor is the *filled-system thermometer*. Temperature variations cause the expansion or contraction of the fluid in the system, which is sensed by the Bourdon spring and transmitted to an indicator or transmitter. Because of the design simplicity, reliability, relatively low cost, and inherent safety, these elements are popular in the process industries. For a more extensive description of these systems the reader is referred to References 2 and 3.

Resistance temperature devices (RTDs) are elements based on the principle that the electrical resistance of pure metals increases with an increase in temperature. Because measurements of electrical resistance can be made with high precision, this also provides a very accurate way to make temperature measurements. The most commonly used metals are platinum, nickel, tungsten, and copper. A Wheatstone bridge is usually used for the resistance, and consequently also the temperature, reading.

Thermistor elements detect very sensitive temperature changes. Thermistors are made of a sintered combination of ceramic material and some kind of semiconducting metallic oxide such as nickel, manganese, copper, titanium, or iron. Thermistors have a very high negative, or sometimes positive, temperature coefficient of resistivity. Some of the advantages are their small size and low cost. Their main disadvantages lie in their nonlinear temperature versus resistance relationship and in the fact that they usually require shielded power lines. Wheatstone bridges are usually used to measure the resistance and, therefore, also temperature.

Table C-4.2 Characteristics of typical temperature sensors

Sensors	Range, °F	Accuracy, °F	Advantages	Disadvantages
Glass-stem thermometer	Practical: −200 to 600 Extreme: −321 to 1100	0.1–2.0	Low cost, simplicity, long life	Difficult to read, only local measurement, no automatic control or recording capability
Bimetallic thermometer	Practical: −80 to 800 Extreme: −100 to 1000	1.0–20	Less subject to breakage, dial reading, less costly than thermal or electrical	Less accurate than glass stem thermometer, changes calibration with rough handling
Filled thermal elements	Practical: −300 to 1000 Extreme: −450 to 1400	0.5–2 % of full scale	Simplicity, no auxiliary power needed, sufficient response times	Larger bulb size than electrical systems and greater minimum spans, bulb to readout distance is maximum of 50–200 ft, factory repair only
Resistance thermometer	−430 to 1800	0.1 (best)	System accuracy, low spans (10°F) available, fast response, small size	Self-heating may be a problem, long-term drift exceeds that of thermocouple, some forms expensive or difficult to mount
Thermocouples	−440 to 5000	0.2 (best)	Small size, low cost, convenient mount, wide range	Not as simple as direct-reading thermometer, cold working on wires can affect calibration, 70°F nominal minimum span
Radiation pyrometer	0 to 7000	0.5–1.0 % of full scale	No physical contact, wide range, fast response, measure small target or average over large area	More fragile than other electrical devices, nonlinear scale, relatively wide span required
Thermistors	−150 to 600	0.1 (best)	Small size, fast response, good for narrow spans, low cost and stable, no cold junction	Very nonlinear response, stability above 600°F is a problem, not suitable for wide spans, high resistance makes system prone to pick up noise from power lines

The last temperature element is the *thermocouple*, probably the best-known industrial temperature sensor. The thermocouple works on a principle discovered by T. J. Seebeck in 1821. The Seebeck effect, or Seebeck principle, states that an electric current flows in a circuit of two dissimilar metals if the two junctions are at different temperatures. The voltage produced by this thermoelectric effect depends on the temperature difference between the two junctions and the metals used. The most common type of thermocouples are platinum–platinum/rhodium alloy, copper–constantan, iron–constantan, chromel–alumel and chromel–constantan. For a more detailed discussion of thermocouples, see References 2 and 3.

C-5   COMPOSITION SENSORS

Another important class of sensors is composition sensors. These sensors are used in environmental and product quality measurement and control. There are many different types of measurement sensors such as density, viscosity, chromatography, pH, and ORP. Because of space limitations, we cannot present these sensors; however, we want to make the reader aware of their importance (see References 2 and 25–28).

C-6   TRANSMITTERS

This section presents an example of a pneumatic transmitter and an example of an electrical transmitter. The objective is to present the reader with the working principles of these typical transmitters. The purpose of a transmitter is to convert the output from the sensor to a signal strong enough to be transmitted to a controller or any other receiving device. Most transmitters are either force-balance types or motion-balance types.

C-6.1   Pneumatic Transmitters

All pneumatic transmitters use a flapper-nozzle arrangement to produce an output signal proportional to the output from the sensor. A pneumatic differential pressure transmitter (see Reference 12), which is a force-balance type transmitter, will be used to illustrate the working principles. This transmitter is shown in Fig. C-6.1.

The twin diaphragm capsule is the sensor. It senses the difference in pressure between the high- and low-pressure sides. Previously, we learned that this type of sensor is used to measure liquid level and flow. The diaphragm is connected to a force bar by a flexure. The force bar is connected to the body of the transmitter by a stainless steel diaphragm. This diaphragm serves as a seal to the measuring cavity and also as a positive fulcrum for the force bar. The top of the force bar is connected by a flexure strap to a range rod. This range rod has a range wheel that also serves as a fulcrum. A feedback bellows and a zero adjustment are located in the bottom part of the range rod. Above the range rod, a flapper-nozzle arrangement and a pneumatic relay are located. As shown in the figure, the flapper is connected to the force bar–range rod combination.

As the diaphragm capsule senses a difference in pressure, this creates a tension or force on the lower end of the force bar. To be more specific, you may assume that the pressure on the high side increases, creating a pulling force on the force bar. This force results in a motion at the outer end of the bar, causing the flapper to move closer to the nozzle. In this case the output of the relay increases and this increases the force that the feedback bellows exerts on the range rod. This force balances the force of the

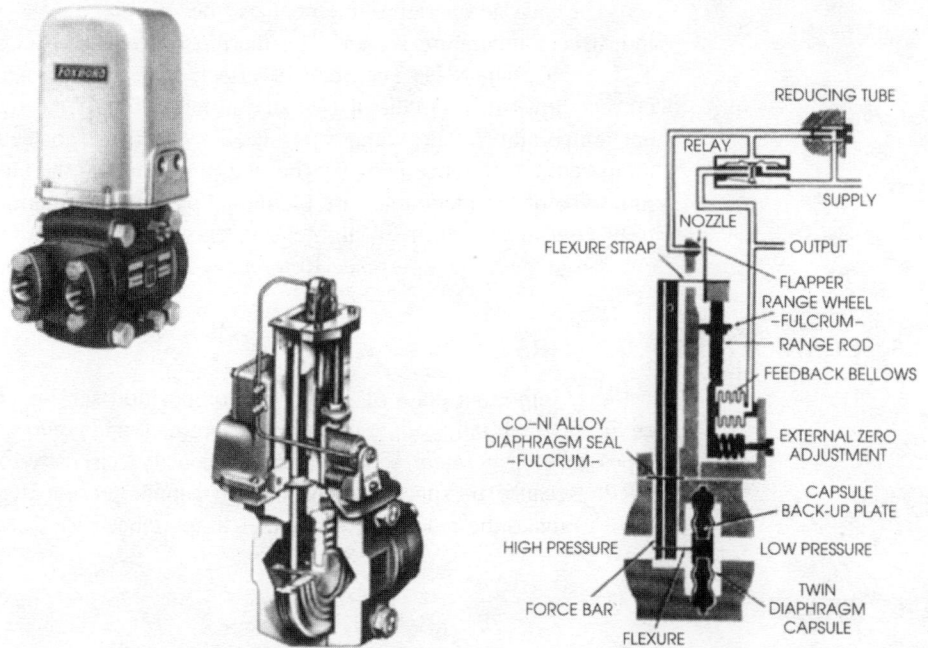

Figure C-6.1 Pneumatic differential pressure transmitter. (Courtesy of Foxboro Co.)

differential pressure across the diaphragm capsule. These balanced forces result in an output signal from the transmitter that is proportional to the difference in pressure.

The recommended supply pressure to most pneumatic instruments is between 20 and 25 psig. This ensures proper performance at the 15-psig output level. The calibration of these instruments requires the adjustment of the zero and span (or range). In the instrument shown in Fig. C-6.1 this is done with the external zero adjustment screw and with the range wheel.

The preceding paragraphs have described the working principle of a typical pneumatic instrument. As we noted at the beginning, all pneumatic instruments use some kind of flapper-nozzle arrangement to produce an output signal. This is a reliable and simple technique that has proved very successful for many years.

C-6.2 Electronic Transmitters

Figure C-6.2 shows a simplified diagram of an electronic differential pressure transmitter (Reference 13). This motion-balance type transmitter will be used to illustrate the working principles of typical electronic instrumentation.

An increase in differential pressure, acting on the measuring element diaphragms, develops a force that moves the lower end of the force beam to the left. This motion of the force beam is transferred to the strain gage force unit thorough the connecting wire. The strain gage force unit contains four strain gages connected in a bridge configuration. Movement of the force beam causes the strain gages to change resistance. This change in resistance produces a differential signal that is proportional to the input differential pressure. This differential signal is applied to the inputs of the input amplifier. One side of the signal is applied to the noninverting input through the zero network. This zero network provides the zero adjustment for the transmitter.

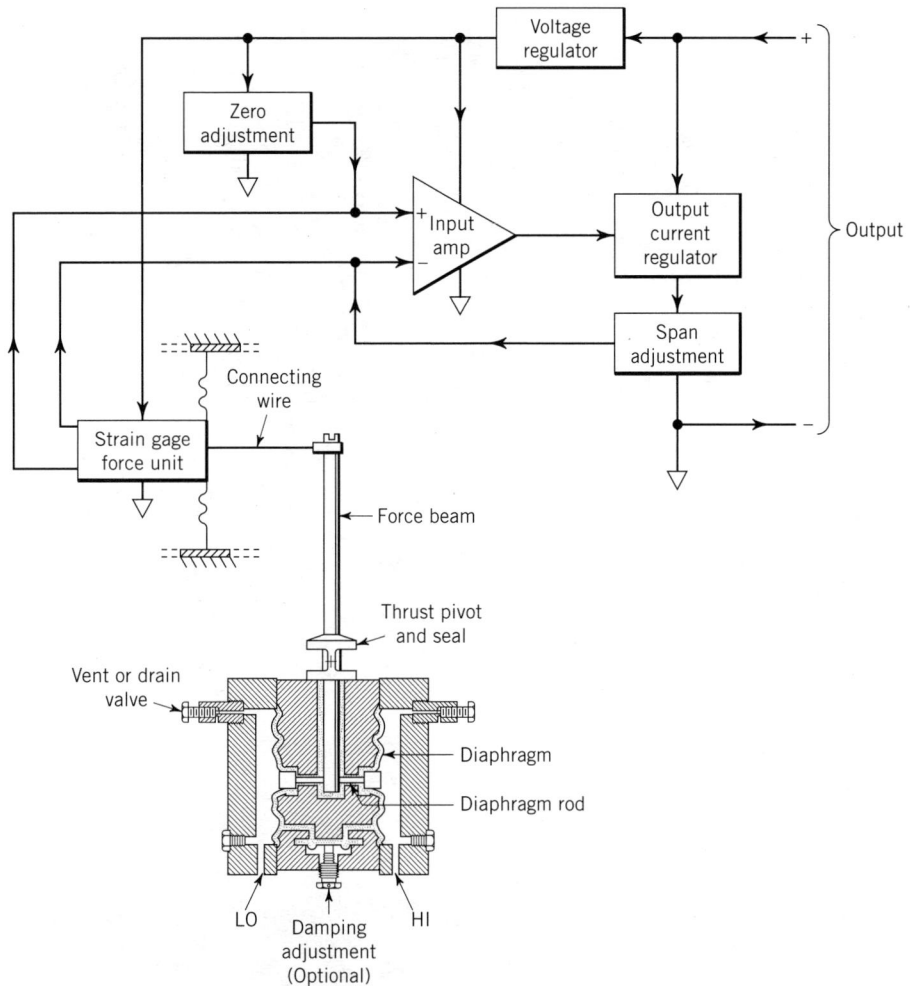

Figure C-6.2 Electronic differential pressure transmitter. (Courtesy of ABB Kent-Taylor.)

The signal from the input amplifier drives the output current regulator. The current regulator controls the transmitter output current through the span network and the output current sense circuit. The span network provides the span adjustment for the transmitter. The signal from the span network is fed back to the input circuit through a buffer amplifier and is used to control the gain of the input circuit. If the transmitter output current increases above 20 mA D.C., the voltage across the output current sense resistor will turn on the output current limiter, which limits the output.

C-7   TYPES OF CONTROL VALVES

There are many different types of control valves on the market (see References 14–20). Consequently, it is difficult to classify them; however, we will classify them into two broad categories: reciprocating stem and rotating stem.

C-7.1   Reciprocating Stem

Figure C-7.1 shows a typical reciprocating stem control valve. This particular valve is called a *single-seated sliding stem globe valve*. *Globe valves* are a family of valves

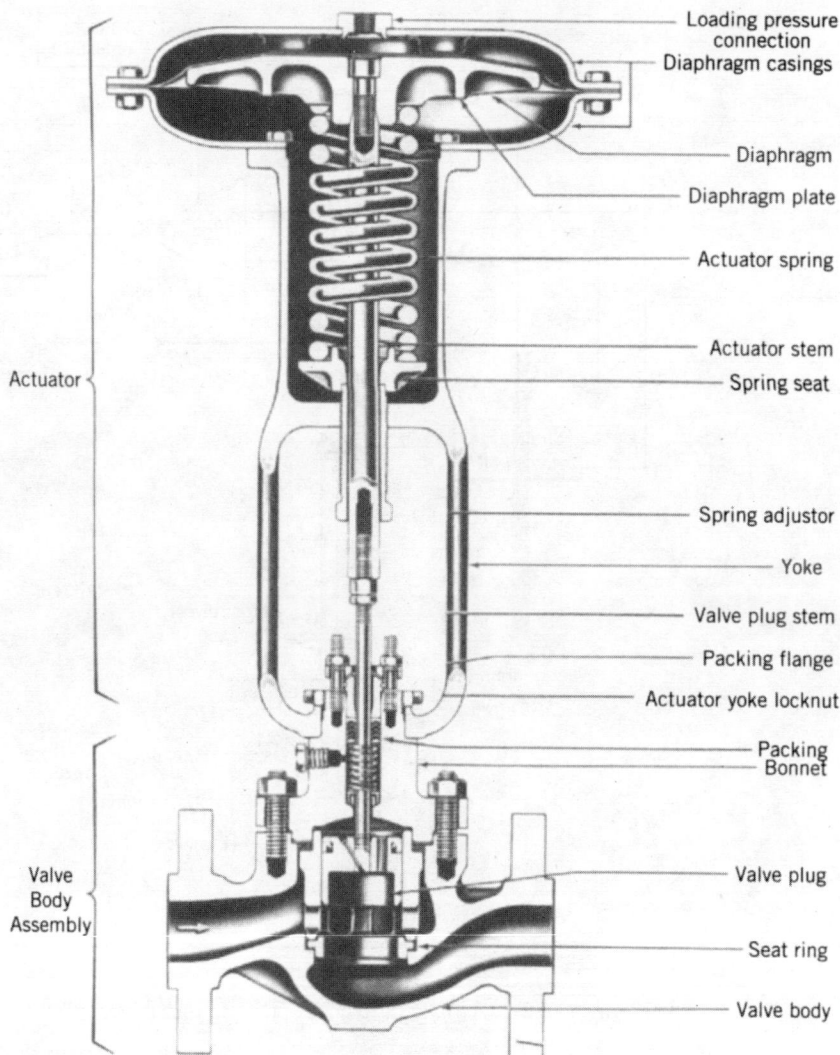

Figure C-7.1 Single-seated sliding stem globe valve. (Courtesy of Fisher Controls.)

characterized by a closure member that travels in a line perpendicular to the valve seat. They are used primarily for throttling purposes and general flow control. Figure C-7.1 also shows in detail the different components of the valve. The valve is shown to be divided into two general areas: the actuator and the body. The actuator is the part of the valve that converts the energy input to the valve into mechanical motion to increase or decrease the flow restriction. Figure C-7.2 shows a *double-seated sliding stem globe valve*. Double-seated valves can handle high process pressure with a standard actuator. However, if tight shut-off is required, single-seated valves are usually used. Double-seated valves tend to have greater leakage when closed than single-seated valves.

Another type of body in common use is the *split-body valve*. This type of body is frequently used in process lines where frequent changes of plug and seat are required because of corrosion.

Cage valves have hollow plugs with internal passages. The valve shown in Fig. C-7.1 is a cage valve.

Travel indicator

Travel indicator scale

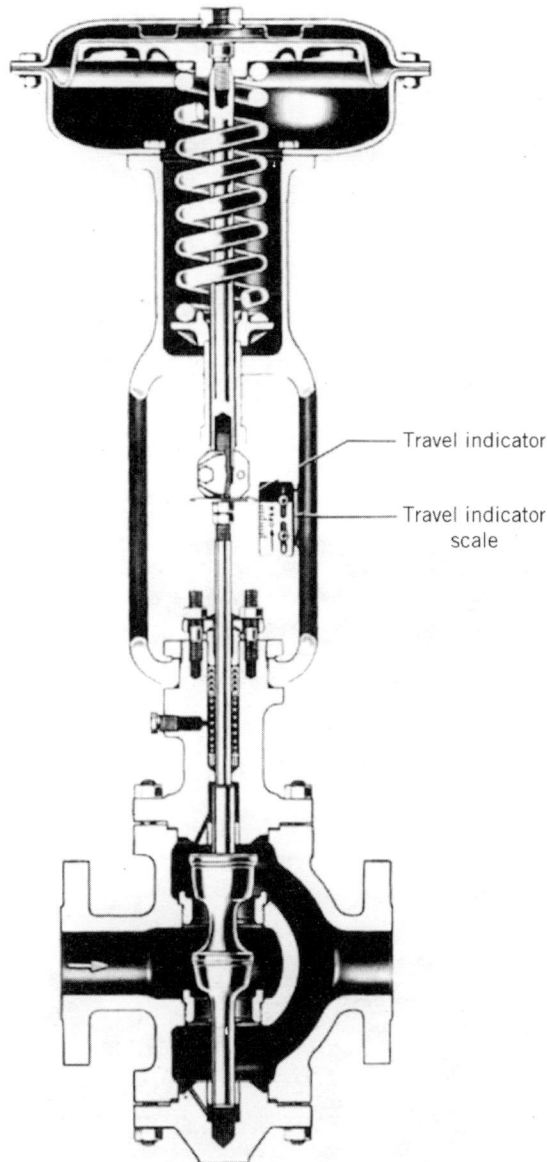

Figure C-7.2 Double-seated sliding stem globe valve. (Courtesy of Fisher Controls.)

Three-way valves are also reciprocating stem control valves. Three-way valves can be either diverging or converging and, consequently, they can either split one stream into two other streams or blend two streams into only one. They are sometimes used for control purposes.

There are some other types of reciprocating stem control valves. Most of them are used in specialized services. Some of these are the *Y-style valve*, which is commonly used in molten metal or cryogenic service. The *pinch or diaphragm valve* consists of some kind of flexure, such as a diaphragm, that can be moved together to open or close the area of flow. They are commonly used for highly corrosive fluids, slurries, and high-viscosity liquids, as well as in some food processing operations. The *gate valve* is another type of reciprocating stem valve. It is mainly used as a block valve, for fully open or fully closed services. It is not used as an automatic valve in throttling services.

(a)

(b)

Figure C-7.3 (*a*) Butterfly valve. (Courtesy of Fisher Controls.) (*b*) Ball valve with positioner. (Courtesy of Masoneilan Division, McGraw-Edison Co.)

C-7.2   Rotating Stem

There are several very popular types of rotating stem valves. One of the most common is the *butterfly valve*, shown in Fig. C-7.3a. These valves consist of a disk rotating about a shaft. They require minimum space for installation and provide high-capacity flow at low cost.

Another common rotating stem valve is the *ball valve* shown in Fig. C-7.3b. Ball valves also provide high-capacity flow at low cost. They are commonly used to handle slurries or fibrous materials. They have low leakage tendency and are small in size.

A very brief introduction to several types of control valves has been presented. However, these are by no means the only control valves, nor are they the only types of valves. There are a great number of valves available to meet the requirements for specialized services as well as safety and other types of regulation.

C-8   CONTROL VALVE ACTUATORS

As previously defined, the actuator is the part of the valve that converts the energy input, either pneumatic or electrical, into mechanical motion to open or close the valve.

C-8.1   Pneumatically Operated Diaphragm Actuators

These are the most common actuators in the process industries. Figure C-8.1 shows a typical diaphragm actuator. These actuators consist of a flexible diaphragm placed between two casings. One of the chambers resulting from this arrangement must be made pressure-tight. The force generated within the actuator is opposed by a "range" spring. The controller air signal goes into the pressure-tight chamber, and an increase or decrease in air pressure produces a force that is used to overcome the force of the actuator's range spring and the forces within the valve body.

The action of the valve, FC or FO, is determined by the actuator. Figure C-8.1a shows a fail-closed or air-to-open valve. Figure C-8.1b shows a fail-open or air-to-close valve. Some valves can also have the action set at body (reversed plug or cage) so that the stem always moves down. That is, in these cases the valve is either FC or FO when the stem moves down.

The size of the actuator depends on the process pressure against which it must move the stem and on the air pressure available. The most common air pressure range is 3 to 15 psig but ranges of 6 to 30 psig and 3 to 27 psig are sometimes also used. These diaphragm actuators are simple in construction and also dependable and economical. Equations for sizing actuators are provided by manufacturers.

C-8.2   Piston Actuators

Piston actuators are normally used when maximum thrust output is required along with fast response. This usually occurs when working against high process pressure. These actuators operate using a high air pressure supply, up to 150 psig. The best designs are double-acting to give maximum thrust in both directions.

C-8.3   Electrohydraulic and Electromechanical Actuators

In this family of actuators, probably the most common one is the solenoid actuator. A solenoid valve can be used to actuate a double-acting piston actuator. By making

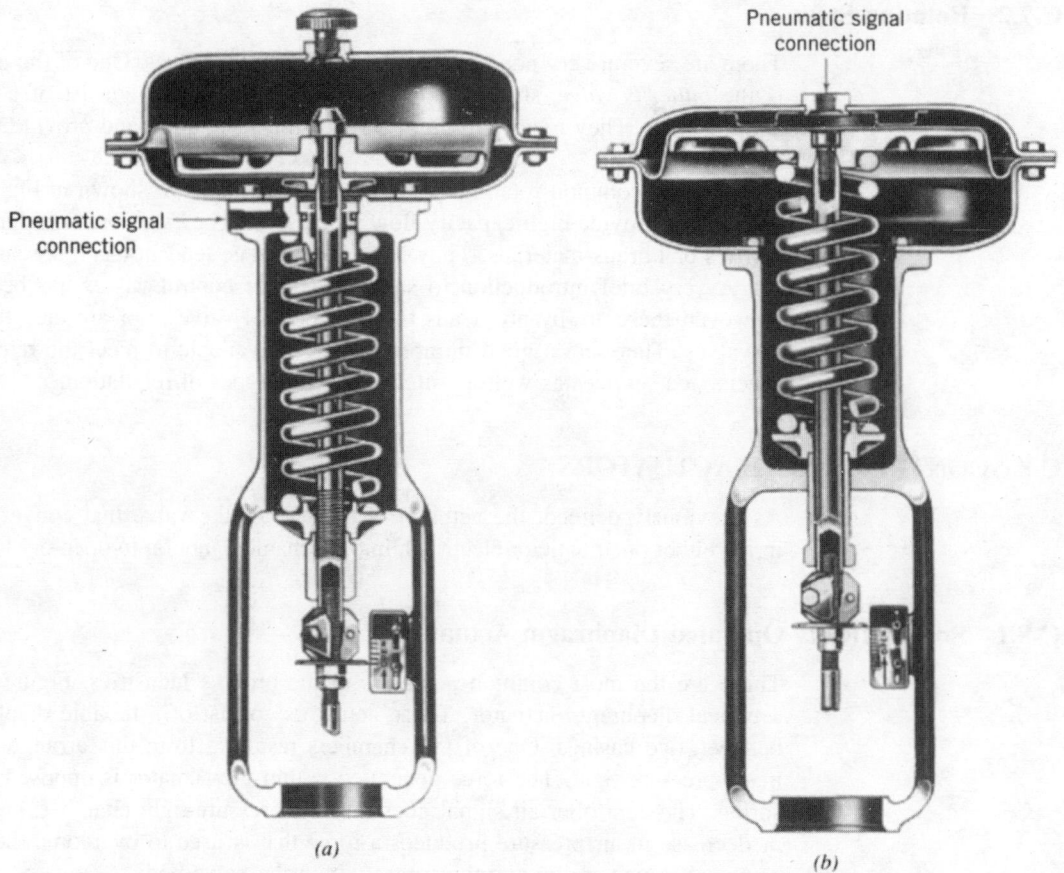

Figure C-8.1 (*a*) Diaphragm actuator: fail-closed, air-to-open. (*b*) Diaphragm actuator: fail-open, air-to-close. (Courtesy of Fisher Controls.)

or breaking an electric current signal, the solenoid switches the output of a connected hydraulic pump to either above or below the actuator piston. Accurate control of valve position can be obtained with this unit. They require electric power to the motor and an electric signal from the controller.

C-8.4 Manual-Handwheel Actuators

These actuators are used where automatic control is not required. They are available for reciprocating stem and rotary stem.

C-9 CONTROL VALVE ACCESSORIES

There are a number of devices, called accessories, that usually go along with control valves. This section presents a brief introduction to some of the most common of these accessories.

C-9.1 Positioners

A positioner is a device that acts very much like a proportional controller with very high gain. Its job is to compare the signal from the controller with the valve stem position.

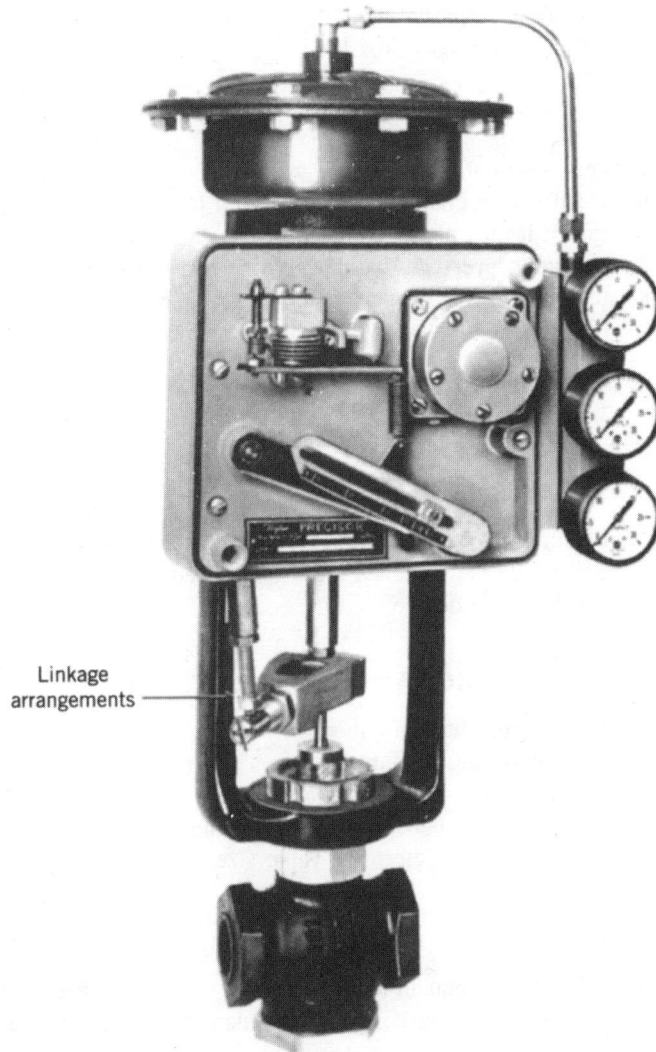

Figure C-9.1 Positioner installed in a valve. (Courtesy of ABB Kent Taylor.)

If the stem is not where the controller wants it to be positioned, the positioner adds or exhausts air from the valve until the correct valve position is obtained. That is, when it is important to position the valve's stem accurately, a positioner is normally used. Figure C-9.1 shows a valve with a positioner. The figure shows the bar-linkage arrangement by which the positioner senses the stem position. Another positioner is shown in Fig. C-7.3*b*.

The use of positioners tends to minimize the effects of:

1. Lag in large-capacity actuators.
2. Stem friction due to tight stuffing boxes.
3. Friction due to viscous or gummy fluids.
4. Process line pressure changes.
5. Hysteresis

Some control loops for which positioners are common are temperature, liquid level, concentration, and gas flow loops.

C-9.2 Boosters

Boosters, also called air relays, are used on valve actuators to speed up the response of the valve to a changing signal from a low-output-capacity pneumatic controller or transducer. It may also be noticed that for fast responding control loops, such as liquid flow or liquid pressure, with which the use of positioners is discouraged, the use of boosters may be the proper choice (Reference 14).

Boosters also have several other possible uses:

1. Amplify a pneumatic signal. Some typical amplification ratios are 1:2 and 1:3.
2. Reduce a pneumatic signal. Typical ratios are 5:1, 3:1, and 2:1.

C-9.3 Limit Switches

Limit switches are mounted on the side of the valves and are triggered by the position of the stem. These switches are usually used to drive alarms, solenoid valves, lights, or any other such device.

C-10 CONTROL VALVES—ADDITIONAL CONSIDERATIONS

This section presents a number of additional considerations to take into account when sizing and choosing a control valve. Thus, this section complements Section 5-2.

Figures C-10.1 show an example of a manufacturer catalog (Masoneilan). Once the C_v coefficient has been calculated using the equations presented in Chapter 5, this figure is used to the determine valve size.

C-10.1 Viscosity Corrections

The valve-sizing equation presented in Chapter 5 does not take into consideration the effect of liquid viscosity in calculating the valve capacity, C_v, coefficient. For

20000 Series

ANSI Class 150-600 (Sch. 40)

Nominal Trim Size	¼	⅜	½	¾	1	1½	2	3	4	6	8	10
Orifice Dia. (in.)	.250	.375	.500	.750	.812	1.250	1.625	2.625	3.500	5.000	6.250	8.000
Valve Size (in.)	Reduced Trim				Full Capacity Trim							
¾	1.7	3.7	6.4	11								
1	1.7	3.7	6.4	11	12							
1½	1.7	3.8	6.6	12	13	25						
2	1.7	3.8	6.7	13	19*	26	46					
3				14		31	47	110				
4						32	49	113	195			
6							53	126	208	400		
8								133	224	415	640	
10									233	442	648	1000

Orifice Diameter 994   Refer to Bulletin 334E

(a)

Figure C-10.1 Example of Masoneilan's valve catalog. (Courtesy of Masoneilan Division, McGraw-Edison Co.)

liquids with the viscosity of water and light hydrocarbons, the viscous effects in valve capacity are negligible. However, for very viscous liquids, the viscous effects can lead to sizing errors.

Masoneilan (Reference 15) proposes to calculate a turbulent C_v and a laminar C_v and then to use the larger value as the required C_v.

Turbulent flow

$$C_v = f\sqrt{\frac{G_f}{\Delta p}}$$

(**C-10.1**)

Laminar flow

$$C_v = 0.072\left(\frac{\mu f}{\Delta p}\right)^{2/3}$$

(**C-10.2**)

where

μ = viscosity, centipoises.

C-10.2   Flashing and Cavitation

The presence of either flashing or cavitation in a control valve can have significant effects on the operation of the valve and on the procedure for sizing it. It is important to understand the meaning and significance of these two phenomena. Figure C-10.2 shows the pressure profile of a liquid flowing through a restriction (possibly a control valve).

To maintain steady-state mass flow, the velocity of the liquid must increase as the cross-sectional area for flow decreases. The liquid velocity reaches its maximum at a point just past the minimum cross-sectional area (the port area for a control valve). This point of maximum velocity is called the *vena contracta*. At this point, the liquid also experiences the lowest pressure. What happens is that the increase in velocity (kinetic energy) is accompanied by a decrease in "pressure energy." Energy is transformed from one form to another.

As the liquid passes the vena contracta, the flow area increases and the fluid velocity decreases and, in so doing, the liquid recovers part of its pressure. Valves such as butterfly valves, ball valves, or most rotary valves have a high-pressure recovery characteristic. Most reciprocating stem valves show a low-pressure recovery characteristic. The flow path through these reciprocating stem valves is more tortuous than through rotary-type valves.

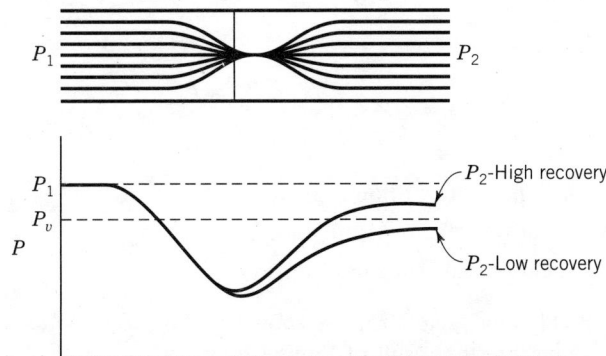

Figure C-10.2 Pressure profile of a liquid across a restriction.

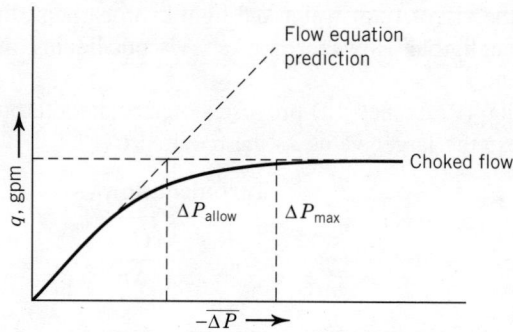

Figure C-10.3 Choked flow condition.

Considering Fig. C-10.2, let us suppose that the vapor pressure of the liquid at the flowing temperature is P_v. When the pressure of the liquid falls below its vapor pressure, some of the liquid starts changing phase from the liquid phase to the vapor phase. That is, the liquid flashes, and can produce serious erosion damage to the valve plug and seat.

Aside from the physical damage to the valve, flashing tends to lower the flow capacity of the valve. As bubbles start forming, this tends to cause a "crowding condition" at the valve, which limits the flow. Furthermore, this crowding condition may get bad enough to "choke" the flow through the valve. That is, beyond this choked condition, increases in pressure drop across the valve will not result in an increased flow. It is important to recognize that the valve-sizing equation of Chapter 5 does not describe this condition. As the pressure drop increases, the equation will predict higher flow rates. This relationship is shown graphically in Fig. C-10.3 along with the choked flow condition.

Notice from this figure that it is important for the engineer to know what maximum pressure drop, $\Delta P_{\max}$, is effective in producing flow. Instead of providing an equation for $\Delta P_{\max}$, manufacturers have chosen to provide an equation for $\Delta P_{\mathrm{allow}}$ and use this term to indicate when choke flow occurs. At higher pressure drops than $\Delta P_{\mathrm{allow}}$, choked flow results. $\Delta P_{\mathrm{allow}}$ is a function of not only the fluid but also of the type of valve. Masoneilan (Reference 15) proposes the following equation:

$$\Delta P_{\mathrm{allow}} = C_f^2 \Delta P_s \tag{C-10.3}$$

and

$$\Delta P_s = P_1 - \left(0.96 - 0.28\sqrt{\frac{P_v}{P_c}}\right) P_v \tag{C-10.4}$$

or, if $P_v < 0.5P_1$,

$$\Delta P_s = P_1 - P_v \tag{C-10.5}$$

where

P_v = vapor pressure of liquid in psia
C_f = critical flow factor (Reference 15)
P_c = critical pressure of liquid in psia

The critical flow factor, C_f, is shown in Fig. C-10.4 for different types of valves. These values are the result of flow tests performed on the valves.

Valve Type	Trim Size	Flow To	C_f (F_L)	K_c*	C_{fr} (F_{LP}) D/d = 1.5 or greater	X_T
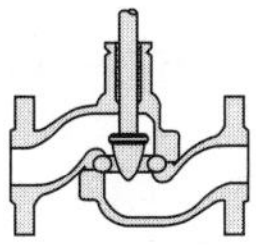 Split body globe valves	A	→ Close Open ←	.80 .75	.51 .46	.77 .72	.54 .47
	B	→ Close Open ←	.80 .90	.52 .65	.80 .89	.54 .68
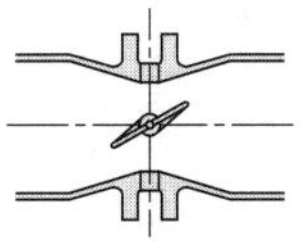 37000 Series	A	Flow in Either Direction	.65	.32	.60	.35
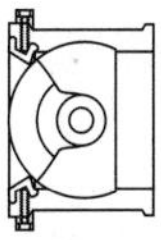 Control ball valve	A	← Open	.60	.24	.55	.30
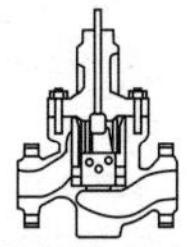 40000 Series balanced	1½" – 4"	→ Close	.94	.71	.87	.74
	6" – 16"	Close →	.92	.68	.89	.71
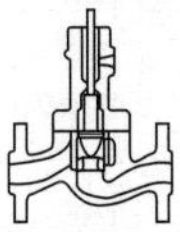 40000 Series unbalanced	A	← Open	.90	.65	.79	.68
	B	← Open	.90	.65	.86	.68
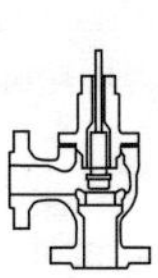 70000 Series	A	→ Close Open ←	.81 .89	.53 .64	.78 .85	.55 .67
	B	→ Close Open ←	.80† .90	.52† .65	.80 .90	.54 .68

(A) Full capacity trim, orifice dia. = .8 valve size.

(B) Reduced capacity trim 50% of (A) and below.

† With Venturi Liner C_f = 0.50, K_c = 0.19.

Note $X_T = 0.84\ C_f^2$

Figure C-10.4 Critical flow factor, C_f, at full opening. (Courtesy of Masoneilan Division, McGraw-Edison Co.)

If the pressure recovery experienced by the liquid is enough to raise the pressure above the vapor pressure of the liquid, then the vapor bubbles will start collapsing, or imploding. This implosion is called *cavitation*. The energy released during cavitation will produce noise, as though gravel were flowing through the valve (Reference 14), and tears away the material of the valve. High-pressure recovery valves, rotary stem valves, tend to experience cavitation more often than low-pressure recovery valves, reciprocating stem.

Tests have shown that for low-pressure recovery valves, such as rotary valves, choked flow and cavitation occur nearly at the same ΔP and, consequently, Eqs. C-10.3 and C-10.6 can also be used to calculate the pressure drop at which cavitation starts. For high-pressure recovery valves, cavitation can occur at pressure drops below ΔP_{allow}. For these types of valves Masoneilan (Reference 15) proposes the following equation

$$\Delta P_{\text{cavitation}} = K_c(P_1 - P_v) \tag{C-10.6}$$

where

K_c = coefficient of incipient cavitation, shown in Fig. C-10.4

Special anticavitation trims are produced by valve manufacturers that tend to increase the K_c term of the valve and, therefore, the pressure drop at which cavitation will occur.

C-11 SUMMARY

The purpose of this appendix is to introduce the reader to some of the most common instrumentation used for process control. The instrumentation shown includes some of the hardware necessary for the measurement of process variables (primary elements) such as flow, pressure, temperature, and pressure. Two types of transmitters are also presented, and their working principles are discussed. Finally, some common types of valves (final control elements) used to take action are presented along with their flow characteristics.

It is impossible to present and discuss in this book all of the details related to the different types of instruments; however, entire handbooks and a very exhaustive collection of articles are available for this purpose. The reader is referred to the fine references given at the end of this appendix. In addition to the many different types of instruments available today, new types of primary elements, transmitters, and final control elements are introduced on the market every month. In the primary elements area new sensors that can measure difficult variables, such as concentration, more exactly, more repeatable, and faster are developed constantly.

In the transmitter area the important word is *smart transmitters*. These are transmitters that with the aid of microprocessors, present the information to the controllers in a more understandable manner. The final control elements present another very active area of research. Not only are pneumatic control valves continually being upgraded, but electric actuators are also being developed and improved to allow interfacing with other electronic components such as controllers and computers. Other final control elements, such as drivers for variable-speed pumps and fans, are continuously being developed and improved. The impetus behind this development is energy conservation. Lack of space prevents a presentation on the feasibility and justification of the use of these variable-speed pumps and fans for flow throttling. The reader is referred to References 21, 22, 23, and 24 for a presentation on this subject.

The last paragraph indicates that there is a lot of research being conducted, principally by manufacturers in the instrumentation area that should result in better measurement and control. This is one reason why process control is such a dynamic field.

REFERENCES

1. RYAN, J. B., "Pressure Control." *Chemical Engineering*, February 3, 1975.

2. LIPTAK, BELA G., ed., *Instrument Engineers' Handbook, Volume I, Process Measurement*. New York: Chilton.

3. PERRY, R., and D. W. GREEN, *Perry's Chemical Engineering Handbook*, 6th Ed., New York: McGraw-Hill.

4. CONSIDINE, D. M., ed., *Handbook of Instrumentation and Controls*. New York: McGraw-Hill, 1961.

5. SMITH, C. L., "Liquid Measurement Technology." *Chemical Engineering*, April 3, 1978.

6. ZIENTARA, DENNIS E., "Measuring Process Variables." *Chemical Engineering*, September 11, 1972.

7. KERN, R., "How to Size Flowmeters," *Chemical Engineering*, March 3, 1975.

8. *Handbook Flowmeter Orifice Sizing*, Handbook No. 10B9000, Warminster, PA: Fischer Porter Co.

9. SPINK, L. K., *Principles and Practice of Flow Meter Engineering*. Foxboro, MA: The Foxboro Co.

10. KERN, R., "Measuring Flow in Pipes with Orifices and Nozzles." *Chemical Engineering*, February 3, 1975.

11. WALLACE, L. M. "Sighting in on Level Instruments." *Chemical Engineering*, February 16, 1976.

12. *Process Control Instrumentation*. Foxboro, MA: The Foxboro Co. Publication 105A-15M-4/71.

13. Taylor Instrument Co., *Differential Pressure Transmitter Manual, 1B-12B215*.

14. *Control Valve Handbook*, Marshalltown, Iowa: Fisher Controls Co.

15. *Masoneilan Handbook for Control Valve Sizing*. Norwood, MA: Masoneilan International, Inc.

16. *Fisher Catalog 10*, Marshalltown, Iowa: Fisher Controls Co.

17. CASEY, J. A., and D. HAMMITT, "How to Select Liquid Flow Control Valves." *Chemical Engineering*, April 3, 1978.

18. CHALFIN, S., "Specifying Control Valves." *Chemical Engineering*, October 14, 1974.

19. KERN, R., "Control Valves in Process Plants." *Chemical Engineering*, April 14, 1975.

20. HUTCHISON, J. W., ed., *ISA Handbook of Control Valves*, Research Triangle Park, NC: Instrument Society of America.

21. FISCHER, K. A., and D. J. LEIGH, "Using Pumps for Flow Control." *Instruments and Control Systems*, March 1983.

22. JARC, D. A., and J. D. ROBECHEK, "Control Pumps with Adjustable Speed Motors Save Energy." *Instruments and Control Systems*, May 1981.

23. PRITCHETT, D. H., "Energy-Efficient Pump Drive." *Chemical Engineering Progress*, October 1981.

24. BAUMANN, H. D., "Control Valve vs. Variable-Speed Pump." *Chemical Engineering*, June 24, 1981.

25. UTTERBACK, V. C. "Online Process Analyzers." *Chemical Engineering*, June 21, 1976.

26. FOSTER, R. "Guideline for Selecting Outline Process Analyzers." *Chemical Engineering*, March 17, 1975.

27. OTTMERS, D. M., *et al.*, "Instruments for Environment Monitoring." *Chemical Engineering*, October 15, 1979.

28. CREASON, S. C., "Selection and Care of pH Electrodes." *Chemical Engineering*, October 23, 1978.

29. YOUNG, A. M., "Coriolis-Based Mass Flow Measurement." *Sensors*, December, 1985.

30. DAHLIN, E., and A. FRANCI, *Practical Applications of a New Coriolis Mass Flowmeter*. Presented at Control Expo '85, Control and Engineering Conference and Exposition, Rosemont, IL, May 21–23, 1985.

Appendix D

Tuning Case Studies

This appendix describes six processes presented in the book's website at www.wiley.com/college/smith to practice the material presented in the book. Process 1, Regenerator Feedback, is used to practice tuning a feedback controller (Chapter 7); Process 2, Regenerator Cascade, is the same Regenerator process but now it is used to practice tuning a two-level cascade control scheme (Chapter 9); Process 3, Paper-Drying Process, is used to practice tuning a two-level cascade control scheme (Chapter 9); Process 4, Scrubber, is used to practice tuning a feedback controller (Chapter 7) and designing a feedforward controller (Chapter 11); Process 4, Mixing Process, is used to tune a feedback controller (Chapter 7); Process 5, Reactor Process, is used to practice selecting the best pairing in a two controlled variables–two manipulated variables process (Chapter 12); and Process 6, Distillation Process, is used to practice tuning controllers and designing a decoupler for an interacting MIMO process (Chapter 12).

As discussed in the chapters, it is necessary to obtain information from the process to tune controllers, to design feedforward controllers, and to decide on the best pairing. The necessary information is in the form of the process response to a step change input signal from the controller, and it is used to model the process by a first-order-plus-dead-time model; all of this is presented in Chapter 7. Unfortunately, there is no recorder to record the process variables in the processes presented in this appendix. Thus, you will have to generate a table of time versus process variable and graph this data. We recommend reading the process variable every 5 s. You should read the variable until a steady state is achieved again. To generate the step change in the controller's output, double-click on the number indicating the output, type the new desired output, and click enter. The charts presented with each process can be easily re-ranged by double-clicking on the low and high numerical in the y-axis and typing the new values.

The program used to develop the processes is Labview, a product of National Instruments. The authors are in debt to Dr. Marco E. Sanjuan and the late Dr. Daniel Palomares for their work in developing this appendix.

PROCESS 1. REGENERATOR FEEDBACK

A catalyst is used in a hydrocarbon reaction. As the reaction proceeds, some carbon deposits over the catalyst, poisoning the catalyst. After enough carbon has deposited it completely poisons the catalyst. At this moment it is necessary to stop the reaction and regenerate the catalyst. This regeneration consists of blowing hot air over the catalyst so that the oxygen in the air reacts with the carbon to form carbon dioxide, and in so doing burns the carbon. Figure D-1 shows the regeneration process. Ambient air

Principles and Practice of Automatic Process Control/Third Edition, by C. A. Smith and A. B. Corripio
ISBN 0-471-66141-4   Copyright © 2006 John Wiley & Sons (Asia) Pte. Ltd.

CATALYST REGENERATOR-R2

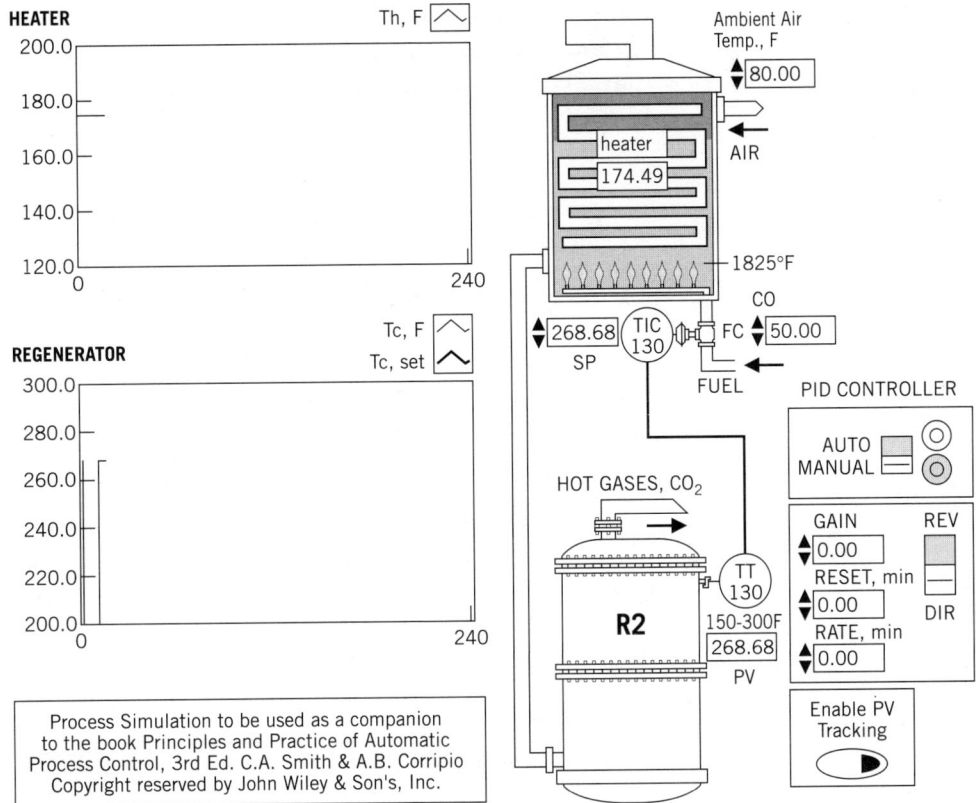

Figure D-1 Catalyst regenerator—feedback control.

is first heated in a small furnace, and then flows to the regenerator, which is full of catalyst. Manipulating the fuel flow controls the temperature in the catalyst bed.

To tune the feedback controller, that is, to find the values of K_C, τ_I, and τ_D, we must first find the process characteristics, process gain (K), time constant (τ), and dead time (t_0). Chapter 7 explains that to obtain the process characteristics a process reaction curve is necessary. To obtain this curve, we first introduce a step change in the controller's output and record the temperature in the regenerator. Unfortunately, there is no recorder to record this temperature. Thus, you will have to generate a table of time versus temperature, and graph this data. We recommend reading the temperature every 5 s. You should read the temperature until a steady state is achieved again. To generate the step change in the controller's output, double-click on the number, type the new desired output, and click enter. You may read again Chapter 7 to review process testing.

Once the process characteristics are obtained and the controller tuned, using the formulas of Chapter 7, then the tuning can be tested by changing the set point, or introducing a disturbance. The temperature of the entering air to the furnace can be changed to induce a disturbance. We recommend changing this inlet temperature and recording the largest deviation from set point. Process 2 is used to implement a cascade control scheme in the same regenerator, and can then compare the control performance given by feedback and cascade control.

PROCESS 2. REGENERATOR CASCADE

A cascade control system is now implemented in the same regenerator. The primary variable is still the temperature in the regenerator, and the secondary variable is the temperature of the air exiting the furnace; the secondary controller manipulates the fuel valve. Figure D-2 shows the process.

Chapter 9 discussed how to tune cascade controllers. The first step is to obtain a process reaction curve for the secondary variable, and for the primary variable. Both curves are generated by the same step change in the secondary controller's output; this is the controller connected to the valve. This time you will have to generate two tables, temperature in the regenerator (primary controlled variable) versus time, and temperature exiting the furnace (secondary controlled variable) versus time. Remember, in taking the data you should read the variables every 5 s after changing the controller's output. From the temperature in the regenerator versus time graph you obtain K_1, τ_1, and t_{01}, and from the temperature exiting the furnace versus time graph you obtain K_2, τ_2, and t_{02}. Using the last set of terms the secondary controller is tuned as a simple feedback controller. Using all the terms and the tuning of the secondary controller, the primary controller is tuned using the formulas given in Chapter 9. Once you have tuned both controllers, you should test the secondary controller first to make sure it works fine by itself; you can set this controller in automatic (not cascade) to test it. Once this is done, you can set the secondary controller in cascade (remote set point) and the primary controller in automatic. Note the modes of the controllers. TIC 130 is not allowed to go into automatic unless TIC 100 is in cascade; this is the correct operation because the primary controller should not be in automatic unless the secondary controller is

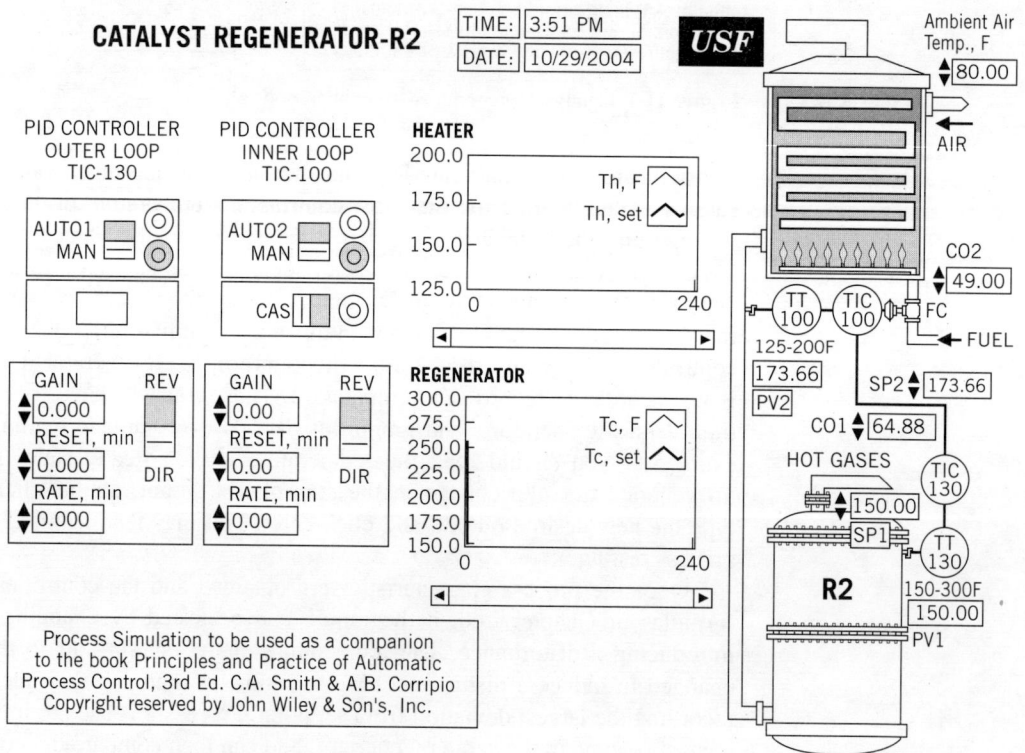

Figure D-2 Catalyst regenerator—cascade control.

ready to receive and follow the signal. Note also that when TIC 100 is in auto and cascade TIC 130 is forced into automatic. Finally, note that (a) when TIC 100 is in automatic (not in cascade) and its set point changes, the output from TIC 130 is forced to change to follow the new set point (output tracking), and (b) when TIC 130 is in manual its set point follows the process variable (set point tracking). As presented in Chapter 9, these tracking options provide a completely "bumpless" transfer in going from manual to automatic operation.

A good test to perform is to change the inlet air temperature to the furnace and record the largest deviation from set point. You can then compare this deviation to the one you obtained under simple feedback control.

PROCESS 3. PAPER-DRYING PROCESS

Figure D-3 shows the schematic of a process to dry paper. Wet paper enters a drying unit where it is dried by blowing hot air through it. Ambient air is heated in a furnace where fuel and air are combusted to provide the energy; from the furnace the air enters the drier. The exit moisture of the paper is the controlled variable; a cascade control scheme controls this moisture. The secondary variable is the temperature of the drying air exiting the furnace. The moisture controller, MIC 210, manipulates the set point to the temperature controller, TIC 220, which in turn manipulates a fail-open fuel valve. The main disturbance to the operation is the moisture of the ambient air entering the furnace.

Chapter 9 discussed how to tune cascade controllers. The first step is to obtain a process reaction curve for the secondary variable, and a process curve for the primary variable. Both curves are generated by the same step change in the secondary controller's output; this is the controller connected to the valve. This time you will have to generate two tables, moisture of the exiting paper (primary controlled variable) versus time, and temperature exiting the furnace (secondary controlled variable) versus time. Remember, in taking the data you should read the variables every 5 s after changing the controller's output. From the moisture of the exiting paper versus time graph you obtain K_1, τ_1, and t_{01}, and from the temperature exiting the furnace versus time graph you obtain K_2, τ_2, and t_{02}. Using the last set of terms the secondary controller is tuned as a simple feedback controller. Using all the terms and the tuning of the secondary controller, the primary controller is tuned using the formulas given in Chapter 9. Once you have tuned both controllers, you should test the secondary first to make sure it works by itself; you can set this controller on AUTO ON and CASCADE OFF to test it. Once this is done, you can set the secondary controller in CASCADE ON. Note the modes of the controllers. MIC 210 is not allowed to go into automatic unless TIC 220 is in cascade; this is the correct operation because, as discussed in Chapter 9, the primary controller should only be in automatic when the secondary controller is ready to receive and follow the signal. Note also that when TIC 220 is AUTO ON and CASCADE ON MIC 210 is forced into automatic (AUTO ON). Finally, note that (a) when TIC 220 is in automatic (not in cascade) and its set point changes, the output from MIC 210 is forced to change to follow the new set point (output tracking), and (b) when MIC 210 is in manual (AUTO OFF) its set point follows the process variable (set point tracking). As presented in Chapter 9, these tracking options provide a completely "bumpless" transfer in going from manual to automatic operation.

PAPER-DRYING UNIT

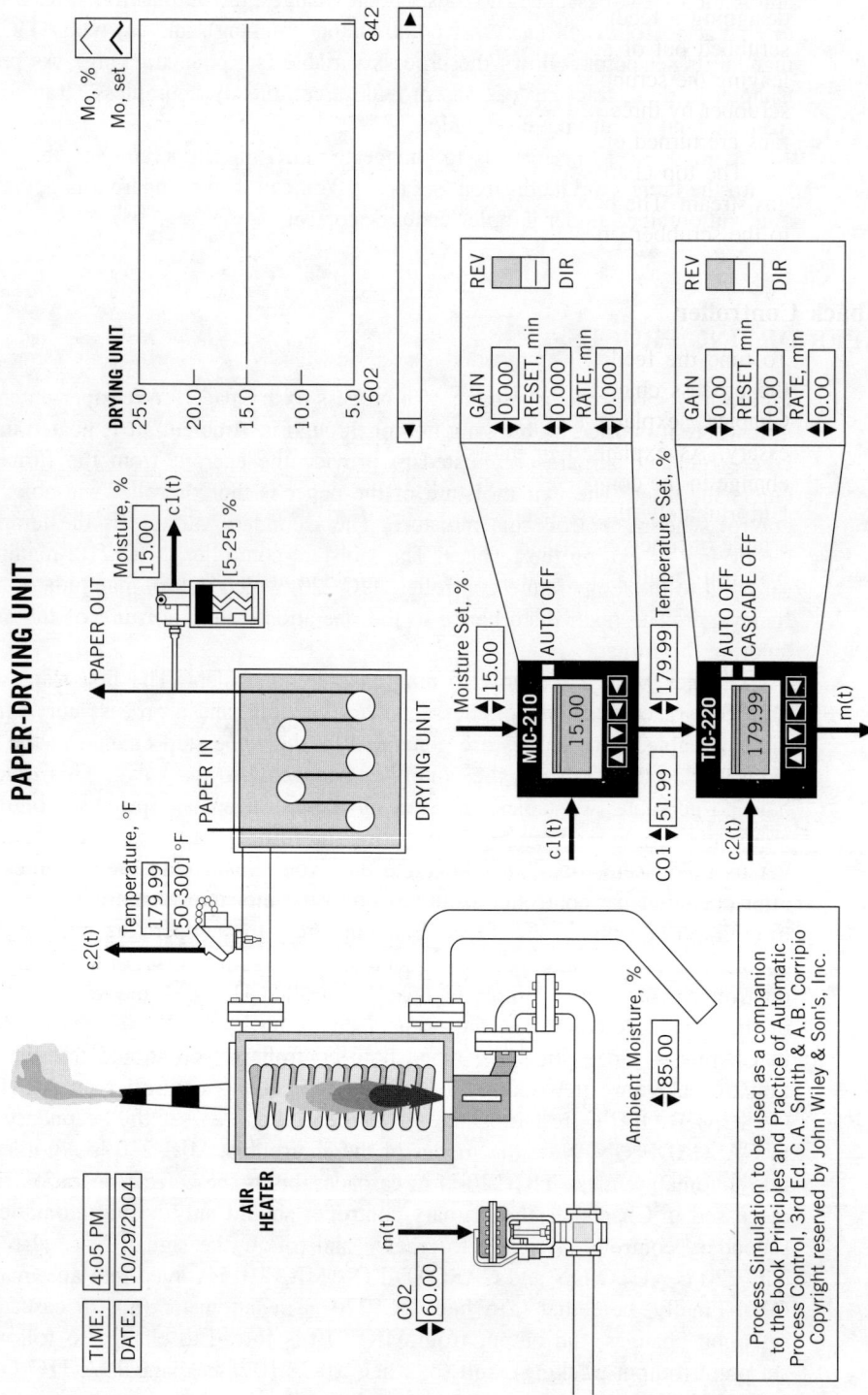

DRYING UNIT

Mo, %
Mo, set

25.0
20.0
15.0
10.0
5.0

602   842

PAPER OUT

Moisture, %
15.00
c1(t)

[5-25] %

PAPER IN

DRYING UNIT

Temperature, °F
179.99
[50-300] °F

c2(t)

AIR HEATER

TIME: 4:05 PM
DATE: 10/29/2004

CO2
60.00

m(t)

Ambient Moisture, %
85.00

Moisture Set, %
15.00

MIC-210
AUTO OFF
15.00
c1(t)
CO1 51.99

GAIN
0.000
RESET, min
0.000
RATE, min
0.000
REV DIR

Temperature Set, %
179.99

TIC-220
AUTO OFF   CASCADE OFF
179.99
c2(t)
m(t)

GAIN
0.00
RESET, min
0.00
RATE, min
0.00
REV DIR

Process Simulation to be used as a companion
to the book Principles and Practice of Automatic
Process Control, 3rd Ed. C.A. Smith & A.B. Corripio
Copyright reserved by John Wiley & Son's, Inc.

Figure D-3 Paper-drying process.

PROCESS 4. HCl SCRUBBER

The process shown in Fig. D-4 is used to practice tuning a feedback controller, and designing a feedforward controller. The process is a scrubber in which HCl is being scrubbed out of air by a NaOH solution. The analyzer transmitter in the gas stream leaving the scrubber has a range of 25 to 150 ppm. The HCl–air mixture is fed to the scrubber by three fans. Fan 1 feeds 50 cfm, and fans 2 and 3 feed 25 cfm each. These fans are turned on/off by simply clicking on the ON/OFF button of each fan.

The top chart shows the set point (in red), and the outlet ppm (in white) of the gas stream. The bottom chart shows the controller's output (in red), and the feed flow to the scrubber (in white).

Tuning the Feedback Controller

To tune the feedback controller, that is, to find K_C, τ_I, and τ_D, we must first find the process characteristics, process gain (K), time constant (τ), and dead time (t_0). Chapter 7 explains that to obtain these characteristics a process reaction curve is necessary. As explained in that chapter, to obtain this curve we first introduce a step change in the controller's output and record the ppm of the NH_3 leaving the scrubber. Unfortunately, there is no recorder to record the ppm. Thus, you will have to generate a table of time versus ppm, and graph this data. We recommend reading the ppm every 5 s. You should read the ppm until a steady state is achieved again. To generate the step

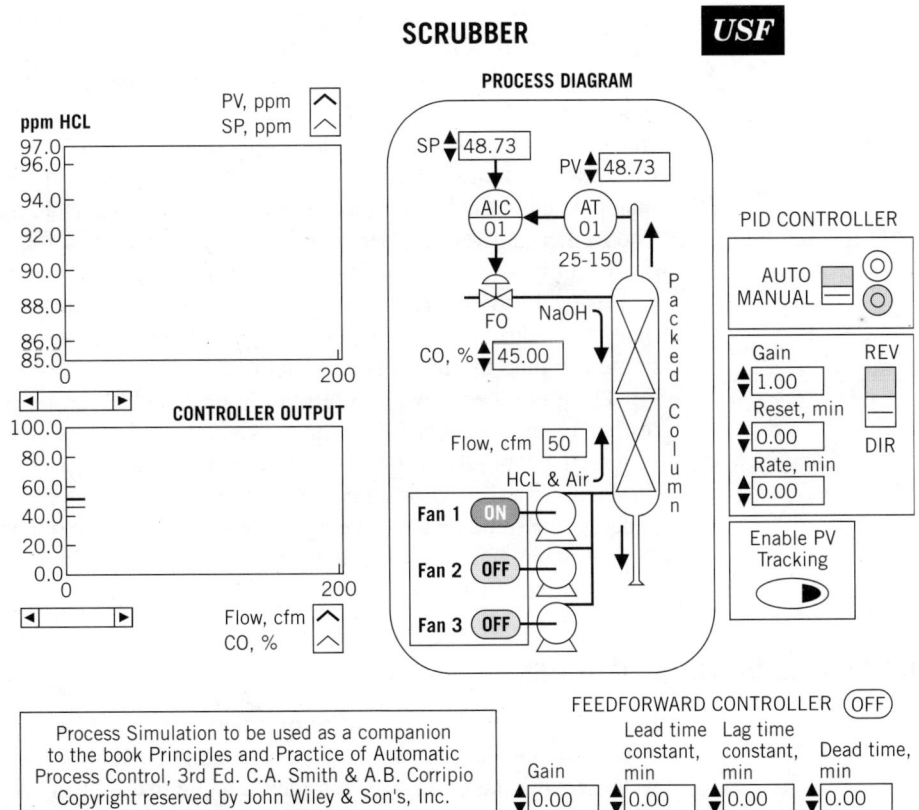

Figure D-4 Scrubber.

change in the controller's output, double-click on the number, type the new desired output, and click enter. You may read again Chapter 7 to review the questions related to this process testing.

Once the process characteristics are obtained and the controller tuned using the formulas of Chapter 7, then the tuning can be tested by changing the set point, or by introducing a disturbance by starting or stopping a fan.

Designing the Feedforward Controller

It is known that for this scrubber unit the input flow can vary much and very often. That is, it is common to have one fan working, for example, fan 2 with a flow of 25 cfm, and suddenly the other two fans turn on to have a total input flow of 100 cfm. When this occurs, the outlet HCl will increase until the feedback controller can take hold by adding more NaOH to bring the HCl back to set point. We must be sure that the HCl ppm do not violate EPA regulations. Thus, under normal operation the set point to the controller must be set low enough to ensure staying within regulation under this upset condition. Although the low set point guarantees not violating the regulation during any upset condition, it costs extra money during normal operation because it requires extra NaOH flow. If we could provide a tighter control, less deviation from set point under upset conditions, then we could raise the set point and thus, save a portion of the extra NaOH.

Feedforward control can provide this tighter control. The idea is to measure the flow entering the scrubber, and if this flow changes, manipulate the NaOH valve. That is, do not wait for a deviation in HCl before taking action. There is a flow sensor-transmitter with a range of 0 to 100 cfm measuring the input flow to the scrubber.

Chapter 11 shows that a way to design a feedforward controller is to use a first-order-plus-dead-time transfer function (K, τ, and t_0) describing how the disturbance (feed to the scrubber in this case) affects the controlled variable (outlet HCl ppm in this case), and another first-order-plus-dead-time transfer function describing how the manipulated variable (the feedback controller's output) affects the controlled variable. The former transfer function is obtained by clicking on/off one of the fans to generate a change in input flow, recording (every 5 s) the HCl ppm, graphing the response curve, and using Fit 3 described in Chapter 7. The latter transfer function was obtained to tune the feedback controller. Once both transfer functions are obtained then Eq. 11-2.5 is used to tune the feedforward controller.

Once the feedforward controller is designed it can be tested. We recommend the testing to proceed the following way. First under feedback control one of the fans is turned on or off and this control performance is used as a baseline to compare the feedforward performance. Then, under steady-state feedforward/feedback control, generate the same disturbance and compare the control obtained with that of feedback. Repeat again but this time add the lead-lag unit. Finally, do the same adding the dead-time compensator if needed.

PROCESS 5. MIXING PROCESS

Figure D-5 shows the schematic of a tank where cold water is mixed with hot water. The valve in the cold water pipe is manipulated to control the temperature of the water leaving the exit pipe. You may assume that a level controller (not shown) maintains perfect level control. The hot water flow and temperature, and the cold water temperature, act as disturbances to the process.

Mixing Tank

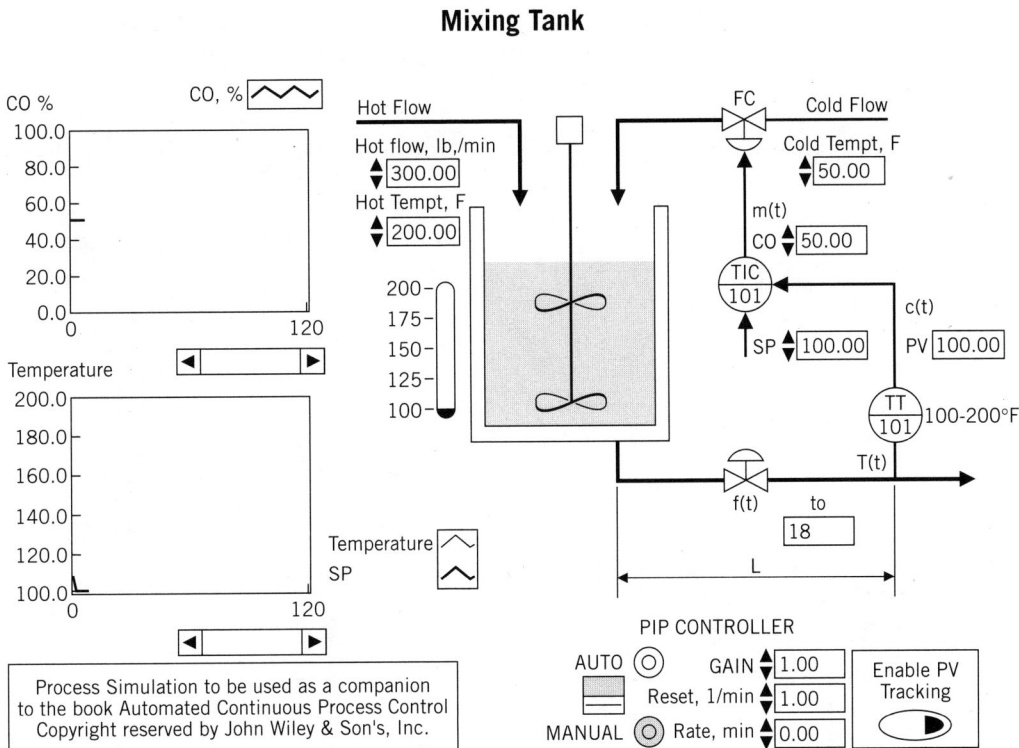

Figure D-5 Mixing process.

PAIR SIGNALS FOR DECENTRALIZED PID REACTOR CONTROL

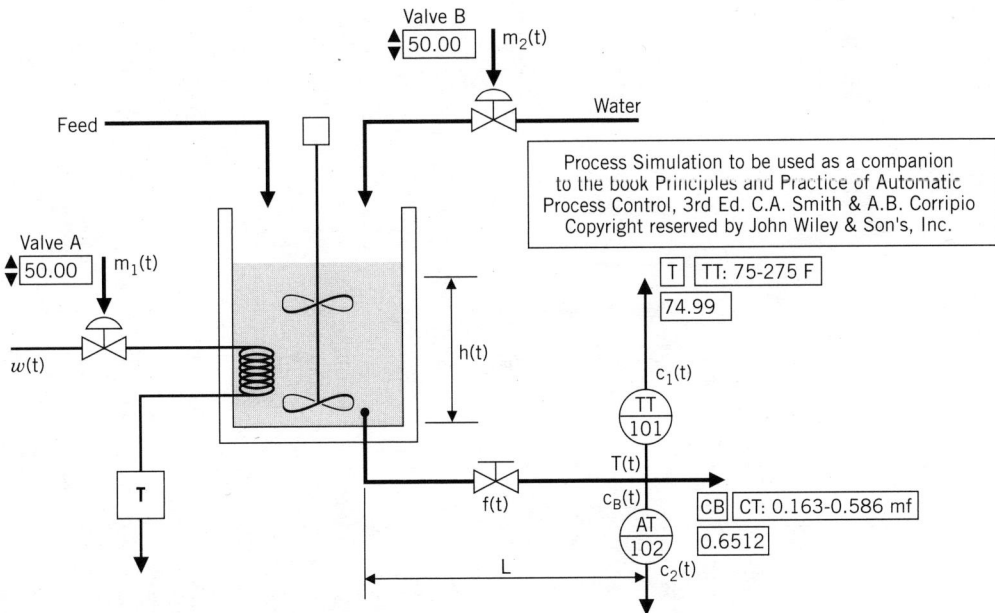

Figure D-6 Reactor process.

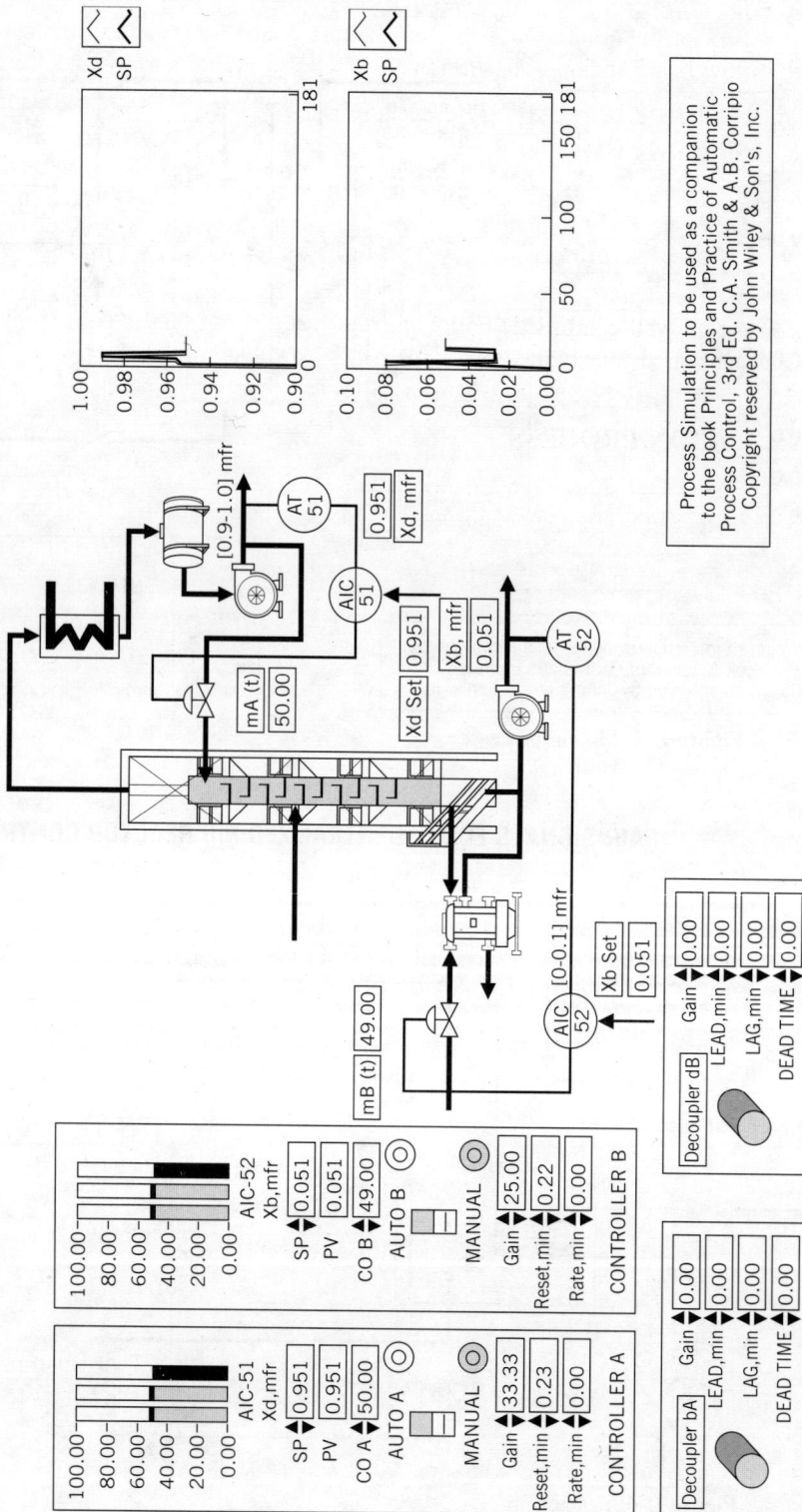

Figure D-7 Distillation process.

Tune the feedback controller by the method presented in Chapter 7, and used in the other processes. Note the units of the reset time. An interesting disturbance is the flow of hot water. Once the temperature controller is tuned, start to decrease the hot flow by 25 lbm/min at a time and watch the control performance.

PROCESS 6. REACTOR PROCESS

Figure D-6 shows a reactor where the endothermic reaction R + water → B takes place. The exit concentration of product B and the exit temperature must be controlled; the figure shows the transmitters and their ranges. The manipulated variables are the steam valve and the water valve. You may assume that the level in the reactor is well-controlled. Following the discussion of Chapter 12, decide on the best controlled–manipulated variable pairing for this process.

PROCESS 7. DISTILLATION PROCESS

Figure D-7 shows a distillation column where it is desired to control the concentration of the light component in the distillate, Xd in mass fraction, and in the bottoms, Xb also in mass fraction. As the figure shows, manipulating the reflux valve controls Xd, and manipulating the steam to the reboiler controls Xb. Using Chapter 12, tune both controllers, and design and implement a decoupler system. As a test, change the set point of one of the controllers and test the response with and without the decoupler.

Appendix E

Operating Case Studies

This appendix presents five cases that integrate the material presented in Chapters 5, 6, and 7. They were written to resemble a possible industrial situation that a junior engineer may face. These cases where contributed by Dr. Marco E. Sanjuan of the Universidad del Norte in Barranquilla, Colombia.

OPERATING CASE STUDY 1: HCL SCRUBBER

You have been hired as a consultant to improve the control system for an air treatment unit. The unit is a scrubber where the HCl concentration in air must be reduced before venting the air to the ambient; Fig. E-1 shows the scrubber. A NaOH solution is used as the scrubbing medium. The main disturbance to the process is the airflow into the scrubber; three fans provide this flow. When the airflow increases, the HCl concentration in the outlet stream also increases. The control system should respond to this disturbance and to set point changes. Usually the set point is 95 ppm. Above this value you are violating federal regulations (actually the maximum allowed is 98 ppm); below you are wasting NaOH.

The transmitter used (AT46) has a range from 25 to 150 ppm, and a time constant of 0.4 min. The sensor has a linear inverse behavior; that is, as the concentration increases the signal decreases, or 0 %TO = 150 ppm. The control valve is a linear fail-open valve. The flow provided by fan 1 is 50 cfm, and fans 2 and 3 provide 25 cfm each; usually only fan 1 works.

In the plant you meet the control engineer, Richard; the plant engineer, Karen; and the operator, Chris. Karen tells you that the controller has been in manual during the past week because the tuning parameters that Richard calculated are not working at all; these tunings are $K_c = 2.8$, $\tau_I = 4.8$ min, *reverse action* (tuning 1).

Chris says that he has been trying to change the tunings a bit with no luck. Karen suggested eliminating the integral mode, using only $K_c = 2.8$, but this did not work either. The current settings are $K_c = 2.5$, *direct action* (tuning 2).

Richard looks quiet and does not say a word. After a while he asks to speak with you in private. Once alone, he says, *"I'm pretty sure about these calculations. I believe that Chris doesn't want the controller to work because he fears that he'd lose his job."* He continues, *"Karen, on the other hand, is an old-fashioned engineer who thinks that humans are more reliable than machines, so if it doesn't work it is because the controller is not suitable for this application."*

(a) You sense a tight atmosphere, and it is not exactly because of the HCl ppm are in excess. Something or someone is wrong. But you need more information. If

---

Principles and Practice of Automatic Process Control/Third Edition, by C. A. Smith and A. B. Corripio
ISBN 0-471-66141-4 Copyright © 2006 John Wiley & Sons (Asia) Pte. Ltd.

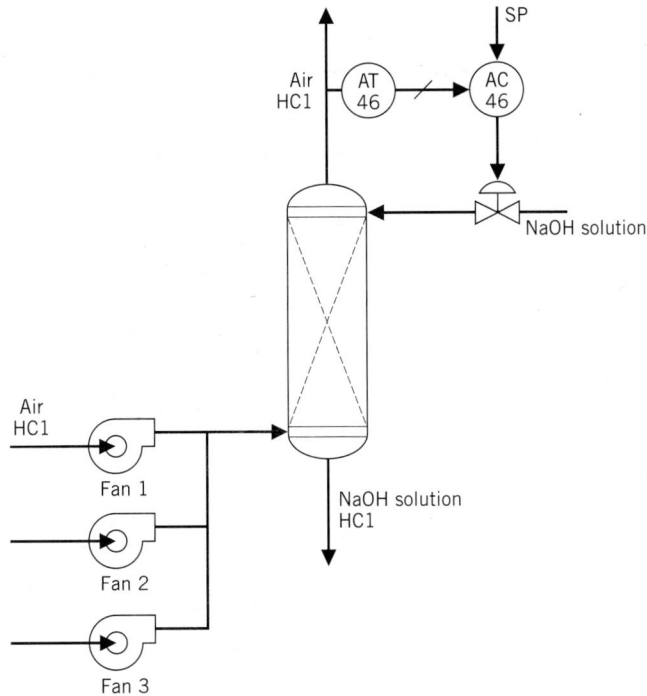

Figure E-1 Process for Operating Case Study 1.

no one is sabotaging the system, and the controller is OK, can you speculate why the different sets of tuning parameters have not worked?

You decide to call Karen, Chris, and Richard for a meeting. This time you want to work out a strategy. When everyone is present, you start your questioning. *"I'm going to need everyone's help to help solve your problem. First of all, why do you need a controller? Why don't you just set a fixed flow rate for the NaOH scrubbing solution?"*

"I can help," Karen says. *"Initially, we used to change the set point every time a new manager was hired. You know, some cared about the environment, others just ignored it. Now, with the new environmental regulations, I don't think it's going to be changed for a while. However, sometimes we have to evacuate more air (with HCl) from the plant, and that is when the other fans (2 and 3) start running. Actually we don't control their start up; someone else in other units does it. In these cases the ppm's can increase a lot, so we have to open the valve."*

"Do you have any information about how much the change is when the fans start running," you ask.

"I'm the expert when it comes to fans disturbing our process." Chris says. *"Yesterday, someone turned fan 2 on at 2:35 P.M. The concentration started changing right away. Before the fan started the concentration was 92.5 ppm and the NaOH flow was 55 gpm (45 %CO). At 2:39 P.M. the concentration was 98.8 ppm and rising. After a while it was stable at 102.5 ppm. Then I decided to start opening the valve."*

"And what happened?" you ask.

"Since Richard was around," Chris says, *"he asked me to record the scrubber behavior from the moment the valve opened and I did it."*

"I have the results here," Richard says. *"I thought you might want to take a look at them."* Richard hands the results to you. The recorded data is shown in Table E-1.

Table E-1   Step test information for Operating Case Study 1

Time	Concentration	Time	Concentration
3:02 PM	102.5	3:13 PM	96.2
3:03 PM	102.5	3:14 PM	95.8
3:04 PM	102.3	3:15 PM	95.4
3:05 PM	101.6	3:16 PM	95.1
3:06 PM	101.1	3:17 PM	94.8
3:07 PM	100.2	3:18 PM	94.6
3:08 PM	99.4	3:19 PM	94.5
3:09 PM	98.5	3:20 PM	94.5
3:10 PM	97.7	3:21 PM	94.4
3:11 PM	97.0	3:22 PM	94.4
3:12 PM	96.5	3:23 PM	94.4

Initial valve signal 45 %CO, final valve signal 35 %CO.

You now think that you have everything you need. However, before you leave, you ask Karen what it means to her that the controller tuning does not work at all. She says that the initial set of parameters got the valve fully open or fully closed in a couple of minutes. The last set of parameters is not that bad. However, when either fan 2 or 3 are turned on, the process never reaches the set point and deviates close to 5.5 ppm from it. *"That,"* she says, *"is not acceptable. Our set point is 95 ppm and federal regulations are 98 ppm, I can live with an offset of 2 ppm but not higher. Also,"* she continues, *"I'd like to get a response where the oscillation amplitude decays fast. For example, if the first overshoot is 2 ppm, the next one should be 0.5 ppm."*

Now you have everything you need.

(b) Draw an overall closed-loop block diagram to illustrate how the system works. Use the actual values (not letters) inside the blocks.

(c) What is the valve maximum flow?

(d) State the reasons why Tuning 1 did not work. Prove your reasons.

(e) State the reasons why Tuning 2 did not work. Prove your reasons.

(f) What tuning (P controller) would you suggest? Why?

(g) What would you say to Chris to eliminate his worries about losing his job if the controller works?

(h) Was Richard right about Chris sabotaging the controller operation?

OPERATING CASE STUDY 2: WATER TREATMENT UNIT

It was a cold day in December. After a long night studying for the Auto Controls final, you drive back home. You cannot forget that you need a 68 in the final to get a C. When you get home, you eat some snacks and go to bed. Just a few minutes after you fall asleep, a weird dream starts taking shape in your head.

Suddenly, you have graduated from college and you are a consultant in Control Engineering, Inc. You are asked to look into some problems at a water treatment unit,

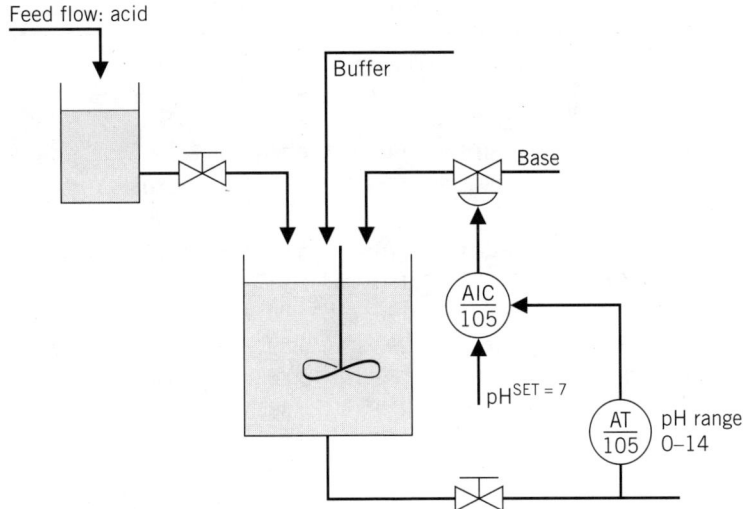

Figure E-2 Process for Operating Case Study 2.

like the one shown in Fig. E-2. A water stream entering the unit (Feed flow) usually has a pH below 7. Before this stream can be delivered to a pond close to the plant, the pH needs to be restored back to 7. The minimum acceptable value is 6.8. However, for safety reasons, the controller set point is 7. The pH control loop installed manipulates a base solution to maintain the outlet pH at the desired value. The sensor-transmitter used in the loop has a range from 0 to 14, and its signal takes 1.5 min to reach steady state.

Rob, the plant manager, tells you that they have been having problems with their controller tuning. *"Sometimes,"* he says, *"it works fine. However, some other times it just goes crazy and produces highly oscillatory behavior. This is costing us a lot of money, because valves are damaged. This year we have had to buy four new control valves."*

When you inquire about how they obtained the tunings, they tell you that they did a step test to identify the process transfer function. *"That sounds good,"* you say, *"but tell me what did you get?"*

They give you Table E-2.

"We even did Process Identification for the feed flow disturbance, but since we realize that we didn't need it for tuning purposes, we just ignore it." Rob says. *"The results are here (Table E-3)."*

"What else have you done?" you wisely ask.

Table E-2 Process identification for pH controller tuning

Initial valve signal	Initial pH	Process gain	Controller gain
52 %CO	7.025	0.74 %TO/%CO	2.64 %CO/%TO
Final valve signal	Final pH	Time constant	Reset time
42 %CO	5.885	7.5 min	7.5 min
		Dead time	Derivative time
		3.2 min	1.6 min

Table E-3   Results for a feed flow change from 166 to 206 gpm

Time, s	pH	Time, s	pH	Time, s	pH	Time, s	pH
0	7.0250	130	6.1745	260	5.7501	390	5.6078
10	7.0250	140	6.1237	270	5.7325	400	5.6023
20	7.0132	150	6.0770	280	5.7163	410	5.5974
30	6.9629	160	6.0340	290	5.7015	420	5.5929
40	6.8821	170	5.9943	300	5.6879	430	5.5888
50	6.7872	180	5.9577	310	5.6755	440	5.5850
60	6.6899	190	5.9239	320	5.6641	450	5.5816
70	6.5965	200	5.8927	330	5.6537	460	5.5786
80	6.5094	210	5.8639	340	5.6441	470	5.5758
90	6.4296	220	5.8373	350	5.6355	480	5.5732
100	6.3568	230	5.8127	360	5.6276	490	5.5709
110	6.2904	240	5.7901	370	5.6203	500	5.5688
120	6.2298	250	5.7693	380	5.6138		

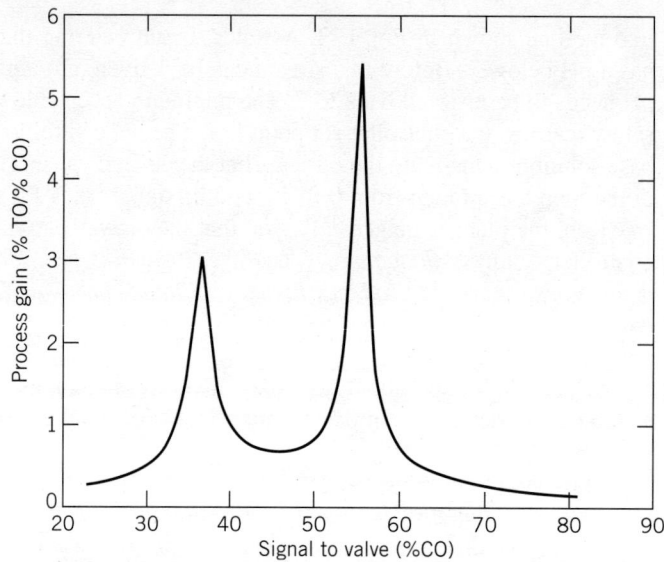

Figure E-3 Process gain versus controller output (%CO).

"Well," says Rob, *"I bought software for process identification for $10,000. We ran it in our computer, and it gave us something like gain, time constant, and dead time versus valve signal."*

"Wow," you exclaim, *"do you have the results?"*

"Sure." replies Rob, *"The time constant and the dead time were sort of constant (didn't change with valve signal). However, the process gain behavior was sort of funny."* And he hands you Fig. E-3.

"Now," you say, *"I have everything I need."*

(a) Draw a closed-loop block diagram indicating inputs, outputs, and transfer functions (using values, not letters). Indicate signals and their units.

(b) Why is their tuning not working?

(c) What is the valve-safe position, and the appropriate controller action?

(d) What is the maximum proportional gain that can be used before getting to unstable behavior?

(e) What should be the appropriate PID tuning if quarter decay ratio based on the ultimate gain and ultimate period were to be used, based on the process identification that they ran first?

(f) What should be the tuning parameters for the PID controller, such that the process never again presents undesired oscillatory behavior? (Even if it means that sometimes it will be to slow.)

OPERATING CASE STUDY 3: CATALYST REGENERATOR

The process shown in Fig. E-4 is a catalyst regenerator unit where hot air removes carbon from solid catalyst. The manipulated variable is the fuel flow and the controlled variable is the regeneration temperature. There is a linear control valve with 120-gpm maximum flow. The steady-state flow is 50 gpm. The temperature sensor has a range from 50 to 275°F. The main disturbance for the system is the ambient air temperature; such temperature is 80°F now.

You decide to identify the process transfer function. To do that, you ask someone to do a step change of 10 %CO to the fuel valve (now it is 60 %), and get the response values to your office. The results are presented in Table E-4. Once you receive the information, you realize that you did not specify the step change sign. The only thing you know is that the fuel control valve is fail-closed. So, you know you can deduce what the sign of the step was, right?

(a) Your first goal is to draw an overall block diagram for the process. Reading records from previous engineers you find the following note:

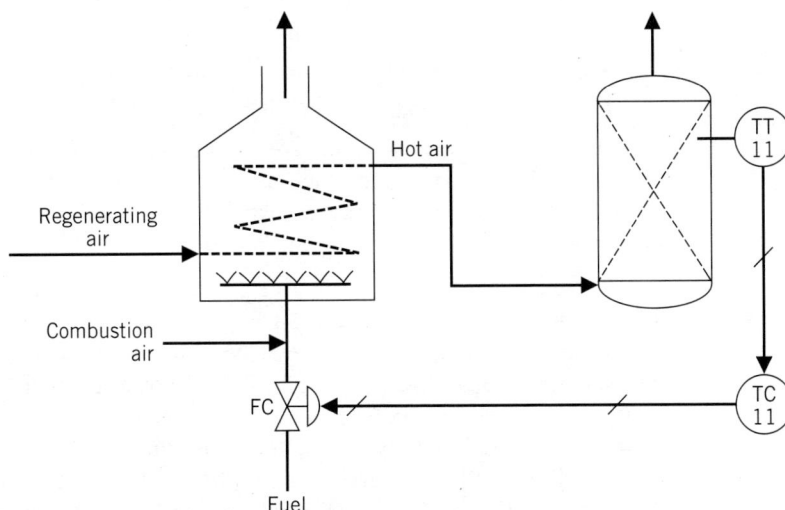

Figure E-4 Process for Operating Case Study 3.

Table E-4   Step test information for Operating Case Study 3

Time, min	Temperature, °F	Time, min	Temperature, °F	Time, min	Temperature, °F	Time, min	Temperature, °F
0.0	159.6	11.0	167.5	22.0	172.7	33.0	174.3
1.0	159.6	12.0	168.3	23.0	172.6	34.0	174.0
2.0	160.2	13.0	168.8	24.0	173.1	35.0	174.6
3.0	160.5	14.0	169.5	25.0	173.4	36.0	174.1
4.0	162.0	15.0	170.1	26.0	173.8	37.0	174.8
5.0	162.2	16.0	170.8	27.0	173.4	38.0	174.2
6.0	163.5	17.0	171.2	28.0	173.8	39.0	174.2
7.0	164.4	18.0	171.3	29.0	173.9	40.0	174.7
8.0	165.1	19.0	172.3	30.0	173.8		
9.0	166.2	20.0	171.8	31.0	173.7		
10.0	167.2	21.0	172.0	32.0	174.0		

As a conclusion, the transfer function between the ambient air temperature and the regenerator temperature sensor is

$$G_D(s) = \frac{0.44e^{-2s}}{4s + 1}$$

Of course, you immediately realize that the gain units are %TO/°F and time units are minutes. This is great!! Now, to prove to everyone that you understand how the system works, you draw an overall block diagram.

(b) Now you remember that the instrument and equipment company wants to install the controller with the factory settings: $K_c = 2.5$, $\tau_I = \infty$, $\tau_D = 0$, *reverse action.* Of course, you do not want to allow them to do it without previously running some calculations. So, you ask yourself the following questions:

- Is reverse acting the appropriate controller action? Why? (You know they will ask you.)

- Even if the controller action is set in the correct way, will the closed-loop process be unstable with the default controller gain (2.5)?

- Can you calculate a controller gain that minimizes the IAE for disturbance changes?

(c) Now you feel that you are the king of the plant! Suddenly, you remember that today the plant manager, the one you just met 2 days ago when you asked for an $8000 raise, is going to be present when they run the new controller. You really want to impress everyone. But how can you do it?

To relax you turn on your TV, and start playing around with the remote control. Then you get to the Weather Channel and listen, "tonight we are going to have an unusual hot night and tomorrow the temperature will be 95°F...blah, blah, high pressure, blah, blah, low pressure...." That power company is going to make money today with all air conditioners running at full speed, eh! But, wait! The temperature is going to increase from 80°F today to 95°F tomorrow! Because the company wants to install a P controller, there will be an offset. *"I bet I can predict the offset,"* you say to yourself, *"that is the way to get that raise!"* Use in your calculation the controller gain you calculated in (b).

OPERATING CASE STUDY 4: BABY BACK RIBS

The semester is over and you have decided to look for a job for the holiday season. *"This is the last small-pay job I'll take,"* you say, *"Once I'm done with Plant Design next semester it will be time for the big dollars."*

You go to Charlie's Bar-BQ and get hired as assistant manager (you know, the person in charge of helping the manager running the place). Once hired, you make sure that the kitchen is fine, that the money is safe, and that the customers are happy. Ah! Customers must be happy. Right? Wrong! There have been complaints lately about how the baby back ribs taste. Some people say they are too cooked, and some say they are too rare. Since you will be here for only a month, you try to ignore the situation.

One day the manager calls you. *"The girl up front says that you had a class in process control. She says you got a B. Is that right?"* he asks.

You nod.

"Well," the manager says, *"I've got a problem for you."*

He tells you that customers have been complaining about the quality of the ribs. You dare to look surprised.

"The cook says that they should cook for 3 hours at 290°F. A change of 4°F doesn't hurt anyone, but more deviation will bring complaints. He says that when we buy the special ones (the expensive ones, you know), they need to be cooked at 310°F. We always keep the oven running, with enough gas to get to 290°F by default. Usually we set the valve to provide 1.5 scfm. But you know what," he says, *"that's a weird valve. You have to enter like a percentage or some ... like that. We spent like 3 hours trying to get the display to show the 1.5 scfm. We had to enter 40 %.*

"The cook has noticed that in cool days," he continues, *"when the temperature outside drops from 90 to 70°F the temperature in the furnace drops to 275°F in about 12 minutes. You can notice how the furnace temperature starts to drop when the outside temperature drops. That happened yesterday. The cook had to put the valve in 80 % (3 scfm) to get the temperature to 290°F. They also have problems when they put additional racks inside. Usually we have 10 racks inside,"* he says. *"But when they put 2 more racks in, the temperature drops to 285°F. The change is not as fast as the other one. First—he states—you don't even notice the temperature change. But after 2 minutes, the change starts. It probably takes 6 minutes to reach the 285°F."*

"I have a question," you say. *"Do you know anything about the sensor?"*

"Sure," he says, *"the ... can measure anything from 200 to 400°F."*

"Wow, I'm impressed," you say. *"Did you calculate the transfer function?"*

"No," he says with surprise. *"I just read the plate behind it."*

"Well," you say, *"I just need to run a test and I'll design a controller to keep the temperature constant."*

"Cool!" he comments. *"Go ahead."*

You run to the kitchen and increase the valve signal to 48 % (from 40 %). You get the results shown in Table E-5. Before you leave the kitchen the cook shouts, *"Hey, thanks for burning the ribs."*

You promise the manager to have the design in 2 hours. Before you leave the manager whispers in your ear, *"By the way, what the ... is a transfer function?"*

(a) Draw an open-loop and closed-loop block diagram with transfer functions included.

(b) Identify the valve (type, safe action, parameters) and the sensor (range).

(c) Tune a P controller.

Table E-5   Step test information for Operating Case Study 4

Time, min	Temperature, °F	Time, min	Temperature, °F
0.0	290.00	11.0	297.62
1.0	290.00	12.0	297.74
2.0	290.72	13.0	297.82
3.0	292.09	14.0	297.88
4.0	293.49	15.0	297.92
5.0	294.68	16.0	297.95
6.0	295.62	17.0	297.96
7.0	296.32	18.0	297.98
8.0	296.83	19.0	297.98
9.0	297.19	20.0	297.99
10.0	297.44		

(d) Will the controller manage the set point change (to 310°F) and the ambient air temperature disturbance?

(e) Tune a PI controller according to Ziegler–Nichols equations.

OPERATING CASE STUDY 5: PAPER DRYING UNIT

A paper processing company Ireland has recently hired you to improve and implement control strategies in their 10-year-old equipment. Once you arrive, you are assigned to the paper-drying unit. This two-part unit has a furnace where air is heated and a drying station where the air circulates on both sides of the paper to remove moisture. If you close your eyes you can imagine the plant but, you would not be able to see Figure E-5 where the plant is depicted.

The plant manager has told you that your first task is to improve and analyze the control system for this unit. Their main problem arises from unexpected changes in the atmospheric air temperature and input paper moisture. They would like to obtain

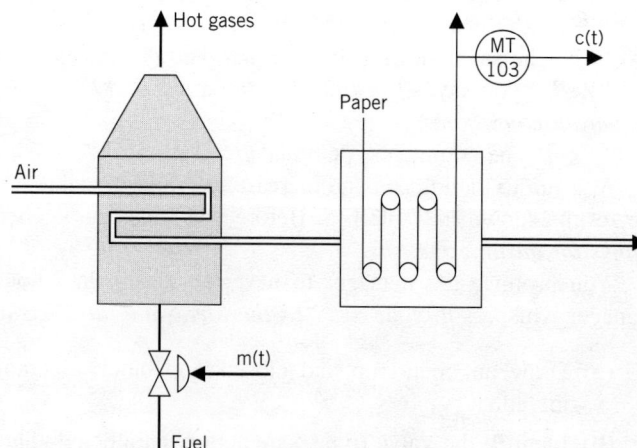

Figure E-5 Process for Operating Case Study 5.

output moisture of 3.5 % with no more than 0.2 % deviation from set point (from 3.3 to 3.7 %).

Your first goal is to gather as much information as possible from the system. So you go to the main control room, and talk the operator into giving you the information you need. First, he tells you that currently there are 120 scfm of fuel going into the system. You read from the screen that the signal to the valve is 34.13 %CO. You also observe that the current output moisture is 3.6 % and the nonscaled value is 52 %. The operator tells you that the minimum moisture that the sensor can read is 1 %, but he does not remember the maximum one. That is OK with you, since you know you can get it.

Then you ask him to increase the signal to the valve by 10 %CO and to start recording the output moisture. You go outside to get a cup of coffee, and when you are back you find the results shown in Table E-6.

After saving the information, you start talking with the operator about what happens when the weather changes. He tells you that when the cold breeze starts blowing suddenly the outside temperature could drop about 8°C. The unit keeps working normally for 3 minutes, but then the output moisture goes from 3.5 to 3.9 % in the next 10 minutes. When you ask about the effect of the input moisture, the operator tells you that there were some records in the drawer, but that he does not know what they indicate.

Looking around you find the following information:

"When we keep the fuel flow constant and are fortunate enough to get a constant ambient temperature, an increase in 2 % in the input moisture (usually 23 %) leads to an increase in 0.25 % in the output moisture. Such a change presents an FOPDT dynamic behavior with time for 28.3 % of change of 4 minutes, and time for 63.2 % of change of 8 minutes."

Now, with all the information you have collected, you are ready to prepare your first analysis for your company following the format presented next.

(a) Draw an open-loop block diagram, indicating the name of the different elements.

(b) Describe the valve (fail-safe position) and sensor (range).

Table E-6 Step change in the signal to the valve

Time	Moisture	Time	Moisture
13:05:00	3.60	13:15:00	3.33
13:06:00	3.60	13:16:00	3.32
13:07:00	3.57	13:17:00	3.32
13:08:00	3.54	13:18:00	3.31
13:09:00	3.50	13:19:00	3.31
13:10:00	3.46	13:20:00	3.31
13:11:00	3.43	13:21:00	3.30
13:12:00	3.40	13:22:00	3.30
13:13:00	3.37	13:23:00	3.30
13:14:00	3.35	13:24:00	3.30

Note: The final flow is 87 scfm.

(c) Draw a closed-loop block diagram, indicating the transfer function inside each block (with numbers), name of the signal and their units.

(d) Find the tuning of a P controller based on 70 % of the ultimate gain.

(e) Will the temperature (8°C) and input moisture (2 %) changes described above be compensated by the control system within the required limits?

Index

Principles and Practice of Automatic Process Control/Third Edition, by C. A. Smith and A. B. Corripio
ISBN 0-471-66141-4   Copyright © 2006 John Wiley & Sons (Asia) Pte. Ltd.